Chinese and Related North American Herbs

Phytopharmacology and Therapeutic Values

Second Edition

Chinese and Related North American Herbs

Phytopharmacology and Therapeutic Values

Second Edition

Thomas S.C. Li, Ph.D.

CRC Press
Taylor & Francis Group
Boca Raton London New York

CRC Press is an imprint of the
Taylor & Francis Group, an **informa** business

CRC Press
Taylor & Francis Group
6000 Broken Sound Parkway NW, Suite 300
Boca Raton, FL 33487-2742

First issued in paperback 2019

© 2009 by Taylor & Francis Group, LLC
CRC Press is an imprint of Taylor & Francis Group, an Informa business

No claim to original U.S. Government works

ISBN-13: 978-1-4200-9415-2 (hbk)
ISBN-13: 978-0-367-38496-8 (pbk)

Library of Congress Cataloging-in-Publication Data

Li, Thomas S. C.
    Chinese and related North American herbs : phytopharmacology and therapeutic values / Thomas S.C. Li. -- 2nd ed.
        p. cm.
    Includes bibliographical references and index.
        ISBN 978-1-4200-9415-2 (hardcover : alk. paper)
        1. Materia medica, Vegetable--North America. 2. Materia medica, Vegetable--China.
    3. Medicine, Chinese. 4. Herbs--Therapeutic use. I. Title.

RS164.L5645 2009
615'.321--dc22   2009028980

Visit the Taylor & Francis Web site at
http://www.taylorandfrancis.com

and the CRC Press Web site at
http://www.crcpress.com

Foreword

Western researchers are increasingly acknowledging the importance of the traditional herbal preparations that have been the mainstay of Eastern medicine for millennia. Indeed, Western society in general is now consuming numerous herbal medicines, and over-the-counter commercial herbals now compete with prescription pharmaceuticals. Given the European origin of Western society, it is not surprising that European plants dominate the medicinal plant industry of the West. However, Asian medicinal plants are now enthusiastically being incorporated into Western medical practice, most particularly Chinese herbs. Unfortunately, while there is an incredible wealth of knowledge about Chinese herbs, most of this information has been unavailable to Western society, and even the accessible information has often been in obscure sources. The famous 15th-century physician Paracelsus taught that the only difference between a medicine and a poison was the dose, so it is critical to know not only what potentially useful chemicals are present in a given plant species, but also the potential for toxicity. Lack of knowledge of both the therapeutic and toxic properties of Chinese medicinal herbs has doubtlessly retarded progress toward developing more effective medications.

Chinese and Related North American Herbs by Dr. Thomas Li represents a milestone in educating Western society about a previously unavailable treasure chest of medicinal knowledge.

This book is an authoritative and comprehensive reference guide to a very large number of significant Chinese medicinal herbs. A gold mine of information is available on their chemical constituents and therapeutic applications. This will be extremely useful to a wide range of healthcare professionals who deal in one way or another with medicinal plants. The current heated debate regarding the comparative values of traditional herbal preparations and physician-prescribed pharmaceuticals should in no way detract from the value of this book, since Chinese medicinal herbs are not simply useful in herbal form, but also have immense potential for contributing to the development of new pharmaceuticals.

Dr. Li's dual presentation of Chinese herbs and their close North American relatives is a stroke of genius. Charles Darwin was one of the first to be puzzled by the fact that many plants of eastern Asia are remarkably similar to many plants of eastern North America. We now know that this phenomenon is due to the existence of an ancient, continuous temperate flora that became separated by geological and climatic changes. Accordingly, many Chinese herbs can be expected to have similar chemicals and similar medicinal values to their North American counterparts, and Dr. Li presents this extremely important information more competently than anyone to date.

Dr. Li, one of the world's leading authorities on medicinal plants, has dedicated many years of effort to acquiring and condensing the information presented in this reference text. He is to be congratulated on this superb and invaluable synthesis.

Ernest Small, Ph.D.
Principal Research Scientist
Eastern Cereal and Oilseed Research Centre
Agriculture and Agri-Food Canada
Ottawa, Canada

Preface

The use of medicinal herbs for treating human disease started in China thousands of years ago. Eighty percent of the world's population is still using traditional medicine, either because they have no access to Western medicine or choose not to use it. Recently, the use of medicinal herbs, especially Chinese herbs and their products, has attracted considerable attention around the world and generated extensive research on their philosophy, principles, and especially the scientific background of the chemical components responsible for their claimed therapeutic value.

Research in Chinese medicinal herbs has been conducted for decades in China, Japan, and Korea and recently in the West. Unfortunately, language barriers and the unreliability of sources and herbal material have hampered progress. A basic scientific understanding of the Chinese herbal preparations is the first step toward building consumer confidence in herbal medications. Proper procedures to eliminate adulteration, contamination, and toxic side effects are also urgently needed to regulate the use of Chinese herbs.

This book is designed to provide researchers with easy access to information on Chinese medicinal herbs compiled from widely scattered sources in the Chinese and Western literature. Table 1A B present current available information on the major constituents and therapeutic values of more than 1800 species of Chinese medicinal herbs. The data are arranged alphabetically by the Latin name followed by the common Chinese and English names. Tables 2 and 3 present data on a total of 700 North American herbs belonging to the same species or genus as Chinese herbs, and a comparison of active ingredients and claimed therapeutic values. Appendices 1, 2, and 3 cross-reference Chinese and scientific names, and major active ingredients and their sources in the Chinese and North American herbs cited in the tables.

The information in this book is primarily for reference and education. It is not intended to be a substitute for the advice of a physician. The uses of medicinal plants described in this book are not recommendations, and the author is not responsible for any liability arising directly or indirectly from the use of information in this book.

Acknowledgments

The author thanks Lynne Boyd and Peggy Watson, librarians, for their efforts in the literature search. I also thank my colleagues, Drs. Tom Beveridge, Cheryl Hampson, Dave Oomah, and Peter Sholberg, for their valuable assistance. Finally, I would like to thank my family for their encouragement.

Contents

Foreword v
Preface vii
Acknowledgments ix

Chapter 1: Chinese Medicinal Herbs: Phytopharmacology and Therapeutic
Values 1

Table 1A : Major Constituents and Therapeutic Values of Chinese
Medicinal Herbs 4

Table 1B: Major Consituents and Therapeutic Values of Chinese Medicinal
Herbs 157

Chapter 2: Phyletic Relationships between Chinese and Western Medicinal
Herbs 175

Table 2: Chinese and North American Medicinal Herbs Belonging to the
Same Species: Major Constituents and Therapeutic Values 176

Table 3: Chinese and North American Medicinal Herbs Belonging
to the Same Genus and Different Species: Major Constituents and
Therapeutic Values 237

References 305

Appendix 1: Chinese and Scientific Names of Chinese Medicinal Herbs 337

Appendix 2: Major Chemical Components and Their Sources in Chinese
Medicinal Herbs 395

Appendix 3: Major Chemical Components and Their Sources in Related
North American Medicinal Herbs 581

Index 641

CHAPTER 1

Chinese Medicinal Herbs
Phytopharmacology and Therapeutic Values

In spite of the great advances of modern scientific medicine, traditional medicine is still the primary form of treating disease for the majority of people in developing countries, including China. Even among those to whom Western medicine is available, the number of people using one form or another of complementary or alternative medicine is rapidly increasing worldwide.[384] In the United States alone, the consumption of medicinal herbs is rising at approximately 15% annually.[385]

Herbal medicine is important to a majority of the world's population, and not only for treating diseases.[33,50,53,58,270,381,382] Many prescription drugs, such as aspirin, codeine, and digoxin, have their origins in herbal medicines.[363] On the basis of global survey data from 1997, about 119 plant-derived compounds of known structure are currently used as prescription drugs.[366,367,369]

With its abundant botanical resources, China has been a pioneer in treating human diseases with medicinal herbs. The medicinal use of herbs in China by tradition has been attributed to a legendary emperor, Shen Nong (3494 BC), who tasted and tested plants and discovered their medicinal properties.[373,389] The recorded use of plants for medicinal purposes in China dates back to 2800 BC.[368] The most comprehensive classic herbal encyclopedia, *Ben Cao Gang Mu*, a description of formulas or prescriptions to treats human diseases, was published in the 16th century by Dr. Li Shizhen (1517–1593 AD).[373,389] This original *materia medica* recorded over 350 crude drugs; since then a great number of drugs and prescriptions have been added.[364,372] In 1958, the year of the "Great Leap Forward," Chairman Mao declared that traditional Chinese medicine (TCM) was a vast "treasure chest" and challenged the Chinese people to validate its efficacy, and to combine the best elements of TCM with modern Western medicine to improves the nation's healthcare delivery system.[50] In 1999, Hong Kong's Chief Executive, Tong Chee-Hwa, announced his intention to develop Hong Kong as a world center for TCM.

In the West, popular demand for and scientific interest in alternative medicine, particularly medicinal herbs, have increased considerably in recent years. The success and acceptance of the Chinese experience have stimulated new research focused on the specific beneficial effects of Chinese herbal medicine.[379,380,391,394,395,396] Some herbs commonly used in Chinese medicine have been studied, and chemical constituents that could represent the therapeutic actions of the herbs have been identified.[372,390] However, numerous mechanisms are likely involved in the various actions of a single herbal medicine. Elucidation of these mechanisms will provide the scientific basis for establishing the efficacy and safety not only of Chinese herbal medicine, but all forms of medicinal herbs.[368]

In China, herbal medicines in the ancient tradition continue to be widely used.[373] These medicines commonly contain ten or more herbs, thereby making it difficult to determine the pharmacological effects of individual drugs incorporated in prescriptions. In modern Western medicine, the use of a single chemical component is preferred in order to avoid drug interactions. In Chinese medical philosophy, therapeutic value and efficiency are increased by combining various herbs and ingredients in one prescription to treats a single disease.[368] A compound prescription often consists of four different functional groups, and each group usually comprises more than one herb. The *principal* provides the principal curative effect; the *adjuvant* helps strengthen the principal effect; the *auxiliary* relieves secondary symptoms or decreases the toxicity of the principal and the *conductant* directs the action of the principal to the target organ or site.[369] There are several logical explanations for the philosophy of mixing several crude extracts to achieve greater benefits. First, crude drugs given in combination may act synergistically. Second, the combination may have unknown interactions that might diminish possible adverse side effects of one or more of the components. Third, the combination may prevent the gradual decline in effectiveness observed when single drugs are given over long periods of time.[368]

Chinese herbal medicine generally uses either the whole plant or crude extracts as medicines, which tend to include a wide range of chemical constituents. Neither the whole plant nor crude extracts deliver highly concentrated medicines.[363] By contrast, conventional Western prescription drugs usually contain a single-molecule active ingredient to treats a single ailment. This practice is more likely to cause side effects than the gentler and less concentrated phytochemicals in traditional herbal medicines.[362,363]

In recent scientific investigations conducted in China, active ingredients have been isolated from herbal preparations. Many studies have focused on the effects of active ingredients both *in vivo* and *in vitro*, and on providing pharmacological data compatible with the modern scientific view.[373] However, it has been suggested that the quality of trials needs substantial improvement in order to promotes evidence-based decision making,[377] and frequently it has not been determined whether actions of isolated compounds shown *in vitro* or in animal studies would be relevant to the doses of herbal medication used in clinical practice.[373] More systematic analysis and testing of Chinese herbs are needed for the development of a standard set of therapeutic agents that may be administered with reliable efficacy and good quality control.[366,378]

In order for Chinese herbal medicine to be accepted as reliable alternative medicine, the safety of medicinal herbs and their efficacy for the treatment of specific diseases must be demonstrated.[368,385,397,400] A first step is establishing reliable sources of ingredients. In addition, the problems of adulteration, contamination, and toxicity must be overcome before Chinese herbal medicine can be accepted as a major medical system in the West. Chinese herbal medications have been reported to be accidentally, or sometimes deliberately, contaminated with heavy metals and/or conventional drugs.[383,392,393] Eleven cases of liver damage were reported following the use of Chinese herbal medicines for skin conditions.[365] One of the herbs used in a weight-reducing pill (*Stephania tetrandra*) was inadvertently replaced in a manufacturing error by Aristolochia fangchi, which is nephrotoxic and carcinogenic.[370,371] In another herbal prescription, *Stephania tetrandra* was incorrectly substituted with *Aristolochia westlandi*, which contains nephrotoxins and Aristolochia acids. Its use caused more than 100 cases of kidney nephropathy.[374] Many herbs used for common purposes contain potentially toxic ingredients and overdoses can also cause problems. An herb commonly used for its anti-inflammatory properties, *Aconitine carmichaeli*, contains aconitine, which causes neurological and cardiac toxicity.[375] The root of licorice (*Glycyrrhiza uralensis*), used in many preparations, is considered safe; however, it contains glycyrrhizic and glycyrrhetenic acids, and large doses can cause hypokalemia and sodium and water retention.[375] *Ginkgo biloba* extract can inhibit platelet aggregation and sometimes cause spontaneous hemorrhaging.[375]

As herbal remedies grow in popularity, it becomes increasingly important to understand potential interactions between herbs and prescription drugs. Many herbs have powerful effects that may be increased or counteracted by pharmaceutical drugs and vice versa.[399] This is equally important to Chinese herbs. However, a major handicap is the lack of sufficient knowledge of chemical components involved in Chinese herbal preparations.

Negative media reports on medicinal use of Chinese herbs have attracted a great deal of attention, especially because the use of these herbs is relatively new to the West. Adulteration, contamination, and toxicity have been found from time to time in medicinal herbs from many parts of the world. However, herbal medicine is still considered comparatively safe. In early 2000, Dr. James Duke of the U.S. Department of Agriculture noted that one quarter to one half of all Americans take herbs or herbal supplements, but only about 40 Americans died from them in the prior year, whereas prescription drugs kill 80,000 to 120,000 people annually.[362] Out of 1,701 patients admitted to a Hong Kong hospital, only three were admitted because of adverse reactions to Chinese herbal drugs.[376] Four percent of 2,695 patients admitted to a Taiwanese hospital had drug-related problems. Herbal medicines ranked third among the categories of medicines responsible for causing adverse effects.[376]

Strong regulations and precise quality control are the best measures for monitoring herbs on the market, especially imported Chinese herbs, and can detect substitution, heavy metal contaminations, and illegally added prescription drugs. In addition, the systematic analysis and testing of Chinese herbs may lead to a greater understanding of the biologically active chemical components that are responsible for the claimed therapeutic values. The level of active ingredients has been used as a standard or marker for the quality of raw plant materials and value-added products in the West.[95] This is an important approach and should be applied to Chinese herbs, since each plant species or variety produces different chemical compounds, with varying medicinal values.

Table 1A Major Constituents and Therapeutic Values of Chinese Medicinal Herbs

Scientific Name	Common Chinese and (English) Name	Major Constituents and (sources)	Therapeutic Values*
Abrus precatorius L.	Siang Si Zi (Prayer beads)	(seed) l-Abrine, precatorine, hypaphorine, cycloarternol, squalene, trigonelline,5-β-cholanic acid.[33,450]	Antiemetic, expectorant, parasiticide.
Abutilon theophrasti Malv. *A. avicennae* Gaertn. Fruct. Sem.	Gou Ma (flowering maple)	(aerial part) Rutin, pentose, pentosan, methylpentosan, uronic acid, methypentose, oil, protein.[48]	Treats dysentery, fevers, a diuretic.
Acacia catechu Willd.	Er Cha (Catechu, Jerusalem thorn)	(peeled branch) d-Catechin, catechutanic acid, epicatechin, gambir-fluorescein, gambirine, mitraphylline, tannin, roxburghine D.[33,450]	Promotes salivation, resolve phlegm, stop bleeding, treats pyogenic infections.
Acacia confusa Merrili	Xiang Si Shu (Black cutch)	(bark) Amino acids.[55]	Externally to stop bleeding, treats snake bite.
Acacia nemu Willd. (Syn. *Albizia julibrissin*)	He Huan Pi (Mimosa)	(bark) Tannins, saponins.[49]	Tonic, stimulant, anthelmintic.
Acalypha australis L.	Tie Xian Cai (Copper leaf)	(aerial part) Acalyphine, tannic acid, gallic acid.[33]	Antibacterial, antiasthmatic, antipyretic, detoxicant, antidysenteric, hemostatic.
Acalypha farnesiana Willd. *A. indica* L.	Jin He Huan Indian Ren Xian (Wild copper leaf)	(whole plant) Acalyphine.[55]	Diuretic, treats diarrhea.

Plant	Name	Constituents	Uses
Acanthopanax gracilistylus Harms. *A. giraldii* (Harms.) Nakai *A. spinosum* Miq.	Wu Jia Pi (Thorny catalpa)	(stem, bark) 4-Methoxysalicylaldehyde, vitamin A, beta-sitosterol, arachic, linoleic acid, essential oil, palmitic acid, diterpene, tannic acid, calcium oxalate, polysaccharides, 6-isoinosine, syringaresinol, diglucoside, l-hexacosene, d-sesamine, triterpene glycosides.[45,50,433,444,481,482,485]	For anodyne, arthritis, backache, beriberi, carminative, antitumor, antipyretic effect, suppressive effect on human lymphocytes *in vitro*, anti-inflammatory
Acanthopanax senticosus (Rupr. et Maxim.) Harms. *A. senticosus* (Rupr. et Maxim.) var. subinermis (Regel) Kitag.	Ci Wu Jia Shao Ci Wu Jia (Siberian gir-seng)	See *Eleutherocossus senticosus*	
Acanthopanax sessiliflorus (Rupr. et Maxim.) Seem.	Duan Geng Wu Jia	(root, bark) l-Sesamin, savinin, acanthosides, syringaresinol, daucosterin, daucosterol.[48]	Diuretic, antiinflammatory.
Acanthopanax trifoliatus (L.) Merr.	San Ye Wu Jia	(leaf) Taraxerol.[54]	Treats cold, cough, neuralgia, rheumatism.
Achillea alpina L. *A. millefolium* L.	Shi Cao (Siberian yarrow)	(aerial part) Alkaloids, essential oils, achillin, flavonoides, betonicine, achilleine, d-camphor, oxalic acids, ether oils, hydroxycinnamic acids, hydrocyanic acids, hydroxybenzoic acids, anthocyanidines, anthraquinones, phytosterines, carotene, coumarins, monoterpene, sesquiterpene glucosides, desacetylmatricarin.[33,222,450,549,568]	Antibacterial; treats menopause, abdominal pain, acute intestinal disorder, wound infection, snakebite.

Table 1A   Major Constituents and Therapeutic Values of Chinese Medicinal Herbs   (continued)

Scientific Name	Common Chinese and (English) Name	Major Constituents and (sources)	Therapeutic Values*
Achyranthes asperia L. var. *indica* L.	Tu Niu Teng (Prickly chaff flower)	(seed) Beta-carotene, thiamine, riboflavin, niacin, saponins, ascorbic acid, protein.[50]	Antispasmodic, diuretic; induces labor, antifertility, anti-inflammatory.
Achyranthes bidentata L.	Huai Niu Teng (Long leaf chaff flower)	(root) Inokosterone, ecdysterone, polysaccharides.[33]	Anticancer.
Achyranthes japonica (Miq.) Nakai	Japan Niu Teng (Japanese chaff flower)	(leaf, root) Calcium oxalate, saponin, oleaolic acid, ecdysterone, inokosterone.[50]	Antirhymatic, anodyne; treats amenorrhea, carbuncles, fever, dystocia, and urinary ailments.
Aconitum balfouri Stapf. *A. carmichaelii* Debeaux *A. chasmanthum* Stapf. *A. deinorrhizum* Stapf. *A. fischeri* Reichb. *A. jaluense* Kom. F. glabrescene (Nakai) Kitag. *A. koreanum* R. Raymund *A. napellus* L. *A. praeparata* Stapf. *A. volubile* Pall. ex Koelle var. oligotrichum (DC) Kitag.	Fu Zi, Wu Tao (Wolfsbane)	(root) Aconitine, hypaconitine, pseudoaconitine, mesaconitine, talatisamine.[33,14,262,450,575] This herb is highly toxic.	Cardiotonic, antinociceptive, anti-inflammatory, analgesic effects.
Aconitum barbatum Persoon *A. austroyunnanense* W. T. Wang.	Xue Shang Yi Zhi Hao	(root) Bullatines, aconitine, talatisamine, vilmorrianines, isotalatizidine.[33,270]	Analgesic effect; relieves pain, activate blood circulation, reduces swelling; curative effect on rheumatism, apoplexy, palsy, fracture.

Aconitum laciniatum Stapf. A. *kusnezoffii* Reichenbach A. *chinense* Paxt. A. *vilmorinianum* Kom. A. *pariculigerum* Nakai	Cao Wu	(root) Hypaconitine, aconitine, aconine, mesaconitine, talatisamine. This herb is highly toxic.[33]	Analgesic, sedative, vagal-stimulation, local anesthetic effect.
Acorus calamus L. var. angustatus Besser A. *gramineus* Ait. A. *tatarinowii* L.	Chang Pu (Sweet flag)	(leaf, root) Acoric acid, beta-asarone, yellow bitter aromatic volatile oil, alpha-pinene, d-camphene, calamene, calamenol, calamenone.[5],[357],[450],[568]	Anticonvulsant, analgesic, aphrodisiac, carminative, contraceptive, dessicant, diaphoretic.
Acronychia pedunculata (L.) Miquel A. *laurifolia* Blume	Jiang Zhen Xiang Sha Tong Mu	(stem) Actronycine, bauerenol, nitroacronycine.[50]	Treats bleeding and pain, heart disease.
Actaea asiatica Hara.	Lai Ye Sheng Ma (Asian baneberry)	(aerial part) *trans*-Aconitic acid. This herb is toxic.[51]	A prophylactic against pestilence, malaria, evil miasma.
Actinidia arguta (Sieb. et Zucc.) Planch ex Miq. A. *chinensis* Planch. A. *japonica* Nakai A. *kolomikta* (Maxim. ex Rupr.) Maxim. A. *polygama* (Sieb. et Zucc.) Planch. ex Maxim.	Mi Hou Tao (Kiwi)	(whole plant) Matatabic acid, iridomyrmecin, actinidine, allomatatabiol, iridomyrmecin, neo-nepetalactone, dihydronepetalactol, matatabiether, isoneomatatabiol, matatabistic acid, neomatabiol, vitamin C, vitamin B.[48],[50],[52],[571]	For esophageal and liver cancers, rheumatoid arthritis, arthralgia, urinary stones, fever.
Adamia chinensis Gard. et Champ. A. *cyanea* Wall. A. *versicolof* Fortune (Syn. *Dichroa febrifuga*)	Chang Shan (Chinese quinine, fever flower)	(root) Alpha-dichroine, beta-dichroine, gamma-dichroine.[49] This herb is toxic.	Antimalarial, antipyretic.

Table 1A   Major Constituents and Therapeutic Values of Chinese Medicinal Herbs   (continued)

Scientific Name	Common Chinese and (English) Name	Major Constituents and (sources)	Therapeutic Values*
Adenophora coronopifolia Fisch. A. *paniculata* Nannf. A. *pereskiaefolia* (Fisch.) G. Don A. *polymorpha* Ledeb. A. *remotiflora* (Sieb. et Zucc.) Miq. A. *stenanthina* (Ledeb.) Kitag. A. *tetraphylla* Mak.	Sha Seng (Bluebell)	(root) Saponins.[33]	Hemolyzes blood cells; stimulates smyocardial contraction; antibacterial.
Adenophora triphylla (Thunb.) DC A. *verticillata* Fisch.	Che Ye Sha Seng Lun Ye Sha Seng (Bellflower)	(root) Inulin, taraxerone, beta-sitosterol, daucosterol, beta-sitosteryl palmitate, lupenone.[53]	Antidotal, aphrodisiac, demulcent, expectorant, restorative, sialogogue, tonic.
Adiantum boreale Presl. A. *capillus-junonis* Rupr. A. *pedatum* L. A. *flabellulatum* L.	Tie Xian Jiu (Black maidenhair) Guo Tan Loan (Maidenhair fern)	(root) Adipedatol, adiantone, hopadiene, isoadiantone, isofernene, fernene, gamma-fernene, filicene, filicenal, fernadiene.[48]	Treats cold and grippe.
Adina rubella Hance. A. *ratemosa* (Sieb. et Zucc.) Miquel	Shui Yang Mei Gen Shui Tuan Hua (Reddish modelwood)	(root, flower) Neucleoside, beta-sitosterol, noreugenin, quinoric acid, saponin, betulinic acid, morolic acid, cincholic acid, stigmasterol.[58]	Astringent, carminative, for dysentery, enteritis, hemorrhage, anticancer.

Adonis brevistyla Franch. A. *chrysocyathus* Hook F. & T. Thoms. A. *vernalis* L.	Fu Shou Cao (Amur adonis)	(aerial part) Cymarol, adonilide, pergularin, corchoroside A, convallatoxin, isoramanone.[33] This herb is toxic.	Treats heart disease and depression; diuretic.
Aesculus chinensis L. A. *indica* Colebr. A. *hippocastanum* L.	Sha Lou Zi (Horse chestnut)	(ripe fruit) Protoescigenine, escigenin, aescine, flavonoid glycosides, aesculine, albumin, fatty oils, amylose, oligosaccharides.[33,450,568]	Promotes circulation, relieves epigastrium pain, promotes digestion.
Agastache rugosa (Fisch. & Mey.) O. Kuntze A. *rugosa* (Fisch. & Mey.) O. Kuntze f. hypoleuca (Maxim.) Hara	Huo Xiang (Chinese giant hyssop)	(leaf) Essential oils, methylchavicol, anethole, anisaldehyde, d-limonene, hexenol, calamene, caryophyllene, p-methoxycinnamaldehyde, d-pinene, beta-pinene, octanol, cymene, linalool, elemene, farnesene.[48,306,558]	Chest congestion, diarrhea, headache, nausea, antipyretic, carminative, febrifuge, stomachic.
Ageratum conyzoides L. A. *houstonianum* Mill.	Sheng Hong Yu (Bastard agrimony)	(leaf, root) Cyanogenic glucoside, coumarin, agerato-chromene, 7-methoxy-2,2-dimethylchromene, beta-caryophyllene.[50]	For digestive disorder, fever, rheumatism, gonorrhea, tetanus, syphilis.

Table 1A Major Constituents and Therapeutic Values of Chinese Medicinal Herbs (continued)

Scientific Name	Common Chinese and (English) Name	Major Constituents and (sources)	Therapeutic Values*
Agrimonia eupatoria L. *A. pilosa* Ledeb. *A. pilosa* Ledeb. var. japonica (Miq.) Nakai *A. pilosa* Ledeb. var. simplex T. Shimizu *A. pilosa* Ledeb. var. viscidula (Bunge) Kom. *A. viscidula* Bunge.	Loan Mao Cao, Xian He Cao (Agrimony)	(whole plant) Agrimophol, agrimols, agrimonine, agrimonolide, cosmosiin, vitamin C, luteolin-7-β-D-glucoside, apigenin-7-β-D-glucoside, vitamin K, tannins, catechin derivatives.[33,48,49,263,568]	Astringent hemostatic in enterorrhagia, hematuria, metrorrhagia, gastrorrhagia, pulmonary, tuberculosis. A cardiotonic, antihemorrhagic, anthelmintic, anti-inflammatory, antimicrobial.
Ailanthus altissima (Mill.) Swingle	Chun Pi (Stinking cedar)	(root, stem, bark) Amarolide, ailanthone, afzelin, syringic acid, vanillic acid, beta-sitosterol, azelaic acid, d-mannitol, amarolide, oleoresin, mucilage.[33,48]	Antidiarrheal; treats dysentery, duodenol ulcers. Astringent, anthelmintic.
Ajuga bracteosa Wallich *A. decumbens* Thunb. *A. pygmaea* A. Gray	Jin Gu Cao Jin Chuang Xian Cao (Bugleweed)	(whole plant) Flavone glucoside, luteolin, tannin, ecdysones cyasterone, ecdysterone, ajugalactone, ajugasterone, beta-sitosterol, ajugasterone, cerotic acid, y-sitosterol, palmitic acid.[33,50,450]	Antitussive, antipyretic, antiphlogistic, antibacterial. Treats bladder ailments, diarrhea, bronchitis, a tonic; stimulant, diuretic.
Akebia quinata (Hoytt.) Decne.	Moo Tune (Chocolate vine)	See *Clematis armandii*	

Species	Common name	Constituents (part)	Activity
Alangium chinense (Lour.) Harms.	Ba Jiao Feng	(root) dl-Anabasine.[33]	Causes myocardial stimulation, increases contractility; may cause fibrillation and increases blood pressure.
Alangium lamarckii Lour.		dl-Anabasine, cephaeline, emetine, psychotrine, tubulosine, ankorine isotubulosine, demethyl-tubulosine, demethylpsychotrin, alangicine, deoxytubulosine, alangimarckine, alamarchine, demethylcephaeline.[33,558]	
Albizia julibrissin Duraz. A. lebbeck (L.) Bentham	Hu Hua Pi (Mimosa tree)	See Acacia nemu	
Aletris formosuna (Hayata) Sasaki A. spicata Franch	Fei Jin Cao (Chinese stargrass)	(root) Stigmasterol, beta-sitosterol, diosgenin.[54]	Antitussive, vermifugal, for ascariasis, marasmus, cough.
Aleurites fordii Hemsl.	You Tong (Candlenut)	(fruit, aerial part, seed) Saponin, alpha-elaeo stearic, oleic acid, palmitic acid, stearic acid, tannins, phytosterols, n-hentriacontane, alpha-amyrin, beta-amyrin, stigmasterol, beta-sitosterolm, campesterol.[50,219]	Analgesic activity. Treats anemia, atrophy, edema, vermicide, oil (toxic internally) for parasitic skin diseases.
Aleurites moluceanu (L.) Willd.	Shi Li (China wood oil)	(bark, seed) Protein, carotene, thiamine.[50] This herb is toxic.	As poultice for fever, headache, swollen joints, and ulcer.
Alisma cordifolia Thunb. A. orientalis (Sam.) Juzep. A. plantago L. A. plantago-aquatica L.	Ze Xie (Water plantain)	(stem, root) Alisol A, alisol B, polysaccharide, alisol monoacetate, sesquiterpenes, triterpenes, glucan, epialisol A (essential oil).[33,451,452,463,464]	Lowers hypercholesteremia, treats hypertriglyceride, immunologic activities; anticomplementary, antiallergic.

Table 1A Major Constituents and Therapeutic Values of Chinese Medicinal Herbs   (continued)

Scientific Name	Common Chinese and (English) Name	Major Constituents and (sources)	Therapeutic Values*
Allamanda cathatica L.	Yuan Xi Huang San	(whole plant) Allamandin.[33]	Treats P-388 leukemia.
Allium chinense Max. A. *odorum* L. A. *sativum* L. A. *tuberosum* Roxb. A. *uliginosum* G. Don	Da Suan (Garlic)	(bulb) Allicin, allistatin, glucominol, neo-allicin, steroidal saponins, polysaccharides, furostanol saponins, proto-isoerubosides, diallyl sulfide.[33,49,438,490,510,569]	Antibacterial, antimutagenic, anticarcinogenesis, carminative, antiarrhythmic; lowers plasma cholesterol and low-density lipoproteins, prevents thrombosis, hypotensive, and vessel protective effect.
Allium fistulusum L. A. *macrostemon* Bunge. A. *tartaricum* Ait.	Jiu Bai (Scallion) Cong (Green onion) Jiu Cai (Scallion)	See *Allium chinense*	
Allium victorialis L. var. platyphyllum (Hult.) Makino	Ge Cong (Serpent garlic)	(whole plant) Methyl allyl disulfide, dially disulfide, methyl allyltrisulfide, l-propenyl sulfonic acid, methyl-l-propenyl disulfide, allyl-l-propenyl disulfide.[48,50]	A diuretic, vermifuge; treats cold.
Alnus japonica (Thunb.) Steudel A. *japonica* (Thunb.) Steudel var. koreana Callier	Ce Yan (Japanese alder)	(leaf, bark) Alpha-amyrin, betulinic acid, glutin-5-en-3-ol, heptacosane, lupenone, taraxerol.[48,50]	Antitumor.
Aloe barbadensis Miller var. chinensis Berger A. *vera* L.	Lu Wen (Aloe)	(aerial part) Aloins, barbaloin, aloe-emodin, polysaccharides.[39b,49,50,108,109,403,450,510,568,56 9]	Laxative, stomachic, emmenagogue.

Alpinia japonica Miq.	Yue Tao (Japanese ginger lily)	(seed) Essential oils, cineole, alpinone, izalpinin, rhamnocitrin, kumatakinin.[56]	Carminative.
Alpinia katsumadai Hayata A. globosum Horan. A. kumatake Mak.	Dou Kou Pi Jiang (Greater galangal)	(whole plant) Kaempferin, galangin, galangol, cineole, citral, carotene, thiamine, riboflavin.[50]	Carminative, stomachic; treats malarial disorders, fluxes, and menstruation.
Alpinia officinarum Hance	Gao Liang Jiang (Lesser galangal)	(rhizome) Galangol, essential oils, cineol, eugenol, pinene, cadinene, methyl cinnamate, sesquiterpene, dioxyflavonol.[49]	As stomachic in chronic enteritis, dyspepsia and gastralgia, carminative, antiperiodic, sialogogue.
Alpinia oxyphylla Miq.	Yi Zhi (Chinese lily ginger)	(fruit) Cincole, zingiberene, zingiberol.[58]	Diuretic, tonic; treats vomiting, and digestive discomfort.
Alpinia speciosa K. Schum.	Shan Jiang (Ginger)	(seed) Zingiberene, zingiberol.[54]	Stomachic.
Alstonia scholaris (L.) R. Br.	Deng Tai Ye (Dita bark)	(leaf) Picrinine, picralinal, echitamine, echitamidine.[33,579,580]	An expectorant, antiphlogistic.
Alternanthera philoxeroides (Mart.) Griseb. A. sessilis (L.) R. Brown	Kong Xin Lian Zi Cao Man Ti Xian	(aerial part) Saponin, coumarin, tannins, falvins.[33]	Treats viral infections, measles, hemorrhagic fever, toxic and icteric hepatitis.
Althaea rosea (L.) Cav.	Shu Kui (Hollyhock)	(shoot, root, seed) Althaeine, dioxybenzoic acid.[50]	As stomachic, regulative, constructive in fevers, dysentery, diuretic.
Amaranthus caudatus L.	Wei Sui Xian (Amaranth)	(leaf) Betaine.[48]	A tonic.

Table 1A Major Constituents and Therapeutic Values of Chinese Medicinal Herbs   (continued)

Scientific Name	Common Chinese and (English) Name	Major Constituents and (sources)	Therapeutic Values*
Amaranthus lividus L. A. blitum Kom. A. viridis L.	Lu Xian (Strawberry blite)	(leaf) Vitamins, protein, thiamine, riboflavin, ascorbic acid.[50]	Treats dysentery and inflammation; vermifuge.
Amaranthus paniculatus L.	Fan Sui Xian	(leaf) Betaine.[48]	Stops bleeding, relieves pain; externally for wounds, broken bones.
Amaranthus tricolor L.	San Se Xian (Jacob's coat, Chinese amaranth)	(leaf) Beta-carotene, thiamine, riboflavin, niacin, ascorbic acid.[50]	Prevent cancer.
Amomum cardamomum L. A. globosum Lour. A. tsao-ko Roxb. A. villosum L. A. xanthioides Wall.	Bai Dou Ku Cao Guo Shan Ren (Siam cardamon, Chinese cardamon)	(seed) d-Borneol, borneol acetate, d-camphor, linalool, nerolidol, terpene.[50]	Treats pyrosis, vomiting, dyspepsia, pulmonary diseases. Antitoxic, antiemetic, carminative, stomachic.
Amorphophallus rivieri Durieu	Mo Yue	(whole plant) Leviduline, levidulinase, mannose.[50] This herb is toxic.	Treats aching bones, eye inflammation, cancer, ulcers, snakebite.
Ampelopsis aconitifolia Bunge.	Cao Bai Ching	(root bark, stem, leaf) Flavonoids, glucosides, amino acids.[48]	Externally as an antiseptic for swollen abscesses.
Ampelopsis brevipedunculata (Maxim.) Trautv.	Ye Pu Tao Teng (Snake grape)		Antitoxic; relieves pain and bleeding; treats arthritis.

Ampelopsis japonica (Thunb.) Mak. *A. bodinieri* (Levl. & Vant.) Rehd. *A. contonensis* (Hook & Arn.) Planch. *A. humulifolia* Bunge.	Bai Lian (Peppervine)	(root bark, stem, leaf) Flavonoids, glucosides, amino acids.[48,60]	Anodyne, astringent anticonvulsive, detoxicant, treats tubercular cervical nodes, hemorrhoidal bleeding. Treats pain of rheumatism.
Anagalis arvensis L.	Hi Lu	(root) Anagalline, anagalligenone, arrenin, cucurbitacins.[55]	Treats snakebite, dog bite; antitoxic.
Ananas comosus (L.) Merrill	Feng Li	(leaf) Ergosterol peroxide, ananasic acid,5-stigmautena-3β,7d-diol, 3,4-dihydroxycinnamic acid, 4-hydroxycinnamic acid, bromelin, vitamins.[57]	Antioxidant activity, for digestion; lowers blood pressure, anticancer.
Andrographis paniculata (Burm. f.) Nees	Chuan Xin Lan (Creat)	(aerial part) Deoxyandrograppholide, andrographolide, neoandrcgrapholide, dehydroandrographolide.[33]	Antibacterial, antipyretic, anti-inflammatory.
Anemarrhena asphodeloides Bunge.	Zhi Mu	(rhizome) Steroidal saponins, mangiferin, isomangiferin, sarsasapogenin, markogenin, neogitogenin.[10,11,33]	Antipyretic, anti-inflammatory, sedative, antibacterial.
Anemone cernua Thunb. *A. pulsatilla* *A. pulsatilla* var. chinensis Bunge.	Bai Tu Own (Pulsatila)	(root) Saponins, protoanemonin.[49]	A cardiac and nervous system sedative, antispasmodic, anodyne in asthma and pulmonary infections, antidiarrheic.

Table 1A Major Constituents and Therapeutic Values of Chinese Medicinal Herbs (continued)

Scientific Name	Common Chinese and (English) Name	Major Constituents and (sources)	Therapeutic Values*
Anemone raddeana Regel *A. rivularis* Buch-Hamilton ex DC *A. rivularis* Buch-Hamilton ex DC var. flore-minore Maxim. *A. vitifolia* (Buch. Ham.) Nakai	Yin Lian Hua, Liang Tao Jian Cao Yu Mei Ye Mian Hua (Anemone)	(rhizome) Raddeanin A, hederasaponin B, raddanoside, ranuneulin, oleanolic acid. [33,48,568]	Antitumor, anti-inflammatory, antirheumatic arthritis.
Anethum graveoleus L.	Shi Luo (Dill)	(fruit, young shoot) Essential oils, d-carvone, dillapiole, limonene, bergapten, umbelliprenin, camphene, dihydrocarvone, dillapiole, dipentene, isomyristicin. [48,50]	Carminative, stimulant.
Angelica amurensis Schischk. *A. anomala* Lallem. *A. dahurica* (Fisch.) Benth. et Hook.	Bai Zhi (Angelica)	(root) Byak-angelicin, byak-angelicol, oxypeucedanine, imperatorin, phellopterin, xanthotoxine, marmesin, scopoletin, marmesin, anomalin, angenomalin, bergapten, imperatoin, pabulenol, isoimperatoin, oxypeucedanin, neobyakangelicol. [33,486]	Antipyretic; treats toothaches, headache; antitumor. Externally for mastitis and wound infection.

Angelica decursiva (Miq.) Franch. et Savat.	Qian Hu	(root) Nodakenin, nodakenetin, decursin, decursidin, umbelliferone, andelin, 3'-angeloyloxy-4'-isovaleroyloxy-3', 4'-dihydroxanthyletin, estragol, umbelliprenin, imperatorin, sioimperatorin, spongesterol, hydroxypeucedanin, decuroside, estragol, spongesterol.[48]	Anodyne, carminative, diuretic, stimulant, suppurative. Treats abscesses, boils, catarrh, cold, coryza, dysmenorrhea, epistaxis, fever.
Angelica grosserrata Maxim.	Fu Shen	(root) Angelic, linoleic, oleic, palmitic, stearic acids.[50]	Antispasmodic, diaphoretic, diuretic. Treats apoplexy, swellings, catarrh, dropsy, headache, leprosy, puerperium.
Angelica polymorpha Max. A. sinensis (Oliv.) Diels	Dan Gui	(root) Vitamin B_{12}, vitamin E, ferulic acid, succinic acid, nicotinic acid, uracil, adenine, butylidenephalide, ligustilide, folinic acid, biotin, polysaccharide.[33,380]	Treats irregular menstruation, anemia, thrombophlebitis, neuralgia, arthritis, chronic nephritis, constrictive aortitis, skin disease such as eczematous dermatitis.
Angelica pubescens Maxim.	Du Huo	(root) Coumarins, bergapten, glabralactone, osthol, angelol, angelic acid, angelicotoxin, byak-angelicin, byak-angelicol, tiglic acid, umbelliferone.[50]	For abscesses, arthritis, cold, epistaxis, headache, toothache, hematochezia, hematurai, lumbago, rheumatism.
Anredera cordifolia (Tenore) Van Steen	Yang Lu Kui	(whole plant) 3-hydroxy-30-horoleana-12, 18-dien-29-oate, larreagenin, ethyl ester, ursolic acid.[57]	Treats boils.

Table 1A   Major Constituents and Therapeutic Values of Chinese Medicinal Herbs   (continued)

Scientific Name	Common Chinese and (English) Name	Major Constituents and (sources)	Therapeutic Values*
Anthriscus aemula (Woron.) Schischk. A. *aemula* (Woron.) Schischk. f. hirtifructa (Ohwi) Kitag. A. *sylvestris* (L.) Hoffm.	Wo Seng (Wild caraway)	(root) Anthricin, deoxypodophyllotoxin, isoanthricin, luteolin, oxalic acids, hydroxycinnamic acids, ether oils, hydroxybenzoic acids, coumarins, anthocyanidines, anthraquinones, phytosterines, carotenes, monoterpene, sesquiterpene glucosides, hydrocyanic acids.[50,222]	Antitumor, glandular tumors, corns, warts.
Antiaris toxicaris (Pers.) Lesch.	Jian Xui Fuan Hou	(seed) Alpha-antiarin, alpha-antioside, convallatoxin, bogoroside, strophalloside, peripalloside.[35]	A cardiotonic, emetic, lactogenic, antipyretic. Treats dysentery.
Apium graveolens L.	Qin Cai (Celery)	(whole plant) Apiin, graveobioside A, graveobioside B.[33,568]	Treats hypertension, hypercholesterolemia.
Apocynum venetum L.	Luo Bu Ma	(leaf, root) Cymarin, strophantidin, k-strophanthin-β, isoquercitrin, quercetin.[33]	Increases myocardial contractility, lowers blood pressure; increases bronchial secretion; diuretic.
Aquilaria agallocha Roxb. A. *sinensis* (Lour.) Gilg.	Chen Xiang (Aloe wood)	(stem wood) Agarospirol, alpha-agarofuran, agarol, beta-agarofuran, benzylacetone, hydrocinnamic acid, hydroagarofuran.[33]	Antiemetic; promotes circulation, relieves pain.

Aquilegia buergeriana Sieb. et Zucc. f. pallidiflora (Nakai) Kitab. A. *buergeriana* Sieb. et Zucc. var. oxysepala (Trautv. et Mey.) Kitam. A. *parviflora* Ledeb.	Xue Jian Chou	(whole plant) Benzylacetone, terpene alcohol, p-methoxybenzylacetone.[48,60]	Treats irregular menstruation, ovarian bleeding, shortness of breath, nausea, pain and gas, chills.
Arachis hypogaea L.	Luo Hua (Peanut, groundnut)	(seed) Amino acids, protein, arachine, globulin, biotin, glycyrrhizin, glucosides, thiamine, riboflavin, niacin, carbohydrate.[48,50]	A cemulcent, nutritive, pectoral, peptic. As an emollient, applied externally for rheumatism.
Aralia chinensis L. A. *cordata* Thunb. var. continentalis (Kitag.) Y. C. Chu A. *elata* (Miq.) Seem. A. *elata* (Miq.) Seem. F. subinermis Y. C. Chu	Jia Mu, Du Huo (Aralia)	(root) Diterpenoids such as (−) pimaradene, (−) kaurene derivatives, l-pimara-8, 15-dien-19-oic acid, aralosides, araligenin, oleanolic acid, beta-taralin, alpha-taralin.[20,48,50]	Carminative, for arthralgia, gastroenteritis, headache, diuretic, antidiabetic, antiseptic.
Araucaria cunninghamii Aitonex Sweet	Na Yang Shan	(shoot) Methyl communate, methyl isocupressate, methyl acetyl-isocupressate, labdadien, diacetate, methyl amentoflavone.[57]	Treats skin diseases.

Table 1A   Major Constituents and Therapeutic Values of Chinese Medicinal Herbs   (continued)

Scientific Name	Common Chinese and (English) Name	Major Constituents and (sources)	Therapeutic Values*
Arctium lappa L.	Niu Bang Chi (Burdock)	(fruit) Arctin, arctigenin, matai-resinol, sesquilignins, stereoisomer, inulin, mucilage, pectin, acetic, butyric, caffeic, chlorogenic, lauric, linoleic, oleic, palmitic, propionic, stearic, tiglic acids, lignans (lappaol).[1,9,450,487,488,489,568]	For dermatitis, tumors, diuretic and arexigenic properties; treats breast cancer, nephritis, antidote, diuretic, antibacterial, anti-inflammatory; relieves sore throat.
Ardisia japonica (Hornst.) Blume	Ai Di Cha or Pin Di Mu (Marlberry)	(whole plant) Bergenine glucoside, essential oil.[33,558]	Antitussive, antiphlegm; promotes blood circulation, hemostatic.
Ardisia quinquegona (Blume) Nakai *A. sieboldii* Miq.	Zhi Jin Niu Shu Gi (Spiceberry)	(leaf, root) Bergenin.[50]	Treats cancer, hepatoma, a diuretic, antidote for poison, antiphlegmatic.
Areca catechu L. *A. hortonsis* Lour.	Bing Lang (Betel nut palm)	(nut) Arecholine, arecholidine, guvacoline, guvacine.[33]	Treats taeniasis.
Arenaria juncea Bieb. *A. juncea* Bieb. var. abbreviata Kitag *A. juncea* Bieb. var. glabra Regel *A. serpyllifolia* L.	Zao Zhui (Thyme-leaved sandwort)	(aerial part) Saponin.[50]	Antitussive, detoxicant, diuretic, febrifuge; treats cough, pulmonary tuberculosis, dysentery.
Arethusa japonica A. Gr.	Ze Lan	(aerial part) Essential oils, tannins.[49]	Diuretic, emmenagogue.

Species	Common name	Constituents	Uses
Arisaema amurense Maxim. A. amurense Maxim. f. purpureum (Nakai) Kitag. A. amurense Maxim. f. serratum (Nakai) Kitag. A. amurense Maxim. f. violaceum (Engler) Kitag. A. consanguineum Mart. A. erubescens (Wall.) Schott. A. heterophyllum Blume A. peninsulae Nakai A. peninsulae Y. C. Chu et D. C. Wu A. thunbergii Blume.	Tian Nan Xing (Arum, serrated arum)	(whole plant) Alkaloids, saponin, benzoic acid.[33,49,144] This herb is highly toxic.	Treats tetanus, spasms, epilepsy, neuralgia. It is a sedative, anticonvulsive, and expectorant.
Aristolochia contorta Bunge. A. kaempferi Willd. A. longa Thunb. A. recurvilabra Hance	Ma Dou Ling	(stem) Aristolochic acid A, aristolochic acid D, aristoloside, magnoflorine, oleanolic acid, beta-sitosterol, hederagenin.[48] This herb is toxic.	Treats pulmonary disorders, antitussive; an expectorant in asthma and bronchitis.
Aristolochia debilis Sieb. et Zucc.	Qing Mu Xiang	(root) Aristolochic acid, debilic acid, magnoflorine, dibilone, cyclanoline, aristolone.[33]	Antihypertensive; lowers heart rate and myocardial contractility, vasodilatation.
Aristolochia manshuriensis Kom.	Mu Tong	See Clematis armandii	
Aristolochia shimadai Hayata	Taiwan Ma Dou Ling	(stem) Aristolochic acid.[54]	Relieves pain; a diuretic, externally for snakebite.

Table 1A   Major Constituents and Therapeutic Values of Chinese Medicinal Herbs   (continued)

Scientific Name	Common Chinese and (English) Name	Major Constituents and (sources)	Therapeutic Values*
Armeniaca ansu (Maxim.) Kostina *A. mandshurica* (Maxim.) Skvortzov *A. sibirica* (L.) Lam. *A. vulgaris* Lam. (Syn. *Prunus armeniaca*)	Xian (Apricot)	(seed) Amygdalin, hydrocyanic acid.[48,49]	Astringent, stomachic, antipyretic.
Arnebia euchroma Forssk.	Zi Cao	(root) Shikonin, acetylshikonin, beta-beta-dimethylacrytoylshikonin, beta-OH-isovalerylshikonin, alkamin-B, beta-di-Me-acrylate.[33,450]	Anti-inflammatory, antiseptic, antibacterial, toothache, eye diseases; a healer of cuts, burns, and wounds.
Artemisia annua L. *A. apiacea* Hance ex Walpers	Qing Guo (Stinking artemisia)	(aerial part) Dihydroartemisinin, artesunate, artemisinin, chloroquine, flavonoids, sesquiterpene.[33,269,476] This herb is mildly toxic.	A schizonticidal agent, antimalarial; treats infections of multidrug-resistant strains of *Plasmodium falciparum*, the cause of human malignant cerebral malaria.

Artemisia argyi Leveille et Variot *A. argyi* Leveille et Variot f. eximia Pamp *A. argyi* Leveille & Variot f. gracilis (Pamp.) Kitag. *A. halodendron* Turez. ex Bess. *A. igniaria* Max. *A. indica* Willd. *A. integrifolia* L. *A. japonica* Thunb. *A. japonica* Thunb. var. manshurica (Kom.) Kitag. *A. keiskeana* Miq. *A. lagocephala* Fisch. ex Bess. *A. lavandulaefolia* DC *A. scoparia* Waldst. & Kitaib. *A. selengensis* Turcz. ex Bess. *A. sieversiana* Ehrh. ex Willd. *A. vulgarts* L.	Ai Ye, Ai Ye You (Artemisia) (Japanese artemisia) (Cottage thatch) (Mugwort)	(aerial part or aerial part oil) Terpinenol-4, β-caryophyllene, artemisia alcohol, linalool, cineol, camphore, borneol, eucalyptol.[33]	Antiasthmatic, antitussive. Treats chronic bronchitis, oral infection, and hypersensitivity.
Artemisia brachyloba Franch.	Shan Guo (Wormwood)	(whole plant) Essential oils, pinene, cinecle, terpene, artemisine, tannins, adenine.[88,394,395] This herb is classified as dangerous by the FDA.[391]	Treats migraine, throat discomfort, malaria.
Artemisia capillaris Thunb.	Yin Chen (Evergreen artemisia)	(shoot) Scoparon, capillene, capillin, capillon, capillarin, capillanol.[33]	A choleretic; treats jaundice, acute infectious hepatitis, gallstone-related illnesses.
Artemisia finita Kitag. *A. frigida* Willd.	Chang Guo Bai Guo	(flower bud) L-beta-santonin, finitin.[48]	Treats intestinal parasites.

Table 1A Major Constituents and Therapeutic Values of Chinese Medicinal Herbs (continued)

Scientific Name	Common Chinese and (English) Name	Major Constituents and (sources)	Therapeutic Values*
Artemisia gmelini Weber ex Stechmann	Bai Lian Guo (Levant wormseed)	(whole plant) Essential oils, borneol, cineole, camphor, azulene, isovaleric acid, umbelliferone, scopoletin, genkwanin.[48]	Treats liver diseases, stops bleeding, arthritis, bronchitis.
Artemisia lactiflora Wallich	Tian Cai	(whole plant) Flavonoid glycoside, coumarin, lactiflorenol, spathulenol, s-guaiazulene, beta-guaienen, *trans*-β-farnesene, *trans*-caryophyllene, limonene, elemene, copaene, myrcene.[57]	Diuretic; regulates menstruation, treats headache, high blood pressure.
Arthraxon hispidus (Thunb.) Makino	Jin Cao	(root, whole plant) Aconitic acid, luteoline, luteolin-7-glucoside, anthraxin, luteolin-monoarabinoside.[48]	For chronic cough and other infections.
Artocarpus altilis (Park.) Fosberg.	Mian Bao Shu	(bark) Triterpenes, beta-amyrin acetate, lupeol acetate.[60]	Poultice for ulcers.
Artocarpus heterophyllus Lam.	Bo Lo Mi (Jackfruit)	(leaf, seed) Caoutchoue, resin, cerotic acid, protein, minerals.[50]	Tonic to treats discomfort from alcohol influence.
Arundinaria graminifolia (D. Don) Hochrentiner	Zhu Ye Lan	(leaf, root) Sitosterol, stigmasterol, campesterol.[50,57]	Antitussive, tonic, anthelmintic, stomachic, carminative.

Arundo donax L. *A. phragmites* L. (Syn. *Phragmites communis*)	Lu Zhu Lu Gen	(root) Glycosides, protein, asparagin.[49]	Stomachic, antiemetic, antipyretic, in acute arthritis, jaundice, pulmonary abscess, food poisoning.
Asarum canadense L. *A. europaeum* L. *A. heterotropoides* Fr. Schmidt var. mandshuricum (Maxim.) Kitag. *A. heterotropoides* Fr. Schmidt var. seouleuse (Nakai) Kitag. *A. sieboldii* Miq.	Xi Xin	(whole plant) Essential oils including ucarvone, safrole, beta-pinene, asoryl-ketone, asariline, chalcone, flavonol glycoside, *trans*-aconitic acid, phenylpropane derivatives.[33,87,453,454,465,466]	Analgesic, sedative, antipyretic, anti-inflammatory.
Asparagus cochinenesis (Lour.) Merr. *A. falcatus* Benth *A. insularis* Hance *A. lucidus* Lindl. *A. officinalis* L.	Tian Men Dong (Asparagus)	(root) Glycolic acid, asparagome, essential oils, methanethiol, (+)-nyasol, asparagine, steroidal, beta-sitosterol, sarsasapogenin, polysaccharide, diosgenin, oleanene derivatives.[50,450,455,456,467,468]	Diuretic, laxative; treats cancer, antitumor, antioxidative activity, neuritis, rheumatism, for parasitic diseases.
Aspidium falcatum Sw. (Syn. *Dryopteris crassirhizoma*)	Guan Zhong (Wood fern)	(whole plant) Filicic acid, tannins, essential oil.[49] This herb is slightly toxic.	Anthelmintic, hemostatic, antidote.
Aster ageratoides L.	Hong Guan Yao	(whole plant) Quercetin, kaempferol.[33]	Antitussive, antiasthmatic; stimulates adrenal cortex.
Aster tataricus L.	Zhi Wen (Purple aster)	(root) Saponins, shionon, quercetin, arabinose.[49,558]	Antitussive, expectorant.
Astilbe longicarpa (Hay.) Hayata *A. chinensis* (Maxim.) Franch. et Sav.	Luo Xing Fu	(whole plant) Astilbin, bergenin, quercetin, 2-hydroxphenylacetic acid.[53]	Antitoxic, against pestilence, malaria, evil miasma.

Table 1A Major Constituents and Therapeutic Values of Chinese Medicinal Herbs (continued)

Scientific Name	Common Chinese and (English) Name	Major Constituents and (sources)	Therapeutic Values*
Astragalus chinensis L.	Sha Yuan Zi (Vetch)	(seed) Astragalin, canavanine, homoserine.[33,436]	Sedative, antibacterial, antiviral, anticarcinogenic effect.
Astragalus complanatus R. Fr. ex Bunge. A. henryi Oliv. A. hoantchy Franch. A. melliotoides Pallas A. membranaceus (Fisch.) Bunge. A. mongholicus Bunge. A. reflexistipulus Franch. A. sinensis L.	Huang Zhi (Yellow vetch, membranous milk vetch)	(root) Gamma-aminobutyric acid, queretin, astragalin, canavanine, coumarin, flavonoid derivatives, saponins, polysaccharide, cycloastrangenol, betaine, rhamnocitrin, isoflavones, astragalosides, formononetin, homoserine, isoliquiritigenin, cosin, kaempferol.[1,33,53,410,411,439,445,448,510,511,552,603]	Hypotensive, antirhinoviral, antitumor, antipyretic, antioxidant effect, diuretic, tonic, an immuno-moderating agent, treats myelosuppression caused by cancer chemotherapy; treats urological tumors.
Atractylis chinensis DC A. lancea Thunb. A. lyrata Sieb. et Zucc. A. ovata Thunb.	Zhang Shu	(root) Essential oils, atractylone, hinesol, atractylodine, atractylol, beta-eudesmol, diacetyl-atractylodiol.[33] This herb is toxic.	Lowers blood sugar; sedative.
Atractylodes chinensis (Bunge.) Koidz. A. chinensis (Bunge.) Koidz. f. simplicifolia (Loes.) Y. C. Chu A. chinensis (Bunge.) Koidz. var. liaotungensis (Kitag.) Y. C. Chu A. japonica Koidz. ex Kitam. A. koreana (Nakai) Kitam. A. lancen (Thunb.) DC A. macrocephala Koidz. A. ovata DC	Bai Zhu	(root) Atractylone, eudesnol, hinesol, bisesquiterpenoid, biatractylolide.[19,258]	Diuretic agent, abdominal and chest tightness, anemia, chills, bronchial cough, diarrhea, CNS suppressing activity.

Atractylodes lancea Bunge.	Cang Zhu	(root) Essential oils, atractylon, atractylol.[49]	As aromatic tonic in chronic gastroenteritis.
Aucklandia costus Falc. *A. lappa* Decne (Syn. *Saussurea lappa*)	Mu Xiang	(root) Saussurine, costulactone, costol, costene, camphene, phellandrene.[49]	Treats asthma; stomachic.
Avena fatua L.	Ye Yen Me (Oats)	(whole plant) Aminoadipic acid, glucovanillin, trigonelline, leucine, isoleucin, threonine, asparaginic acid, oxylysin, beta-sitosterol, aconitic acid, avenasterol, secalose, erucic acid, xanthophyllepoxyd.[48]	Stops bleeding; a tonic.
Azalea japonica A. Gray *A. mollis* Blume *A. pontica* var. sinensis Lindl.	Yang Zhi Zu (Azalea)	See *Rhododendron sinensis*	
Azolla imbricata (Roxb.) Nakai	Man Jiang Hong	(whole plant) Luteolinidin 5-glucoside, aesculetin, caffeic acid.[57]	Treats cough, arthritis, eczema, swelling; diuretic.
Baphicanthus cusia (Nees.) Bremek.	Ban Lan or Da Qing Ye	(leaf, root) Indirubin, indigo, indo-brown, indo-yellow, isoindigo, lacerol, tryptanthrin.[53]	Antidotal, febrifugal; treats fever, epidemic mumps, erysipelas, rashes, sore throat.
Bauhinia championi Bentham *B. variegata* L.	Jiu Hua Teng (Orchid tree)	(bark, sepal) Kaempferol-3-galactoside, daempferol-3-rutinoside, protein, flavonoids, carbohydrates, stigmasterol, beta-sitosterol, beta-p glucophyranoside.[50,450]	Astringent, tonic; treats scrofula, skin ailments, leprosy, ulcers, and diarrhea.

Table 1A Major Constituents and Therapeutic Values of Chinese Medicinal Herbs (continued)

Scientific Name	Common Chinese and (English) Name	Major Constituents and (sources)	Therapeutic Values*
Belamcanda chinensis (L.) DC *B. panctata* Moench.	She Gan (Blackberry lily, leopard lily)	(root) Tectoridin.[50]	Antipyretic, antifungus, analgesic, detoxicant, stomachic. Externally for boils, cancer, contusions, swellings.
Benincase cerifera Savi. *B. hispida* Cogn.	Don Gua (Gourd melon)	(fruit, seed) Palmitic acid, stearic acid, linoleic acid, thiamine, riboflavin, niacin, ascorbic acid.[50]	Diuretic, laxative; treats diabetes, dropsy, rhinitis.
Berberis amurensis Rupr. *B. poiretii* Schneid. *B. sibirica* Pall. *B. soulieana* C. K. Schneid.	Xiao Yeh (Chinese barberry)	(root) Berberine, berbamine, palamatine, jatrorrhizine, oxycanthine.[33]	Antibacterial; promotes leukocytosis; choleretic.
Betula mandshurica (Regel) Nakai *B. platyphylla* Suk.	Bai Hua (White birch tree)	(bark, tree sap) Betuloside, betulafolienetriol, betulafolienetetraol, betulin.[48,50]	Anticancer, mammary carcinoma.
Bidens bipinnata L. *B. parviflora* Willd.	Kuei Chen Gao (Black jack)	(whole plant) Flavonoids, essential oils.[48]	Treats bug bites, diarrhea, snakebite.
Bidens pilosa L. var. minor (Blume) Sheff.	Sien Feng Cao (Bur marigold)	(leaf) Polyacetylenes (it is phototoxic), phenytheptatriyne.[50]	Antibiotic; treats bug bites, diarrhea, snakebite. Bactericidal, fungicidal.
Bidens tripartita L.	Lang Ba Cao (Water hemp)	(whole plant) Luteolin, butin, buteine, coumarin, dihydroxycoumarin, scopoletin, umbelliferone.[48]	Treats chronic dysentery, heart ailments, eczema.
Bignonia grandiflora Thunb. *B. chinensis* Lam. (Syn. *Campsis chinensis*)	Zi Wei Hua (Trumpet vine)	See *Campsis chinensis*	

Biota chinensis Hort. *B. orientalis* L. (Syn. *Platycladus orientalis*, *Thuja orientalis*)	Ce Bai Ye	(twig) Quercitrin, pinipicrin, thujone, essential oils.[33]	Hemostatic; shorten blood clotting time, antitussive.
Bistorta lapidosa Kitag. (Syn. *Polygonum lapidosum*)	Shi Sheng Yu	See *Polygonum lapidosa*	
Bletilla hyacinthina R. Br. *B. striata* (Thunb.) Reichb.	Bai Ji (Amethyst orchid)	(tuber) Gelatin, essential oil, stilbenoids, blespirol, blestrianol, phenanthrene glucosides, bisphenanthrene theres.[33, 434,491,492,493,494,495]	Hemostatic; promotes leukocyte and platelet aggregation. Treats hematuria, blood splitting, primary hepatic carcinoma, antimicrobial.
Blumea balsamifera (L.) DC var. microcephala Kitumura	Ai Na Xian (Blumea camphor)	(leaf, shoot) Essential oils, borneol, camphor, cineole, limonene, palmitic acid, myristic acid, sesquiterpene alcohol, dimethy ether, cineole, limonene, pyrocatechic tannins.[48,53]	Treats itch, sores, wounds. A stomachic, sudorific, tonic, diaphoretic, anticatarrhal.
Blumea hieraciifolia (D. Don) DC	Tu Er Cao (Camphor)	(whole plant)[56] No information is available in the literature.	Treats pneumonia, water in the lung, diarrhea, snakebite.
Blumea lacera (Burm. f.) DC	Hong Tu Cao	(leaf) Carotene, coniferyl alcohol, angelic acid, vitamin C, cineole, citral, fenchone, camphor.[48,56]	Insectifuge, vermifuge: treats cholera, eczema, fever, itch, scurvy.
Blumea riparia (Blume) DC var. megacephala Randeria	Sha Hong Fan Cao	(root) No information is available in the literature.	Treats headache; relieves colic.

Table 1A Major Constituents and Therapeutic Values of Chinese Medicinal Herbs (continued)

Scientific Name	Common Chinese and (English) Name	Major Constituents and (sources)	Therapeutic Values*
Boehmeria densiflora Hooker et Arnott	Mu Yu Ma (Ramie)	(leaf, root) Beta-carotene, thiamine, lignin, riboflavin, niacin, ascorbic acid, protein.[48]	Astringent, antiabortifacient, drooling, demulcent, diuretic, resolvent, uterosedative, antihemorrhagic, styptic.
Boehmeria nivea Gaudich. *B. tenacissima* Gaudich.	Yu Ma Gen (Grass cloth plant, ramie)	See *Urtica tenacissima*	
Boenninghausenia albiflora (Hook.) Meisn.	Chou Chie Cao	(aerial part) Daphnoretin.[59]	Treats malaria.
Bougainvillea brasiliensis Raeusch *B. glabra* Choisy var. sanderiana Hort.	Jiu Chung Ko	(flower, stem) Betanidin, isobeturudin, 6-O-β-sophoraside, 6-O-rhamnosyl cophoroside.[54]	Treats liver infection; regulates menses.
Brassica alba (L.) Rabenh. *B. juncea* (L.) Czern. et Coss.	Bai Jie Zi (Indian mustard)	(seed, young shoot) Sinigrin, myrocin, sinapic acid, sinapine, potassium myronate, mustard oil, allyl isothiocyanate, behenic acid, erucic acid, benzyl isothiocyanate, eicosenic acid.[48,50]	Relieves bladder inflammation, hemorrhage, abscesses, lumbago, rheumatism, stomach disorders.
Brucea javanica (L.) Merrill *B. sumatrana* Roxb.	Ya Dan Zi (Kosam seed)	(fruit) Bruceines, bruceolide, brusatol, oleic acid, yatanoside.[33,510] This herb is toxic.	Antiamebial, anticancer, antiprotozoan.
Buddleia formosana Hatushima *B. madaguscariensis* Hance *B. officinalis* Maxim.	Bei Pu Jiang Mi Meng Hua (Butterfly bush)	(flower bud) Buddleoglycoside.[33,49,53]	Improves visual acuity, prescribed as ophthalmic in nyctalopia, asthenopia, cataract.

Species	Constituents	Properties
Bupleurum chinense DC *B. falcatum* L. *B. scorzoneraefolium* Willd.	(root) Triterpenoid saponins, sapogenins, saikosaponins, bupleuran, lignin-like polyphenolic substances, L-arabinose, D-glucose, arabinan polymer. [21,22,33,242,259,266,441,510]	Relieves tightness,; antitumor antipyretic, inflammation of inner organs; treats chronic hepatitis, nephrosis syndrome, autoimmune diseases, antiulcer, immunopharmacological activities.
Buxus harlandii Hance	(leaf) Cyclovirobuxine D, buxarmine E, cycloprotobuxine C, buxpiine K. [58]	Improves blood circulation, enhances heart muscle, regulates heartbeat, treats hepatitis, arthritis.
Buxus microphylla Sieb. et Zucc.	(root) Cyclovivobuxine C and D, buxtamine E, cycloprotobuxamine A and C, buxtauine, buxpiine. [58]	Treats heart conditions; a detoxicant.
Caesalpinia decapetula (Roth.) Alston	(seed) Volatile oil, bonducin, saporin, glycosides. [60]	Astringent, anthelmintic, antipyretic, antimalarial.
Caesalpinia pulcherrima Swartz	(flower, leaf, seed) Alkaloid, gallic acid, resins, tannins. [60] This herb is toxic.	Febrifuge, stomachic, diuretic, astringent, anticholeric.
Caesalpinia sappan L.	(heartwood) Brasilin, tetraacetylbrazilin, proesapanin A, essential oils, tannic acid, gallic acid, saponin. [33,49,50,621]	Activates blood flow, removes blood stasis, reduces swelling; against human cancer cells.
Calamus margaritae Hance	(root) [50] No information is available in the literature.	Antidysenteric, antibilious, hypotensive, to treats liver infections.

Table 1A   Major Constituents and Therapeutic Values of Chinese Medicinal Herbs   (continued)

Scientific Name	Common Chinese and (English) Name	Major Constituents and (sources)	Therapeutic Values*
Calendula officinalis L.	Jin Tsan Jiu (Marigold)	(whole plant) Arnidiol, carotenes, calenduline, cerylalcohol, flavoxanthin, lycopene, oleanolic acid, inulin, rebixanthin, violaxanthin. tocopherol, salicylic acid.[50,87]	Treats bleeding gums, bleeding piles, for amenorrhea, bruises, cholera, cramps, eruption, fevers, flu.
Callicarpa formosana Rolfe.	Tu Hung Hua (Callicarpa)	(flower, root)[54] No information is available in the literature.	Diuretic, for arthritis and nerve pain, gonorrhea, and emmenagogue. Externally applied as a styptic to wounds.
Callicarpa macrophylla L.	Zi Zhu Cao	(leaf, root) Tannins, flavone, resin.[33]	Hemostatic, constricting the blood vessels, antibacterial; treats tubercular bleeding.
Callicarpa nudiflora Hook & Am.	Luo Hua Zi Zhu	(leaf) Tannins.[33]	Treats suppurative skin infections and burns.
Calloglossa lepieurii (Mont.) J. Ag.	Mei She Chao, Zhe Gu Cai	(whole plant) Alpha-kainic acid, digeneaside.[33]	Inhibits the myocardium and causes a drop in blood pressure.
Caltha palustris L. var. membranacea Turcz. C. palustris L. var. sibirica Regel	Luo Ti Cao (Marsh marigold)	(whole plant) Anemonin, protoanemonin, choline, hellebrin, cevadine, berberine, scopoletin, saponin, umbelliferone, isorhamnetin, xanthophyllepoxyl.[48,50]	Antirheumatic, antitumor.
Calystegia hederacea Willich ex Roxb. C. japonica Choisy iu Zoll.	Da Wan Hua (Ivy bindweed, Japanese bindweed)	(root, flower) Kaempferol, kaempferol-3-rhamnoglucoside, columbin, palmatine.[48,50]	Diuretic; stimulates kidney secretions.

Camellia bohea Griff. *C. sinensis* (L.) Kuntze *C. theifera* Griff. *C. viridis* Link.	Cha (Tea)	See *Thea sinensis*	
Camellia japonica L.	Sha Cha Hua (Camelia)	(flower bud) Camelliagenins, d-catechol, l-epicatechol, leucoanthocyanin, arabinose, camellin, rhamnose, theasaponin.[49,50]	For hemoptysis, epistaxis, gastrointestinal hemorrhage, metrorrhagia.
Campanula gentianoides Lam. *C. giauca* Thunb. *C. grandiflora* Jacq. (Syn. *Platycodon* *grandiflorum*)	Jie Geng	(root) Saponins, inulin, platycodigenin.[49]	As an expectorant.
Campanula glomerata L. f. canescens (Maxim.) Kitag. *C. glomerata* L. var. dahurica Fisch. ex Ker-Gawl. *C. punctata* Lam.	Feng Lin Cao	(whole plant) Quercetin, isorhamnetin, kaempferol, hyperoside, isoquercetin, trifolin, chlorogenic acid, methyl caffeate, coumaroylquinic acid.[48]	For throat infection, headaches
Campsis adrepens Lour. *C. chinensis* Voss. *C. grandiflora* (Thunb.) Loiseleur (Syn. *Bignonia grandiflora*)	Zhu Wei (Chinese trumpet creeper)	(flower) Protein, dextrose, cyanidin-3-rutinoside.[48]	As emmenagogue. Treats amenorrhea, dysmenorrhea, leucorrhea, menorrhagia, metrorrhagia.
Camptotheca acuminata Decne.	Xi Shu (Happy tree)	(fruit) Camptothecine, venoterpine, hydroxyleamptothecin, methoxyl- camptothecin, irinotecan, 10-hydroxycamptothecin.[33,457,458,469] This herb is toxic.	Treats breast cancer, carcinoma of the stomach, rectum, colon and bladder, chronic leukemia.

Table 1A Major Constituents and Therapeutic Values of Chinese Medicinal Herbs   (continued)

Scientific Name	Common Chinese and (English) Name	Major Constituents and (sources)	Therapeutic Values*
Canarium album Raeusch. *C. sinense* Rumph.	Gan Lan (Chinese olive)	(seed) Beta-carotene, thiamine, riboflavin, niacin, ascorbic acid.[50]	Antiphlogistic, astringent in pharngitis.
Canavalia gladiata (Jacq.) DC *C. ensiformis* (L.) DC	Dao Dou (Broad bean)	(seed) Canavaline, canavanine, urease, gibberelin A_{21}, gibberelin A_{22}, canavalia gibberelin I, canavalia gibberelin II.[33]	A tonic, bactericidal, fungicidal, stomachic.
Cannabis chinensis Del. *C. sativa* L.	Huo Ma Ren or Da Ma Ren (Hemp)	(fruit, seed) Vitamin B_1, vitamin B_2, muscarine, choline, trigonelline, l(d)-isoleucine betaine, cannabinol, tetra-hydrocannobinol, cannabidiol.[33,450,558]	Purgative; stimulates intestinal mucosa causing an increase in secretions and peristalsis.
Capsella bursa-pastoris (L.) Medicus	Jie Cai (Shepherd's purse)	(whole plant) Bursic acid, alkaloids, vitamin A, choline, citric acid.[33]	Hemostatic, antihypertensive, chyluria, nephritis, edema, hematuria.
Caragana franchetiana Koma *C. intermedia* Kuang *C. microphylla* Lam. *C. sinica* Lam.	Jin Gi Er (Chinese caragana)	(root) Alkaloids, glucosides.[33]	Antihypertensive, anti-inflammatory.
Cardamine leucantha (Tausch.) O. E. Schulz. *C. lyrata* Bunge.	Sui Mi Jie	(root, leaf, seed) Erucic acid, linolenic acid, linoleic acid, oleic acid, sinigroside, lecithine, myrosinase.[60]	Treats abdominal pain, antidysenteria.
Carduus acaulis Thunb. *C. crispus* L. *C. japonicus* Franch. (Syn. *Cirsium japonicum*)	Xiao Ji (Plumeless thistle)	(leaf, stem) Essential oils, glycosides, bitter principle.[49]	Hemostatic.

Carpesium abrotanoides L. *C.athunbergianum* Sieb. et Zucc.	He Shi or Tian Min Qing (Starwort)	(whole plant, fruit) Essential oils, inlin.[49]	Ascariasis, enterobiasis, taeniasis, antiphlogistic in pharyngitis, tracheitis, laryngitis.
Carthamus tinctorius L.	Hong Jua (Safflower)	(flower) Cartharmin, neocarthamin, safflower yellow, quinochalone, safflomin A.[33,558]	Promotes blood circulation, removes blood stasis, and restores normal menstruation.
Carum carvi L.	Ye Hao (Caraway)	(fruit, aerial part) Essential oil, d-carvone, coumarin, chromone, polyacetylene, herniarin, scopoletin, umbelliferone, d-limonene, phytosterols.[48,50,250,450]	Carminative; treats stomach pain.
Cassia alata L.	Dui Ye Dou (Ringworm bush)	(whole plant) Fatty acids, aloe-emodin, rhein chrysarobin, chrysophanic acid, oxymethyl anthraquinone, rutin, isochaksine, quercetin.[48,450,510,621]	Improves night vision, migraines; astringent, purgative.
Cassia angustifolia Vahl.	Fan Xie Ye	(leaflet) Sennosides, aloe-emodin, dianthrone glucoside, rhein monoglucoside, rhein, kaempferin, myricyl alcohcl, anthraquinone derivative.[3,510]	Purgative, laxative, cathartic.
Cassia nomame (Sieb.) Honda *C. obtusifolia* L. *C. tora* L.	Jue Ming Zi (Sicklepod)	(seed) Anthraquinones such as emodin, chrysophanol, physcion, rhein aurantio-obtusin, obtusifolin, chryso-obtusin, naphthopyrones, obtusin, aurantio-obtusin rubrofusarin, nor-rubrofusarin, toralacton.[33,621]	Purgative; treats ophthalmia, hypercholesterolemia, vaginitis.

Table 1A Major Constituents and Therapeutic Values of Chinese Medicinal Herbs (continued)

Scientific Name	Common Chinese and (English) Name	Major Constituents and (sources)	Therapeutic Values*
Cassia occidentalis L. C. torosa Cav.	Wang Jiang Nan (Coffee senna, sicklepod)	(seed, root) Anthraquinones, torosachrysone, n-methylmorpholine, apigenin, galactomannan, cassiollin, xanthorin, dianthronic heteroside, helminthosporin.[4,33,496]	Mild purgative; lowers blood pressure; antioxidative, antiasthmatic, antitoxic, antimalarial, antibacterial, anthraquinones, and hepatoprotective activities.
Cassia siamea Lam.	Tie Dao Mu	(leaf, flower, fruit) Chrysophanic acid, chrysarobin, oxymethyl anthraquinone.[60]	A tonic to relieves stomach pains.
Cassytha filliformis L.	Pan Chan Teng (Dodder laurel)	(stem) Cassyfiline, cassythidine, galactitol, cassythine, laurotetanine.[50]	Diuretic, for gonorrhea, kidney problems.
Castanea crenuta Sieb. et Zucc. C. mollissima Blume	Japan Su (Chestnut)	(flower, stem bark) Quercetin, urea, protein, beta-carotene, riboflavin, thiamine, ascorbic acid, niacin.[48,50]	Treats diarrhea, poisoned wounds, lacquer poisoning; astringent.
Catharanthus roseus (L.) G. Don	Chang Chun Hua (Madagascar periwinkle)	(whole plant) Vinblastine, vincristine, carosine, vinrosidine, lenrosine, lenrosivine, rovidine, perivine, perividine, vindolinine, pericalline.[33] This herb is toxic.	Anticancer in chronic lymphocytic leukemia and Hodgkin's disease, in acute lymphocytic leukemia.
Caulophyllum robustum Maxim.	Wei Yan Xian	(root) Magnoflorine, taspine, methylcytisine, alpha-lupanine, cauloside, hederagonin.[48]	Treats arthritis, wounds; regulates menstruation.

Celastrus alatus Thunb. *C. striatus* Thunb. (Syn. *Evonymus alatus*)	Wei Mao (Bittersweet)	See *Evonymus alatus*
Celosia argentea L. var. cristata Bth. *C. cristata* L.	Ji Guan Hua (Quail grass or Cockscomb)	(stem, leaf, flower) Celosiaol, nicotinic acid.[48]	Treats high blood pressure, itchiness, arthritis pain.
Celosia argentea L. *C. margariacea* L.	Cao Jue Ming (Quail grass)	(stem, leaf) Guijaverin, hyperoside, quercitin, isoquercitrin.[50]	Insecticidal.
Celtis bungeana Blume *C. sinensis* Pers.	Po Shu (Hackberry)	(bark) Essential oils.[48]	For dyspepsia, poor appetite, shortness of breath, swollen feet.
Centaurium meyeri (Bunge.) Druce	Ai Lei	(whole plant) Bitter glycoside, ophelic acid, chiretta.[60]	Treats headache, fever, and infections.
Centella asiatica (L.) Urb.	Ji Xue Cao (Gotu kola)	(whole plant) Asiaticoside, madecassoside, brahmoside, brahmissoside, glucoside asiaticoside, sitosterol, tannin, vallarine, pectic acid, resin.[33,450,510]	Antibacterial; lowers blood pressure; antipyretic, diuretic, detoxicant.
Centipeda minima (L.) A. Braun. et Ascherison	Shi Wu Tou (Centipeda)	(whole plant) Essential oil, myriogynine, alkaloids, glycosides, saponin.[60]	Antidotal; treats conjunctivitis, piles, malaria.
Cephalanoplos segetum	Xiao Ji (Field thistle)	(aerial part) Alkaloids, choline, saponins.[33]	Hemostatic, cardiac stimulation.

Table 1A   Major Constituents and Therapeutic Values of Chinese Medicinal Herbs   (continued)

Scientific Name	Common Chinese and (English) Name	Major Constituents and (sources)	Therapeutic Values*
Cephalotaxus fortunei Hook. *C. oliveri* Mast. *C. qinensis* (Rehd. et Wils.) Li	San Jian Shan (Plum yew)	(branch) Cephalotaxine, harringtonine, epicephalotaxine, epiwilsonine, demethylcephalotaxine, wilsonine, cephalotaxinone.[33] This herb is toxic.	Treats malignant tumors.
Cephalotaxus wilsoniana Hayata	Taiwan Cu Fei	(shoot) Cephalotaxine, cephalotaxinone, acetycophalotaxine, wilsonine, demethylcephalotaxine, epicephalotaxin, harringtonine, hormoharringtonine, c-3epi-wilsonine, hydroxyeephalotaxine, isoharringtonine.[56]	Antitumor, anticancer. Treats lymphatic gland swelling, improves digestion; an insecticide.
Chaenomeles japonica (Thunb.) Lind. *C. sinensis* Koch. *C. speciosa* (Sweet) Nakai	Japan Mu Gua Xuan Mu Gua	(fruit) Vitamin C, malic acid, tartaric acid, citric acid, hydrocyanic acid.[49]	Treats arthralgia, diarrhea, cholera, gout, arthritis.
Chamaenerion angustifolium (L.) Scop. *C. angustifolium* (L.) Scop. f. pubescens (Hausskn.) Kitag.	Liu Lan	(whole plant) Crataegolic acid, penta-o-galloyl-β-d-glucose, maslinic acid, chanerol, cerylalcohol.[48]	Regulate menstruation, improves breast milk production. Externally for wounds; stops bleeding.
Changium smyrnioides Wolff.	Min Dong Seng	(root)[50] No information is available in the literature.	Tonic for lungs, stomach; antiemetic, bechic.

Species	Name / Constituents	Uses
Chelidonium album L. C. hybridum L. C. majus L. C. serotinum L.	Bai Qu Cai (Celandine poppy) (whole plant) Chelidonine, chelidocystatin, protopine, stylopine, allocryptopine, chelerythrine, sparteine, coptisine.[33,256,449,497]	Anodyne, analgesic, diuretic, antitussive, detoxicant, anticancer. Treats abdominal pain, peptic ulcers, chronic bronchitis, and whooping cough.
Chenopotium ambrosiodes L.	Chou Xing (Lambs quarter) (leaf) Volatile oil, ascaridol, geraniol, saponin, 1-limonene, p-cymene, d-camphor, kaemferol-7 shamnoside, ambroide.[60,450]	An anthelmintic to treats ascarids, ancylostomiasis, vermifuge, carminative.
Chimaphila umbellata (L.) W. Barton	Mei Li Cao (whole plant) Arbutin, ursolic acid, homoarbutin, chimaphilin, isohomoarbutin, hyperin, avicularin, kaempferol, renifolin, beta-amyrin, ericolin, andromedotoxin, chinic acid.[48]	Diuretic; relieves stomach, tooth and after-birth pains; antifungal.
Chloranthus glubra (Thunb.) Nakia C. oldhnami Solms.	Jiu Jie Cha Si Ye Lian (Chloranthus) (leaf, stem) Essential oils, flavonoids, pelargonidin-3-rhamncsylglucoside.[54]	Treats bone fractures, vomiting, contusions, lung infection; an astringent. Antitumor; improves immune system, relieves arthritis pain.
Chrysanthemum boreale (Makino) Makino C. indicum L. C. lavandulaefolium (Fisch.) Mak. C. procumbens Lour. C. tripartium Sw.	Ye Jiu Hua (Chrysanthemum) (flower, petal) Alpha-pinene, limonene, carvone, cineol, camphor, borneol, chrysanthrnin, yejuhualactone, chrysanthemaxanthin.[33]	Antibacterial, relieves headache, insomnia, and dizziness due to high blood pressure.
Chrysanthemum cinerriaefolium Visiont	Chu Gu Jiu (Chrysanthemum) (flower) Essential oil, adenine, choline, stachydrine.[60]	Used as insecticides.

Table 1A Major Constituents and Therapeutic Values of Chinese Medicinal Herbs (continued)

Scientific Name	Common Chinese and (English) Name	Major Constituents and (sources)	Therapeutic Values*
Chrysanthemum jucundum Nakai & Kitag. C. *koraiense* Nakai C. *morifolium* Ramat. C. *sinense* Sabine.	Jiu Hua (Chrysanthemum)	(flower) Bornol, chrysanthemin, camphor, stachydrine, choline, acacetin-7-rhamnoglucoside, cosmosiin, acacetin-7-glucoside, diosmetin-7-glucoside, adenine.[33]	Antipyretic, antitoxin, remedy for common cold, headache, dizziness, red eye, swelling, hypertension.
Cibotium barometz (L.) J. Smith	Hie Quin Cao (Lamb of Tartary)	(root) Palmitic acid, linoleic acid.[50,230]	A tonic, digestive, laxative, analgesic in rheumatism, lumbago, myospasm.
Cimicifuga dahurica (Turcz.) Maxim. C. *foetida* L. C. *heracleifolia* Kom. C. *racemosa* (L.) Nutt. C. *ussuriensis* Oettingen	Sheng Ma (Stinking bugbane)	(rhizome) Ferulic acid, isoferulic acid, cimigenol, khellol, aminol, cimifugenol, cimitin.[33,570,586]	Induces diaphoresis; promotes skin eruption.
Cinnamomum aromaticum Nees. C. *cassia* Presl.	Gui Zhi (Cinnamon)	(twig, bark) Cinnamic aldehyde, cinnamyl acetate, cinnamic acid, eugenol, phellandrene, phenylpropyl alcohol, coumarin, cinnamic aldehyde, orthomethylcoumaric aldehyde.[33,49,254,435,510,567]	Antibacterial, vasodilatation, aromatic stomachic, astringent, tonic, analgesic; stimulates human lymphocytes to proliferate.
Cinnamomum camphora (L.) J. S. Presl.	Chang Shu (Cinnamon)	(root, branch, leaf) d-camphor, eucalyptole, cineole, pinene, aromadendrene, cumaldehyde, pinocarveol, 1-acetyl-4-isopropylidenecyclopentene.[33,53]	Stimulates nervous system; relaxes gastrointestinal muscle contractions.

Cinnamomum zeglanicum Blume	Ceylon Rou Gui (Ceylon cinnamon)	(bark) Cinnamic aldehyde, p-cymene, hydrocinnamic aldehyde, pinene, benzaldehyde, cuminic aldehyde, nonylic aldehyde, eugenol, caryophyllene, l-phellandrine, methyl-n-amyl ketone, l-linalool.[60]	Stimulant to digestion, respiration, and circulation.
Cirsium albescens Kitamura *C. brevicaule* A. Grey *C. littorale* Max. *C. maakii* Max. *C. segetum* Bunge. *C. setosum* (Willd.) Bieb. *C. vlassovianum* Fisch. ex DC	Xiao Ji	(whole plant) Essential oil, rutin, acacetin-7-rhomnoglucoside, protocatechuic acid, caffeic acid, chlorogenic acid.[48,49]	Hemostat, diuretic; stops bleeding; externally for wound infections.
Cirsium chinense Gardn. et Champ. *C. japonicum* DC	Chinese Ji Da Ji (Thistle)	(leaf, stem) Alpha-amyrin, beta-amyrin, beta-sitosterol, stigmasterol, taraxsteryl acetate, inulin, labenzyme, pectolinarin.[50,60]	Hemostat, diuretic; treats intestinal bleeding caused by ulcers; externally for abscesses and scabies.
Cissampelos pareira L.	Xi Sheng Teng (Ice vine)	(plant) Cissampareine, hayatine, pelosine, isoquinoline, hayatinine, berberine, dl-beheerine, dl-curine, D-guereitol, d-isochondrocendrine, hayatidine, cissamine, menisnine, reserpine, cissampeline.[33,450,558]	Blockade of NMJ depolarization. Used externally on wound surfaces to relieves pain.
Cistanche deserticola Y. C. Ma	Rou Chon Wun (Broomrape)	(whole plant) Boschnialactone, boschniakine, neoboschnialactone, echinocoside.[33,244,593]	Antioxidant activity, antisenile, immunopharmacological effect;; stimulates hypothalamus-pituitary system; increases memory power, sex function.

Table 1A   Major Constituents and Therapeutic Values of Chinese Medicinal Herbs   (continued)

Scientific Name	Common Chinese and (English) Name	Major Constituents and (sources)	Therapeutic Values*
Citrullus anguria Duch. *C. edulis* Spach. *C. lanatus* Matsumura & Nakai *C. vulgaris* Schrad.	Xi Gua (Watermelon)	(fruit, seed) Cucurbitacins, carprylic acid, capric acid, lauric acid, myristic acid, palmitic acid, stearic acid, oleic acid, linoleic acid, sterol, citrulline.[50,351]	For alcohol poisoning, diabetes, nephritis, sore throat, stomatitis; demulcent.
Citrus aurantium (Christm.) Swingle var. amara	Suan Cheng (Bitter orange)	(unripe fruit) Synephrine, N-methyltyramine, flavones including tangeratin and nobiletin.[33]	Treats indigestion, relieves abdominal distension, ptosis of the anus or uterus.
Citrus deliciosa Tenore *C. nobilis* Lour.	Jiu Pi (Orange)	(fruit skin) Vitamin A, B and C, hesperidin, limonene, citral, methyl anthranilate.[49]	Stomachic, digestant, expectorant, antitussive, antiemetic.
Citrus reticulata Blanco *C. reticulata* Blanco. var. chachiensis	Jiu Hong, Chen Pi (Orange)	(external layer of pericarp) Citral, geraniol, linalool, methylanthranilate, stachydrine, putrescine, apyrocatechol, naringin, poncirin, hesperidin, neohespiridin, nobiletin.[33,568]	Expectorant, antitussive; treats indigestion; an antiemetic agent.
Clematis armandii Franch. *C. heracleifolia* DC *C. heracleifolia* DC var. davidiana (Decaisue ex Verlot) O. Kuntze	Mu Tong (Clematis)	(stem) Aristolochic acid, saponin akebin, triterpenoids.[25,33]	Diuretic, antibacterial.

Clematis chinensis Retz. C. *florida* Thunb. C. *hexapetala* Pall. C. *hexapetala* Pall. f. *longiloba* (Freyn) S. H. Li et Y. H. Huang C. *minor* Lour. C. *sinensis* Lour. C. *terniflora* DC	Wei Ling Xian (Clematis)	(root) Anemonin, anemonol, saponins.[33,49,246]	Analgesia, diuresis, carminative, diuretic, anti-inflammatory; treats arthritis, backache, and headaches.
Clematis intriicata Bunge. C. *mandshurica* Rupr.	Tie Xian Lian (Clematis)	(whole plant) Clematoside A, oleanolic acid.[48]	Relieves arthritis pain and related infections.
Cleome spinosa Jacquin C. *gynandra* L. C. *viscosa* L.	Xi Yang Bai Hua Cai Xiang Tian Huang (Spiderwisp)	(seed) Cleomin, lactone, tannins, volatile oils.[50]	Treats dysentery, gonorrhea, malaria; rheumatoid arthritis.
Clerodendrum cyrtophyllum Turcsaninow	Da Qing (Clerodendrum)	(leaf, root) Indirubin, ingigo, tryptanthrin, isatan B, glucobrassicin, 3-indolylmethylgluco-sinolate, neoglucobrassicin, isoindigo, indican, lacerol.[53]	Antipyretic, detoxicant, diuretic, preventitive for epidemic meningitis.
Clerodendrum fragrans Ventenat	Chou Mu Lee	(stem, root) 24beta-methylcholesta-5, 22E,25-trien-3beta-ol, clerosterol, 24alpha-ethyl-5alpha-cholest-22E-en-3beta-ol, 22E-dehydroclerosterol.[48,196,232]	Strengthens weak leg muscles, skin trouble, and smallpox.
Clerodendrum paniculatum L. var. albiflorum (Hemsl.) Hsieh.	Bai Long Chuan Hua	(root) 24beta-epimer poriferasterol, 24alpha-epimer stigmasterol.[56]	For gonorrhea, skin diseases; diuretic; regulates menses.

Table 1A   Major Constituents and Therapeutic Values of Chinese Medicinal Herbs   (continued)

Scientific Name	Common Chinese and (English) Name	Major Constituents and (sources)	Therapeutic Values*
Clerodendrum trichotomum Thunb. C. *spicatus* (Thunb.) C. Y. Wu C. *trichotomum* Thunb. var. ferrugineum Nakai	Chou Wu Tong (Hairy clerodendrum)	(leaf, stem, root) Glycosides clerodendrin, acacetin-7-glucurono-(1,2)-glucuronide, clerodendrin, mesoinositol, clerodolone, apigenin-7-diglucuronide, friedelin, epifriedelin.[33,48,71]	Treats hypertension, arthritis pain.
Clinopodium chinense (Benth.) O. Kuntze. C. *gracile* (Benth.) O. Kuntze. C. *polycephalum* (Benth.) C. *umbrosum* (Bleb.) C. Koch.	Duan Xue Liu Guang Feng Lun Cai Feng Lun Cai	(whole plant) Dydimin, hesperidin, siosakuranetin, apigenin, ursolic acid.[48]	Hemostatic; stimulates uterine contractions; antibacterial.
Clivia miniata Lindley	Jun Zi Lian	(whole plant) Clividine, miniatine, lycorine.[57]	Anticancer, antitumor.
Cnidium monnieri (L.) Cusson	She Cheung Zi	(whole plant) Archangelicin, columbianetin, O-acetylcolumbianetin, O-isovaleryl columbianetin, cnidiadin, cnidimine, l-pinene, l-camphen.[33]	A trichomonicidal agent, antiascariac, andantifungal.
Cocculus diversifolius Miq. C. *thunbergii* DC	Fang Ji, Japan Han Fang Ji, Qing Teng	(root) Sinomenine, disinomenine, sinoacutine, isosinomenine, sinactine, tuduranine, michelalbine, acutumine, acutumidine.[33]	Similar to morphine but less potent. Sedative, antitussive, anti-inflammatory.

Cocculus laurifolius DC C. *sarmentosus* DC C. *trilobus* (Thunb.) DC	Mu Fang Ji Japan Mu Fang Ji	(root) Magnoflorine, triboline, homotrilobine, isotrilobine. normenisarine, coclobine, cocculolidine, trilobamine.[33]	Analgesic effect; can reduce swelling, relieves arthritis pain and neuralgia, treats pulmonary and cardiac edema.
Codonopsis lanceolata (Sieb. et Zucc.) Trautv.	Yang Lu (Bellflower)	(whole plant) Apigenin, luteolin, alpha-spinasterol, stigmastenol, oleanolic acid, echinocystic acid, albigenic acid.[48]	Treats Lung abscesses, stimulates milk flow, and treats amenorrhea.
Codonopsis pilosula (Franch.) Nannfeldt C. *tangshen* Oliv. C. *ussuriensis* (Rupr. et Maxim.) Hemsl.	Dong Seng	(root) Taraxeryl acetate, friedelin, n-butyl allophanate, inulin, sucrose, amino acids, stigmasterol, spinastrerol, methyl palmitate, taraxerol, triterpenoids, delta-spinasterol, delta-7-st gmasternol, perlolyrine glucopyranosides.[48,253,470,471 380,568,580]	For amnesia, anorexia, asthma, cachexia, cancer, impotence, insomnia, palpitations, hypotensive and vasorelaxant activities.
Coix agrestis Lour. C. *chinensis* Tod. C. *lachryma* L. C. *lachryma-jobi* L. var. ma-yuen (Roman) Stapf.	Yi Yi (Rosary beads)	(seed, root) Coixenolide, coixol, protein, myristic acid, palmitic acid, stearic acid, oleic acid, linoleic acid, polysaccharides, triglycerides, phospholipids, benzoxazinones, adenosine.[48,50,239,437,472,,473,474,475]	For intestinal or lung cancers and warts, antitumor, antirheumatic, diuretic, refrigerant.
Commelina communis L.	Ya Zhi Cao (Day flower)	(aerial part) Awobanin, commelin, flavocommelitin.[33]	Antibacterial, antipyretic, diuretic, antiedematic.
Commiphora myrrha Engler	Mo Yao (Myrrh)	(stem) From gum resin, essential oils including myrcene, alpha-camphorene, Z-guggulsterol, guggulsterol, makulor, cembrene.[33,568]	Stimulates blood flow, relieves pain, and promotes tissue regeneration.

Table 1A   Major Constituents and Therapeutic Values of Chinese Medicinal Herbs   (continued)

Scientific Name	Common Chinese and (English) Name	Major Constituents and (sources)	Therapeutic Values*
Conioselinum univittatum Turcz.	Gong Chong	(root) Essential oil.[49]	Emmenagogue, sedative.
Convallaria keiskei Miq.	Ling Lan	(whole plant) Convallatoxin, convalloside, convallamarin, convallatoxol.[33] This herb is toxic.	Treats heart disease, detoxifies the liver.
Convolvulus arvensis L.	Tian Xuan Hua (Bindweed)	(whole plant) Quercetin, kaempferol, caffeic acid, beta-methylaesculetin.[48,568]	Improves blood circulation, relieves pain and itchiness.
Conyza canadensis (L.) Cronq.	Xiao Fei Peng	(aerial part) Essential oils, matricaria ester, dehydromatricaria ester, linoleyl acetate, limonene, linalool, centaur X, dephenyl methane-2-carboxylic acid, cumulene, O-benzylbezoic acid.[48]	Relieves swelling, itchiness, treats intestine and liver infection; a detoxicant; externally for skin eczyma, wounds, pain caused by arthritis, toothache.
Coptis chinensis Franch. C. japonica Makino C. teeta Wall.	Huang Lian (Gold thread)	(root) Berberine, coptisine, urbenine, worenine, palmatine, jatrorrhizine, columbamine, lumicaerulic acid.[33,60,248,510,578,588] This herb is toxic.	Antiarrhythmic, antibacterial, antiviral, antiprotozoal, anticerebral ischemic.

Species	Name (Part)	Constituents	Uses
Corchorus capsularis L. *C. olitorius* L.	Huang Ma Sha Ma (Jute)	(leaf, flower) Glycosides, capsularin, corchorin, corchoritin, aglycone, strophanthidin, digitoxigenin, coroloside, glucoevatromonoside, erysimoside, olitoriside, linolic acid, corchoroside, helveticoside, corchotoxin, oleic acid, palmitic acid, stearic acid.[60,498,499,500]	Treats dysentery, consumptive cough, epistaxis, bladder diseases. Inhibitory effect on lipopolysaccharide-induced NO production in cultured mouse peritoneal macrophages.
Cordyceps sinensis Link.	Dong Chong Xia Chao	(fruit body) 2'-deoxyadenosine, adenosine, sterols, saccharides, protein, cordycepin, d-mannitol.[33,91,92,401,402,413,414,459] Lead poisoning was reported.[238]	Antisenescence, hypolipidemic, endocrine, antitumor, antiatherosclerotic and sexual function-restorative activities. Treats respiratory, renal, liver and cardiovascular diseases, antileukemic cells, hyposexuality, and hyperlipidemia.
Coriandrum sativum L.	Hu Sui (Coriander)	(leaf) Acetone, borneol, coriandrol, cymene, decanal, decanol, decylic aldehyde, dipentene, geraniol, rutin, limonene, linalool, malic acid, nonanal, oxalic acid, phellandrene, tannic acid, terpinene, terpinolen, umbelliferone, scopoletun, coumarins, quercetin, kaepferal, aflatoxins.[50,450,568,574]	Eruptions of pox and measles.
Coriolus versicolor (L. ex Fr.) Quel.	Yun Chih	Polysaccharides, polysaccharopeptide.[415,416,417,418]	Antimetastatic effect, antilung cancer, tumor inhibition, against immunodeficiency virus.
Cornus alba L. *C. kousa* Hance *C. macrophylla* Wallich	Si Zhao Hua Jian Zi Mu (Dogwood)	(bark, shoot, leaf) Quercitol, kaempferol, dihydroxyglutamic acid, phenethylamine.[48]	Astringent, antimalarial; treats arthritis, backache, diabetes, hepatitis, malaria, metrorrhagia, cancer.

Table 1A   Major Constituents and Therapeutic Values of Chinese Medicinal Herbs   (continued)

Scientific Name	Common Chinese and (English) Name	Major Constituents and (sources)	Therapeutic Values*
Cornus officinalis Sieb. et Zucc.	Shan Zhu Yu (Dogwood)	(sarcocarp) Morroniside, 7-O-methyl-morroniside, sworoside, loganin, longiceroside, tannic acid, resin, tartaric acid, cornin, gallic acid, malic acid.[33,60]	Diuretic; treats dysmenorrhea, excessive menstruation, impotency, backache, dizziness.
Cornus walteri Wangerin	Korean Si Zhao Hua (Korean cornel)	(leaf, fruit) Fatty acid, loganin, linolenic acid.[48,53]	An astringent.
Corydalis ambigua Cham. et Schlecht. var. amurensis Maxim. C. repens Mandl. et Muehld. var. watnabei (Kitag.) Y. C. Chu C. ternata (Nakai) Nakai C. turtschaninovii Besser Bess f. yanhusa C. yanhusuo W.T.Wang ex Z. Y. Su et C. Y. Wu	Yan Hu Suo Korean Yan Hu Suo	(tuber) d-corydaline, corydalis, dl-tetrahydropalmatine, crybulbine, tetrahydrocoptisine, corydalamine, tetrahydrocolumbamine, protopine, alpha-allocryptopine, coptisine, dehydrocorydaline, columbamine, dehydrocorydalmine.[33,558] Toxic if overdosage.	Analgesic, sedative, hypnotic, synergistic;;increases coronary blood flow.
Corydalis decumbens (Thunb.) Pers.	Xia Tian Wu	(aerial part) Protopine, bulbocapnine, d-tetrahydropalmatine.[33,240] Toxic if overdosage.	Relieves pain after bone fractures; antihypertensive, antirheumatic.
Corydalis incisa (Thunb.) Pers. C. bungeana Turcz.	Chuan Duan Chang Cao Di Ding Zi Jing (Corydalis)	(whole plant) Protopine, pallidine, sinocecatine, corynoline, isocorynoline, coptisine, corycavine, acetylcorynoline, corynoloxin, coreximine, reliculine, corydamine, scoulerine.[33,50]	For rectal prolapse, abscesses, hemorrhoids.

Corylus heterophylla Fisch. ex Besser. C. mandshurica Maxim. ex Rupr. C. mandshurica Maxim. ex Rupr. f. brevituba (Kom.) Kitag.	Zhen (Filbert)	(seed) Beta-carotene, thiamine, riboflavin, niacin, ascorbic acid.[50]	To improves appetite; a digestive.
Costus specious (Koen.) Smith	Bi Qao Jiang (Crepe ginger)	(whole plant) Diogenin, tigogenin, corticosteroids, 3-(4-hydroxyphenyl)-2 (E)-propenoate.[50,194,450]	For fever, anasarca, asthma, bronchitis, cholera, antifungal.
Cotinus coggygria Scop.	Huang Lu (Smoke tree)	(leaf, twig) Myricetin, myricitrin, fisetin, fustin.[33]	Antipyretic.
Cotyledon fimbriatum Turcz. C. malacophylla Pall.	Zuo Yie He Cao	(whole plant)[50] No information is available in the literature.	Treats tumors, for dysentery, hemostatic; stops intestinal bleeding.
Crataegus cuneata Sieb. et Zucc. C. chlorusarca Maxim. C. dahurica Koehne ex Schneid. C. maximowiczii Schneid. C. pentagyna Waldst. et Kit. C. pinnatifida Bunge. C. sanguinea Pall.	Shan Zha (Hawthorn)	(unripe or ripe fruit) Flavonoids, quercetin, hyperoside, l-epicatechin, d-catechin, saponins, chlorogeric acid, caffeic acid, citric acid, crataegolic acid, corosolic acid, maslinic acid, ursolic acid.[13,33,231]	Cardiotonic agent; treats hypercholesterolemia, angina pectoris, hypertension.
Crocus sativus L.	Shi Hong Hua (Saffron)	(root) Crocetin, crocetin geniobiose glucose ester, crocetin di-glucose ester, crotin, lycopene beta-carotene.[33,450]	Ameliorating effect on ethanol-induced impairment of learning and memory.

Table 1A   Major Constituents and Therapeutic Values of Chinese Medicinal Herbs   (continued)

Scientific Name	Common Chinese and (English) Name	Major Constituents and (sources)	Therapeutic Values*
Crotalaria mucronata Desv.	Zhu Shi Tou (Rattlebox)	(whole plant) Mucronatine, mucronatinine, retroresine, usaramine, nilgirine, vitexin, vitrexin-4-O-xyloside, apigenin.[33]	Treats frequent urination in children, edema, chronic diarrhea, pelvic infections.
Crotalaria sessiliflora L.	Ye Bai He (Narrow-leaved rattlebox)	(whole plant) Monocrotalines.[33] This herb is toxic.	Anticancer.
Croton cascarilloides Raeushel *C. tiglium* L.	Ba Dou (Croton)	(seed) Croton resin, phorbol, crotonic acid, crotin, crotonoside.[33,144] This herb is very toxic.	Purgative, wound healing properties.
Cryptotaenia japonica Hasskarl *C. canadensis* (L.) DC	Japan Liu Shan Ya Er Qin	(whole plant) Cryptotaenen, kiganen, kiganol, petroselic acid, isomesityl oxide, mesityl oxide, methyl isobutyl ketone, *trans*-beta-ocimene, terpinolene.[48,50]	For diarrhea, dysmenorrhea, rheumatism, tubercular glands.
Cucumis melo L.	Gua Di (Cantaloupe)	(pedicel) Melotoxin, cucurbitacin B, cucurbitacin E, sterol.[33,351]	Produce vomiting for drug intoxication, treats toxic and chronic hepatitis and cirrhosis of the liver.
Cucumis sativus L.	Huang Gua (Cucumber)	(leaf, fruit, seed) Arginine, caffeic acid, chlorogenic acid, cucurbitacins, fructose, galactose, isoquercitrin, mannose, 2,6-nonadienol, rutin, linoleic acid, oleic acid, palmitic acid, stearic acid, sterol.[50,351]	Diuretic, purgative, vermifuge; pulp can be used for burns, scalds, and skin ailments.

Cucurbita moschata Duch. var. melonaeformis (Carr.) Makino *C. pepo* L.	Nan Gua Zi (Winter crookneck squash)	(seed) Cucurbitine, sterol.[33,235,351,568]	Treats taeniasis.
Cunninghamia lanceolata (Lamb.) Hooker	Shan (China fir)	(stem) Borneol, camphene, cineole, citrene, limonene, phellandrine, pinene, terpineol, essential oils.[50]	For lacquer poisoning, chronic ulcers, cholera, flatulence.
Curculigo capitulata (Lour.) O. Kuntze *C. ensifolia* R. Br. *C. malabarica* Labill. *C. orchiodes* Gaertn. *C. stams* Labill.	Da Xian Mao Xian Mao (Black musli)	(rhizome) Calcium oxalate, resin, tannins.[50]	Improves immunicity, stimulates endocrine system.
Curcuma longa L. *C. domestica* L.	Yu Jin (Turmeric)	(tuber) l-curcamene, sesquiterpene, camphor, camphene, curmarin, curzernone, curzenene, curcumol, furanodienone, furanodiene, zederone, curcolone, diol, procurcumenol, curdione, curcumin.[33,398,460,510,568]	Anti-inflammatory, antitumor, anti-infectious properties, antioxidative activity. Activates blood flow, removes blood stasis.
Curcuma pallida Lour. *C. phaeocoulis* Val.	E Zhu Peng Wo Mao	(rhizome) Volatile oil, cineole, camphene, zingiberene, borneol, camphor, curcumin, zedoarin, gum, resin.[60]	Stomachic, carminative.

Table 1A   Major Constituents and Therapeutic Values of Chinese Medicinal Herbs   (continued)

Scientific Name	Common Chinese and (English) Name	Major Constituents and (sources)	Therapeutic Values*
Curcuma zedoaria (Christ.) Roscoe C. aromatica Salisk. C. kwangsiensis A. Lee	E Zu (Wild turmeric)	(rhizome) Curzerenone, curzerene, zederone, zerumbone, furanodiene, curdione, furanodienone, curculone, diol, procurcumenol, curcumin, turmerone, zingiberene, 3-(4-hydroxyphenyl-2 (E)-propenoate.[33,192,194] This herb is toxic.	Inhibits mutagenesis and tumor promotion; anti-inflammatory, antitumor, anti-infectious, antifungal, anti-HIV.
Cuscuta australis R. Brown	Dou Tu Si (Dodder)	(seed, aerial part) Carotenoids, alpha-carotene-5, 6-epoxide, taraxanthin, lutein.[48]	For fever, constipation; diuretic.
Cuscuta chinensis Lam. C. europaea L. C. japonica Choisy C. lupuliformis Krocker	Tu Si Zi (Dodder)	(seed) Cuscutalin, bergenin, cuscutin, amarbelin, cholesterol, campesterol, beta-sitosterol, stigmasterol, beta-amyrin.[48,568]	Improves immunity, increases blood sugar metabolism.
Cyathula prostrate (L.) Blume	Bei Xian	(leaf, root) Ecdysterone.[50]	Laxative; for dysentery, rheumatism, syphilis.
Cycas revoluta Thunb.	Tie Shu (Sago palm)	(leaf) Sotelsulflavone, hinokiflavone, amentoflavone.[33]	Promotes blood circulation.
Cydonia sinensis Thou. (Syn. *Chaeomeles sinensis*)	Xuan Mu Gua (Quince)	(fruit) Vitamin C, malic acid, tartaric acid, citric acid, hydrocyanic acid.[49]	As astringent in diarrhea, analgesic in arthralgia, gout, cholera.

Botanical name	Common name	Part / Constituents	Uses
Cymbidium hyacinthinum Sm. *C. striatum* Sw. (Syn. *Bletilla hyacinthina*)	Bai Ji	(root) Mucilage, essential oil, glycogen.[49]	For stomachache, venereal disease. Externally as emollient for burns and skin disorders.
Cymbopogon citratus (DC) Stapf.	Ning Meng Sian Mao (West Indian lemongrass, citronella)	(leaf, root) Elemicin, cymbopogonol, citral, dipentene, methylheptenone, beta-dihydropseudoionone, linalool, methylheptenol, alpha-terpineol, geraniol, nerol, farnesol, caprylic, citrogellol, citronellal, decanal, farnesal, isovaleric, geranic, citronellic.[50,60,568]	Treats blood in the urine, fever; antiseptic, preservative.
Cymbopogon distans (Nees ex Steud.) J. F. Watson *C. goeringii* (Steud.) A. Camus *C. nardus* Rendle	Yun Xian Cao, Xian Mao	(aerial part) Piperitone, essential oils.[33,192]	Antagonizes muscle contraction; antitussive, antibacterial.
Cynanchum atratum Bunge. *C. auriculatum* Royle	Bai Way	(stem, root) Cynanchol, cynanchin, cynanchocerin.[50,267] This herb cause abortion in sows.	For fever, leucorrhea, nephritis, tuberculosis, antipyretic, diuretic.
Cynanchum japonicum Moore et Decne.	Bai Chen	(root)[50] No information is available in the literature.	Antitussive, expectorant; for bad cold with cough and discomfort in the chest, asthmatic breathing, and acute bronchitis.
Cynanchum paniculatum L.	Xu Chang Qing	(root) Paeonol, paeonin, tomentogenin, deacylcynanchogenin, sarcostin, deacylmetaplexigenin.[33,50]	Sedative, analgesic, effect on the cardiovascular system, lowers plasma cholesterol level.
Cynodon dactylon (L.) Persoon	Tie Xian Cao	(root) Beta-sitosterol.[50]	Anticancer, depurative, diuretic, emollient.

Table 1A Major Constituents and Therapeutic Values of Chinese Medicinal Herbs (continued)

Scientific Name	Common Chinese and (English) Name	Major Constituents and (sources)	Therapeutic Values*
Cynoglossum divaricatum stemphan	Dao Ti Hu (Cynoglossum, hound's tongue)	(leaf) Potassium nitrate.[96]	A diuretic.
Cynomorium coccineum L. *C. songarium* L.	Su Yang (Juniper)	(stem) Anthocyanin, beta-sitosterol, palmitic acid, ursolic acid, daucosterol, catechin, naringenin-4'-O-pyranogluoside, succinic acid.[53,215]	Improves immunity, stimulates endocrine system; aphrodisiac, spermatopoietic.
Cyperus brevifolius (Rottb.) Hassk. *C. difformis* L. *C. glomeratus* L. *C. iria* L.	Sha Cao	(rhizome) Allelopathic essential oils, terpenes, alpha-cyperone, beta-selinene, alpha-humulene.[60,197,198]	A vermifuge, antidote, remedy for dysentery, alleviate stress, sedative.
Cyperus rotundus L.	Xiang Fu (Nut grass)	(tuber) Essential oils, alpha-cyperene, beta-cyperene, alpha-cyperol, beta-cyperol, cyperoone, patchoulenone, kobusone, capadiene, epoxyquaine, rotundone, rotunol, terpenes, olealonic acid, beta-sitosterol, pinene, sesquiterpenes.[33,450]	Treats dysmenorrhea, menstrual irregularities.
Cypripedium guttatum Swartz *C. macranthum* Swartz *C. macranthum* Swartz f. albiflorum Y. C. Chu *C. pubescens* Willd.	Shao Lan	(root, flower) Flavonoids, phenols, sterols, vitamin C.[48]	Diuretic; improves blood circulation, relieves pain.

Species	Chinese Name	Constituents	Uses
Cyrtomium falcatum (L. f.) Presel.	Quan Yuan Guan Zhong	(root) Flavonoid, cyrtomin, astragalin, isoquercitrin.[57]	Treats cold, fever, dizziness due to high blood pressure, insomnia.
Cytisus scoparius (L.) Link.	Jin Que Hua (Scotch broom)	(root) Sparteine, sarothamine, genisteine, scoparin.[60,568]	As a fomentation for bruises, a remedy for coughs, colds.
Daemonorops draco Blume.	Xue Jie	(resinous secretion from fruits) Amorphous dracoresene, amorphous dracoalban, benzoic acid, cinnamic acid, resin.[49]	Astringent, hemostatic, anticancer, for cancerous sores.
Daemonorops margaritae (Hance) Beccari	Huang Teng	(aerial plant) Dracoalban, dracoresene, dracoresinotannol, benzolacetic ester.[60]	Astringent.
Daphne fortunei Lindl. *D. genkwa* Sieb. et Zucc.	Yuan Hua (Fish poison)	(flower) Genkwanin, yuanhuacine, apigenin, hydroxygenkwanin, yuanhuatine, yuanhuadine, genkwadaphnin, 12-benzoxydaphnetoxin.[33,53,144,235] This herb is toxic.	Induces abortion, treats chronic bronchitis, malaria, cutaneous infections.
Daphne giraldii Nitsche *D. gurakduu* Nitsche *D. retusa* Hemsl. *D. tangutica* Maxim.	Zu Si Ma (Mezerum)	(root bark) Daphnetins.[33]	Analgesic, anti-inflammatory, antibacterial.
Daphne koreana Nakai	Chang Bai Rui Xiang (Daphne)	(root, stem) Daphnetins.[33]	Treats angina pectoris, arthritis.
Daphnidium myrrha Sieb. et Zucc. *D. strychnifolius* Sieb. et Zucc.	Wu Yao	See *Lindera strychnifolia*	

Table 1A Major Constituents and Therapeutic Values of Chinese Medicinal Herbs (continued)

Scientific Name	Common Chinese and (English) Name	Major Constituents and (sources)	Therapeutic Values*
Datura alba Nees. *D. fastuosa* L. var. alba Clark *D. innoxia* Mill. *D. metel* L. *D. stramonium* L. *D. tatula* L.	Man Tu Luo (Jimsonweed)	(leaf, seed, flower) Scopolamine, hyoscyamine, daturodiol, daturolone, hyoscine.[33,144,450,558] This herb is toxic.	Spasmolytic, analgesic, antiasthmatic, antirheumatic agent. A general anesthetic for major operations.
Datura suaveolens Humb. & Bonpl. ex Wlld.	San Hu Shu	(leaf, seed) 1-hyoscyamine, scopolamine, atropine, anisodine, anisodamine.[53]	Antispasmodic, bronchodilator.
Daucus carota L. subsp. sative Hoffm.	Nan He Chi (Carrot)	(whole plant) Carotenes, lycopene, phytofluere, umbelliferone, alpha-pinene, camphene, myrcene, daucol, alpha-phellandrene, bisabolene, luteolin-7-glucoside, daucine, pyrrolidine, geraniol, citronellol, carotol, citral, caryophyllene, p-cymene, asarone, daucosterol, petroselinic acid.[48,568]	For chronic dysentery, worms; carminative, diuretic, emmenagogue; lowers blood sugar, prevents cancer, diabetes, dyspepsias and gout.
Delonix regia (Boj.) Raf.	Feng Huan Mu	(bark, leaf) Gum, saponin, alkaloid.[61]	Febrifuge.
Delphinium grandiflorum L.	Cui Que	(root, whole plant) Methyllcaconitine.[48]	Emetic, cathartic.
Dendrobium nobile Lindl.	Shi Dou (Orchid)	(stem) Dendrobine.[50,558]	Analgesic, hyperglycemic, hypotensive, hypothermic.
Desmodium microphyllum (Thunb.) DC	Sui Me Jie	(whole plant) Kaempferitrin.[48]	Antitoxic; relieves diarrhea, cough, pain, snakebite.

Desmodium pulchellum (L.) Benth.	Pai Qian Chao	(aerial part) Bufotenine, nigerine, donoxime.[33]	Antimalarial, antipyretic, antischistosomiasis.
Desmodium triforum (L.) DC	San Dian Jir Cao	(whole plant) Potassium oxide, silicic acid, tannins.[60]	For dysentery, antirheumatic, antipyretic, jaundice, gonorrhea. Externally for wounds, abscesses, ulcers.
Desmodium triquetrum (L.) DC	Hu Lu Cao	(leaf) Potassium oxide, silicic acid, tannins.[50,60]	A tonic for dyspepsia, hemorrhoids, infantile spasms; insecticide, vermicide.
Dianthus barbatus L. var. asiaticus Nakai *D. superbus* *D. oreadum* Hance	Qu Mai (Carnation)	(aerial part) Dianthus saponin, essential oils, eugenol.[33]	Antipyretic, diuretic. Treats urinary tract infections, relieves strangury.
Dichroa cyanitis Miq. *D. febrifuga* Lour. *D. latifolia* Miq.	Chang Shan (Chinese quinine, fever flower)	(root) Dichroines, dichroidine, 4-quinazolone, dichrins.[33,558] This herb is toxic.	Antiamebial, antipyretic; for use against chicken malaria.
Dicranopteris linearis (Burm. f.) Under.	Mang Ji	(leaf, stem) Quercitrin, afzelin, nonacosane, heptacosane, nonacosan-10-one, nonacosan-10-ol.[54]	Anthelmintic, a poultice for fever; improves blood circulation; diuretic.
Dictamnus albus L. subsp. dasycarpus (Turcz.) Winter *D. dasycarpus* Turcz.	Bai Xian Pi (Fraxinella)	(root bark) Dictamnine, skimmianine, saponins, preskinnianine, choline, fragarine, aurapten, bergapten, isomaculosindine, limonin, obakinone, fraxinellone, psoralen, trigonelline.[50,60]	Antifungal, antipyretic, antiseptic, antitussive, sedative, emmenagogue, tonic.

Table 1A   Major Constituents and Therapeutic Values of Chinese Medicinal Herbs   (continued)

Scientific Name	Common Chinese and (English) Name	Major Constituents and (sources)	Therapeutic Values*
Digitalia purpurea L. *D. sanguinalis* (L.) Scop. *D. sanguinalis* (L.) Scop. var. ciliaris (Retz.) Parl.	Mao Di Huang (Foxglove)	(whole plant) Digitoxigenin, gitoxigenin, gitanin, gitaloxigenin, digitoxin, gitoxin, gitaloxin, digicoside, strospeside, digipurin, digicirin, digifolein, digitonin, purpureal glycosides.[60,87]	For gonorrhea, scleroses of the breast.
Dioscorea batatus Decaisue	Shu Yu (Yam)	(root) Allantoin, arginine, d-abscisin, mannan, phytic acid, diosgenin, protein, glycosides, triterpene glucosides.[48,461,462]	Antitumor, sore throat, swelling, food poisoning, goiter, hernia, purulent inflammations.
Dioscorea bulbifera L.	Huang Yao Zi (Potato yam)	(rhizome) Saponins, dioscorecin, iodine, dioscoretoxin, saponins, diosgenin, diosbulbin, tannins, campesterol, beta-sotpsterols, stigmasterol, diosbulbines.[33,48]	Treats cancer, goiter.
Dioscorea cirrhosa L. *D. hispida* Dennst. *D. japonica* Thunb.	Shu Liang (Dyeing yam)	(tuber) Tannins, mucus.[33]	Hemostatic; increases platelet aggregation, increases uterine contraction.
Dioscorea nipponica Makino	Chuan Shan Long (Japanese yam)	(root) Dioscin, diosgenin, trillin, 25-D-spirosta-3,5-diene.[33,53]	Antiinflammatory, antitussive, expectorant, antiasthmatic.
Dioscorea opposita Thunb.	Shan Yao (Chinese yam)	(leaf, tuber, root) Allantoin, arginine, choline, glutamine, leucine, tyrosine, diosgenin, sinodiosgenin.[50,568]	leaf juice for snakebite, root for asthma, cachexia, cough, debility, diarrhea, neurasthenia, polyuria; tuber is anthelmintic.

Diospyros chinensis Blume D. costata Carr. D. kaki L. D. lotus L. D. roxburgii Carr.	Shi Zi (Varnish persimmon, date plum)	(stem bark, fruit) Betulinic acid, acetylcholine, choline, shibuol.[50]	Astringent, stomachic; treats diarrhea, enterorrhagia, hemorrhoids; antifebrile, antivinous, demulcent.
Dipsacus asper Wall.	Xu Duan (Teazel)	(root) Essential oil, alkaloid lamine.[50]	Increases the leukocyte count, prevents spontaneous abortion.
Dodonaea viscosa (L.) Jacquin	Che Sang Zi	(leaf, bark) Alkaloid, glucoside, tannins, resins.[60]	Remedy for fever, astringent to treats eczema.
Dolichos lablab L.	Bai Ben Dou (Hyacinth bean)	(flower, seed) Glucokinin, plant insulin, tryptophane, arginine, lysine, tyrosine.[62]	Treats menorrhagia, leucorrhea, metritis.
Draba nemorosa L.	Ting Li Zi	(fruit stalk, seed) Allyl sinapic oil.[60]	As an expectorant, diuretic in chronic trachitis, asthma, pleurisy, hydrothorax.
Draceana graminifolia L. (Syn. Liriope spicata)	Mo Men Dong	(rhizome) Mucilage, dracorubin, dracorhodin, nordracorubin.[49,53]	Antitussive, expectorant, emollient.
Dracocephalum integrifolium L.	Quao Ye Ging Lan	(aerial plant) Essential oil, flavone glucoside.[33]	Antitussive, antiasthmatic, antiphlogistic, antibacterial.
Drosera anglica Hudson D. burmunni Vahl. D. rotundifolia L.	Mao Gao Cai (Sundew)	(whole plant) Citric acid, malic acid.[57]	Treats dysentery, scrofula, and malaria.
Dryobalanops aromatica Gaertn. D. camphora Colebr.	Loan Now Xiang (Borneo camphor)	(kernel of the fruit) Borneol, camphene, terpineol, sesquiterpene.[60] This herb is toxic.	A tonic and aphrodisiac; cataracts; reduces swelling. Externally for mucous membrane of the nose, eyes, throat, and on piles.

Table 1A   Major Constituents and Therapeutic Values of Chinese Medicinal Herbs   (continued)

Scientific Name	Common Chinese and (English) Name	Major Constituents and (sources)	Therapeutic Values*
Dryopteris crassirhizoma Nakai	Guan Zhong	(rhizome) Filmarone, filicic acid, diplotene, albaspididin, flavaspidin, fernene, dryocrassin.[33]	Anthelmintic, an insecticide, antitumor.
Dryopteris laeta (Kom.) Christ. *D. filix-mas* (L.) Schott.	Mian Ma Guan Zhong	(rhizome) Dryocrassin, filicic acid, filicin, paraaspidin, deaspidin, albaspidin, oleoresin, filmarone, flavaspidic acid, resin, diploptene.[50,53,60]	Anthelmintic to treats tapeworm, hemorrhage, hookworm, influenza. Externally treats leucoderma.
Duchesnea indica (Andr.) Focke.	She Mei (Mock strawberry)	(whole plant) Emodin, chrysophanic acid, phytosterol, volatile oil, calcium.[60]	Insecticide, antidote; treats whitlow, burns, snakebite.
Dysosma pleiantha (Hance) Woodson	Pa Jiao Lian	(rhizome) Podophyllotoxin, astragalin, peltatin, etoposide, hyperin, deoxypodophyllotoxin.[33]	Treats condyloma acuminata, exophytic warts.
Echinops dahuricus Fisch. *E. gmelini* Ledeb. *E. grijsii* Hance *E. latifollus* Tausch. *E. sphacrocephalus* Miq.	Lou Lu (Globe thistle)	(root, flower stalk) Echinopsine.[48]	Anthelmintic, galactagogue, depurative; treats tumors, swellings, leucorrhea, and gout.
Eclipta alba Hassk. *E. marginata* Boiss. *E. prostrata* (L.) L. *E. thermalis* Bunge.	Fang Kui (Pink plant)	(aerial part) Alkaloids, nicotine, ecliptine.[60]	Leaves heated or crushed in oil are applied to keep the hair black and to encourage its growth. Astringent, hemostatic, tonic.

Eclipta erecta L.	Mo Han Lian (Eclipta)	(aerial part) Essential oils, tannic acid, saponin, wedolactone, nicotine, ecliptine, demethylwedolactone, alpha-tertiary methanol, stigmasterol, beta-amyrin.[33,450]	Hemostatic effect, antimyotoxic, antihemorrhagic.
Elaeagnus formosana Nakai	Tiawan Hu Tin Chi	(root)[56] No information is available in the literature.	For arthritis pain, throat swelling, cough, bleeding, menses, stomachache.
Elaeagnus glabra Thunb.	Teng Hu Tin Chi (Elaeagnus)	(leaf, bark) Flavonoids, epigallocatechin, amino acids.[56]	Antifungal, antibacterial properties; relieves pain swelling, treats hepatitis, gastritis.
Elaeagnus oldhumii Maixmowicz	Yi Wu	(root) Sitosterol, muslinic acid, sitosteryl glucopyranosid, arjunolic acid.[56]	Treats arthritis pain, asthma.
Elaeagnus pungens Thunb. *E. umbellata* Thunb.	Hu Tin Chi	(root, leaf, fruit) Harman, tetrahydroharman, dihydroharman, 2-methyl-1,2,3, 4-tetrahydro-β-carboline, caffeic acid, chlorogenic acid, catechin, neochlorogenic acid, epicatechin.[48]	Treats coughs, watery diarrhea; an astringent in hemoptysis.
Elephantopus elatus L. *E. grandiflorus* Smith	Di Dan Tou	(leaf) Elaeocarpid, saponin.[60] This herb is toxic.	Treats bilious attacks, staphylococcus.
Elephantopus molis H. B. K.	Mao Liang Cai (Hairy elephant's foot)	(whole plant) Elephantin, deoxyelephantcpin, isodexyelephantopin, elephantopin, molephantin, motephantinin, phantomolin, dotriacontanol, epifriedelinol, lupeol, lupeol acetate.[58]	A tonic, diuretic; treats swellings, diarrhea.

Table 1A   Major Constituents and Therapeutic Values of Chinese Medicinal Herbs   (continued)

Scientific Name	Common Chinese and (English) Name	Major Constituents and (sources)	Therapeutic Values*
Elephantopus scaber L.	Tian Ja Cai (Rough elephant's foot)	(whole plant) A bitter principle, a glycosidic compound.[50,52]	Diuretic, tonic, vermifuge, for diarrhea, dysentery, leucorrhea.
Elettaria cardamomum Maton.	Yi Zhi Zi (Cluster cardamom)	(seed) Phytosterol, palmitic acid, oleic acid, linoleic acid, p-cymene, camphene, d-limonene, myrcene, alpha-phellandrene, pinene, sabinene, terpinene, thujone, cineole, camphor citral, linalol, citronellal, dl-borneol, citronellol, geraniol, terpineol, sabinene.[50]	Carminative, emmenagogue, stimulant, stomachic, tonic. Treats ague, cachexia, dyspepsis, enuresis, gastralgia, nausea, spermatorrhea.
Eleutherococcus senticosus (Rupr. ex Maxim.) Maxim.	Ci Wu Jia (Siberian ginseng)	(bark, root) Eleutherosides, beta-sitosterol glucoside, l-sesamen, syringareinol.[7,33,396,568]	Central nervous system activating and antistress action.
Elsholtzia argyi Lev. E. cristata Willd. E. feddei Lev. E. souliei Lev. E. splendens Nakai	Xiang Xu (Aromatic madder)	(whole plant) Essential oils, elsholtzia ketone, elsholtzianic acid, furylmethyl ketone, furylpropyl ketone, furylisobutyl ketone, furane, pinene, terpene.[49]	Stomachic, carminative, diuretic.
Emilia sonchifolia (L.) DC	Zi Bei Cao (Red tasselflower)	(leaf) Alkaloids.[63]	For dysentery, phthisis, coughs; a detoxicant, diuretic, febrifuge.
Entada phaseoloides (L.) Merrill.	Guo Gang Long	(stem) Entageric acid.[33]	Antirheumatic, promotes collateral flow, relieves blood stasis.

Ephedra distachya L. E. equisetina Bunge. E. intermedia Schrenk ex Mey. E. monosperma Gmel. ex. Mey. E. sinica Stapf.	Ma Huang (Ma Huang)	(aerial part) l-ephedrine, l-methylephedrine, l-norephedrine, methylephedrine, d-pseudoephedrinem, ephedrine, d-N-methylpseudoephedrine, norpseudoephedrine.[30,33,352,510,558,568] This herb is toxic.	Treats asthma, sympathomimetic action; relieves headache, body ache and coughing, and lowers fever by increasing perspiration.
Epidendrum monile Thunb.	Shi Dou	See Dendrobium nobile	
Epidendrum striatum Thunb, E. tuberosum Lour.	Bai Ji	See Bletilla hyacinthina	
Epilobium amurense Haussku. E. hirsutum L. E. palustre L. E. tanguticum (L.) Hausskn.	Liu Ye Cai	(hair of the seed, shoot) Anthocyanin.[50,358]	A tonic, galactagogue, stomachache, dropsy. seed hairs are applied as a styptic.
Epimedium brevicorum Maxim. E. koreanum Nakai E. macranthum Moore et Decne.	Jin Yang Huo	(aerial part) Icariin, noricariin, korepimedoside A, korepimedoside B, icariine, des-O-methyl-licariine, magnoflorine, epimedoside A, polysaccharides.[33,48]	Dilates the coronary vessels and increases the coronary flow by reducing vascular resistance.
Equisetum arvense L. E. hyemale L. E. ramosissimum Desf.	Mo Ja Chac (Horsetail)	(whole plant) Equisetonin, equisetrin, articulain, isoquereitrin, galuteolin, populnin, kaempferol-3,7-diglucoside, astragalin, palustrine, gossypitrin, 3-methoxypyridine, herbacetrin.[48,568]	Antihemorrhagic, anodyne, carminative, diaphoretic, diuretic.
Erigeron canadensis L. E. annuus (L.) Persoon	Canada Pon Yi Nian Pon (Fleabane)	(leaf) Essential oils, gallic acid, tannic acid, limonene, dipentene, methylacetic acid, terpeneol, lacnophyllum, matricaria, dehydromatricaria, erigeron, hexahydromatricaria.[50]	For hemorrhage, diarrhea, dysentery, internal hemorrhage of typhoid fever.

Table 1A Major Constituents and Therapeutic Values of Chinese Medicinal Herbs (continued)

Scientific Name	Common Chinese and (English) Name	Major Constituents and (sources)	Therapeutic Values*
Eriobotrya japonica Linkdl.	Pi Pa Yie (Loquat)	(leaf, flower, fruit) Levulose, sucrose, malic acid, citric acid, tartaric acid, succinic acid, amygdalin, crytoxanthin, carotenes, phenyl ethyl alcohol pentosans, essential oils.[50]	Antitussive, expectorant, treats bronchitis, cough, fever, nausea, externally applied to epistaxis, smallpox, ulcers.
Eriocaulon sieboldianum Stend. *E. buegerianum* Koern.	Ke Jing Cao	(whole plant)[50] No information is available in the literature.	Antiphlogistic, diuretic, febrifuge, ophthalmic.
Erycibe henryi Prain *E. aenea* Prain	Ding Gong Teng	(leaf, stem, root) Scopoline, erycbelline, scopoletin.[56]	leaf poultices applied to sores and to the head to treats headache, arthritis, swelling, pain.
Erysimum amurense Kitag. var. bungei (Kitag.) Kitag. *E. cheiranthoides* L.	Tang Jie	(root, leaf, shoot) Erysimoside, erysimosol, erucic acid, canescein, erychroside, helveticosol, erythriside, corchoroside A, erysimotoxin.[35,48]	Treats cold and cold-related infections, sore throat, dizziness.
Erythrina corallodendron L. *E. indica* Lam. *E. variegata* L.	San Hu Ci Tong Hai Tong Pi (Indian coral tree)	(leaf, stem bark) Alkaloids.[50] This herb is toxic.	Anthelmintic, antisyphilitic, laxative, analgesic in arthritis, neuralgia, rheumatism.
Erythroxylum coca Lam.	Guo Ko Yi	(leaf) l-cocaine, cinnamylococaine, alpha-trevilline, beta-trevilline, ecgonine, benzoylecgonine.[33,568]	For local anesthetic; has vasoconstriction effect.

Eucalyptus robusta Sm.	Da Ye An (Swamp mahogany)	(leaf) Essential oils, cineol, thymol, gallic acid.[33]	Antibacterial, antimalarial. Externally treats Trichomonas vaginalis.
Euchresta japonicum Benth.	Shan Duo Gen	(stem)[50] Lupin alkaloid, (+)-5, 17-dehydromatrine N-oxide, (−)-12-cytisineacetic acid, euchrestaflavanones.[199,200,201]	A disinfectant, for asthma, bronchitis, cancer, congestion, fever, snakebite; aphrodisiac.
Eucommia ulmoides D. Oliver	Du Zhong	(bark) Pinoresinol-di-β-D-glucoside, resin, aucubin, ajugoside, reptoside, harpagide acetate, encommiol.[33,558]	Improves liver and kidney function, lowers blood pressure.
Eugenia aromatica Baill. E. caryophyllata (L.) Thunb. E. ulmoides Oliv.	Ding Xian (Clove tree)	(flower bud) Rhamnetin, eugenitin, kaempferol, oleanolic acid, isoeugenitin. Bark: ellagic acid, beta-sitosterol, mairin. Essential oil: ugenol, humulene, acetyleugenol, chavicol, alpha-caryophylline, beta-caryophylline, ylangene.[33,568]	Stimulates gastric secretions, increased in digestion; and a dispelling of gases. Antibacterial, antifungal.
Euonymus alatus (Thunb.) Sieb. E. alatus (Thunb.) Sieb. var. apterus Regel E. bungeanus Maxim. E. maackii Rupr.	Wei Mao (Thimble tree)	(young branch, leaf, fruit) Quercetin, dulcite, epifriedelinol, friedelin, fatty acid, alatamine, kaempferol glucosides, wilfordine, resin, sesquiterpene alkaloids.[33,337,338,339]	Regulates blood flow, relieves pain, eliminates stagnant blood, and treats dysmenorrhea.
Eupatorium chinense L. var. simplicifolium (Malcino) Kitam. E. lindleyanum DC E. japonicum Thunb.	Zi Lan (Thorowort)	(whole plant) Cumarin, D-cumaric acid, lactones, rhymohydroquinone, volatile oil, euparin, bornyl acetate, dimethyl thymohydroquinone, linalool.[48]	Sedative in disturbances of pregnancy and puerperium. Carminative, diuretic, vermifuge.

Table 1A Major Constituents and Therapeutic Values of Chinese Medicinal Herbs (continued)

Scientific Name	Common Chinese and (English) Name	Major Constituents and (sources)	Therapeutic Values*
Eupatorium formosanum L.	Taiwan Pai Lan	(whole plant) Sesquiterpine lactones, eupatolide, eupaformonin, eupaformosanin, michelenolide, costunolide, parthenolide, santamarine.[33,501]	Anticancer.
Eupatorium odoratum L.	Pei Lan	(seed) Eupatol, lupeol, beta-amyrin, salvigenin, isosakuranetin, odoratin, aromatic acids, anisic acid.[50]	Anodyne, hemostat, spasmolytic, vermifuge.
Euphorbia antiquorum L.	Huo Yu Jin (Fleshy spurge)	(whole plant) Friedelaun-3-ol, alpha-taraxerol, beta-amyrin, cycloarternol, euphol, alpha-euphorbol.[33]	Diuretic.
Euphorbia coraroides Thunb. *E. lasiocaula* Boiss. *E. lunulata* Bunge. *E. pallasii* Turcz. *E. pekinensis* Rupr. *E. sampsoni* Hance *E. sieboldiana* Moore et Decne.	Da Ji (Peking spurge)	(root) Euphorbon, euphorbias, butyric acid, calcium malate, calcium oxalate, vitamin C.[48,50,558]	Diuretic, emetic, emmenagogue, purgative.
Euphorbia esula L. *E. helioscopia* L.	Ze Qi or Di Jin Cao (Spurge)	(whole plant) Phasin, tithymalin, helioscopiol, butyric acid, euphorbine, phasine, saponin.[33,50]	Diuretic, febrifuge, vermifuge.

Euphorbia hirta L.	Da Fei Yang Cao (Asthma herb)	(stem) Camphol, leucocyanidol, quercitol, quercitrin, rhamnose, euphorbon, chlorophenolic acid, taraxerol, taraxerone, gallic acid.[50]	For asthma, bronchitis; externally for athlete's foot.
Euphorbia humifusa Willd.	Deng Qing Cao (Wolf's milk)	(aerial part) Camphol, euphorbon, gallic acid.[48]	Antibacterial, detoxicant against diphtheria toxin.
Euphorbia kansui Lion.	Qian Jin Zi, Gan Suei	(root) Alpha-euphol, tirucallol, alpha-euphorbol, kansuinine.[33,144] This herb is very toxic.	Diuretic, expectorant; for ascites, constipation, dysuria, hydrothorax.
Euphorbia lathyrus L. *E. lucorum* Rupr. *E. resinfera* Berger *E. thymifolia* L.	Xu Sui Zi Da Ji Ru Zi Shu Xiao Fei Yang Cao (Caper spurge, petroleum plant)	(seed) Euphorbiasteroid, betulin, 7-hydroxylathyrol, lathyrol diacetate benzoate, lathyrol diacetate nicotinate, euphol, euphorbol, euphorbetin, esculetin, daphnetin.[33,53,144] This herb is very toxic.	Diuretic to remove edema; eliminate blood stasis and resolved masses; antitumor.
Euryale ferox Salisb.	Qian Shi (Water lily)	(seed) Protein, starch.[53]	Treats diarrhea, spontaneous emission, and leucorrhagia.
Evodia lepta (Spreng.) Merrill. *E. triphylla* DC	San Ya Ko	(root, leaf) Amino acids.[55]	For arthritis, chickenpox, fever, hemorrhoids, itch, infectious hepatitis.
Evodia rutaecarpa (Juss.) Berth	Wu Zhu Yu (Evodia)	(fruit) Alkyl methyl quinolone alkaloids, evodiamine, limonin, evocarpine, rutaecarpine, N-methyl anthranilic acid, evodol, hydroxyevodiamine, N-methylanthranflamide, N,N-dimethyl-5-methoxytryptamine, dehydroevodiamine.[32,33,237]	Antiemetic, analgesic; lowers blood pressure; antibacterial.

Table 1A Major Constituents and Therapeutic Values of Chinese Medicinal Herbs (continued)

Scientific Name	Common Chinese and (English) Name	Major Constituents and (sources)	Therapeutic Values*
Evonymus alatus Regel E. subtriflorus Blume E. thunbergianus Blume	Wei Mao	(twig)[50] No information is available in the literature.	For analgesic, emmenagogue, purgative in female disorders.
Fagopyrum esculentum Moench. F. cymosum (Trev.) Meisn. F. sagittatum Gilib.	Qiao Mai (Buckwheat)	(seed, leaf, stem) Rutin, quercetin, caffeic acid, rutin, orientin, homoorientin, vitexin, saponaretin, cyanidin, fagopyrin, flavanol, fagomine, alanine, leucoanthocyanin. seeds contain amylase, linamarase, maltase, phosphatides, protease, quercitol, rhamnose, urease.[48,50,421,446,450]	For colic and diarrhea; stops cold sweats, tumor inhibition, treats lung cancer.
Fagopyrum tataricum (L.) Gaertn.	Ku Qiao Mai (India wheat)	(whole plant) Rutin, flavones.[48]	For stomachache, leg pain; a digestive.
Ferula assa-foetida L. F. bungeana Kitag.	A Wei (Asafetida)	(gum or resin) Vanillin, asarensinotannol, ferulic acid, farnesiferols.[33]	Anthelmintic; treats ascites, dysentery, malaria.
Fibraurea recisa Hance	Huang Teng	(stem) Palmatine, jatrorrhizine, fibramine, fibraminine, fibralactone, sterol.[33]	Antipyretic, detoxicant; treats tonsillitis and pharyngitis.
Ficus awkeotsang Makino	Ai Yu Zi	(root, leaf, stem) Resin, glucose, fructose, gum, protein, fat.[55]	Treats arthritis, joint discomfort.

Ficus carica L.	Wu Hua Go (Fig)	(leaf, fruit) Bergaptin, cerotinic acid, ficusin, glutamine, papain, pepsin, psoralen, guaiaxulene, amyrin, lupeol, retin, octacosane, guaiacol, quercitin, rhamnose, sitosterol, tyrosine, urease.[50,55,502]	For stomachache; externally for swollen piles, corns, warts. Fruit is a laxative, digestive, anthelmintic, hypolipidaemic, and hypotriglyceridaemic activities.
Ficus pumila L. F. inicrocarpa L.	Bi Li Go Rong Shu (Creeping fig)	(whole plant) Latex.[50]	Carbuncle, dysentery, hematuria, piles, hernia, bladder inflammation.
Firmiana simplex (L.) W. F. Wight	Wu Tong (Chinese parasol tree)	(leaf, seed, bark, root) Betaine, choline, beta-amyrin, beta-amyrin acetate, rutin, lupenone, heutriacontane, octacosanol, beta-sitosterol.[33]	Detoxicant, smooths lung function, increases appetite.
Flagellaria indica L.	India Bian Teng	(leaf, fruit) Alkaloidal substances, cyanogetic glycosides, emilsin-like enzyme.[60]	Astringent, vulnerary, diuretic; treats pox.
Foeniculum officinale All. F. vulgare Mill.	Xiao Hui Xiang (Fennel)	(fruit) Anethol, d-fenchone, anisaldehyde, methylchavicol.[33,87]	Restore normal functioning of the stomach.
Forsythia suspensa (Thunb.) Vahl.	Lian Qiao (Forsythia)	(leaf, fruit, root) Phillyrin, rutin, taraxasteryl palmitate and acetate, bigelovin, dihydrobigelovin.[50,260,558]	Febrifuge, for cancer, carbuncle, chickenpox, antiphlogistic, diuretic, emmenagogue, antiemetic, laxative, antipyretic.
Fortunella crassifolia Swingle F. japonica (Thunb.) Swin. F. margarita (Lour.) Swin.	Jin Gan Yuan Jin Gan Jin Ju (Kumquat)	(whole plant) Glucosides, galactose, essential oil, pentosane, vitamin C.[60]	Antiphlogistic, antivinous, carminative, deodorizing, stimulant.

Table 1A   Major Constituents and Therapeutic Values of Chinese Medicinal Herbs   (continued)

Scientific Name	Common Chinese and (English) Name	Major Constituents and (sources)	Therapeutic Values*
Fragaria indica Andr.	She Mei (Mock strawberry)	See *Duchesnea indica*	
Fraxinella dictamnus Moench	Bai Xian Pi	See *Dictamnus albus*	
Fraxinus bungeana DC *F. chinensis* Roxb. *F. floribunda* Bunge. *F. obovata* Blume *F. ornus* L. var. bungeana Hance *F. rhynchophylla* Hance	Zhen Pi (Chinese ash)	(bark) Fraxin, aesculin.[33] This herb is toxic.	Antibacterial, analgesic, anti-inflammatory.
Fritillaria anheunensis Chen et Yin *F. collicola* Hance *F. maximowiczii* Freyn *F. roylei* Hook *F. thunbergii* Miq. *F. ussuriensis* Maxim. *F. verticillata* Willd.	Bei Mu (Fritillaria)	(bulb) Fritilline, fritillarine, verticine, verticinine, peimine, peiminine, peimisine, peiniphine, peimidine, peimilidine, propeimin, puqiedinone, isosteroidal alkaloids.[33,243,503]	Causes bronchodilatation and inhibition of mucosal secretions. Antitussive; stimulates uterine and intestinal contractions.
Galium bungei Stead. *G. spurium* L. *G. verum* L. var. leiocarpum Ledeb. *G. verum* L. var. trachycarpum DC	Si Ye Lu Zhu Yin Yin Peng Zi Cao (Bedstraw)	(rhizome) Alisarin, rubrierythrinic acid, purpurin.[60]	Treats rheumatism, jaundice, menstrual difficulties, epistaxis, hemorrhages.

		(whole body) Ergosterol, fungal lysozyme, amino acids, proteinase, organic acids, polysaccharides, adenosine, triterpenoids. [33,41,403,404,407,268]	Improves immune system, reduce cholesterol, treats blood pressure, prevent blood clot, regulates blood circulation, antitumor, antiviral. Treats hepatitis, hyperlipemia, angina pectoris, chronic bronchitis, leucopenia.
Ganoderma lucidum (Polyporaceae)	Ling Zhi		
Gardenia angusta (L.) Merrill. G. *jasminoides* Ellis.	Shan Zhi (Cape jasmine)	(fruit, flower, bark) Gardenin, alpha-crocetin, volatile oil, chlorogenin, glycosides, mannit. [64]	Emetic, stimulant, febrifuge, diuretic, hemostatic, antihemorrhagic, emmenagogue.
Gardenia florida L. G. *grandiflora* Sieb. et Zucc. G. *maruba* Sieb. G. *pictorum* Hassk. G. *radicans* Thunb.	Zhi or Zhi Zi (Gardenia)	(fruit) Gardenoside, shanzhiside, cardoside. [33]	Antipyretic, choleretic, sedative, hypnotic, anticonvulsant, antibacterial, anthelmintic properties.
Gastrodia elata Blume G. *elata* Blume f. pallens (Kitag.) Tuyama	Tian Ma	(root) Vanillyl alcohol, vanillin, vitamin A, gastrodin. [33]	Anticonvulsive, sedative, analgesic effect.
Gaultheria leucocarpa f. var. cumingiana (Vidal) Sleumer	Bai Zhu Shu	(leaf) Methylsalicylate, salicylic acid. [60]	Treats rheumatism; an antiseptic.
Gelidium amansii (Lamx.)	Qiong Zhi	(isolated mucous substance) Agarose, agaropectin, taurine. [33]	A mild laxative in the treatment of chronic constipation.
Gelsemium sempervirens (L.) Ait. G. *elegans* Benth.	Gou Min (Jessamine)	(root, stem) Gelsemine, gelsemidine, kcumine, sempervirine, kouminine, kouminicine, douminidine. This herb is highly toxic. [33,46,50]	For caked breast, perspiring feet, skin eruptions, wounds.

Table 1A   Major Constituents and Therapeutic Values of Chinese Medicinal Herbs   (continued)

Scientific Name	Common Chinese and (English) Name	Major Constituents and (sources)	Therapeutic Values*
Gentiana algida Pall. G. barbata Froel. G. manshurica Kitag. G. olivieri DC G. scabra Bunge. G. squarrosa Ledeb. G. triflora Pall.	Long Dan (Gentian)	(root) Gentiopicrin (or gentiopicroside), indoid compounds such as geniposide and gardenoside. Saponins, gentianine.[16,17,33]	For arthritis, cancer, carbuncle, cold, conjuctivitis, diarrhea, gastritis, neuralgia.
Gentiana dahurica Fisch. G. lutea L. G. macrophylla Pall.	Jue Chuang (Gentian)	(root) Gentianine, gentianidine, gentianol.[33]	Treats rheumatism and fever; antipyretic, anti-inflammatory, antihypersensitivity, and antihistaminic effects.
Geranium dahuricum DC G. eriostemon Fisch. ex DC G. eriostemon Fisch. ex DC f. hypoleucum (Nakai) Y. C. Chu G. eriostemon Fisch. ex DC f. megalanthum (Nakai) Y. C. Chu G. sibiricum L. G. wilfordi Maxim. G. wlassowianum Fisch. ex Link	Lao Huan Cao (Geranium)	(aerial part) Kaempferitrin, gallic acid, quercetin, succinic acid, tannins.[48,50,65]	Astringent, for diarrhea, endometritis, nervous diseases, numbness of limbs, pains, rheumatism. It helps circulation and strengthens bones and tendons.
Geum aleppicum Jacquin G. aleppicum Jacquin f. glabricaule (Juzepczuk) Kitag.	Shui Yang Mei (Avens)	(whole plant) Flavones, fatty acids, eugenol, gein, geoside.[48]	Treats bleeding, bug bite, convulsive disorder, fevers, irritability, obstinate skin diseases.

Ginkgo biloba L.	Yin Xing (Ginkgo)	(leaf, seed kernel) Kaempterol-3-rhamnoglucoside, gibberellin, cytokinin, ginkgolic acid, ginkgol, bilobal, ginnol, rutin, ginkgolides, querretin, quercitrin, ginkgetin, isoginketine, bilobetin, isorhamnetin, shikimic acid, D-glucaric acid, anacardic acid, sesquiterpene, diterpenes, beta-sterol.[33,48,422,450,510,396,58,603]	Antitussive, antiasthmatic; anodyne, treats coronary artery disease, angina pectoris, hypercholesterolemia, Parkir son's disease; inhibits the growth of human cancer cell lines.
Glechoma hederacea L. var. grandis (A. Gray) Kudo *G. longituba* (Nakai) Kuprijan.	Jin Qian Cac (Ground ivy)	(aerial part) l-pinocamphone, l-menthone, isomenthone, l-pulegone, alpha-pinene, beta-pinene, 1,8-cineol, isopinocamphone, limonene, menthol, alpha-terpineol, linalool, p-cymene.[48]	Febrifuge, anodyne, treats earache, arthritis, fever, toothache, diuretic, decocgulant.
Gleditsia horrida Willd. *G. sinensis* Lam. *G. xylocarpa* Hance	Zao Ci (Chinese honey locust)	(leaf, fruit, seed) Saponin, arabinon, gleditsin, fisetin, fustin.[50,558]	Anthelmintic, febrifuge; treats cough, cysertery, flatulence, rectal prolapse, stroke, throat numbness, tetanus; emtic.
Glehnia littoralis F. Schmidt et Miq.	Bei Za Seng (Beech silve-top)	(leaf, root) Stigmasterol, beta-sitosterol, imperatorin, psoralen, osthenol-7-o-β-gentiobioside, petroselenic acid, petroselidinic acid, polyine, polysaccharides, falcalindiol, anthocyanin, furanocoumarin.[53,84,477,478,479,480,383,484,384]	Anthelmintic, for chronic bronchitis, cough and hoarseness, antiproliferative activities, antimycobacterial, immunosuppressive activities.

Table 1A　Major Constituents and Therapeutic Values of Chinese Medicinal Herbs　(continued)

Scientific Name	Common Chinese and (English) Name	Major Constituents and (sources)	Therapeutic Values*
Glycine max (L.) Merrill G. *soja* Sieb. & Zucc.	Da Dou Ye Da Dou (Soybean)	(seed) Protein, isoflavone derivatives, genisteine, daidzein, riboflavin, thiamine, niacin, pantothenic acid, choline.[33,67,568]	Phytoestrogenic; elevates the vasomotor system, prevents cancer; a potent inhibitor of protein tyrosine kinase.
Glycosmis cochinchinensis Pierre G. *pentaphylla* (Retz.) Correa.	Xiao Shan Ju	(leaf, root) Glycosmine, skimmianine, glycosminine, glycosine.[66]	Treats coughs, inflammation. An appetite enhancer.
Glycyrrhiza pallidiflora Maxim. G. *uralensis* Fisch. ex DC	Gan Cao (Licorice)	(outer cortex of root) Glycyrrhiza, triterpenoid saponin, flavonone glucoside, liquiritin, aglycone, liquiritigenin, chalcone glucose, isoliquiritin, aglycone, isoliquiritigenen, glycyrrhizic acid, beta-glycyrrhetinic acid.[1,33,355,356,510,567,568]	Anti-inflammatory, anticonvulsant, carminative, antidote, antitumor, antispasmodic, antiulcer.
Gnaphalium affine L. G. *arenarium* Thunb. G. *confusum* DC G. *javanum* DC G. *luteo-album* L. var. multiceps Hook G. *multiceps* Wall. G. *ramigerum* DC G. *tranzschelii* Kirpicznikov G. *uliginosum* L.	Shu Qu Cao (Cudweed)	(whole plant) Fat, resin, phytosterol, essential oils, carotene, vitamin B_1.[48,49,50,449]	Remedy for lung disease; antifebrile, antimalarial; reduces blood pressure and stomach and intestinal ulcers. Externally for wounds, against cancer.

Gomphrena globosa L.	Qian Ri Hong (Globe amaranth)	(flower) Saponins, betacyamines, gomphrenin, amaranthin, isoamaranthin.[33]	Treats chronic bronchitis, whooping cough.
Gossypium herbaceum L.	Mian Zi Soo or Mian Hua Gen (Cotton)	(root) Gossypol, hemigossypol, 6,6'-dimethoxylgossypol, aflatoxin B (in seed), methoxylhemigosipol, acetovanillone, hirsutrin (in leaf).[33]	Antitussive; treats bronchitis.
Gynostemma pentaphyllum (Thunb.) Makino	Joe Koo Lan	(root) Panaxatriol, panaxadiol, saponin, glypenosides, sterol.[33,34,349,350,351]	Regulating effect on lymphocyte transformation, protective effect against myocardial and cerebral ischemia; relaxes isochemic heart ventricles.
Gynura bicolor DC	Mu Er Cao (Velvet plant)	(whole plant) Flavonoids.[54]	Improves blood circulation, stops bleeding; a detoxicant, relieves swelling, cough with blood.
Gynura japonica Mak. *G. pinnatifida* Vanniot *G. segetum* Merr.	San Qi (Canton tusanchi)	(root, leaf) Saponins.[-9]	Hemostat, furnculosis, hemorrhage, hemorrhea. Externally for bruises and wounds, insect bites, snakebites.
Haemanthus multiflorus Mart. ex Willd.	Huo Qin Hua	(bulb) Haemanthramine, haemanthridien.[57] This herb s toxic.	Detoxicant, relieves swelling.
Hedera rhombea (Miq.) Bean *H. helix* L.	Chang Chun Ton	(leaf) Hederin, hederaic acid, tannic acid, oleic acic.[50,568]	For cough, headache; diaphoretic, emmenagogue.
Hedychium coronarium Koen.	Shan Ren (Ginger lily)	(flower, rhizome) Sesquiterpenes, phenols, aldehyde, ketone, 1,8-cineole, camphene, beta-pinene.[60,195]	Stimulant.

Table 1A Major Constituents and Therapeutic Values of Chinese Medicinal Herbs (continued)

Scientific Name	Common Chinese and (English) Name	Major Constituents and (sources)	Therapeutic Values*
Hedyotis corymbosa (L.) Lamarck.	Shui Xian Cao	(seed) Borneol, bornyl acetate, l-camphor, linalool, nerolidol.[49]	Stomachic, mouthwash to relieves toothache, as a poultice to heal wounds, small sores.
Hedyotis diffusa Willd.	Bai Hua She Shi Chao	(leaf) Acyl flavonol di-gycoside, iridoid glucosides, anthraquinone, essential oils, p-vinylphenol, p-vinylguaiacol, linalool.[50,202,203,204,205,206]	Immunopotentiation activity; treats tumors, antibacterial, antipyretic, detoxicant, diuretic, anticancer, externally applied as lotion.
Hemerocallis flava L.	Huang Hua Xuan Cao (Day lily)	(root) Protoveratrine, jervine, pseudojervine.[60]	Sternutative, anthelmintic, evacuant properties.
Hepatica asiatica Nakai	Xi Shin (Liver leaf)	(root) No information is available in the literature.	Anodyne, antifebrile; for angina and sunstroke, local application in smallpox ulcerations.
Heracleum dissectum Ledeb. *H. lanatum* Michx.	Niu Fang Feng	(root) Oxalic acids, hydroxycinnamic acids, hydroxybenzoic acids, coumarins, anthocyanidines, anthraquinones, phytosterines, carotenes, ether oils, monoterpene, sesquiterpene glucosides, hydrocyanic acids, xanthotoxin, coumarin, bergapten.[222,223,450]	Relieves headache, toothache, hematuria, gonorrhea, itching skin, swelling; removes corns from the feet.
Hibiscus chinensis DC *H. rhombifolius* Cav. *H. syriacus* L. *H. trionum* L.	Mu Jin Chuan Jin Pi (Shrubby althea)	(bark) Saponarin.[33]	Treats dysentery, diarrhea, jaundice, eczema, tinea, and scabies. Antiphlogistic.

Species	Name	Constituents	Uses
Hibiscus mutabilis L.	Fu Rong Yie (Cotton rose)	(leaf, flower) Isoquercitrin, hyperoside, rutin, quercetin-4-glucoside, quercetin, quercimeritrin.[50]	Applied to swellings, burns, ulcers. Internally lung ailments, cough, dysuria, menorrhagia.
Hibiscus rosa-sinensis L. *H. rhombifolius* Cav.	Zhu Jin Chuan Jin Pi (Rose of China)	(leaf, flower) Protein, thiamine, riboflavin, niacin, cyandidin-3-sophoroside.[50]	Used as poultice on cancerous swellings and mumps.
Hibiscus sabdariffa L.	Luo Sheng Kui	(leaf, flower, stem bark) Saponin, saponaretin, vitexin.[50]	Stomachic, diuretic, expectorant, hematochezia, gas, vertigo.
Hieracium umbellatum L.	Shan Liu Jiu	(whole plant) Vitamin C, tannic acid.[48]	Relieves pain, bladder infection, diarrhea.
Hierochloe odorata (L.) Beauv.	Mao Xian	(root, flower head) Coumarin, coumarinic acid-β-glucoside.[48,568]	Relieves internal bleeding, kidney infection.
Hippeastrum hybridum Hortorum	Shi Suan Hua	(bulb) Lycorine, lycoramine, tazettine, galanthamine. This herb is toxic.	Detoxicant, relieves swelling, induces vomiting.
Hippophae rhamnoides L.	Sha Ji (Sea buckthorn)	(seed, fruit, leaf) Cryptoxanthin, harman, harmol, hemin, isorhamnetin, lycopene, serotonin, isorhamnetin-3-mono-beta-D-glucoside, polyphenols, fatty acids flavonoid, essential oils, tannins, quercitin, vitamin C, vitamin E, beta-carotenoid.[50,450,568]	Improves resistance to infection, skin irritation and eruption; treats heart disease; oil for cosmetic use.

Table 1A   Major Constituents and Therapeutic Values of Chinese Medicinal Herbs   (continued)

Scientific Name	Common Chinese and (English) Name	Major Constituents and (sources)	Therapeutic Values*
Hordeum vulgare L.	Mai Ya (Barley)	(germinated seed) Enzymes such as invertase, amylase, proteinase, vitamin B, vitamin C, maltose, dextrose.[33]	Improves digestion of carbohydrates and protein.
Houttynia cordata Thunb.	Yu Xing Cao (Fishwort)	(aerial part) Essential oil, houttuynium, decanoylacetaldehyde, quercitrin, isoquercitrin.[33]	Antibacterial, antiviral, analgesic, hemostatic, antitussive.
Hovenia dulcis Thunb.	Zhi Bei Zi (Japanese raisin tree)	(stem bark) Ebelin lactone, hovenosides, potassium malate, potassium nitrate.[27,50]	For rectal diseases, constipation, infantile convulsions; antispasmodic, febrifuge.
Hoya carnosa (L. F.) R. Brown	Yu Dei Mei	(leaf) Condurangin, hoyin, phytosterindigitonid.[50]	To hasten maturation of anthrax and furuncles.
Humulus lupulus L.	She Ma (Hop)	(female flower, unripe fruit) Humulone, resin, lupulone, choline asparaginer, lupulin, isohumulone, isovaleric acid.[33,450,568] This herb is toxic.	Inhibits the growth of tubercle bacillus and arrests stuberculosis.
Humulus scandens (Lour.) Merr.	Lu Cao	(aerial part) Humulone, lupulone, asparagine, choline, luteolin.[33]	Inhibits tubercle bacillus; antipyretic, diuretic.
Hydnocarpus anthelmintica Pierre *H. castaneus* H. F. & Th.	Da Feng Zi (Krabao oil tree, chaulmoogra)	(seed) Hydrocarpus oil, hydnocarpic acid, chaulmoogric acid, gorlic acid.[33,558]	Anthelmintic.

Species	Chinese name	Parts / Constituents	Uses
Hydrangea macrophylla (Thunb.) Seringe	Xiu Qiu (Sugar-leaf hydrangea, French hydrangea)	(leaf, flower, root) Febrifugin, hydrangeic acid, hydrangenol, rutin.[50]	Antimalarial, antitussive, diuretic.
Hymenocallis speciosa Salisbury	Shui Gui Jiao	(bulb) Lycorin.[60] This herb is toxic.	As a vulnerary.
Hyoscyamus bohemicus F. W. Schmidt / *H. niger* L.	Liang Shi (Henbane)	(root, leaf) Alkaloid, hyoscyamine, hyoscine, scopolomine, hyoscypierin, choline, mucilage, albumin.[60,144,450] This herb is toxic.	Antispasmodic activity.
Hypericum attenuatum Choisy / *H. ascyron* L. / *H. japonicum* Thunb. / *H. perforatum* L. / *H. sumpsonii* Hance	Jin Si Tao or Di Er Cao / Guan Ye Lean Qiao / Tian Bao Cao (St. John's wort)	(whole plant) Quercetin, quercitrin, isoquercitrin, sarolactone, hypericin, usigtoercin, protohypericin, kaempferol.[33,53,87,265,450,396,568]	Antipyretic, antibacterial; detoxicant, treats acute icteric hepatitis, lowers blood pressure, dysmenorrhea, gonorrhea, skin ailments.
Hypericum triquetrifolium Turra. / *H. chinensis* L.	Jin Ci Tau (St. John's wort)	(whole plant) Hypericin, pseudohypericin, hyperin.[33]	Antidepressant, anti-HIV, antitumor.
Hyperzia serrata (Thunb.) Trev.	Shi Shan	(root) Huperzine A, isovanilhyperzine A.[33]	Active cognition enhancer, treats senile dementia including Alzheimer's disease.
Hyssopus ocymifolius Lam.	Xiang Xu	See *Elsholtzia cristata*	
Ilex chinensis Sims	Shi Ji Qing (Wintergreen holly)	(leaf) Protocatechuic acid, protocatechuic aldehyde, ursolic acid, tannic acid.[33]	Treats angina pectoris, thrombophlebitis, extremity ulceration.
Ilex pubescens Hook & Arn.	Mao Dong Qing (Holly)	(leaf, root) Flavone, ursolic acid, scopoletin, 3,4-dihydroxyacetophenone, hydroquinone, vomifliol.[33]	Treats angina pectoris, acute myocardial infarction, central angiospastic retinitis, cerebral thrombosis, thrombophlebitis.

Table 1A Major Constituents and Therapeutic Values of Chinese Medicinal Herbs (continued)

Scientific Name	Common Chinese and (English) Name	Major Constituents and (sources)	Therapeutic Values*
Illicium verum Hook f.	Ba Jiao Hui Xiang (Star anise)	(fruit) Anethol, anisaldehyde, safrole, anisic ketone.[33]	Capable of warming the viscera and expelling cold.
Illicium lanacedatum A. S. Smith	Hong Hui Xiang Gen (Japanese star anise)	(fruit) Anisatin, neoanisatin, shikimic acid, pseudoanisatin.[33] This herb is toxic.	
Impatiens balsamina L. *I. noli-tangere* L. *I. textori* Miq.	Tou Gu Cao or Feng Xian Hua (Garden balsam)	(whole plant) Gentisic acid, ferulic acid, p-coumaric acid, sinapic acid, caffeic acid, scopoletin, lawsone.[33,144] This herb is toxic.	Treats arthritis, relieves pain.
Imperata arundinaceae Cyrill. *I. cylindrica* Beauv.	Bai Mao (Thatch grass)	(root) Fernenol, arundoin, arborinol, arborinone, glutinol, cylindrin, simiarenol.[33]	Diuretic, hemostatic, antibacterial.
Inula britannica L. *I. japonica* Thunb. *I. linariaefolia* Turcz. *I. linariaefolia* Turcz. f. simplex Kom. *I. salsoloides* (Turcz.) Ostenfeld	Xuan Fu Hua (Elecampane)	(aerial part, including flower head) Inusterol A, taraxasterol, inusterol B, inulicin, flavone, caffeic acid, chlorogenic acid, isoquercitrin, quercetin, taraxasteryl palmitate, bigelovin, dihydrobigelovin.[48,50,260]	Discutient, vulnerary, anti-emetic, carminative, diuretic, deobstruent; treats ascites, bronchitis, cancer, chest congestion.
Iphigenia indica Bak. (Syn. *Tulipa edulis*)	Shan Ci Kodfghvb	(aerial part) Colchicine, colchicine amide, N-formyl-N-deacetylcolchine, cornigerine, P-lumicolchicine.[50]	Antitumor activity against hepatoma, lymphosarcoma.

Ipomoea barbata Both. *I. caerulea* Koeh. *I. hederacea* Jacq. *I. triloba* Thunb		See *Pharbitis hederacea*
Ipomoea cairica (L.) Sweet	Qian Niu (Sweet potato)	Purgative.
Iris aqyatuca Forskal *I. buatatas* (L.) Lamarck. *I. dichotoma* Pallas	Wu Zhao Jin Long (Cairo morning glory)	(flower) Muricatin A, beta-sitosterol.[50]
	Rong Cai Gan Su She Gan (Iris)	(root, whole plant) Tectoridin, iricin, flavone.[48]	Treats lung diseases, cough, pneumonia, uneasy breathing.
Iris lactea Pall. subsp. chinensis (Fisch.) Kitag.	Wu Gan	(seed, flower, leaf, root) Iridin, irigenin, irisflorentin.[60]	Astringent, diuretic, hemostatic; remedy for hemorrhage, postpartum difficulties.
Iris pallasii Fisch.	Ma Lan Zi (North China iris)	(seed) Irisquinone.[33]	Treats cancer, hepatoma, lymphatic sarcoma.
Isatis chinensis (Thunb.) Nakai *I. chinensis* (Thunb.) Nakai var. graminifolia (Ledeb.) H. C. Fu *I. tinctoria* L.	Ban Lan Gen	(leaf, root) Quercetin, kaempferol, stachyose, manneotetrose, lupeose, cicerose, isatan, indoxyl-5-ketogluc?nate.[50]	Antiviral, antibacterial; increase blood flow, improves microcirculation, and lowers blood pressure.
Isatis indigotica Fortune ex Lindley *I. oblongata* DC	Da Qing	(leaf) Indican, meoglucobrassicin, isatan B, indigo, glucobrassicin.[33]	Antibacterial, antipyretic, anti-inflammatory, choleretic.
Jasminum mesnyi Hance *J. nudiflorum* Lindley	Ying Chun Hua (Jasmine)	(leaf) Syringin, jasmiflorin, jasmipierin, mannose, tannins.[60]	Diaphoretic.
Jasminum samba (L.) Aiton	Mo Li Hua (Arabian jasmine)	(flower, root) Formic acid, benzoic acid, acetic acid, anthranil acid, sesquiterpene, sesquijasmine.[60] This herb (root) is toxic.	Sedative, anesthetic, vulnerary properties. For congestive headache, lactifuge.

Table 1A   Major Constituents and Therapeutic Values of Chinese Medicinal Herbs   (continued)

Scientific Name	Common Chinese and (English) Name	Major Constituents and (sources)	Therapeutic Values*
Jatropha gospiifolia L. var. delgans Muel. *J. curcas* L.	Hong Ma Feng Shu (Sweet cassava)	(seed) Phytotoxin, curcin, curcasin, arachidic, linoleic acid, myristic acid, oleic acid, palmitic acid, stearic acid.[50] The herb (seed) is toxic.	seed oil emetic, laxative, purgative; treats skin ailments.
Jatropha podagrica Hooker	San Hu You Tong	(stem bark) Tetramethylpyrazine, steroids, n-hexacosane, beta-amirine, lupeol palmitate, beta-sitosterol, rutin, flavonoids, quercetin, apigenin, vitexin, isovitexin.[57,207,208,568] This herb is toxic.	Detoxicant, hypotensive, neuromuscular and cardiovascular actions; antibacterial; relieves swelling, pain; externally treats snakebite, infection.
Juglans mandshurica Maxim. *J. regia* L.	Hu Tao Ren (English walnut)	(seed) Alpha-hydrojuglone-4-β-D-glucoside, jugone, juglanin.[33]	Nourishes and invigorates the lungs and kidneys.
Juncus communis Mey.	Den Xin (Bulrush)	(whole plant) Arabinose, xylose.[68]	Antilithic, pectoral, discutient, refrigerant, diuretic, depurative, sedative.
Juncus effusus L.	Deng Xin Cao (Common rush)	(whole plant) Tripeptide, r-glutamyl-valyl-glutamic acid, apigenin, juglandic acid, juglonone, barium, luteolin-7-glucoside, luteolinidin, oxalic acid, arsenic, vitamins.[48,450]	Diuretic, sexually transmitted diseases, anti-inflammation.

Juniperus rigida Sieb. et Zucc. *J. rigida* Sieb. et Zucc. f. modesta (Nakai) Y. C. Chu	Tu Soon (Juniper)	(fruit) Alpha-pinene, myrcene, carene, limonene, p-cymene, beta-elemene, caryophyllene, humulene, borneol, r-cadinene, terpinene, citronellol, anethole.[48]	Hemorrhage; treats hemoptysis, inflammation, kidney infection, arthritis joint infection.
Justicia gendarussa L. *J. procumbens* L.	Qin Jiu	See *Gentiana macrophylla*	
Kadsura japonica (L.) Dunal	Nan Wu We Zi (Kadsura)	(vine) Kadsuric acid, kadsurin, kadsurarin A, germacrene.[33,227]	Against hepatitis B. Relieves pain; a detoxicant; improves blood circulation, relieves arm and leg numb feelings.
Kaempferia galanga L.	Shan Na (Galanga)	(rhizome) Borneol, camphor, cineol, ethyl alcohol.[49]	Stomachic, carminative, stimulant.
Lactuca raddeana Maxim. *L. indica* L. *L. sativa* L.	Shan Wo Ju (Lettuce)	(seed) Pectic compound, oxalic acid, malic acid, citric acid, ceryl alcohol, ergosterol, vitamin E.[50]	Anodyne, lactogogue; for genital swelling, hemorrhoids, lumbago.
Laggera alata (D. Don) Schultz- Bip ex Oliver	Lu Er Jin	(whole plant) Flavonoid glycoside, phenols, amino acid, essential oil.[56]	Detoxicant, relieves swelling; treats fever, cough, hepatitis.
Laminaria angusta Kjellim. *L. cichorioides* Miyabe. *L. japonica* Aresch. *L. longipedalis* Okam. *L. religiosa* Miyabe.	Kun Bu or Hai Dai	(thallus) Iodine, potassium, calcium, amino acids, laminarin, laminine, algin.[33]	Improves thyroid function, corrects the malignant vicious cycle effect of iodine deficiency, and lowers blood pressure.

Table 1A   Major Constituents and Therapeutic Values of Chinese Medicinal Herbs   (continued)

Scientific Name	Common Chinese and (English) Name	Major Constituents and (sources)	Therapeutic Values*
Lappa communis Coss et Germ. *L. edulis* Sieb. *L. major* Gaerth. *L. minor* DC (Syn. *Arctium lappa*)	Niu Bang, Zong Shi	(seed, root) seed: arctin, arctigenin, gobosterin, essential oil, fatty oil. root: Inulin, lappine, lappatin, resin, essential oil, tannins.[49]	Diuretic, antipyretic, expectorant, antiphlogistic in throat infections, pneumonia, scarlet fever, measles, smallpox, syphilis.
Lawsonia inermis L.	Zhi Jia Hua (Henna)	(flower) Alpha-ionone, beta-ionone, gallic acid, lawsone.[50]	Antibiotic, antitumor, anthelmintic, astringent, bactericidal, fungicidal, sedative.
Ledebouriella divaricata Hiroe.	Fang Feng	(root) Essential oils, alcohol derivatives, organic acids.[33,603,606]	Antipyretic, analgesic, antibacterial; treats migraine headache, common cold, and rheumatoid arthritis.
Ledum palustre L. subsp. decumbens (Aiton) Hulten	Tu Xian	(leaf, shoot) Alpha-pinene, camphene, sabinene, myrcene, alpha-phellandrene, beta-pinene, limonene, quinene, isothujene, ascaridol, arbutin ericolin.[48]	Treats cough, asthma; lowers blood pressure; antifungal.
Lemmaphyllum microphyllum Presl.	Jing Mian Cao	(whole plant) Vitamins, luteolin-7-β-D-glucopyranoside, flavonoids, d-apiose, protein, resin.[48]	A poultice for animal bites, itchiness; a lotion for smallpox; relieves headache.
Lemna minor L. *L. perpusilla* Torrey	Qing Ping (Duckweed)	(aerial part) Luteolin-7-beta-D-glucopyranoside.[50]	For circulation, measles, swollen feet; depurative, diuretic, soporific.

Leonurus heterophyllus Sweet L. japonicus Houttuyn. L. macranthus Maxim. L. mongolicus V. Kreczet. et Kupr. L. pseudo-macranthus Kitag.	Yi Mu Cao (Motherwort)	(aerial part) Leonurine, stachydrine, vitamin A, leonaridine, leonurinine, fatty oils.[33,558]	Stimulates uterine contractions, respiratory system, proliferation of T. cells, skeletal muscles.
Leonurus sibiricus L. L. sibiricus L. f. albiflorus (Nakai et Kitag.) G. Y. Wu	Chung Way Bai Hua Yi Mu Cao (Siberian motherwort)	(seed) Essential oil, leonurine.[49]	Emmenagogue, diuretic, vasodilator.
Lepidium apetalum Willd. L. virginicum L.	Do Xing Cao Bei Mei Do Xing Cao	(seed) Isothiocyanates.[50]	Antibacterial, cardiotonic.
Lespedeza cuneata G. Don	Ye Guan Men (Perennial lespedeza)	(whole plant) Beta-sitosterol, pinitol, flavoroid.[33]	Antitussive, antiasthmatic, antiphlogistic, antibacterial.
Leucaena leucocephala (Lam.) De Wit	Yin He Huan	(leaf, seed) Leucanol, leucaenine, phenolic compounds, condensed tannins.[60,228] This herb is toxic.	Anthelmintic; for diabetes; an emollient, emmenagogue.
Ligusticum chuanziang Hort.	Chuan Xiang	(rhizome) Tetramethylpyrazine, perlolyrine, leucylphenylalanine anhydride, cnidilide, neocnidilide, ligustilide, acetylsalicylic acid, phthalide, benzoquinone.[33,226,419,420]	Promotes blood flow, remove blood stasis; and relieves pain.
Ligusticum jeholense (Nakai et Kitag.) Nakai et Kitag. L. pyrenacum Couan. L. sinense Oliv. L. tenuissimum (Nakai) Kitag.	Gao Ben	(root) Nothosmyrnol, coumarin, chromone, polyacetylene.[33,259]	Induces diaphoresis; for gout; an anodyne, emmenagogue, sedative.

Table 1A Major Constituents and Therapeutic Values of Chinese Medicinal Herbs (continued)

Scientific Name	Common Chinese and (English) Name	Major Constituents and (sources)	Therapeutic Values*
Ligustrum lucidum Mill. *L. japonicum* Thunb.	Nu Zhen Zi (Wax tree)	(fruit) Nuzhenide, oleanolic acid, ursolic acid.[33,445]	Increase leukocyte count; a cardiac tonic, diuretic; treats urological tumors.
Lilium brownii F.E. Brown var. viridulum Baker *L. concolor* Salisb. var. buschianum (Ledeb.) Baker *L. concolor* Salisb. var. partheneion (Sieb. & De Vries) Baker *L. dauricum* Ker-Gawler *L. distichum* Nakai ex Kamibayashi *L. japonicum* Thunb. *L. lancifolium* Thunb. *L. pumilum* DC	Bai He (Lily, Star lily)	(bulb) Protein, colchicine.[49]	Relieves coughing, ease anxiety, improves digestion, treats anxiety, apprehension; carminative, sedative, gynecologic disorders.
Linaria vulgaris Miller subsp. sinensis (Bebeaux) Hong	Liu Chun Yu	(aerial part) Peganine, linarin, pectolinarin, neolinarin, flavones, pectolinarigenin, linaracrine, linarezine, phytosterine.[48]	Diuretic; treats headache, dizziness, heart condition. Externally treats burns, skin diseases.
Lindera akoensis Hayata	Nei Don Zi	(leaf)[55] No information is available in the literature.	Treats wounds.
Lindera communis Hemsley	Xian Ye Shu	(fruit) Fatty acids.[55]	Relieves swelling, pain, bleeding; treats infection.

Species	Chinese Name	Constituents	Uses
Lindera glauca (Sieb. et Zucc.) Blume	Bai Ye Diao Zhang	(fruit) Essential oils, cineole, limonene, caryophyllene, bornylautate, fatty acids, camphene, beta-pinene.[55]	Carminative properties; treats arthritis joint pain.
Lindera megaphylla Hemsley	Da Xian Ye Shu	(root, seed) Essential oils.[55]	Promotes sweating, treats wounds.
Lindera obtusiloba Blume f. villosa (Blume) Kitag.	Nei Don Zi	(bark) Campesterol, linderol, capric acid, lauric acid, myristic acid, linderic acid, dodecen-4-oic acid, oleic acid, tetradecen-4-oic acid, tsudzuic acid, linoleic acid.[48]	Reduces swelling, pain.
Lindera strychnifolia Vill.	Wu Yao	(fruit, root, seed) Essentail oils, lindestrene, liderane, linderoxide, linderalactone, isolinderalactone, isolinderoxide, lindestreolide, neolinderalactone, isofuranogermacrene, linerene.[33]	Improves circulation, relieves pain, abdominal distention, fever.
Linum stelleroides Planch. *L. usitatissimum* L.	Ya Ma (Flax)	(whole plant) Fatty acids, geranylgeraniol, cholesterol, campesterol, orientin, stigmasterol, avenasterol, vitexin cycloartenol, eikosanol, leucine, valine, linamarin, lotaustralin.[48,568]	For diarrhea, sensitive skin, itchiness, loss of hair.
Liquidambar acerifolia Max. *L. formosana* Hance *L. maximowiczii* Miq.	Fon Xian Chi (Liquid amber)	(bark, leaf, root) Balsam (resin), cinnamic alcohol, cinnamic acid, l-borneol, camphene, dipentane, terpene.[60,69]	Analogous, externally as antiphlogistic and astringent in skin diseases, antihemorrhagic.

Table 1A   Major Constituents and Therapeutic Values of Chinese Medicinal Herbs   (continued)

Scientific Name	Common Chinese and (English) Name	Major Constituents and (sources)	Therapeutic Values*
Liriope graminifolia Bak. *L. platyphylla* Wang & Tang *L. spicata* Lour.	Mai Men Dong	(root) Mucilage.[49] This herb is used to produce Ophiopogon.[60]	Antitussive, expectorant, emollient.
Litchi chinensis Sonn.	Li Chi (Lychee)	(leaf, fruit, seed) Citric acid, vitamins A, B, C, sugar, amino acids, lysine, leucine, valine, alanine, glutamic acid, serine, proline, asparagic acid, theronine, arginine, lysine, beta-phenethyl alcohol.[49,50,70]	Remedy for gland enlargement, tumors; treats bites of poisonous animals. Astringent, analgesic in gastralgia, colic, orchitis.
Lithospermum erythrorhizon Sieb. et Zucc. *L. officinalis* var. erythrorhizon Sieb. et Zucc.	Zhu Cao (Groomwell)	(root) Quinonoid, alkannan, acetylshikonin, shikonin, lithospermin, dihydroshikonin, cycloshikonin.[1,69]	Ointment to treat wounds and burns, antitumor, antipyretic, regulating blood circulation, diuretic, purgative, remedy for smallpox.
Litsea cubeba Lour.	Shan Cong Zi (Cubebs)	(fruit) Citral, linalool, laurotetanine.[33]	Treats chronic bronchitis and bronchial asthma, protects hypersensitization shock.
Livistona chinensis (Jacq.) R. Br. ex Mart.	Kui Shu Zi	(shoot, leaf, seed) Triglyceride.[54]	Anticancer, treats tuberculosis, regulates menses; externally for bites or stings.
Lobelia chinensis L. *L. pyramidalis* Wallich. *L. sessilifolia* Lambert	Ban Bian Lian Sha Gen Cai (Lobelia)	(whole plant) Lobeline, lobelanine, lobelanidine, isolobelamine. (Lobeline has been approved by the FDA to curb the tobacco habit).[33,50,71,558] This herb may be toxic.	Diuretic; increase respiration via stimulation of carotid chemoreceptors. Treats snakebites, insecticide; reduces swelling, depurative, antirheumatic, antisyphilitic.

Lonicera acuminata Wallich L. apodonta Ohwi L. brachypoda DC L. chinensis Wats. L. confusa Miq. L. flexuosa Thunb. L. hypoglauca Miq. L. japonica Thunb. L. japonica var. chinensis Bak. L. maackii (Rupr.) Maxim.	Rui Ye Ren Dong Duan Geng Ren Dong Jin Yin Hua Ren Dong (Honeysuckle)	(flower bud, whole plant) Luteolin, inositol, loganin, lonicerin, syringin, saponin, tannins, chlorogenic acid, luteolin-7- rhamnoglucoside.[33,48,55,600]	Antibacterial, cytoprotective, antilipemic, antiphlogistic.
Lophanthus chinensis Walp. L. rugosus Fisch. et Mey. (Syn. Agastache rugosa)	Huo Xiang	(leaf) Essential oils.[49]	Carminative, stomachic.
Lophatherum gracile Brongn.	Dan Zhu Ye	(aerial part) Arundoin, cylindrin, friedelin.[33]	Antipyretic, diuretic, antibacterial.
Loranthus parasiticus L. yadoriki Sieb. et Zucc.	Song Ji Shang (Mistletoe)	(leaf, stem) Saponins including avicularin, quercetin.[40]	Treats angina pectoris, cardiac arrhythmia, hypertension. Ointment to treats frostbite.
Loropetalum chinense (R. Br.) D. Oliver	Ji Mu (Strap flower)	(plant) Flavone, quercitrin, isoquercitrin.[33]	Antipyretic, a detoxicant, hemostatic; treats angina pectoris, bronchitis, bleeding, alimentary indigestion.
Luffa aegyptiaca Mill. L. cylindrica Roem. L. faetida Sieb. et Zucc. L. petola Ser.	Se Gua (Luffa sponge)	(fruit fibers) Xylose, mannosan, galactan, saponins, acetic acid, valeric acid, pinenes, limonene, cineole, sterol, menthone, linalool, bourbonene, caryophyllene, menthol, carvone, vitamins A, B, C.[49,50,351]	Hemostatic, analgesic in enterorrhagia, dysentery, metrorrhagia, orchitis, hemorrhoids.

Table 1A   Major Constituents and Therapeutic Values of Chinese Medicinal Herbs   (continued)

Scientific Name	Common Chinese and (English) Name	Major Constituents and (sources)	Therapeutic Values*
Lupinus luteus L.	Yu Shan Dou	(whole plant) Lupinine, lupinidin, rechts-lupinine, d-lupaine.[56] This herb is toxic.	Diuretic, insecticidal; treats fever, respiratory difficulties.
Lycium barbarum L. *L. megistocarpum* Dun. *L. ovatum* Loisel. *L. trewianum* G. Don *L. turbinatum* Loisel.	Gou Gi, Gou Qi Zi (Ningxia wolfberry)	(fruit) Betaine, zeaxanthin, physalein, carotenes, nicotinic acid, vitamin C.[33,447,568]	Increases leukocyte count; anticancer, increase immunity, stimulation of tissue development.
Lycium chinense Miller	Di Gu Pi (Matrimony vine)	(root bark) Cinnamic acid, betaine, peptides, acyclic diterpene glycosides, polysaccharide, kukoamines.[33]	Lowers blood sugar and blood pressure; antipyretic; stimulates uterine contraction, antibacterial.
Lycopersicon esculentum L.	Fan Qie (Tomato)	(root, leaf) Protein, vitamin A, thiamine, nicotinic acid, riboflavin.[50]	Relieves toothache; insecticide, laxative.
Lycopodium annotinum L. *L. cernum* L. *L. compianatum* L.	Shan Ye Man Shi Song Jin Gu Cao (Devil's powder) Di Shua Zi (Ground cedar)	(whole plant) Clavatine, lycopodine, complanatine, alpha-obscurine, serratenediol, tohogenol.[48]	Relieves numb feeling, arthritis pain, sexually transmitted disease.

Botanical name	Chinese name	Constituents (part)	Uses
Lycopodium clavatum L. var. nipponicum Nakai *L. obscurum* L. *L. selago* L. *L. serratum* Thunb.	Shen Jin Cao (Running pine, staghorn clubmoss, princess pine)	(whole plant) Lycopodine, lycodoline, clavatine, fawcetine, clavoloninine, azelaic acid, clavatoxine, fawcetimine, deacetylfawcetine, nicotine, vanillic acid, ferulic acid, alpha-onocerin, lycoclavanol, lycoclavanin, lycopodine.[33,45,568]	Relieves the rigidity of muscles and joints; treats arthritis and dysmenorrhea.
Lycopus fargesii Herter *L. lucidus* Turcz. *L. lucidus* Turcz. f. hirtus (Regel) Kitag. *L. maackianus* (Maxim.) Makino *L. parviflorus* Maxim. *L. ramosissimus* (Makino) Makino var. japonicus (Matsum et Kudo) Kitam. *L. veitchii* Christ.	Shan Ye Shi Song Shi Song Yu Shan Shi Song (Shining water horehound)	(aerial part) Resin, lycopose, raffinose, glucose, stachyose.[48]	For abdominal distention, abscesses, congestive edema, blood extravasation.
Lycoris aura (L'Her.) Herb. *L. longituba* Y. Han et Fan *L. radiata* (L'Her.) Herb.	Shi Suan (Amaryllis)	(rhizome) Galanthamine, lycoremine, lycorine, lycoramine, lycorenine, tazettine, pseudolycorine, dihydrolycorine, homolycorine, lycoricidine, lycoricidinol.[33]	As a cholinesterase inhibitor, lowers blood pressure, stimulates secretion from the pituitary gland, and increases antidiuretic hormone secretion.
Lygodium japonicum Swartz. *L. flexuosum* (L.) Sw.	Hai Jin Sha Teng (Climbing fern)	(leaf with or without sporangia) Fatty oil.[49]	Diuretic, antirheumatic against venereal diseases, disorder of the urinary tract.
Lyonia ovalifolia (Wall.) Drude.	Nan Zhu	(leaf, fruit) Andromedotoxin, lyoniols.[60] This herb is highly toxic.	A tonic.

Table 1A Major Constituents and Therapeutic Values of Chinese Medicinal Herbs (continued)

Scientific Name	Common Chinese and (English) Name	Major Constituents and (sources)	Therapeutic Values*
Lysimachia barystachys Bunge. *L. christinae* Hance *L. clethroides* Duby *L. davurica* Ledeb. *L. davurica* Ledeb. f. latifolia Korsh.	Jin Qian Chao (Loosestrife)	(whole plant) Essential oils, l-pinocamphone, l-menthone, l-pinene, limonene, 1,8-cineol, p-cymene.[33]	Diuretic; a choleretic agent, antibacterial.
Lysionotus pauciflorus D. Don	Shi Diao Lan	(aerial part) Organic acids, flavones, nevadersin.[33]	Antibacterial, antitussive; lowers blood pressure.
Lythrum salicaria L. *L. salicaria* L. var. glabrum Ledeb.	Qian Qu Cai (Purple loosestrife)	(aerial part) Tannins, salicarin, chlorogenic acid, cyanidin-3-monogalactoside, ellagic acid, malvidin, malvin, orientin, vitexin.[50,72,568]	Astringent, styptic, treats bacillary dysentery.
Machilus thunbergii Sieb. et Zucc.	Hong Nan	(bark, root) dl-N-noramepavine, quercetin, N-norarmepavine, reticuline, lignoceric acid, dl-catechol.[57]	Removes eczema, treats spleen and stomach disease, asthma.
Macleaya cordata (Willd.) R. Br.	Bo Lou Hui	(whole plant including root, fruit) Sanguinarine, oxysanguinarine, ethoxysanguinarine, protopine, alpha-allocryptopine, bocconine, chelerythrine, coptisine, herberinecorysamine, bocconoline, ethoxychelerythrine, chelilutine, chelirubine.[33] This herb is toxic.	Antiplasmodial; treats vaginal trichomonas, antibacterial.

Macrocarpium officinalis (Sieb. et Zucc.) Nakai	Shan Zhu Yu	(fruit) Tannic acid, resin, tartaric acid, cornin, gallic acid, malic acid.[60]	A tonic, astringent, diuretic, antilithic, anthelimintic, febrifuge.
Maesa japonica (Thunb.) Moritzi	Du Jing Sha	(aerial part) Maesaguinone.[50]	Emetic, febrifuge, resolvent, styptic; for fever, malaria.
Maesa perlarius (Lour.) Merrill.	Sha Gui Hua	(leaf, root) Alkaloids, quinonic substance.[73]	Crushed leaves are bound over broken bones or to treat measles. root is diuretic, stomachic.
Maesa tenera Mez.	Taiwan Sha Gui Hua	(whole plant) Maesaquinone.[58]	Stomachache, hepatitis, lowers cholesterol level, treats cold, headache.
Magnolia biloba Cheng *M. denudata* Desr. *M. discolor* Vent. *M. liliflora* Desr. *M. purpurea* Curt.	Xin Yi, Mu Lan (Red magnolia)	(flower bud, leaf) flower: eugenol, safrole, citrol, anethol. leaf: salicifoline, citral, magnocurarine.[33] Essential oils, safrole, anethole, estragole. cineol, eugenol.[49]	Relieves nasal congestion, sinusitis, rhinitis, coryza, headache, vertigo.
Magnolia coco (Lour.) DC *M. fortunei* (Lindl.) Fedde	Ye He Hua Gong Lao Mu	(bud, flower, stalk) Alkaloid.[74]	Febrifuge, stimulant, tonic; treats chronic rheumatism.
Magnolia grandiflora L.	Yang Yu Lan	(bark, flower bud) Magnocurarine, salicifoline, fatty acids, volatile oil.[56]	A tonic; treats malaria, high blood pressure, headache.
Magnolia hypoleuca Diels. *M. officinalis* Rehd. et Wils. *M. japonica* (Thunb.) DC	Hou Po Huang Bai Mu (Magnolia)	(bark) Alkaloids, magnocurarine, magnoflorine, beta-eudesmol, neo-lignans, magnolol, konokiol, liriodenine, crytomeridiol.[6,33] This herb may cause kidney failure.[392] *Magnolia officinalis* bark is toxic.	Central nervous system depressant action, sedative, anticonvulsant, muscle relaxation.

Table 1A　Major Constituents and Therapeutic Values of Chinese Medicinal Herbs　(continued)

Scientific Name	Common Chinese and (English) Name	Major Constituents and (sources)	Therapeutic Values*
Mahonia japonica DC	Gou Gu	(leaf, root, stem, seed) Berberine, jatrorhizine, palmatine.[97,255]	Antipyretic, backache, cough, dysentery, fever.
Mallotus japonicus (Thunb.) Muell.	Ye Wu Tong	(stem, leaf) Resin, tannins, alkaloids.[60]	Treats lumbar pain, stomachache; crushed leaves are applied to tumors and swellings.
Mallotus paniculatus (Lam.) Muell.-Arg.	Bai Bao Zi	(stem leaf) Amino acids.[57]	To cleanse wounds.
Mallotus repandus (Willd.) Muell.	Gong Xian Teng	(stem, leaf) Mallorepine, bergenin, repandusinin, repandusinic acids, mallotinin.[340,341]	An insecticide; relieves itching, anti-inflammatory.
Malva chinensis Mill. *M. pulchella* Berhn. *M. sylvestris* L. *M. verticillata* L.	Dong Kui Zi (Chinese mallow)	(whole plant) l-arabinose, l-rhamnose, d-galacturonic acid.[75,87]	Treats stomach and intestinal disorders, to make labor easier; laxative; treats gonorrhea, congestion, constipation.
Manihot esculenta Crantz.	Shu Shu	(root) Hydrocyanic acid.[76] This herb is toxic.	To dress ulcerous sores.
Marsdenia tenacissima (Roxb.) Wight. et Arn.	Tong Guan Teng	(stem) Saponins, marsdeoreophisides, metaploxigenin, sarcostin.[33]	Antiasthmatic, hypotensive, antibacterial.
Matricaria chamomilla L.	Yang Gan Jiu (Matricary)	(flower head, leaf) Volatile oil, azulene, isoamyl, isobutyl, angelic acid, tiglic acid, anthelmic acid, tannins, malic acid.[77,87]	Carminative, diaphoretic.

Species	Chinese name	Constituents	Uses
Matteuccia struthiopteris (L.) Todaro	Jia Gou Ju	(root) Ponasterone A, ecdysterone, palmitic acid, astragalin, caffeic acid, chlorogenic, vanillic acid, p-hydroxybenzoic, p-coumaric, ferulic, protocatechuic, beta-sitosterol, campesterol, filicin, stigmasterol, pterosterone.[48,217]	Tonic; lowers blood pressure.
Maytenus diversifolia Hou. M. confertiflorus Luo & Chen.	Ci Luo Shi	(leaf, bark) Dulcitol, maytansine, succinic acid, syringic acid, 3-oxykojie acid, loliolide.[50]	Antitumor; bark is used for cancer of the liver and stomach.
Maytenus serrata (Hochst. ex A. Rich) Wilcz. M. hookeri Loes.	Mei Deng Mu	(fruit, bark, rhizome) Maytansine, maytanprine, maytanbutine, maytanvaline, maytanacine, maytansinol.[33]	Treats lung cancer, breast and ovarian cancer, acute lymphocytic leukemia, colon carcinoma, kidney carcinoma.
Medicago falcata L. M. lupulina L. M. polymorpha L. M. ruthenica (L.) Ledeb. M. sativa L.	Mu Xu (Alfalfa)	(whole plant) Lucernol, sativol, coumesterol, formonetin, daidzein, tricin, citrulline, canaline, dicoumarol, methylene-bishydroxy-coumarin, medicagemic acid, ononitol, petunidin, malvidin, delphinidin, linalool, myrcene, limonene.[68,568]	Depurative, diuretic, stomachic, treats intestinal and kidney disorders, kidney stones, poor night vision.
Melaleuca leucadendra L.	Bai Qian Ceng (Cajeput)	(leaf) Cajuputol, cineole, eucalyptol, lignin, melaleucin, pinene, terpinol, l-limonene, dipentene, nerolidiol, sesquiterpenes, azulene, sesquiterpene alcohols, valeraldehyde, benzaldehyde, betulinic acid, oleanolic acid, ursolic acid, quercimeritrin, isoquercitrin, gallic acid.[50]	Treats dropsy, oil is for gout; inhaled to treats colds, rhinitis, embrocation against rheumatism. It is an anodyne, antiseptic.

Table 1A   Major Constituents and Therapeutic Values of Chinese Medicinal Herbs   (continued)

Scientific Name	Common Chinese and (English) Name	Major Constituents and (sources)	Therapeutic Values*
Melasma arvense (Benth) Handel- Maxxetti	Hei Shuo	(whole plant) Musaenoide, aucubin.[57]	Treats child's whitish tongue; diuretic.
Melia japonica G. Don M. toosendan L. M. azedarach L.	Chuan Lian or Ku Lian Ku Lian Pi or Ku Lian Chi (Chinaberry tree)	(stem, root bark) Toosendanin, nimbin, kulinone, methylkulonate, melianol, gedunin, melianodiol, melianotriol, melialactone, azadarachtin, nimbolins, fraxinella, palmitic acid, lauric acid, valerianic acid, butyric acid, stearic acid, cycloencalenol.[33,49,144] This herb is toxic.	Treats intestinal parasite; antibacterial, anthelmintic.
Melilotus alba Medicus M. indica (L.) All. M. suaveolens Ledeb.	Be Han Cao (Clover)	(whole plant) Hydroxycinnamic acid, coumarinic acid, 4-hydroxycinnamic acid, cumaric acid, umbelliferone, scopoletin, melilotoside, melilotic acid, beta-D-glucosyloxy, dicumarol, chlogogenic acid, caffeic acid, melilotic acid.[48]	Anticoagulant; treats bowel complaints, infantile diarrhea. A bactericide.
Melochia corchonifolia L.	Ye Lu Kui	(leaf) Trifalin, melocorin, hibifolin.[57]	As poultice to treats sores, swelling, and pain in the abdomen. Also treats vomiting.
Menispermum dauricum L.	Ye Dou Gen, Shan Dou Gen (Moonseed)	(root) Dauricine, daurinoline, l-stepharine, dauricoline, acutumine, tetrandrine, dauricinoline, stepholidine, magnoflorine, menisperine, sinomenine.[33,505]	Antiarrhythmic, analgesic effect; relieves headache, insomnia.

Menispermum dauricum DC f. pilosum (Schneider) Kitag. (Syn. *Cocculus diversifolius*)	Fang Chi (Siberian moonseed)	(whole plant) Acutumine, acutuminine, dauricine, disinomenine, magnoflorine, menispermine, sinomenine, stepharine, tetrandrine.[50]	Antitumor, cytotoxic; alleviates skin allergies, antirheumatic, anticancer against esophageal cancer.
Mentha arvensis L. *M. dahurica* Fisch. ex Benth. *M. haplocalyx* Briq. *M. sachalinensis* (Briq.) Kudo *M. sachalinensis* (Briq.) Kudo f. arguta (Kitag.) Y. C. Chu	Bo Hoo (Peppermint)	(aerial parts) Menthol, menthone, menthyl acetate.[33]	Stimulates gastrointestinal tract motility and central nervous system, dilates peripheral blood vessels, and increases sweat gland secretion.
Menyanthes trifoliata L.	Shui Cai (Bogbean)	(whole plant) Aromadendrine, betulinic acid, cadinene choline, gentiatibetin, cineole, dihydrofoliamenthin, foliamenthin, gentialutine, loganin, gentianine, gentiatibetine, invertin, gurjuncene, meliatin, menthiafolin, menyanthin, secologanin, alpha-spinasterol, stigmast-7-enol, trifolioside.[50]	Antitumor; increases gastric secretions, as cathartic, cholagogue, narcotic, sedative, tonic, vermifuge.
Michelia alba DC *M. figo* DC	Bai Yu Lan Han Xiao Hua (White champac)	(flower bud) Acetic acid, linalool, michelabine, methylethylacetic ester, methyl eugenol, oxoushinsunine, salicifoline, ushinsunine.[50]	For sapremia following miscarriage.
Milletia reticulata Bentham. *M. taiwaniana* (Matsum.) Hayata	Ji Xue Teng Lu Teng	(leaf) Rotenone, anhydroderrid.[60]	Insecticide.
Mimosa arborea Thunb.	Han Xiou Cao	See *Albizia julibrissin*	

Table 1A   Major Constituents and Therapeutic Values of Chinese Medicinal Herbs   (continued)

Scientific Name	Common Chinese and (English) Name	Major Constituents and (sources)	Therapeutic Values*
Mimosa invisa Mart. et Colla *M. pudica* L.	American Han Xiou Cao Han Xiou Cao	(whole plant) Minosine.[78] This herb is toxic if overdosed.	Treats neurosis, trauma wounds, and hemoptysis. It has a tranquilizing effect.
Momordica charantia L.	Ku Gua (Bitter melon)	(seed) Anti-HIV protein MAP 30, sterol.[33,351,408,423]	For immune disorders and common infections. Capable of inhibiting infection of HIV-1 in T. lymphocytes and monocytes, antitumor.
Momordica cylindrica L.	Si Gua	See *Luffa cylindrica*	
Momordica grosvenori Swingle	Luo Han Guo	(fruit) Esgoside.[33]	An expectorant; control coughing.
Morinda citrifolia L. *M. officinalis* L.	Je Shu Ba Ji Tian	(root) Dihydroxy methyl anthraquinone, glucoside morindin, rubichloric acid, alizarin, alpha-methyl ether, rubiadin-l-methyl ether, tannins, morindadiol, masperuloside, soranjudiol, nordamnacanthal.[50,424]	Treats beriberi, cancer, lumbago, cholecystitis; increases leukocyte count, and stimulates endocrine system.
Morinda parvifolia Bartling	Xiao Ye Yang Jiao Teng	(root) Methanolic, morindaparvin-a, alizarin-l-methyl ether.[50]	Against p-388 lymphocytic leukemia growth (*in vivo*), cytotoxic, antileukemic.
Morus alba L. *M. constantinopolitana* Poir. *M. indica* L.	Sang Zhi or Sang Gen Bai Pi (Mulberry tree)	(young twig) Morin, dihydromorin, maclurin, dihydrokaempterol, mulberrin, 2,4,4', t-tetrahydroxybenzophenone, mulberrochromene, cyclomulberrochromene.[33]	Antirheumatic, antihypertensive, diuretic; removes obstructions of the intestinal tract.

Murraya paniculata (L.) Jack	Jiu Li Xiang (Orange jessamine)	(leaf, young branches) l-cadinene, methylanthranilate, bisabolene, beta-caryophyllene, geraniol, carene,5-guaizulene, osthol, paniculatincoumurrayin.[33]	Relieves pain, remove toxic substances, an antispasmodic; antagonizes muscular spasms.
Musa paradisiaca L. var. sapientum O. Ktze.	Xiang Jiao (Banana)	(root, trunk juice, fruit, flower) Serotonin, norepinephrine.[60]	Carbuncles all kinds of tumors, swellings, measles, headache with fever and sunburn. Stimulates the smooth muscle of the intestine, treats certain forms of heart collapse.
Mussaenda parviflora Miq.	Yu Ye Jin Hua	(leaf, root) Triterpinoid.[79]	Treats malarial fever.
Myrica rubra (Lour.) Sieb. et Zucc.	Gou Mei (Chinese strawberry)	(fruit) Myricetin.[33]	Treats gastric pain, diarrhea, dysentery.
Myristica fragrans Houtt.	Rou Dau Kou (Nutmeg)	(seed) Lauric acid, myristic acid, stearic acid, hexadeceonic acid, oleic acid, linoleic acid, amylodextrins, pectins, resins, campherene, cymene, dipentene, eugenol, geraniol, isoeugenol, linalool, myristicin, pinene, safrole, terpineol.[50] Volatile oil from this herb may be toxic.	For hysteria, hypochondria, agoraphobia, laughter, cramps, crying jags, dysmenorrhea, amnesia.
Nandina domestica Thunb.	Nan Tian Zhu (Sacred bamboo)	(fruit, bark, leaf) Domesticine, nandinine, cyanic acid, nandazurine, berberine.[49] This herb is toxic.	Antitussive.
Narcissus tazetta L. var. chinensis Roem.	Shui Shai (Polyanthus narcissus)	(bulb) Lycorine, tazettine, narcitine.[49,60] This herb is toxic if overdosed.	Antiphlogistic; analgesic for boils, abscesses, mastitis.

Table 1A   Major Constituents and Therapeutic Values of Chinese Medicinal Herbs   (continued)

Scientific Name	Common Chinese and (English) Name	Major Constituents and (sources)	Therapeutic Values*
Nardostachys jatamansi DC	Ga Song Xiang (Spikenard)	(root) Essential oil, jatamansic acid, sesquiterpene.[49,80,82]	Aromatic stomachic, sedative, antispasmodic.
Nauclea rhynchophylla Miq. *N. sinensis* Oliv.	Gou Teng	(stem, spine) Rhynchophylline, isorhynchophylline.[49,72]	Lowers blood pressure, paralyzes sympathetic nerve ending; sedative, antispasmodic in infantile nervous disorders.
Nelumbium nelumbo Druce.	Lian Zi Xin (Lotus)	(plumule) Liensinine, isoliensinine, neferine, lotusine, methyl-corypalline, demethyl-coclaurine.[33]	Tranquilizing and antihypertensive.
Nelumbium nuciferum Gaertner *N. speciosum* Willd.	Lian, He Ye (East Indian lotus)	(leaf) Nuciferine, roemerine, anonaine, O-nornuciferine, liriodenine, anneparine, dihydronuciferine, pronuciferine, N-methylcoclaurine, N-methylisococlaurine.[33]	Relaxing effect on smooth muscles, increases essential body energies.
Neoalsomitra integrifoliola (Cogn.) Hutch	Bang Chui Hui	(stem) Cucurbitacin B, iso-cucurbitacin B, carotenoids.[56]	Laxative for diarrhea; treats intermittent fever, hepatitis, thyroid gland swelling. Used as a wash for contusions.
Nepenthes raffsiana Masilus	Zhu Long Cao (Pitcher plant)	(root, stem) Flavonoids, anthraquinoids, amino acids, phenols.[57]	As a poultice to treats stomachache and dysentery. Internally to treats remittent fever.
Nephelium longana Camb. *N. lappaceum* L.	Ron Yen Raw Hong Mao Dan (Rambutan)	(aerial part, kernel) Glucose, sucrose, tartaric acid, vitamins A, B, saponins, tannins.[49]	Nutrient tonic in neurasthenia, insomnia, styptic.

Species	Chinese name	Part used / Constituents	Uses
Nerium indicum Mill.	Jia Zhu Tao (Indian oleander)	(leaf, stem, flower, root) Oleandrin (toxic), oleandrose, neriodorin, nerioderin, karabin, scopoletin, scopoline, neriodin, ursolic acid, adynerin. [33,450]	Treats psychosis, congestive heart failure; analgesic, emmenagogue.
Nervilia purpurea (Hayata) Schltr.	Yi Dian Hong	(whole plant) Cyclonerviol, cyclomonerviol, stigmasterol, dihydrocyclonerviol, ergosterol, epibrassicasterol, nervisterol, cyclonervilol. [58]	As a protective medicine postpartum; treats throat infection, pneumonia, high blood pressure, diabetes.
Nicotiana tabacum L.	Yan Cao (Tobacco)	(leaf) Nicotine, nicotimine, nicoteine, nicotelline. [60] This herb is toxic.	Treats soreness in the joints, numbness, hemicrania, poisonous snakebites; insecticide, antidysenteric, emetic.
Nothosmyrnium japonicum Miq.	Hao Mu	(root) Essential oil, nothosmyrnol, dimethoxyallylbenzene. [49]	Cerebral sedative, analgesic, antispasmodic.
Notopterygium incisium Ting	Giang Huo	(root) Notoptero, isoimperatorin, falcarindiol, essential oils, limonene. [53]	Antirheumatic; for arthritis, cold, excessive sweating.
Nuphar japonicum DC *N. pumilum* (Timm) DC	Japan Pin Peng Cao (Yellow pond lily)	(seed, root) Nupharamine, sitosterol, palmitic acid, oleic acid. [74,75]	For digestive organs; increases body strength.
Nymphaea tetragona Georgi *N. tetragona* Georgi var. crasifolia (Hand. Mazz.) Y. C. Chu	Shui Lian (Pigmy water lily)	(flower, leaf, root) Amino acids. [48]	A cooling lotion to apply to eruptive fevers, treats colic, gonorrhea, lowers blood pressure.
Oenothera biennis L. *O. odorata* Jacq.	Shan Zhi Ma Ri Jian Cao	(seed oil, root) 6,9,12-octadecatrienoic acid. [48,87,568]	Lowers cholesterol, regulate heartbeat, and treats arthritis.

Table 1A   Major Constituents and Therapeutic Values of Chinese Medicinal Herbs   (continued)

Scientific Name	Common Chinese and (English) Name	Major Constituents and (sources)	Therapeutic Values*
Oenothera javanica (Bl) DC	Shui Jin	(seed, leaf) Bis (2-ethyl butyl) phthalate, n-butyl-2-ethyl butyl phthalate, diethyl phthalate, myrcene, alpha-pinene, terpinolene, limonene, beta-pinene, alpha-terpinene, persicarin, petroscelinic acid.[48,50]	For plethora, cholera, dysuria, fever, hematuria, influenza, jaundice, metrorrhagia; antivinous, hemostat. Externally for abscesses, cancerous swelling, snakebite.
Oenothera terythrosepala Borbus	Da Ri Jian Cao	(root) Linolenic acid.[53]	
Oldenlandia chrysotricha L. O. corymbosa L.	Shi Da Chuan Shui Xian Cao	(leaf) Biflorine, biflorone, gamma-sitosterol, stigmasterol, ursolic acid, oleanolic acid.[50]	Ferbrifuge, for hepatomegaly, lymphadenitis, neophasia, splenomegaly.
Oldenlandia diffusa L.	Bai Hua Shi Shi Cao	(aerial part) Asperuloside, palderoside, desacetylasperuloside, beta-sitosterol, stigmasterol, ursolic acid, oldenlandoside.[33]	Treats malignant tumors, hepatomas, hepatomegaly, cancer of the cervix, esophagus, stimulates reticuloendothelial system.
Omphalia lapidescens Baill.	Lei Wan	(sclerotium)[50] Glucan, OL-2.[209]	Antitumor; treats ascariasis, taeniasis, ankylostomiasis.
Onychium japonicum (Thunb.) Kunze.	Japan Jin Fen Ju	(spores, aerial part) Kaempferol-rhamnoside.[56]	To relieves chest and abdominal pains; stop bleeding; diuretic, detoxicant, intestinal infection.
Ophioglossum japonicus (Thunb.) Ker-Gawl.	Mai Dong	(root) Beta-sitosterol.[69]	Depurative; a tonic, thirst-quenching or sialagogue; treats phthisis.

Ophioglossum vulgatum L.	Yi Zhi Jian (Adder's tongue)	(whole plant) 3-O-methylquercetin.[33]	As a hemostatic, abscesses; treats gangrene. Externally treats snakebite.
Ophiopogon gracilus Kunth. *O. longifolius* Decne. *O. spicatus* Ker-Gawler	Mai Meng Dong (Lilyturf)	See *Liriope spicata*	
Ophiopogon japonicus Wall.	Yan Jie Cao or Mar Dong (Japanese lilyturf)	(root) Beta-sitosterol, stigmasterol, ophiopogenins, polysaccharides, kaempferol-3-glucosylgalactoside.[33,50]	Antitussive, expectorant, emollient, anticancer. Smoothes lung functions, stops coughing.
Ophiorrhiza japonica Blume *O. mungos* L.	Japan She Gen Cao She Gen Cao	(whole plant) Resin, alkaloid, beta-sitosterol, 5 alpha-ergost-en-3 beta-ol, 5 alpha-ergost-8(-14)-en-3B-ol, tannates, hydrogen cyanide.[50]	For circulatory and pulmonary ailments.
Oplopanax elatus (Nakai) Nakai	Ci Seng	(stem, root) Essential oil, echinopanacene, n-caprylaldehyde, echinopanacol, oplopanaxosides, flavonoids.[48,50,72,354]	A remedy and tonic for progressive emaciation.
Orchis latifolia L.	Hong Men Lan	(whole plant) Alkaloids.[48,359]	Sialagogue; treats anemia.
Oryza sativa L.	Nou Me (Rice)	(whole plant) Isoleucine, leucine, lysine, phenylalanine, tyrosine, amino acids, methioine, threonine, tryptophane, valine.[50]	For dysentery, gout, rheumatism, hemorrhoids; an astringent, anhydrotic, anuria.
Osmanthus fragrans (Thunb.) Lour.	Mu Gui, Gui Hua (Cassia tree)	(flower) Beta-phellandrene, osmane, nerol, methyl-laurate, methylmyristate, methylpalmitate, uvaol.[33]	Reduces phlegm, removes blood stasis.

Table 1A   Major Constituents and Therapeutic Values of Chinese Medicinal Herbs   (continued)

Scientific Name	Common Chinese and (English) Name	Major Constituents and (sources)	Therapeutic Values*
Osmunda japonica L.	Zi Kee Guan Zhong (Japanese royal fern)	(whole plant) Ponasterone A, ecdysterone, custeodysine, ecdysone.[33]	Anthelmintic; treats inflammation of the salivary glands.
Oxalis corriculaza L. *O. corymbosa* DC	Sha Jiang Cao (Wood sorrel)	(leaf) Oxalate, vitamin C, calcium, citric acid, malic acid, tartaric acid.[50]	Antidote to arsenic and mercury; for bruises, clots, diarrhea, fever, influenza, snakebite, urinary infections.
Oxyria digyna (L.) Hill	Gao Shan Liao	(whole plant) Protein, fat, ash, carbohydrate, retinol, mineral elements.[48,210]	For hepatitis.
Pachyrhizus thunbergianus Sieb. et Zucc.		See *Pueraria thunbergiana*	
Paeonia albiflora Pall. *P. edulis* Salisb. *P. japonica* (Makino) Miyabe. et Takeda *P. lactiflora* Pall. *P. lactiflora* Pall. var. trichocarpa (Bunge.) Stern *P. moutan* Sims. *P. officinalis* L.	Bai Shao, Shao Yao (Peony, tree peony)	(root) Benzoic acid, paeoniflorin, oxypaeoniflorin, albiflorin, benzoyl paeoniflorin, acetylsalicylic acid.[14,15,226,5][10,87,380,568,571]	Carminative, antispasmodic, analgesic, sedative.
Paeonia obovata Maxim. *P. mourian* Sim. *P. suffruticosa* Anders. *P. veitchii* Lynch.	Mu Dan Pi (Tree peony)	(root bark) Paeonol, paeonoside, paeonin, pelargonin, paeonolide, astragalin (paeoniflorin contained in *P. mourian*).[1,2,33,87]	Sedative, antipyretic, analgesic actions.

Panax ginseng C. A. Meyer *P. pseudoginseng* Wall.	Ren Seng (Ginseng)	(root) Triterpenoid, quinquenosides, ginsenosides, oleanolic acid, panaxynol, beta-elemene, spemine, putrescine, spermindine.[26,53,510,511,568,571,573]	A stimulant, tonic, expectorant.
Panax japonicum C. A. Meyer	Zhu Je Seng (Japanese ginseng)	(rhizome) Saponins including chikusetsa saponin II and chikusetsa saponin IV, ginsenoside R$_0$.[25,33]	A stimulant, tonic, expectorant.
Panax notoginseng (Burk) F. H. Chen	Tian Qi	(root) Ginsenosides, panaxatriol, panaxadiol, dencichine, saponins flavonoids.[53,425,568]	A stimulant, tonic expectorant, anti-inflammatory.
Panax quinquefolium L.	Xi Yang Seng (American ginseng)	(root) Ginsenosides, phytosterols.[26]	Stimulating effect on central nervous system; antifatigue.
Panax zingiberensis C. Y. Wu & K. M. Feng	San Qi	(root) Saponins including arasaponins, panaxadiol, panaxatriol.[33,568] This herb is mildly toxic.	Arrests bleeding, removes blood stasis, and relieves pain. Treats angina pectoris, hemorragic diseases.
Papaver amurense (N. Busch) N. Busch ex Tolmatchev. *P. nudicaule* L. *P. radicatum* Rottb. var. pseudoradicatum (Kitag.) Kitag.	Ying Su (Poppy)	(whole plant) Amurine, amuroline, amuroine, coptisine, nudaurine, muramine, nudicaulin.[48]	For cough, headache, intestinal infection, blood in the urine, stomach ulcer.
Papaver rhoeaes L.	Li Chun Hua (Corn poppy)	(flower, root) Rhoeadine, rhoeagenine.[72,87]	For jaundice, as a gargle or ingested as bechic.

Table 1A   Major Constituents and Therapeutic Values of Chinese Medicinal Herbs   (continued)

Scientific Name	Common Chinese and (English) Name	Major Constituents and (sources)	Therapeutic Values*
Papaver somniferum L.	Yu Mei Ku (Opium poppy)	(whole plant) Berberine, codeine, papaverine, isocorypalmine, laudanine, magnoflorine, meconine, 6-methylocodine, morphine, narcotine, pseudomorphine, rhoeadine, sanguinarine, beta-sitosterol, stigmasterol, thebaine, zanthaline.[50,87]	Antitussive, antispasmodic, analgesic, astringent, narcotic; treats chronic enteritis, diarrhea, enterorrhagia, headache, toothache, asthma.
Paracyclea insularis Kudo et Yamamoto	Tu Fang Ji	(root) Insulanoline, insularine, iso-chondrodendrine.[58]	Treats headache, throat pain, arthritis pain; externally for snakebite.
Paracyclea ochiaiana Kudo et Yamamoto	Taiwan Tu Fang Ji	(stem) Insularine.[58]	Relieves pain caused by arthritis, headache.
Parechites adnascens Hance *P. thunbergii* A. Gray	Luo Shi	See *Trachelospermum jasminoides*	
Parietaria micrantha Ledeb.	Qiang Cao	(root, vine) Protocatechuic acid.[50]	For fractures, hemorrhage, lumbago, myalgia, numbness, renitis, rheumatism.
Paris polyphylla Smith *P. quadrifolia* L.	Zao Xiu (Himalayan paris)	(root) Alpha-paristyphnin, diosgenin glycosides, furostanol, spirostanol saponins.[50,506]	Antispasmodic, antiinflammatory, febrifuge, antitumor.
Parnassia palustris L.	Mei Hua Cao	(whole plant) Kaempferol, hyperin.[48]	Treats diarrhea, liver infection, cough.

Parthenocissus tricuspidata Planchon.	Di Jin (Boston ivy)	(root, stem) Cyanidin, lysopine, octopinic acid, fatty acids.[48]	Treats arthritis, stomachic, headache, blood in the stool.
Patrinia heterophylla Juss.	Mu Tou Hui	(root) Essential oils.[48]	Stimulates circulation, eliminates blood stasis in cancers of the blood and cervix.
Patrinia scabiosaefolia Fisch ex Link.	Ye Huang Hua or Bai Jiang Cao	(leaf) Essential oils, patrinoside, isopatrinene.[33,50]	Antidote, astringent, anodyne. Treats insomnia caused by neurasthenia or acute infections.
Pedicularis resupinata L. *P. resupinata* L. f. pubescens Kom. *P. resupinata* L. f. ramosa Kom.	Ma Xian Gao (Lousewort)	(whole plant) Alpha-amyrin, beta-amyrin, betulinic acid, cholesterol, kaempferol.[50,218,329]	Used in fever, leucorrhea, rheumatism, sterility, urinary difficulties, anti-inflammation, dryness of the mouth, tongue, and tinnitus.
Pericamylus formosanus Diels	Peng Lai Teng (Salt vine)	(whole plant) Narcotic alkaloid, mucilage.[50]	Antirheumatic.
Perilla frutescens (L.) Britt. *P. ocymoides* L. *P. ocymoides* L. var. crispa Benth. *P. polystachya* D. Don (Syn. *Elsholtzia cristata*) *P. arguta* Benth	Xiang Xu (Perilla) Zi Su	(leaf) Perilladehyde, l-perilla, aldehyde, apigenin, luteolin, limonene, beta-caryophyllene, alpha-bergamotene, linalool, 3-p-coumarylglycoside-5-glucoside of cyanidin, 7-caffeyl-glucosides of apigenin and luteolin, anthocyanins.[33,50,249,264]	Antibacterial, antitussive, stomachic, antiseptic.
Periploca sepium Berge.	Xiang Jia Pi (Silk vine)	(root bark) Steroid glycosides, carenolide, periplocin, pregnenes (low toxicity).[29,33]	Antirheumatic, cardiotoxic.
Persicaria amphibia (L.) S. F. Gray	Liang Xi Liao	(whole plant) Hyperoside, avicularin, quercetin, kaempferol, quercimeritrin, luteolin-7-glucoside.[48]	Treats diarrhea.

Table 1A   Major Constituents and Therapeutic Values of Chinese Medicinal Herbs   (continued)

Scientific Name	Common Chinese and (English) Name	Major Constituents and (sources)	Therapeutic Values*
Persicaria hydropiper (L.) Spach.	Shui Liao	(whole plant) Tadeonal, polygodiol, isotadeonal, confertifolin, polygonone, persicarin, quercetin, rhamnazin, quercitrin, quercimeritrin, hyperin, leucoanthocyanins.[48]	Antitoxin, insecticide; relieves itchiness, stops bleeding.
Persicaria orientalis (L.) Spach.	Hong Cao	(fruit, whole plant) Orientin, orientoside, vitexin, plastoquinone.[48]	Treats arthritis, relieves swelling, diuretic.
Petasites japonicus F. Schmidt	Feng Dou Cai	(flower, root, leaf) Beta-sitosterol, beta-carotene, thiamine, riboflavin, niacin, ascorbic acid.[50]	For colds, asthma, cough, dyspnea, tuberculosis.
Peucedanum decursivum Max.	Qian Hu (Hogfennel)	(root) Glycoside nodakenin.[49]	Analgesic, antipyretic, antitussive; treats headache, bronchitis, asthma, pertussis.
Peucedanum formosanum Hayata	Taiwan Qian Hu	(root) Anomalin, coumarin, peuformosin.[56]	Cooling function, relieves pain, cough; treats cold, headache.
Peucedanum japonicum Thunb. *P. praeruptorum* L. *P. rubricaule* Shan et Shch.	Fang Kui	(root) Nodakenetin, nodakenin, decursidin, umbelliferone, pencordin, qianhucocumarin, rubricauloside.[33,247]	Antitussive, expectorant.

Pharbitis diversifolia Lindl. *P. hederacea* Choisy *P. nil* (L.) *P. triloba* Miq.	Qian Niu Zi (Morning glory)	(seed) Glycoside pharbitin, gibberellin, pharbilic acid.[33,144] This herb is toxic.	Potent purgative; purges parasites, ascaris, and taenia. Treats constipation, edema.
Phaseolus angularis (Willd.) W. F. Wight *P. lunatus* L. *P. radiatus* L. *P. vulgaris* L.	Jin Jia Dou (Black gram)	(seed, leaf) Alpha-globuline, beta-globulin, fatty acids, vitamins A, B, and B_2, robinin, kaempferol-3-robinobiosido-7-rhamnoside.[48]	Diuretic, for abscesses, beri-beri, dysentery, sores, swelling.
Phellodendron amurense Rupr. *P. amurense* Rupr. f. molle (Nakai) Y. C. Chu *P. chinensis* Schneid	Huang Bai (Cork tree)	(bark) Berberine, palmatine, candicine, phellodendrine, obacunone.[33,58,568]	Antibacterial; stimulates the phagocitic activity of leukocytes, against dysentery.
Photinia serrulata Lindl.	Shi Nan Ye (Photinia)	(leaf) Hydrocyanic acid, tannins.[49]	Tonic and stimulant in neurasthenia, impotence, spermatorrhea, amenorrhea, infecundity.
Phragmites communis Trin.	Lu Gen (Reed)	(root) Glycosides, protein, asparagin.[49]	As stomachic, antiemetic, antipyretic. Treats arthritis, jaundice, pulmonary abscess.
Phyllanthus emblica L.	Ye Gan Zi	(whole plant) Chebulinic acid, mucic acid, alpha-leucodelphinidin, vitamin C.[33,307,568]	For conjuctivitis, diarrhea, abdominal tumors, nephritis, urogenital ailments.
Phyllanthus simplex Ketziu	Dan Ye Xiz Zhu	(leaf) Alpha-ketoglutaric acid, simplexine.[57,211,212]	Treats sore eyes, hepatitis, mammary gland infection.

Table 1A Major Constituents and Therapeutic Values of Chinese Medicinal Herbs   (continued)

Scientific Name	Common Chinese and (English) Name	Major Constituents and (sources)	Therapeutic Values*
Phyllanthus urinaria L. *P. niruri* Li *P. reticulatus* Poiret	Ye Xia Zhu Zhu Zi Cao	(fruit, leaf) Phyllanthine, phyllantidine. In leaf, phyllanthin, hypophyllanthin, niranthin, nirtetralin, phylteralin.[33]	Treats coughing, promotes digestion and secretion.
Phyllanthus virgatus Forster	Xi Ye Zhu Chi Cao	(whole plant) Vitamin C, amino acids.[57]	Treats heptitis, cold, fever, blood vomiting, diarrhea.
Phyllodium pulchellum (L.) Desvaux.	Pai Qian Shu	(whole plant) Bufotenine, N,N-dimethyltryptamine, N,N-dimethyltryptamine oxide, framine, physcim-l-gluco-rhamnoside.[54]	Diuretic; relieves swelling, treats cold, pain, and regulates menses.
Phyllostachys bambusoide Sieb. et Zucc. *P. nigra* Munro. var. henonis Mak.	Chu Ye (Timber bamboo)	(leaf, shoot) Benzoic acid, silica, potassium hydroxide, aluminum oxide, iron oxide, calcium.	Antipyretic, hematuria, sedative, antiemetic, antispasmodic in catarrh.
Physalis alkekengi L. var. francheti (Mast.) Makino	Jin Deng Long (Chinese lantern)	(calyx, fruit) Physanols, physalien, zeaxanthin, glycolic acid, cryptoxanthin, physoxanthin, mutaxanthin, auroxanthin, physalin A, B, and C, luteolin, tigloidine, physalins, hystonin.[33,4887,568]	Antibacterial; stimulates myocardial contraction, causes vasoconstriction, uterine contraction.
Physalis angulata L.	Ku Zhi (Groundcherry)	(fruit, leaf) Hystonin.[60] Overdose may cause dizziness.	Antifebrile, laxative, diuretic; causes uterine contractions.

Physochlaina infundibularis Kuang.	Hua Shan Seng	(root) Hyoscyamine, scopolamine, scopoletin, scoplin.[33] This herb is toxic.	A cholinergic blocking agent, relaxing effect on bronchial muscles.
Phytolacca acinosa Roxb. *P. americana* L. *P. japonica* Makino *P. kaempferi* A. Gray *P. octandra* Bge. *P. pekinensis* Hance	Shang Lu (Pokeberry)	(root) Phytolacine, phytolaccatoxin, oxyristic acid, jaligonic acid, saponins.[33,144,568] This herb is toxic.	Antitussive, diuretic, antibacterial, anti-inflammatory.
Picrasma quassioides (D. Don) Benn. *P. quassioides* (D. Don) Benn. f. *dasycarpa* Kitag.	Ku Shu	(stem bark) 2,4-dichloro-6-aminopyridine, 4,5-dimethoxycanthin-6-one, 2,6-dimethoxy-p-benzo-quinone, methyl nigakinone, picrasmin, nigakihemiacetal A, nigakilactone A, nigakinone, quassin.[50]	Treats fever, stomachache.
Picrorhiza kurroa Royle.	Hu Huang Lain	(root) Cathartic acid, picrorhizin, kurrin, aglycone, kutkin, vanillic acid.[49,450]	Antipyretic, stomachic.
Pileostegia viburnoides Hooker f. Thomson	Ching Mian Hua	(whole plant) Lunularic acid, abscisic acid, quercetin, leucocyanidin.[57]	Treats arthritis.
Pimela alba Lour.	Gan Lan	See *Canarium album*	
Pimpinella thellungiana Wolff *P. thellungiana* Wolff var. *tenuisecta* Chu	Hui Qin (Aniseed)	(leaf, root, seed) Ilungianin A, ilungianin B.[50,220]	A stimulant, anodyne, hypotensive; treats choleraic infections and flatulence.
Pinellia ternata (Thunb.) Breit. *P. tuberifera* Tenore	Ban Xia	(tuber) l-ephedrine, choline, amino acids.[33,144] This herb is toxic.	Antiemetic, antitussive, and antidote for strychnine intoxication.

Table 1A   Major Constituents and Therapeutic Values of Chinese Medicinal Herbs   (continued)

Scientific Name	Common Chinese and (English) Name	Major Constituents and (sources)	Therapeutic Values*
Pinus bungeana Zucc. ex Endl. *P. densiflora* Sieb. et Zucc. *P. koraiensis* Sieb. et Zucc. *P. sylvestris* L. var. mongolica Litv. *P. sylvestris* L. var. sylvestriformis (Takenouchi) Cheng et C. D. Chu *P. tabulaeformis* Carr.	Song Ta (Pine)	(cone) Essential oil, limonene, pinitol.[33]	Antitussive, antiasthmatic, antibacterial.
Piper cubeba L.	Bi Cheng Qie (Cubeb, tailed pepper)	(unripe fruit) Cubebin, dipentene, cadinene, cineol, carene, camphene, pinene, sabinene, azulene, terpineol.[49]	Urinary antiseptic, stomachic, carminative.
Piper longum L.	Bi Ba (Indian long pepper)	(fruit) Volatile oil, piperine,	Antipyretic, carminative, aromatic stomachic, analgesic in gastralgia, flatulence, headache.
Piper nigrum L.	Hu Jiao (Black pepper)	(fruit) Piperine, chavicine, piperamine, piperonal, dihydrocarveol, cryptone, caryophyllene. This herb may cause irritation to the system.[33,45]	Anticonvulsive, sedative.
Pistacia lentiscus L.	Ru Xiang	(resin) Masticinic acid, masticonic acid, masticoresene, fisetin, fustin, gallic acid, quercetin, taxifolin.[49,50]	Antitumor, antitussive, analgesic, sedative in gastralgia, cardiodynia, mastitis, peptic ulcer.

Pittosporum tobira (Thunb.) Aiton	Hi Tong	(bark) Dihydroterpene, heptane.[60]	Treats dysentery and rheumatism.
plantago asiatica L. P. depressa Willd. P. exaltata Horn. P. loureiri Roem. et Schult. P. major L. P. major L. var. asiatica DC	Che Chen Zi (plantain)	(seed) d-xylose, l-arabinose, d-galacturonic acid, l-rhamnose, plantasan, plantenolic acid, plantagin, homoplantagin, aucubin, ursolic acid, hentriacontane.[48,510,568]	Diuretic, expectorant, intestinal infection, diarrhea caused by bacteria.
Platycladus orientalis (L.) Franco.	Ce Bai Ye	See Thuja orientalis	
Platycodon autumnalis Decne. P. grandiflorum DC P. sinensis Lam.	Jie Geng (Kikio root)	(root) Platycodigenin, polygalacic acid, platycodigenic acids, platyconin, prosapogenin, betulin, 3-O-β-glucosylplatycodigenin, platycodosides, spinasterols, platycodonin.[23,33,48,510,558]	An expectorant, antitussive, analgesic.
Plumbago zeylanica L.	Bai Hua Teng (Ceylon leadwort)	(root) Plumbagin, glucose, fructose, protease, invertase, plumbagin, naphthaquinone, slliptinone, 3-chloroplumbagin.[50,450]	Bactericidal, antifertility.
Plumeria rubra L.	Hong Je Dan Hua (Frangipani)	(leaf, stem bark, flower) Agoniadin, cerotinic acid, fulvoplumierin, lupeol, plumieric acid, plumieride, quercetin, pectins, plumieric acid, cerotic acid, acetyl lupeol, essential oils, geraniol, citronellol, farnesol, phenylethyl alcohol, linalool, kaempferol, aldehydes, ketones.[50]	Inhibits the tuberculosis bacterium; fungicidal, stimulant, emmenagogue, febrifuge, purgative; for dropsy, herpes; and venereal infections.

Table 1A   Major Constituents and Therapeutic Values of Chinese Medicinal Herbs   (continued)

Scientific Name	Common Chinese and (English) Name	Major Constituents and (sources)	Therapeutic Values*
Podocarpus macrophyllus (Thunb.) D. Don	Luo Han Song (Southern yew)	(stem bark, leaf, root, fruit) Pinene, camphene, cadinene, podocarpene, neocryuptomerin, kaurene, ecdysterone, ponasterone, makisterones, hinokiflavone, sciadopitysin, podocarpus flavones, macrephyllic acid, podototarin, totarol.[54]	For ringworms, blood disorders; tonic for heart, kidneys, lungs, stomach.
Podophyllum peltatum L. *P. pleianthum* Hance	Ba Jian Lian (Podophylium)	(rhizome) Podophyllotoxin, flavonoids, desoxypodophyllotoxin.[5,87,589] This herb is toxic.	An antitumor agent.
Pogostemon cablin Benth.	Huo Xiang (Patchouli)	(branch, leaf) Essential oils.[60]	Antiseptic, for abdominal pain, cold, diarrhea.
Polyanthus narcissus (Syn. *Narcissus tazetta*)	Shui Shai Gen	See *Narcissus tazetta*	
Polygala japonica Houtt *P. sibirica* L. *P. tatarinowii* Regel.	Su Cao (Milkwort)	(rhizome, bark) Saponins, tenuidine, tenuifolin (prosenegenin).[28,33]	Stimulates bronchial secretions; antibacterial.
Polygala tenuifolia Willd.	Yuan Zhi	(root) Onjisaponin A, onjisaponin B.[24,578]	Sedative; strengthens nervous system.

Species	Common name / (part) / Constituents	Actions
Polygonatum chinense Kunth. *P. cirrhifolium* Royle. *P. macropodium* Turez. *P. officinale* All. *P. sibiricum* Delar. ex Redoute *P. stenophyllum* Maxim. *P. odoratum* (Mill.) Druce var. pluriflorum (Miq.) Ohwi f. ovarifolium Y. C. Chu *P. vulgare* Desf.	Huang Ging (Solomon's seal) Jiang Sun (root, stem) Convallarin, convallamarin, steroidal saponin POD-II, beta-sitosterol, mucilage.[49,245]	Stimulates the appetite, increase peristalsis, slows the heart and raises the arterial tension, slow and deepens respiration, purgative.
Polygonum aviculare L. *P. aviculare* L. var. vegetum Ledeb. *P. lapidosa* Kitag. *P. manshuriensis* Komarov *P. vivipara* (L.) S. F. Gray	Bian Xu (Knot grass) (aerial part) Avicularin, tannins, vitamin E, mucilage, gallic, caffeic, guercitrin, chlorogenic, osalic, silicle, p-coumaric acids, d-catechol, leucoanthocyanins.[33,60,450]	Treats urethritis, lithiasis, and chyluria. Anti-inflammatory, against dysentery and parotitis, an antiascardiasis agent.
Polygonum bistorta L.	Cao He Che (Snakeweed, bistort) (stem, root) Iodine, oxalic acids, coumarins, hydroxycinnamic acids, ether oil, hydroxybenzoic acids, hydrocyanic acids, anthocyanidines, carotenes, anthraquinones, phytosterines, monoterpene, sesquiterpene glucoside, caffeic acid, quercimeritrin, avicularin, gallic acid, protocatechuic acid.[50,221,222,223,224,568]	Diuretic, laxative, hemostatic, antifebrile.
Polygonum cuspidatum Siebold & Zucc.	Hu Chang (Japanese knotweed) (stem, root) Polygonin, glucofragulin, emodin, polydatin, flavonoids.[33]	Treats hypercholesterol.
Polygonum hydropiper L.	La Lian (Water pepper) (whole plant) Persicarin, rhamnazin, isotadeonal, quercimeritrin, tadeonal.[33]	Improves indigestion, treats dysentery and enteritis.

Table 1A Major Constituents and Therapeutic Values of Chinese Medicinal Herbs (continued)

Scientific Name	Common Chinese and (English) Name	Major Constituents and (sources)	Therapeutic Values*
Polygonum multiflorum Thunb. *P. chinensis* L.	He Shou Wu Huo Tan Mo Cao (Hill buckwheat)	(root, stem, leaf) Chrysophenol, emodin, emodin methyl ester, rhein, glycoside rhaphantin, lecithin, parietin, chrysophanic acid, anthron. [33,49,54,442]	A laxative, detoxicant for boils. Treats neurosis, neurasthenia, insomnia, hypercholesterolemia.
Polygonum orientale L.	Shui Hong Cao (Prince's feather)	(whole plant) Orientin, vitexin, isovitexin, isoorientin, plastoquinone-9. [33]	Antibacterial.
Polygonum perfoliatum L. *P. tinctorium* Lour.	Gang Ban Gui Ban Lan Geng (Chinese indigo)	(leaf) Flavonoids, amino acids, organic acids, sugar, indican, emodin, chrysophanol, protein. [56]	Cooling property; relieves swelling, blood circulation, detoxicant, diarrhea. Juice is dropped into the ear to cure deafness.
Polyporus umbellatus (Pers.) Fr.	Zhu Ling	(dried fungus) Ergosterol, biotin, protein. [33]	Diuretic; stimulates the immune system; anticancer.
Poncirus trifoliata Rafin	Gou Gi (Trifoliate orange)	(fruit) Poncirin, limonin, imperatorin, bergapten, neohesperidin, citrifoliol, myrcene, camphene, gamma-terpinene. [33]	Treats gastric pain, constipation, and prolapse of the uterus or rectum.
Pongamia pinnata (L.) Pierre ex Merrill.	Shui Huang Pi	(bark) Behenic acid, gamatin, kaempferol, kanugin, karanjin, pinnatin, pongapin. [50]	Antiseptic.
Populus alba L. *P. davidiana* Dode *P. tomentosa* Carr.	Yin Bai Yang (White poplar)	(leaf, stem bark) Salicin, populin, benzoyl salicin, tannins, erisin, salicinase, salicortin, tremulacin, salireposide. [50]	Depurative, for colic, eczema, herpes, labialis, fever, dysuria; antiseptic, antiperiodic.

Species	Chinese name	Constituents	Medicinal effects
Poria cocos (Polyporaceae) (Syn. Sclerotium cocos)	Fu Ling	(fungus body) Pachymic acid, tumulosic acid, eburicoic acid, pinicolic acid, pachymarose.[33,567,568]	A diuretic, cardiotonic, it has a tranquilizing effect, lowers blood sugar levels, it is an antibacterial and anticancer.
Portulaca grandiflora Hooker	Song Ye Mo (Purslane)	(whole plant) Portulal, betacyanin, betanin, betanidin.[55]	Treats throat swelling and pain, externally for burns, wounds and infections.
Portulaca oleracea L.	Ma Chi Xian (Purslane)	(aerial part) Potassium salts, catecholamines, norepinephrine, dopamine, vitamin A, vitamin B, magnesium.[33,49]	Antibacterial, diuretic; causes vasoconstriction, stimulates uterine and intestinal smooth muscle contraction.
Portulaca pilosa L.	He Que She	(aerial part) Tannins, phosphates, magnesium, iron, aluminum, manganese, calcium, potassium, sodium, urea.[60]	Antihemorrhagic, antiscorbutic, vulnerary properties.
Potentilla bifurca L. var. canescens Bong. et Mey. P. bifurca L. var. glabrata Lehm. P. chinensis Seringe P. discolor Bunge. P. fragariodes L. P. fragariodes L. var. major Maxim. P. freyaiana Bornmuller P. kleiniana Wight & Arnott var. robusta (Franch. & Savat.) Kitag.	Wei Ling Cai (Wolfstooth, cinquefoil)	(leaf) D-catechin.[50] This herb is toxic.[60]	Antibacterial, antiplasmodium, smoothes muscle relaxation, gynecological bleeding.
Poterium officinale Benth.	Di Yu (Ground elm)	See Sanguisorba officinalis	

Table 1A Major Constituents and Therapeutic Values of Chinese Medicinal Herbs (continued)

Scientific Name	Common Chinese and (English) Name	Major Constituents and (sources)	Therapeutic Values*
Primula sieboldii E. Morren var. patens (Turcz.) Kitag. *P. asiatica* Nakai *P. asiatica* Nakai f. albiflora (Koidz.) Kitag. *P. asiatica* Nakai f. lilacina (Nakai) Kitag. *P. vulgaris* L.	Yin Cao	(whole plant) Primulagenin A, aegicerin, protoprimulagenin A.[48,568]	Relieves cough, throat infection.
Prunella vulgaris L.	Xia Ku Chao (Heal-all)	(leaf) Caffeic acid, d-camphor, cyanidin, delphinidin, d-fenchone, hyperoside, oleanolic acid, rutin, ursolic acid.[48,450,568]	Antibacterial, antipyretic, cardiac tonic, diuretic, anticancer.
Prunus armeniaca L.	Xing Ren (Apricot)	(kernel) Amygdalin, prunasin, fatty acids, mandelonitrile (enzyme amygdalase can hydrolyze amygdalin to produce cyanic acid).[33,53]	Stimulates respiratory center reflexively and produces a tranquilizing effect.
Prunus domestica L. *P. glandulosa* Thunb. *P. japonica* Thunb.	Yu Lee Ren (Dwarf flowering cherry)	(leaf, fruit) Amygdalin, citric acid, fatty acids.[53]	Diuretic, laxative.
Prunus mume (Sieb.) Sieb. et Zucc.	Wu Mai (Black plum)	(fruit) Prudomenin, malic acid, succinic acid, citric acid, tartaric acid, amygdalin.[33,53]	Treats biliary ascariasis and hookworm.
Prunus padus L.	Chou Lee	(fruit, leaf) Hyperin, quercetin-3-galacto-xylo-glucoside, nonacosane, beta-sitosterol, lupeol, amygdalin, fatty acids.[48]	Treats diarrhea, cough.

Species	Common name	Constituents (part)	Uses
Prunus persica (L.) Batsch.	Tou Ren (Peach)	(leaf, flower, fruit) Malic acid, citric acid, octalactone, leucoanthocyanins, tannins, hexalactone, hectalactone, benzyl alcohol, nonalactone, decalactone, ethanol, hexanol, acetadehyde, benzaldehyde, acetic acid, pentanoic acid, hexanoic acid.[50]	Astringent, febrifuge, parasiticide, diuretic, sedative, vermifuge.
Pseudostellaria heterophylla (Miq.)	Tai Zi Shen (Gorden latch)	(root) Fructose, starch.[48,380]	A tonic for lung disease; used as an appetizer.
Psidium guajava L.	Fan Shi Lui (Guava)	(fruit) Avicularin, guaijaverin, arabinose ester, amritoside, crataegolic acid, luteioic acid, argamolic acid.[33]	Treats dysentery and acute gastrointestinal inflammation.
Psoralea corylifolia L.	Bu Gu Zi (Scuffy pea)	(fruit) Psoralen, angelicin, psoralidin, coryfolin, bavachinin, isobavachin, corylifolinin, d-backuchiol.[33]	Coronary vasodilating effect; increases the myocardial contraction; antibacterial, anticancer.
Psychotria rubra (Lour.) Poir. *P. serpens* L.	Jiu Jie Mu Ling Bi Long (Red psychctria)	(leaf, stem) Alkaloids.[83]	A remedy for contusions; relieves pain of bruises; externally for swellings.
Pteris cretica L. *P. ensiformis* Burmann *P. multifida* Poir *P. vittata* L. *P. wallichiana* Agardh.	Feng Wei Cao (Brake)	(whole plant) Starch, filicic acid, tannins.[84]	Treats arthritis, dysentery, diarrhea.
Pterocarya stenoptera DC	Feng Yang (Wing nut)	(stem, leaf, bark) Salicylic acid, kino-tannic acid, pyrocatechine acid, protocatechinic acid.[48,60]	Diuretic; used on wounds and ulcers, hemorrhage, suppuration.

Table 1A   Major Constituents and Therapeutic Values of Chinese Medicinal Herbs   (continued)

Scientific Name	Common Chinese and (English) Name	Major Constituents and (sources)	Therapeutic Values*
Pueraria lobata (Willd.) Ohwi. *P. pseudo-hirsuta* Tang.	Ge Gen (Kudzu)	(root) Isoflavones, daidzin, diadzin-4, 7-diglucoside, daidzein, puerarin, xylopurarin, robinin, kaempferol-rhamnoside, fatty acids. [12,33,48,558,568]	Antispasmodic, hypotensive; and stabilizes blood pressure, treats angina pectoris.
Pueraria montana (Lour.) Merrill. *P. thunbergiana* Benth.	Shan Ge	(root, leaf) Glutamic acid, butyric acid, asparagin, adenine. [49]	Antipyretic, refrigerant.
Pulsatilla ambigua Turcz. ex Pritz. *P. cernua* (Thunb.) Bercht. et Opiz. *P. chinensis* (Bunge.) Regel *P. chinensis* (Bunge.) Regel var. kissii (Mandl.) S. H. Li et Y. H. Huang	Bai Tou Went (Pasque flower)	(root) Protoanemonin, anemonin, okinalin, okinalein, ranunculin, saponins, triterpenoids. [33,426]	Antiamebial, antibacterial; treats amebic dysentery, anticancer.
Punica granatum L.	Shi Liu Pi (Pomegranate)	(pericarp, root bark) Pelletierine, isopelletierine, methylisopelletierine, methyl-pelletierine, pseudopelletierine, gallotannic acid, sitosterol, ursolic acid, maslinic acid, elegic and gallic acid. [33,450] This herb is toxic.	Purges intestinal parasites.
Pyrethrum cinerariifolium (L.) Trev. *P. sinense* DC	Chu Chong Jiu	(flower) Essential oil, adenine, choline, stachydrine. [49,568]	Sedative, refrigerant in headache, influenza.

Botanical name	Common name	Constituents	Medicinal uses
Pyrola decorata P. incarnata Fisch. ex DC P. japonica Klenze ex Alefeld P. renifolia Maxim. P. rotundifolia L.	Lu Xian Cao (Wintergreen)	(whole plant) Arbutin, homoarbutin, isohomoarbutin, chimaphillin, monotropin.[33]	Antibacterial, antiarrhythmic; lowers blood pressure; hemostatic effect.
Pyrrosia adnascens (Sw.) Ching	Shu Long	(frond) Amygdalin, tannins, formic acid, tartaric acid, arbutin.[56,60,225]	Treats burns, a remedy for dysentery, diuretic, nerve pain.
Pyrrosia lingua (Thunb.) Farwell P. petiolosa (Chris.) Ching P. sheareri (Baker) Ching	Shi Wei (Felt fern)	(leaf) Isomangiferin, diplotene.[33]	Diuretic; treats urinary tract infections and urolithiasis.
Quercus acutissima Carr. Q. aliena Blume var. acutiserrata Maxim. ex Wenzig Q. dentata Thunb. Q. liaotungensis Koidz. Q. mongolica Fisch. ex Turcz. Q. variabilis Blume	Li Shu (Oak)	(stem bark, root) Lignin, cellulose, protein, pentosan, galactan.[56]	Promotes absorption of tuberculous nodules, remedy for diarrhea, hypertrophy of the gastrointestinal tract, root makes a cleansing dressing for foul sores.
Quisqualis grandiflora Miq. Q. indica L. Q. longifolia Presl. Q. loureiri G. Don. Q. pubescens Burm. Q. sinensis Lindl.	Shi Jiun Zi (Rangoon creeper)	(fruit) Quisqualic acid, trigonelline.[33,235]	Treats internal parasites.
Rabdosia lasiocarpus (Hayata) Hara	Mao Guo Yan Ming Cao	(whole plant) Terpenes, oridonin, rubescensins, 5-fluorouracil.[50] This herb is toxic.	For carcinomas of esophagus and stomach; antiarthritic, antidotal, febrifuge.

Table 1A Major Constituents and Therapeutic Values of Chinese Medicinal Herbs (continued)

Scientific Name	Common Chinese and (English) Name	Major Constituents and (sources)	Therapeutic Values*
Rabdosia rubescens Hora	Dong Ling Cao	(aerial part) Rubescensine B, oridonin, tannic acid, ponicidine, essential oils.[33] This herb is toxic.	Treats esophageal cancer, malignant cancer.
Ranunculus chinensis Bunge.	Hui Hui Suan	(whole plant) Protoanemonin, anemonin, ranunculin.[48]	Relieves swelling, asthma, liver disorders, toothache, night blindness.
Ranunculus japonicus Thunb. *R. sarmentosa* Adams	Mao Liang (Japanese radish)	(whole plant) Anemonin, protoanemonin.[50]	Antitumor, sedative, bactericidal against bacillae of diphtheria, staphylococcus.
Ranunculus sceleratus L.	Shi Long Nei (Ground mulberry)	(whole plant) Ranunculin, anemonin, 5-hydroxytryptamine, seratonin, protoanemonin, pyrogallol tannins.[48,50] This herb is toxic.	Relieves swelling, pain; antitoxin; treats lymphatic gland disorders, antirheumatic.
Ranunculus ternatus Thunb.	Mao Zhua Chao	(leaf) Tannins, phenolic acids, volatile phenols, nonvolatile terpenic compounds, volatile carbonyl and S-containing compounds.[60,223]	Treats abscesses.
Raphanus sativus L.	Cai Fu (Radish)	(leaf, flower, seed, root) Raphanin.[50,568]	For asthma, cough, diarrhea, dysentery, eruptive fevers; bactericidal, antitumor.
Rauvolfia verticilata (Lour.) Baill.	Luo Fu Mu	(root) Reserpine, rescinnamine, beta-sitosterol, aricine, vellosimine, peraksine, serpentine, robinin.[33,39,510]	Treats hypertension, psychosis, schizophrenia.
Rehmannia chinensis Fisch. *R. glutinosa* (Gaertn.) Libosch.	Di Huang (Chinese foxglove)	(root) Catalpol, campesterol, rehmannin, polysaccharide.[16,33,558]	Lowers blood sugar; immuno-antitumor activity.

Species	Common name	Constituents	Uses
Rhamnus davurica Pall. R. davurica Pall. var. nipponica Makino R. parvifolia Bunge.	Shu Li (Buckthorn)	(fruit, root, stem bark) Emodin, chrysophanol, kaempferol, rhamnodiastase, aloe-emodin.[48,308] This herb is slightly toxic.	Insecticidal; treats respiratory infection, cough, improves bowel movement.
Rhaponticum uniflorum Ludl.	Lour Lu	(root) Lactones, flavonoids, essential oils.[48]	Febrifuge, an emmenagogue, antidysenteric.
Rheum officinale Baill. R. koreanum Nakai R. palmatum L. R. tanguticum Maxim R. undulatum L.	Tai Huang (Rhubarb)	(rhizome) Anthraquinones, chrysophanol, emodin, physcion, aloe-emodin, rhein, chrysophenol, rheum tannic acid, gallic acid, calechin, bianthraquinonyl, sennosides (*R. undulatum* also contains rhaponticin).[1,33,236,510,558,567,568] This herb may be toxic.	Potent laxative, antibacterial, anthelmintic, anticancer; stimulates the large intestine and increase the movement of luminal contents toward the anus, resulting in defecation. Antispasmodic, choleretic, hemostatic, diuretic; lowers blood pressure, lowers cholesterol level.
Rhodea japonica Roth.	Won Nian Qing	(leaf, rhizome) Rhodexin A, B, C, and D.[33] This herb can cause vasoconstriction.	Improves heart muscle; used as an emetic, antibacterial.
Rhodiola elongata (Ledeb.) Fisch. & Meyer	Hong Gin Tian	(root) p-tyrosol, rhodioloside, flavonoids.[48]	A tonic; improves heart muscle; aphrodisiac.
Rhododendron anthopogon G. Don	Lie Xiang Du Juan (Rhododendron)	(leaf) Essential oils, saponins, quercetin, gossypetin.[33]	Antitussive, antiasthmatic.
Rhododendron dauricum DC	Man Shan Hong (Daurian rhododendron)	(leaf) Germacrone, flavonoid, farreol, feriol, quercetin, myricetin, anromedotoxin, rhodotoxin.[33]	Antitussive, antiasthmatic.

Table 1A   Major Constituents and Therapeutic Values of Chinese Medicinal Herbs   (continued)

Scientific Name	Common Chinese and (English) Name	Major Constituents and (sources)	Therapeutic Values*
Rhododendron molle (Blume) G. Don	Ba Li Ma (Chinese azalea, yellow azalea)	(fruit) Rhomotoxin.[37,144] This herb is toxic.	Treats tachycardia, palpitations, hypertension.
Rhododendron mucronatum G. Don	Bai Du Juan Hua (white azalea)	(flower) Essential oil, germacrone, farreol, grayanotoxin, gossypetin, azaleatin, 5-methyl kaempferol, 5-methyl myricetin, syringic acid, dihydroquercetin, coumarins, phenolic acid, p-hydroxybenzoic acid, protocatechuic acid, vanillic acid.[48]	Treats cough, asthma, headache, respiratory infection.
Rhododendron sinensis Sw.	Yang Zhi Zu (Chinese rhododendron)	(flower) Andromedotoxin, veratrine alkaloids.[49] This herb is toxic.	Sedative, analgesic, anesthetic in rheumatism.
Rhus chinensis Mill. *R. cotinus* L. *R. javanica* L. *R. osbeckii* Decne.	Wu Bei Zi (Chinese sumach)	(leaf) Gallotannic acid, gallic acid, resin, polysaccharides.[33,144] This herb is toxic.	Treats chronic intestinal infections, hematochezia, protoptosis, skin infections, bleeding wounds.
Rhus semialata Murr.	Po Yen (Sumac)	(nutgalls on leaves) Tannins.[49]	As an astringent, styptic; treats diarrhea, hemorrhage.
Rhus verniciflua Stokes	Gan Qi (Lacquer tree)	(exudation of the bark) Resinous oil urushiol.[49] This herb may be toxic.	As hemolytic, emmenagogue, vermifuge.
Ribes mandshurica (Maxim.) Kom. *R. mandshurica* (Maxim.) Kom. f. subglabrum (Kom.) Kitag.	Shan Ma Zi	(fruit) Citric acid, malic acid, organic acids.[48]	Treats cold.

Ricinus communis L.	Bi Ma Zi (Castor bean)	(seed) Ricinine, ricinolein, olein acid, stearin acid, isoricinoleic acid, cytochrome C, castor oil.[33,87,427,450]	Cathartic, tumor inhibition.
Rorippa indica (L.) Hiern. R. *islandica* (Oeder) Borbas R. *montana* (Wall) Small	Han Cai (Nasturtium)	(whole plant) Alpha-phenylethylisothiocyante, gluconasturtin, rorifone, rorifamide.[33,235]	Antitussive, expectorant, diuretic, detoxicant.
Rosa acicularis Lindl. R. *amygdalifolia* Ser. R. *davurica* Pall. R. *davurica* Pall. var. *alpestris* (Nakai) Kitag. R. *koreana* Kom. R. *laevigata* Michx. R. *maximowicziana* Regel	Jin Jing Zi (Climbing rose)	(flower, fruit, root) Vitamins, teteracylic triterpene acids, flavonoids, ethyl beta-fructopyranoside, methyl-3-O-beta-glucopyranosyl-gallate, gallocatechin, epigallocatechin, epicatechin gallate, catechin, epicatchin, fatty acids.[48,251]	Stop vomiting blood, stomachache; relieves pain caused by nervous system, menstruation.
Rosa chinensis Jacq. R. *indica* Lindl.	Yue Je Hua (Tea rose)	(leaf, fruit, flower bud) Essential oils.[49]	For arthritis, boils, cough, hematuria, rheumatoid joint pains, circulatory stimulant.
Rosa multiflora Thunb.	Chen Wei (Multiflora rose, seven sisters rose)	(leaf, fruit, seed) Ascorbic acid, multiflorin, quercetol, kaempferol-3-glucoside, catechin.[50]	Anodyne, diuretic, laxative.
Rosa rugosa Thunb.	Mei Gui Hua (Rose)	(flower bud) Essential oils, l-citronellol, citral, geraniol, nerol, eugenol, cyanin, n-phenylethyl alcohol, citro, nonyl aldehyde, l-linalool, l-p-menthene, nonacosane, menthene, paeonidin, bensaldehyde, phenylacetic acid, rosenoxide.[48,50]	Promotes blood circulation, treats abscesses, blood diseases, dyspepsia, hematemesis, hepatitis, stomachache.
Rubia akane Nakai	Hong Gen Cao	(root) Alizarin, rubierythrinic acid, purpurin.[85]	Treats rheumatism.

Table 1A   Major Constituents and Therapeutic Values of Chinese Medicinal Herbs   (continued)

Scientific Name	Common Chinese and (English) Name	Major Constituents and (sources)	Therapeutic Values*
Rubia chinensis Regel & Maack *R. cordifolia* Thunb. *R. cordifolia* L. f. *pratensis* (Maxim.) Kitag. *R. mungista* Roxb. *R. sylvatica* (Maxim.) Nakai	Qian Cao (Madder)	(root) Rubierythrinic acid, alizarin, purpurin, pseudopurpurin, munjistin.[33,49]	Hemostatic; shorten the blood clotting time; antibacterial, antitussive; stimulates uterine contractions.
Rubus coreanus Miq. *R. crataegifolius* Bunge. *R. matsumuranus* Levelle & Vaniot *R. matsumuranus* Levelle & Vaniot var. concolor (Kom.) Kitag. *R. saxatilis* L.	Fu Pen Zi (Briar rose)	(fruit, root) Beta-sitosterol, stigmasterol, campesterol, cholestanol, ursolic acid, flavonoids.[48]	Diuretic, aphrodisiac; level infection, joint infection caused by arthritis.
Rubus parvifolius L.	Hong Mei Xiao	(root, stem) Flavonoids.[48]	Treats fever, throat pain, blood vomiting, liver and intestinal infections.
Rumex acetosa L. *R. acetosella* L. *R. amurensis* Fr. Schm. *R. aquaticus* L. *R. gmelini* Turcz. *R. longifolius* DC *R. maritimus* L. *R. marschallianus* Rehb. *R. stenophyllus* Ledeb. var. *ussuriensis* (A. Los.) Kitag. *R. thyrsiflorus* Fingerh.	Suan Mo (Garden sorrel)	(whole plant) Vitexin, quercetin-3-galactoside, violaxanthin, vitamin C, emodin, chrysophanein, chrysophanol, nepodin, hyperin, physcion.[48,50,568]	Homeopathically for cramps, hemorrhage, sore throat, esophagitis, diuretic; treats blood vomiting.

Rumex crispus L. R. japonicus	Yang Ti Gen (Dock)	(root) Chrysophanein, nepodin.[48]	Treats ovarian bleeding, eczema, tuberculosis, sexually transmitted diseases.
Rumex patientia L. var. callosus Fr. Schm.	Tu Tai Huang	(root) Chrysophanol, emodin, physcion, sloeemodin, anthranol, emodin-monomethylether.[33]	Hemostatic; treats thrombopenia and uterus.
Sagittardia sagittifolia L.	Ci Gu (Arrow head)	(whole plant) d-raffinose, d-stachyose, d-verbascose, d-fructose, d-galactose, glucose, asparagine, vitamin B.[48]	Bruised leaves for bug bite, foul sores, scrofulous ulcers, antilactogogue.
Salix babylonica L. S. matsudana Koidz. S. microstachya Turcz. ex Trautv.	Liu Ye (Weeping wil ow)	(leaf, root) Saligenin glucoside, iodine, pyrocaledol, saponins.[33]	Artigoiter, antibacterial; treats tubercule bacilli.
Salsola collina Pall.	Zhu Mao Chao	(whole plant) Salsoline, salsolidine, betaina.[33]	Treats hypertension at an early stage.
Salvia chinensis Benth. S. pogonocalyx Hance S. przewalskii Maxim. S. miltiorrhiza Bunge.	Shi Jian Chuan	(rhizome) Scutellarin, danshenols.[60,440,507]	Treats abdominal pain, arthritis, inflammation, metrorrhagia, uteritis, women's diseases; treats nasopharyngeal carcinoma.
Salvia coccinea L.	Zhu Chun Hua	(whole plant) Saluianin.[56]	Stops bleeding, cooling effect; stimulates sweating, relieves swelling.
Salvia miltiorrhiza Bunge.	Tan Seng (Red-rooted sage)	(root) Tanshinone, cryptotanshinone, isocryptotanshinone, mitirone, tanshinol, salviol, acetylsalicylic acid.[33,226,235,428,429]	Treats angina pectoris, cerebral atherosclerosis, diffusive intravascular clotting, thrombophlebitis; antioxidant.

Table 1A   Major Constituents and Therapeutic Values of Chinese Medicinal Herbs   (continued)

Scientific Name	Common Chinese and (English) Name	Major Constituents and (sources)	Therapeutic Values*
Salvia plebeia R. Brown	Li Zhi Cao	(aerial part) Flavonoids, homoplantaginin, hispidulin, eupafolin, essential oils.[48]	Diuretic, vermifuge, astringent.
Sambucus coreana Kom. & Klob. Alisova S. latipinna Nakai S. manshurica Kitag. S. peninsularis Kitag. S. sieboldiana (Miq.) Blume ex Graebner var. miquelii (Nakai) Hara S. williamsii Hance	Jie Gu Mu (Elder)	(leaf, flower, stem, root bark) Chlorogen acid.[60]	Diaphoretic, diuretic, carminative; treats arthralgia, fever.
Sambucus formosana Nakai	Hu Gu Xiao	(leaf) Alpha-amyrin palmitate.[56]	Detoxicant; stops swelling; diuretic; relieves pain.
Sanguisorba officinalis L. S. grandiflora (Maxim.) Makino S. officinalis L. S. officinalis L. f. latifoliata (Liou et C. Y. Li) Y. C. Chu S. officinalis L. var. longa Kitag. S. officinalis L. var. longa Kitag. f. dilutiflora Kitag. S. parviflora (Maxim.) Takeda S. x tenuifolia Fisch. ex Link	Zi Yu (Burnet)	(root) Oxalic acids, hydroxycinnamic acids, hydroxybenzoic acids, coumarins, anthocyanidines, anthraquinones, phytosterines, carotenes, ether oils, monoterpene, sesquiterpene glucosides, Zi Yu glucoside I, hydrocyanic acids, Zi Yu glucoside II, sanguisorbin A, sanguisorbin B, sanguisorbin C.[33,222]	Astringent effect to stop diarrhea and relieves chronic intestinal infection, duodenal ulcer and bleeding. Externally for eczema.
Sansevieria trifosciate Prain	Hu Wei Lan (Snake plant)	(leaf) Abamagenin, haemolytic sapogenin, organic acids.[50]	leaf juice for earache. Treats itchiness.

Species	Chinese name (parts)	Constituents	Uses
Santalum album L. *S. myrtifolium* Roxb. *S. verum* L.	Tan Xian (Sandalwood)	(heartwood) Alpha-santalol, beta-santalol, alpha-santalene, beta-santalene, santene, alpha-santenone, alpha-santenol, santalone, santalic acid, teresantalic, isovaleraldehyde, teresantalol, tricycloekasantal, santalin, deoxysantalin, sinapyl aldehyde, caniferyl aldehyde, syringic aldehyde.[33,568]	Treats stomachache.
Sapindus mukorossi Gaertner	Wu Huan Shu (Bodhi seeds)	(flower, fruit, seed, root) Saponin, mukorosside.[60,450]	For conjuctivitis, eye diseases; removes freckles and suntan.
Sapium sebiferum (L.) Roxb. *S. discolor* Mueller-Arg.	Wu Jiu Shan Jiu (Chinese tallow tree, Chinese vegetable tallow)	(leaf, root bark) Xanthoxylin, corilagin, sebiferic acid, lauric acid, margaric acid, palmitic acid.[33,50]	Antihypertensive activity; for constipation, poisoning, skin diseases.
Saponaria officinalis L. *S. vaccaria* L. (Syn. *Vaccaria segetalis*)	Wang Bu Liu Xing (Cow herb)	(seed, root) Saponarin.[65,568] This herb is contraindicated in pregnancy.	For abscesses, furuncles, ulcers, scabies, mastitis, lymphangitis. root is used to treats syphilis, glandular and chronic skin disease.
Sarcandra glabra (Thunb.) Nakai	Shong Jie Fong	(whole plant) Glucosides, essential oils, fumaric acid, succinic acid.[33,508,509]	Treats malignant tumors.
Sargassum pallidum (Harv.) Setch.	Hai Zao (Seaweed)	(seaweed) Iodine, alginic acid, algin, iron, potasium.[33]	Antigoiter, anticoagulant.
Sargentodoxa cuneata L.	Hong Teng	(leaf, stem) Acetylsalicylic acid.[50,226]	Antibacterial, antipyretic; activates blood flow.

Table 1A　Major Constituents and Therapeutic Values of Chinese Medicinal Herbs　(continued)

Scientific Name	Common Chinese and (English) Name	Major Constituents and (sources)	Therapeutic Values*
Saururus chinensis (Lour.) Baillon	San Bai Cao (Lizard's tail)	(whole plant) Quercitrin, isoquercitrin, avicularin, hyperin, amino acids.[55]	To clean abscesses; antimalarial, diuretic, depurative, eliminative, parasiticide.
Saussurea japonica (Thunb.) DC S. *japonica* (Thunb.) DC f. alata (Chen) Kitag. S. *japonica* (Thunb.) DC var. maritima Kitag. S. *lappa* Clarke	Mu Ziang (Custus)	(root) Saussurine, phene, phellandrene.[49]	As a stomachic.
Schisandra arisanensis Hayata S. *sphenanthera* Rehd.	Taiwan Wu Wei Zi	(stem)[56] Schisantherin A, B, C, D, E.[235]	For blood vomiting, pain caused by cold, overtiredness, wounds.
Schisandra chinensis (Turcz.) Baill.	Wu Wei Zi (Chinese magnolia vine)	(fruit, kernel) Schizandrin, deoxyschizandrin, schizandrol, schizandrer.[8,33,558,568]	Antitussive, a tonic. A tendency to lowers SGPT caused by hepatitis.
Schizonepeta multifida (L.) Briquet S. *tenuifolia* (Benth.) Briquet	Jing Jie	(aerial part, spikes) Essential oils, d-menthone, d-limonene, campesterol, stigmasterol, beta-sitosterol, hesperidin.[33,214,602]	Diaphoresis, anti-inflammatory, analgesic, antipyretic, antispasmodic, antidiabetic, lowers body temperature, increases blood coagulation; anticonvulsive.

Scopalia dulcis L.	Tian Zhu Cao	(whole plant) Amellin, dulciol, hexacosanol, mannitol, beta-sitosterol, mannitol, tannins, hexacosanol, salicylic acid, scopanol, dulcilone, tetulinic acid, ifflaionic acid.[50,86]	A cough remedy; induces labor; used as an opium substitute. Therapeutic action in diabetes in some reports.
Scopolia tangutica Max.	San Long Zhi	(root) Hyoscyamine, scopolamine, anisodamine, anisodine.[33,42]	Treats shock caused by acute infectious diseases, cerebral thrombosis, acute spinal cord inflammation.
Scrophularia buergeriana Miq. *S. kakudensis* Franch var. latisepala (Kitag.) Kitag. *S. ningpoensis* Hemsl. *S. oldhami* Oliv. *S. puergeriana* Miq.	Xuan Seng (Figwort)	(root) Scrophularin, iridoid glycosides, 8-(O-methyl-p-coumaroyl)-harpagide, harpagoside, essential oils, flavonoids, p-methoxylcinnamic acid.[33]	Lowers blood pressure and blood sugar. A health strengthening agent.
Scutellaria baicalensis Georgi *S. grandiflora* Adams *S. lanceolaria* Miq. *S. macrantha* Fisch. *S. rivulararis* Benth. *S. viscidula* Bunge.	Huang Qin (Skullcap)	(root) Baicalein, baicalin, wogonin, beta-sitosterol, wognoside, 7-methoxy-baicalein, 7-methoxynorwogonin, skullcap flavones.[33,257,430,432,558,568,603,371,576,578]	Antibacterial, antiviral, an antioxidant, antipyretic, anti-inflammatory, antitumor, antineoplastic.
Scutellaria formosana Brown	Taiwan Huang Qin	(whole plant) Berberine, baicalin.[54,233,234]	Relieves swelling, pain, treats cold, wounds, liver infection.
Securinega suffruticosa (Pall.) Rehd.	Yi Ye Chan	(leaf, flower, twig) Securinine, allosecurinine, securinol, dihydrosecuritinine, securitinine, phyllantidine.[33] This herb may be toxic.	Treats infantile paralysis, neurasthenia, neuroparalysis.

Table 1A Major Constituents and Therapeutic Values of Chinese Medicinal Herbs (continued)

Scientific Name	Common Chinese and (English) Name	Major Constituents and (sources)	Therapeutic Values*
Securinega virosa (Roxb.) Pax & Hoffmann	Bai Yin Shu	(leaf, root) Virosine, norsecurinine, dihydrosecurinine, virosecurinin, viroallosecurinine, norsecurinine, fluggein.[56]	Leaves used as a maturative, a detergent; it has antibiotic activity. root to treats teeth and gum disease.
Sedum aizoon L.	Jing Tian San Qi	(whole plant) Sedoflorin, sedocaulin, sedocitrin, sedoheptulose, arbutin.[33,48]	Hemostatic, removes blood stasis.
Sedum erythrostichum Miq. S. *kamtschaticum* Fisch. S. *verticillatum* L.	Jing Tian (Stonecrop)	(whole plant) Sedoheptulose, sarmentosin.[48,235]	Detoxicant; relieves swelling, stop bleeding and pain.
Sedum formosanum N. E. Brown	Taiwan Fo Jia Cao	(whole plant) Triterpenes, amyrenone, amyrenol.[54]	Treats diabetes, relieves swelling. pain, diarrhrea, and aids digestion.
Sedum lineare Thunb.	Fo Jia Cao (Linear stonecrop)	(whole plant) Sedoheptose, glucose, fructose. This herb is slightly toxic.[50]	Applied locally to burns and scalds; treats throat infection, diabetes.
Sedum sarmentosum Bunge.	Chui Pen Chao or Jing Tian (Stringy stonecrop)	(whole plant) Sarmentoslin, dihydro-N-methyl-isopelletierine, N-methyl-2-(β-OH-propyl) piperidine, N-methyl-isopelletierine, dl-methylisopelletierine, dihydroisopelletierine.[33,48,50]	Antipyretic, detoxicant, diuretic; treats hepatitis.
Selaginella involvens (Sw.) Sprengel S. *doederieninii* Heironyus	Shi Juan Bai Shi Shang Bai	(whole plant) Alkaloids, trehalose, d-glucose.[55]	A febrifuge, antihemorrhagic, detoxicant in molar cancer, for cough, gravel, rectum, blood thinning property, amenorrhea.

Selaginella tamarisina (Beauv.) Spring	Juan Bai	(whole plant) Soteisuflavone, amentoflavone, apigenin, trehalose, hinokiflavone, isocryptomerin, sotetsuflavone.[33,48]	Treats hematurai, dysmenorrhea, stops postpartum bleeding.
Semiaquilegia adoxoides (DC) Mak.	Tian Kui Zi	(root)[50,60] No information is available in the literature.	For scabby skin, urinary disorders; an insecticide.
Senecio argunensis Turcz. S. nemorensis L. S. scandens Buch-Ham ex D. Don	Qian Li Guang (Ragwort)	(aerial part) Lavoxanthin, macrophylline, cynarin, chlorogenic acid, chrysanthemaxanthin, sarracine.[33,48]	Antibacterial, antiplasmodial, treats acute bacterial dysentery and bronchitis.
Senecio campestris (Retz.) DC	Gou Shi Cac (Dog's tongue)	(whole plant) Alkaloids.[48] This herb may be toxic.	Depresses leukemia; detoxicant, diuretic, insectisidic.
Senecio cannabifolius Lessing	Huan Hun Cao (Groundsel)	(whole plant) p-hydroxyacetophenone, arbutin.[48]	Treats heart disease, respiratory infection, sexually transmitted diseases.
Senecio vulgaris L.	European Qian Li Guang (German ivy)	(aerial part) Senecionine, inulin.[58]	Used in ointment on hemorrhoids and swellings; lowers blood pressure; laxative.
Sesamum indicum L.	Wu Ma (Sesame)	(seed) Olein acid, linolein acid, palmitine acid, stearin acid, myristic acid, sesamin, sesamol, pentosan, phytin, lecithin, choline, calcium oxalate, chlorogenic acid, vitamin A, vitamin B.[49,568]	A nutrient, laxative, hyperchlorhydria; a lenitive in scybalous constipation; as a nutrient tonic in degenerative neuritis, neuroparalysis.
Sesbinia grandiflora (L.) Persoon	Da Hua Tian Qing	(bark, root) Agathin, xanthoagathin.[57]	A tonic, antipyretic; for gastric troubles, colic with diarrhea, and dysentery.

Table 1A Major Constituents and Therapeutic Values of Chinese Medicinal Herbs (continued)

Scientific Name	Common Chinese and (English) Name	Major Constituents and (sources)	Therapeutic Values*
Sesbinia javanica (L.) Persoon	Tian Qing	(whole plant) Pentosan, d-galactose, d-mannose.	Diuretic, detoxicant; stops pain.
Sesbinia sesbin (L.) Merrill.	Indian Tian Qing	(root, leaf, bark, seed)[57] Saponins, triterpene glycosides, steroid glycosides, glycoalkaloids, kaempferol trisaccharide.[342,343,344]	Diuretic, irregular menses; externally for bug bites, antitumor.
Silene jenisseensis Willd. S. jenisseensis Willd. f. dasyphylla (Turcz.) Schischk. S. jenisseensis Willd. f. parviflora (Turcz.) Schischk. S. jenisseensis Willd. f. setifolia (Turcz.) Schischk. S. jenisseensis Willd. var. oliganthella (Nakai ex Kitag.) Y. C. Chu S. jenisseensis Willd. var. viscifera Y. C. Chu	Han Mai Bin Cao	(root) 6,8-di-C-galactopyranosylapigenin, 6-C-galactopyranosyl-isoscutellarein, essential oil.[84]	For fever, kala-azar, malaria.
Silybum marianum (L.) Gaertn.	Shui Fei Ji	(fruit) Silybin, silymarin, silydiamin, silyckristin, dehydrosilybin, silybinomer.[33,568,569]	Maintains normal functioning of the liver, promotes the regeneration of injured hepatic cells, and increases glycogenesis and nucleic acid metabolism.
Sinapis alba L.	Bai Jie (Mustard)	(leaf) Arachidic acid, erucic acid, lignoceric acid, linoleic acid, myrosinase, phosphatase, sinalbine.[50]	Carminative; toothache; seal for eruptions and ulcers.

Sinomenium acutum (Thunb.) Rehd. et Wils. S. diversifolium Diels.	Japanese Fuag Ji, Qing Teng	See Cocculus diversifolius	
Sinomenium acutum var. cinereum	Qing Feng Teng	(stem) Sinomenine, disinomenine, magnofloine, acutumine, sinactine, isosinomenine, tuduranine, sinoacutine.[33]	Analgesic, anti-inflammatory; lowers blood pressure.
Smilacina japonica A. Gray	Lu Yao	(root)[48] No information is available in the literature.	For arthritis; relieves swelling and pain; aphrodisiac; regulates monthly period; breast gland infection.
Smilax china L. S. nipponica Miq. subsp. manshurica Kitag. S. riparia DC subsp. ussuriensi (Regel) Kitag. S. sieboldii Miq.	Tu Gu Ling (China root)	(root) Crystalline saponin smilacin, tannins, resin, tigogenin, neotigogenin, laxogenin[48,49]	As alternative, diuretic in syphilis, gout, skin disorders, rheumatism.
Solanum aculeatissimum Jacquin	Xiao Ying Qie	(fruit) Solasonine, beta-solamargine, solasurine.[55]	For cough, asthma, diuretic, pain.
Solanum biflorum Loureiro	Hong Si Xian	(whole plant) Glycoside alkaloids, steroid alkaloid glycosides.[55,360,361]	Detoxicant; for cough, swelling, dog bites.
Solanum capsicastrum Link.	Mao Dong San Hu	(leaf) Solanocapsine.[55]	With cooling effect, relieves swelling, treats liver inflammation.
Solanum incanum L.	Huang Shui Jia	(root) Beta-sitosterol, D-glucose, ursolic acid, alkaloids, solasodine, solamargine.[55]	Treats liver inflammation, lymphatic gland; a detoxicant.

Table 1A   Major Constituents and Therapeutic Values of Chinese Medicinal Herbs   (continued)

Scientific Name	Common Chinese and (English) Name	Major Constituents and (sources)	Therapeutic Values*
Solanum indicum L.	Niu Zi Qie (Indian nightshade)	(root, leaf, fruit) Diosgenin, solanidine, solanine, solasodine, alkaloids, carbohydrases, maltase, saccharase, melibiase.[50]	Antidote for poison, for urinary disease.
Solanum lyratum Thunb. S. *melongena* L.	Bai Ying (Eggplant)	(root, leaf, flower, fruit) Trigonelline, stachydrine, choline, solanine, nasunin, shisonin, delphinidin-3-monoglucoside, adenine, imidazolylothylamine, solasodine, arginine glucoside.[48]	For arthritis, respiratory disorder, swelling, cough, diarrhea, blood in the urine.
Solanum nigrum L.	Long Kui (Black nightshade)	(whole plant) Solanigrines, saponin, riboflavin, nicotinic acid, vitamin C.[33]	Antibacterial, diuretic; treats mastitis, cervicitis, chronic bronchitis, dysentery.
Solanum pseudo-capsicum L.	Dong San Hu	(root) Solanocapsine.[55]	A detoxicant; relieves pain. Treats tuberculosis, pneumonia.
Solanum verbascifolium L.	Shan Yan Cao (Tobacco nightshade, turkey berry)	(root) Solasonine.[54]	Treats dysentery, intestinal pain, and fever.
Solidago canadensis L.	N. Am. Yi Zhi Huang Hua	(whole plant) Cadinene, quercitrin.[57,87]	Antibacterial, treats infection, stops bleeding, throat swelling.
Solidago dahurica (Kitag.) Kitag. S. *pacifica* Juzepczuk. S. *virgaurea* L.	Yi Zhi Huang Hua (Golden rod)	(whole plant) Caffeic acid, chlorogenic acid, cyanidin-3-glucoside, flavonoids, astragalin, cyanidin-3-gentiobioside, kaempferol-rhamno glucoside, hydroxycinnamic acid, quinic acid, polygalacic acid.[48,50]	Decoagulant, carminative; for bladder ailments, cholera, diarrhea, dysmenorrhea.

Sonchus arvensis L. S. oleraceus L.	Ju Shi Cai (Sow thistle)	(whole plant) Inositol, lactucerol, mannitol, taraxasterol, palmitic acid, stearic acid, tartaric acid, lactucerols.[50,585]	Used as an insecticide, asthma, bronchitis, cough, ophthalmia, insomnia, pertussis, swellings; and tumors.
Sophora flavescens Ait. S. alopecurosides L.	Ku Seng Gu Dong Zi	(root) d-oxymatrine, d-sophoranol, cytisine, l-anagyrine, l-baptifoline, l-methylcytisine, trifolirhizin, d-matrine, norkurarinone, kuraridin.[36]	Anthelmintic, antipruritic; treats irregular heart beat, eczema, acute dysentery, trichomoniasis.
Sophora japonica L.	Huai Hua (Japanese pagoda tree)	(flower bud) Rutin, sophoradiol, genisteine, sophoricoside, sophorabioside, sophoraflavonoloside, isorhammetin.[33,252]	Antihemostatic, increases capillary resistance and decreases capillary fragility and permeability.
Sophora subprostrata Chun et T. Chen	Shan Dou Gen	(root) Matrine, oxymatrine, anagyrine, methylcytisine, sophoranone, sophoranochromene, sophoradin, daidzein.[33] This herb is toxic.	Against tubercle bacilli; treats intractable ulcerative colitis; antiarrhythmic, anticancer; promotes leukocytosis.
Sophora tomatosa L.	Ling Nan Huai	(seed, leaf, root) Cytisine (sophorine).[88]	For diarrhea, cholera, colic, dysentery.
Sorbus alnifolia (Sieb. & Zucc.) K. Koch S. alnifolia (Sieb. & Zucc.) K. Koch var. lobulata Rehd. S. amurensis Koehne S. pohuashanensis (Hance) Hedl. var. manshuriensis (Kitag.) Y. C. Chu.	Shui Yu	(stem, bark, fruit) Fatty acids, starch, essential oils, flavonoids, isochlorogenic acid, parasorbic acid.[48]	For stomach infection and ache, swellings, cough, vitamin deficiencies.

Table 1A   Major Constituents and Therapeutic Values of Chinese Medicinal Herbs   (continued)

Scientific Name	Common Chinese and (English) Name	Major Constituents and (sources)	Therapeutic Values*
Spatholobus suberectus Dunn.	Ji Xue Teng	(stem) Friedelin, taraxerone.[33]	Slow the heart rate, lowers blood pressure.
Sphenomeris chusana (L.) Copel.	Wu Ju	(whole plant) Sphenone A, phenanthrene-1, 4-quinone.[60,229]	Treats feverish conditions and bladder difficulties.
Spilanthes acmella (L.) Murray S. *acmella* L. var. oleracea Clarke	Tian We Cao Liu Shen Cao	(whole plant) Alpha-amyrenol, beta-amyrenol, myricyl, stigmasterol, sitosteryl-o-β-d-glucoside, spilanthol.[58]	Treats aphrodisiac; depurative, diuretic, ophthalmic, tonic.
Spiraea salicifolia L. S. *salicifolia* L. var. grosseserrata Liou & Liou fil. S. *salicifolia* L. var. oligodonta Yu	Xiu Xian Jiu (Bridal wreath)	(whole plant) Flavonoids, carotenoids, vitamin C, alkaloids, seed oil.[33]	Diuretic; treats cough, pain, monthly period, constipation.
Spirodela polyrhiza Schleid.	Fu Ping (Duckweed fern)	(whole plant) Apigenin-7-O-glucoside, apigenin-8-C-glucoside.[48,50]	Carminative, diaphoretic, diuretic.
Stachys chinensis Bunge. ex Benth. S. *baicalensis* Fisch. ex Benth. S. *baicalensis* Fisch. ex Benth. var. angustifolia Honda S. *japonica* Miq.	Shui Su (Camphor mint)	(whole plant) Coumarin, alkaloids, stachydrine chloride.[48]	Treats cold, influenza.
Stauntonia hexaphylla Dence S. *chinensis* Bunge.	Ye Mu Gua	(fruit, stem, root) Stauntonin.[50,440]	Antirheumatic, diuretic; treats nasopharyngeal carcinoma.

Species	Name	Constituents	Uses
Stellaria alsine Grimm var. undulata (Thunb.) Ohwi	Tian Peng Cao (Starwort)	(whole plant)[50] No information is available in the literature.	For colds, pimples, snakebite, traumatic injuries. It is a carminative, lactagogue.
Stellaria media (L.) Cyrillo	Fan Lu (Chickweed)	(whole plant) r-linolenic acid, octadecatetraenoic acid.[48]	A postpartum depurative, emmenagogue, lactagogue; promotes circulation, treats mucus disorder. Externally for rheumatic pains, ulcers, wounds.
stemona japonica (Bl.) Miq. S. tuberosa Lour.	Bai Bu Dei Ye Bei Bu	(root) stemonine, isotemonidine, stemonidine, protostemonine.[33,50,558,568,570]	Suppress excitation of the respiratory center and inhibits the cough reflex. Antitubercular, antibacterial, antifungal.
Stephania cepharantha Diels.	Jin Xian Diao Wu Gui	(root) Cepharanthine, isotetrandrine, cycleanine, cepharanoline, berbamine, cepharamine, homoaromoline.[33,43]	A diuretic, antiphlogistic, antirheumatic, analgesic, anti-inflammatory.
Stephania hernendifolia (Willd.) Walp.	Qian Jin Teng	(root) dl-tetrandrine, fangchinoline, 4-dementhyl-hasubanonine, isochondrodendrine, hernandine, stephisoferuline, hernandoline, hernandolinol, 3-O-demethylhernandifoline.[33]	Treats nephritic edema, urinary tract infection, rheumatic arthritis, sciatic neuralgia.
Stephania japonica (Thunb.) Miers.	Qian Jin Teng	(root) Stephanine, protostephanine, epistephanine, hypoepistephanine, homostephanoline, metaphanine, prometaphanine, hasubanonine, insularine, cyclanoline, steponine stephanoline, stepinonine.[33]	Treats nephritic edema, urinary tract infection, rheumatic arthritis, sciatic neuralgia.

Table 1A   Major Constituents and Therapeutic Values of Chinese Medicinal Herbs   (continued)

Scientific Name	Common Chinese and (English) Name	Major Constituents and (sources)	Therapeutic Values*
Stephania sinica DC	Hua Qian Jin Teng	(root) l-tetrahydropalmatine, stepharotine, stepharine, tuduranine.[33]	Analgesic effect; treats stomachache, neuralgia, toothache.
Stephania tetrandraq Moore	Fang Ji or Han Fang Ji	(root) d-tetrandrine, fangchinotine, cyclanoline.[33,38] This herb may cause kidney failure.[392]	A diuretic, antiphlogistic, antirheumatic, analgesic, anti-inflammatory.
Stevia rebaudiana (Bertoni) Hemsl.	Tian Jiu (Stevia)	(stem, leaf) Stevioside, steviolbioside, rebaudiosides, austroinulin.[57,568]	Treats diabetes, tonic; lowers blood pressure.
Strophanthus divaricatus (Lour.) Hook. et Arn.	Yang Guo Nau	(seed) Divaricoside, divostroside, sinoside, sinostroside, caudoside, caudostroside, sarmutoside.[33] This herb is toxic.	Cardiac stimulating action causing an increase of myocardial contractility; slows the heartbeat.
Strychnos nux-vomica L.	Fan Mu Pen (Strychnine)	(seed) Strychnine, monomeric tertiary indole alkaloids, brucine.[50,144,504] This herb is highly toxic.	Treats neurasthenia, aphrodisiac, vasomotor stimulation; regulates blood pressure, treats nerve diseases.
Strychnos pierriana L.	Ma Qian Zi	(seed) Strychnine, brucine, vomicine, pseudostrychnine, pseudobrucine, novacine.[33]	Increases central nervous system reflex stimulation.
Styrax suberifolus Hook. et Arnott.	Hong Pi	(root, leaf)[55] No information is available in the literature.	Stomachache, pain caused by arthritis.

Styrax tonkinensis Pierre. *S. benzoin* Dryand	An Xi Xian (Styrax)	(leaf) Sumaresinolic acid, coniferyl cinnamate, styracin, vanillin, alpha-phenylpropyl cinnamyl cinnamate, balsamic acid.[33,50]	As an aromatic stimulant; for aphrodisiac, an astringent.
Swertia diluta (Turcz.) Benth. et Hook. f. *S. mileensis* L.	Qing Ye Dan	(whole plant) Oleanolic acid.[33]	Treats acute icteric hepatitis.
Swertia pseudochinensis Hara	Dang Yao	(whole plant) Swertiamarin, swertisin, methyl-bellidifolin, homoorentin, methyl-swertianin, isovitexin, bellidifolin, decussatin, swertifrancheside.[33]	Choleretic; improves hepatic function. Treats acute icteric hepatitis, chronic liver disease.
Syringa dilatata Nakai *S. oblata* Lindley *S. oblata* Lindley var. alba Hort. ex Rehd. *S. reticulata* (Blume) Hara var. mandshurica (Maxim.) Hara *S. suspensa* Thunb. (Syn. *Forsythia suspensa*) *S. vulgaris* L.	Lian Qiao	(bark, fruit) Syringin, 3,4-dihydroxyphenethyl alcohol, saponons, phillyrin.[49]	Antipyretic, antiphlogistic in infectious fevers, suppurative inflammation, phlegmon, variola, erysipelas, measles.
Syzygium aromaticum (L.) Merr. & Perry	Ding Xian (Clove)	(clove bud) Phytosterols, campesterol, crataegol acid, sitosterols, stigmasterol, niacin, ascorbic acid.[50]	Antiemetic, carminative, stimulant; treats diarrhea, halitosis, nasal polyps, uterine fluxes, sterility, toothache.
Syzygium cuminii (L.) Skeels	Hei Nan Pu Tao	(bark, leaf) Betulinic acid, eugianin, friedelin, epifriedelanol, beta-sitosterol, acetyl oleanolic acid, ellagic acid, myricetin, cyanidin rhamno-glucoside, petunidin glucoside, maluidin glucoside, jambolin.[57]	Cooling effect; relieves itchiness, stops bleeding, infection, diarrhea.

Table 1A   Major Constituents and Therapeutic Values of Chinese Medicinal Herbs   (continued)

Scientific Name	Common Chinese and (English) Name	Major Constituents and (sources)	Therapeutic Values*
Tagetes erecta L.	Chou Fu Yong (Marigold)	(leaf, flower) Alpha-terthienyl, d-limonene, l-linalool, tagetone, n-nonyl aldehyde.[50]	Treats sores and ulcers, cold, conjuctivitis, cough, mastitis, mumps.
Tagetes patula L.	Wan Shou Jiu (French marigold)	(whole plant) Tagetone, linalool, limonene, linalylacetate, ocimene, patuletin, patulitrin, cyanidin diglycoside, quercetagetin, quercetagetin, helenien, polythienyls.[50] This herb is toxic.	For coughs and dysentery.
Talinunm triangulare Willd.	Tu Ren Shen	(root)[60] No information is available in the literature.	A tonic for general weakness; treats inflammation, swelling.
Tamarindus indicus L.	Luo Huang Zi (Tamarind)	(stem, fruit) Tannins, beta-amyrin, campesterol, beta-sitosterol, palmitic acid, oleic acid, linoleic acid, eicosanoic acid, arabinose, xylose, galactose, glucose, uronic acid, pectins, mucilage, vitamin B.[60,216]	Diuretic, purgative; for liver disorders, inappetence, digestion, and hypoglycemic, hypocholesterolemic properties.
Tamarix juniperina Bunge.	Shen Liu (Tamarisk)	(young shoot, flower, gum) Quercetin-monomethylether.[48]	Treats cold, blood vomiting, respiratory infection.
Taraxacum formosanum Kitamura	Taiwan Pu Gong Ying	(aerial part) Taraxasterol, choline, inulin, pectins.[54]	Cure for swollen breasts; a diuretic; treats fever, tracheatis, hepititis, tonsillitis.

Taraxacum mongolicum Hand-Mazz. *T. sinicum* Kitag.	Pu Gong Ying (Mongolian dandelion)	(aerial part) Taraxasterol, taraxerol, taraxol, taraxacerin, taraxacin, cryptoxanthin, zeaxanthin, lutein antheraxanthin, violaxanthin, neoxanthin, myristic acid, lauric acid, palmitic acid, stearic acid, beta-sitosterol, beta-amyrin, cysteic acid, cysteine, cystine, serine, glycine, asparagine, lysine, alanine.[33,48,607]	Antibacterial, antispirochetic, antiviral; a choleretic agent.
Taraxacum officinale G. H. Weber ex Wigg.	Western Pu Gong Ying (Dandelion)	(root) Inulin, essential oils, choline, hydroxycinnamic acids, carotenes, ether oils, monoterpene, oxalic acids, hydrocyanic acids, sesquiterpene glucosides, flavonoids, hydroxybenzoic acid, coumarins, anthocyanidines, anthraquinones, phytosterines, squalene, cerylic alcohol, arabinose, vitamins A, B, C.[88,222,450,87,213,568,570]	Sudorific, stomachic, tonic; a remedy for sores, boils, ulcers, abscesses, snakebites.
Taxus cuspidata Sieb. et Zucc. *T. chinensis* (Pily) Rehd. *T. yunnanensis* Cheng et L. K. Fu	Zhu Shan, Huang Dao Shan (Yew tree)	(bark, leaf) Taxol, baccatin, cephalomannine, 10-deacetylbaccatin, yunnanxana, abeotaxanes, taxinine E.[33]	Antineoplastic, anticancer; treats ovarian carcinoma.
Tephrosia purpurea Persoon	Hui Mao Dou	(root) Rotenone, degueline, tephrosin, rutin, quercetin glucoside.[57]	Used as a cordial and a stomachic; a deobstruent, emmenagogue.
Terminalia chebula Retz.	He Zi (Myrobalans)	(leaf, fruit) Chebulic acid, fatty oil, tannins, ellagic acid, chebulinic acid.[49,450]	An astringent in diarrhea, enterorrhagia, metrorrhagia, metritis, leukorrhea.

Table 1A   Major Constituents and Therapeutic Values of Chinese Medicinal Herbs   (continued)

Scientific Name	Common Chinese and (English) Name	Major Constituents and (sources)	Therapeutic Values*
Tetragonia tetragonoides (Pall.) O. Kuntz.	Fan Xing	(leaf, stem)[60] Phosphatidylcholine, phosphatidyl-ethanolamine, phosphatidyl-serine, phosphatidyl-inositol, tetragonin, trigonelline, choline, adenine.[56]	A remedy for carcinoma; treats ventriculi; stomach ulcers, leukemia.
Thalictrum aquilegifolium L. var. sibiricum Regel & Tiling *T. baicalense* Turcz. *T. baicalense* Turcz. f. levicarpum Tamura *T. fauriel* Hayata *T. petaloideum* L. *T. petaloideum* L. var. supradecompositum (Nakai) Kitag. *T. simplex* L. *T. simplex* L. var. affine (Ledeb.) Regel *T. simplex* L. var. brevipes Hara *T. squarrosum* Steph. ex Willd. *T. thunbergii* DC	Tang Song Cao (Meadow)	(root) Flavonoids, fetidine, phetidine, thalfoetidine, thalpine, thalphinine, rhalidasine, hernandezine, thelic simidine, coptisine, oxypurpureine, berbamine, isotetrandrine, alpha-allocryptopine, oxycanthine, isothalidenzine, glaucine, berberine, palmatine, jatrorrhizine, protopine, cryptopine, thalidezine.[48,53]	Anticancer activity; treats fever, nausea, thirst, hemorrhages, and conjunctivitis.
Thalictrum foetidum L.	Taiwan Tang Song Cao	(whole plant) Thalfoetidine, thalpine, thalphinine, fetidine, flavonoid glycoside, saponin, cardiac glucoside, berberine, magnoflorine, palmitine, jatrorrhizine.[56]	Lowers blood pressure, treats hepatitis, cold, arthritis, intestinal infection.
Thalictrum ichangense Lecoyer ex Oliver *T. glandulissimum* L.	Ma Wei Lian	(rhizome) Berberine, palmatine, jatrorrhizine, talictrine, thalidasine, thalicarpine, saponaretin.[33]	Antibacterial; treats influenza, childhood fevers, measles, malaria.

Thea assamica Mast T. bohea L. T. cantoniensis Lour. T. chinensis Sims. T. cochinchinensis Lour. T. sinensis L. T. viridis Link.	Cha (Tea)	(leaf) Caffeine, theophylline, tannic acid, theobromine, xanthine, polyphenols.[33,47,405,406,409]	Diuretic effect, increases renal blood flow, stimulates central nervous system; antitumor; prevents lung cancer.
Thesium chinense Turcz.	Bai Rui Cao	(whole plant) Flavonoids, mannitol.[48]	Breast gland, lung, throat, tonsil infections, fever caused by cold; relieves swelling.
Thevetai peruviana (Pers.) K. Schum.	Huang Hua Jia Zhu Tao	(seed, flower, leaf) Thevetin A and B, theveside, peruvosides, vertiaflavone, theviridoside.[33]	Tranquilizing effect; treats congestive heart failure.
Thlaspi arvense L.	Jin Moa (Field pennycress)	(aerial part) Sinigrin, fatty acids, essential oil, myrocin, myrosinase.[48]	For ophthalmia, lumbago; an antidote, antipyretic; improves circulation, diaphoretic.
Thuja chinensis Hort. T. koraiensis Nakai T. orientalis L. (Syn. Biota orientalis, Platycladus orientalis)	Ce Bai Ye (Oriental arborvitae)	(seed kernel, young leaf) Thujene, thujone, fenchone, pinene, caryophyllene, aromadendrin, quercetin, myricetin, hinokiflavone, amentoflavone.[48,353]	Antipyretic, astringent, diuretic; for dysmenorrhea, epistaxis, gonorrhea, metrorrhagia.
Thymus amurensis Klokov T. disjunctus Klokov T. kitagawianus Tscherneva T. komarovii Sergievskaja T. przewalskii (Kom.) Nakai T. quinquecostatus Celakovsky	Di Jiao	(aerial part) Scutellarein heteroside, luteolin-7-glucoside, apigenin, volatile oils, carvacrol, p-cymene, p-terpinene, alpha-terpineol, zingiberene, borneol, ursolic acid, thymol.[48]	Treats high blood pressure, stomachache, intestinal infection, cough, digestion, diarrhea.

Table 1A Major Constituents and Therapeutic Values of Chinese Medicinal Herbs (continued)

Scientific Name	Common Chinese and (English) Name	Major Constituents and (sources)	Therapeutic Values*
Thymus vulgaris L.	She Xiang Cao (Thyme)	(aerial part) Tymol, terpinen-4-ol, pinenes, camphene, myrcene, alpha-phellandrene, limonene, 1,8-cineol, p-cymene, linalool, linalyl acetate, bornyl acetate, alpha-terpinyl acetate, alpha-terpineol, borneol, citral, geraniol, carvacrol.[50,510,568]	Anthelmintic, antispasmodic, carminative, diaphoretic, sedative. Treats bronchitis, cancer, diarrhea, gastritis, rheumatism, skin ailments.
Tilia amurensis Rupr. *T. mandshurica* Rupr. & Maxim. *T. mongolica* Maxim.	Zi Duan (Linden)	(flower, stem, leaf) Flavonoids, essential oils.[48]	Promotes sweating, bactericidal; treats cold, kidney infection, throat infection.
Tinnevelly senna O. Kuntz. (Syn. *Cassia angustifolia*)	Fan Xie Ye	See *Cassia angustifolia*	
Trachelospermum jasminoides Lam.	Luo Shi (Star jasmine)	(stem) Tracheloside, nortracheloside, matairesinoside.[33]	Relieves muscle rigidity, removes blood stasis, stops bleeding.
Trachycarpus wagnerianus Beccari *T. fortunei* H. Wendl.	Zong Lu	(seed) Mannosan, galactan, saccharose, tannins.[49]	An astringent, hemostatic.
Trapa bispinosa Roxb.	Ling (Water chestnut)	(fruit) Ergostatetraen, dihydrostigmast, beta-sitosterol, amylose, protein.[48]	Treats stomach ulcer, diarrhea, breast, ovary, gullet cancer.
Tribulus terrestsis L.	Ci Hi Li (Calthrop)	(fruit) Glycosides tribuloside, astragalin, harmane, harmine.[33]	Anticonvulsive, a spasmolytic agent.

Trichosanthes kirilowii Maxim. *T. uniflora* Hao	Gua Lou, Tian Hua Fen (Chinese snakegourd)	(root, seed) Trichosanthin, polysaccharides, saponin, organic acids, resin, protein (TAP29).[33,261,558] This herb is highly toxic.	Treats pectoris and acute mastitis. Antitussive, as an expectorant, anti-HIV activity.
Trifolium pratense L. *T. repens* L.	Che Zhou Cao (Red clover)	(whole plant) Phytoestrogens, genisteine, iodine, daidzein, formononetin.[33,48,221]	Stimulating effect on female reproductive organs.
Trigonella foenum-graecum L.	Wu Ru Ba (Fenugreek)	(seed) Trigonelline, saponins, flavone derivatives including vitex, saponaretin, isoorientin, vitexin-7-glucoside.[33,568]	Reduces plasma cholesterol levels; supports hepatic and renal functions.
Trillium camschatcense Ker-Gawler	Yan Ling Cao	(root) Trillin, trillarin, diosgenin, cyasterone, ecdysterone.[48]	Improves blood circulation, detoxicant; treats headache, high blood pressure; stops bleeding.
Tripterygium hypoglaucum (Levl.) Hutch.	Shan Hai Tor or Zi Jin Pi	(whole plant) Alkaloides, triptolide.[33]	Anti-inflammatory, antiswelling.
Tripterygium wilfordii Hook. f.	Lei Gong Teng (Yellow vine)	(root) Tripchlorolide, celastrol, triptein, wilfordine, triptophenolide, triptonide, triptolide, triptdiolide, triptolidenol, tripchlorolide, triptolide, tripdiolitonide, trihydroxytriptolide, triptolide.[33,241,390,431,443] This herb is toxic with adverse effects on gastrointestinal tract.	Antifertility effect on males, anti-inflammatory, antitumor; treats cancer, antirheumatoid arthritis, suppressive effects.
Triticum vulgare Vill.	Foo Shao Ma (Wheat)	(kernel) Protein, fat, carbohydrate, vitamins A, B, E, G.[49]	Sedative, antipyretic in night sweats, insomnia.

Table 1A   Major Constituents and Therapeutic Values of Chinese Medicinal Herbs   (continued)

Scientific Name	Common Chinese and (English) Name	Major Constituents and (sources)	Therapeutic Values*
Tulipa edulis Bak. *T. gesneriana* L.	Shan Ci Ko Yu Jin Xian (Tulip)	(bulb) Colchicine, alkaloids, starch.[48]	Relieves swelling, lymphatic gland infection, throat infection.
Tussilago farfara L.	Kuan Dong Hua (Colts foot)	(flower bud) Faradiol, rutin, hyperin, saponins, taraxanthin, tannins, essential oil.[33,568]	Antitussive, expectorant, antiasthmatic; stimulates the medullary center and slowly raise blood pressure.
Typha angustata Bory et Chaub. *T. angustifolia* L. *T. davidiana* (Kronfeld) Hand. Mazz. *T. latifolia* L. *T. minima* Hoppe *T. orientalis* Presl. *T. przeqalskii* Skv.	Pu Huang (Bulrush)	(pollen) Isothamnetin, alpha-typhasterol, oligosaccharides.[33]	Treats hypercholesteremia, angina pectoris, exudative eczema, postdelivery bleeding; stops bleeding in hematemesis and hematuria.
Typhonium divaricatum (L.) Decaisne	Li Tou Cao	(leaf, tuber)[50,144] This herb is toxic. Overdose causes numbness or nausea.	An expectorant, rubefacient; used for cough and pulmonary disorders.
Typhonium giganteum Engl.	Du Jiao Lian	(whole plant) Uracil, succinic acid, tyrosine, valine, linolein, dipalmiin.[48]	Antispasmodic, carminative; for apoplexy, headache, paralysis.
Ulmus campestris L. *U. macrocarpa* Hance *U. pumila* L.	Yu Bai Pi (Siberian elm, Chinese elm)	(leaf) Butyric acid, capric acid, lipase, hexylenaldehyde, phlobaphene, phytosterol, sitosterol.[50]	For urinary calculi; diuretic, febrifuge.

Species	Chinese name	(Part) Constituents	Uses
Uncaria hirsuta Havil U. rhynchophylla Miq. (Syn. Nauclea sinensis Oliv.)	Gou Teng (Gambir)	(stem) Rhynchophylline, corynoxeine, iso-rhynchophylline, isocorynoxeine, corynantheine, hirsutine, hirsuteine.[33]	A sedative, anticonvulsive; lowers blood pressure, it has a triphasic effect. Treats childhood epilepsy.
Uraria crinita Desvaux U. lagopodiodes (L.) Dexvaux	Hu Li Wei Tu Wei Cao	(leaf, root) Vitexin, vitexin-7-0-glucoside, orientin-7-0-glucoside, saponartin-4'-0-glucoside.[56]	Treats hemorrhoids, dysentery, diarrhea, cough, pain, arthritis, irregular menses.
Urena procumbens L.	Fan Tian Hua (Duck foot)	(leaf, twig) Phenols, flavonoid glycoside, amino acids.[57]	Treats rheumatism, toothache.
Urtica angustifolia Fisch. ex Hornem. U. cannabina L. U. cannabina L. f. angustiloba Chu U. lobata L. U. tenacissima Roxb. U. urens L. U. utillis Hort. (Syn. Boehmeria nivea)	Yu Ma (Chinese nettle)	(root) Chlorogenic acid, alkaloids, 5-hydroxytryptamine, protein, fat, carbohydrate, ash, fabric.[48,49,304,568]	Diuretic, tonic, stomachache, arthritis.
Urtica laetevirens Maxim.	Shi Mu Zi	See Urtica angustifolia	
Usnea diffracta Dill. ex Adans. U. longissima Acharius	Lao Jium Xiu Song Lo	(whole plant) Barbatic acid, usnic acid, diffractaic acid, ramalic acid, lichenin.[54]	Treats throat discharges, malaria, headache, cough; a detoxicant.
Vaccaria segetalis (Neck.) Garcke V. pyramidata Medic.	Wang Bu Liu Xing Liu Xing Zi (Cow cockle)	(seed) Vacsegoside, vaccaroside, gypsogenin, vaccarin.[33]	Activates blood flow, promotes milk secretion, and treats amenorrhea and breast infections.

Table 1A Major Constituents and Therapeutic Values of Chinese Medicinal Herbs (continued)

Scientific Name	Common Chinese and (English) Name	Major Constituents and (sources)	Therapeutic Values*
Vaccinium bracteatum Thunb. *V. vitis-idaea* L.	Wu Fan Shu (Mountain cranberry)	(leaf) 6-o-acetyl-arbutin, arbutin, avicularin, 2-o-caffeoylarbutin, d-catechol, l-epicatechol, d-gallocatechol, hyperin, hyperoside, sioquercitrin, salidroside, tannins, ursone.[50]	For gonorrhea.
Valeriana alternifolia Bunge. *V. alternifolia* Bunge. var. stolonifera Baranov & Skv. *V. alternifolia* Bunge. var. stolonifera Baranov & Skv. F. angustifolia (Kom.) Kitag. *V. amurensis* P. Smiru. ex Kom. *V. fauriei* Briq. *V. fauriei* Briq. var. dasycarpa Hara *V. subbipinnatifolia* A. Baranow.	Jiao Cao	(root) Bornyl isovalerate, isovaleric acid, borneol, camphene, pinene, d-terpineol, l-limonene, pyrryl-methyl ketone, alpha-fenchene, myrcene, phellandrene, l-caryophyllene, erpinene, terpinolene, eremophilene, selinene, cadinene, valerianol, valerenone, myrtenol, bisabolene, chatinine, caffeic acid.[48,510]	Antispasmodic, aphrodisiac, emmenagogue, stimulant, sudorific, backache, cramps, influenza, nausea, numbness.
Veratrum dahuricum (Turcz.) Loes *V. maackii* Regel *V. nigrum* L.	Li Lu (Mountain onion)	(rhizome) Jervine, pseudojervine, rubijervine, tienmulimine, tienmuliminine, zygadenine, germine.[33] This herb may cause mucosal irritation.	Lowers blood pressure, slows heart rate, antibacterial. It has an insecticidal effect.
Veratrum formosanum Loesener	Taiwan Li Lu	(root) Protoveratrine, jervine, alkaloids, veratramine.[55]	Lowers blood pressure, stops vomiting; antifungal, a stimulant.

Species	Chinese name	Constituents	Uses
Verbena officinalis L. / V. oxysepalum Turcz.	Ma Bian Cao (Vervain)	(aerial part) Verbenalir, verbenalol, adenosine, tannins, essential oils.[33]	Antiplasmodial, antibacterial, antitoxin, anti-inflammatory.
Vernonia andersonii C. B. Carke / V. cinerea (L.) Less. / V. patula (Ait.) Merr.	Ban Jiu Jiu Shang Han Cao Xian Xia Hua	(leaf, root) Triterpinoid, alkaloid, saponin.[89]	As restorative, febrifuge and antidiarrheic; treats colic, stomachache.
Veronica anagallis-aquatica L. / V. anagallis-aquatica L. f. pumila Kitag.	Shui Ku Shi (Speedwell)	(root) Aucubin.[50]	For fever; a gargle for throat ailments, stomatitis.
Veronica linariaefolia Pall. ex Link / V. linariaefolia Pall. ex Link subsp. dilatata (Nakai et Kitag.) Hong	Shui Man Chin	(whole plant) Cordycepic acid, flavonoids.[48]	For windpipe infection, blood vomiting, relieves pain; detoxicant.
Veronica sibirica L. / V. sibirica L. f. glabra (Nakai) Kitag. / V. undulata Wallich	Po Po Na	(whole plant) Mannitol, veronicastroside, inteolin-7-β-neohesperidoside, luteolin, 7-β-glucopyranoside, ancubin, arbutin.[48]	Relieves swelling, stops bleeding, and treats cold, cough.
Viburnum sargenti Koehne f. glabrum Kom. / V. sargenti Koehne f. intermedium (Kom.) Kitag. / V. sargenti Koehne var. puberulum (Kom.) Kitag.	Shan Teng Zi (Snowball)	(whole plant) Scopoletin, aesculetin, caffeic acid, citric acid, malic acid, chlorogenic acid, isochlorogenic acid, essential oil, kaempferol-3-glucoside, beta-amyrin, beta-sitosterol, paeonin.[48]	For blood circulation, swelling; detoxicant; relieves itchiness, arthritis.
Vicia faba L.	Cam Dou (Horseean)	(aerial part) Betulin, furraric acid, kaempferol.[50]	Antitumor.

Table 1A   Major Constituents and Therapeutic Values of Chinese Medicinal Herbs   (continued)

Scientific Name	Common Chinese and (English) Name	Major Constituents and (sources)	Therapeutic Values*
Viola acuminata Ledeb. V. alisoviana Kiss V. alisoviana Kiss f. candida (Kitag.) Takenouchi V. alisoviana Kiss f. intermedia (Kitag.) Takenouchi V. collina Bess. V. dissecta Ledeb. V. dissecta Ledeb. f. pubescens (Regel) Kitag. V. mandshurica W. Becker V. patrini DC ex Ging. V. prionantha Bunge. V. verecunda A. Gray	Jin Cai (Violet)	(whole plant) Saturated acids, cerotic acid, unsaturated acids, hydrocarbons, alcohols.[43;48]	Mucilaginous, emollient, suppurative inflammations, abscesses, ulcers.
Viscum album L. subsp. coloratum Kom. V. album L. subsp. coloratum Kom. f. rubroaurantiacum (Makino) Kitag. V. coloratum (Kom.) Nakai	Hu Ji Shang (Asiatic mistletoe)	(leaf, stem) Oleanolic acid, beta-amyrin, fatty acids, mesoinositol, flavoyadorinin, homoflavoyadorinin, lupeol, myristic acid, agglutinins, alkaloids, quercitol, querbrachitol, quencetine, acetylcholine, choline, histamine, tyramine, vitamins E and C.[33,450]	Antihypertensive; prolongs the life of patients with late-stage stomach cancer.
Vitex chinensis Miller V. jeguaod L.	Mu Jing	(leaf) Essential oils, beta-caryophyllene, caryophyllene oxide.[33]	Antitussive, antiasthmatic, antibacterial.

Vitex nequndo L.	Huong Jing (Five-leaved chaste tree)	(leaf, fruit, root) Essential oil, phenolic derivatives, cineol acid, pinere acid, dipentene, citronellol, geraniol, eugenol, camphene, delta-3-carene, tannic acid, nishindine, hydrocotylene, glucononitol, hydroxybenzoic acid, iridoidglycoside-nishindaside, negundoside, aucubin, agnuside, casticin, orientin, isoorientin [50,450]	An astringent, sedative, for cholera, eczema, gravel, anxiety, convulsions, cough, headache, vertigo.
Vitex trifolia L. var. simplicifolia Cham. *V. rotundifolia* L. f.	Mu Jing Chi Dan Ye Mu Jing (Indian privet, seashore vitex)	(fruit, leaf, shoot) Camphene, pinene, vitricine, terpenylacetate, aucubin, agnuside, casticin, orientin, isoorientin, luteolin-7-glucoside, vitexicarpin, casticin, flavons. [48,50]	For fever; analgesic sedative; promotes beard growth, breast cancer.
Vitis amurensis Rupr. *V. vinifera* L.	Shan Pu Tao (Wine grape)	(leaf, fruit) Malic acid, tartaric acid, racemic acid, oxalic acid [50]	For abortion, cholera, dropsy, nausea.
Wahlenbergia marginata (Thunb.) A. DC	Xi Ye Sha Seng	(root) Alkaloids. [55]	Treats pulmonary disorder, skin eruption; stops sweating.
Wikestroemia indica C. A. Meyer	Liao Ge Wang	(root) Wikstroemin, hydroxygenkwanin, daphnetin, acidic resin. [33,558]	Antibacterial.
Wisteria sinensis (Sims) Sweet	Zi Teng (Chinese wisteria)	(seed, bark) Toxic glycosides, toxic resin. [60] This herb is toxic.	Diuretic.
Woodwardia japonica (Lif.) Sm.	Gou Ji Guan Zhong (Chain fern)	(rhizome) Inokosterone, woodwardic acid, woodorien [33]	Antiviral, against herpes simplex virus type-1.

Table 1A Major Constituents and Therapeutic Values of Chinese Medicinal Herbs (continued)

Scientific Name	Common Chinese and (English) Name	Major Constituents and (sources)	Therapeutic Values*
Xanthium chinense Mill. X. japonicum Widder X. mongolicum Kitag. X. sibiricum Patr. ex Widd. X. strumarium L.	Cang Er (Cocklebur)	(fruit) Xanthinin, xanthumin, xanthanol, isoxanthanol, strumaroside, tetrahydroxy flavone, caffeic acid, dicaffeoxylquinic acid.[33,48]	Antibacterial, antitussive, respiratory stimulating effect; lowers blood pressure and blood sugar levels.
Xanthoxylum piperitum DC	Chuan Jian	(fruit) Essential oils, phellandrene, limonene, citronellol, geraniol, and sanshol in fruit; sesquiterpene lactones-xanthatin, limonene in seed; saponin, citral, citronellol, geraniol in leaf; berberine, xanthoxylinin root.[49,50]	Diaphoretic properties, prophylactic against hydrophobia, used as a diuretic, stomachic, carminative, stimulant; resolves inflammatory swellings, it is a sedative.
Zanthoxylum ailanthoides Sieb. et Zucc.	Shi Zhu Yu	(aerial part) Essential oils, methyl n-nonylketone, isopimpinellin, dictamine, skimmianine, magnoflorine, laurifoline.[64,94]	Treats chills, influenza, sunstroke, indigestion.
Zanthoxylum bungeanum Maxim.	Shan Hua Jiao (Szechuan pepper)	(fruit, leaf, seed) Essential oils, limonene, cumic alcohol, linalool, myrcene, benzene tert-butyl, sabinene, rerpinenol, piperitone, beta-gurjunene, alpha-piene, geraniol, estragole, cadinene, clovene.[53]	Anthelmintic, aromatic, astringent, carminative, emmenagogue, stimulant, sudorific.

Zanthoxylum nitidum (Roxb.) DC	Shuang Mian Ci (Shiny bramble)	(root) Nitidine, oxynitidine, vitexin, 6-ethoxy-chelerythrin, diosmin, oxynitidine, oxycheleryhrine, N-desmethylchelerythrine, skimmiarine.[33,50,53]	Analgesic, anodyne, antitumor against leukemia, carminative, detoxicant; increases blood flow.
Zanthoxylum schinifolium Sieb. et Zucc.	Hua Jiao (Pepper)	(pericarp) Estragol, citronellol, phellandrene, xanthoxylene, skimmianine, magnoflorine, xanthoplanine, dictamnine, bergapten, berberine, esculetin dimethyl ether.[33,48,53]	Treats ascaris, relieves abdominal pain caused by ascariar obstruction.
Zea mays L.	Yu Mi Xu (Corn)	(leaf, flower, root, seed) Carbohydrate, beta-carotene, thiamine, riboflavin, niacin, ascorbic acid.[50]	For dropsy, diabetes mellitus, hypertension, epistaxis, menorrhagia, cancers, tumors, warts.
Zephyranthes candida Herbert	Cong Lan (White zephyr lily)	(aerial part) Lycorine, haemanthidien, nerinine, taxettin.[50]	For convulsion; hepatitis.
Zephyranthes carinata Herbert	Jiu Lan	(leaf, bulb) Alkaloids, lycorine.[63,93]	To relieves fever; used as as poultice for abscesses.
Zingiber officinale Roscoe	Sheng Jiang (Ginger)	(root) Essential oils, zingiberol, zingiberene, phellandrene, camphene, citral, linalool, methylheptenone, nonylaldehyde, d-borneol, gingerol.[53,398,510,87,568]	Anti-inflammatory, antitumor; stimulates gastric secretion.

Table 1A Major Constituents and Therapeutic Values of Chinese Medicinal Herbs (continued)

Scientific Name	Common Chinese and (English) Name	Major Constituents and (sources)	Therapeutic Values*
Zingiber zerumbet Smith	Qiu Jiang (Ginger)	(rhizome)[55,60] 3δ,4δ-O-diacetylafzelin, zerumbone, zerumbone epoxide, curcuminoids diferuloylmethane, feruloyl-p-coumaroylmethane, di-p-coumaroylmethane, essential oils, alpha-humulene.[192,193,195] These compounds are cytotoxic.	A tonic, stimulant, depurative; to treats asthma, stomachache, antimicrobial properties. This plant plays an important role in masculine rituals and it makes women sterile.
Ziziphus jujuba Mill. Z. spinosa	Suan Zao or Suan Zao Ren (Jujube, Chinese date)	(seed) Saponins, betulinic acid, betulic acid, betulin, jujuboside A, jujuboside B, sanjoinines, daechu alkaloids.[1,33,44,53]	For insomnia, neurasthenia, and irritation.

* This information should not be used for the diagnosis, treatment, or prevention of diseases in humans. The information contained herein is in no way intended to be a guide to medical practice or a recommendation that herbs be used for medicinal purposes. The information is presented here mainly for educational purposes and should not be used to promotes the sale of any product or replace the services of a physician.

Table 1B Major Constituents and Therapeutic Values of Chinese Medicinal Herbs

Scientific Name	Common Chinese and (English) Name	Major Constituents and (sources)	Therapeutic Values
Achillea millefolium L.	Yang Shi Cao (Yarrow)	(plant) HCN, betaine, choline, stachydrine, trigonelline, apigenin, rutin. [513,514,521,532,534,538,548,568]	Dry herb used for bleeding, hemorrhoidal and menstrual. Fresh herb used for sores, snakebite, dogbite and other wounds.
Acorus calamus L.	Bai Chang (Sweet flag calamus)	(leaf) Acoric acid, hallucinogenic, fungicidal, insecticidal, hypotensive, β-asarone, phenylpropane derivatives, sesquiterpene ketones, shyobunones, acorns. [510,512,513,525,552,542,544,547,568,610,611]	Anticonvulsant, lotion for leprous, pustular sores, chest congestion, anorexia, antirheumatic, hypotensive, sedative, fever, gastritis, arthritis, cancer, convulsions, diarrhea epilepsy, dyspepsia
Actinidia chinensis Planch	Mi Hou Tao (Chinese gooseberry)	(leaf, fruit, root) Vitamin B, C, phosphorus, potassium, calcium, iron, minerals, 2α-hydroxyursolic acid. [510,511,513,612]	Fruit is used for quenchin thirst gravel, astringent, anti-scorbutic.
Adonis amuresis Regel & Radde	Fu Shou Cao (Amur adonis)	(whole plant with root) Adonilide, fukujusone, umbelliferone, scopoletin, cymarin, cymarol, cymarigenin, corchoroside A, couvallatoxin, k-strophanthin-β, somalin, lineolone, isolineolone. [510,512,555,557]	Diuretic, tranquilizer, congestive heart failure. t has direct-action on heart muscle: causes contraction.
Agerodum conyzoldes L.	Sheng Hong Ji (Bastard agrimony)	(leaf) Cyanogenic, glucoside, coumarin, ageratochromene, β-caryophyllene. [510,514,613]	leaf, vulrerary for abscesses, boils, bruises bites, itch, sores, swellings, digestive
Alstonia scholaris (L) R. Br.	Xiang Pi Mu (Dita bark)	(leaf) α-amyrin, β-sitosterol, α-amyrin acetate, Scholarisine A campesterol, echitamidine. [513,515,518,563]	leaf tips with roast coconut for stomatitis.

Table 1B   Major Constituents and Therapeutic Values of Chinese Medicinal Herbs (continued)

Scientific Name	Common Chinese and (English) Name	Major Constituents and (sources)	Therapeutic Values
Amaranthus tricolor L.	Yan Lai, Hong (Jacob's coat, Chinese amaranth)	(leaf) β-carotene, thiamine, riboflavin, ascorbic acid, vitamin A & C. [510,547,554,513]	root decoction with pumpkin used to control hemorrhage following abortion.
Andrographis paniculata (Burmf.) Nees	Chuan Xin Lian (Creat)	(leaf) Andrographan, andrographolide, andrographon, 12-di-dehydroandrog-rapholide. [510,512,514,521,535,585,614,5 13,568]	Juice used for diarrhea, fever, poulticed onto swollen legs or feet, itch and female disorders.
Anethum graveolens L.	Shi Luo Zi (Dill)	(seed) Bergapten, camphene, corvine, dihydrocarvone, dillapiole, dipentene, isomyristicin, limonene, monoterpene, phellandrene, pinene, umbelliprenin. [510,537,538]	seeds are used for carminative, stimulant.
Angelica gigas Maxim	Du Huo (unknown)	(root) Decursin, nodakenin, imperatorin, nodakenetin, coumarins, umbelliferone, bergapten, glasbra lactone, osthol,5 [10]	Decoction root emmenagogue for abscesses, arthritis, cold, epistaxis, headache, lumbago, hematochezia, hematuria, rheumatism, swellings, toothache.
Angelica sinensis (Oliv.) Diels	Dang qui (Angelica)	(root) Coumarins, bergapten, glabralactone, osthol, angelic acid, angelicotoxin, byak-angelicin, byak-angelicol [512,524,526,527, 28,530,531,535,538, 42,559,513]	Analgesic, deobstruent, emmenagogue, sedative; used for anemia, boils, constipation.
Aralia mandschurica (Rupr. & Maxim) seem	Ci Lao Ya (Manchurian aralin)	(root bark) Araloside A, B, C. including oleanolic acid, glucuronic acid. [514]	root bark used for stimulates the central nervous system, restores appetite, memory, vigor; vitality.
Areca catechu L.	Bing Lang (Betelnut palm)	(fruit) Arecoline [527,530,536,513]	Cholinergic, stimulating the neurons, beriberi, canker, diarrhea, dysentery, dyspepsia.

Species	Chinese name	Constituents	Uses
Asclepias curassavica L.	Lian Sheng Gui Zi hua (West Indian ipecac)	(flower) Asclepiadin, calotropin, asclepogenin, coroglaucigenin, uzarigenin, asclepin [514-515].	Juice vermifuge for gonorrhea, powdered and mixed into a past; The root is used to spread on sores.
Berberis amurensis Rupr.	Xiao Bo (Amur barberry)	(root) Berberine. [515,544,513]	Antirheumatic, anticancer.
Blumea balsamifera (L.) DC.	Ai Na Xiang (Camphor plant)	(leaf) Borneol, camphor, cineole, limonene, palmitic-myristic acid. [510,513,514,520,521,543,550,551,615]	Decocted dry leaves used for itchy sores;and wounds. Pland is used for stomachic, sudorific, diaphoretic, anti-catarrhal.
Boschniakia rossica (Cham & Schlecht).	Cao Cong Rong (unknown)	(plant) Boschniakinic acid, boschniakine, boschnialactone, actinidine. [510,593,594]	A tonic, used to treats impotence and sterility decoction an tipyretic for dysmenorrheal, stimulates hormone secretion.
Botrychium strictum Underw	Yin Dijue (Moonwort fern)	(root, whole plant) Luteolin [510]	It is used for stomachic, cancer, tonic and vulnerary are used for cancer, consumption, diarrhea, ophthalmic, phthisis, ruptures, snakebite, sores and wounds.
Cacalia hastate L.	Shan Jian Chai (Cacalia)	(whole plant above the ground, leaf) Hastanecine, potassium, tartrate [510]	Young leaves-raw, cooked or used as a flavoring.
Cajanus indicus L.	Shan Tou Ken (Pigeon pea)	(root, fruit, stem, leaf) Potassium, hastanecine, tartrate [510,532,586,587]	It is an anthelmintic, sedative, expectorant, and with vulenary properties
Cajanus cajan L.	Mu Dou (Pigeon pea)	(leaf) Arginine, cystine, histidin, isoleucine, lysine, methionine, phenylalanine [510,532,535,542]	Leaves used for dysentery, gingivitis, mouthwash, parturition, toothache.
Carex kobomug Ohwi	Shin-Ts'ao (spontaneous grain)	(fruit, whole plant) Fiber. [510]	Prevent nausea, anorexia, produces bodily strength.
Carpesium abrotanoides L.	Tian Ming Jing (Starwort)	(stem) Hentriacotane, essential oil, inulin. [510,521,513]	Juice applied for bug bites. Fruit is anathematic. seed is laxative, pectoral, vermifuge.

Table 1B   Major Constituents and Therapeutic Values of Chinese Medicinal Herbs (continued)

Scientific Name	Common Chinese and (English) Name	Major Constituents and (sources)	Therapeutic Values
Carum carvi L.	Zang, Hui, Xiang (Caraway)	(seed) Acetaldehyde, acetylinic compound, carveol, carvone, dihydrocarveol, falearindione, isohydro carrveol. [510,532,519,568]	Caraway seeds are carminative.
Catalpa ovata G. Don	Zi Bai Pi (Chinese Catawba)	(plant) p-coumaric acid, ferulic acid, p-hydroxylbenzoyl, catalposide, isoferulic, catalpalactone. [510,515,518,541]	Twigs used for dropsy and kidney ailments, beriberi, peritonitis.
Catharanthus roseus (L.) G. Don.	Chang Chun Hua (Madagascar periwinkle)	(plant) Alkaloids including catharanthine, leurosine sulphate, lochnerine, vindoline, tetrahydroalstonine, vindolinine. [510,513,515,539,540,541,568]	Astringent, bechic, depurative, diuretic, emmenagogue, anti-cancer.
Chenopodium album L.	Hui-Hsien (Lambs quarters)	(whole plant, seed) Olcanolic acid, ferulic acid, vanillic acid, carnaubic acid, nonacosane, oleyl, alcohol, betaine. [510,532,562,568]	It has insecticidal properties, used in cases of insect stings and bites. Expressed juice in freckles and sunburn. seeds are for anthelmintic and remedy.
Cicuta virosa L.	Yeh-Chin-Ts'ai (Radish)	(root, stem) Coniine, cicutine, conicine, conhydrine, pseudo-conhydrine, N-methyl coniine, cicutoxin, camphene, cumaldehyde, limonene, cicutol, cuminaldehyde, cymene, α-terpinene, pinene, myrcene, α-pinene. [510,513,519,52]	Leaves applied to bug bites, sunstroke, swollen feet, stem juice applied to freckles and sunburn.
Clerodendrum trichotomum Thumb.	Chou Wu Tong (Hairy clerodendron)	(leaf, new stem) Clerodendrin, apigen in-7-diglucuronide acacetin, fricdelin, epifriedelin, acacetin, 7-β-D-glucurono-β-D-glucuronide, cleroden dronin A. clerodendrin A Picein, epifriedelinol, clerodolone, clerodone, clerosterol. [510,512,526,531,535]	Leaves are used externally for dermatitis, hypertension, and rheumatoid arthritis (internally), hypotensive, sedative.

Coriandrum sativum L.	Hu Sui (Coriander)	(leaf) Acetone, borneol, decanal, decanal, cariandrol, cymene, decanol, decylic, aldehyde, linalool, malic acid, oxalic acid, phellandrene. [510,513,514,515,535,536,538,568]	Hastens the eruptions of pox and measles. Fruit for dysentery, measles, hemorrhoids.
Cotinus coggygria Scop	Huang Lu (Smoke tree)	(stem) Fisetine, fustine, myricetol, quercitol, gallic acid, myricetin, sulfuretin. [510,515,513]	Yellow wood used as cholagogue, febrifuge, and eye ailments.
Curculigo orohiodes Gaertn.	Xian Mao (Black musli)	(root) Calcium oxalate, resin, tannin. [510,513,524,535]	root for arthritis, blenorrhea, cachexia, enuresis, impotency, weak kidneys.
Cyanchum glaucescens Decaisne C. *atratum* Bge.	Bai Qian (White stem)	(root) Cynanchol, cynanchin, cynanchocerin. [510,511,512,524,526,535]	Used for asthma, cold, cough, dyspnea, sore throat; antitussive, expectorant.
Daucus carota L.	Hu Lo Po (Carrot)	(seed, root, fruit, whole plant) Carotenes, lycopene, phytofluere, umbelliferone, lycopene, camphene, myrcene, α-phellandrene, bisabolene [513,514,516,521,525,542,568,616]	root is considered to be beneficial to the digestive tract, increasing the appetite and acting as a carminative. The seeds are used in chronic dysentery.
Descurania Sophia (L) Webb ex Prantl.	Ting Li Zi (flaxweed)	(flower, seed) Linolenic, linoleic, oleic-, erucic-, palmitic-, and stearic acid. Allyl-, benzyl-and propenyl-isothiocyanate. [510,512,561]	Antiscorbutic, astringent, used for cough, dyspnea, dysuria, edema, excess sputum.
Dianthus chinensis L.	Shih-Chu (Rainbow pink)	(flower, whole plant) Eugenol, phenylethylalcohol, benzyl benzoate, benzyl salicylate, methyl salicylate. [512,514,526]	Diuretic, vulnerary, abortifacient, relieves opacities of the corneas alleviate fluxes, promotes the growth of fair, and used in the treatment of gravel, amenorrhea, resolvent for incipient abscesses.
Dysosma pleiantha (Hance) Woodson	Ba Jiao Lian (Chinese mayapple)	(root) Astragalin, hyperin, deoxypodophyllotoxin, hyperin, kaempferol, podophyllotoxin, quercetin. [510,513,526,531,544]	Rhizome antirheumatic, antisceptic for syphilis. Externally used as a liniment, for snakebite.

Table 1B   Major Constituents and Therapeutic Values of Chinese Medicinal Herbs (continued)

Scientific Name	Common Chinese and (English) Name	Major Constituents and (sources)	Therapeutic Values
Eauisetum palustre L.	Mu Zei (Common horsetail)	(whole plant, stem) Palustrine, palustridine, nicotine, aconitic acid, methyl, sulforn, thymine.[510,511,521,513]	Diaphoretic, diuretic, expectorant, febrifuge, hemostat; used for colds, conjunctivitis, dysentery, edema, enterorrha-epipnora, gia, and fever.
Eclipta prostrate L.	Mo Han Lian (Eclipta)	(leaf) Estrogenic acitivty, thiophene activity.[510,511,512,521,524,535,551,582,547,568]	Astringent for hemorrhage, eye ailments, enter-orrhagia, hematemesis, hepatitis, anti-proliferative activity of triterpenoids.
Eleutherococcus senticosus Maxim	Wu Jia Pi (Siberian ginseng)	(plant) Saponins.[526,617,618]	Bronchitis, heart ailments, rheumatism, improves appetite, effects on arousal and performance energy metabolism, cardiovascular system.
Elshoitzia ciliate (Thunb) Hylander	Xiang Ru (Aromatic madder)	(whole plant with flower) Elsholiziaketon, naginataketone, iso-butyl, iso-valcrate, α,-β-naginatene[510,512,519]	Sudorific, used for cold, dropsy, nausea, stomachache, and typhoid.
Emilia sonchifolia (L.) D.C.	Yang Ti Cao (Red tassel flower)	(leaf) Carbohydrate, fiber.[510,535,554,513.]	Leaf tea for dysentery, plant for detoxicant, diuretic febrifuge, refrigerant, sudorific.
Equisetum arvense L.	Wen Jing (Horsetail)	Trehalase, articulatin, equistic acid, equisetine, equisetrin, galuteolin, gossypitrin, isoquercitrin, silicilic acid.[513]	Antihemorrhagic, andyne carminative, diaphoretic, fever, gonorrhea, hepatitis, stomach and urinary disorders.
Equisetum arvense L.	Qi Zhou Yi Zhi Hao (Horsetail)	(plant) Equisetonin, equisetrin, articulain, isoqueictrin, galutcolin, populnin.[510,515,520,542,554,513,568]	Antihemorrhagic, anodyne, carminative, diaphoretic, diuretic.
Erigeron Canadensis L.	Qi Zhou Yi Zhi Hao (Horseweed)	(plant) Protein, carbohydrate, essential oil, gallic and tannic acid.[536,553,554,513]	Used in folk remedies for bronchitis, catarrh, cystitis, diarrhea, dropsy, dysentery, eczema, gonorrhea, hemorrhage, lungs, metrorrhagia, parturition, piles, renosis, ringworm.

Eupatorium odoratum L.	Fei Ji Cao (Siam weed)	(plant) Ceryl alcohol, sequiterpine, eupatol, trihydric alcohol, triterpene, lupeol, β-amyrin. [511,513,514,535,554,583]	This plant is used for anodyne, hemostat, nervine, spasmolytic, vermifuge.
Ferula assafoetida L.	A wei (Asafetida)	(plant) Asaresinotannol, axulene, bassorine, ferulic acid, pinene, umbelliferone, asaresinotannol, farnesiferol A, B, C. [510,511,513,524,537,538]	Alexeritic, alterative, anthelmintic, antispasmodic, laxative, sedative, stomachic, vermifuge.
Filifolium sibiricum (L.) Kitam	Xian Ye Chiu (unknown)	(whole plant, flower) Filifolin, eriodictyol, 3,6-dimethoxy-quercetagetin. [510,512,516,519]	The plant is used in folk remedies for cancer and tumors. It is an alterative, artiacid, diuretic, antispoismodic, astringent.
Guelden staedtia Maxim	Di Ding (Maritima)	(whole plant, flower) Psyllcstearyl alcohol, Soyasapogenol B, E. [510,512,542]	Used for gall disorder, scrofula, syphilis. plant is also used for appendicitis, dermatitis, epistaxis, snakebite, and sloughing ulcers.
Gymnadenia conopsea (L.) R. Brown.	Shou Jao Sun (unknown)	(root) Kaempferol-3-β-glycoside-7-β-glycoside, astragalin, quercetin-3-β-glycoside-7-β-glycoside, isoquercitrin, piperonal, methyl vanillin. [516,519,591]	The compounds isolated were evaluated for activity in *in vitro* assays for acetylcholine, esterase and monoamine. It also has oxidase inhibitory activities.
Helianthus annuus L.	Xiang Ri Kui Zi (Sunflower)	(flower) Beta-carotene, thiamine, riboflavin, niacin, ascorbic acid. [568]	Chaff and receptacle boiled with inner ear of pig for ringing of ear.
Hemerocallis fulva L.	Yuan Cao Gen (Yellow daylily) (flower)	(root) Umbelliferone, bergapten, isobergapten, quercetin, kaempferol, rutin. [510,513,515,531,559,592]	It is used as an anodyne in headache, in fluenza, toothache, and vertigo. It is also used for poor memory, melancholy and agitation, indigestion, and asthma. Externally for healing and rheumatic pain and palpitations.

Table 1B   Major Constituents and Therapeutic Values of Chinese Medicinal Herbs (continued)

Scientific Name	Common Chinese and (English) Name	Major Constituents and (sources)	Therapeutic Values
Hemerocallis minor Miller	Xuan Caogen (Yellow day lily)	(root) Friedelin, colchicines, vitamin A, B, C. chrysophanol, β-sitosterol. asparagines, hemerocallone, chrysophanol, mi-hem erocallin, hemerocallin, rhein, heptacosane, trebalase [510,515,531,559]	Considered anodyne, antiemetic, antispasmodic, depurative, febrifuge, sedative.
Heteropappus altaicus (Willd.) Novopokr.	A Er Tai Zi Wan (unknown)	(whole plant) Demethyl, nobiletin. [510]	It is used for poisoning, feverish conditions.
Holarrhenia antidysenterica Wall	Zhi xie mu pi (Conessi bark)	(stem bark) Conamine, cencuressine, corssidine, conessimine, conkurchin, dihydro-conessine, 3-epiconamine. [510,515,520,521]	stem bark is used for dysentery, fever, stomachic.
Hypecoum erectum L.	Jiao Wei Xiang (unknown)	(root, whole palnt) Protopeine, sanguinarine, chelirubine, chelerythrine, coptisine. [519,520,524]	It is used for anti-bacteria, antipyretic, calm cough, abdominal pain. It is also used to treats cold and grippe, diarrhea.
Inula britannica L.	Xuan Fu Hua (Chinese elecampane)	(leaf, plant) Inulin, flavone, caffeic acid, chlorogenic acid, isoquercitrin, quercetin. [510,511,512,513,535,555]	Discutient, vulnerary, alternative, carminative, deobstruent, depurative, diuretic, hematic, lavative, stomachic.
Inula helenium L.	Tu Mu Xiang (Elecampane)	(root) Alantolactone, helenin. [513,514,520,521,526,542,538,524,556,568]	Anthelmintic, antiseptic, used for asthma, bronchitis, cancer, cholera, diarrhea, dysentery.
Jeffersonia dubia (Maxim) Benth et Hook f.	Xian Huang Lian (Asian twinleaf)	(root, stem) It has anti-tumor compound berberine. [510,520,542]	It is aromatic, emetic, diuretic, expectorant, stimulant' and tonic. plant is used for syphilis' rheumatism, scarlet fever, spasms, sores, sore throat. root used for fever. Decoction a collyrium, stomachic, substituted for coptis.

Kalopanzx septemlobus (Thunb.) Koidz.	Ci Qiu Shu Pi (unknown)	(stem wood, root) Glucan, pectic substances, kalopanax sapenin A Kalosaponin sapenin B, Kalosaponin, kalotoxin, trachitin. [510]	Infusion leaf, a stomachic tea. root expectorant, wook decoction for skin diseases.
Koelreuteria paniculata Laxm	Luan Hua (unknown)	(flower, root) Koelreuteria A, B. [510,511]	Carminative, stomachic, peptic, emmenagogue, and cholagogue properties.
Kyllingia brevifolia Rottb.	Chin-nin-Ts'ao (Duck weed)	(whole plant) Volatile oil. [510,511,520]	plant is dried and burned to produce a smoke to drive away all sorts of parasitic insects.
Lamium album L.	Hsii-Tuan (Teazel)	(flower head). Isoquercitrine, kaemp-ferol-3-glucoside. [510,515,520,521,568]	It is considered capable of nine joining together broken bones. A tonic in exhausting diseases, wounds, tumors, fractures, and ruptured tendons, suppression of the secretion of milk dysmenorrhoea, hemorrhage.
Lamium amplexi-caule L.	Bao Gai, Cao (Henbit)	(plant) Ipolamiide, iridoid, lamiide, lamiol, lamioside. [510,520,521]	An excitant, febrifuge. It is used for influenza.
Lathyrus pratensis L.	Yeh-Wan-Tou (Beach pea)	(whole plant) Quercetin, kaempferol, trifolin, isoquercitrin, orientin, isoorientin. [510,519]	It is used as a pot-herb, and upon prolonged used. It is said to be very nourishing and to greatly benefit the intestinal tract. A tonic to the urinary organs.
Levisticum officinale Koch	Dang Gui (Lovage)	(plant) Phthalides, terpenoids, volatile acids, coumarins, β-sitosterol [537,538,568]	It is said to treats female diseases.

Table 1B   Major Constituents and Therapeutic Values of Chinese Medicinal Herbs (continued)

Scientific Name	Common Chinese and (English) Name	Major Constituents and (sources)	Therapeutic Values
Ligularia fischeri Ledeb.	Hu Lu Qi Cho (unknown)	(root) 1(10) eremophilen-11-ol, 3β-hydroxyeremophile mophilenolide, furanoeremophilane, petasalbin, 1β, 10-β-epoxyfuranoeremophilane, 2-hydroxymethyl prop-2-enoate, isopentenic acid, ligularone, liguloxide, liguloxidol acetate, liguloxidol, epiliguloxide.[510,521]	
Ludwigia prostrate Roxy	Shui Ding Xiang (unknown)	(seed, whole plant) Triethyl chebulate.[510,520]	An astringent, purgative, vermifuge. seeds are used for whooping cough, myalgia, and toothache
Lycoris aurea Herb.	Da Yi Zhi Jian (Golden spider lily)	(root) Galanthamine, homolycorine, lycoramine, lycorenine, lycorine, pseudoly corine, tazettine.[510,513,515,518]	Crushed bulb poulticed onto burns, scalds, and ulcers.
Malachium aquaticum L. Fries.	E Chang Cao (unknown)	(whole plant) Cyclolaudenol.[510,515]	Leaves used for patients suffering appendicitis, beriberi or carcinoma, ventriculi. Leaves steeped in water are applied to aching, bones, bruises, and vaginal catarrh. It is also used for fistulae, swellings, and tumors.
Mangifera indica L.	Mang Guo (Mango)	(leaf) Ambolic acid, ambonic acid, arabinan, carotenoides, m-digalic acid, galacturonan.[510,516]	Ashes from this plant are used for burns and scalds, skin ailments, asthma, and cough. seed used for diarrhea, leucorrhea.

Marsilea quadrifolia L.	Ping (European water clder)	(leaf) β-carotene, thiamine, riboflavin, niacin, ascorbic acid.[510,511,513,535]	Juice drunk for snakebite, applied to abscesses, dermatitis, ulcers; diuretic, febrifuge, decoction for fever and swelling.
Metaplexis japonica (Thunb)	Lo-Mo (unknown)	(root, fruit shall, whole plant, seeds) Benzoylramanone, metaplexigenin, isoramanone, sarcostin, gagaminin, dibenzoylgagaimol deacylmetaplexigenin, deacylcynanchogenin, pergularin, utendin, ester A. Cinnamoyl nicotinoyl-7-dehydroxygagaimol.[510,515,596]	It is tonic and constructive. The crushed seeds are applied to wounds and ulcers as an astringent and hemostatic remedy. It can also be applied to all sorts of insect bites. This plant has escharotic properties.
Miscanthus sinensis Andress	Mang Jing (Miscanthus)	(flower, root, and stem) Prunin, miscanthoside.[510,512,597]	Juice from young stem to disperse poison, dissolve blood clots, dissipate extravasated blood, and remove inflammation. This plant is a diuretic and refrigerant.
Mosla chinensis Maxim	Shi Xiang Rou (Chinese mosla)	(plant) Borned, carvacrol, cymene, linalool, alphathujene.[510,515]	It is a diuretic, sudorific. plant applied for cold, diarrhea, edema, fever, and headache.
Nelumbo nucifera Gaertner, Fruct et Sem.	Ho (East Indian lotus)	(root, stem, seed, leaf) Galuteolin, isoquercit, rine, nulumbine, raffinose, p-hydroxybenzyl-6,7-dibydroxy-1,2,3,4-tetrahydro-isoquinoline, nuciferine.[511,512,524,526,515,521,535,547,559]	It is considered to be nutritious;stomachic, tonic, increasing the mental faculties and quieting the spirits. leaf is considered to be antiebrile, antihemorrhagic, constructive to the blood.
Nerium indicum Mill	Jia Zhu Tao (Indian oleander)	(leaf) Dambonitol, deacetylo-leandrin, digitoxigenin, gitoxigenin, karabin neriantin, neriorcorin.[510,511513,518,521,535.]	Leaves cardiotonic, diaphoretic, diuretic, emetic, expectorant, applied externally for bruises, bug bites, fungus, maggots, swelling.
Nymphoides peltate S. G. Gmelin	Xing Cai (Floating heart)	(whole plant) Rutin, peltatoside, β-vicianosyl-3-quercetin, isoquercitrin.[510,515,598]	Bruised plant for burns, fevers, rodent, ulcers, snake bites, swellings, diuretic, febrifuge, refrigerant.

Table 1B　Major Constituents and Therapeutic Values of Chinese Medicinal Herbs (continued)

Scientific Name	Common Chinese and (English) Name	Major Constituents and (sources)	Therapeutic Values
Oenanthe javanica Blume DC	Shui qin (Water dropwort, Chinese celery)	(root, stem, whole plant) Petroselinic acid, α-pinene, myrcene, terpinolene, diethyl-phthalata, n-butyl-2-phthalata, n-butyl-2-ethyl butylphthalate, persicarin.[510,511,512,515,535,537,554,563]	seed for plethora, plant prescribed for fever, hematuria, influenza, jaundice, metrorrhagia, antivinous, hemostat.
Ophioglossum thermale Kom	Ping Er Xiao Cao (Adder's tongue)	(whole plant) 3-0-methylquer-cetin-7-0-diglucoside-4'-0-glucoside.[510,520,521]	plant used for abscesses, bad teeth, gangrene; cooked with pock for a depurative, lymphoderopathy.
Orobanche caerulescens Stephan	Jou Tsung Jung (Broom rape)	(root, whole plant) Orobanchin, chlorogenic acid.[511,535]	It is used in spermatorrhea, menstrual difficulties, gonorrhea and all forms of difficulties of the genital organs.
Orostachys fimbriatus Turcz	Wa Song (Chinese hens-and chickens)	(whole plant) Sedoheptulosan, isopropylidene, sedoh eptulosan.[510,524]	Hemostat for dogbite, dysentery, dysmenorrheal, enterror-rhagia gravel, powdered for boils, piles, swellings, wounds. Prescribed as an expectorant and diuretic in chronic trachitis asthma, pleurisy, hydrothorax edematous beriberi.
Patrinia villosa (Thunb.)	Bai Jiang (unknown)	(root, stem, whole plant with root) Villoside, morroniside, loganin, sesquiterpenes, patrinene, isopatrinene [510,511,524,535,560,566,.601]	Crushed leaves applied to abscesses and boils. Whole plant decocted; antiphlogistic, decoagulant, diuretic, febrifuge, resolvent, It is also used for abscesses, appendicitis, dropsy, dysentery, enteritis, fever, inflammation.
Paulownia tometosa (Thunb.)	Tong Pi (Princess tree)	(stem skin) Ursolic acid, matteucinol, polyphenols, paulownin, isopaulownin, d-sesamin, d-asarinin, paulownioside, catalpinoside, syringin.[510,511,514,520,52 1,542,564.]	Leaves decoction for foul ulcers, promotes hair growth and restore its color. Inner bark for high fever, delirium, astringent, vermicide, alopecia, delirium, typhoid, ulcers.

Perilla frutescens (L.) britt	Bai Su Zi (Perilla)	(leaf, seed) Apigenin, luteolin, 3-p-coumarylglycoside-5-glucoside 7-caffeyl-glucosides anthocyanins. [511,512,526,561,565,513,605]	This plant is used for an emollient, stomachic, tonic, antiseptic, antitussive, diaphoretic, pectoral.
Periploca sepium Bunge	Xiang Jia Pi (Silk vine)	Cardioactive glycosides, periplocin, periplocymarin, glycosides G and K, 4-methoxy salicylaldehyde. [510,511,512,513,515,518]	stem bark is a cardiotonic, it is used for rheumatic and bone pains. It has repellent activity against the olive weevil. Antinematodal activity.
Plumeria rubra L. var. acutifolia (Poir) Ball	Ji Dan Hua (Frangi pani)	(leaf, stem) Agoniadin, cerotinic acid, fulvoplumierin, lupeol, plumieric acid, plumierde. [510,513,521]	Poulticed onto swellings, stem latex poisonous, purgative, and rub efficient.
Phryma leptostachya L.	Pai Chiang (Soy sauce)	(whole plant) Phrymarolin – I, Phrymarolin – II. Leptostachyol acetate. [510,520,532,535,604]	The seeds are nutritious. It is prescribed in rheumatism, seminal losses, asthma, obstinate coughs.
Pinus madshurica Rupr.	Sung (Chinese pine)	(seed, leaf) Pinacene, lambertianic acid, lamber-tianic methylate folene. Sabinene, myrcene, dipentene, β-phellandrene, r-terpinene, p-cymene, 4-epiisocembrol, agathodie nediol. [510,511,526,536]	It is carminative, antifebrile. It is also beneficial to the tendons, eyes, and ears. Externally for skin eruptions, old ulcers, indolent wounds, vomiting, cold, rheumatism, toothache.
Potamogeton perfoliatus L.	Yu (Purslane)	(whole plant, root) Lutein, violaxanthin, neoxanthin. [510]	A tonic, giving brightness in medicine, acuteness to the hearing; antifebrile, diuretic.
Prunella asiatica Nakai	Hsia-Ku-ts'ao (Heal-all)	(fruit, whole plant) Prunelin, ursolic acid, stachyose. [510,511,512,515,520,524,526,530,542,560]	The stalk and leaves are the parts used. The drug is considered as cooling, used in fevers, antirheumatic, alterative, and tonic remedy.
Prunella vulgaris L.	Xia Ku Cao (Self-heal)	(leaf, flower) Caffeic acid, d-camphor cyaniding, delphinidin, d-fen chone, hyperuside, oleanolic acid, ursolic acid. [510, 511,512,513,515,520,524,526,530,542,560,568]	Alterative, antipyretic, diuretic for gout, scrofula, rheumatism. In florescence for cancer, boils, conjunctivitis, scrofula.

Table 1B Major Constituents and Therapeutic Values of Chinese Medicinal Herbs (continued)

Scientific Name	Common Chinese and (English) Name	Major Constituents and (sources)	Therapeutic Values
Pteridium aquilinum (L.) Kuhu	Jue (Bracken)	(root, stem, whole plant) 2-5-7-trimethl-1-oxoin, dan-6-acetic acid, ponasterone, crustecdysone, hexenal, indan, isoquercitrin, taeniafuge, antivitamin B, K, aspidinol, filicic acid, hydrogen cyanide, pteridine, tannin. [510,524,542,560]	Young shoot is used for diuretic, refrigerant, vermifuge; root used in tincture in wine for rheumatism.
Rhododendron aureum Georgi	Tu Chiian (Azalea)	(leaf, flower) Geraniol, azalein, andromedotoxin, ericolin. [510,515]	The flowers are used as a sedative in rheumatism, neuralgias, contractions, and bronchitis.
Rhododendron micranthum Turcz	Shih-nan (Azalea)	(leaf) Germacron, grayanotoxin, romedotoxin, p-hydroxybenzoic acid, protocatechuic acid, vanilic acid, syringic acid. [531,566]	Leaves are used in medicine, bitter and slightly poisonous. It strengthens the kidneys, cures internal injury and weakness. It is also prescribed in fevers, colds, and intestinal worms.
Rhododendron molle (Bl) G. Don	Nao Yang Hua (Chinese azelea flower)	(flower) Andromedotoxin, pesticidal for maggots, mosquito larvae, oncomelania snails. [510,515,524,535]	Applied externally for arthritis, caries, itch, maggots and traumatic, analgesic, anesthetic, sedative in rheumatism.
Rhus chinensis Mill	Yan Fu Zi (Chinese sumach)	(leaf) Gallic acid, penta-m-pdigalloyl-βglucose. [510,511,513,515,524,529,535,513,584]	Leaves and roots decocted for hemoptysis, inflammations, laryngitis, snakebite, stomachache, and traumatic fracture, anti-HIV-1 acitivities.
Rhus verniciflua Strokes	Qan Qi (Chinense lacquer tree)	(leaf) Urushio, fisetin, fustin, dibasic acids. Eicosanedicarboxylic acid, dioxybenzol, galactose, sorbose, urvshiol. [510,511,513,521,524]	leaf used for wasting diseases and intestinal parasites. seeds are used for dysentery. plant is resin emmenagogue, hemolytic, stimulant, tonic, vermifuge for amenorrhea coughs and ecchymoses.

Species	Common name	Constituents (part)	Uses
Saposhnikovia divaricata (Turcz) Schischk	Fang Feng (unknown)	(root) Psoralen, bergapten, imperatorin, phellopterin, deltoin, marmesin, nodakenetin, hamaudol, 3'-0-acetylhamaudol, hamaudol, 3'-0-angeloyl, hamaudol, ledebouriellol, cimifugin.	(root) Antidote to aconite, analgesic, antipyretic. It is used for arthritis, chills, headache, influenza, numbness, rheumatism, and tetanus.
Saussurea lappa (Clarke)	Mu xiang (Costus)	(root) Aplotaxene, camphene, alpha-cos-tene, beta-costene, costol, kushtin, phelladrene, costunolide.[511,512,513,524,542]	It is used for abdominal pain, asthma, cancer, cholera, nausea, rheumatism, stomachache, and tenesmus.
Senecio scandens Buch-hami	Quan Li Guang (German ivy, ragwort)	(leaf) Penicillin, streptomycin, purulent, appendicitis.[510,511,512,513,515,535,542]	For eye ailments, stem and leaves decocted for abscesses, boils, dermatitis, eye ailments, piles.
Siegesbeckia orientallis L.	Xi Xian (Divine herb)	(root, plant) Salicylic acid, darutoside, aglucones, isodarutigenals B and C.[510,524,526,535,542,511,512.]	Root is used for an analgesic; antirheumatic; used externally for abscesses, boils; and ulcers.
Siegesbeckia pubescens Makino	Chu Kao Mu (Herbe de flacq)	(whole plant, leaf) Darutin-bitter, manool, sclareol, 16,17-dihydroxy-16β-kauran-19-oic acid.[510,511,512,521,524,526,535,608]	The leaves are used as a tonic; treats cancerous sores.
Solidago pacifica Juzepczuk	Lung Kuei (Golden rot)	(whole plant, flower, seed) Cerylalcohol, kautschuk, mannitol, linositol.[510,511,521,535,554]	It is used in hemorrhages, wounds, menstrual disorders, cholera, diarrhea and hemorrhage from the bladder in children.
Solidago virgaurea L.	Yi Zhi Huang Hua (Aaron's rod)	(seed) Astragalin, caffeic-, chlorogenic-, hydroxycinnamic-, quinic-acids. Kaempferol-rhamnoside, quercetin.[510,511,513,521,535,554,568]	Decoagulant, carminative for bladder ailments, cholera, diarrhea; dysmenorrheal, hemorrhages, and wounds.
Sonchus arvensis L.	Niu She Tou (Perennial sowthistle)	(plant) Inositol, lactucerol, mannitol, taraxasterol, palmitic acid, steanic acids.[510,515,520,521,542,513]	Insecticide, nonenolides and cyrochalasins, with phytotoxic activity against cirsium avense.
Sonchus oleraceus L.	Ku Cai (Annual sowthistle)	(leafs, stem) β-carotene, thiamine, riboflavin, niacin, ascorbic acid.[510,515,520,521,543,513]	Latex used for warts. plant juice is a powerful hydragogue, cathartic.

Table 1B Major Constituents and Therapeutic Values of Chinese Medicinal Herbs (continued)

Scientific Name	Common Chinese and (English) Name	Major Constituents and (sources)	Therapeutic Values
Spinacia oleracea L.	Po-ts'ai (Spinach)	(whole plant, seed) α-tocopherol, 6-hydroxymethyllumazin, xanthophylls, neo-β-carotene V, α-spinasterol, 7-stigmastenol, cholesterol, cholesterol, patuletin, spinacetin [510,515,520]	The herbage with the root is regarded as a cooling, carminative, antivinous, thirst-relieving vegetable. (no special medicinal uses are noted)
Tagetes erecia L.	Wan Shou Ju (Marigold)	(leaf, flower) α-terthienyl, d-limonene, l-linalool, tagetone, n-nonyl aldehyde. [510,525,513]	Leaf to treats sores and ulcers. flowers for cold, conjunctivitis, cough, mastitis, mumps, sore eyes.
Tagetes patula L.	Knog Que Cao (French marigold)	(leaves, stem) Essential oil, tagetone, linalool, limonene, linalyl-lacetate, patuletin, patulitrin, cyaniding, diglycoside. [521,538,554,513]	Whole herb is powdered or decocted for coughs and dysentery.
Taraxacum mongolicum Hand.-Mazz. Monogr. Tarax	P'u Kung Ying (Dandelion)	y-aminobutyric acid, valine, leucine, β-sitosterol, β-amyrin, cysteic acid, glycine, asparagines, lysine, alanine, threonine, glutamine, tataxasterol, taraxerol, taraxol, taraxanthin, cryptoxanthin, cryptoxan-thin-epoxide, zeaxanthin, lutein, antheraxanthin, myristic acid, behenic acid, linolenic acid, choline, steraric acid. [510,511,512,542,524,525,526,530,535,557,513,607]	The tender shoots are tonic and alterative. It is prescribed in all sorts of abscesses and swellings, carious teeth, and snakebites.
Torilis japonica (Houtt.) DC prodr.	He Shi (unknown)	(fruit) Cadinene, torilene, petroselinic acid, toriliol, torilolone, torilin, myristin, olein. [510,543,609]	Fruit is used for expectorant, tonic, dysentery, fever, hemorrhoids, leucorrhea, skin disease, spasm, vaginal swelling.

Trachycarpus excelsa Wendl.	Zong Lu Zi (Windmill palm)	(flower, seed) Leucoanthocyanins. [511,524,535]	Flower used for fluxes, hemorrhage. seed is an stringent, hemostat; plant ashes used for epistaxis, gonorrhea, hematochezia, hemoptysis, hematemesis, metrorrhagia, venereal diseases.
Trachelospermum jasminoides (Lindl.) Lem	Luo Shi Teng (Star jasmine)	(leaf) Arctiin, cymarose, dambonitol, glucoside, matairesinoside, nortracheloside, β-sitosterol, tracheloside. [510,511,515,524,535,513]	Leaf restorative, tonic. It is used for carcinomatous growths, gonorrhea, sciatice, snakebites.
Trapa bicornis Osbeck	Ling (Water chestnut)	(flower) Dihydrostigmast, sitosterol. [510]	A tonic, flower used for astringent in fluxes. fruit for fever, sunstroke. It is anti-cancer, anti-pyretic.
Trapa manshurica Flerov	Ling (Water chestnut)	(fruit meal, shall, fruit stem). 22-ergostate traen-3-one, 22-dihydrostigmast-4, en-3,6-dione, β-sitosterol, amylase. [511]	Flower used for astringent, influxes. Fruit used for fever, sunstroke. It is anti-cancer, antipyretic, tonic.
Trigonotis peduncularis Trevir	Chi Chang Ts'ao (unknow)	(shole plant) Delphinidin-3,5-diglucoside.	It is used in medicine as a diuretic and as an emollient application in wounds. A bland remedy in diarrhea and the dysenteries of children.
Xanthium sibiricum Patr. Ex widd.	Cang er (Cocklebur)	(leaf, flower) Hypoglycemic activity. [510,511,512,515,520,524,535,555,,513]	Astringent, hemostat, tranquilizer. flower used for colds, stem used for astringent, cooling, hemostat, sedative.
Zephyranthes candida Herb.	Gan Feng, Cao (White zephyr lily)	(plant) Lycorine, nerinine, haemanthidine, tazettin. [513,515]	Useful in convulsions and hepatitis.

CHAPTER **2**

Phyletic Relationships between Chinese and Western Medicinal Herbs

The recognition that active ingredients extracted from native herbs may have potential utility in modern medicine has given new incentive to worldwide efforts to conserve vulnerable populations of wild plant species.[50] During the past decade, market demand for Chinese herbs around the world has increased sharply. And these resources have been avidly sought as raw material by drug processors in the East as well as the West. As a result of mass collecting, many natural habitats and the plant communities which they sustain have been decimated and some species are threatened or have become scarce.

Many Chinese herbal species currently are unavailable commercially in North America. However, hundreds of Chinese medicinal herbs commonly used in China also can be found in natural habitats in North America. The majority have a phyletic relationship with either the same species or the same genus of Chinese herb (see Tables 2 and 3). Some of the principal ingredients in Chinese herbs can be extracted from related plant species in the West. Thus, it may be possible to substitute Chinese herbs with more readily available herbs in the West. Moreover, these North American plants are, or can be, cultivated, harvested, and processed under proper management that will ensure their safety, quality, and efficacy.

The information presented herein is intended for use by biologists, chemists, and the interested layman as a guide to the Chinese medicinal plant resources and their uses.

Table 2 Chinese and North American Medicinal Herbs Belonging to the Same Species: Major Constituents and Therapeutic Values

Scientific Name	Source	Major Constituents	Therapeutic Values*
Abrus precatorius L.	China	L-abrine, precatorine, squalene, hypaphorine, trigonelline, cycloarterenol, 5-β-cholanic acid.[33]	Antiemetic, an expectorant, parasiticide.
	N.A.	Seeds: Abrin, anthocyanins, indole alkaloids. Root and leaves: glycyrrhizin, abrin. This herb is toxic.[100]	Seeds: A contraceptive, aborufacient; treats chronic conjuctivitis. Leaves: treats asthma, bronchitis.
Acacia catechu Willd.	China	d-catechin, epicatechin, gambir-fluorescein, gambirine, mitraphylline, roxburghine D.[33]	Promotes salivation, resolves phlegm, stops bleeding, and treats pyogenic infections.
	N.A.	Tannins, mucilage, flavonoids, resins.[100]	An astringent, clotting agent, it helps reduces excess mucus in the nose, the large bowel, or vagina. It treats eczema, hemorrhages, diarrhea, and dysentery.
Acalypha indica L.	China	Acalyphine.[55]	Diuretic; treats diarrhea.
	N.A.	Acalyphine, resin, tannins, volatile oil, cyanogetic glucoside, triacetonamine, querbrachitol, hydrociannic acid.[100]	A diaphoretic, expectorant, laxative. Leaves used as an anthelmintic.
Achillea millefolium L.	China	Alkaloids, essential oils, flavonoides, achillin, betonicine, achilleine, d-camphor, desacetylmatricarin.[33]	Antibacterial, treats menopause, abdominal pain, acute intestinitis, wound infection, snakebite.
	N.A.	Achilleine, tannins, cineole, chamazulene, sesquiterpene, lactones, menthol, camphor, sterols, triterpenes.[98,99,100,101,102,103]	Reduces fever, anti-inflammatory; treats common cold, diarrhea, dysentery, hypertension, and gastrointestinal complaints.

Achyranthes bidentata L.	China	Inokosterone, ecdysterone, polysaccharides. [33]	Anticancer.
	N.A.	Triterpenoid saponins. [99]	Treats canker sores, toothache, bleeding gums and nosebleeds. Invigorate blood flow, stimulates menstruation, and ease menstrual pain.
Aconitum carmichaelii Debeaux	China	Aconitine, hypaconitine, mesaconitine, talatisamine. [33]	A cardiotonic.
	N.A.	Aconitine, malonic acid, caffeic acid, hypaconitine, mesaconitine, neoline, napelline, benzol-aconitine. [100,102,104]	For congestive heart failure.
Aconitum napellus L.	China	Aconitine, hypaconitine, mesaconitine, talatisamine. [33] This herb is highly toxic.	A cardiotonic.
	N.A.	Aconitine, malonic acid, caffeic acid, hypaconitine, mesaconitine, neoline, napelline, benzol-aconitine. [100,102,104] This herb is toxic.	Heart and nerve sedative, anticarcinogenic; reduces fever.
Acorus calamus L. *A. gramineus* Ait.	China	Acoric acid. [50]	Anticonvulsant, analgesic, aphrodisiac, carminative, contraceptive, dessicant, diaphoretic.
	N.A.	Acoric acid, asarone, linalool, palmitic acid, methylamine, saponin, mucilage, sesquiterpenes. [99,100,103,105]	Used as a panacea. It is antibacterial, antifungal, antiseptic, antiamebic, antiprotozoal, a vermifuge. Treats digestive upset, fevers.
Actinidia polygama (Sieb. et Zucc.) Planch ex Maxim.	China	Matatabic acid, iridomyrmecin, actinidine, allomatatabiol, iridomyrmecin, neo-nepetalactone, dihydronepetalactol, matatabiether, isoneomatatabiol, matatabistic acid, neomatabiol, vitamin C, vitamin B. [48,50,52]	Used for escophageal and liver cancers, rheumatoid arthritis, arthralgia, urinary stones, fever.
	N.A.	Actindine. [100]	For colic, rheumatism.

Table 2 Chinese and North American Medicinal Herbs Belonging to the Same Species: Major Constituents and Therapeutic Values (continued)

Scientific Name	Source	Major Constituents	Therapeutic Values*
Adiantum capillus-junonis Rupr.	China	Adipedatol, adiantone, hopadiene, isoadiantone, isofernene, fernene, gamma-fernene, filicene, filicenal, fernadiene.[48]	Treats cold and grippe.
	N.A.	Rutin, isoquercitin, terpenoids, adiantone, tannin, mucilage.[99]	Treats coughs, bronchitis, excess mucus, sore throat, chronic nasal congestion.
Adonis vernalis L.	China	Cymarol, corchoroside A, convallatoxin, adonilide, isoramanone, pergularin.[33] This herb is toxic.	Treats heart disease and central depression, diuretic.
	N.A.	Adonitoxin.[99] This herb is toxic.	Treats venereal disease, heart disorders, sedative.
Aesculus hippocastanum L.	China	Protoescigenine, escigenin, oligosaccharides, amylose.[33]	Promotes circulation, relieves epigastrium pain, and promotes digestion.
	N.A.	Aescin, citric acid, resin, saponin, tannin, uric acid, quercetin, kaempferol, flavonoids, coumarin derivatives.[99,100]	Antipyretic, antithrombin, antiexudative. Treats lymphatic congestions, cerebral and pulmonary edema, crural ulcer and hemorrhoidal complaints.
Agrimonia eupatoria L.	China	Agrimophol, agrimols, agrimonine, agrimonolide, luteolin-7-β-D-glucoside, apigenin-7-β-D-glucoside, cosmosiin, vitamin c, vitamin K, tannin.[33,48,49]	An astringent hemostatic in enterorrhagia, hematuria, metrorrhagia, gastrorrhagia, pulmonary, tuberculosis. A cardiotonic.
	N.A.	Tannins, coumarins, flavonoids, luteolin, polysaccharides.[99,106]	Heal wounds and encourages clot formation; treats diarrhea, used as a tonic for digestion.

Species	Origin	Constituents	Uses
Ailanthus altissima (Mill.) Swingle	China	Amarolide, ailanthone, afzelin, syringic acid, vanillic acid, beta-sitosterol, azelaic acid, d-mannitol, amarolide, oleorsin, mucilage.[33,48]	Antidiarrheal; treats dysentery, duodenol ulcers. Astringent, anthelmintic, deobstruent.
	N.A.	Quassinoids, ailanthone, quassin, alkaloids, flavonols, tannins.[99]	Antimalarial, against cancerous cells; counter worms, excessive vaginal discharge, gonorrhea, malaria, antispasmodic, cardiac depressant.
Aleurites moluceanu (L.) Willd.	China	Saponin, alpha-elaeo stearic acid, oleic acid, palmitic acid, stearic acid, tannins, phytosterols.[50]	Treats anemia, atrophy, edema; vermicide, oil (toxic internally) for parasitic skin diseases.
	N.A.	Oleostearic acid, hydrociannic acid, tannin, linolenic acid, cleic acid, linoleic acid, protein, thiamine.[100] This herb is toxic.	Laxative, stimulant, and sudorific.
Allium sativum L. A. fistulosum L. A. tuberosum Rottl.	China	Allicin, allistatin, glucominol, neo-allicin, steroid saponins, polysaccharides, proto-isoerubosides, diallyl sulfide.[33,49,510]	Antibacterial, antimutagenic, anticarcinogenic, carminative, abtuarrhythmic, lowers plasma cholesterol and low-density lipoproteins, prevents thrombosis, hypotensive and vessel protective effect.
	N.A.	Alliin, iodine, diallyl trisulfide, 2-vinyl-4h-1,3-dithin, ajoene, linoleic acid, diallyl disulfide, scordinins, selenium.[98,99,107,511]	Reduces serum cholesterol, lowers blood pressure, and platelet aggregation. It is an anticancer, antimicrobial, antithrombotic.
Aloe barbadensis Miller A. vera L.	China	Aloins, barbaloin, aloe-emodin.[49,50,510]	Laxative, stomachic, emmenagogue.
	N.A.	Aloin isobarbaloin, aloeresin A, B, aloesin glycone, aloesone, emodin, chrysophanic acid, 1,8-dihydroxy-anthracene derivatives, barbaloin, anthaquinone glycosides.[99,100,108,109,510,511]	Purgative, eupeptic, and cholagogue effect. It is a laxative and cathartic. Juice from leaves used for cuts and other skin problems.

**Table 2 Chinese and North American Medicinal Herbs Belonging to the Same Species:
Major Constituents and Therapeutic Values (continued)**

Scientific Name	Source	Major Constituents	Therapeutic Values*
Alstonia scholaris (L.) R. Br.	China	Picrinine, picralinal, echitamine, echitamidine.[33]	An expectorant, antiphlogistic.
	N.A.	Alkaloids, rescrpine.[99] This herb is toxic.	Treats malarial fever, antispasmodic, lowers blood pressure, reduces high blood pressure.
Anagalis arvensis L.	China	Anagalline, anagalligenone, cucurbitacins, arrenin.[55]	Treats snakebite, dog bite, antitoxic.
	N.A.	Saponins, anagalline, tannins, cucurbitacins.	Diuretic, sweat-inducing and expectorant properties.
Ananas comosus (L.) Merrill	China	Ergosterol peroxide, ananasic acid, 5-stigmautena-3β,7d-diol, 3,4-dihydroxycinnamic acid, 4-hydroxycinnamic acid, bromelin, vitamins.[57]	Antioxdant activity, for digestion, lowers blood pressure; anticancer.
	N.A.	Bromelain, citric acid, vanillin, methyl-n-propyl ketone, valerianic acid, malic acid, isocaproic acid, acrylic acid.[100]	Unripe fruits improves digestion, increase appetite, and relieves dyspepsia. Ripe fruits reduces excessive gastric acid.
Anemone pulsatilla L.	China	Saponins, protoanemonin.[49]	A cardiac and nervous sedative, antispasmodic, anodyne in asthma and pulmonary infections, antidiarrheic.
	N.A.	Ranunculin, tannin, resin, saponin, anemonin, delphinidin, pelargonidin glycosides, beta-amyrin, beta-sitosterol.	Treats asthma, bronchitis, catarrh, diarrhea, rheumatism, and warts. It is an alterative, antidotal, diuretic, and emmengague.

Species	Region	Constituents	Uses
Anethum graveoleus L.	China	Essential oils, d-carvone, dillapiole, limonene, bergapten, umbelliprenin, camphene, dihydrocarvone, dillapiole, dipentene, isomyristicin.[48,50]	Carminative, stimulant.
	N.A.	Carvone, limonene, flavonoids, coumarins, xanthones, triterpenes.[99,100,107]	Used for infant colic, cough, cold, and flu remedies. It relieves digestive disorders.
Angelica polymorpha Max.	China	Vitamin B_{12}, vitamin E, ferulic acid, succinic acid, nicotin c acid, uracil, adenine, butylidenephalide, ligustilide, folinic acid, biotin, polysaccharide.[33]	Treats irregular menstruation, anemia, thrombophlebitis, neuralgia, arthritis, chronic nephritis, constrictive aoritis, and skin diseases such as eczematous dermatitis.
	N.A.	Butylphthalide, cadinene, carvacrol, n-dodecanol, isosafrole, linoleic acid, palmitic acid, safrole, sesquiterpene, sesquiterpenic alcohol, n-tetradecanol.[100]	Immunosuppressive activity; treats hay fever, asthma, and atopic dermatitis. Analgesic, deobstruent, emmenagogue, sedative.
Apium graveolens L.	China	Apiin, graveobioside A, graveobioside B.[33]	Treats hypertension, hypercholesterolemia.
	N.A.	Limonene, coumarins, apiin, oleic, linoleic, palmitic, paliloleic, petroselinic, petroselaidic, stearic, myristic, and myristoleic acids, bergapten.[99,102,110]	It is a carminative and antirheumatic.
Arctium lappa L.	China	Arctin, arctigenin, matai-resinol, sesquilignins, stereoisomer.[1,9]	For dermatitis, tumors; antibacterial, relieves sore throat.
	N.A.	Inulin, mucilage, tannins, resin, arctin, arctic acid, arctiol, dehydrofukinone.[99,100]	For rheumatism, gout, and lung disease. It is a laxative, diuretic, and perspiration inducer.
Areca catechu L.	China	Arecholine, arecholidine, guvacoline, guvacine.[33]	Treats taeniasis.
	N.A.	Arecoline, arecaine, arecaidine, arecolidine, isoguvacine, guvacine, givacoline, tannins, palmitic acid, stearic acid, myristic acid, lauric acid, margaric acid, nonadecanoid, heneicosanic acid.[100]	A breath sweetening masticatory; treats abdominal tumor; an astringent, stomachic, stimulant, and anthelmintic.

Table 2   Chinese and North American Medicinal Herbs Belonging to the Same Species: Major Constituents and Therapeutic Values   (continued)

Scientific Name	Source	Major Constituents	Therapeutic Values*
Arisaema consanguineum Mart.	China	Alkaloids, saponin, benzoic acid.[33,49] This herb is highly toxic.	Treats tetanus, spasms, epilepsy, neuralgia. Sedative, anticonvulsive, an expectorant.
	N.A.	Triterpenoid saponins, benzoic acid.[100]	Treats chest problems. Externally, fresh rhizome for skin ulcers.
Artemisia annua L.	China	Dihydroartemisinin, artesunate, artemisinin, chloroquine.[33] This herb is mildly toxic.	A schizonticidal agent, antimalarial.
	N.A.	Abrotamine, artemisinin, vitamin A.[99]	Treats fever, headaches, dizziness, and tight-chested sensation.
Artemisia vulgarts L.	China	Terpinenol-4, β-caryophyllene, artemisia alcohol, linalool, cineol, camphore, borneol, eucalyptol.[33]	Reduces or stop menstrual bleeding. Antiasthmatic, antitussive. Treats chronic bronchitis, oral infection, and hypersensitivity.
	N.A.	Cineole, thujone, ascorbic acid, thiamine, inulin, resin, tannin.[100,102]	Improvess appetite, digestive function, and absorption of nutrients. Antiseptic, a uterine stimulant.
Asarum canadense L.	China	Essential oils including ucarvone, safrole, beta-pinene, asoryl-ketone, asariline.[33]	Analgesic, sedative, antipyretic, anti-inflammatory.
	N.A.	Pinenes, delta-linalool, borneol, terpineol, arislolochic acid.[107]	Treats asthma, sore throats, stomach cramps, recurrence of herpes lesions.
Asparagus officinalis L.	China	Glycolic acid, asparagome, essential oils, methanethiol.[50]	Diuretic, laxative; treats cancer, neuritis, rheumatism, for parasitic diseases.
	N.A.	Asparagosides, asparagine, flavonoids.[99]	Diuretic, for rheumatic conditions, sedative.

Aster tataricus L.	China	Saponins, shionon, quercetin, arabinose.[49]	Antitussive, expectorant.
	N.A.	Coumarins, polyacetylenes, terpenoids, flavonoids, phenylpropanoids, saponins.[185]	A stimulant, expectorant herb for the bronchial system. Treats tuberculosis.
Astragalus membranaceus (Fisch.) Bunge.	China	Gamma-aminobutyric acid, astragalin, canavanine, coumarin, flavonoid derivatives, saponins, polysaccharide, cycloastrangenol, betaine, rhamnocitrin, saponin, astragalosides, formononetin, homoserine, isoliquiritigenin, kaempferol, quereetin, cosin.[1,33,53,510]	Hypotensive, antirhinoviral, antitumor, antipyretic, diuretic, tonic, an immuno-moderating agent.
	N.A.	Asparagine, calcyosin, formononetin, astragalosides, kumatakenin, sterols.[99,511]	An energy tonic, for excessive sweating; relieves fluid retention; immune stimulant; treats uterine bleeding.
Atractylodes macrocephala Koidz.	China	Atractylone, eudesnol, hinesol.[19]	Diuretic agent, abdominal and chest tightness, anemia, chills, bronchial cough, diarrhea, CNS suppressing activity.
	N.A.	Atractylol, lactones atractylenolide II and III.[99]	As a tonic, strengthens the spleen, relieves fluid retention, excessive sweating, diarrhea, and vomiting.
Benincase hispida Cogn.	China	Palmitic acid, stearic acid, linoleic acid, thiamine, riboflavin, niacin, ascorbic acid.[50]	Diuretic, laxative; treats diabetes, dropsy, renitis.
	N.A.	Saponins, guaridine.[99]	Fruit has an anticancerous effect.
Bidens tripartita L.	China	Luteolin, butin, buteine, coumarin, dihydroxycoumarin, scopoletin, umbelliferone.[48]	Treats chronic dysentery, heart ailments, eczema.
	N.A.	Flavonoids, xanthophylls, volatile oil, acetylenes, sterols, and tannins.[99]	As an astringent, diuretic, to treats bladder and kidney problems. Staunches blood flow, for uterine hemorrhage.

Table 2 Chinese and North American Medicinal Herbs Belonging to the Same Species: Major Constituents and Therapeutic Values (continued)

Scientific Name	Source	Major Constituents	Therapeutic Values*
Biota orientalis L.	China	Quercitrin, pinipicrin, thuzone, essential oils.[33]	Hemostatic; shortens blood clotting time. Antitussive.
	N.A.	Alpha-thujone, fenchone, beta-thujone, sabinen, beyerene, bornyl acetate, camphor, borneol, sesquiterpenes, lignans, flavonoids.[185,186] This herb is toxic.	Against amebas, parasites, bacteria, fungi, and viruses.
Blumea balsamifera (L.) DC	China	Essential oils, borneol, camphor, cineole, limonene, palmitic acid, myristic acid, sesquiterpene alcohol, dimethyl ether, cineole, limonene, pyrocatechic tannin.[48,53]	Treats itch, sores, wounds. A stomachic, sudorific, tonic, diaphoretic, anticatarrhal.
	N.A.	Camphor, cinnamon.[105]	Externally for joint and muscle pain. Used as an inhalant for bronchial and nasal congestion.
Brassica alba (L.) Rabenh. *B. juncea* (L.) Czern. et Coss.	China	Sinigrin, myrocin, sinapic acid, sinapine, potassium myronate, mustard oil, allyl isothiocyanate, behenic acid, erucic acid, benzyl isothiocyanate, eicosenic acid.[48,50]	Relieves bladder inflammation, hemorrhage, abscesses, lumbago, rheumatism, stomach disorders.
	N.A.	Mustard-oil glycosides.[147]	Antibiotic effects. A pungent, stimulant; improvess digestion and circulation.
Bupleurum falcatum L.	China	Triterpenoid saponins, sapogenins, saikosaponins.[21,22,33,510]	Relieves tightness; antipyretic, inflammation of inner organs.
	N.A.	Bupleurumol, triterpenoid saponins, flavonoids, saikosides.[99]	A tonic, anti-inflammatory, antiviral; protects liver.

Species	Origin	Constituents	Uses
Calendula officinalis L.	China	Arnidiol, carotin, calenduline, cerylalcohol, flavoxanthin, lycopene, oleanolic acid, inulin, rebixanthin, violaxanthin, tocopherol, salicylic acid.[50]	Treats bleeding gums, bleeding piles; for amenorrhea, bruises, cholera, cramps, eruption, fevers, flu.
	N.A.	Carotenoids, saponins, flavonoids, phytosterols, mucilage, triterpenes, resin.[99,100]	Anti-inflammatory; heals wounds, bed sores, ulcers, and skin rashes.
Camellia sinensis (L.) Kuntze	China	Caffeine, theophylline, tannic acid, theobromine, xanthine.[33,47]	Diuretic effect: increases renal blood flow, stimulates central nervous system.
	N.A.	Methylxanthines, caffeine, purine, polyphenols, ascorbic acid, beta-carotene, thiamine, niacin, theophylline.[100,111]	Antioxidant with stimulating effects.
Cannabis sativa L.	China	Vitamin B_1, vitamin B_2, muscarine, choline, trigonelline, l(d)-isoleucine betaine, cannabinol, tetra-hydrocannobinol, cannabidiol.[33]	Purgative; stimulates intestinal mucosa causing an increase in secretions and peristalsis.
	N.A.	Tetrahydro-cannabinols, thiamine protein. Seeds contain choline, inositol, xylose, phytosterols, trigonelline.[100,102]	Induce euphoria and exhilaration; sedative, antispasmodic.
Capsella bursa-pastoris (L.) Medicus	China	Bursic acid, alkaloids, vitamin A, choline, citric acid.[33]	Hemostatic, antihypertensive, chyluria, nephritis, edema, hematuria.
	N.A.	Amine choline, acetylcholine, bursine, histamine, flavonoids, polypeptides, tyramine.[99,102]	Controls internal bleeding, profuse menstruation.
Carthamus tinctorius L.	China	Cartharmin, neocartharmin, safflowers yellow, quinochalone, safflomin A.[33]	Promotes blood circulation, removes blood stasis, and restores normal menstruation.
	N.A.	Carthamone, lignans, vitamin E, polysaccharides.[112]	Reduces fever by inducing perspiration, it has a laxative effect.

Table 2   Chinese and North American Medicinal Herbs Belonging to the Same Species: Major Constituents and Therapeutic Values (continued)

Scientific Name	Source	Major Constituents	Therapeutic Values*
Carum carvi L.	China	Essential oil, d-carvone, d-limonene, phytosterols.[48,50]	Carminative; treats stomach pain.
	N.A.	Carvone, limonene, flavonoids, polysaccharides.[99,107,113]	Relieves gas pains; antispasmodic and carminative.
Cassia angustifolia Vahl.	China	Fatty acids, aloe-emodin, rhein chrysarobin, chrysophanic acid, oxymethyl anthraquinone.[48,510,511]	Improvess night vision, migraines; astringent, purgative.
	N.A.	Anthraquinone, beta-sitosterol, rhein, dianthrone glucosides, sennosides A, B, naphthalene glycosides, aloe-emodin, mucilage.[99,100,510,511]	Laxative, stimulant, anticancer, cathartic.
Catharanthus roseus (L.) G. Don	China	Vinblastine, vincristine, carosine, vinrosidine, lenrosine, lenrosivine, rovidine, perivine, perividine, vindolinine, pericalline.[33] This herb is toxic.	Anticancer in chronic lymphocytic leukemia and Hodgkin's disease, in acute lymphocytic leukemia.
	N.A.	Alkaloids, tannins, saponins, pectin, oleoresin, aldehydes, dimeric indole alkaloids, vinblastine, sesquiterpenes.[100,114,315]	Treats diabetes, leukemia, reduces blood pressure, Hodgkin's disease, hypotensive, sedative and tranquillizing, anticancer.
Centella asiatica (L.) Urb.	China	Asiaticoside, madecassoside, brahmoside, brahmissoside.[33,510]	Antibacterial, lowers blood pressure; antipyretic, diuretic, detoxicant.
	N.A.	Triterpenoid, saponins, oleic acid, vellarin, hydrocotyline, sitosterol, asiatic, madecassic, madasiatic acids, asiaticoside.[99,100,115,116,511]	Treats skin disease, leprosy; antipyretic, detoxicant, diuretic, antirheumatic, mild diuretic, sedative, and peripheral vasodilator.

Species	Origin	Constituents	Uses
Chaenomeles speciosa (Sweet) Nakai	China	Vitamin C, malic acid, tartaric acid, citric acid, hydrocyanic acid.[49]	Treats arthralgia, diarrhea, cholera, gout, arthritis.
	N.A.	Calcium, iron, magnesium, potassium, dodium.[334]	Anti-inflammatory, antispasmodic, a circulatory and digestive stimulant; treats rheumatism, arthritis, cramps.
Chamaenerion angustifolium (L.) Scop.	China	Crataegulic acid, penta-o-galloyl-β-d-glucose, maslinic acid, chanerol, cerylalcohol.[48]	Regulate menstruation, improves breast milk production. Externally for wounds, stops bleeding.
	N.A.	3-O-β-D-glucuronide, mucilage, tannins, flavones.[103,117]	Treats skin irritation and burns, infused flowers for gargle for sore throat and laryngitis. It is an anti-irritant and used as a mild sunscreen, inhibiting microbial growth.
Chelidonium majus L.	China	Chelidonine, protopine, stylopine, allocryptopine, chelerythrine, sparteine, coptisine.[33]	Anodyne, analgesic, diuretic, antitussive, detoxicant. Treats abdominal pain, peptic ulcers, chronic bronchitis, and whooping cough.
	N.A.	Isoquinoline alkaloids, allocryptopine, berberine, chelidonine, sparteine.[99]	Analgesic, antispasmodic; lowers blood pressure. A mild sedative.
Chenopodium ambrosiodes L.	China	Volatile oil, ascaridol, geraniol, saponin, 1-limonene, p-cymene, d-camphor.[60]	An anthelmintic to treats ascarids, ancylostomiasis; vermifuge, carminative.
	N.A.	Ascaridole, saponins, myrcene, geraniol.[99,100,107]	Anthelmintic.
Chimaphila umbellata (L.) W. Barton	China	Arbutin, ursolic acid, homoarbutin, chimaphilin, isohomoarbutin, hyperin, avicularin, kaempferol, renifolin, beta-amyrin, ericolin, andromedotoxin, chinic acid.[48]	Diuretic, relieves stomach, tooth and after-birth pains; antifungal.
	N.A.	Hydroquinones (arbutin), flavonoids, triterpene, methyl salicylate, tannins.[99]	An astringent, tonic, and diuretic. An infusion for urinary tract problems.

Table 2   Chinese and North American Medicinal Herbs Belonging to the Same Species: Major Constituents and Therapeutic Values   (continued)

Scientific Name	Source	Major Constituents	Therapeutic Values*
Chrysanthemum cinerriaefolium (Trevir.) Vis.	China	Essential oil, adenine, choline, stachydrine.[60]	Used as an insecticide.
	N.A.	Pyrethrins, cinerins, palmitic, linoleic acid, sesquiterpene lactones.[100,107,118]	Externally used as a contact insecticide.
Cimicifuga foetida L.	China	Ferulic acid, isoferulic acid, cimigenol, khellol, aminol, cimifugenol, cimitin.[33]	Induces diaphoresis, promotes skin eruption.
	N.A.	Triterpene glycosides, actein, tannins, cimicifugoside, isoflavones, isoferulic acid, salicylic acid, resin.[99,100,103,119,120]	Promotes menstrual flow, antirheumatic, expectorant, sedative. Treats inflammatory arthritis, high blood pressure, whooping cough, and asthma.
Cimicifuga racemosa (L.) Nutt.	China	Ferulic acid, isoferulic acid, cimigenol, khellol, aminol, cimifugenol, cimitin.[33]	Induces diaphoresis, promotes skin eruption.
	N.A.	Triterpene glycosides, isoflavones, isoferulic acid, resin, salicylates, sterols, methylcytisine, cimicifugin, actein.[100,120]	Treats rheumatism, neuralgia, diarrhea, bronchitis, measles, whooping cough, tuberculosis, high blood pressure, migraine headaches, arthritis; relieves depression and suppresses hot flashes.
Cinnamomum camphora (L.) J. S. Presl.	China	d-camphor, eucalyptole, cineole, pinene, aromadendrene, cumaldehyde, pinocarveol, 1-acetyl-4-isopropylidenecyclopentene.[33,53,510]	Stimulates nervous system, relaxes gastrointestinal muscle contractions.
	N.A.	Camphor, safrole, eugenol, terpineol, lignans.[99,100,511]	Carminative, antispasmodic.

Cinnamomum cassia Presl.	China	Cinnamic aldehyde, cinnamyl acetate, cinnamic acid, eugenol, phellandrene, phenylpropyl alcohol, coumarin, orthomethylcoumaric aldehyde.[33,49]	Antibacterial, vasodilatation, aromatic stomachic, astringent, tonic, analgesic, stimulant.
	N.A.	Camphor, camphene, dipentene, limonene, phyllandrene, pinene, cinnamalcehyde.[99,100,119]	Carminative, antispasmodic, antiseptic, and antiviral.
Cinnamomum zeglanicum Blume	China	Cinnamic aldehyde, p-cymene, hydrocinnamic aldehyde, pinene, benzaldehyde, cuminic aldehyde, nonylic aldehyde, eugenol, caryophyllene, l-phellandrine, methyl-n-amyl ketone, l-linalool.[60]	Stimulant to digestion, respiration, circulation.
	N.A.	Cinnamaldehyde, eugenol, tannins, coumarins, mucilage.	A stimulant, carminative, antispasmodic, antiseptic, and antiviral. It is a sedative, analgesic; reduces blood pressure and fevers.
Cissampelos pareira L.	China	Cissampareine, hayatine, hayatinine, dl-beheerine, dl-curine, D-guereitol, d-isochondrodendrine, hayatidine, cissamine, menisnine.[33]	A blockade of NMJ depolarization. Used externally on wound surfaces to relieves pain.
	N.A.	Cissampeline.[105]	A potent muscle relaxant.
Citrullus vulgaris Schrad.	China	Cucurbitacins, carprylic acid, capric acid, lauric acid, myristic acid, palmitic acid, stearic acid, oleic acid, linoleic acid, citrulline.[50]	For alcoholic poisoning, diabetes, nephritis, sore throat, stomatitis; demulcent.
	N.A.	Citrulline, arginine.[99]	Increases flow of urine and cleanses the kidneys. Treats hepatitis, bronchitis, asthma.

Table 2 Chinese and North American Medicinal Herbs Belonging to the Same Species: Major Constituents and Therapeutic Values (continued)

Scientific Name	Source	Major Constituents	Therapeutic Values*
Citrus aurantium (Christm.) Swingle	China	Synephrine, N-methyltyramine, flavones including tangeratin and nobiletin.[33]	Treats indigestion, relieves abdominal distension, ptosis of the anus or uterus.
	N.A.	Coumarins, bioflavonoids, mucilage, vitamins A, B, C, benzoic acid, cinnamic acid, coumarins, carotenoids.[99,121]	Antiseptic, antirheumatic, antibacterial, antioxidant.
Clerodendrum trichotomum Thunb.	China	Glycosides clerodendrin, acacetin-7-glucurono-(1,2)-glucuronide, clerodendrin, mesoinositol, clerodolone, apigenin-7-diglucuronide, friedelin, epifriedelin, friedelin.[33,48,71]	Treats hypertension, arthritis pain.
	N.A.	Clerodendrin acacetin, mesoinositol.[99]	Lowers blood pressure, eases joint pain, numbness, and paralysis.
Cnidium monnieri (L.) Cusson	China	Archangelicin, columbianetin, O-acetylcolumbianetin, O-isovaleryl columbianetin, cnidiadin, cnidimine, l-pinene, l-camphen.[33]	A trichomonicidal agent, antiascariac, antifungal.
	N.A.	Pinene, camphene, bornyl isovalerate, isoborneol.[99]	Antifungal; treats vaginitis and vaginal discharge.
Codonopsis pilosula (Franch.) Nannfeldt C. *tangshen* Oliv.	China	Taraxeryl acetate, friedelin, n-butyl allophanate, inulin, sucrose, amino acids, stigmasterol, spinasterol, methyl palmitate, taraxerol.[48]	For amnesia, anorexia, asthma, cachexia, cancer, impotence, insomnia, palpitations.
	N.A.	Triterpenoid saponins, sterins, perlolyrin, alkenyl, polysaccharides, alkenyl glycoside, tangshenoside.[99,122]	An adaptogen, stimulant, and tonic.

Commiphora myrrha Engler	China	From gum resin, essential oils including myrcene, alpha-camphorene, Z-guggulsterol, guggulsterol, makulor, cembrene.[33]	Activate blood flow, relieves pain, and promotes tissue regeneration.
	N.A.	Gum, acidic polysaccharides, resin.[99]	A stimulant, antiseptic, astringent, and expectorant. It is anti-inflammatory, antispasmodic, and carminative.
Conyza canadensis (L.) Cronq.	China	Essential oils, mrtaicaria ester, dehydromatricaria ester, linoleyl acetate, limonene, linalool, centaur X, diphenyl methane-2-carboxylic acid, cumulene, O-benzylbezoic acid.[48]	Relieves swelling, itchiness, treats intestine and liver infection; a detoxicant, externally for skin eczema, wounds, pain caused by arthritis, toothache.
	N.A.	Limonene, terpineol, linalool, tannins, flavonoids, terpenes.[99]	For gastrointestinal problems such as diarrhea and dysentery. Treats bleeding hemorrhoids, bladder problems, gonorrhea.
Coptis chinensis Franch. C. *teeta* Wall	China	Berberine, coptisine, urbenine, worenine, palmaline, jatrorrhizine, columbamine, lumicaerulic acid.[33,60] This herb is toxic.	Antiarrhythmic, antibacterial, antiviral, antiprotozoal, anticerebral ischemic.
	N.A.	Isoquinline, berberine, coptisine, worenine.[99]	Antibacterial, amebicidal, and antidiarrheal.
Coriandrum sativum L.	China	Acetone, borneol, coriandrol, cymene, decanal, decanol, decylic aldehyde, dipentene, geraniol, limonene, linalool, malic acid, nonanal, oxalic acid, phellandrene, tannic acid, terpinene, terpinolene.[50]	Eruptions of pox and measles.
	N.A.	Linalool, proteins, vitamin C, alpha-pinene, terpinene.[99,107]	A digestive tonic, carminative, and sedative.
Cornus officinalis Sieb. et Zucc.	China	Morroniside, 7-O-methyl-morroniside, sworoside, loganin, longiceroside, tannic acid, resin, tartaric acid, cornin, gallic acid, malic acid.[33,60]	Diuretic; treats dysmenorrhea, excessive menstruation, impotency, backache, dizziness.
	N.A.	Iridoid glycosides, verbenalin, saponins, tannins.[100,102,123]	Mild effect on the involuntary nervous system, which governs the digestive system.

Table 2 Chinese and North American Medicinal Herbs Belonging to the Same Species: Major Constituents and Therapeutic Values (continued)

Scientific Name	Source	Major Constituents	Therapeutic Values*
Corydalis yanhusuo W.T. Wang ex Z.Y. Su et C.Y. Wu	China	d-corydaline, corydalis, dl-tetrahydropalmatine, crybulbine, alpha-allocryptopine, tetrahydrocoptisine, corydalamine, tetrahydrocolumbamine, protopine, coptisine, dehydrocorydaline, columbamine, dehydrocorydalmine.[33] Overdosage is toxic.	Analgesic, sedative, hypnotic, synergistic; increases coronary flow.
	N.A.	Corydalis, corydaline, leonticine, tetrahydropalmatine, protopine.[99]	Analgesic, antispasmodic, sedative.
Crocus sativus L.	China	Crocetin, crocetin geniobiose glucose ester, crocetin di-glucose ester.[33]	Ameliorating effect on ethanol-induced impairment of learning and memory.
	N.A.	Crocine glycosides, beta-carotene, phytoene, phytofluene, pinene, safranal, cineole.[100,107]	Saffron stomachic, antispasmodic, emmenagogue properties.
Croton tiglium L.	China	Croton resin, phorbol, crotonic acid, crotin, crotonoside.[33] This herb is very toxic.	Purgative.
	N.A.	Croton oil.[105] Oil is carcinogenic, can be fatal.	For constipation, dysentery, biliary colic, intestinal obstructions, food poisoning, malaria, mastitis. Externally for warts, dermatitis, abscesses, boils.
Cryptotaenia canadensis (L.) DC	China	Cryptotaenen, kiganen, kiganol, methyl isobutyl ketone, petroselic acid, isomesityl oxide, *trans*-beta-ocimene, terpinolene.[48,50]	For diarrhea, dysmenorrhea, rheumatism, tubercular glands.
	N.A.	Volatile oils.[105]	A stimulant.

Cryptotaenia japonica Hasskarl	China	Cryptotaenen, kiganen, kiganol, petroselic acid, isomesityl oxide, mesityl oxide, methyl isobutyl ketone, *trans*-beta-ocimene, terpinolene.[48,50]	For diarrhea, dysmenorrhea, rheumatism, tubercular glands.
	N.A.	Apiole, myristicin, pinene, apiin, flavonoids, phthalides, coumarins.[99,100,107]	Diuretic, stomachic, carminative, irritant, and emmenagogue properties.
Cucumis sativus L.	China	Arginine, caffeic acid, chlorogenic acid, cucurbitacins, fructose, galactose, isoquercitrin, mannose, 2,6-ncnadienol, rutin, linoleic acid, oleic acid, palmitic acid, stearic acid.[50]	Diuretic, purgative, vermifuge; pulp can be used for burns, scalds, and skin ailments.
	N.A.	Palmitic acid, stearic acid, linoleic acid, oleic acid.[187,188]	Internally for blemished skin, heat rashes, tapeworm. Externally for sunburn, scalds, sore eyes, and conjunctivitis.
Curcuma aromatica Salisbury	China	Curzerenone, curzenene, furanodiene, furanodienone, zederone, curculone, curcumenol, procurcumenol, curcumadiol, curdione, curcumin, turmerone, zingiberene.[33,510] This herb is toxic.	Inhibits mutagenesis and tumor promotion;anti-inflammatory. antitumor, anti-infectious, anti-HIV.
	N.A.	Curcuminoids, essential oils.[99]	Biliary disorder, anti-inflammatory, sedative.
Curcuma longa L.	China	l-curcamene, sesquiterpene, camphor, camphene, curmarin, curzernone, curzenene, curcumol, furanodienone, furanodiene, zederone, curcolone, curcumadiol, procurcumenol, curdione, curcumin.[33,510]	Anti-inflammatory, antitumor, anti-infectious properties, antioxidative activity, active blood flow; removes blood stasis.
	N.A.	Volatile oil, zingiberen, turmerone, curcumin, resin.[99,511]	Stimulates secretion of bile; antibacterial, anti-inflammatory; relieves stomach pain; antioxidant.

Table 2   Chinese and North American Medicinal Herbs Belonging to the Same Species: Major Constituents and Therapeutic Values   (continued)

Scientific Name	Source	Major Constituents	Therapeutic Values*
Cuscuta chinensis Lam.	China	Cuscutalin, bergenin, cuscutin, amarbelin, cholesterol, campesterol, beta-sitosterol, stigmasterol, beta-amyrin.[48]	Improves immunity, increases blood sugar metabolism.
	N.A.	Flavonoids, hydroxycinnamic acid, bergenin.[102]	Remedy for kidney disorder and liver disease, laxative.
Cymbopogon citratus (DC) Stapf. C. nardus Rendle	China	Elemicin, cymbopogonol, citral, dipentene. Methylheptenone, beta-dihydropseudoionone, linalool, methylheptenol, alpha-terpineol, geraniol, nerol, farnesol, caprylic, citrogellol, citronellal, decanal, farnesal, isovaleric, geranic, citronellic.[50,60]	Treats blood in the urine, fever, antiseptic, preservative.
	N.A.	Citral, citronellal.[100,107,117,124]	Treats digestive problems, relieves cramping pains.
Cyperus rotundus L.	China	Essential oils, alpha-cyperene, beta-cyperene, alpha-cyperol, beta-cyperol, cyperoone, patchoulenone, kobusone, capadiene, epoxyquaine, rotundone, rotunol.[33]	Treats dysmenorrhea, menstrual irregularities.
	N.A.	Fixed oil known as tiger nut oil.[99]	A digestive tonic. Promotes urine production and menstruation.
Cytisus scoparius (L.) Link.	China	Sparteine, sarothamine, genisteine, scoparin.[60]	As a fomentation to bruises, a remedy for coughs, colds.
	N.A.	Sparteine, scoparoside, flavone.[100,125]	Diuretic, cathartic.

Daphne genkwa Sieb. et Zucc.	China	Genkwanin, yuanhuacine, apigenin, hydroxygenkwanin, yuanhuafine, yuanhuadine, 12-benzoxydaphnetoxin, yuanhuatine, genkwadaphnin.[33,53] This herb is toxic.	Induces abortion, treats chronic bronchitis, malaria, cutaneous infections.
	N.A.	Daphnetoxin, mezerein, mucilage, tannins.[99] This herb is toxic.	An abortifacient, alterative, carcinogenic, diuretic, purgative, stimulant, sudorific.
Datura innoxia Mill. *D. metel* L. *D. stramonium* L.	China	Scopolamine, hyoscyamine, daturodiol, daturolone.[33] This herb is mildly toxic.	A spasmolytic, analgesic, antiasthmatic, antirheumatic agent. A general anesthetic for major operations.
	N.A.	Tropane alkaloids (hyoscyamine, hyoscine), flavonoids, withanolides, coumarins, tannins.[100]	Treats asthma, coughs, fevers, skin conditions.
Daucus carota L.	China	Carotenes, lycopene, phytofluere, umbelliferone, alpha-pinene, camphene, myrcene, alpha-phellandrene, daucol, bisabolene, luteolin-7-glucoside, citral, daucine, pyrrolidine, geraniol, caroto, citronellol, caryophyllene, p-cymene, asarone, daucosterol, petroselinic acid.[48]	For chronic dysentery, worms; carminative, diuretic, emmenagogue, lowers blood sugar, prevents cancer, diabetes, dyspepsia, and gout.
	N.A.	Thiamine, nicotinic acid, phytin, lipids, carotenes, vitamin B complex, vitamin C.[100]	Anthelmintic, diuretic.
Dictamnus albus L.	China	Dictamine, skimmianine, saponins, preskinnianine, choline, fragarine, aurapten, bergapten, isomaculosindine, limonin, obakinone, fraxinellone, psoralen, trigonelline.[50,60]	Antifungal, antipyretic, antiseptic, antitussive, sedative, emmenagogue, tonic.
	N.A.	Estragol, anethole, dictamnin.[99] This herb is toxic.	Stimulates the muscles of the uterus; antispasmodic.

Table 2   Chinese and North American Medicinal Herbs Belonging to the Same Species: Major Constituents and Therapeutic Values (continued)

Scientific Name	Source	Major Constituents	Therapeutic Values*
Digitalia purpurea L.	China	Digitoxigenin, gitoxigenin, digitonin, gitaloxigenin, digitoxin, gitoxin, gitanin, gitaloxin, digicoside, strospeside, digipurin, digicirin, digifolein, purpureal glycosides.[60]	For gonorrhea, sclerosis of the breast.
	N.A.	Purpurea-glycosides A and B, digoxin, digitoxin, caffeic acid, lanatoside, choline, saponins, chlorogenic acid.[100]	Improving blood circulation to the kidneys, it has a cardiologic effect.
Dioscorea batatus Decaisue	China	Allantoin, arginine, d-abscisin, mannan, phytic acid, diosgenin, protein.[48]	Sore throat, swellings, food poisoning, goiter, hernia, purulent inflammations.
Dioscorea opposita Thunb.	N.A.	Steroidal saponins.[126]	Strengthens a weak digestion, improvess appetite, it has a hormonal effect. It counters excessive sweating, frequent urination, and chronic thirst.
	China	Allantoin, arginine, choline, glutamine, leucine, tyrosine, diosgenin, sinodiosgenin.[50]	Leaf juice for snakebite, root for asthma, cachexia, cough, debility, diarrhea, neurasthenia, polyuria, tuber is anthelmintic.
	N.A.	Steroidal saponins, albuminoides, diosgenin, progestoron, sapogenin.[126]	Hormonal effect;treats vaginal discharge; diuretic and anti-inflammatory properties.
Dodonaea viscosa (L.) Jacquin	China	Alkaloid, glucoside, tannin, resins.[60]	Remedy for fever, astringent to treats eczema.
	N.A.	Tannin.[105]	Internally for fever, externally for pain relief of toothache, sore throat, wounds, and stings.
Drosera rotundifolia L.	China	Citric acid, malic acid.[57]	Treats dysentery, scrofula, and malaria.
	N.A.	Naphthaquinones, enzymes, flavonoids, volatile oil.[99]	Antimicrobial, antispasmodic; relaxes the muscles of the respiratory tract.

Dryobalanops aromatica Gaertn.	China	Borneol, camphene, terpineol, sesquiterpene.[60] This herb is toxic.	A tonic and aphrodisiac, cataracts, and reduces swelling. Externally for mucous membrane of the nose, eyes, throat, and on piles.
	N.A.	Camphor oil, d-borneol.[105]	Internally for fainting, convulsions associated with high fever, cholera, pneumonia. Externally for rheumatism, ringworm, abscesses, boils, cold sores, mouth ulcers.
Dryopteris filix-mas (L.) Schott.	China	Dryocrassin, filicic acids, paraaspidin, deaspidin, albaspidin, oleoresin, filmarone, filicin, flavaspidic acids, filicin, resin albaspidin, diploptene.[50,53,60]	Anthelmintic to treats tapeworm, hemorrhage, hookworm, influenza. Externally to treats leucoderma.
	N.A.	Oleo-resin, triterpenes, alkanes, volatile oil, resins.[99] This herb is toxic.	Treats tapeworms.
Eclipta alba Hassk. *E. prostrata* (L.) L.	China	Alkaloids, nicotine, ecliptine.[60]	Leaves heated or crushed in oil are applied to keep the hair black and to encourage its growth. Astringent, hemostatic, tonic.
	N.A.	Saponins, alpha-terthienylmethanol, ecliptine.[99]	Prevents premature graying of the hair, staunches bleeding especially from uterus.
Elettaria cardamomum Maton.	China	Phytosterol, palmitic acid, oleic acid, linoleic acid, p-cymene, camphene, d-limonene, myrcene, alpha-phellandrene, pinene, sabinene, terpinene, thujene, cineole, camphorm citral, linalol, citronellal, dl-borneol, citronellol, geraniol, terpineol, sabinene.[50]	Carminative, emmenagogue, stimulant, stomachic, tonic. Treats ague, cachexia, dyspepsis, enuresis, gastralgia, nausea, spermatorrhea.
	N.A.	Borneol, camphor, pinene, humulene, caryophyllene, carvone, eucalptole, terpinene, sabinene.[99]	Eases stomach pain; carminative, antispasmodic; and digestive stimulant.

Table 2   Chinese and North American Medicinal Herbs Belonging to the Same Species: Major Constituents and Therapeutic Values   (continued)

Scientific Name	Source	Major Constituents	Therapeutic Values*
Eleutherococcus senticosus (Rupr. ex Maxim.) Maxim.	China	Eleutherosides, beta-sitosterol glucoside, l-sesamen, syringareinol.[7,33]	Central nervous system activating and anti-stress action.
	N.A.	Eleutherosides, lignans, coumarins, phenylproparnoids, isofraxin, pectin, triterpenoid saponins, resins, glycans, polysaccharides.[99,100,127,128]	An adaptogen, tonic, stimulant; protects the immune system.
Entada phaseoloides (L.) Merrill.	China	Entageric acid.[33]	Antirheumatic; promotes collateral flow, relieves blood stasis.
	N.A.	Saponins.[99]	Treats female sterility, indigestion, and as a painkiller.
Ephedra distachya L.	China	l-ephedrine, l-methylephedrine, l-norephedrine, methylephedrine, d-pseudoephedrinem, d-N-methylpseudoephedrine.[30,31,33] This herb is toxic.	Treats asthma, sympathomimetic action, relieves headache, body ache, and coughing, lowers fever by increasing perspiration.
	N.A.	Alkaloids, ephedrine, l-ephedrine, d-pseudoephedrine.[106,129,511]	Treats fevers, relieves kidney pain, asthma, nose and lung congestions, hay fever, and as a hypertensive aid.
Ephedra sinica Stapf.	China	l-ephedrine, l-methylephedrine, l-norephedrine, methylephedrine, d-pseudoephedrinem, d-N-methylpseudoephedrine.[30,31,33,510]	Treats asthma, sympathomimetic action; relieves headache, body ache, and coughing, lowers fever by increasing perspiration.
	N.A.	Protoalkaloids, tannins, saponin, flavone, volatile oil, ephedrine, l-ephedrine, d-pseudoephedrine.[99,106,511]	Increases sweating, dilates bronchioles, stimulant, diuretic, and raises blood pressure.

Equisetum arvense L.	China	Equisetonin, equisetrin, articulain, isoquereitrin, galuteolin, populnin, kaempferol-3,7-diglucoside, astragalin, palustrine, gossypitrin, herbacetrin, 3-methoxypyridine.[48]	Antihemorrhagic, anodyne, carminative, diaphoretic, diuretic.
	N.A.	Silicic acid, trace of nicotine, equisitine, silicates.[99,100,102]	Treats bleeding wounds; antibiotic, for oral infection, antidiaphoretic.
Equisetum hyemale L.	China	Equisetonin, equisetrin, articulain, isoquereitrin, galuteolin, populnin, kaempferol-3,7-diglucoside, astragalin, palustrine, gossypitrin, herbacetrin, 3-methoxypyridine.[48]	Antihemorrhagic, anodyne, carminative, diaphoretic, diuretic.
	N.A.	Silicic acid, silicates, flavonoids, phenolic acid, nicotine, sterols.[100]	Regeneration of connective tissue, clotting agent; astringent effect on genitourinary system.
Erigeron canadensis L.	China	Essential oils, erigeron, tannic acid, limonene, dipentene, methylacetic acid, terpeneol, lacnophyllum, matricaria, dehydromatricaria, gallic acids, hexahydromatricaria.[50]	For hemorrhage, diarrhea, dysentery, internal hemorrhage of typhoid fever.
	N.A.	Limonene, terpineol, linalool, tannins, flavonoids, terpenes.[99]	As an astringent, for gastrointestinal problems, bleeding hemorrhoids.
Eriobotrya japonica Linkdl.	China	Levulose, sucrose, malic acid, citric acid, tartaric acid, succinin acid, amygdalin, crytoxanthin, carotenes, phenyl ethyl alcohol pentosans, essential oils.[50]	Antitussive, expectorant; treats bronchitis, cough, fever, nausea, externally applied to epistaxis, smallpox, ulcers.
	N.A.	Volatile oil, flavonoids, resin.	Treats tracheitis, bronchitis, and asthma.
Erythroxylum coca Lam.	China	l-cocaire, cinnamylococaine, alpha-trevilline, beta-trevilline, ecgonine, benzoylecgonine.[33]	For local anesthetic, has a vasoconstriction effect.
	N.A.	Cocaine, nicotine, benzoylecgonine, cinnamylcocaine, ecgonine, methyl salicylate.[100]	An esthetic, aphrodisiac, stimulant.

Table 2   Chinese and North American Medicinal Herbs Belonging to the Same Species: Major Constituents and Therapeutic Values (continued)

Scientific Name	Source	Major Constituents	Therapeutic Values*
Eugenia caryophyllata (L.) Thunb.	China	Esssential oils, eugenol, humulene, acetyleugenol, chavicol, alpha-caryophylline, beta-caryophylline, ylangene. Flowers bud: Rhamnetin, kaempferol, oleanolic acid, eugenitin, isoeugenitin. Bark: ellagic acid, beta-sitosterol, mairin.[33]	For nausea, vomiting, hiccups, stomach chills, impotence, therapeutic, antiherpes simplex virus.
	N.A.	Sesquiterpenes, eugenol, tannins, gum.[130,131]	For gastroenteritis, intestinal parasites. Externally for toothache and insect bites.
Euphorbia hirta L.	China	Camphol, leucocyanidol, quercitol, quercitrin, rhamnose, euphorbon, gallic acid, chlorophenolic acid, taraxerol, taraxerone.[50]	For asthma, bronchitis, externally for athlete's foot.
	N.A.	Flavonoids, terpenoids, alkanes, phenolic acids, shikimic acid, choline.[99]	For bronchial asthma, mildly sedative; treats intestinal amebiasis.
Euphorbia lathyrus L.	China	Euphorbiasteroid, betulin, 7-hydroxylathyrol, lathyrol diacetate benzoate, lathyrol diacetate nicotinate, euphol, euphorbol, euphorbetin, esculetin, daphnetin.[33,53]	Diuretic to remove edema; eliminates blood stasis and resolves masses; antitumor.
	N.A.	Fixed oil, resin, euphorbone.[99,100] This herb is toxic.	Depilatory, removes corns.
Fagopyrum esculentum Moench	China	Rutin, quercetin, caffeic acid, orientin, homoorientin, vitexin, saponaretin, cyanidin, leucoanthocyanin. Seeds contain amylase, linamarase, maltase, phosphatides, protease, quercitol, rhamnose, urease.[48,50]	For colic and diarrhea, stops cold sweats.
	N.A.	Bioflavonoids (rutin).[99]	Antioxidant; strengthens the inner lining of blood vessels.

Ferula assa-foetida L.	China	Vanillin, asarensinotannol, ferulic acid, farnesiferols.[33]	Anthelmintic; treats ascites, dysentery, malaria.
	N.A.	Disulphides, resin, gum, sesquiterpenoid coumarins, foetidin.[99]	An expectorant, for digestive problems, bronchitis, bronchial asthma, whooping cough; lowers blood pressure and thins the blood.
Ficus carica L.	China	Bergaptin, cerotinic acid, ficusin, glutamine, papain, pepsin, psoralen, guaiaxulene, amyrin, lupeol, octacosane, guaiacol, quercitin, rhamnose, rutin, sitosterol, tyrosine, urease.[50,55]	For stomachache, externally for swollen piles, corns, warts. Fruit is laxative, digestive.
	N.A.	Glucose, flavonoids, vitamins, enzymes.[99]	A gentle laxative effect; treats tumors, swellings, and gum abscesses.
Foeniculum vulgare Mill.	China	Anethol, d-fenchone, anisaldehyde, methylchavicol.[33]	Restores normal functioning of the stomach.
	N.A.	Anethole, fenchone.[100,107]	Antispasmodic.
Forsythia suspensa (Thunb.) Vahl.	China	Phillyrin, rutin.[50]	Febrifuge, for cancer, carbuncle, chickenpox, antiphlogistic, diuretic, emmenagogue, laxative, antipyretic.
	N.A.	Forsythin, vitamin P.[132]	Antiseptic, remedy for colds, flu, sore throats, and tonsillitis.
Fraxinus ornus L.	China	Fraxin, aesculin.[33] This herb is toxic.	Antibacterial, analgesic, anti-inflammatory.
	N.A.	Coumarins, flavonoids, tannins, volatile oil.[99]	A laxative for children and pregnant women.
Fritillaria verticillata Willd.	China	Fritilline, fritillarine, verticine, verticinine, peimine, peimirine, peimisine, peiniphine, peimidine, peimidine, propeimin, verticine, verticinine.[33]	Causes bronchodilatation and inhibition of mucosal secretions. Antitussive; stimulates uterine and intestinal contractions.
	N.A.	Alkaloids, peimine.[99]	Affects the parasympathetic nervous system.

Table 2 Chinese and North American Medicinal Herbs Belonging to the Same Species: Major Constituents and Therapeutic Values (continued)

Scientific Name	Source	Major Constituents	Therapeutic Values*
Galium verum L.	China	Alisarin, rubrierythrinic acid, purpurin.[60]	Treats rheumatism, jaundice, menstrual difficulties, epistaxis, hemorrhages.
	N.A.	Asperuloside, flavonoids, alkanes, anthraquinones.[107]	A diuretic, for skin problems.
Gardenia angusta (L.) Merr.	China	Gardenin, alpha-crocetin, chlorogenin, volatile oil, mannit, glycosides.[64]	Emetic, stimulant, febrifuge, diuretic, hemostatic, antihemorrhagic, emmenagogue.
	N.A.	Volatile oil, gardenin crocin, geniposide.[99]	For fever, irritability and restlessness, insomnia, urination, and jaundice. Treats cystitis, headaches, difficulty in breathing.
Gelsemium sempervirens (L.) Ait.	China	Gelsemine, gelsemidine, koumine, sempervirine, kouminine, kouminicine, douminidine.[33,46,50] This herb is highly toxic.	For caked breast, perspiring feet, skin eruptions, wounds.
	N.A.	Gelsemine, gelsedine, iridoids, coumarins, tannins.[99] This herb is toxic.	A sedative, antispasmodic; treats neuralgia, facial nerve pain. Externally treats intercostal neuralgia.
Gentiana lutea L. *G. macrophylla* Pall.	China	Gentianine, gentianidine, gentianol.[33]	Treats rheumatism and fever; antipyretic, anti-inflammatory; antihypersensitivity and antihistaminic effects.
	N.A.	Gentianine, gentianindine.[99]	Stimulates digestion.
Gentiana scabra Bunge.	China	Entiopicrin (or gentiopicroside), saponins, geniposide, gardenoside, gentianine.[16,17,33]	For arthritis, cancer, carbuncle, cold, conjuctivitis, diarrhea, gastritis, neuralgia.
	N.A.	Gentianine, gentianindine.[99]	Stimulates digestion.

Species	Source	Constituents	Uses
Geum aleppicum Jacq.	China	Flavones, fatty acids, eugenol, gein, geoside.[48]	Treats bleeding, bug bite, convulsive disorder, fevers, irritability, obstinate skin diseases.
	N.A.	Phenolic glycosides, eugenol, tannins, sesquiterpene lactone.[99]	Treats fever, stomach and intestinal complaints, diarrhea, and reduces bleeding, inflammation, and hemorrhoids.
Ginkgo biloba L.	China	Kaempterol-3-rhamnoglucoside, gibberellin, cytokinin, ginkgolic acid, ginkgol, bilobal, ginnol, ginkgolides, querretin, quercitrin, ginkgetin, rutin, isoginketine, bilobetin, isorhametin, shikimic acid, D-glucaric acid, anacardic acid.[33,48,510,511]	Antitussive, antiasthmatic, anodyne, treats coronary artery disease, angina pectoris, hypercholesterolemia, Parkinson's disease.
	N.A.	Ginkgoside A, B, C, J, and M, flavonoids, bilobalide, sciadopitysin, ginkgetin, isoginkgetin, bilobetin, carotenoids, 4'-0-methylpyridoxine.[133,134,135,136,137,311,510,511]	Treats dementia and cerebral insufficiency, relieves asthma, and treats cerebral disorders.
Glechoma hederacea L.	China	l-pinocamphone, l-menthone, 1,8-cineol, isomenthone, l-pulegone, alpha-pinene, beta-pinene, isopinocamphone, limonene, menthol, alpha-terpineol, linalool, p-cymene.[48]	Febrifuge, anodyne, treats earache, fever, toothache, diuretic, decoagulant; treats arthritis.
	N.A.	Glechomine, tannins, flavonoids, resins, saponins, sesquiterpene.[99]	For mucous (respiratory) problems, glue ear, lung congestion, urine retention.
Glycine max (L.) Merr.	China	Protein, isoflavone derivatives, genistein, daidzein, riboflavin, thiamin, niacin, pantothenic acid, choline.[33,67]	Phytoestrogenic; elevates the vasomotor system, prevents cancer; a potent inhibitor of protein tyrosine kinase.
	N.A.	Lecithin, globuline, glycine, mineral, daidzine astrogen, caffeic acid, choline, coumestrol, tocopherol, saponins, phytic acid, isoflavones, protein, fatty acid, vitamins, carbohydrates, and fiber.[130,138]	Prevents arteriosclerosis and coronary heart disease; an astringent; treats hypercholesterol, a starting source of stigma sterol.

Table 2   Chinese and North American Medicinal Herbs Belonging to the Same Species: Major Constituents and Therapeutic Values   (continued)

Scientific Name	Source	Major Constituents	Therapeutic Values*
Glycyrrhiza uralensis Fisch. ex DC	China	Glycyrrhiza, triterpenoid saponin, flavonone glucoside, liquiritin, aglycone, liquiritigenin, chalcone glucose, isoliquiritin, aglycone, isoliquiritigenen.[1,33,510,511]	Anti-inflammatory, anticonvulsant, calminative, antidote. Antispasmodic, antiulcer.
	N.A.	Triterpene saponins, chalcones flavonoids, isoflavonoids.[99,312,511]	Sweet-tasting tonic; treats sore throats, wheezing, coughs, canker sores, peptic ulcer, and gastritis.
Gnaphalium uliginosum L.	China	Fat, resin, phytosterol, essential oils, carotene, vitamin B₁.[48,49,50]	Remedy for lung disease; antifebrile, antimalarial, reduces blood pressure and stomach and intestinal ulcers. Externally for wounds.
	N.A.	Volatile oil, tannins.[99]	An astringent, antiseptic, decongestant.
Gossypium herbaceum L.	China	Gossypol, hemigossypol, 6,6'-dimethoxylgossypol, aflatoxin B (in seed), methoxylhemigosipol, acetovanillone, hirsutrin (in leaf).[33]	Antitussive; treats bronchitis.
	N.A.	Gossypol, flavonoids.[99] This herb is toxic.	As a labor-inducing agent, promotes abortion or onset of menstruation. Gossypol causes infertility in men.
Hibiscus rosa-sinensis L.	China	Protein, thiamine, riboflavin, niacin, cyandidin-3-sophoroside.[50]	Used as poultice on cancerous swellings and mumps.
	N.A.	Mucilage, citric, malic, tartaric acids, hibiscus acid, thiamine, gossypetin, anthocyanin, myristic acid, palmitic acid.[100,107]	Soothing effect on mucous membranes that line the respiratory and digestive tracts. Seeds used for cramps, flowers as an astringent.
Hibiscus sabdariffa L.	China	Saponin, saponaretin, vitexin.[50]	Stomachic, diuretic, expectorant, hematochezia, gas, vertigo.

	N.A.	Mucilage, citric, malic, tartaric acids, hibiscus acid, thiamine, gossypetin, anthocyanin, myristic acid, palmitic acid.[100,107]	Soothing effect on mucous membranes that line the respiratory and digestive tracts. Seeds used for cramps, flowers as an astringent.
Hierochloe odorata (L.) Beauv.	China	Coumarin, coumarinic acid-β-glucoside.[48]	Relieves internal bleeding, kidney infection.
	N.A.	Coumarin, massoilactone, lactone.[103]	Treats cough, sore throat, venereal infection, bleeding after childbirth, chapped or wind-burned skin, and eye irritations.
Hippophae rhamnoides L.	China	Cryptoxanthin, harman, harmol, hernin, isorhamnetin, lycopene, serotonin, isorhamnetin-3-mono-beta-D-glucoside, fatty acids, flavonoid, essential oils, tannins, quercitin, vitamin C, vitamin E, beta-carotenoid.[50]	Improves resistance to infection, skin irritation and eruption; treats heart disease, oil for cosmetic use.
	N.A.	Carotenoid, flavonoid, essential oil, fatty acids, tannins, quercitin, provitamin A, vitamins C, B complex, and E.[102,139,140141]	Improves resistance to infection, skin irritation, and eruptions. Treats heart conditions, good source of vitamins C and E.
Hordeum vulgare L.	China	Enzymes such as invertase, amylase, proteinase, vitamin B, vitamin C, maltose, dextrose.[33]	Improves digestion of carbohydrates and protein.
	N.A.	Hordenine, gramme.[99]	For minor infections of diarrhea; treats fever.
Humulus lupulus L.	China	Humulone, resin, lupulone, isohumulone, isovaleric acid.[33] This herb is toxic.	Inhibits the growth of tubercle bacillus and arrests tuberculosis.
	N.A.	Humulone, lupulone, humulene, alpha, beta-acids, polyphenols, steroids, resins, tannins.[103,142,143]	Sedative effect, hypnotic, stomachic, diuretic. Against gram-positive organisms and tuberculosis.

Table 2 Chinese and North American Medicinal Herbs Belonging to the Same Species: Major Constituents and Therapeutic Values (continued)

Scientific Name	Source	Major Constituents	Therapeutic Values*
Hyoscyamus niger L.	China	Alkaloid.[60] This herb is mildly toxic.	Antispasmodic activity.
	N.A.	Tropane alkaloids, hyoscyamine, hyoscine.[105] Overdose can be toxic.	For asthma, whooping cough, motion sickness. Externally for neuralgia and dental and rheumatic pain.
Hypericum perforatum L.	China	Quercetin, quercitrin, isoquercitrin.[33,53]	Antipyretic, antibacterial, detoxicant effect; treats acute icteric hepatitis, lowers blood pressure, dysmenorrhea, gonorrhea, skin ailments.
	N.A.	Hypericin, hyperoside, rutin, quercitin, chlorogenic acid, pseudohypericin, flavonoids.[99,100,102]	Antidepressant, anti-inflammatory, diuretic, antiseptic, and astringent properties.
Illicium verum Hook f.	China	Anethol, anisaldehyde, safrole, anisic ketone.[33]	Warming the viscera, expelling cold; relieves pain.
	N.A.	Anethole, methyl chavicol, safrole.[100]	Antibacterial, stimulant, diuretic and digestive properties, for rheumatism, back pain, hernias.
Impatiens balsamina L.	China	Gentisic acid, ferulic acid, p-coumaric acid, sinapic acid, caffeic acid, scopoletin, lawsone.[33]	Treats arthritis, relieves pain.
	N.A.	Balsaminones, 2-methoxy-1, 4-naphthoquinone, saponins, quercitin, kaempferol derivatives, balsaminasterol, parinaric acid, hosenkosides.[302]	Remedy for rashes, pain caused by insect bites, anti-inflammation.
Inula britannica L. I. japonica Thunb.	China	Inusterol A, taraxasterol, inusterol B, inulicin, flavone, caffeic acid, chlorogenic acid, isoquercitrin, quercetin.[48,50]	Discutient, vulnerary, carminative, deobstruent, diuretic; treats ascites, bronchitis, cancer, chest congestion.
	N.A.	Volatile oil, flavonoids, phenolic acids, triterpenes, taraxasterol.[99]	An expectorant. For bronchitis, wheezing, chronic coughing, chest complaints.

Species	Source	Constituents	Uses
Isatis tinctoria L.	China	Quercetin, kaempferol, stachyose, manneotetrose, lupeose, cicerose, isatan, indoxyl-5-ketogluconate.[50]	Antiviral, antibacterial; increase blood flow, improves microcirculation, and lowers blood pressure.
	N.A.	No information is available in the literature.	For meningitis, encephalitis, mumps, influenza, erysipelas, heat rash, sore throat.[335]
Jatropha gospiifolia L.	China	Phytotoxin, curcin, curcasin, arachidic, linoleic acid, myristic acid, oleic acid, palmitic acid, stearic acid.[50] This herb (seed) is toxic.	Seed oil emetic, laxative, purgative; treats skin ailments.
	N.A.	Jutrophine, emetic, purgative oil, diterpene jatrophone, isovitexin, resins, isophytosterol, tannin, cyanidin, apigenin, histamine.[145,146] This herb is toxic.	A folk remedy for cancer. Treats asthma, constipation, diabetes, diarrhea. It is a disinfectant, laxative. Externally applied to piles and burns.
Juglans regia L.	China	Alpha-hydrojuglone-4-β-D-glucoside, jugone, juglanin.[33]	Nourishes and invigorates the lungs and kidneys.
	N.A.	Tannin, juglandin, juglone, hydrojuglone.[147]	Astringent, hemostatic, anti-inflimmatory, antispasmodic, antiphlogistic, and mild sedative.
Juniperus rigida Sieb. et Zucc.	China	Alpha-pinene, myrcene, carene, limonene, p-cymene, beta-elemene, caryophyllene, humulene, r-cadinene, terpinene, borneol, citronellol, anethole.[48]	Hemorrhage; treats hemoptysis, inflammation, kidney infection, arthritic joint infection.
	N.A.	Myrcene, sabinene, alpha-pinene, beta-pinene, cineole, tinnins, diterpenes, resin.[99] This herb is potentially toxic.	A tonic, diuretic, antiseptic, for cystitis; relieves fluid retention.
Kaempferia galanga L.	China	Borneol, camphor, cineol, ethyl alcohol.[49]	Stomachic, carminative, stimulant.
	N.A.	n-pentadecane, ethyl cinnamate, ethyl-p-methoxycinnamate, carene, camphene, borneol, p-methoxystyrene.[100]	Carminative, diuretic, expectorant, pectoral, stimulant.

Table 2 Chinese and North American Medicinal Herbs Belonging to the Same Species: Major Constituents and Therapeutic Values (continued)

Scientific Name	Source	Major Constituents	Therapeutic Values*
Lawsonia inermis L.	China	Alpha-ionone, beta-ionone, gallic acid, lawsone.[50]	Antibiotic, antitumor, anthelmintic, astringent, bactericidal, fungicidal, sedative.
	N.A.	Coumarins, naphthaquinones, lawsone, flavonoids, sterols, tannins.[99]	As a gargle for sore throats, treats diarrhea, dysentery. An astringent; prevents hemorrhaging.
Ledum palustre L.	China	Alpha-pinene, camphene, sabinene, myrcene, alpha-phellandrene, beta-pinene, limonene, quinene, isothujene, ascaridol, arbutin ericolin.[48]	Treats cough, asthma; lowers blood pressure; antifungal.
	N.A.	Coumarins, naphthaquinones, lawsone, flavonoids, sterols, tannins.[100]	A gargle for sore throat, for diarrhea, dysentery. Prevents hemorrhaging, promotes menstrual flow.
Lemna minor L.	China	Luteolin-7-beta-D-glucopyranoside.[50]	For circulation, measles, swollen feet; depurative, diuretic, soporific.
	N.A.	Arginine, lysine, iron manganese.[102]	For fever, skin disease, rash, and water retention.
Lepidium virginicum L.	China	Isothiocyanates.[50]	Antibacterial, cardiotonic.
	N.A.	Vitamin C.[99]	Treats poison ivy symptoms, vitamin C deficiency, diabetes, expels intestinal worms.
Ligustrum lucidum Mill.	China	Nuzhenide, oleanolic acid, ursolic acid.[33]	Increases leukocyte count; a cardiac tonic, diuretic.
	N.A.	Essential oil, phthalides, terpenoids.[100]	prevents bone marrow loss, treats acquired immune deficiency syndrome, respiratory tract infections, hypertension, Parkinson's disease, and hepatitis.
Linaria vulgaris Miller	China	Peganine, linarin, pectolinarin, neolinarin, flavons, pectolinarigenin, linaracrine, linarezine, phytosterine.[48]	Diuretic; treats headache, dizziness, heart conditions. Externally treats burns, skin diseases.
	N.A.	Linarin, sterols, sugars, tannins, mucilage.[99]	Treats jaundice, chronic constipation, skin disease.

Linum usitatissimum L.	China	Fatty acids, geranylgeraniol, cholesterol, campesterol, orientin, stigmasterol, avenasterol, vitexin cycloartenol, eikosanol, leucine, valine, linamarin, lotaustralin.[48]	For diarrhea, sensitive skin, itchiness, loss of hair.
	N.A.	Linseed oil, linoleic acid, linolenic acid, stearic acid, oleic acid, mucilage, linamarin.[99]	Relieves constipation; demulcent, laxative. Externally as a poultice for boils, burns.
Lithospermum erythrorhizon Sieb. et Zucc. *L. officinale* L.	China	Quinonoid, alkannan, acetylshikonin, shikonin, lithospermin, dihydroshikonin, cycloshikonin.[1,69]	Ointment to treat wounds and burns; antitumor, antipyretic; regulate blood circulation; diuretic, purgative, remedy for smallpox.
	N.A.	Lithospermic acid.[100]	Used as a form of birth control, prevents gonadotrophin from stimulating ovaries in lab mice.
Lobelia chinensis L.	China	Lobeline, lobelanine, lobelanidine, isolobelamine (lobeline has been approved by the FDA to curb the tobacco habit).[33,50,71] This herb may be toxic.	Diuretic, increases respiration via stimulation of carotid chemoreceptors. Treats snakebites, insecticide; reduces swellings; depurative, antirheumatic, antisyphilitic.
	N.A.	Stictic acid, stictinic acid, fatty acids, mucilage, tannins.[99]	Expectorant, tonic. For congested mucus, increases appetite.
Lonicera japonica Thunb.	China	Luteolin, inositol, lonicerin, loganin, syringin, saponins, tannin, chlorogenic acid, luteolin-7-rhamnoglucoside.[33,48,55]	Inhibits tuberculosis bacillus and counters infection.
	N.A.	Volatile oil, tannins, salicylic acid.[99,102]	Diuretic, antispasmodic; relieves gout, kidney stones, coughs; as a gargle for sore throats, canker sores.
Lophanthus rugosus Fisch. et May	China	Essential oils.[49]	Carminative, stomachic.
	N.A.	Volatile oils.[105]	Antibacterial, stimulates the digestive system, relaxes spasm; and lowers fever.

Table 2 Chinese and North American Medicinal Herbs Belonging to the Same Species: Major Constituents and Therapeutic Values (continued)

Scientific Name	Source	Major Constituents	Therapeutic Values*
Luffa aegyptiaca Mill. L. cylindrica Roem.	China	Xylose, mannosan, galactan, saponins, acetic acid, valeric acid, pinenes, limonene, cineole, menthone, linalool, bourbonene, caryophyllene, menthol, carvone, vitamins A, B, C.[49,50]	Hemostatic, analgesic in enterorrhagia, dysentery, metrorrhagia, orchitis, hemorrhoids.
	N.A.	Xylan, xylose, galactan.[99]	Treats pain in the muscles, joints, chest, and abdomen.
Lycium barbarum L.	China	Betaine, zeaxanthin, physalein, carotine, nicotinic acid, vitamin C.[33]	Increases leukocyte count, increase immunity, stimulates tissue development.
	N.A.	Betaine, beta-sitosterol. Berry has physalin, carotene, vitamins B_1, B_{12}, C. Root has cinnamic acid, psyllic acid.[99]	Berry: treats high blood pressure; a tonic to protect liver, menopausal complaints. Root: treats chronic fevers, lowers blood pressure, internal hemorrhage, tuberculosis.
Lycium chinense Miller	China	Cinnamic acid, betaine, peptides, acyclic diterpene glycosides, polysaccharide, kukoamines.[33]	Lowers blood sugar and blood pressure, antipyretic, stimulates uterine contractions; antibacterial.
	N.A.	Betaine, beta-sitosterol.[99]	Treats high blood pressure, menopausal complaints.
Lycopersicon esculentum L.	China	Protein, vitamin A, thiamine, nicotinic acid, riboflavin.[50]	Relieves toothache; insecticide, laxative.
	N.A.	Carotene, thiamine, nicotinic acid, riboflavin, folic acid, pantothenic acid, biotin, glutamic acid, serine, glycine, aminobutyric acid, globulin, amino acids.[100]	An antiseptic, aperient, depurative, digestive, pectoral, a folk remedy for asthma.
Lycopodium annotinum L.	China	Clavatine, lycopodine, complanatine, alpha-obscurine, lycopodine, serratenediol, tohogenol.[48]	Relieves numb feeling, arthritis pain, sexually transmitted disease.

	N.A.	Lycopodine, polyphenols, flavonoids, triterpenes.[99]	Diuretic, sedative, antispasmodic. Treats chronic urinary complaints.
Lycopodium clavatum L. *L. obscurum* L.	China	Lycopodine, lycodoline, clavatine, fawcetine, clavolonine, fawcetimine, deacetylfawcetine, clavatoxine, nicotine, vanillic acid, ferulic acid, azelaic acid, alpha-onocerin, lycoclavanol, lycoclavanin, lycopodine.[33,48]	Relieves the rigidity of muscles and joints, treats arthritis and dysmenorrhea.
	N.A.	Lycopodine, dihydrolycopodine, resins, myristic acid, polyphenols, flavonoids, triterpenes.[99]	A diuretic for kidney and bladder complaints.
Lythrum salicaria L.	China	Tannin, salicarin, chlorogenic acid, cyanidin-3-monogalactoside, ellagic acid, malvidin, malvin, orientin, vitexin.[50,72]	Astringent, styptic; treats bacillary dysentery.
	N.A.	Tannin, triacylglycerols, salicarin, vitexin.[99]	Lowers serum cholesterol, glucose, and triglyceride levels, and antiatherosclerotic action. Relieves diarrhea; gargle for sore throat; cleans wounds.
Magnolia liliflora Desr.	China	Flower: eugenol, safrole, citrol, anethol. Leaf: salicifoline, magnocurarine.[33] Essential oils, citral, safrole, anethole, estragole, cineol, eugenol.[49]	Relieves nasal congestion, sinusitis, rhinitis, coryza, headache, vertigo.
	N.A.	Volatile oil, magnocurarine.[99]	Relieves cramping pain and flatulence, for abdominal distension, indigestion, loss of appetite, vomiting, diarrhea.
Magnolia officinalis Rehd. et Wils.	China	Tannin, salicarin, chlorogenic acid, cyanidin-3-monogalactoside, ellagic acid, malvidin, malvin, orientin, vitexin.[50,72]	Astringent, styptic; treats bacillary dysentery.
	N.A.	Volatile oil, magnocurarine.[99]	Relieves cramping pain and flatulence, for abdominal distension, indigestion, loss of appetite, vomiting, diarrhea.

Table 2 Chinese and North American Medicinal Herbs Belonging to the Same Species: Major Constituents and Therapeutic Values (continued)

Scientific Name	Source	Major Constituents	Therapeutic Values*
Manihot esculenta Crantz.	China	Hydrocyanic acid.[76] This herb is toxic.	To dress ulcerous sores.
	N.A.	Cyanogenic glycosides.[99]	Treats scabies, diarrhea, dysentery.
Matricaria chamomilla L.	China	Volatile oil, azulene, isoamyl, isobutyl, angelic acid, tiglic acid, anthelmic acid, tannin, malic acid.[77]	Carminative, diaphoretic.
	N.A.	Flavonoid, glycosides, tannins, luteolin, n-coumaric acid, herniarin, cynaroside, umbelliferone, alpha-bisabolol, azulene, anthemidin, luteolin, coumarins.[99,100,107]	Antispasmodic for relieving cramps, nervous digestive upsets, insomnia; antiallergenic.
Matteuccia struthiopteris (L.) Todaro.	China	Ponasterone A, ecdysterone, pterosterone, filicin.[48]	Tonic; lowers blood pressure.
	N.A.	Palmitic acid, astragalin, caffeic acid, chlorogenic, p-coumaric, oleoresins, p-hydroxybenzoic, vanillic, stigmasterol, protocatechuic, beta-sitosterol, ferulic, campesterol.[148]	Expels parasites, treats inflammation of lymphatic glands.
Medicago sativa L.	China	Lucernol, sativol, coumesterol, formonetin, daidzein, tricin, citrulline, canaline, dicoumarol, methylene-bishydroxy-coumarin, medicagemic acid, ononitol, petunidin, myrcene, malvidin, delphinidin, linalool, limonene.[48]	Depurative, deobstruent, diuretic, stomachic; treats intestinal and kidney disorders, kidney stones, poor night vision.
	N.A.	Isoflavones, coumarins, alkaloids, vitamins, porphyrins, stachydrine, l-homostarchydrine.[100,102]	For menstruation and menopause.

Species	Region	Constituents	Uses
Melaleuca leucadendra L.	China	Cajupputol, terpinol, l-pinene, aldehydes.[189]	Against rheumatism and pain in the joints.
	N.A.	Terpenoids, cineole, beta-pinene, alpha-terpineol.[99]	Antiseptic, treats cold, sore throats, coughs, chest infections.
Melia azedarach L.	China	Toosendanin, nimbin, kulinone, methylkulonate, melianol, gedunin, melianodiol, melianotriol, melialactone, azadarachtin, nimbolins, fraxinella, palmitic acid, lauric acid, valerianic acid, butyric aicd, stearic acid, cycloencalenol.[33,49] This herb is toxic.	Treats intestinal parasite; antibacterial, anthelmintic.
	N.A.	Meliacins, triterpenoid bitters, tannins, flavonoids.[100]	For hemorrhoids, malaria, peptic ulcers, intestinal worms. Antifungal, antiviral, anti-inflammatory, antibacterial.
Melilotus alba Medik.	China	Hydroxycinnamic acid, coumarinic acid, 4-hydroxycinnamic acid, cumaric acid, umbelliferone, scopoletin, mellilotoside, mellilotic acid, beta-D-glucosyloxy, dicumarol, chlogogenic acid, caffeic acid, melilotic acid.[48]	Anticoagulant; treats bowel complaints, infantile diarrhea. A bactericide.
	N.A.	Flavonoids, coumarins, resin, tannins, volatile oil, dicoumarol.[102]	Relieves varicose veins and hemorrhoids, reduces the rash of phlebitis and thrombosis.
Mentha arvensis L. M. haplocalyx Briq.	China	Menthol, menthone, menthyl acetate.[33]	Stimulates gastrointestinal tract motility and central nervous system, dilates peripheral blood vessels. Increases sweat gland secretion.
	N.A.	Menthol, menthone, menthyl acetate, camphene, limonene, terpenoids.[99]	Treats colds, sore throats, sore mouth.

Table 2   Chinese and North American Medicinal Herbs Belonging to the Same Species: Major Constituents and Therapeutic Values   (continued)

Scientific Name	Source	Major Constituents	Therapeutic Values*
Menyanthes trifoliata L.	China	Aromadendrine, betulinic acid, cadinene, choline, gentiatibetin, cineole, dihydrofoliamenthin, foliamenthin, gentialutine, loganin, gentianine, gentiatibetine, invertin, gurjuncene, meliatin, menthiafolin, menyanthin, secologanin, alpha-spinasterol, stigmast-7-enol, trifolioside.[50]	Antitumor, increases gastric secretions, as cathartic, cholagogue, narcotic, sedative, tonic, vermifuge.
	N.A.	Iridoid glycosides, flavonol glycosides, coumarins, phenolic acids, sterols, triterpenoids, tannins.[99]	Stimulates digestive secretions, treats fluid retention, scabies, and fever.
Mimosa pudica L.	China	Mimosine.[78] This herb is toxic if overdosed.	Treats neurosis, trauma wound, and hemoptysis. It has a tranquilizing effect.
	N.A.	Nigerine (N, N-dimethyltryptamine).[100]	An astringent; cures fatigue, fortifies the uterus.
Momordica charantia L.	China	Anti-HIV protein MAP 30.[33]	For immune disorders and common infections. Capable of inhibiting infection of HIV-1 in T. lymphocytes and monocytes.
	N.A.	Fixed oil, insulin-like peptide, mormordin, charantin, mormordicine.[99]	Treats diabetes, ulcers, urinary stones; a stomach tonic; induces menstruation.
Morus alba L.	China	Morin, dihydromorin, maclurin, dihydrokaempterol, mulberrin, 2,4,4′,t-tetrahydroxybenzophenone, mulberrochromene, cyclomulberrochromene.[33]	Antirheumatic, antihypertensive, diuretic; removes obstructions of the intestinal tract.
	N.A.	Flavonoids, anthocyanins, artocapin, vitamins A, B$_1$, B$_2$, and C.[99]	An expectorant; helps coughing up of mucus.

Species	Region	Constituents	Uses
Myristica fragrans Houtt.	China	Lauric acid, myristic acid, stearic acid, hexadeceonic acid, oleic acid, linoelic acid, amylodextrins, pectins, resins, campherene, cymene, dipentene, eugenol, geraniol, isoeugenol, linalool, myristicin, pinene, safrole, terpineol.[50] Volatile oil from this herb may be toxic.	For hysteria, hypochondria, agarophobia, laughter, cramps, crying jags, dysmenorrhea, amnesia.
	N.A.	Safrole, myristicin, lauric acid, oleic acid, stearic acid, hexadecenoic acid, linoleic acid, d-camphene.[98,130]	For diarrhea, dysentery, vomiting, abdominal distention, indigestion, and colic.
Narcissus tazetta L.	China	Lycorine, tazettine, narcitine.[49,60] Toxic if overdosed.	Antiphlogistic, analgesic for boils, abscesses, mastitis.
	N.A.	Acetylated alkaloids, lectins.[149,150,151]	
Nicotiana tabacum L.	China	Nicotine, nicotimine, nicotelline.[60] This herb is toxic.	Treats soreness in the joints, numbness, hemicrania, poisonous snakebites; insecticide, antidysenteric, emetic.
	N.A.	Alkaloids, nicotine, volatile oil.[99] Nicotine is toxic.	A good insecticide. No longer used medicinally.
Oenothera biennis L.	China	6,9,12-octadecatrienoic acid.[48]	Lowers cholesterol, regulates heartbeat; and treats arthritis.
	N.A.	Linoleic acid, linolenic acid, phenolics, flavonoids, tannins.[103,118,152]	Treats asthma, arteriosclerosis, multiple sclerosis, atopic eczema, schizophrenia, diabetic neuropathy, cardiovascular diseases; antitumor.
Oxyria digyna (L.) Hill	China	Protein, fat, mineral elements.[48]	For hepatitis.
	N.A.	Protein, fat, ash, carbohydrate, retinol, mineral elements.[210]	Used as nutrient food.

Table 2 Chinese and North American Medicinal Herbs Belonging to the Same Species: Major Constituents and Therapeutic Values (continued)

Scientific Name	Source	Major Constituents	Therapeutic Values*
Paeonia albiflora Pall.	China	Benzoic acid, paeoniflorin, oxypaeoniflorin, benzoyl paeoniflorin, albiflorin. [14,15]	Carminative, antispasmodic, analgesic, sedative.
	N.A.	Monoterpenoid glycosides, paenoiflorin, albiflorin, benzoic acid, pentagalloyl glucose. [153]	Antispasmodic, tonic, astringent, analgesic.
Paeonia lactiflora Pall.	China	Benzoic acid, paeoniflorin, oxypaeoniflorin, benzoyl paeoniflorin, albiflorin. [14,15,510]	Carminative, antispasmodic, analgesic, sedative.
	N.A.	Monoterpenoid glycosides, benzoic acid, albiflorin, paeonol, astragalin, palmitic acid, gallotannin, pentagallotannin, beta-sitosterol, benzoic acid, myoinositol, pentagalloyl glucoside. [99,153,511]	Antispasmodic, tonic, astringent, analgesic, sedative, anti-inflammatory, prophylactic effect on stress ulcer and hypotension.
Paeonia officinalis L.	China	Benzoic acid, paeoniflorin, oxypaeoniflorin, benzoyl paeoniflorin, albiflorin. [14,15]	Carminative, antispasmodic, analgesic, sedative.
	N.A.	Glycosides, tannins, anthocyanidin, peregrinine, paeonine. [99,147]	Antispasmodic, diuretic, sedative properties.
Paeonia suffruticosa Andr.	China	Paeonol, paeonoside, paeonin, pelargonin, paeonolide, astragalin (paeoniflorin contained in *P. mourian*). [1,2,33]	Sedative, antipyretic, analgesic actions.
	N.A.	Monoterpenoid glycosides, benzoic acid. [99,153,154]	Antispasmodic, tonic, astringent, analgesic.
Panax ginseng C. A. Meyer	China	Triterpenoid, quinquenosides, ginsenosides, oleanolic acid, panaxynol, beta-elemene, spermine, putrescine, spermindine. [26,53,510]	A stimulant, tonic, expectorant.
	N.A.	Ginsenosides, acetylenic compounds, polysaccharides, panaxosides. [103,125,140,141,155,156,314,511]	A stimulant, tonic, adaptogen, diuretic, stomachic agent, carminative, aphrodisiac, healing properties; provides energy, retards the aging process.

Species	Origin	Constituents	Uses
Panax quinquefolium L.	China	Ginsenosides, phytosterols.[26]	Stimulation effects on central nervous system, antifatigue.
	N.A.	Ginsenosides, acetylenic compounds, polysaccharides, panaxosides.[125,140,141,155,156,193]	A stimulant, tonic, adaptogen, aphrodisiac, healing properties; provides energy, retard the aging process. American ginseng may lowers the blood pressure.
Papaver rhoeaes L.	China	Rhoeadine, rhoeagenine.[72]	For jaundice, as a gargle, or ingested as a bechic.
	N.A.	Thebaine, oripavine, morphine, codeine.[99,100]	Mild sedative to induce sleep in babies, ease cough, relieves pain; narcotic analgesic, antitussive.
Papaver somniferum L.	China	Berberine, codeine, papaverine, isocorypalmine, laudanine, magnoflorine, meconine, 6-methylocodine, morphine, narcotine, pseudomorphine, rhoeadine, sanguinarine, beta-sitosterol, stigmasterol, thebaine, zanthaline.[50]	Antitussive, antispasmodic, analgesic, astringent, narcotic; treats chronic enteritis, diarrhea, enterorrhagia, headache, toothache, asthma.
	N.A.	Morphine, narcotine, codeine, papaverine, meconic acid, albumin, mucilage, sugars, resin, wax.[99]	Sedates or suppresses nervous system activity, pain, and coughs.
Paris quadrifolia L.	China	Alpha-paristyphnin, diosgenin glycoside.[50]	Antispasmodic, anti-inflammatory, febrifuge.
	N.A.	Paradin, paridol, paristyphnine, l-asparagine, citric acid, pectin.[100] Overdose of this herb is toxic.	For bronchitis, cramps, gout, neuralgia, rabies, tumors, ulcers.
Perilla frutescens (L.) Britt.	China	l-perilla, aldehyde, apigenin, luteolin, 3-p-coumarylglycoside-5-glucoside of cyanidin, 7-caffeyl-glucosides of apigenin and luteolin, anthocyanins.[33,50]	Antibacterial, antitussive, stomachic, antiseptic.
	N.A.	Protein, flavone glycosides, shishonin, anthocyanin, perillanin chloride, aldehyde antioxine, citral, l-limonene, alpha-pinene.[100] This herb may be toxic.	Antispasmodic, diaphoretic, sedative; treats pulmonary and uterine disorders.

Table 2 Chinese and North American Medicinal Herbs Belonging to the Same Species: Major Constituents and Therapeutic Values (continued)

Scientific Name	Source	Major Constituents	Therapeutic Values*
Phaseolus vulgaris L.	China	Alpha-globuline, beta-globulin, fatty acids, vitamins A, B, and B_2, robinin, kaempferol-3-robinobiosido-7-rhamnoside.[48]	Diuretic; for abscesses, beriberi, dysentery, sores, swelling.
	N.A.	Allantoin, sugars, leucine, tyrosine, arginine, inositol.[99]	The pods are diuretic, stimulating urine flow, and flushing toxins from the body.
Phellodendron amurense Rupr. *P. chinensis* Schneid	China	Berberine, palmatine, candicine, phellodendrine, obacunone.[33]	Antibacterial; stimulates the phagocyte activity of leukocytes, against dysentery.
	N.A.	Isoquinoline alkaloids (berberine), sesquiterpene lactones, sterols.[157]	Treats diarrhea, dysentery, jaundice, vaginal infection, skin conditions.
Phragmites communis Trin.	China	Glycosides, protein, asparagin.[49]	A stomachic, antiemetic, antipyretic. Treats arthritis, jaundice, pulmonary abscess.
	N.A.	Protein, carbohydrate, crude fiber, minerals.[190]	For fevers, vomiting, coughs, urinary tract infections.
Phyllostachys nigra Munro.	China	Benzoic acid, silica, potassium hydroxide, aluminum oxide, iron oxide, calcium.	Antipyretic, hematuria, sedative, antiemetic, antispasmodic in catarrh.
	N.A.	No information is available in the literature.	Diuretic; lowers fever, treats lung infections with cough and phlegm.
Physalis alkekengi L.	China	Physanols, physalien, zeaxanthin, glycolic acid, cryptoxanthin, physoxanthin, mutaxanthin, auroxanthin, physalin A, B, and C, luteolin, tigloidine, physalines, hystonin.[33,48]	Antibacterial; stimulates myocardial contraction, causes vasoconstriction, uterine contraction.
	N.A.	Physalin, vitamin C, alkaloids, flavonoids, sterols.[99]	Diuretic; treats kidney and urinary disorders.

Species	Origin	Constituents	Uses
Phytolacca acinosa Roxb.	China	Phytolacine, phytolaccatoxin, oxyristic acid, jaligonic acid, saponins.[33]	Antitussive, diuretic, antibacterial, anti-inflammatory.
	N.A.	Triterpenoid saponins, lectins, proteins, resin, mucilage.[99] This herb is toxic.	Anti-inflammatory, antiviral; treats rheumatic and arthritic conditions, respiratory tract infections.
Phytolacca americana L.	China	Phytolacine, phytolaccatoxin, oxyristic acid, jaligonic acid, saponins.[33]	Antitussive, diuretic, antibacterial, anti-inflammatory.
	N.A.	Caryophyllen, isobetanine, isoprebetanine.[99,100]	Treats catarrh, dyspepsia, granular conjunctivitis, and rheumatism.
Pinus sylvestris L.	China	Essential oil, limonene, pinitol.[33]	Antitussive, antiasthmatic, antibacterial.
	N.A.	Alpha-pinene, beta-pinene, delta-limonene.[98,99]	Mild antiseptic effect, essential oil for asthma, respiratory infections, digestive disorder.
Piper cubeba L.	China	Cubebin, dipentene, cadinene, cineol, carene, camphene, pinene, sabinene, azulene, terpineol.[49]	Urinary antiseptic, stomachic, carminative.
	N.A.	Volatile oil, cubebin, piperidine, resin.[99]	Antiflatulent, antiseptic; relieves digestive problems.
Piper longum L.	China	Volatile oil, piperine.	Antipyretic, carminative, aromatic stomachic, analgesic in gastralgia, flatulence, headache.
	N.A.	Piperine, volatile oil, protein, l-phyllandrene, caryophyllene.[100]	Stimulant effect on digestive and circulatory system.
Piper nigrum L.	China	Piperine, chavicine, piperamine, piperonal, dihydrocarveol, cryptone, caryophyllene. This herb may cause irritation to the system.[33,45]	Anticonvulsive, secative.
	N.A.	Piperine, volatile oil, protein, l-phyllandrene, caryophyllene.[100]	Stimulant effect on digestive and circulatory system.

Table 2   Chinese and North American Medicinal Herbs Belonging to the Same Species: Major Constituents and Therapeutic Values   (continued)

Scientific Name	Source	Major Constituents	Therapeutic Values*
Pistacia lentiscus L.	China	Masticinic acid, masticonic acid, masticoresene, fisetin, fustin, gallic acid, quercetin, taxifolin. [49,50]	Antitumor, antitussive, analgesic, sedative in gastralgia, cardiodynia, mastitis, peptic ulcer.
	N.A.	Alpha-masticoresin, beta-masticoresin, alpha-pinene, tannins, masticin, mastic acid. [99]	As an expectorant for bronchial troubles and coughs; treats diarrhea.
Plantago asiatica L.	China	d-xylose, l-arabinose, d-galacturonic acid, l-rhamnose, plantasan, plantenolic acid, plantagin, homoplantagin, aucubin, ursolic acid, hentriacontane. [48,510]	Diuretic, expectorant, intestinal infection, diarrhea caused by bacteria.
	N.A.	Mucilage, linoleic, oleic, palmitic acid, fiber. [100,124,511]	Demulcent, laxative, antidiarrheal.
Plantago major L.	China	Xylose, galacturonic acid, rhamnose, plantasan, plantenolic acid, plantagin, homoplantagin, aucubin, ursolic acid. [48]	Diuretic, expectorant, intestinal infection, diarrhea caused by bacteria.
	N.A.	Aucubin, mucilage, carotene, tannin, chlorogenic acid. [100,102]	Expectorant, emollient, demulcent, vulneraria, and astringent, soothing effects.
Pogostemon cablin Benth.	China	Essential oils. [60]	Antiseptic, for abdominal pain, cold, diarrhea.
	N.A.	Sesquiterpenes patchoulol, bulnesene. [99]	Aphrodisiac, antidepressant, antiseptic.
Polygonatum odoratum (Mill.) Druce	China	Convallarin, convallamarin, mucilage. [49]	Stimulates the appetite, increases peristalsis, slows the heart and raise the arterial tension, slows and deepens respiration, and purgative.
	N.A.	Saponins, flavonoids, vitamin A. [99]	Prevents excessive bruising and stimulates tissue repair. An astringent, treats tuberculosis.

Species	Origin	Constituents	Uses
Polygonum aviculare L. P. viviparum L.	China	Avicularin, caffeic acid, tannin, chlorogenic acid, vitamin E. [33,60]	Treats urethritis, lithiasis, and chyluria. Against dysentery and parotitis, an antiascardiasis agent.
	N.A.	Tannins, flavonoids, polyphenols, silicic acid, mucilage. [99,102]	With astringent and diuretic properties. Treats diarrhea, hemorrhoids, expels worms.
Polygonum bistorta L.	China	Iodine, oxalic acids, coumarins, gallic acid, hydroxycinnamic acids, ether oil, carotin, hydroxybenzoic acids, hydrocyanic acids, anthocyanidines, anthraquinones, phytosterines, caffeic acid, monoterpene, sesquiterpenen glucoside, avicularin, quercimeritrin, protocatechuic acid. [50,221,222,223,224]	Diuretic, laxative, hemostatic, antifebrile.
	N.A.	Chrysophanic acid, anthraquinones, lecithin. [99]	Mild sedative; nourishes the blood; a tonic.
Polygonum hydropiper L.	China	Persicarin, rhamnazin, isotadeonal, quercimeritrin. tadeonal. [33]	Improves indigestion, treats dysentery and enteritis.
	N.A.	Chrysophanic acid, anthraquinones, lecithin. [99]	Mild sedative; nourishes the blood; a tonic.
Polygonum multifolrum Thunb.	China	Chrysophenol, emodin, emodin methyl ester, rhein, glycoside rhaphantin, lecithin, parietin, chrysophanic acid, anthron. [33,46,54]	A laxative, detoxicant for boils. Treats neurosis, neurasthenia, insomnia, hypercholesterolemia.
	N.A.	Tannins, flavonoids, polyphenols, silicic acid, mucilage. [99,102]	With astringent and diuretic properties. Treats diarrhea, hemorrhoids, expels worms.
Populus alba L.	China	Salicin, populin, benzoyl salicin, tannin, erisin, salicinase, salicortin, tremulacin, salireposide. [50]	Depurative, for colic, eczema, herpes, labialis, fever, dysuria, antiseptic, antiperiodic.
	N.A.	Flavoncids, flavones, flavonols, flavanones, coumaric acid, cinnamic acid, terpenoids. [190]	Treats diabetes, high blood pressure, asthma.

Table 2   Chinese and North American Medicinal Herbs Belonging to the Same Species: Major Constituents and Therapeutic Values   (continued)

Scientific Name	Source	Major Constituents	Therapeutic Values*
Poria cocos (Polyporaceae)	China	Pachymic acid, tumulosic acid, eburicoic acid, pinicolic acid, pachymarose.[33]	A diuretic, cardiotonic, it has a tranquilizing effect, lowers blood sugar levels; it is antibacterial and anticancer.
	N.A.	Beta-pachyman, beta-pachymanase, pachymic acid.[99]	For urinary ststem, stress-related anxiety, tension headaches, palpitations, and difficulty in sleeping.
Portulaca oleracea L.	China	Potassium salts, catecholamines, norepinephrine, dopamine, vitamin A, vitamin B, magnesium.[33,49]	Antibacterial, diuretic; causes vasoconstriction, stimulates uterine and intestinal smooth muscle contraction.
	N.A.	Mucilage, calcium.[99]	Treats urinary and digestive problems. It has mild antibiotic effect.
Poterium officinale Benth.	China	Zi Yu glucoside I, Zi Yu glucoside II, sanguisorbin A, sanguisorbin B, sanguisorbin C.[33]	Astringent effect to stop diarrhea and relieves chronic intestinal infection, duodenal ulcer, and bleeding. Externally for eczema.
	N.A.	Tannins, sanguisorbic acid, dilactone, gum.[99]	Slow blood flow; treats heavy periods and uterine hemorrhage; externally for hemorrhoids, burns, wounds, and eczema.
Primula vulgaris Huds.	China	Primulagenin A, aegicerin, protoprimulagenin A.[48]	Relieves cough, throat infection.
	N.A.	Triterpenoid saponins, flavonoids, phenols, tannins, volatile oil.	Internally for bronchitis, respiratory tract infections, insomnia, anxiety, rheumatic disorders.

Prunella vulgaris L.	China	Caffeic acid, d-camphor, cyanidin, delphinidin, d-fenchone, hyperoside, oleanolic acid, rutin, ursolic acid.[48]	Antibacterial, antipyretic, cardiac tonic, diuretic, anticancer.
	N.A.	Tannins, saponins, aucubin, vitamins B, C, and K, caffeic acid, ursolic acid, betulinic acid, deanolic acid.[99,102]	Astringent, anti-inflammatory, hemostatic; gargle for sore throat; cleans wounds.
Prunus armeniaca L.	China	Amygdalin, prunasin, fatty acids, mandelonitrile (enzyme amygdalase can hydrolyze amygdalin to produce cyanic acid).[33,53]	Stimulates respiratory center reflexively and produces a tranquilizing effect.
	N.A.	Amygdalin, prussic acid, cyanogenic glycoside, lactrile, hydrocyanic acid.[99] Kernel is toxic.	Treats coughs, asthma, wheezing, and excessive mucus, constipation. Treats cancer.
Prunus domestica L.	China	Amygdalin, citric acid, fatty acids.[53]	Diuretic, laxative.
	N.A.	Cyanogenic glucosides.[336]	For constipation; a laxative.
Prunus mume Siebold & Zucc.	China	Prudomenin, malic acid, succinic acid, citric acid, tartaric acid, amygdalin.[33,53]	Treats biliary ascariasis and hookworm.
	N.A.	Laetrile, cyanide, beta-carotene, thiamine, ascorbic acid, malic acid, citric acid, oligopeptides, polysaccharide.[158,159]	Internally for chronic coughs, externally for fungal skin infections, warts, improving blood fluidity; has immunochemical characterization.
Prunus persica (L.) Batsch.	China	Malic acid, citric acid, octalactone, leucoanthocyanins, tannins, hexalactone, hectalactone, benzyl alcohol, nonalactone, decalactone, ethanol, hexanol, acetadehyde, benzaldehyde, acetic acid, pentanoic acid, hexanoic acid.[50]	Astringent, febrifuge, parasiticide, diuretic, sedative, vermifuge.
	N.A.	Essential oils.[105]	For gastritis, coughs, whooping cough, bronchitis.

Table 2   Chinese and North American Medicinal Herbs Belonging to the Same Species: Major Constituents and Therapeutic Values   (continued)

Scientific Name	Source	Major Constituents	Therapeutic Values*
Psoralea corylifolia L.	China	Psoralen, angelicin, psoralidin, coryfolin, bavachinin, isobavachin, corylifolinin, d-backuchiol.[33]	Coronary vasodilating effect; increases the myocardial contraction; antibacterial, anticancer.
	N.A.	Psoraline, isopsorlin, bavachin.[99]	Treats impotence, premature ejaculation.
Pueraria lobata (Willd.) Ohwi.	China	Isoflavones, daidzin, diadzin-4, 7-diglucoside, daidzein, puerarin, xylopurarin, robinin, kaempferol-rhamnoside, fatty acids.[12,33,48]	Antispasmodic, hypotensive, and stabilizing blood pressure; treats angina pectoris.
	N.A.	Daidzin, diadzein, isoflavonoids, puerarin, sterol.[99]	For colds, influenza, feverish illness, thirst in diabetes, externally for snakebite.
Pueraria thunbergiana Benth.	China	Glutamic acid, butyric acid, asparagin, adenine.[49]	Antipyretic, refrigerant.
	N.A.	Isoflavonoids, puerarin, daidzein, sterols.[99]	For muscle aches, headache, dizziness due to high blood pressure.
Pulsatilla chinensis (Bunge.) Regel	China	Protoanemonin, anemonin, okinalin, okinalein, ranuneulin, saponins.[33]	Antiamebial, antibacterial; treats amebic dysentery.
	N.A.	Lactone, protoanemonin, anemonin, pulsatoside, anemonol.[99]	Antibacterial, as an irritant.
Punica granatum L.	China	Pelletierine, isopelletierine, methyl-pelletierine, methylisopelletierine, pseudopelletierine, tannic acid, granatin.[33] This herb is toxic.	Treats intestinal parasties; antibacterial.
	N.A.	Pelletierene alkaloids, elligatannins, triterpenoids.[99] This herb is toxic.	For tapeworm infestation.

Pyrethrum cinerariifolium (L.) Trev.	China	Essential oil, adenine, choline, stachydrine.[49]	Sedative, refrigerant in headache, influenza.
	N.A.	Pyrethrins, cinerins, palmitic, linoleic acid, sesquiterpene lactones.[100,107,118]	Externally used as a contact insecticide.
Pyrola rotundifolia L.	China	Arbutin, homoarbutin, isohomoarbutin, chimaphillin, monotropin.[33]	Antibacterial, antiarrhythmic; lowers blood pressure, hemostatic effect.
	N.A.	Flavonoid glycosides, chimpahilin, sesquiterpenes, arbutin, ursolic acid.[186]	Anti-inflammatory; relieves pain, improves myocardial circulation.
Raphanus sativus L.	China	Raphanin.[50]	For asthma, cough, diarrhea, dysentery, eruptive fevers; bactericidal, antitumor.
	N.A.	Glucosinolates, arginine, histidine, vitamins A, B, and C.[102]	Leaf is diuretic, laxative, root for hemorrhoids.
Rehmannia glutinosa (Gaertn.) Libosch.	China	Catalpo, campesterol, rehmannin, polysaccharide.[16,33]	Lowers blood sugar, immuno-antitumor activity.
	N.A.	Phytosterols, β-sitosterol, stigmasterol, mannitol, rehmannin.[99]	Preventing poisoning and liver damage. Treats blood pressure, fever.
Rheum officinale Baill. *R. palmatum* L. *R. tanguticum* Maxim.	China	Anthraquinones, chrysophanol, emodin, physcion, aloe-emodin, rhein, chrysophenol, rheum tannic acid, gallic acid, catechin, bianthraquinonyl, sennosides.[1,33,510,511] This herb may be toxic.	Potent laxative, antibacterial, anthelmintic, anticancer; stimulates the large intestine and increases the movement of luminal contents toward the anus, resulting in defecation. Antispasmodic, choleretic, hemostatic, diuretic, lowers blood pressure, lowers cholesterol level.
	N.A.	Cinnamic acid, gallic acid, emodin, rhein, rhein anthrones, catechin, anthraquinone compounds, tannin, calcium oxalate.[99,100,107,510,511]	Treats diarrhea, stimulates appetite, chronic constipation; laxative, cathartic.

Table 2 Chinese and North American Medicinal Herbs Belonging to the Same Species: Major Constituents and Therapeutic Values (continued)

Scientific Name	Source	Major Constituents	Therapeutic Values*
Ricinus communis L.	China	Ricinine, ricinolein, olein, stearin, isoricinoleic acid, cytochrome C.[33]	Cathartic.
	N.A.	Ricinoleic acid, ricin, ricinine, lectins.[99] The seeds are toxic.	Laxative, prompting a bowel movement.
Rosa acicularis Lindl.	China	Vitamins, gallocatechin, epigallocatechin, epicatechin gallate, catechin, epicatchin, fatty acids.[48]	Stop vomiting blood, stomachache, relieves pain caused by nervous system, menstruation.
	N.A.	Vitamins, malic acid, citric acid, pectin, geraniol, l-citronellol.[160]	A tonic, astringent, diuretic, laxative.
Rosa rugosa Thunb.	China	Essential oils, l-citronellol, citral, geraniol, nerol, eugenol, cyanin, n-phenylethyl alcohol, citrol, nonyl aldehyde, l-linalool, l-p-menthene, nonacosane, menthene, bensaldehyde, phenylacetic acid, rosenoxide, paeonidin.[48,50]	Promotes blood circulation, treats abscesses, blood diseases, dyspepsia, hematemesis, hepatitis, stomachache.
	N.A.	Vitamins, malic acid, citric acid, pectin, geraniol, l-citronellol.[160]	A tonic, astringent, diuretic, laxative.
Rubus coreanus Miq.	China	Beta-sitosterol, stigmasterol, campesterol, cholestanol, ursolic acid, flavonoids.[48]	Diuretic, aphrodisiac; treats liver infection, joint infection caused by arthritis.
	N.A.	Tannins, organic acids, vitamin C.[138]	An astringent, antiseptic, diuretic.

Rumex acetosella L. R. *aquaticus* L.	China	Vitexin, quercetin-3-galactoside, violaxanthin, vitamin C, emodin, chrysophanein, chrysophanol, nepodin, hyperin, physcion.[48,50]	Homeopathically for cramps, hemorrhage, sore throat, esophagitis, diuretic; treats blood vomiting.
	N.A.	Oxalates, anthraquinone (chrysophanol, emodin, physcion), phenol, physcion, tannic acid.[100,102,118]	Antiseptic, laxative, rheumatic pains.
Rumex crispus L.	China	Chrysophanein, nepodin.[48]	Treats ovarian bleeding, eczema, tuberculosis, sexually transmitted diseases.
	N.A.	Anthraquinones, nepodin, emodin, chrysophanol, tannins, oxalates, volatile oil.[99]	Mild laxative; stimulates bile flow, as a cleansing.
Salvia coccinea L.	China	Saluianin.[56]	Stop bleeding, cooling effect, stimulates sweating, relieves swelling.
	N.A.	Thujone.[107,161]	Treats fever; an antiseptic, astringent.
Sanguisorba officinalis L.	China	Zi Yu glucoside I, Zi Yu glucoside II, sanguisorbin A, sanguisorbin B, sanguisorbin C.[33]	Astringent effect to stop diarrhea and relieves chronic intestinal infection, duodenal ulcer, and bleeding. Externally for eczema.
	N.A.	Tannins, sanguisorbic acid, dilactone, gum.[99]	To slow blood flow, treats uterine hemorrhage.
Santalum album L.	China	Alpha-santalol, beta-santalol, alpha-santalene, beta-santalene, santene, alpha-santenone, alpha-santenol, santalone, santalic acid, teresantalic, isovaleraldehyde, teresantalol, tricycloekasantal, santalin, deoxysantalin, sinapyl aldehyde, caniferyl aldehyde, syringic aldehyde.[33]	Treats stomachics.
	N.A.	Dihydro-β-agarofuran, curcumin, sesquiterpene hydrocarbons, dendrolasin, santalols.[8,100]	Internally for genitourinary disorder, fever, sunstroke; externally for skin disorder.

Table 2   Chinese and North American Medicinal Herbs Belonging to the Same Species: Major Constituents and Therapeutic Values (continued)

Scientific Name	Source	Major Constituents	Therapeutic Values*
Saponaria officinalis L.	China	Saponarin.[65] This herb is contraindicated in pregnancy.	For abscesses, furuncles, ulcers, scabies, mastitis, lymphangitis. Root is used to treats syphilis, glandular and chronic skin disease.
	N.A.	Saponins, resin, sapogenin, sterol, trace of volatile oil.[162]	As an expectorant, bronchitis, coughs, asthma, rheumatic and arthritic pain.
Saussurea lappa Clarke	China	Saussurine, phene, phellandrene.[49]	As a stomachic.
	N.A.	Terpenes, sesquiterpenes, aplotaxene, sausarine, resin.[99]	Depresses the parasympathetic nervous system.
Schisandra chinensis (Turcz.) Baill.	China	Schizandrin, deoxyschizandrin, schizandrol, schizandrer.[8,33]	Antitussive, a tonic. A tendency to lowers SGPT caused by hepatitis.
	N.A.	Lignans, phytosterols, vitamins C, E.[99]	Tonic, adaptogenic; protects liver.
Schizonepeta tenuifolia (Benth.) Briquet	China	Essential oils, d-menthone, d-limonene.[33]	Diaphoresis, lowers body temperature; anticonvulsive, increase blood coagulation.
	N.A.	Menthone, limonene.[99]	To alleviate skin boils and itchiness; treats fever and chills.
Scrophularia ningpoensis Hemsl.	China	Scrophularin, iridoid glycosides, 8-(O-methyl-p-coumaroyl)-harpagide, harpagoside, essential oils, flavonoids, p-methoxylcinnamic acid.[33]	Lowers blood pressure and blood sugar. A health strengthening agent.
	N.A.	Aucubin, harpagoside, acetyl harpagide, flavonoids, phenolic acid.[99]	Antiarthritic; treats infections and to clears toxicity.

Plant	Origin	Constituents	Uses
Scutellaria baicalensis Georgi S. macrantha Fisch.	China	Baicalein, baicalin, wogonin, beta-sitosterol, wognoside, 7-methoxy-baicaleir, 7-methoxynorwogonin, skullcap flavones.[33]	Antibacterial, antiviral, antipyretic, anti-inflammatory, antitumor.
	N.A.	Scutellarin, baicalin, baicalein, wogonin, benzoic acid, catapol, tannins, beta-sitosterol, camphesterol, stigmasterol.[99,102,163]	Sedative and antispasmodic, prevents epileptic seizures; antiallergic.
Senecio vulgaris L.	China	Senecionine, inulin.[58]	Used in ointment on hemorrhoids and swellings, lowers blood pressure; laxative.
	N.A.	Volatile oil, seneciphyline, jacoline, pyrrolizidine, senecionine, tinnins, resin.[164]	As a poultice ointment or location to relieves pain and inflammation.
Sesamum indicum L.	China	Olein acid, linolein acid, palmitin acid, stearin acid, myristin acid, sesamin, sesamol, pentosan, phytin, lecithin, choline, calcium oxalate, chlorogenic acid, vitamin A, vitamin B.[49]	A nutrient, laxative, hyperchlorhydria, a lenitive in scybalous constipation; as a nutrient tonic in degenerative neuritis, neuroparalysis.
	N.A.	Phenol, lignan, oleic acid, linoleic acid, protein, vitamins B, E, folic acid.[165,166]	An antioxidant, antitumor, antimitotic, antiviral, prevents breast cancer. Internally for premature hair loss and graying, strengthens bones and teeth.
Silybum marianum (L.) Gaertn.	China	Silybin, silymarin, silydiamin, silychristin, dehydrosilybin, silybinomer.[33]	Maintain normal functioning of the liver, promotes the regeneration of injured hepatic cells, and increases glycogenesis and nucleic acid metabolism.
	N.A.	Flavonolignans, silibinin, silymarin.[167,168,169]	Treats hepatitis, cirrhosis, regeneration of diseased liver, liver poisoning, digestion.
Sinapis alba L.	China	Arachidic acid, erucic acid, lignoceric acid, linoleic acid, myrosinase, phosphatase sinalbine.[50]	Carminative, toothache; seals eruptions; and ulcers.
	N.A.	Mustard oil.[191]	Stimulant; promotes urination. Mustard plasters for rheumatism, arthritis, chest congestion, aching backs, sore muscles.

Table 2   Chinese and North American Medicinal Herbs Belonging to the Same Species: Major Constituents and Therapeutic Values   (continued)

Scientific Name	Source	Major Constituents	Therapeutic Values*
Smilax china L.	China	Crystalline saponin smilacin, tannin, resin, tigogenin, neotigogenin, laxogenin. [48,49]	As alternative, diuretic in syphilis, gout, skin disorders, rheumatism.
	N.A.	Steroidal saponins, phytosterols (beta-sitosterol), starch, resin, sarsapic acid, minerals. [99]	Anti-inflammatory and cleansing. Relieves skin eczema, psoriasis, itchiness.
Solanum aculeatissimum Jacquin	China	Solasonine, beta-solamargine, solasurine. [55]	For cough, asthma, diuretic, pain.
S. melongena L.	N.A.	Proteins, carbohydrates, vitamins A, B_1, B_2, and C. [99]	Lowers blood cholesterol level, regulate high blood pressure.
Solanum nigrum L.	China	Solanigrines, saponines. [33]	Antibacterial, diuretic, treats mastitis, cervicitis, chronic bronchitis, dysentery.
	N.A.	Linoleic acid, palmitic acid, stearic acid, sitosterol, diosgenin, tigonenin, solanine, chaconine, solasodine, solasonine, solamargine. [145] This herb is toxic.	Remedy for tumors and cancer, diuretic, treats eye diseases, fevers, hydrophobia. It is a laxative, emollient, anti-inflammatory.
Solidago canadensis L.	China	Cadinene, quercitrin. [57]	Antibacterial; treats infection, stops bleeding, throat swelling.
	N.A.	Tannins, saponins, polygalic acid, cariaester, inulin, salicylic acid. [100]	Alleviate intestinal gas, relieves fever.
Solidago virgaurea L.	China	Caffeic acid, chlorogenic acid, cyanidin-3-glucoside, flavonoids, astragalin, cyanidin-3-gentiobioside, kaempferol-rhamno glucoside, hydroxycinnamic acid, quinic acid, polygalacic acid. [48,50]	Decoagulant, carminative, for bladder ailments, cholera, diarrhea, dysmenorrhea.
	N.A.	Tannins, saponins, polygalic acid, cariaester, inulin, salicylic acid. [100]	For urinary infections, chronic excess mucus, skin diseases, influenza, whooping cough.

Species	Origin	Constituents	Uses
Sophora japonica L.	China	Rutin, sophoradiol, genistein, sophoricoside, sophorabioside, sophoraflavono-oside.[33]	Increase capillary resistance and decreases capillary fragility and permeability.
	N.A.	No information is available in the literature.	For internal hemorrhage, hypertension and poor peripheral circulation.[345]
Stellaria media (L.) Cyrillo	China	r-linolenic acid, octadecatetraenoic acid.[48]	A postpartum depurative, emmenagogue, lactagogue, promotes circulation, treats mucus disorder. Externally for rheumatic pains, ulcers, wounds.
	N.A.	Triterpenoid saponins, vitamin C, coumarins, flavonoids, linolenic acid, octadecatetraenic acid.[99,100]	Treats internal and external inflammations, irritated skin.
Strychnos nux-vomica L.	China	Strychnine.[50] This herb may be toxic.	Treats neurasthenia, aphrodisiac, vasomotor stimulation, regulate blood pressure.
	N.A.	Indole alkaloids, strychnine, loganin, chlorogenic acid.[99] This herb is toxic.	A stimulant for the nervous system, a homeopathic remedy for digestive problems; sensitivity to cold, and irritability.
Syringa suspensa Thunb. S. vulgaris L.	China	Syringin, 3,4-dihydroxyphenethyl alcohol, saponors, phillyrin.[49]	Antipyretic, antiphlogistic in infectious fevers, suppurative inflammation, phlegmon, variola, erysipelas, measles.
	N.A.	Lilacin, ligustrin, lignans, hydroxyphenylethanol glycosides.[102,170]	Tonic, neurotrophic, adaptogenic, immune stimulating, antimicrobial from leaves.
Syzygium aromaticum (L.) Merr. & Perry	China	Phytosterols, campesterol, crataegol acid, sitosterols, stigmasterol, niacin, ascorbic acid.[50]	Antiemetic, carminative, stimulant; treats diarrhea, halitosis, nasal polyps, uterine fluxes, sterility, toothache.
	N.A.	Sesquiterpenes, volatile oil, eugenol, tannins, gum.[130,131,314]	Internally for gastroenteritis and intestinal parasites, externally for toothache and insect bites.

Table 2   Chinese and North American Medicinal Herbs Belonging to the Same Species: Major Constituents and Therapeutic Values (continued)

Scientific Name	Source	Major Constituents	Therapeutic Values*
Tagetes erecta L.	China	Alpha-terthienyl, d-limonene, l-linalool, tagetone, n-nonyl aldehyde.[50]	Treats sores and ulcers, cold, conjuctivitis, cough, mastitis, mumps.
	N.A.	Limonene, linalool.[171]	Treats rheumatism.
Tagetes patula L.	China	Tagetone, linalool, limonene, linalylacetate, ocimene, patuletin, patulitrin, cyanidin diglycoside, quercetagetin, quercetagetrin, helenien, polythienyls.[50] This herb is toxic.	For coughs and dysentery.
	N.A.	Essential oils, tagetone, limonene, linalool, ocimene, linalyl acetate, thiophenes.[171]	Treats rheumatism; externally for boils, carbuncles, earache.
Taraxacum officinale G. H. Weber ex Wigg.	China	Inulin, essential oils, choline, cerylic alcohol, arabinose, vitamins A, B, C.[88]	Sudorific, stomachic, tonic, a remedy for sores, boils, ulcers, abscesses, snakebites.
	N.A.	Taraxacin, taraxerol, taraxasterol, inulin, gluten, gum, choline, levulin, pulin, tannins, provitamin A, vitamins B, C.[103,172,173]	Tonic, diuretic; stimulates appetite, digestion; treats fever, insomnia, jaundice, eczema, rheumatism, and arthritis.
Terminalia chebula Retz.	China	Chebulic acid, fatty oil, tannin, ellagic acid.[49]	An astringent in diarrhea, enterorrhagia, metrorrhagia, metritis, leukorrhea.
	N.A.	Anthraquinones, tannins, chebulic acid, resin.[99]	Laxative, astringent; improves bowel regularity.

Thevetai peruviana (Pers.) K. Schum.	China	Thevetin A, B, theveside, peruvosides, vertiaflavone, theviridoside.[33]	Tranquilizing effect; treats congestive heart failure.
	N.A.	Caoutchouc, resin, palmitic acid, stearic acid, arachidic acid.[100] This herb is very toxic.	Used for skin ailments.
Thlaspi arvense L.	China	Sinigrin, fatty acids, essential oil, myrosin, myrosinase.[48]	For ophthalmia, lumbago; an antidote, antipyretic; improves circulation, diaphoretic.
	N.A.	Amine choline, acetylcholine, bursine, histamine, flavonoids, polypeptides, tyramine.[99,102]	Controls internal bleeding, profuse menstruation.
Thymus vulgaris L.	China	Tymol, terpinen-4-ol, pinenes, camphene, myrcene, alpha-phellandrene, limonene, 1,8-cinole, p-cymene, linalool, linalyl acetate, bornyl acetate, alpha-terpinyl acetate, alpha-te pineol, borneol, citral, geraniol, carvacrol.[50]	Anthelmintic, antispasmodic, carminative, diaphoretic, sedative. Treats bronchitis, cancer, diarrhea, gastritis, rheumatism, skin ailments.
	N.A.	Thymol, tannins, carvacrol, sapanins, apigenin, luteolin.[99,100,107]	Antispasmodic, antitussive; relieves coughing.
Tribulus terrestsis L.	China	Glycosides tribuloside, astragalin, harmane, harmine.[33]	Anticonvulsive, a spasmolytic agent.
	N.A.	Sitosterol, tannins, saponins, tribulusamide A and B, n-*trans*-feruloyltyramine, terrestriamide, n-*trans*-coumaroyltyramine.[174]	Estrogenic properties, antiandrogenic action; reduces benign prostate hyperlasia (BPH).
Trichosanthes kirilowii Maxim.	China	Trichosanthin, polysaccharides, saponin, organic acids, resin, protein (TAP29).[33] This herb is toxic.	Treats pectoris and acute mastitis. Antitussive, as an expectorant, anti-HIV activity.
	N.A.	Trichosanic acid.[100]	No information is available in the literature.

Table 2 Chinese and North American Medicinal Herbs Belonging to the Same Species: Major Constituents and Therapeutic Values (continued)

Scientific Name	Source	Major Constituents	Therapeutic Values*
Trifolium pratense L.	China	Phytoestrogens, genistein, daidzein, formononetin.[33,48]	Stimulating effect on female reproductive organs.
	N.A.	Tannins, phenolic glycosides, p-coumaric acid, silicic acid, caffeic acid, salicylic acid.[100,102]	Remedy for sore throat, colds, coughs, bronchitis, diarrhea, chronic skin disease.
Trigonella foenum-graecum L.	China	Trigonelline, saponins, flavone derivatives including vitex, saponaretin, isoorientin, vitexin-7-glucoside.[33]	Reduces plasma cholesterol levels, supports hepatic and renal functions.
	N.A.	Protein, linoleic, oleic, linolenic and palmitic acids, trigonelline, choline, coumarin, nicotinic acid.[100,117,175]	Reduces total cholesterol and triglycerides without affecting the HDL, reduces blood sugar.
Tussilago farfara L.	China	Faradiol, rutin, hyperin, saponins, taraxanthin, tannin, essential oil.[33]	Antitussive, expectorant, antiasthmatic; stimulates the medullary center and slowly raises blood pressure.
	N.A.	Mucilage, sterols, pigments, inulin, gallic, malic, tartaric acids, tannins, pyrrolizidine alkaloids.[99,100]	Expectorant, demulcent, astringent, anti-inflammatory.
Typha angustifolia L. *T. latifolia* L.	China	Isothamnetin, alpha-typhasterol, oligosaccharides.[33]	Treats hypercholesteremia, angina pectoris, exudative eczema, postdelivery bleeding, stops bleeding in hematemesis and hematuria.
	N.A.	Isorhamnetin, pentacosane, phytosterols.[99] Do not use during pregnancy.	Treats angina.

Species	Origin	Constituents	Uses
Urtica urens L.	China	Chlorogenic acid, alkaloids, 5-hydroxytryptamine, protein, fat, carbohydrate, ash, fabric.[48,49]	Diuretic, tonic, stomachache; arthritis.
	N.A.	Stigmast-4-3-one, stigmasterol, beta-sitosterol, polysaccharides, aretylcholine, serotonin, quercitin, histamine, choline, glucoquinone.[99,102,176,304]	Treats benign prostatic hyperplasia; hair tonic and growth stimulation, used in antidandruff shampoo.
Vaccinium vitis-idaea L.	China	6-o-acetyl-arbutin, arbutin, avicularin, 2-o-caffeoylarbutin, d-catechol, l-epicatechol, d-gallocatechol, hyperin, hyperoside, sioquercitrin, salidroside, tannin, ursone.[50]	For gonorrhea.
	N.A.	Anthocyanosides, hippuric acid, vitamins A and C.[103,177,178,179,180]	Treats urinary infection and stones. Juice has antioxidant value.
Verbena officinalis L.	China	Verbenalin, verbenalol, adenosine, tannin, essential oils.[33]	Antiplasmodial, antibacterial, antitoxin, anti-inflammatory.
	N.A.	Ververin, verbenalin, volatile oil, alkaloids, mucilage, tannins.[99]	A tonic, mild sedative; stimulates bile secretion.
Viscum album L.	China	Oleanolic acid, beta-amyrin, mesoinositol, flavoyadorinin, homoflavoyadorinin, lupeol, myristic acid, agglutinins, alkaloids, quercito, querbrachitol, vitamins E and C.[33]	Antihypertensive; prolongs the life of patients with late stage stomach cancer.
	N.A.	Galactoside-specific lectin, lignans, viscotoxin, choline, alkaloids, resin, acetylcholine, protein, flavonoids, caffeic acid, viscin, carotenoids.[99,100,181]	Lowers blood pressure, stimulates heart action, and treats arteriosclerosis.
Vitis vinifera L.	China	Malic acid, tartaric acid, racemic acid, oxalic acid.[50]	For abortion, cholera, dropsy, nausea.
	N.A.	Linoleic, oleic, palmitic, and stearic acids, flavonoids, malic acid, anthocyanins, tartaric, tannins, monoterpene glycosides.[95,182]	Antioxidant, internally for varicose veins, excessive menstruation, menopausal syndrome, hemorrhage, and hypertension.

Table 2　Chinese and North American Medicinal Herbs Belonging to the Same Species: Major Constituents and Therapeutic Values　(continued)

Scientific Name	Source	Major Constituents	Therapeutic Values*
Zea mays L.	China	Carbohydrate, beta-carotene, thiamine, riboflavin, niacin, ascorbic acid.[50]	For dropsy, diabetes mellitus, hypertension, epistaxis, menorrhagia, cancers, tumors, warts.
	N.A.	Saponins, fatty acids, tannins, resin, maysin, essential oil, thiamine, mucilage.[99]	Treats cystitis, urethritis, prostatitis, urinary stones.
Zingiber officinale Roscoe	China	Essential oils, zingiberol, zingiberene, phellandrene, camphene, citral, linalool, methylheptenone, nonylaldehyde, d-borneol, gingerol.[53]	Anti-inflammatory; stimulates gastric secretion.
	N.A.	Volatile oil, gingerol, shogaols, l-zingiberene.[99,107,183,184]	Carminative, circulatory stimulant, anti-inflammatory, antiseptic.
Ziziphus jujuba Mill.	China	Saponins, betulinic acid, betulic acid, betulin, jujuboside A, jujuboside B, sanjoinines, daechu alkaloids.[1,33,44,53]	For insomnia, neurasthenia, and irritation.
	N.A.	Saponins, flavonoids, sugars, mucilage, vitamins A, B, and C.[99]	Improves muscular strength, weight gain; increases stamina.

*This information should not be used for the diagnosis, treatment, or prevention of diseases in humans. The information contained herein is in no way intended to be a guide to medical practice or a recommendation that herbs be used for medicinal purposes. The information is presented here mainly for educational purposes and should not be used to promotes the sale of any product or replace the service of a physician.

Table 3 Chinese and North American Medicinal Herbs Belonging to the Same Genus and Different Species: Major Constituents and Therapeutic Values

Source	Scientific Name	Major Constituents	Therapeutic Values*
China	*Abutilon theophrasti* Malv. *A. avicennae* Gaertn. Fruct. Sem.	Rutin, pentose, pentosans, uronic acid, methylpentosans, methypentose, oil, protein.[48]	Treats dysentery, fevers; a diuretic.
N.A.	*Abutilon indicum* (L.) Sweet	Mucilage, tannins, asparagine.[99]	For bronchitis, skin conditions such as boils and ulcers, threadworms.
China	*Actaea asiatica* Hara	*trans*-Aconitic acid. This herb is toxic.[51]	A prophylactic against pestilence, malaria, evil miasmas.
N.A.	*Actaea rubra* (Ait.) Willd. *A. alba* L.	Resin, *trans*-aconitic acid, protoanemonoid compound.[102]	Treats headache, insomnia, melancholy, and convulsions.
China	*Adenophora coronopifolia* Fisch. *A. paniculata* Nanuf. *A. pereskiaefolia* (Fisch.) G. Don *A. polymorpha* Ledeb. *A. remotiflora* (Sieb. et Zucc.) Miq. *A. stenanthina* (Ledeb.) Kitag. *A. tetraphylla* Mak.	Saponins.[33]	Hemolyze blood cells, stimulates myocardial contraction; antibacterial.
China	*Adenophora triphylla* (Thunb.) A. DC *A. verticillata* Fisch.	Inulin, taraxerone, beta-sitosterol, daucosterol, beta-sitosteryl palmitate, lupenone.[181]	Antidotal, aphrodisiac, demulcent, expectorant, restorative, sialogogue, tonic.
N.A.	*Adenophora stricta* Miq.	No information is available in the literature.	Treats dry coughs, chronic bronchitis, tuberculosis.
China	*Adonis chrysocyathus* Hook F. & T. Thoms. *A. brevistyla* Franch.	Cymarol, corchoroside A, convallatoxin, adonilide, isoramanone, pergularin.[33] This herb is toxic.	Treats heart disease and central nervous system, depression; diuretic.

Table 3   Chinese and North American Medicinal Herbs Belonging to the Same Genus and Different Species: Major Constituents and Therapeutic Values   (continued)

Source	Scientific Name	Major Constituents	Therapeutic Values*
N.A.	*Adonis vernalis* L.	Cardiac glycosides, adonitoxin.[99]	For heart conditions such as irregular beat, low blood pressure.
China	*Agastache rugosa* (Fisch. & Mey.) O. Kuntze	Essential oils, methylchavicol, anethole, anisaldehyde, d-limonene, hexenol, calamene, beta-pinene,	Chest congestion, diarrhea, headache, nausea, antipyretic, carminative, febrifuge, stomachic.
	A. rugosa (Fisch. & Mey.) O. Kuntze f. hypoleuca (Maxim.) Hara	p-methoxycinnamaldehyde, d-pinene, octanol, cymene, linalool, elemene, caryophyllene, farnesene.[48]	
N.A.	*Agastache anethrodora* L. *A. foeniculum* L.	Methylchavicol, anerhole, anisaldehyde.[99,306]	Relieves abdominal distention, nausea, vomiting.
China	*Ailanthus altissima* (Mill.) Swingle	Amarolide, ailanthone, afzelin, syringic acid, vanillic acid, beta-sitosterol, azelaic acid, d-mannitol, amarolide, oleorsin, mucilage.[33,48]	Antidiarrheal; treats dysentery, duodenol ulcers. Astringent, anthelmintic.
N.A.	*Ailanthus glandulosa* Desf.	Quassinoids, ailanthone, quassin, alkaloids, flavonols, tannins.[99]	To counter worms, excessive vaginal discharge, gonorrhea, malaria, asthma, antispasmodic, cardiac depressant.
China	*Ajuga bracteosa* Wallich *A. decumbens* Thunb. *A. pygmaea* A. Gray	Flavon glucoside, luteolin, ecdysones cyasterone, ecdysterone, ajugalactone, ajugasterone, ajugasterone.[33,50]	Antitussive, antipyretic, antiphlogistic, antibacterial. Treats bladder ailments, diarrhea, bronchitis.
N.A.	*Ajuga reptans* L.	Indoid glycosides (harpagide).[99]	An astringent, mild analgesic, laxative.
China	*Akebia quinata* (Hoytt.) Decne.	Aristolochic acid, saponin akebin, triterpenoids.[25,33]	Diuretic, antibacterial.

N.A.	Akebia trifoliata (Thunb.) G. Koidz.	No information is available in the literature.	Controls infection, stimulates the circulatory and urinary systems. Diuretic properties.[345]
China	Aletris formosuna (Hayata) Sasaki A. spicata Franch.	Stigmasterol, beta-sitosterol, diosgenin.[54]	Antitussive, vermifugal, for ascariasis, marasmus, cough.
N.A.	Aletris farinosa L.	Steroidal saponins, diosgenin, volatile oil, resin.[99]	For gynecological problems during menopause. Treats loss of appetite, indigestion, flatulence, and bloating.
China	Alnus japonica (Thunb.) Steudel	Alpha-amyrin, betulinic acid, glutin-5-en-3-ol, heptacosane, lupenone, taraxerol.[48,50]	Antitumor.
N.A.	Alnus crispus (Ait.) Pursh A. glutinosa (L.) Gaertn.	Tannins, resins, phlobaphenes, flavone glycoside, alnulin, taraxerol, protoalnulin, beta-sitosterol.[100,102]	As an astringent, reduces inflammation and internal hemorrhage.
China	Alpinia japonica Miq.	Essential oils, cineole, alpinone, izalpinin, rhamnocitrin, kumatakinin.[56]	Caraminative.
	Alpinia globasum Horan. A. katsumadai Hayata A. kumatake Mak.	Kaempferin, galangin, galangol, cineole, citral, carotene, thiamine, riboflavin.[50]	Caraminative, stomachic; treats malarial disorders, fluxes, and menstruation.
	Alpinia officinarum Hance	Galangol, essential oils, cineol, eugenol, pinene, cadinene, methyl cinnamate, sesquiterpene, dioxyflavonol.[49]	As stomachic in chronic enteritis, dyspepsia, and gastralgia, carminative, antiperiodic, sialogogue.
	Alpinia oxyphylla Miq.	Cincole, zingiberene, zingiberol.[58]	Diuretic, tonic; treats vomiting, and digestive discomfort.
	Alpinia speciosa K. Schum.	Zingiberene, zingiberol.[54]	Stomachic.

Table 3   Chinese and North American Medicinal Herbs Belonging to the Same Genus and Different Species: Major Constituents and Therapeutic Values   (continued)

Source	Scientific Name	Major Constituents	Therapeutic Values*
N.A.	*Alpinia galanga* Miq.	Volatile oil, alpha-pinene, cineole, linalool, sesquiterpene lactones, galangol, galangin.[99]	A stimulant, carminative; prevents vomiting, antifungal.
China	*Althaea rosea* (L.) Cav.	Althaeine, dioxybenzoic acid.[50]	As stomachic, regulative, constructive in fevers, dysentery, diuretic.
N.A.	*Althaea officinalis* L.	Mucilage, asparagine, pectin, flavonoids.[99,177]	For antitussive, bronchitis, asthma, stomach disorder.
China	*Amaranthus caudatus* L.	Betaine.[48]	A tonic.
	Amaranthus blitum Kom. *A. lividus* L. *A. virdis* L.	Vitamins, protein, thiamine, riboflavin, ascorbic acid.[50]	Treats dysentery and inflammation; vermifuge.
	Amaranthus tricolor L.	Beta-carotene, thiamine, riboflavin, niacin, ascorbic acid.[50]	Prevents cancer.
N.A.	*Amaranthus hypochondriacus* L.	Tannins, a red pigment.[99]	An astringent; reduces blood loss, treats diarrhea.
China	*Amomum cardamomum* L. *A. globosum* Lour. *A. tsao-ko* Roxb. *A. villosum* L.	d-borneol, borneol acetate, d-camphor, linalool, nerolidol, terpene.[50]	Treats pyrosis, vomiting, dyspepsia, pulmonary diseases, dyspepsia. Antitoxic, antiemetic, carminative, stomachic.
N.A.	*Amomum xanthioides* Soland ex. Maton.	No information is available in the literature.	Carminative, diuretic; stimulates appetite, relieves indigestion and controls nausea and vomiting.[345]

China	*Anemone cernua* Thunb. A. *raddeana* Regel A. *rivularis* Buch-Hamilton ex DC A. *vitifolia* (Buch-Ham.) Nakai	Raddeanin A, hederasaponin B, raddanoside, ranuneulin, oleanolic acid.[33,48]	Antitumor, anti-inflammatory, antirheumatic arthritis.
N.A.	*Anemone hepatica* (DC.) Ker- Gawl. A. *patens* L. A. *pulsutilla* L.	Lactone protoanemonin (anemonin), triterpenoid saponins, tannins, volatile oil.[99]	For cramps, menstrual problems, distress, spasmodic pain of the reproductive system.
China	*Angelica amurensis* Schischk. A. *anomala* Lallem. A. *dahurica* (Fisch.) Benth. et Hook.	Byak-angelicin, byak-angelicol, oxypeucedanine, imperatorin, phellopterin, xanthotoxine, marmesin, scopoletin, marmesin, scopoletin, anomalin, angenomalin, bergapten.[33]	Antipyretic: treats toothache, headache. Externally for mastitis and wound infection.
	Angelica decursiva (Miq.) Franch. et Savat.	Nodakenin, nodakenetin, decursin, decursidin, umbelliferone, andelin, 3'-angeloyloxy-4'-isovaleroyloxy-3', 4'-dihydroxanthyletin, estragol, estragol, umbelliprenin, imperatorin, decuroside, sioimperatorin, spongesterol, hydroxypeucedanin.[48]	Anodyne, carminative, diuretic, stimulant, suppurative. Treats abscess, boils, catarrh, cold, coryza, dysmenorrhea, epistaxis, fever.
	Angelica grosserrata Maxim. A. *pubescens* Maxim.	Angelic, linoleic, oleic, palmitic, stearic acids.[50]	Antispasmodic, diaphoretic, diuretic. Treats apoplexy, swellings, catarrh, dropsy, headache, leprosy, puerperium.
	Angelica sinensis (Oliv.) Diels	Vitamin B_{12}, vitamin E, ferulic acid, succinic acid, nicotinic acid, uracil, adenine, butylidenephalide, ligustilide, folinic acid, biotin, polysaccharide.[33]	Treats irregular menstruation, anemia, thrombophlebitis, neuralgia, arthritis, chronic nephritis, constrictive aoritis, skin disease such as eczematous dermatitis.

Table 3   Chinese and North American Medicinal Herbs Belonging to the Same Genus and Different Species: Major Constituents and Therapeutic Values   (continued)

Source	Scientific Name	Major Constituents	Therapeutic Values*
N.A.	Angelica archangelica L.	Angelicide, brefeldin A, ligustilide, n-butyldenephthalide, phyllandrene, tinnins, valeric acid, ferulic acid, lactones, limonene, courmarin.[98,99,100,107,272]	Stimulates blood circulation, regulate menstruation, stimulates appetite, and alleviates coughs and pain.
China	Anthriscus aemula (Woron.) Schischk. A. sylvestris (L.) Hoffm.	Anthricin, deoxypodophyllotoxin, isoanthricin, luteolin.[50]	Antitumor, glandular tumors, corns, warts.
N.A.	Anthriscus cerefolium (L.) Hoffm.	Volatile oil, coumarins, flavonoids.[99]	To settle digestion; lowers blood pressure; a diuretic. Externally, juice for wounds, eczema, and abscesses.
China	Apocynum venetum L.	Cymarin, strophantidin, k-strophanthin-β, isoquercitrin, quercetin.[33]	Increases myocardial contractility, lowers blood pressure, and increases bronchial secretion; diruetic.
N.A.	Apocynum androsaemifolium L.	Glucoside apocynamarin, a bitter principle cymarin, apocynein, apocynin, volatile oils, fixed oils, caoutchouc.[100]	For rheumatism, scrofula, and syphilis.
China	Aquilaria agallocha Roxb. A. sinensis Kitam.	Agarospirol, alpha-agarofuran, agarol, beta-agarofuran, benzylacetone, hydrocinnamic acid, hydroagarofuran.[33]	Antiemetic; promotes circulation, relieves pain.
N.A.	Aquilaria flavescens S. Wats.	Hydrocyanic acid.[118] This herb is highly toxic.	Externally for skin diseases.

China	*Aquilegia buergeriana* Sieb. et Zucc. *A. parviflora* Ledeb.	Benzylacetone, terpene, p-methoxybenzylacetone.[48,60]	Treats irregular menstruation, ovary bleeding, shortness of breath, nausea, pain and gas, chills.
N.A.	*Aquilegia vulgaris* L.	Delphinidin-3,5-diglucoside, lipase, nitryl-glycoside, capronic acid, palmitic acid, oleic acid, linoleic acid.[100]	
China	*Aralia chinensis* L. *A. cordata* Thunb. *A. elata* (Miq.) Seem.	Diterpenoids such as (−) pimaradene, (−) kaurene derivatives, l-pimara-8, 15-dien-19-oic acid, aralosides, araligenin, oleanoic acid, beta-taralin, alpha-taralin.[20,48,50]	Carminative, for arthralgia, gastroenteritis, headache; diuretic, antidiabetic, antiseptic.
N.A.	*Aralia catechu* L. *A. nudicaulis* L. *A. racemosa* L.	Arctiin, tannins, diterpene acids, glucoside, volatile oil.[99,102]	Treats rheumatism, asthma, coughs.
China	*Arenaria juncea* Bieb. *A. serpyllifolia* L.	Saponin.[50]	Antitussive, detoxicant, diuretic, febrifuge;treats cough, pulmonary tuberculosis, dysentery.
N.A.	*Arenaria rubra* (Wahlenb.) Sm.	Resin.[346]	Relaxes muscle walls of the urinary tubules and bladder. Treats kidney stones, acute, and chronic cystitis.
China	*Aristolochia contorta* Bunge. *A. kaempferi* Willd. *A. longa* Thunb. *A. recurvilabra* Hance.	Aristolochic acid A, aristolochic acid D, aristoloside, magnoflorine, oleanoic acid, beta-sitosterol, hederagenin.[48] This herb is toxic.	Treats pulmonary disorders; antitussive, an expectorant in asthma and bronchitis.
China	*Aristolochia debilis* Sieb. et. Zucc.	Aristolochic acid, debilic acid, magnoflorine, dibilone, cyclancline, aristolone.[33]	Antihypertensive; lowers heart rate and myocardial contractility, vasodilation.

Table 3   Chinese and North American Medicinal Herbs Belonging to the Same Genus and Different Species: Major Constituents and Therapeutic Values   (continued)

Source	Scientific Name	Major Constituents	Therapeutic Values*
China	*Aristolochia manshuriensis* Kom.	Aristolochic acid, saponin akebin, triterpenoids.[25,33]	Diuretic, antibacterial.
	Aristolochia shimadai Hayata	Aristolochic acid.[54]	Relieves pain, a diuretic; externally for snakebite.
N.A.	*Aristolochia clematitis* L. *A. serpentaria* L.	Aristolochic acids, volatile oil, tannins.[99]	Treats wounds, sores, snakebite; taken after childbirth to prevents infection, heal ulcers, treats asthma and bronchitis.
China	*Armeniaca ansu* (Maxim.) Kostina *A. mandshurica* (Maxim.) Skvortzov. *A. sibirica* (L.) Lam. *A. vulgaris* Lam.	Amygdalin, hydrocyanic acid.[48,49]	An astringent, stomachic, antipyretic.
N.A.	*Prunus americana* Marsh.	Amygdalin, cyanogenic glycoside, laetrile, hydrocyanic acid, tannins.[99]	Treats cancer, coughs, asthma, and wheezing.
China	*Artemisia apiacea* Hance ex Walpers	Dihydroartemisinin, artesunate, artemisinin, chloroquine.[33] This herb is mildly toxic.	A schizonticidal agent, antimalarial.
	Artemisia argyi Leveille & Vaniot *A. halodendron* Turez. ex Bess. *A. igniaria* Max. *A. indica* Willd. *A. integrifolia* L. *A. japonica* Thunb. *A. keiskeana* Miq. *A. scoparia* Waldst. & Kitaib. *A. selengensis* Turcz. ex Bess.	Terpinenol-4, β-caryophyllene, artemisia alcohol, linalool, cineol, camphore, borneol, eucalyptol.[33]	Antiasthmatic, antitussive. Treats chronic bronchitis, oral infection, and hypersensitivity.

China	*Artemisia capillaris* Thunb.	Scoparon, capillene, capillin, capillon, capillarin, capillanol.[33]	A choleretic; treats jaundice, acute infectious hepatitis, gallstone-related illnesses.
	Artimisia finita Kitag. *A. frigida* Willd.	L-beta-santonin, finitin.[48]	Treats intestinal parasites.
	Artimisia gmelini Weber ex Stechmann	Essential oils, borneol, cineole, camphor, azulene, isovaleric acid, umbelliferone, scopoletin, genkwanin.[48]	Treats liver diseases, stops bleeding, arthritis, bronchitis.
	Artimisia lactiflora Wallich	Flavonoid glycoside, coumarin, lactiflorenol, spathulenol, s-guaiazulene, beta-guaienen, *trans*-β-farnesene, *trans*-caryophyllene, limonene, elemene, copaene, myrcene.[57]	Diuretic; regulates menstruation, treats headache, high blood pressure.
N.A.	*Artemisia absinthium* L.	Absinthol, tannins, thujyl alcohol, flavonoids, phenolic acid, lignins.[99,102]	Anthelmintic.
	A. dracunculus L.	Estragole, phelandrine, methyl chavicol, iodine, rutin, tannins, flavonoids, coumarins.[99]	Diuretic, appetite stimulant.
	A. tridentata Nutt.	Furanoid, pentane, volatile oil.[99]	Aromatic, bug repellent.
China	*Aspidium falcatum* Sw.	Filicic acid, tannin, essential oil.[49] This herb is slightly toxic.	Anthelmintic, hemostatic, antidote.
N.A.	*Aspidium filix-mis* (L.) Schott.	Oleo-resin, triterpenes, alkanes, volatile oil, resins.[99]	Treats tapeworms.
China	*Aster ageratoides* L.	Quercetin, kaempferol.[33]	Antitussive, antiasthmatic; stimulates adrenal cortex.

Table 3   Chinese and North American Medicinal Herbs Belonging to the Same Genus and Different Species: Major Constituents and Therapeutic Values   (continued)

Source	Scientific Name	Major Constituents	Therapeutic Values*
N.A.	*Aster tataricus* L.	Monoterpenes, sesquiterpenes, diterpenes, triterpenes, saponins, flavonoids, coumarins.[271]	Stops bleeding, treats pinkeye (conjunctivitis).
China	*Astragalus chinensis* L.	Astragalin, canavanine, homoserine.[33]	Sedative, antibacterial, antiviral.
	A. complanatus R. Fr. ex Bunge. *A. henryi* Oliv. *A. hoantchy* Franch. *A. melilotoides* Pallas. *A. mongholicus* Bunge. *A. reflexistipulus* Franch. *A. sinensis* L.	Gama-aminobutyric acid, astragalin, canavanine, coumarin, flavonoid derivatives, saponins, polysaccharide, cycloastrangenol, betaine, rhamnocitrin, saponin, astragalosides, formononetin, homoserine, isoliquiritigenin, kaempferol, quereetin, cosin.[1,33,53]	Hypotensive, antirhinoviral, antitumor, antipyretic, diuretic, tonic, an immuno-moderating agent.
N.A.	*Astragalus americana* Bunge.	Asparagine, calcyosm, sterols, formononetin, kumatakenin.[99]	Improves immune system, lowers blood pressure.
China	*Atractylodes lancea* Bunge. *A. chinensis* DC *A. japonica* Koidz. ex Kitam. *A. koreana* (Nakai) Kitam. *A. lancea* Bunge. *A. ovata* DC	Atractylone, eudesnol, hinesol.[19]	Diuretic agent, abdominal and chest tightness, anemia chills, bronchitis cough, diarrhea, CNS suppressing activity.
N.A.	*Atractylodes macrocephala* Koidz.	Atractylol, lactones, atractylenolide II and III.[99]	Protects liver, to relieves fluid retention, excessive sweating, diarrhea, vomiting.

China	Avena fatua L.	Aminoadipic acid, glucovanillin, trigonellin, leucin, isoleucin, threonin, asparaginic acid, oxylysin, beta-sitosterol, aconitic acid, avenasterol, secalose, erucic acid, xanthopyhllepoxyd.[48]	Stops bleeding; a tonic.
N.A.	Avena sativa L.	Proteins, vitamin B complex, saponin, carotenes.[102,138]	Antidepressant; heals skin disorders.
China	Belamcanda panctata Moench.	Tectoridin.[50]	Antipyretic, antifungus, analgesic, detoxicant, stomachic. Externally for boils, cancer, contusions, swellings.
N.A.	Belamcanda chinensis (L.) DC	Belamcandaquinones A and B, isoflavones, tectoridin, iridin, iridals, tectorigenin, irigenin, irisflorentin.[318,319,320]	Treats throat disorders, stimulates the mucous membrane of the throat.
China	Berberis amurensis Rupr. B. poiretii Schneid. B. sibirica Pall. B. soulieana C. K. Schneid.	Berberine, berbamine, palamatine, jatrorrhizine, oxycanthine.[33]	Antibacterial; promotes leukocytosis, choleretic.
N.A.	Berberis aquifolium L.	Berberine, protoberberine alkaloids, oxyberberine, magnoflorine, columbamine.[100,273,274]	For eczema, gall bladder disorder, chronic hepatitis B, gastritis, diarrhea, antisporiasis.
	Berberis vulgaria L.	Berberine, tannins, resin, berbamine, berberubine.[99,100]	Improves liver function; antiseptic and antidiarrhea.
China	Betula mandshurica (Regel) Nakai B. platyphylla Suk.	Betuloside, betulafolienetriol, betulafolienetetraol, betulin.[48,50]	Anticancer, mammary carcinoma.
N.A.	Betula lenta L. B. pendala Roth. B. verrucosa J. F. Ehrh.	Saponins, hyperoside, tannins, gallic acid, methyl salicylate, essential oil.[102]	For headaches, rheumatic pain; anti-inflammatory.

Table 3   Chinese and North American Medicinal Herbs Belonging to the Same Genus and Different Species: Major Constituents and Therapeutic Values   (continued)

Source	Scientific Name	Major Constituents	Therapeutic Values*
China	*Bidens bipinnata* L. *B. parviflora* Willd. *B. pilosa* L. var. minor (Blume) Sheff.	Flavonoids, essential oils. [48]	Treats bug bites, diarrhea, snakebite.
N.A.	*Bidens tripartita* L. *B. connata* Muhl.	Flavonoids, xanthophylls, volatile oil, acetylenes, sterols, tannins. [99]	An astringent, diuretic.
China	*Bignonia chinensis* Lam. *B. grandiflora* Thunb.	Protein, dextrose, cyanidin-3-rutinoside. [48]	As emmenagogue. Treats amenorrhea, dysmenorrhea, leucorrhea, menorrhagia, metrorrhagia.
N.A.	*Bignonia catalpa* (L.) Karst.	Catalpine, oxylenzoic acid, protocatechetic acid. [99]	Treats asthma, whooping cough, spasmodic coughs.
China	*Blumea hieraciifolia* (D. Don) DC	No information is available in the literature.	Treats pneumonia, water in the lung, diarrhea, snakebite.
	Blumea lacera (Burm. f.) DC	Carotene, coniferyl alcohol, angelic acid, vitamin C, cineole, citral, fenchone, camphor. [48,56]	An insectifuge, vermifuge; treats cholera, eczema, fever, itch, scurvy.
	Blumea riparia (Blume) DC var. megacephala Randeria	No information is available in the literature.	Treats headache, relieves colic.
N.A.	*Blumea balsumifera* (L.) DC	Flavonoids, sesquiterpene lactones, camphor. [316,317,345]	Carminative, vermifuge, disphoretic; an expectorant.

China	*Buxus harlandii* Hance	Cyclovirobuxine D, buxanmine E, cycloprotobuxine C, buxpiine K.[58]	Improves blood circulation, enhance sheart muscle, regulate heartbeat; treats hepatitis, arthritis.
	Buxus microophylla Sieb. et Zucc.	Cyclovirobuxine C and D, buxtamine E, cycloprotobuxamine A and C, buxtauine, buxpiine.[58]	Treats heart conditions; a detoxicant.
N.A.	*Buxus sempervirens* L.	Steroidal alkaloids, alpha-tocopherol.[125,275]	Used for recurrent fevers, rheumatism, intestinal parasites.
China	*Caesalpinia decapetula* (Roth.) Alston	Volatile oil, bonducin, saponin, glycoside.[60]	Astringent, anthelmintic, antipyretic, antima arial.
	Caesalpinis pulcherrima Swartz	Alkaloid, gallic acid, resins, tannins.[60] This herb is toxic.	Febrifuge, stomachic, diuretic, astringent, anticho.leric.
	Caesalpinis sappan L.	Brasilin, tetraacetylbrazilin, proesapanin A, essential oils, tannic acid, gallic acid, saponin.[33,49,50]	Activate blood flow, removes blood stasis, reduces swelling; against human cancer cells.
N.A.	*Caesalpinia ascendens* L. C. *bonducella* L. C. *sylvatica* L.	Fix oil, bonducin, tannins.[99]	Treats fever; aphrodisiac.
China	*Caltha palustris* L. var. sibirica Regel	Anemonin, protoanemonin, choline, hellebrin, cevadine, berberine, scopoletin, saponin, umbelliferone, isorhamnetin, xanthophyllepoxyl.[48,50]	Antirheumatic, antitumor.
N.A.	*Caltha leptosepala* DC	No information is available in the literature.	Diaphoretic, emetic, expectorant. Diuretic, laxative, antitumor activity.[347]

Table 3   Chinese and North American Medicinal Herbs Belonging to the Same Genus and Different Species: Major Constituents and Therapeutic Values   (continued)

Source	Scientific Name	Major Constituents	Therapeutic Values*
China	Calystegia hederacea Willich ex Roxb. C. japonica Choisy iu Zoll.	Kaempferol, kaempferol-3-rhamnoglucoside, columbin, palmatine.[48,50]	Diuretic; stimulates kidney secretions.
N.A.	Calystegia sepium (L.) R. Br.	Lectin, calystegins.[324,325]	Glycosidase inhibitor.
China	Campanula glomerata L. f. canescens (Maxim.) Kitag. C. giauca Thrunb. C. grandiflora Jacq. C. punctata Lam.	Quercetin, isorhamnetin, kaempferol, hyperoside, isoquercetin, trifolin, chlorogenic acid, methyl caffeate, coumaroylquinic acid.[48]	For throat infection, headache.
N.A.	Campanula rotundifolia L. C. palustris L.	Lutein-7-primveroside, luteolin-7-0-beta-D-glucopyranosil, rhamnetin-3-0-beta-D-galactoside, esculetin, caffeic, n-coumaric, ferulic acids.[302]	For faintness or a weak heart, stops bleeding, reduces swelling.[274]
China	Cardamine leucantha (Tausch.) O. E. Schulz.	Erucic acid, linolenic acid, linoleic acid, oleic acid, sinigroside, lecithine, myrosinase.[60]	Treats abdominal pain; antidysenteria.
N.A.	Cardamine pratensis L.	Minerals, vitamin C.[274]	Stimulates appetite, eases indigestion, cough remedy.
China	Carduus acaulis Thunb. C. crispus L. C. japonicus Franch.	Essential oils, glycoside, bitter principle.[49]	Hemostatic.

N.A.	Carduus benedita L.	Lignins, sesquiterpene lactones (cnicin), volatile oil, polyacetylenes.[98,99]	Stimulates the secretions of the stomach, intestines, gallbladder.
	Carduus marianus L.	Flavonlignans (silymarin), polyacetylenes.[99]	Protect the liver, stimulates secretion of bile, and increases breast-milk production. An antidepressant.
China	Cassia alata L.	Fatty acids, aloe-emodin, rhein chrysarobin, chrysophanic acid, oxymethyl anthraquinone.[48]	Improves night vision, migraines, purgative, astringent.
	Cassia angustifolia Vahl.	Sennosides, aloe-emodin, dianthrone glucoside, rhein monoglucoside, rhein, kaempferin, myricyl alcohol, anthraquinone derivative.[33]	Purgative, laxative, cathartic.
	Cassia nomame (Sieb.) Honda C. obtusifolia L.	Anthraquinones such as emodin, chrysophanol, physcion, rhein aurantioobtusin, obtusifolin, chryso-obtusin, naphthopyrones, obtusin, aurantio-obtusin rubrofusarin, nor-rubrofusarin, toralacton.[33]	Purgative; treats ophthalmia, hypercholesterolemia, vaginitis.
	Cassia occidentalis L. C. tora L. C. torosa Cav.	Anthraquinones, torosachrysone, N-methylmorpholine, apigenin, galactomannan, cassiollin, xanthorin, dianthronic heteroside, helminthosporin.[4,33]	Mild purgative; lowers blood pressure; antibacterial, antiasthmatic, antitoxic.
	Cassia siamea Lamark	Chrysophanic acid, chrysarobin, oxymethylanthraquinone.[60]	A tonic to relieves stomach pains.
N.A.	Cassia senna L.	Anthraquinone, beta-sitosterol, rhein, dianthrone glucosides, sennosides A, sennosides B, naphthalene glycosides, aloe-emodine, mucilage.[99,100]	Laxative, stimulant, cathartic, anticancer, cathartic.
China	Castanea crenuta Sieb. et Zucc. C. striatus Thunb.	Quercetin, urea, protein, beta-carotene, riboflavin, thiamine, ascorbic acid, niacin.[48,50]	Treats diahrrea, poisoned wounds, lacquer poisoning; astringent.

Table 3 Chinese and North American Medicinal Herbs Belonging to the Same Genus and Different Species: Major Constituents and Therapeutic Values (continued)

Source	Scientific Name	Major Constituents	Therapeutic Values*
N.A.	*Castanea sative* Mill.	Tannins, plastoquinones, mucilage.[99]	Treats whooping cough, bronchitis, sore throat.
China	*Caulophyllum robustum* Maxim.	Magnoflorine, taspine, methylcytisine, alpha-lupanine, cauloside, hederagonin.[48]	Treats arthritis, wounds; regulates menstruation.
N.A.	*Caulophyllum thalietroides* (L.) Michx.	Caulophylline, caulosaponin, methylcytisine, anagyrine, steroidal saponins, laburnine, magnoflorine.[99,103,276]	Antispasmodic, diuretic, antirheumatic; promotes menstrual flow, induce abortion.
China	*Celtis bungeana* Blume *C. sinensis* Pers.	Essential oils.[48]	For dyspepsia, poor appetite, shortness of breath, swollen feet.
N.A.	*Celtis australis* L.	Tannins, mucilage.[99]	Reduces heavy menstrual flow, intermenstrual uterine bleeding.
China	*Centaurium meyeri* (Bunge.) Druce	Bitter glycoside, ophelic acid, chiretta.[60]	Treats headache, fever, and infections.
N.A.	*Centaurium erythraea* Rafn.	Secoiridoid glucosides, xanthones, benzophenone, swertiamarin, gentiopicroside.[305,321,322,323]	Antipyretic, antidiabetic.
China	*Chrysanthemum boreale* (Makino) Makino *C. jucundum* Nakai & Kitag. *C. koraiense* Nakai. *C. procumbens* Lour. *C. sinense* Sabine	Alpha-pinene, limonene, carvone, cineol, camphore, borneol, chrysanthinin, yejuhualactone, chrysanthemaxanthin.[33]	Antibacterial; relieves headache, insomnia, and dizziness due to high blood pressure.
China	*Chrysanthemum cinerriaefolium* Visiont	Essential oil, adenine, choline, stachydrine.[60]	Used as an insecticide.

N.A.	*Chrysanthemum parthenium* (L.) Benn.	Camphor, tannins, mucilage, sesquiterpene lactone.[277]	Treats fevers, migraine, arthritis, colds, indigestion, diarrhea, hysteria.
	Chrysanthemum vulgare L.	Thujone, borneol, camphor.[98,102]	Antispasmodic, vermifuge, emmenagogues.
China	*Clematis armandii* Franch. *C. heracleifloia* DC	Aristolochic acid, saponin akebin, triterpenoids.[25,33]	Diuretic, antibacterial.
	Clematis chinensis Retz. *C. florida* Thunb. *C. hexapetala* Pall. *C. minor* Lour. *C. sinensis* Lour. *C. terniflora* DC	Anemonin, anemonol, sapcnins.[33,49]	Analges a, diuresis, carminative, diuretic; treats arthritis, backache, headaches.
	Clematis intricata Bunge. *C. mandshurica* Rupr.	Clematoside A, oleanolic acid.[48]	Relieves arthritis pain and related infections.
N.A.	*Clematis vitalba* L. *C. virginiana* L.	Protoanemonin, saponins.[99] This herb is toxic.	Analgesic; relieves pain to arthritic joints, diuretic, counters urinary problems.
China	*Clinopodium chinense* Benth. *C. gracile* (Benth.) O. Kuntze. *C. polycephalum* Benth. *C. umbrosum* (Bleb.) C. Koch.	Dydimin, hesperidin, sicsakuranetin, apigenin, ursolic acid.[48]	Hemostatic, stimulates uterine contractions; antibacterial.
N.A.	*Clinopodium acinos* L.	No information is available in the literature.	
China	*Commiphora myrrha* Engler	From gum resin, essential oils including myrcene, alpha-camphorene, Z-guggulsterol, guggulsterol, makulor, cembrene.[33]	Stimulates blood flow, relieves pain, promotes tissue regeneration.

Table 3   Chinese and North American Medicinal Herbs Belonging to the Same Genus and Different Species: Major Constituents and Therapeutic Values   (continued)

Source*	Scientific Name	Major Constituents	Therapeutic Values*
N.A.	*Commiphora molmol* Engl. ex Tschirch. C. *myrrha* Engler	Gum, acidic polysaccharides, resin.[99]	Antiseptic, astringent, expectorant, anti-inflammatory, antispasmodism, carminative.
China	*Convallaria keiskei* Miq.	Convallatoxin, convalloside, convallamarin, convallatoxol. This herb is toxic.[33]	Treats heart disease, detoxifies the liver.
N.A.	*Convallaria majalis* L. C. *sepium* L.	Cardiac glycosides, cardenolides, convallotoxin, convalloside, convallatoxol, flavonoid glycosides.[99]	Affect in heart failure, regulate heart beat, and lowers blood pressure.
China	*Convolvulus arvensis* L.	Quercetin, kaempferol, caffeic acid, beta-methylaesculetin.[48]	Improves blood circulation, relieves pain and itchiness.
N.A.	*Convolvulus jalapa* L.	Resin, convolvulin.[99] Large dose can cause vomiting.	Elimination of profuse watery stools.
China	*Coptis japonica* Makino	Berberine, coptisine, urbenine, worenine, palmaline, jatrorrhizine, columbamine, lumicaerulic acid.[33,60] This herb is toxic.	Antiarrhythmic, antibacterial, antiviral, antiprotozoal, anticerebral ischemic.
N.A.	*Coptis groenlandica* Salisb. C. *trifolia* (L.) Salisb.	Isoquiniline alkaloids, berberine, coptisine.[99]	For indigestion and stomach weakness. Treats peptic ulcers. A mouthwash, lotion for canker sores.

China	Cornus alba L. C. kousa Hance. C. macrophylla Wallich	Quercitol, kaempferol, phenethylamine, dihydroxyglutamic acid.[48]	Astringent, antimalarial; treats arthritis, backache, diabetes, hepatitis, malaria, metrorrhagia, cancer.
	Cornus officinalis Sieb. et Zucc.	Morroniside, 7-O-methyl-morroniside, sworoside, loganin, longiceroside, tannic acid, resin, tartaric acid, cornin, gallic acid, malic acid.[33,60]	Diuretic treats dysmenorrhea, excessive menstruation, impotency, backache, dizziness.
	Cornus walteri Wangerin	Fatty acid, loganin, linolenic acid.[48,53]	An astringent.
N.A.	Cornus canadensis L.	Cornine, cornic acid, quercitin, tannins.[102]	Decreases inflammation, pain, fever.
	Cornus florida L.	Verbenalin, saponins, tannins, resin, gallic acid, malic acid, tartalic acid, tannic acid.[100,123]	An astringent, tonic, and hemostatic.
China	Corydalis ambigua Cham. et Schlecht. var. amurensis Maxim. C. repens Mandl. et Muehld. C. ternata (Nakai) Nakai C. turtschaninovii Besser.	d-Corydaline, corydalis, columbamine, dl-tetrahydropalmatine, crybulbine, tetrahydrocoptisine, dehydrocorydaline, corydalamine, tetrahydrocolumbamine, protopine, alpha-allocryptopine, coptisine, dehycrocorydalmine.[33] Toxic if overdosage.	Analgesic, sedative, hypnotic, synergistic; increases coronary blood flow.
	Corydalis bungeana Turcz.	Protopine, pallidine, sinocecatine, corynoline, isocorynoline, coptisine, corycavine, acetylcorynoline, corynoloxin, coreximine, reliculine, corydamine, scoulerine.[33,50]	For rectal prolapse, abscesses, hemorrhoids.
	Corydalis decumbens (Thunb.) Pers.	Protopine, bulbocapnine, d-tetrahydropalmatine.[33] Toxic if overdosage.[33]	Relieves pain after bone fractures.;
N.A.	Corydalis solida (L.) DC	Corydalis, corydaline, leonticine, tetrahydropalmatine, protopine.[99]	An analgesic, antispamodic, sedative.

Table 3   Chinese and North American Medicinal Herbs Belonging to the Same Genus and Different Species: Major Constituents and Therapeutic Values (continued)

Source	Scientific Name	Major Constituents	Therapeutic Values*
China	*Corylus heterophylla* Fisch. ex Besser. *C. mandshurica* Maxim. ex Rupr.	Beta-carotene, thiamine, riboflavin, niacin, ascorbic acid.[50]	To improves appetite; a digestive.
N.A.	*Corylus avellana* L. *C. cornuta* Marsh. *C. rostrata* Marsh. *C. americana* Marsh.	Tannins, essential oil, ferric oxide, beta-sitosterol.[102]	For coughs, colds, diuretic, prostaglandin inhibition, anti-inflammation.
China	*Crataegus cuneata* Sieb. et Zucc. *C. chlorusarca* Maxim. *C. dahurica* Koehne ex Schneid. *C. maximowiczii* Schneid. *C. pentagyna* Waldst. et Kit. *C. pinnatifida* Bunge. *C. sanguinea* Pall.	Flavonoids, quercetin, hyperoside, l-epicatechin, d-catechin, saponins, chlorogenic acid, caffeic acid, citric acid, crataegolic acid, maslinic acid, ursolic acid.[13,33]	Cardiotonic agent; treats hypercholesterolemia, angina pectoris, hypertension.
N.A.	*Crataegus laevigata* *C. monongyna* Jacq. *C. oxyacantha* L.	Flavonoid glycosides, procyanidins, catechins, triterpenoid acid, pectins, amygdalin, proanthocyanidins, emulsin, tartaric acid, tannins, crataegus acid, rutin, coumarins, quercitin, amines.[99,100,231,278,279]	Therapeutic treatment of heart insufficiency, hypotensive, coronary blood supply, arrhythmia.
China	*Cucurbita moschata* Duch. *C. pepo* L.	Cucurbitine.[33]	Treats taeniasis.
N.A.	*Cucurbita maxima* L.	Linoleic acid, oleic acid, cucurbitacins, vitamins.[99]	Against tapeworms in pregnant women and in children; treats nephritis, urinary problems.

China	Cuscuta australis R. Brown	Carotenoids, alpha-carotene-5,6-epoxide, taraxanthin, lutein.[48]	For fever, constipation; diuretic.
	Cuscuta chinensis Lam. C. europaea L. C. japonica Choisy C. lupuliformis Krocker	Cuscutalin, bergenin, cuscutin, amarbelin, cholesterol, campesterol, beta-sitosterol, stigmasterol, beta-amyrin.[48]	Improves immunity, increases blood sugar metabolism.
N.A.	Cuscuta epithymum Murr.	Flavonoids, hydroxycinnamic acid, bergegin.[102]	For kidney disorder, liver disease.
China	Cydonia sinensis Thou.	Vitamin C, malic acid, tartaric acid, citric acid, hydrocyanic acid.[49]	As astringent in diarrhea, analgesic in arthralgia, gout, cholera.
N.A.	Cydonia oblonga Mill.	Tannins, pectin, mucilage, cyanogenic glycosides, amygdalin, fixed oil, tannins.[99]	For diarrhea, mouthwash; gargle to treats canker sores, gum problems, and sore throat.
China	Cymbopogon citratus (DC) Stapf. C. goeringii (Steud.) A. Camus	Elemicin, cymbopogonol, citral, caprilic, dipentene, methylheptenone, linalool, geranic, methylheptenol, nerol, alpha-terpineol, geraniol, farnesol, citrogellol, decanal, citronellal, farnesal, beta-dihydropseudoionone, isovaleric, citronellic.[50,60]	Treats blood in the urine, fever, antiseptic, preservative.
	Cymbopogon distans (Nees ex. Steud.) J. F.Watson C. nardus Rendle	Piperitone.[33]	Antagonizes muscle contraction; antitussive, antibacterial.
N.A.	Cymbopogon citratus (DC ex Nees) Stapf. C. martinii (Roxb.) Wats. C. winterianus Jowitt	Volatile oil, citral, citronellal.[100,107,117,124]	Treats digestive problems, relieves cramping pains.

Table 3   Chinese and North American Medicinal Herbs Belonging to the Same Genus and Different Species: Major Constituents and Therapeutic Values   (continued)

Source	Scientific Name	Major Constituents	Therapeutic Values*
China	*Cynoglossum divaricatum* Stemphan	Potassium nitrate.[96]	A diuretic.
N.A.	*Cynoglossum officinale* L.	No information is available in the literature	
China	*Cyperus difformis* L. *C. glomeratus* L.	Allelopathic essential oils, terpenes, alpha-cyperone, beta-selinene, alpha-humulene.[60,197,198]	A vermifuge, antidote; remedy for dysentery.
	Cyperus rotundus L.	Essential oils, alpha-cyperene, beta-cyperene, alpha-cyperol, beta-cyperol, cyperoone, patchoulenone, kobusone, capadiene, epoxyquaine, rotundone, rotunol.[33]	Treats dysmenorrhea, menstrual irregularities.
N.A.	*Cyperus esculentus* L. *C. brevifolius* (Rottb.) Hassk.	Fixed oil (chufa, tiger nut oil).[99]	A digestive tonic; promotes urine production and menstruation.
China	*Cypripedium guttatum* Swartz *C. macranthum* Swartz.	Flavonoids, phenol, sterols, vitamin C.[48]	Diuretic; improves blood circulation, relieves pain.
N.A.	*Cypripedium calceolus* L. *C. pariflorum* var. pubescens *C. calceolus* L. var. pubescens	Cypripedin, tannic acid, gallic acid.[100]	Treats headache, nervousness; anodyne, antispasmodic, sedative.
China	*Daphne fortunei* Lindl. *D. giraldii* Nitsche *D. koreana* Nakai	Genkwanin, yuanhuacine, apigenin, 12-benzoxydaphnetoxin, genkwadaphnin, hydroxygenkwanin, yuanhuadine, yuanhuatine.[33,53] This herb is toxic.	Induces abortion, treats chronic bronchitis, malaria, cutaneous infections.

N.A.	Daphne mezereum L. D. genkwa Sieb. et Zucc.	Diterpenes (daphnetoxin, mezerein), mucilage, tannins.[99] This herb is toxic.	As an external counterirritant only, for rheumatic joints.
China	Desmodium microphyllum (Thunb) DC	Kaempferitrin.[48]	Antitoxic; relieves diarrhea, cough, pain, snakebites.
	Desmodium pulchellum (L.) Benth.	Bufotenine, nigerine, donoxime.[33]	Antimalarial, antipyretic, antischistosomiasis.
	Desmodium triforum (L.) DC	Potassium oxide, silicic acid, tannin.[60]	For dysentery; antirheumatic, antipyretic, jaundice, gonorrhea. Externally for wounds, abscesses, ulcers.
	Desmodium triquetrum (L.) DC	Potassium oxide, silicic acid, tannin.[50,60]	A tonic for dyspepsia, hemorrhoids, infantile spasms; insecticide, vermicide.
N.A.	Desmodium gangeticum (L.) DC	Volatile oil, alkaloid.[99]	Improves appetite and digestion, treats dysentery and hemorrhoids.
China	Dianthus barbatus L. var. asiaticus Nakai D. oreadum Hance D. superbus L.	Dianthus saponin, essential oils, eugenol.[33]	Antipyretic, diuretic. Treats urinary tract infections, relieves strangury.
N.A.	Dianthus caryophyllus L.	Eugenol, benzyl benzoate, methyl salicylate.[99]	Treats kidney stones, urinary tract infections, blood in the urine.
China	Dipsacus asper Wall.	Essential oil, alkaloid lamire.[50]	Increases the leukocyte count, prevents spontaneous abortion.
N.A.	Dipsacus fullonum L.	Inulin, scabioside.[99]	Diuretic, sweat-inducing and stomach-soothing properties.

Table 3　Chinese and North American Medicinal Herbs Belonging to the Same Genus and Different Species: Major Constituents and Therapeutic Values　(continued)

Source	Scientific Name	Major Constituents	Therapeutic Values*
China	*Dolichos lablab* L.	Glucokinin, plant insulin, tryptophane, arginine, lysine, tyrosine.[62]	Treats menorrhagia, leucorrhea, metritis.
N.A.	*Dolichos pruriens* L.	No information is available in the literature.	
China	*Dryopteris crassirhizoma* Nakai *D. laeta* (Kom.) Christ.	Filmarone, filicic acid, diplotene, albaspididin, flavaspidin, fernene, dryocrassin.[33]	Anthelmintic, an insecticide, antitumor.
N.A.	*Dryopteris filix-mas* (L.) Schott.	Oleo-resin, filicin, triterpenes, alkanes, volatile oil.[99]	For tapeworms.
China	*Ephedra distachya* L. *E. equisetina* Bunge. *E. intermedia* Schrenk ex Mey. *E. monosperma* Gmel. ex Mey. *E. sinica* Stapf.	l-ephedrine, l-methylephedrine, l-norephedrine, methylephedrine, d-pseudoephedrinem, d-N-methylpseudoephedrine.[30,31,33] This herb is toxic.	Treats asthma, sympathomimetic action, relieves headache, body ache, and coughing, and lowers fever by increasing perspiration.
N.A.	*Ephedra nevadensis* Wats.	Pseudoephedrine, l-ephedrine, d-pseudoephedrine.[100]	A decongestant and asthma remedy, for hypertension, hay fever.
China	*Epilobium amurense* Hausskn. *E. hirsutum* L. *E. palustre* L.	No information is available in the literature.	A tonic, galactagogue, stomachache, dropsy. Seed hairs are applied as a styptic.
N.A.	*Epilobium angustifolium* L.	3-O-β-D-glucuronide, mucilage, tannins.[103,117]	Treats skin irritation and burns; gargle for sore throat, laryngitis.
N.A.	*Epilobium parviflorum* Schreb.	Flavonoids, sitosterol, gallic acid derivatives.[147]	Antiphlogistic.

China	*Epimedium koreanum* Nakai *E. brevicorum* Maxim. *E. macranthum* Moore et Decne. *E. tanguticum* (L.) Hausskn.	Icariin, noricariin, korepimedoside A, korepimedoside B, icariine, des-O-methyl-licariine, magnoflorine, epimedoside A, polysaccharides. [33,48]	Dilate the coronary vessels and increases the coronary flow by reducing vascular resistance.
N.A.	*Epimedium sagittatum* Jack.	No information is available in the literature. [345]	Internally for asthma, bronchitis, cold or numb extremities, arthritis, lumbago, impotence, premature ejaculation, high blood pressure.
China	*Erysimum amurense* Kitag. var. bungei (Kitag.) Kitag. *E. cheiranthoides* L.	Erysimoside, erysimosol, erucic acid, canescein, erychroside, helveticosol, erythriside, corchoroside A, erysimotoxin. [35,48]	Treats cold and cold-related infections, sore throat, dizziness.
N.A.	*Erysimum officinale*	No information is available in the literature.	
China	*Erythrina corallodendron* L. *E. indica* Lam. *E. variegana* L.	Alkaloids. [50] This herb is toxic.	Anthelmintic, antisyphilitic, laxative, analgesic in arthritis, neuralgia, rheumatism.
N.A.	*Erythrina centaurium* Lour.	Secoiridoids. [99]	
China	*Eucalyptus robusta* Sm.	Essential oils, cineol, thymol, gallic acid. [33]	Antibacterial, antimalarial. Externally treats *Trichomonas vaginalis.*
N.A.	*Eucalyptus citriodora* Hool. *E. globulus* Labill.	Cineole, eucalyptol, caffeic acid, coumaric acid, gallic acid, gentisic acid, hydroxybenzoc acid, syringic acid, vanillic acid. [99,100]	Externally for athlete's foot, dandruff, herpes, and an inhalation for fevers and asthma.
China	*Euonymus alatus* (Thunb.) Sieb. *E. bungeanus* Maxim. *E. maackii* Rupr.	Quercetin, dulcite, epifriedelinol, friedelin, resin, fatty acid. [33]	Regulates blood flow, relieves pain, eliminate stagnant blood, and treats dysmenorrhea.

Table 3   Chinese and North American Medicinal Herbs Belonging to the Same Genus and Different Species: Major Constituents and Therapeutic Values (continued)

Source	Scientific Name	Major Constituents	Therapeutic Values*
N.A.	*Euonymus atropurpureus* Jacq.	Cardienolides, cardiac glycosides, asparagine, sterols, tannins.[99]	A gallbaldder remedy with laxative and diuretic properties. Treats biliousness, liver problems, eczema, constipation.
China	*Eupatorium chinense* L. var. simplicifolium (Malcino) Kitam. E. formosanum L. E. japonicum Thunb. E. lindleyanum DC E. odoratum L.	Sesquiterpine lactones, eupatolide, eupaformonin, eupaformosanin, michelenolide, costunolide, parthenolide, santamarine.[33]	Anticancer.
N.A.	*Eupatorium perfoliatum* L.	Sesquiterpene lactones (eupafolin), polysaccharides, flavonoids, diterpenes, sterols, volatile oil.[100]	Immunostimulant. Relieves common cold, stimulates resistance to viral, bacterial infection.
China	*Fagopyrum esculentum* Moench. F. sagittatum Gilib.	Rutin, quercetin, caffeic acid, orientin, homoorientin, vitexin, saponaretin, cyanidin, leucoanthocyanin. Seeds contain amylase, linamarase, maltase, phosphatides, protease, quercitol, rhamnose, urease.[48,50]	For colic and diarrhea; stops cold sweats.
	Fagopyrum tataricum (L.) Gaertn.	Rutin, flavones.[48]	For stomachache, leg pain; a digestive.
N.A.	*Fagopyrum tutricum* (L.) Gaertn. F. esculentum Moench.	Bioflavonoids, rutin.[99]	Strengthens the inner lining of blood vessels.
China	*Fragaria indica* Andr.	Emodin, chrysophanic acid, phytosterol, volatile oil, calcium.[60]	Insecticide, antidote, treats whitlow, burns, snakebite.

N.A.	Fragaria vesca L.	Tannins, vitamin C, pectin, citric acid, malic acid.[102]	Stimulates appetite; antidyspeptic.
China	Fraxinus bungeana DC F. chinensis Roxb. F. floribunda Bunge. F. obovata Blume. F. ornus L. var. bungeana Hance F. rhynchophylla Hance.	Fraxin, aesculin.[33] This herb is toxic.	Antibacterial, analgesic, anti-inflammatory.
N.A.	Fraxinus americana L. F. excelsior L. F. ornus L.	Coumarins, flavonoids, tinnins, volatile oil.[99]	A tonic, astringent, laxative, diuretic; treats fevers.
China	Galium bungei Stead. G. spurium L. G. verum L. var. leiocarpum Ledeb.	Alisarin, rubrierythrinic acid, purpurin.[60]	Treats rheumatism, jaundice, menstrual difficulties, epistaxis, hemorrhages.
N.A.	Galium aparine L.	Iridoid valepotriates, polyphenolic acids, anthraquinones, tannins.[99,107]	For vitamin C deficiency.
China	Gaultheria leucocarpa f. var. cumingiana (Vidal) Sleumer	Methylsalicylate, salicylic acid.[60]	Treats rheumatism; an antiseptic.
N.A.	Gaultheria procumbens L.	Methylsalicylate.[100]	Antiseptic, carminative, diuretic.
China	Gelidium amansii Lamx.	Agarose, agaropectin, taurine.[33]	A mild laxative in the treatment of chronic constipation.
N.A.	Gelidium cartilagineum L.	Polysaccharides, agarose, agaropectin, mucilage.[99]	Laxative, stimulating bowel activity and elimination of feces.

Table 3   Chinese and North American Medicinal Herbs Belonging to the Same Genus and Different Species: Major Constituents and Therapeutic Values   (continued)

Source	Scientific Name	Major Constituents	Therapeutic Values*
China	*Geranium dahuricum* DC G. *eriostemon* Fisch. ex DC G. *sibiricum* L. G. *wilfordi* Maxim.	Kaempferitrin, gallic acid, quercetin, succinic acid, tannin.[48,50,65]	Astringent, for diarrhea, endometritis, nervous diseases, numbness of limbs, pains, rheumatism. It helps circulation and strengthens bones and tendons.
N.A.	*Geranium macrorrhizum* L. G. *robertianum* L. G. *maculatum* L.	Tannins.[99]	Treats stomach disorder; aphrodisiac, colitis, peptic ulcer.
China	*Geum aleppicum* Jacquin	Flavones, fatty acids, eugenol, gein, geoside.[48]	Treats bleeding, bug bite, convulsive disorder, fevers, irritability, obstinate skin diseases.
N.A.	*Geum urbanum* L.	Phenolic glycosides (eugenol), tannins, volatile oil, sesquiterpene lactone, cnicin.[99]	Treats mouth, throat, and gastrointestinal tract disorders. For peptic ulcers, irritable bowel syndromes.
China	*Glycyrrhiza pallidiflora* Maxim. G. *uralensis* Fisch. ex DC	Glycyrrhiza, triterpenoid saponin, flavonone glucoside, liquiritin, aglycone, liquiritigenin, chalcone glucose, isoliquiritin, aglycone, isoliquiritigenen.[1,33]	Anti-inflammatory, anticonvulsant, calmative, antidote, antispasmodic, antiulcer.
N.A.	*Glycyrrhiza glabra* L.	Glycorrhizin, mucilage, flavonoids, glycyrrhetinic acid, saponin, glabridin, tannic acid, 2-β-glucuronosyl, glucuronic acid.[99,100,107,280,281,312]	Antiulcerative; treats stomach, duodenal ulcers, anti-inflammatory, antiallergic, antihepatitis.
China	*Hedera rhombea* (Miq.) Bean	Hederin, hederaic acid, tannic acid, oleic acid.[50]	For cough, headache; diaphoretic, emmenagogue.

N.A.	Hedera helix L.	Tannins, hederin, aglycone, iodine, beta-elemone, elixen, hederacoside E, hederacoside C, germacrene B.[100]	An expectorant with antispasmodic and cardiac actions.
China	Hepatica asiatica Nakai	No information is available in the literature.	Anodyne, antifebrile, for angina and sunstroke, local application in smallpox ulcerations.[33]
N.A.	Hepatica nobilis Gars.	No information is available in the literature.	For bronchial and digestive complaints, and liver and gall bladder disorders.[345]
China	Heracleum dissectum Ledeb.	This herb is used in the same way and as a substitute for Angelica (ferulic acid, succinic acid, nicotinic acid, uracil, adenine, butylidenephalide, ligustilide, folinic acid, biotin, polysaccharide),[33] with less effect.[60]	Relieves headache, toothache, hematuria, gonorrhea, itching skin, swellings; remove corns from the feet.
N.A.	Heracleum maximum Barr. H. lanatum Michx. H. sphondylium L.	Sphondin, psoralen, heraclein, glutamine, essential oil.[102]	For headaches, poor memory, melancholy, agitation, indigestion, and asthma.
China	Hieracium umbellatum L.	Vitamin C, tannic acid.[48]	Relieves pain, bladder infection, diarrhea.
N.A.	Hieracium pilosella L.	Coumarin, umbelliferone, flavonoids, caffeic acid.[99]	Antifungal. Relaxes the muscles of the bronchial tubes, stimulates the cough reflex; and reduces mucus.
China	Hydnocarpus anthelmintica Pierre H. castaneus H. F. & Th.	Hydnocarpus oil, hynocarpic acid, chaulmoogric acid, gorlic acid.[33]	Anthelmintic.
N.A.	Hydnocarpus kurzii (King) Warb.	No information is available in the literature. This herb may cause vomiting, dizziness, and breathing difficulties.[345]	For leprosy, scabies, eczema, psoriasis, scrofula, ringworm, and intestinal worms.[345]

Table 3   Chinese and North American Medicinal Herbs Belonging to the Same Genus and Different Species: Major Constituents and Therapeutic Values   (continued)

Source	Scientific Name	Major Constituents	Therapeutic Values*
China	*Hydrangea macrophylla* (Thunb.) Seringe	Febrifugin, hydrangeic acid, hydrangenol, rutin.[50]	Antimalarial, antitussive, diuretic.
N.A.	*Hydrangea arborescens* L.	Flavonoids, cyanogenic glycoside, saponins, hydrangein, tannin.[99,100]	Treats kidney and bladder stones.
China	*Hyoscyamus bohemicus* F. W. Schmidt	Alkaloid.[60] This herb is mildly toxic.	Antispasmodic activity.
N.A.	*Hyoscyamus niger* L.	Tropane alkaloids, hyoscyamine, hyoscine.[99]	A sedative, painkiller, antispasmodic.
China	*Hyssopus ocymifolius* Lam.	Essential oils, elsholtzia ketone, elsholtzianic acid, furylmethyl ketone, furylpropyl ketone, furylisobutyl ketone, furane, pinene, terpene.[49]	Stomachic, carminative, diuretic.
N.A.	*Hyssopus officinalis* L.	Pinene, limonene, pinecamphene, hesperidin, tannins, terpenes.[99,107]	Treats respiratory problem, coughs, sore throat, hoarseness, asthma, bronchitis.
China	*Ilex chinensis* Sims.	Protocatechuic acid, protocatechuic aldehyde, ursolic acid, tannic acid.[33]	Treats angina pectoris, thrombophlebitis, extremity ulceration.
	Ilex pubescens Hook & Am.	Flavone, ursolic acid, 3,4-dihydroxyacetophenone, scopoletin, hydroquinone, vomifiliol.[33]	Treats angina pectoris, acute myocardial infarction, central angiospastic retinitis, cerebral thrombosis, thrombophlebitis.
N.A.	*Ilex aquifolium* L. *I. paraguensis* st. Hil.	Triterpenoids, salicylic acid, caffeine, isophthalic acid.[147]	Relieves menstrual cramps, calms nervous stomach.

	Species	Compounds	Uses
China	Impatiens balsamina L. / I. noli-tangere L. / I. textori Miq.	Gentisic acid, ferulic acid, p-coumaric acid, sinapic acid, caffeic acid, scopoletin, lawsone. [33]	Treats arthritis, relieves pain.
N.A.	Impatiens pallida Nutt. / I. capensis Meerb.	Lawsone, seed oil contains alpha-spinasterol, beta-ergosterol. [302]	Remedy for rashes, pain caused by insect bites, anti-inflammatory.
China	Inula britannica L. / I. japonica Thunb. / I. linariaefolia Turcz. / I. salsoloides (Turcz.) Ostenfeld.	Inusterol A, taraxasterol, inusterol B, inulicin, flavone, caffeic acid, chlorogenic acid, isoquercitrin, quercetin. [48,50]	Discutient, vulnerary, carminative, deobstruent, diuretic; treats ascites, bronchitis, cancer, chest congestion.
N.A.	Inula helenium L.	Inulin, resin, mucilage, helenalin, dammaranedienol. [99]	For asthma, chest cold, stomach ulcers, antitussive, diuretic, antiseptic.
China	Ipomoea barbata Both. / I. hederacea Jacq. / I. triloba Thunb.	Glycoside pharbitin, gibberellin, pharbilic acid. [33] This herb may be toxic.	Potent purgative; purged parasites, ascaris, and taenia. Treats constipation, ecema.
	Ipomoea cairica (L.) Sweet	Muricatin A, beta-sitosterol. [50]	Purgative.
N.A.	Ipomoea purga (Wender) Hayne	Convolvulin. [99]	Elimination of profuse watery stools.
China	Iris aqyatuca Forskal / I. buatatas (L.) Lamarck. / I. dichotoma Pallas	Tectoridin, iridin, flavon. [48]	Treats lung diseases, cough, pneumonia, uneasy breathing.
	Iris lactea Pall. subsp. chinensis (Fisch.) Kitag.	Iridin, irigenin, irisflorentin. [60]	Astringent, diuretic, hemostatic; remedy for hemorrhage, postpartum difficulties.
	Iris pallasii Fisch.	Irisquinone. [33]	Treats cancer, hepatoma, lymphatic sarcoma.
N.A.	Iris versicolor L. / I. pseudacorus L.	Triterpenoids, salicylic acid, isophthalic acid, alpha-phytosterol, myricyl alcohol. [100]	Relieves menstrual cramps, calms nervous stomach.

Table 3 Chinese and North American Medicinal Herbs Belonging to the Same Genus and Different Species: Major Constituents and Therapeutic Values (continued)

Source	Scientific Name	Major Constituents	Therapeutic Values*
China	*Isatis chinensis* (Thunb.) Nakai	Quercetin, kaempferol, stachyose, manneotetrose, lupeose, cicerose, isatan, indoxyl-5-ketogluconate.[50]	Antiviral, antibacterial; increases blood flow, improves microcirculation, and lowers blood pressue.
	Isatis indigotica Fortune ex Lindley *I. oblongata* DC	Indican, isatan B, indigo, glucobrassicin, meoglucobrassicin.[33]	Antibacterial, antipyretic, anti-inflammatory, choleretic.
N.A.	*Isatis tinctoria* L.	No information is available in the literature.[345]	For meningitis, encephalitis, mumps, influenza, erysipelas, heat rash, sore throat, abscesses, and swellings.
China	*Jasminum mesnyi* Hance *J. nudiflorum* Lindley	Syringin, jasmiflorin, jasmipierin, mannose, tannin.[60]	Diaphoretic.
	Jasminum samba (L.) Aiton	Formic acid, benzoic acid, acetic acid, anthranil acid, sesquiterpene, sesquijasmine.[60] This herb (root) is toxic.	Sedative, anesthetic, vulnerary properties. For congestive headache, lactifuge.
N.A.	*Jasminum grandiflorum* L. *J. officinale* L.	Essential oil, isoquercitrin, ursolic acid, 2-3,4-dihydroxyphenyl-ethanol.[282,283]	Treats high fever, sunstroke, cancer, and Hodgkin's disease.
China	*Juniperus rigida* Sieb. et Zucc.	Alpha-pinene, myrcene, carene, limonene, p-cymene, beta-elemene, caryophyllene, humulene, r-cadinene, terpinene, borneol, citronellol, anethole.[48]	Hemorrhage;treats hemoptysis, inflammation, kidney infection, arthritis joint infection.
	Juniperus rigida Sieb. et Zucc. f. modesta (Nakai) Y. C. Chu	Alpha-pinene, myrcene, carene, limonene, p-cymene, beta-elemene, caryophyllene, humulene, r-cadinene, terpinene, borneol, citronellol, anethole.[48]	Hemorrhage; treats hemoptysis, inflammation, kidney infection, arthritis joint infection.

N.A.	*Juniperus communis* L. *J. horizontalis* Moench. *J. sabina* L.	Resin, pinene, borneol, inositol, juniperin, limonene, cymene, terpinene.[100,102,107]	For dropsy, bladder and kidney disorders, rheumatic pain.
China	*Justicia gendarussa* L. *J. procumbens* L.	Gentianine, gentianidine, gentianol.[33]	Treats rheumatism and fever; antipyretic, anti-inflammatory, antihypersensitivity, and antihistaminic effects.
N.A.	*Justicia adhatoda* L.	Alkaloids, volatile oil.[99]	For bronchitis, tuberculosis.
China	*Lactuca raddeana* Maxim. *L. indica* L. *L. sativa* L.	Pectic compound, oxalic acid, malic acid, citric acid, ceryl alcohol, ergosterol, vitamin E.[50]	Anodyne, lactogogue, for genital swelling, hemorrhoids, lumbago.
N.A.	*Lactuca serriola* L.	Sesquiterpene lactones, lactucopicrin, lactucerin, flavonoids, coumarins.[99]	Sedative, for excitability in children; treats coughs, lowers the libido; relieves pain.
China	*Laminaria angusta* Kjellium *L. japonica* Aresch. *L. religiosa* Miyabe.	Iodine, potassium, calcium, amino acids, laminarin, laminine, algin.[33]	Improves thyroid function, corrects the malignant vicious cycle effect of iodine deficiency, and lowers blood pressure.
N.A.	*Laminaria digitata* (Hudss.) Lank. *L. longicruris* Lank. *L. saccharine* (L.) Lank.	Phenols, polysaccharides, iodine.[284,285,286,287,288]	Treats iodine deficiency; antibiotic, promotes hormone production.
China	*Leonurus heterophyllus* Sweet *L. japonicus* Houttuyn. *L. macranthus* Maxim. *L. mongolicus* V. Kreczet. et Kupr.	Leonurine, stachydrine, leonaridine, leonurinine, vitamin A, fatty oils.[33]	Stimulates uterine contractions, respiratory system, proliferation of T. cells, skeletal muscles.
	Leonurus sibiricus L. *L. sibiricus* L. f. albiflorus (Nakai et Kitag.) G.Y. Wu et H.W. Li	Essential oil, leonurin.[49]	Emmenagogue, diuretic, vasodilator.

Table 3   Chinese and North American Medicinal Herbs Belonging to the Same Genus and Different Species: Major Constituents and Therapeutic Values   (continued)

Source	Scientific Name	Major Constituents	Therapeutic Values*
N.A.	*Leonurus cardiaca* L.	Leonurin, leonuride, pyrogallol, catechins, choline, saponins.[100]	Emmenagogue, cardiologic, astringent, antispasmodic, hypotensive.
China	*Ligusticum chuanziang* Hort.	Tetramethylpyrazine, perlolyrine, leucylphenylalanine anhydride, cnidilide, neocnidilide, ligustilide.[33]	Promotes blood flow, removes blood stasis, and relieves pain.
China	*Ligusticum jeholense* (Nakai et Kitag.) Nakai et Kitag. *L. sinense* Oliv. *L. tenuissimum* (Nakai) Kitag.	Nothosmyrnol.[33]	Induce diaphoresis, for gout, an anodyne, emmenagogue, sedative.
N.A.	*Ligusticum scoticum* L.	Phthalides, terpenoides, essential oil.[99]	Prevents bone marrow loss, treats acquired immune deficiency syndrome, respiratory tract infections, hepatitis, hypertension, Parkinson's disease.
China	*Ligustrum japonicum* Thunb. *L. lucidum* Mill.	Nuzhenide, oleanolic acid, ursolic acid.[33]	Increases leukocyte count; a cardiac tonic, diuretic.
N.A.	*Ligustrum vulgare* L.	Essential oil, phthalides, terpenoides.[100]	Prevents bone marrow loss, treats acquired immune deficiency syndrome, repiratory tract infections, hypertension, Parkinson's disease, and hepatitis.
China	*Lilium japonicum* Thunb. *L. lancifolium* Thunb. *L. pumilum* DC *L. concolor* Salisb.	Protein, colchiceine.[49]	Relieves coughing, eases anxiety, improves digestion, treats anxiety, apprehension, carminative, sedative; gynecologic disorders.

N.A.	*Lilium candidum* L.	No information is available in the literature.
China	*Lindera akoensis* Hayata *L. obtusiloba* Blume f. *villosa* (Blume) Kitag.	Campesterol, linderol, capric acid, lauric acid, myristic acid, linderic acid, dodecen-4-oic acid, tetradecen-4-oic acid, tsudzuic acid, oleic acid, linoleic acid.[48]	Treats wounds, reduces swelling, pain.
	Lindera communis Hemsley	Fatty acids.[55]	Relieves swelling, pain, bleeding, treats infection.
	Lindera glauca (Sieb. et Zucc.) Blume.	Essential oils, cineole, caryophyllene, bornylautate, camphene, beta-pinene, limonene, fatty acids.[55]	Carminative properties; treats arthritis joint pain.
	Lindera megaphylla Hemsley	Essential oils.[55]	Promotes sweating, treats wounds.
	Lindera strychnifolia Vill.	Essential oils including lindestrene, liderane, linderene, linderalactone, isolinderalactone, isolinderoxide, lindestreolide, isofuranogermacrene, linderoxide, neolinderalactone.[33]	Improves circulation, relieves pain, abdominal distention, fever.
N.A.	*Lindera benzoin* (L.) Blume	No information is available in the literature.	
China	*Liquidambar acerifolia* Max. *L. formosana* Hance *L. maximowiczii* Miq.	Balsam (resin), cinnamic alcohol, cinnamic acid, l-borneol, camphene, dipentane, terpene.[60,69]	Analogous, externally as antiphlogistic and astringent in skin diseases, antihemorrhagic.
N.A.	*Liquidambar orientalis* Mill. *L. styraciflua* L.	Levant storax: cinnamic acid, cinnamyl cinnamate, phenylprepyl cinnamate, triterpene acid.[99]	Internally for strokes, infantile convulsions, coma, heart disease, and pruritus.

Table 3   Chinese and North American Medicinal Herbs Belonging to the Same Genus and Different Species: Major Constituents and Therapeutic Values (continued)

Source	Scientific Name	Major Constituentst	Therapeutic Values*
China	*Lobelia chinensis* L. *L. pyramidalis* Wallich. *L. sessilifolia* Lambert	Lobeline, lobelanine, lobelanidine, isolobelamine. Lobeline has been approved by the FDA to curb the tobacco habit.[33,50,71] This herb may be toxic.	Diuretic; increases respiration via stimulation of carotid chemoreceptors. Treats snakebites, insecticide, reduces swelling; depurative, antirheumatic, antisyphilitic.
N.A.	*Lobelia inflata* L.	Lobeline, lobelidiol, lobelanidine, carboxylic acid.[99,100,289]	Respiratory stimulant, antispasmodic; inducec vomiting.
	Lobelia siphilitica L.	Alkaloids.[154]	Treats syphilis.
N.A.	*Lobelia pulmonaria* L.	d-Usnic acid, thamnolic, polysaccharides, anthraquinones.[154]	Stimulates immune system; antitumor, cancer.
China	*Lonicera acuminata* Wallich *L. apodonta* Ohwi *L. brachypoda* DC *L. chinensis* Wats. *L. hypoglauca* Miq.	Luteolin, inositol, lonicerin, loganin, syringin, saponins, tannin, chlorogenic acid, luteolin-7-rhamnoglucoside.[33,48,55]	Antibacterial, cytoprotective, antilipemic, antiphlogistic.
N.A.	*Lonicera caerulea* L. *L. caprifolium* L.	Sorbitol, inositol, limonic acid, malic acid, citric acid, tannins, salicylic acid.[102]	Hypotensive, sedative, antipyretic.
China	*Loranthus parasiticus* (L.) Merr. *L. yadoriki* Sieb. et Zucc.	Saponins, including avicularin, quercetin.[40]	Treats angina pectoris, cardiac arrhythmias, hypertension. Ointment to treats frostbite.
N.A.	*Loranthus europaeus* L.	Flavonoids, kaempferol, quercetin.[328]	

China	*Lycium barbarum* L. *L. chinense* Miller	Betaine, zeaxanthin, physalein, carotine, nicotinic acid, vitamin C.[33]	Increases leukocyte count, increases immunity, stimulation of tissue development.
N.A.	*Lycium pallidum* L.	Betaine, beta-sitosterol, physalin, cinnamic acid, psyllic acid, carotene.[99]	Treats blood pressure, menopausal complaints, chronic fevers, internal hemorrhage, tuberculosis.
China	*Lycopus fargesii* Herter *L. lucidus* Turcz. *L. obscurum* L. *L. phlegmaria* L. *L. veitchii* Christ.	Resin, lycopose, raffincse, stachyose, glucose.[48]	For abdominal distention, abscesses, congestive edema, blood extravasation.
N.A.	*Lycopus virginicus* L.	Phenolic acids, caffeic derivatives, chlorogenic derivatives, ellagic acids.[99]	Treats overactive thyroid gland; an astringent to reduces the production of mucus.
China	*Lysimachia barystachys* Bunge. *L. christinae* Hance *L. clethroides* Duby *L. davurica* Ledeb.	Essential oils, l-pinocamphone, l-menthone, l-pinene, limonene, 1,8-cineol, p-cymene.[33]	Diuretic, a choleretic agent, antibacterial.
N.A.	*Lysimachia vulgaris* L.	Saponins, flavonoids, tannins, benzoquinene.[99]	Treats gastrointestinal conditions such as diarrhea, dysentery; stops bleeding.
China	*Mahonia japonica* DC	Berberine, jatrorhizine.[97]	Antipyretic, backache, cough, dysentery, enteritis, fever.
N.A.	*Mahonia aquifolium* (Lindl.) Don	Berberine, protoberberine alkaloids, oxyberberine, magnoflorine, columbamine.[100,273,274]	Treats eczema, gall bladder disorder, chronic hepatitis B, gastritis, diarrhea, antisporiasis.

Table 3 Chinese and North American Medicinal Herbs Belonging to the Same Genus and Different Species: Major Constituents and Therapeutic Values (continued)

Source	Scientific Name	Major Constituents	Therapeutic Values*
China	*Malva chinensis* Mill. *M. pulchella* Berhn. *M. verticillata* L.	l-arabinose, l-rhamnose, d-galacturonic acid.[75]	Treats stomach and intestinal disorders, to make labor easier; laxative; treats gonorrhea, congestion, constipation.
N.A.	*Malva rotundifolia* L. *M. sylvestris* L.	Flavonol glycosides, gossypin-3-sulfate, mucilage, tannins, anthocyanin, malvin.[99]	A demulcent, a poultice to reduces swelling and draw out toxins. Internally to reduces gut irritation, laxative effect.
China	*Marsdenia tenacissima* (Roxb.) Wight. et. Arn.	Saponins, marsdeoreophisides, metaploxigenin, sarcostin.[33]	Antiasthmatic; hypotensive, antibacterial.
N.A.	*Marsdenia condurango* R. Br.	Condurangogenins, volatile oil, phytosterols.[99]	Stimulates stomach secretions. A digestive tonic.
China	*Melilotus alba* Medicus *M. indica* (L.) All. *M. suaveolens* Ledeb.	Hydroxycinnamic aicd, coumarinic acid, 4-hydroxycinnamic acid, cumaric acid, umbelliferone, scopoletin, mellilotoside, melilotic acid, beta-D-glucosyloxy, dicumarol, chlogogenic acid, caffeic acid, melilotic acid.[48]	Anticoagulant, treats bowel complaints, infantile diarrhea. A bactericide.
N.A.	*Melilotus arvensis* L.	Flavonoids, coumarins, resin, tannins, volatile oil, dicoumarol.[102]	Help varicose veins and hemorrhoids, reduces the rash of phlebitis and thrombosis.
	Melilotus officinalis Lamk.	Flavonoids, resin, tannins, coumarins, hydroxycoumarin, hydrocoumarin.[99,107]	Reduces the risk of phlebitis and thrombosis; sedative, antispasmodic.
China	*Melochia corchonfolia* L.	Trifalin, melocorin, hibifolin.[57]	Poultice to treats sores, swelling, and pain in the abdomen. Also treats vomiting.

N.A.	*Melochia tomentosa* L.	Melovinone, melosatin D, stigmasterol, beta-sitosterol, beta-sitosterol, beta-D-glucoside, octacosanol. [326,327]	Tumorigenic properties.
China	*Menispermum dauricum* L.	Acutumine, acutuminine, dauricine, disinomenine, magnoflorine, menispermine, sinomenine, stepharine, tetrandrine. [50]	Antitumor, cytotoxic; alleviates skin allergies, antirheumatic, anticancer against esophageal cancer.
N.A.	*Menispermum canadense* L.	Dauricine, tetrandrine, viburnito, acutumine, acutomidine, daurinoline, N-desmethyldauricine, magnoflorine. [100] Fruits are toxic if eaten in quantity.	Cyanogenetic, diuretic, laxative, nervine, stomachic.
N.A.	*Menispermum palmatum* L.	Pulegone, pinenes, limonene, lauric acid, myristic acid, palmitic acid, beta-methyl-adipic acid, phenol, cresols, eugenal. [100]	For uterine tumors, uterine fibroids, indurations of the uterus.
China	*Mentha arvensis* L. M. *dahurica* Fisch. ex Benth. M. *haplocalyx* Briq. M. *sachalinensis* (Briq.) Kudo	Menthol, menthone, methyl acetate. [33]	Stimulates gastrointestinal tract motility and central nervous system, dilate peripheral blood vessels. Increases sweat gland secretion.
N.A.	*Mentha pulegium* L.	Pulegone, isopulegone, menthol, terpenoids. [99]	Digestive tonic; relieves flatulence and colic.
N.A.	*Mentha spicata* L. M. x *piperita* L.	Menthol, menthone, isomenthone, pinene, myrcene, limonene, cineole, cymene, terpinene, carvone, luteolin. [99,100,107,130]	Carminative, stomachic, mild antispasmodic, expectorant, antiseptic, and local anesthetic properties.

Table 3   Chinese and North American Medicinal Herbs Belonging to the Same Genus and Different Species: Major Constituents and Therapeutic Values   (continued)

Source	Scientific Name	Major Constituents	Therapeutic Values*
China	*Mimosa arborea* Thunb.	Tannin, saponins.[49]	Tonic, stimulant, anthelmintic.
	Mimosa invisa Mart. et Colla	Minosine.[78] This herb is toxic if overdosed.	Treats neurosis, trauma wound, and hemoptysis. It has a tranquilizing effect.
N.A.	*Mimosa hostilis* Benth.	Nigerine.[100]	An astringent, cure fatigue.
China	*Morinda citrifolia* L. *M. officinalis* L.	Dihydroxy methyl anthraquinone, glucoside morindin, rubichloric acid, alizarin, alpha-methyl ether, rubiadin-l-methyl ether, tannin, morindadiol, soranjudiol, masperuloside, nordamnacanthal.[50]	Treats beri-beri, cancer, lumbago, cholecystitis, increases leukocyte count, stimulates endocrine system.
	Morinda parvifolia Bartling	Methanolic, morindaparvin-A, alizarin-l-methyl ether.[50]	Against p-388 lymphocytic leukemia growth (*in vivo*), cytotoxic, antileukemic.
N.A.	*Morinda didyma* L. *M. fistulosa* L. *M. punctata* L.	Morindin, vitamin C.[99]	Treats impotence and premature ejaculation in men, infertility.
China	*Murraya paniculata* (L.) Jack	L-Cadinene, methylanthranilate, carene, bisabolene, paniculatincoumurrayin, 5-guaizulene, osthol, beta-caryophyllene, gerariol.[33]	Relieves pain, remove toxic substances; an antispasmodic; antagonizes muscular spasms.
N.A.	*Murraya koenigii* (L.) K. Spreng.	Glycoside (koenigin), volatile oil, tannins.[99]	Increases digestive secretions, relieves nausea, indigestion, and vomiting. Treats diarrhea and dysentery.
China	*Myrica rubra* (Lour.) Sieb. et Zucc.	Myricetin.[33]	Treats gastric pain, diarrhea, dysentery.

N.A.	Myrica cerifera L. M. penxylvanica Lois.	Triterpenes, flavonoids, tannins, phenols, resins.[99]	Increases circulation, stimulates perspiration.
China	Nardostachys jatamansi DC	Essential oil, jatamansic acid, sesquiterpene.[49,80]	Aromatic stomachic, sedative, antispasmodic.
N.A.	Nardostachys grandiflora DC	No information is available in the literature.[345]	For nervous indigestion, insomnia, depression, and tension headaches.
China	Nelumbium nelumbo Druce.	Liensinine, isoliensinine, neferine, lotusine, methyl-corypalline, demethyl-coclaurine.[33]	Tranquilizing and antihypertensive.
	Nelumbium nuciferum Gaertner N. speciosum Willd.	Nuciferine, roemerine, anonaine, O-nornuciferine, lirodenine, anneparine, dihydronuciferine, pronuciferine, N-methylcoclaurine, N-methylisococ aurine.[33]	Relaxing effect on smooth muscles, increases essential body energies.
N.A.	Nelumbium officinale L.	No information is available in the literature.[345]	For hemorrhage, nosebleed, excessive menstruation, hypertension.
China	Nerium indicum L.	Oleandrin (toxic), oleandrose.[33]	Treats psychosis, congestive heart failure, analgesic, emmenagogue.
N.A.	Nerium oleander L.	Oleandrin, neriin, folinerin, rosagenin, cornerin, pseudocuramine, rutin, cortenerin, oleandomycin.[100] This herb is highly toxic.	Cardiac, cardiotonic, cyanogenetic, diuretic, emetic, emmenagogue, insecticidal, parasiticide, purgative, sternuatory, stimulant.
China	Nymphaea tetragona Georgi	Amino acids.[48]	A cooling lotion to apply to eruptive fevers; treats colic, gonorrhea, lowers blood pressure.

Table 3   Chinese and North American Medicinal Herbs Belonging to the Same Genus and Different Species: Major Constituents and Therapeutic Values   (continued)

Source	Scientific Name	Major Constituents	Therapeutic Values*
N.A.	*Nymphaea alba* L.	Tinnins, nupharine, nymphaeine, resin.[99]	Astringent, cardiologic and antispasmodic properties, a proprietary medicine to reduces sexual drive.
China	*Oplopanax elatus* (Nakai) Nakai	Essential oil, echinopanacene, n-caprylaldehyde, echinopanacol, oplopanaxosides, flavonoids.[48,50,72]	A remedy and tonic for progressive emaciation.
N.A.	*Oplopanax horridus* (Sm.) Miq.	Sesquiterpene.[103,290]	Hypoglycemic effects; reduces serious implications caused by diabetes such as kidney and heart disease. Treats arthritis, rheumatism, stomach and digestive problems.
China	*Orchis latifolia* L.	No information is available in the literature.	Sialagogue; treats anemia.
N.A.	*Orchis mascula* L.	Mucilage.[99]	Treats diarrhea, irritated gastrointestinal tracts in children.
China	*Oxalis corriculaza* L. *O. corymbosa* DC	Oxalate, vitamin C, calcium, citric acid, malic acid, tartaric acid.[50]	Antidote to arsenic and mercury, for bruises, clots, diarrhea, fever, influenza, snakebite, urinary infections.
N.A.	*Oxalis acetosela* L.	No information is available in the literature. It is toxic in large quantities.[348]	An astringent, diuretic; treats fevers and urinary problems.

China	*Papaver amurense* (N. Busch) N. Busch ex Tolmatchev. *P. nudicaule* L.	Amurine, amuroline, amuroine, coptisine, nudaurine, muramine, nudicaulin.[48]	For cough, headache, intestinal infection, blood in the urine, stomach ulcer.
	Papaver rhoeaes L.	Rhoeadine, rhoeagenine.[72]	For jaundice, as a gargle, or ingested as bechic.
	Papaver somniferum L.	Berberine, codeine, papaverine, isocorypalmine, laudanine, magnoflorine, meconine, 6-methylocodine, morphine, narcotine, pseudomorphine, rhoeadine, sanguinarine, beta-sitosterol, stigmasterol, thebaine, zanthaline.[50]	Antitussive, antispasmodic, analgesic, astringent, narcotic; treats chronic enteritis, diarrhea, enterorrhagia, headache, toothache, asthma.
N.A.	*Papaver bracteatum* Lindl.	Thebaine, oripavine, morphine, codeine.[99,100]	Mild sedative to induce sleep in babies, ease cough, relieves pain; narcotic analgesic, antitussive.
China	*Parietaria micrantha* Ledeb.	Protocatechuic aicd.[50]	For fractures, hemorrhage, lumbago, myalgia, numbness, renitis, rheumatism.
N.A.	*Parietaria judaica* L.	Flavonoids, tannins.[99]	A diuretic, demulcent, laxative. Restorative action on the kidneys, for nephritis, pyelitis, kidney stones, renal colic, cystitis, and edema.
China	*Pedicularis resupinata* L.	Alpha-amyrin, beta-amyrin, betulinic acid, cholesterol, kaempferol.[50,218]	Used in fever, leucorrhea, rheumatism, sterility, urinary difficulties, anti-inflammation, dryness of the mouth, tongue, and tinnitus.
N.A.	*Pedicularis palustris* L. *P. canadensis* L.	Alkaloids, phenyl-propanoid glycosides, iridoid glucosides.[303]	Treats swelling internally, coughs, uterine spasms, with antioxidant property.

Table 3   Chinese and North American Medicinal Herbs Belonging to the Same Genus and Different Species: Major Constituents and Therapeutic Values (continued)

Source	Scientific Name	Major Constituents	Therapeutic Values*
China	*Peucedanum decursivum* Max.	Glycoside nodakenin.[49]	Analgesic; antipyretic, antitussive, treats headache, bronchitis, asthma, pertussis.
	P. formosanum Hayata	Anomalin, coumarine, peuformosin.[56]	Cooling function; relieves pain, cough, treats cold, headache.
	P. japonicum Thunb. *P. praeruptorum* L.	Nodakenetin, nodakenin, decursidin, umbelliferone, pencordin, qianhucocumarin.[33]	Antitussive, expectorant.
N.A.	*Peucedanum graveolens* L.	Volatile oil (carvone), flavonoids, coumarins, xanthones, triterpenes.[99]	Relieves intestinal spasms and cramps, increases milk production by nursing mothers.
China	*Phragmittes communis* Trin.	Glycosides, protein, asparagin.[49]	As stomachic, antiemetic, antipyretic. Treats arthritis, jaundice, pulmonary abscess.
N.A.	*Phragmites australis* (Cav.) Trin.	Glycosides, protein, asparagin, ferulic acid, colxol, tricin, asparamide, coniferaldehyde, syringaldehyde, 4-hydroxyinnamic acid, vanillic acid, 4-hydroxybenzaldehyde, 2,5-dimethoxypara-quinone, polysaccharide, serotonin, tricin.[302]	For toothache, earache; remedy for hiccoughs, seafood poisoning, parched throat with fever, acute bronchitis with mucus, acute gastritis with vomiting, urinary tract infections, blood or stones in urine, eruptive fevers like measles and chickenpox.

Region	Species	Constituents	Uses
China	Physalis alkekengi L. var. francheti (Mast.) Makino	Physanols, physalien, zeaxanthin, glycolic acid, cryptoxanthin, luteolin, physoxanthin, mutaxanthin, tigloidine, auroxanthin, physalin A, B, and C, physalines, hystonin.[33,48]	Antibacterial; stimulates myocardial contraction, cause vasoconstriction, uterine contraction.
	Physalis angulata L.	Hystonin.[60] Overdose may cause dizziness.	Antifebrile, laxative, diuretic, causing uterine contractions.
N.A.	Physalis franchetti L. P. pubescene L.	Flavonoids, plant sterols, vitamins A and C, alkaloids.[100,310]	A diuretic for urinary and arthritic problems including kidney and bladder stones, fluid retention, and gout.
China	Picrasma quassioides (D. Don) Benn.	2,4-Dichloro-6-aminopyridine, 4,5-dimethoxycanthin-6-one, 2,6-dimethoxy-p-benzo-quinone, methyl nigakinone, picrasmin, nigakihemiacetal A, nigakilactone A, nigakinone, quassin.[50]	Treats fever, stomachache.
N.A.	Picrasma excelsa (Sw.) Planch.	Quassinoid (quassin), alkaloids, coumarin (scopoletin), vitamin B_1.[99]	Strengthen digestive systems, increases bile flow, secretion of salivary juices, and stomach acid production.
China	Pimpinella thellungiana Wolff.	Ilungianin A, Ilungianin B.[50,220]	A stimulant, anodyne, hypotensive; treats choleraic affections and flatulence.
N.A.	Pimpinella anisum L.	Anethole, creosol, coumarin, acetylinic, flavonoids, fatty oil, protein.[99]	Antispasmodic, carminative, diuretic; relieves gas pain.
China	Pinus bungeana Zucc. ex Endl. P. densiflora Sieb. et Zucc. P. koraiensis Sieb. et Zucc. P. sylvestris L. var. mongolica Litv. P. tabulaeformis Carr.	Essential oil, limonene, pinitol.[33]	Antitussive, antiasthmatic, antibacterial.

Table 3 Chinese and North American Medicinal Herbs Belonging to the Same Genus and Different Species: Major Constituents and Therapeutic Values   (continued)

Source	Scientific Name	Major Constituents	Therapeutic Values*
N.A.	Pinus albicaulis Engelm. P. contorta Dougl. ex. Loud. P. mugo Turra var. pumilio P. palustris Mill. P. strobus L.	Bishomophinolenic acid, resins, mallol, borneol acetate, tannins, vitamin A, vitamin C, galactose, alpha-pinenes, beta-pinenes, anthocyanin.[8,102]	Relieves fever, bronchial and nasal congestion, improves blood flow. Anthocyanin from bark has antioxidant activity, inhibits the enzymes that cause inflammation.
China	Plantago asiatica L. P. depressa Willd. P. exaltata Horn. P. loureiri Roem. et Schult. P. major L.	d-Xylose, l-arabinose, d-galacturonic acid, l-rhamnose, plantasan, plantasan, plantenolic acid, plantagin, homoplantagin, aucubin, ursolic acid, hentriacontane.[48]	Diuretic, expectorant, intestinal infection, diarrhea caused by bacteria.
N.A.	Plantago psyllium L.	Mucilage, linoleic acid, oleic acid, palmitic acid.[100,154]	Demulcent, laxative, antidiarrhea.
China	Platycladus orientalis (L.) Franco.	Thujene, thujone, pinene, myricetin, caryophyllene, aromadendrin, quercetin, hinokiflavone, fenchone, amentoflavone.[48]	Antipyretic, astringent, diuretic, for dysmenorrhea, epistaxis, gonorrhea, metrorrhagia.
N.A.	Platycladus occidentalis L.	Catechin, gallocatechin, afzelechin, epicatechin, epigallocatechin, epiafzalechin, procyanidins, flavones, myricetin, 3-0-glucoside, neothujic acid, podophyllotoxin type lignins.[303]	An expectorant for bronchial catarrh accompanied by heart weakness. Treats skin problems, vaccination, and menstruation.
China	Polygala japonica Houtt. P. sibirica L. P. tatarinowii Regel	Saponins, tenuidine, tenuifolin (prosenegenin).[28,33]	Stimulates bronchial secretions; antibacterial.
	Polygala tenuifolia Willd.	Onjisaponin A, onjisaponin B.[24]	Sedative; strengthens nervous system.

Origin	Species	Constituents	Uses
N.A.	Polygala senega L.	Triterpenoid saponins, phenolic acids, polygalitol, methyl salicylate, sterols.[99,103,291,292]	Treats rattlesnake bite, cough, bronchitis, asthma
	Polygala vulgaris Thunb.	Triterpenoid saponins, volatile oil, gaultherin, mucilage.[99]	Treats respiratory disorders such as chronic bronchitis, bronchial asthma, convulsive coughs. A diuretic.
China	Polygonatum chinense Kunth. P. cirrhifolium Royle. P. macropodium Turez. P. odoratum (Mill.) Druce var. pluriflorum (Miq.) Ohwi f. P. officinale All. P. ovarifolium Y. C. Chu P. sibiricum Delar. ex Redoute P. stenophyllum Maxim. P. vulgare Desf.	Convallarin, convallamarin, mucilage.[49]	Stimulates the appetite, increases peristalsis, slows the heart and raises the arterial tension, slows and deepens respiration; purgative.
N.A.	Polygonatum multiflorum (L.) All. P. biflorum (Walt.) Elliott	Saponins, flavonoids, vitamin A.[99]	A poultice to stimulates tissue repair. Treats tuberculosis, accelerates healing.
China	Populus alba L. P. davidiana Dode. P. tomentosa Carr.	Salicin, populin, benzoyl salicin, tannin, erisin, salicinase, salicortin, tremulacin, salireposide.[50]	Depurative, for colic, eczema, herpes, labialis, fever, dysuria, antiseptic, antiperiodic.
N.A.	Populus balsamifera L. P. candicans L.	Flavonoids, phenolic glycoside.[102]	Antiseptic, for sore throats, dry irritable coughs.
	Populus tremuloides Michx.	Salicin, populin, tannins.[99]	Reduces fever, relieves pain, anti-inflammatory.

Table 3   Chinese and North American Medicinal Herbs Belonging to the Same Genus and Different Species: Major Constituents and Therapeutic Values   (continued)

Source	Scientific Name	Major Constituents	Therapeutic Values*
China	Potentilla bifurca L. P. chinensis Seringe P. discolor Bunge. P. fragariodes L. P. freyaiana Bornmuller	D-Catechin.[50] This herb is toxic.[60]	Antibacterial, antiplasmodium; smoothes muscle relaxation, gynecological bleeding.
N.A.	Potentilla anserina L.	Ellagitannins, flavonoids, choline.[99]	Gargle for sore throats, remedy for diarrhea.
	Potentilla erecoa (L.) Rauschel. P. tormentilla (L.) Rauschel.	Tannins, catechins, ellagitannins, phlobaphene.[99]	Gargle for throat infections, mouthwash for canker sores and infected gums.
China	Primula sieboldii E. Morren var. patens (Turcz.) Kitag. P. vulgaris L.	Primulagenin A, aegicerin, protoprimulagenin A.[48]	Relieves cough, throat infection.
N.A.	Primula veris L.	Triterpenoid saponins, flavonoids, phenols, tannins, volatile oil.[99]	For bronchitis, respiratory tract infections, insomnia, anxiety, rheumatic disorder.
China	Pulsatilla ambigua Turcz. ex Pritz. P. chinensis (Bunge.) Regel	Protoanemonin, anemonin, okinalin, okinalein, ranuneulin, saponins.[33]	Antiamebial, antibacterial; treats amebic dysentery.
N.A.	Pulsatilla vulgaris Mill.	Lactone protoanemonin, triterpenoid saponins, tannins, volatile oil.[99]	For cramps, menstrual problems, distress. Treats spasmodic pain of the reproductive system.

China	Quercus acutissima Carr. / Q. dentata Thunb. / Q. liaotungensis Koidz. / Q. mongolica Fisch. ex Turcz. / Q. variabilis Blume	Lignin, cellulose, protein, pentosan, galactan.[56]	Promotes absorption of tuberculous nodules, remedy for diarrhea, hypertrophy of the gastrointestinal tract, root makes a cleansing dressing for foul sores.
N.A.	Quercus robur L.	Tannins, cutins, suberins.[99]	Treats sore throat, tonsillitis; an astringent.
China	Ranunculus chinensis Bung.	Protoanemonin, anemonin, ranunculin.[48]	Relieves swelling, asthma, liver disorders, toothache, night blindness.
	Ranunculus japonicus Thunb. / R. sarmentosa Adams	Anemonin, protoanemonin.[50]	Antitumor, sedative, bactericidal against bacillae of diphtheria, staphylococcus.
	Ranunculus sceleratus L.	Ranunculin, anemonin, 5-hydroxytryptamine, seratonin, protoanemonin, pyrogallol tannin.[48,50] This herb is toxic.	Relieves swelling, pain; antitoxin; treats lymphatic gland disorders; antirheumatic.
	Ranunculus ternatus Thunb.	Tannins, phenolic acids, volatile phenols, non-volatile terpenic compounds, volatile carbonyl.[60,223]	Treats abscesses.
N.A.	Ranunculus ficaria L.	Anemoni, tannins, saponins, volatile oil.[99]	Diuretic, anti-inflammatory; a tonic for digestive system, kidney, and urinary stones.
N.A.	Ranunculus occidentalis Nutt.	Anemonin.[102]	A stimulant, externally to relieves chronic sciatica.
China	Rauvolfia verticilata (Lour.) Baill.	Reserpine, beta-sitosterol, aricine, vellosimine, peraksine, serpentine, robinin.[33,39]	Treats hypertension, psychosis, schizophrenia.

Table 3 Chinese and North American Medicinal Herbs Belonging to the Same Genus and Different Species: Major Constituents and Therapeutic Values (continued)

Source	Scientific Name	Major Constituents	Therapeutic Values*
N.A.	*Rauvolfia serpentina* (L.) Benth.	Indole alkaloids, reserpine, rescinnamine, ajmaline, yohimbine.[99]	Regulates heartbeat, treats high blood pressure and anxiety. Sedative and depressant effect on sympathetic nervous system.
China	*Rhamnus davurica* Pall. *R. parvifolia* Bunge.	Emodin, chrysophanol, kaempferol, rhamnodiastase, aloe-emodin.[48,308] This herb is slightly toxic.	Insecticidal; treats respiratory infection, cough, improves bowel movement.
N.A.	*Rhamnus catharticus* L. *R. frangula* L. *R. purshianus* L.	Anthraquinone glycosides, phenolic flavonols, pectin, vitamin C, glucofrangulin A, B, frangulin A, B, emodin, chrysophanol, physcion.[100,103,107]	Laxative, diuretic, constipating, astringent, antibacterial, purgative, digestive complaints.
China	*Rheum koreanum* Nakai *R. officinale* Baill. *R. palmatum* L. *R. undulatum* L.	Anthraquinones, chrysophanol, emodin, physcion, aloe-emodin, rhein, chrysophenol, rheum tannic acid, gallic acid, catechin, bianthraquinonyl, sennosides (*R. undulatum* also contains rhaponticin).[1,33] This herb may be toxic.	Potent laxative, antibacterial, anthelmintic, anticancer; stimulates the large intestine and increases the movement of luminal contents toward the anus, resulting in defecation. Antispasmodic, choleretic, hemostatic, diuretic; lowers blood pressure, lowers cholesterol level.
N.A.	*Rheum tanguticum* L.	Cinnamic acid, gallic acid, emodin, rhein, rhein anthrones, catechin, anthraquinone compounds, tannin.[99,100,107]	Treats diarrhea, stimulates appetite, chronic constipation; laxative, cathartic.
China	*Rhodiola elongata* (Ledeb.) Fisch. & Meyer	p-Tyrosol, rhodioloside, flavonoids.[48]	A tonic; improves heart muscle; aphrodisiac.

Region	Species	Constituents	Uses
N.A.	Rhodiola rosea (L.) Scop.	Rhodioloside, flavanol glycosides.[103,293]	Improves learning and memory and reduces stress, anticancer; stimulates the central nervous system.
China	Rhododendron sinensis Sw.	Andromedotoxin, veratrine alkaloids.[49] This herb is toxic.	Sedative, analgesic, anesthetic in rheumatism.
	Rhododendron anthopogon G. Don	Essential oils, saponins, quercetin, gossypetin.[33]	Antitussive, antiasthmatic.
	Rhododendron dauricum DC	Germacrone, flavonoid, farrerol, feriol, quercetin, myricetin, anromedotoxin, rhodotoxin.[33]	Antitussive, antiasthmatic.
	Rhododendron molle (Blume) G. Don	Rhomotoxin.[37] This herb is mildly toxic.	Treats tachycardia, palpitations, hypertension.
	Rhododendron mucronatum G. Don	Essential oil, germacrone, farreol, grayanotoxin, gossypetin, azaleatin, 5-methyl kaempferol, 5-methyl myricetin, syringic acid, dihydroquercetin, coumarins, phenolic acid, p-hydroxybenzoic acid, protocatechuic acid, vanillic acid.[48]	Treats cough, asthma, headache, respiratory infection.
N.A.	Rhododendron maximum L.	No information is available in the literature. Large quantity may be toxic.	Used as a tonic for the kidneys and itchiness.
China	Rhus chinensis Mill.	Gallotannic acid, gallic acid, resin, wax, polysaccharides.[33]	Treats chronic intestinal infections, hematochezia, protoptosis, skin infections, bleeding wounds.
	Rhus semialata Murr.	Tannin.[49]	As an astringent, styptic; treats diarrhea, hemorrhage.
	Rhus verniciflua Stokes	Resinous oil urushiol.[49] This herb may be toxic.	As a hemolytic, emmenagogue, vermifuge.

Table 3   Chinese and North American Medicinal Herbs Belonging to the Same Genus and Different Species: Major Constituents and Therapeutic Values (continued)

Source	Scientific Name	Major Constituents	Therapeutic Values*
N.A.	*Rhus radicans* L. *R. glabra* L. *R. toxicodendron* L.	Toxicodendrol, urushiol, 3-n-pentadecylcatechol.[102]	Sympathetic stimulant; restores nerve function, facial neuritis, ulcerated sores on the lips, mouth, and nasal membrane.
China	*Ribes mandshurica* (Maxim.) Kom.	Citric acid, malic acid, organic acids.[48]	Treats colds.
N.A.	*Ribes nigrum* L.	Anthocyanosides, antiprotease, tannins, vitamins B₁, B₂, C, P, citric acid, pectin.[102]	Diuretic and diaphoretic properties; for urinary infection, rheumatism, and diarrhea.
	R. lacustre (Pers.) Poir.	Anthocyanosides.[102]	Infusion of leaves to lessen the pain associated with female menstrual cycle.
China	*Rorippa indica* (L.) Hiern. *R. islandica* (Oeder) Borbas	Alpha-phenylethylisothiocyante, gluconasturtin, rorifone, rorifamide.[33]	Antitussive, expectorant, diuretic, detoxicant.
N.A.	*Rorippa nasturtium-aquaticum* (L.) Hayek.	Raphanolide, raphanol, diastase, ferment, gluconasturin, bitters, essential oils, phenyl ethyl, vitamins, niacin.[303]	A blood builder, antidyskratic diuretic activities, lymphatic and digestive cleansing; treats prostate irritation, vaginal pruritis, chronic skin irritations.

China	Rosa acicularis Lindl.	Vitamins, gallocatechin, epigallocatechin, epicatechin gallate, catechin, epicatchin, fatty acids.[48]	Stop vomiting blood, stomachache; relieves pain caused by nerve system, menstruation.
	Rosa chinensis Jacq.	Essential oils.[49]	For arthritis, boils, cough, hematuria, rheumatoid joint pains, circulatory stimulant.
	Rosa multiflora Thunb.	Ascorbic acid, multiflorin, quercetol, kaempferol-3-glucoside, catechin.[50]	Anodyne, diuretic, laxative.
	Rosa rugosa Thunb.	Essential oils, l-citronellol, citral, geraniol, nerol, eugenol, cyanin, n-phenylethyl alcohol, citrol, nonyl aldehyde, l-linalool, l-p-menthene, nonacosane, menthene, bensaldehyde, phenylacetic acid, rosenoxide, paeonidin.[48,50]	Promotes blood circulation, treats abscesses, blood diseases, dyspepsia, hematemesis, hepatitis, stomachache.
N.A.	Rosa canina L. / R. damascena Mill. / R. gallica L.	Malic acid, citric acids, pectin, geraniol, citronellol, vitamins C, B complex.[102,107,160]	Astringent, mild diuretic and laxative effect. Excellent source of vitamin C when it's fresh.
China	Rubia akane Nakai	Alizarin, rubrierythrinic acid, purpurin.[85]	Treats rheumatism.
	Rubia chinensis Regel & Maack / R. cordifolia Thunb. / R. cordifolia L. f. pratensis (Maxim.) Kitag. / R. mungista Roxb. / R. sylvatica (Maxim.) Nakai	Rubierythrinic acid, alizarin, purpurin, pseudopurpurin, munjistin.[33,49]	Hemostatic; shorten the blood clotting time; antibacterial, antitussive; stimulates uterine contractions.
N.A.	Rubia tinctorum L.	Anthraquinone derivatives, ruberythric acid, alzarin, purpurin, indoid, asperuloside, resin, calcium.[99]	Treats kidney and bladder stones.

Table 3   Chinese and North American Medicinal Herbs Belonging to the Same Genus and Different Species: Major Constituents and Therapeutic Values   (continued)

Source	Scientific Name	Major Constituents	Therapeutic Values*
China	*Rubus coreanus* Miq.	Beta-sitosterol, stigmasterol, campesterol, cholestanol, ursolic acid, flavonoids.[48]	Diuretic, aphrodisiac, liver infection, joint infection caused by arthritis.
	Rubus parvifolius L.	Flavonoids.[48]	Treats fever, throat pain, blood vomiting, liver and intestine infection.
N.A.	*Rubus chamaemorus* L.	Tocopherol, benzoic acid, salicylic acid, ascorbic acid, vitamin C.[102,275]	Laxative, tonic; treats cough and fever.
	Rubus fruiticosus L.	Tannins, organic acids, vitamin C.[154]	Mild astringent, antiseptic, antifungal, diuretic and tonic properties.
	Rubus idaeus L.	Tannins, vitamin C, anthocyanins, pectin, flavonoids, gallic acid.[99]	Treats diarrhea; antispasmodic.
China	*Rumex acetosa* L. *R. stenophyllus* Ledeb. var. *ussuriensis* (A. Los.) Kitag.	Vitexin, quercetin-3-galactoside, violaxanthin, vitamin C, emodin, chrysophanein, chrysophanol, nepodin, hyperin, physcion.[48,50]	Homeopathically for cramps, hemorrhage, sore throat, esophagitis, diuretic; treats blood vomiting.
	Rumex crispus L.	Chrysophanein, nepodin.[48]	Treats ovarian bleeding, eczema, tuberculosis, sexually transmitted diseases.
N.A.	*Rumex obtusifolia* L.	Oxalates, anthraquinones, phanol, physcion, tannic acid.[100,102,118]	Antiseptic, laxative, rheumatic pains.
China	*Salix babylonica* L. *S. matsudana* Koidz. *S. microstachya* Turcz. ex Trautv.	Saligenin glucoside, iodine, pyrocaledol, saponins.[33]	Antigoiter, antibacterial; treats tubercule bacilli.

	Species	Constituents	Properties/Uses
N.A.	*Salix alba* L. *S. discolor* Muhlenb.	Salicin, tannins, phenolic, flavonoid, glycosides, salicortin, triandrin.[99,102]	Antipyretic, diaphoretic, antirheumatic, analgesic.
China	*Salvia chinensis* L. *S. pogonocalyx* Hance *S. przewalskii*	Scutellarin.[60]	Treats abdominal pain, arthritis, inflammations, metrorrhagia, uteritis, women's diseases.
	Salvia coccinea L.	Saluianin.[56]	Stop bleeding, cooling effect; stimulates sweating, relieves swelling.
	Salvia miltiorhiza Bunge.	Tanshinone, cryptotanshinone, isocryptotanshinone, miltirone, tanshincl, salviol.[33]	Treats angina pectoris, cerebral atherosclerosis, diffusive intravascular clotting, thrombophlebitis.
	Salvia plebeia R. Brown	Flavonoids, homoplantaginin, hispidulin, eupafolin, essential oils.[48]	Diuretic, vermifuge, astringent.
N.A.	*Salvia clevelandii* (A. Gray) Greense *S. divinorum* Epl. & Jutiva	No information is available in the literature.[100]	Emetic, hallucinogenic, psychotropic.
	S. officinalis L.	Thujone, borneol, cineole, camphor, salvin, tannin, fumaric acid, malic acid, oxalic acid.[100,107,161]	Carminative; lowers fever; antiseptic, antifungal, astringent, diuretic, antidiarrheal, antispasmodic.
Chinea	*Sambucus coreana* Kom. & Klob. Alisova *S. latipinna* Nakai *S. manshurica* Kitag. *S. peninsularis* Kitag. *S. sieboldiana* (Miq.) Blume ex Graebner var. miquelii (Nakai) Hara *S. williamsii* Hance	Chlorogen acid.[60]	Diaphoretic, diuretic, carminative; treats arthralgia, fever.

Table 3   Chinese and North American Medicinal Herbs Belonging to the Same Genus and Different Species: Major Constituents and Therapeutic Values (continued)

Source*	Scientific Name	Major Constituents	Therapeutic Values*
China	*Sambucus formosana* Nakai	Alpha-amyrin palmitate.[56]	Detoxicant; stops swelling; diuretic; relieves pain.
N.A.	*Sambucus nigra* L. *S. canadensis* L.	Flavonoids, phenolic, triterpenes, sterols, cyanogenic glycosides, vitamins A, C.[99]	Increases sweating; diuretic, anti-inflammatory.
	Sambucus racemosa L.	Rutin, tannins, cyanogenic, glucans, baldrianic acid.[102]	Antiulcer, antimutagens, anticoagulant.
China	*Sargassum pallidum* (Harv.) Setch.	Odine, alginic acid, algin, iron, potasium.[33]	Antigoiter, anticoagulant.
N.A.	*Sargassum officinalis* L. *S. fusiforme* L.	No information is available in the literature.[345]	For goiter, tuberculosis of lymph nodes, cysts, bronchitis, edema, hydocele.
China	*Saussurea japonica* (Thunb.) DC	Saussurine, phene, phellandrene.[49]	As a stomachic.
N.A.	*Saussurea lappa* Clarke	Terpenes, sesquiterpenes, aplotaxene, saussurine, resin.[99]	Depresses the parasympathetic nervous system. It has tonic, stimulant, and antiseptic properties.
China	*Scutellaria baicalensis* Georgi	Baicalein, baicalin, wogonin, beta-sitosterol, wognoside, 7-methoxy-baicalein, 7-methoxynorwogonin, skullcap flavones.[33]	Antibacterial, antiviral, antipyretic, anti-inflammatory, antitumor.
	Scutellaria formosana Brown	Berberine, baicalin.[54,233,234]	Relieves swelling; pain, treats cold, wounds, liver infection.

N.A.	*Scutellaria lateriflora* L.	Scutellarin, baicalin, baicalein, wogonin, benzoic acid, catalpol, tannins, beta-sitosterol, camphesterol, stigmasterol.[99,102,163]	Sedative and antispasmodic; prevents epileptic seizures, tonic; antispasmodic, antiallergic.
China	*Sedum aizoon* L.	Sedoflorin, sedocaulin, sedocitrin, sedoheptulose, arbutin.[33,48]	Hemostatic; removes blood stasis.
	Sedum formosanum N. E. Brown	Triterpenes, amyrenone, amyrenol.[54]	Treats diabetes, relieves swelling, pain, digestion, diarrhea.
	Sedum lineare Thunb.	Sedoheptose, glucose, fructose.[50] This herb is slightly toxic.	Applied locally to burns and scalds; treats throat infection, diabetes.
	Sedum sarmentosum Bunge.	Sarmentoslin, dihydro-N-methyl-isopelletierine, N-methyl-2-(β-OH-propyl) piperidine, N-methyl-isopelletierine, dl-methylisopelletierine, dihydroisopelletierine.[33,48,50]	Antipyretic, detoxicant, diuretic; treats hepatitis.
	Sedum erythrostichum Miq. *S. kamtschaticum* Fisch *S. verticillatum* L.	Sedoheptulose.[48]	Detoxicant, relieves swelling; stops bleeding and pain.
N.A.	*Sedum acre* L.	Sedacrine, n-methyl anabasine, sedinine, sedacryptine, flavanol glycosides.[102,254]	Insomnia, depressant, hemorrhoidal pain; treats excessive menstrual flow during menopause.
China	*Senecio arguensis* Turcz.	Lavoxanthin, macrophylline, cynarin, chlorogenic acid, chrysanthemaxanthin, sarracine.[33,48]	Antibacterial, antiplasmodial; treats acute bacterial dysentery and bronchitis.
	S. campestris (Retz.) DC	Alkaloids.[48] This herb may be toxic.	Depresses leukemia; detoxicant, diuretic, insectisidic.

Table 3   Chinese and North American Medicinal Herbs Belonging to the Same Genus and Different Species: Major Constituents and Therapeutic Values* (continued)

Source	Scientific Name	Major Constituents	Therapeutic Values*
China	S. cannabifolius Lessing	p-Hydroxyacetophenone, arbutin.[48]	Treats heart disease, respiratory infection, sexually transmitted diseases.
	S. vulgaris L.	Senecionine, inulin.[58]	Used in ointment on hemorrhoids and swellings, lowers blood pressure; laxative.
N.A.	Senecio aureus L.	Seneciphyline, jacoline, pyrrolizidine, senecionine, tannins, resin.[164]	A poultice, ointment, or lotion to relieves pain and inflammation.
China	Silene jenisseensis Willd.	6,8-di-C-Galactopyranosylapigenin, 6-C-galactopyranosyl-isoscutellarein, essential oil.[84]	For fever, kala-azar, malaria.
N.A.	Silene ocaulis L. S. virginica L.	Spinasterol, ecdysterones, 22-dihydrospinasterol, 2-(6'-cinnamoyl) glucosido-methyl-4H-pyran-4-one.[302]	Anabolic, tonic, adaptogenic effects.
China	Smilacina japonica A. Gray	No information is available in the literature.	For arthritis, relieves swelling and pain; aphrodisiac, regulates monthly period, breast gland infection.
N.A.	Smilacina stellata (L.) Desf.	No information is available in the literature.	
China	Smilax china L. S. nipponica Miq. subsp. manshurica Kitag. S. riparia DC subsp. ussuriensi (Regel) Kitag. S. sieboldii Miq.	Crystalline saponin smilacin, tannin, resin, tigogenin, neotigogenin, laxogenin.[48,49]	As alternative, diuretic in syphilis, gout, skin disorders, rheumatism.

		Chemical Constituents	Medicinal Uses
N.A.	Smilax aristolochiifolia Mill.	Steroidal saponins, phytosterols (beta-sitosterol), starch, resin, sarsapic acid, minerals.[99]	Anti-inflammatory; relieves eczema, psoriasis, and itchiness. Treats rheumatism, rheumatoid arthritis, and gout. It has a progesterogenic action.
China	Solanum aculeatissimum Jacquin	Solasonine, beta-solamargine, solasurine.[55]	For cough, asthma, diuretic, pain.
	Solanum biflorum Loureiro	No information is available in the literature.	Detoxicant, for cough, swelling, dog bites.
	Solanum capsicastrum Link.	Solanocapsin.[55]	With cooling effect, relieves swelling, treats liver inflammation.
	Solanum incanum L.	Beta-sitosterol, D-glucose, ursolic acid, alkaloids, solasodine, solamargine.[55]	Treats liver inflammation, lymphatic gland; a detoxicant.
	Solanum indicum L.	Diosgenin, solanidine, solanine, solasodine, alkaloids, carbohydrases, maltase, saccharase, melibiase.[50]	Antidote for poison, for urinary disease.
	Solanum lyratum Thunb. S. melongena L.	Trigonelline, stachydrine, choline, solanine, nasunin, shisonin, delphinidin-3-monoglucoside, adenine, imidazolylothylamine, solasodine, arginine glucoside.[48]	For arthitis, respiratory disorder, swelling, cough, diarrhea, blood in the urine.
	Solanum nigrum L.	Solanigrines, saponins.[33]	Antibacterial, diuretic; treats mastitis, cervicitis, chronic bronchitis, dysentery.
	Solanum pseudo-capsicum L.	Solanocapsine.[55]	A detox cant; relieves pain. Treats tuberculosis, pneumonia.
	Solanum verbascifolium L.	Solasonine.[54]	Treats cysentery, intestinal pain, and fever.

Table 3   Chinese and North American Medicinal Herbs Belonging to the Same Genus and Different Species: Major Constituents and Therapeutic Values (continued)

Source	Scientific Name	Major Constituents	Therapeutic Values*
N.A.	*Solanum dulcamara* L.	Steroidal alkaloids, solasodine, soldulcamaridine, steroidal saponins, tannins.[99]	Treats eczema, itchiness, psoriasis, and warts. It relieves asthma, chronic bronchitis, and rheumatic conditions.
	Solanum tuberosum L.	Vitamins A, B_1, B_2, C, and K, minerals, atropine alkaloids.[99]	Potato juice treats peptic ulcers, relieves pain and acidity.
N.A.	*Solanum xanthocarpum* L.	Steroidal alkaloids (solanocarpine).[99]	Treats gas and constipation, throat and gum disorder. It is an anticongestive.
China	*Sorbus alnifolia* (Sieb. & Zucc.) K. Koch	Fatty acids, starch, essential oils, flavonoids, isochlorogenic acid, parasorbic acid.[48]	For stomach infection and ache, swellings, cough, vitamin deficiencies.
N.A.	*Sorbus aucuparia* L.	Tannins, vitamin C, pectin, organic acids.[102]	Astringent for hemorrhoids and diarrhea, source of vitamin C.
China	*Spiraea salicifolia* L. S. *salicifolia* L. var. grosseserrata Liou & Liou fil. S. *salicifolia* L. var. oligodonta Yu	Flavonoids, carotenoids, vitamin C, alkaloids, seed oil.[33]	Diuretic, treats cough, pain, monthly period, constipation.
N.A.	*Spiraea ulmaria* L.	Salicylates, flavonol glycosides, heliotropin, vanillin, tannins.[99,100,118]	Laxative; treats headache.
China	*Stachys chinensis* Bunge. ex Benth. S. *baicalensis* Fisch. ex Benth. S. *baicalensis* Fisch. ex Benth. var. angustifolia Honda S. *japonica* Miq.	Coumarin, alkaloids, stachydrine chloride.[48]	Treats cold, influenza.

N.A.	Stachys officinalis (L.) Trev.	Tannins, stachydrine, betonicine, betaine, choline.[99,100,107]	Stops bleeding from open wounds; antispetic.
China	Strophanthus divaricatus (Lour.) Hook. & Arn.	Divaricoside, divostroside, sinoside, sinostroside, caudoside, caudostroside, sarmutoside.[33] This herb is toxic.	Cardiac stimulating action causing an increases of myocardiac contractility; slow the heartbeat.
N.A.	Strophanthus gratus (Wallich & Hook. ex Benth.) Ball. S. kombe L.	Cardiac glycosides[99]	Treats snakebite, delay blood clotting. A mild heart tonic, improves heart efficiency.
China	Styrax tonkinensis Pierre	Sumaresinolic acid, coniferyl cinnamate, styracin, vanillin, alpha-phenylpropyl cinnamyl cinnamate, balsamic acid.[33,50]	As an aromatic stimulant, for aphrodisiac, an astringent.
China	Styrax suberifolus Hook. et Arnott.	No information is available in the literature.	Stomachache, pain caused by arthritis.
N.A.	Styrax benzoin Dryander	Cinnamic, benzoic, sumaresinolic acid esters, benzoic acid, benzaldehyde, vanillin.[99]	Antiseptic, astringent. Externally for wounds and ulcers, internally to settle cramps; stimulates coughing, disinfects the urinary tract.
China	Swertia diluta (Turcz.) Benth. et Hook. f.	Swertiamarin, swertisin, methyl-bellidifolin, homoorentin, methyl-swertianin, isovitexin, bellidifolin, decussatin, swertifrancheside.[33]	Choleretic; improves hepatic function. Treats acute icteric hepatitis, chronic liver disease.
China	Swertia pseudochinensis Hara	Swertiamarin, swertisin, methyl-bellidifolin, homoorentin, methyl-swertianin, isovitexin, bellidifolin, decussatin, swertifrancheside.[33]	Choleretic; improves hepatic function. Treats acute icteric hepatitis, chronic liver disease.
N.A.	Swertia chirata L.	Xanthones, indoids, amarogentin, alkaloids, flavones.[99]	A tonic, antimalarial; stimulates appetite, ease stomach pain, reduces fever.

Table 3   Chinese and North American Medicinal Herbs Belonging to the Same Genus and Different Species: Major Constituents and Therapeutic Values   (continued)

Source	Scientific Name	Major Constituents	Therapeutic Values*
China	*Syzygium cuminii* (L.) Skeels	Betulinic acid, eugianin, friedelin, epifriedelanol, beta-sitosterol, acetyl oleanolic acid, ellagic acid, myricetin, cyanidin rhamno-glucoside, petunidin glucoside, maluidin glucoside, jambolin.[57]	Cooling effect; relieves itchiness, stop bleeding, infection, diarrhea.
N.A.	*Syzygium aromaticum* (L.) Merr.	Sesquiterpenes, eugenol, tannins, gum.[314]	For gastroenteritis and intestinal parasites.
China	*Tagetes erecta* L. *T. patula* L.	Alpha-terthienyl, d-limonene, l-linalool, tagetone, n-nonyl aldehyde.[50]	Treats sores and ulcers, cold, conjuctivitis, cough, mastitis, mumps.
N.A.	*Tagetes minuta* L. *T. lucida* Cav.	Coumarin derivatives, resin, gallic acid, tannins, glucose, pectin, gum.[107]	For diarrhea, indigestion, nausea, externally for smooth muscles, scorpion bites, and to remove ticks.
China	*Taxus cuspidata* Sieb. et Zucc. *T. chinensis* (Pily) Rehd. *T. cuspidata* Sieb. et Zucc. *T. yunnanensis* Cheng et L. K. Fu	Taxol, baccatin, cephalomannine, 10-deacetylbaccatin, yunnanxana, abeotaxanes, taxinine E.[33]	Antineoplastic, anticancer; treats ovarian carcinoma.
N.A.	*Taxus x media* Rehd. *T. brevifolia* Nutt.	Taxol, resin.[103,295,296]	Treats cancer, gout, and rheumatism, arthritis.
China	*Tephrosia purpurea* Person	Rotenone, deguelline, tephrosin, rutin, quercetin glucoside.[57]	Used as a cordial and a stomachic, an emmenagogue.
N.A.	*Tephrosia virginiana* (L.) Pers.	Deguelin, dehydrorotenone, rotenone, tephrosin.[100]	For alopecia, cholecystosis, cough, syphilis, bladder trouble.

	Species	Constituents	Uses
China	Thalictrum aquilegifolium L. var. sibiricum Regel & Tiling; T. baicalense Turcz.; T. baicalense Turcz. f. levicarpum Tamura; T. fauriel Hayata; T. petaloideum L.; T. petaloideum L. var. supradecompositum (Nakai) Kitag.; T. simplex L.	Flavonoids, fetidine, phetidine, thalfoetidine, thalpine, thalphinine, thalidasine, hernandezine, thelic simidine, oxypurpureine, berbamine, isotetrandrine, oxyacanthine, isothalidenzine, glaucine, berberine, palmatine, jatrorrhizine, coptisine, protopine, cryptopine, alpha-allocryptopine, thalidezine.[48,53]	Anticancer activity; treats fever, nausea, thirst, hemorrhages, and conjunctivitis.
N.A.	Thalictrum ichangense Lecoyer ex Oliver; T. glandulissimum DC	Berberine, palmatine, jatrorrhizine, talictrine, thalidasine, thalicarpine, saponaretin.[33]	Antibacterial; treats influenza, childhood fevers, measles, malaria.
N.A.	Thalictrum dasycarpum Fisch. & Ave-Lall.; T. occidentale A. Gray	Thalidasine, thalicarpine, thalisopavine, corypalline, norargemonine, trans-5, cis-9-octadecadienoic acid, dasycarponin, bis-norargemonine, L-laudanidine, trans-5-hexadecenoic acid.[302,333]	A tumor inhibitor, relieves dizziness and ear problems.
China	Thuja chinensis Hort.; T. koraiensis Nakai; T. orientalis L.	Thujene, thujone, fenchone, myricetin, caryophyllene, aromadendrin, quercetin, pinene, hinokiflavone, amentoflavone.[48]	Antipyretic, astringent, diuretic, for dysmenorrhea, epistaxis, gonorrhea, metrorrhagia.
N.A.	Thuja occidentalis L.	Thujone, flavonoids, wax, mucilage, tannins.[99]	Antiviral; treats warts and polyps. It induces menstruation.
China	Thymus amurensis Klokov; T. disjunctus Klokov; T. kitagawianus Tscherneva; T. komarovii Sergievskaja; T. przewalskii (Kom.) Nakai; T. quinquecostatus Celakovsky	Scutellarein heteroside, luteolin-7-glucoside, apigenin, volatile oils: carvacrol, p-cymrene, p-terpinene, alpha-terpineol, zingiberene, borneol, ursolic acid, thymol.[43]	Treats high blood pressure, stomachache, intestinal infection, cough, digestion, diarrhea.

Table 3 Chinese and North American Medicinal Herbs Belonging to the Same Genus and Different Species: Major Constituents and Therapeutic Values (continued)

Source	Scientific Name	Major Constituents	Therapeutic Values*
N.A.	Thymus capitatus L. T. citriodorus (Pers.) Schreb. T. praecox Opiz. T. pulegiodes L. T. serpyllum L. T. vulgaris L.	Thymol, tannins, carvacrol, saponins, apigenin, lutolin.[99,100,107]	Antispasmodic, antitussive; relieves coughing.
China	Tilia mandshurica Rupr. & Maxim. T. amurensis Rupr. T. mongolica Maxim.	Flavonoids, essential oils.[48]	Promotes sweating, bactericidal, treats cold, kidney infection, throat infection.
N.A.	Tilia cordata Mill. T. europaea L.	Mucilage, tannins, flavonoid, caffeic acid, taraxerol, tiliadine, vanillin, phytosterols, mucilage.[99,100]	Diaphoretic, antispasmodic, diuretic, mild sedative.
China	Trifolium pratense L.	Phytoestrogens, genistein, daidzein, formononetin.[33,48]	Stimulating effect on female reproductive organs.
N.A.	Trifolium incarnatum L.	Flavonoids, salicylic acid.[154]	Treats skin conditions, spasmodic coughs.
	Trifolium pratense L.	Tannins, phenolic glycosides, p-coumaric acid, silicic acid, caffeic acid, salicylic acid.[100,102]	Remedy for sore throat, colds, coughs, bronchitis, diarrhea, chronic skin conditions.
China	Trillium camschatcense Ker-Gawler	Trillin, trillarin, diosgenin, cyasterone, ecdysterone.[48]	Improves blood circultation; detoxicant; treats headache, high blood pressure; stops bleeding.

Region	Species	Constituents	Uses
N.A.	*Trillium erectum* L.	Saponins (trillin), tannin, resin, fixed oil, volatile oil.[99]	For heavy menstrual or intermenstrual bleeding; treats bleeding associated with uterine fibroids.
China	*Ulmus campestris* L. *U. macrocarpa* Hance *U. pumila* L.	Butyric acid, capric acid, hexylenaldehyde, lipase, phlobaphene, phytosterol, sitosterol.[50]	For urinary calculi, diuretic, febrifuge.
N.A.	*Ulmus rubra* Muhl. *U. procera* L.	Tannins, mucilage, cholesterol, campesterol, beta-sitosterol, pentoses.[99,100]	Astringent, demulcent, anti-inflammatory.
China	*Uncaria hirsuta* Havil *U. rhynchophylla* Miq.	Rhynchophylline, corynoxeine, iso-rhynchophylline, isocorynoxeine, corynantheine, hirsutine, hirsuteine.[33]	A sedative, anticonvulsive; lowers blood presure, it has a triphasic effect. Treats childhood epilepsy.
N.A.	*Uncaria gambir* (Hunter) Roxb.	Rhyncophylline, corynoxeine, hirsutine, isorhyncophylline, nicotinic acid, catechin.[297,298,299,313]	Lowers blood pressure, protects the liver from infection. An astringent.
China	*Vaccinium bracteatum* Thunb. *V. vitis-idaea* L.	6-o-acetyl-arbutin, arbutin, avicularin, 2-o-caffeoylarbutin, d-catechol, l-epicatechol, d-gallocatechol, hyperin, hypercside, sioquercitrin, salidroside, tannin, ursone.[50]	For gonorrhea.
N.A.	*Vaccinium macrocarpon* Ait.	Anthocyanosides, hippuric acid, vitamins C, A.[103,177,178,180,300]	Prevents urinary infection and stones; an antioxidant, effect on clogged heart arteries.
N.A.	*Vaccinium myrtilloides* Michx. *V. myrtillus* L. *V. oreophilum* Rydb.	Tannins, arbutin, iridoids, insulin, anthocyanosides, myrtocyan.[103,300,301]	Strengthening cardiovascular system; improves vision, treats diabetes, digestive disorder, urinary disorder; an antioxidant.

Table 3 Chinese and North American Medicinal Herbs Belonging to the Same Genus and Different Species: Major Constituents and Therapeutic Values (continued)

Source	Scientific Name	Major Constituents	Therapeutic Values*
China	Valeriana alternifolia Bunge. V. amurensis P. Smiru. ex Kom. V. fauriei Briq. V. fauriei Briq. var. dasycarpa Hara V. subbipinnatifolia A. Baranow	Bornyl isovalerate, isovaleric acid, borneol, camphene, pinene, d-terpineol, l-limonene, pyrryl-α-methyl ketone, alpha-fenchene, myrcene, phellandrene, l-caryophyllene, erpinene, terpinolene, eremophilene, selinene, cadinene, valerianol, valerenone, myrtenol, bisabolene, chatinine, caffeic acid.[48]	Antispasmodic, aphrodisiac, emmenagogue, stimulant, sudorific, backache, cramps, influenza, nausea, numbness.
N.A.	Valeriana officinalis L.	Essential oil, valtrate, valepotriates, bornyl esters, alkaloids, isovaltrate.[99,100]	Sedative for nervous disorders, antispasmodic.
China	Veratrum dahuricum (Turcz.) Loes V. formosanum Loesener	Jervine, pseudojervine, rubijervine, tienmulilmine, tienmulilminine, zygadenine, germine.[33] This herb may cause mucosal irritation.	Lowers blood pressure, slows heart rate; antibacterial. It has an insecticidal effect.
N.A.	Veratrum viride Ait.	Steroidal, alkaloids, chelidonic acid.[99]	Lowers blood pressure, dilates the peripheral blood vessels.
China	Veronica anagallis-aquatica L.	Aucubin.[50]	For fever, a gargle for throat ailments, stomatitis.
	Veronica linariaefolia Pall. ex Link	Cordycepic acid, flavonoids.[48]	For windpipe infection, blood vomiting; relieves pain; detoxicant.
China	Veronica sibirica L. V. undulata Wallich	Mannitol, veronicastroside, inteolin-7-β-neohesperidoside, luteolin, 7-β-glucopyranoside, ancubin, arbutin.[48]	Relieves swelling, stops bleeding, and treats cold, cough.
N.A.	Veronica officinalis L.	Tannins, essential oil, aucuboside, vitamin C, flavonoids, acetopenone glucoside.[99]	Diuretic, expectorant.

China	*Viburnum sargenti* Koehne f. glabrum Kom.	Scopoletin, aesculetin, caffeic acid, citric acid, malic acid, chlorogenic acid, isochlorogenic acid, essential oil, kaempferol-3-glucoside, beta-amyrin, beta-sitosterol, paeonin.[48]	For blood circulation, swelling; detoxicant;relieves itchiness, arthritis.
N.A.	*Viburnum opulus* L. *V. prunifolium* L.	Hydroquinones, coumarins, tannins, resin.[99,102]	Antispasmodic, sedative, an astringent.
China	*Viola acuminata* Ledeb.	Saturated acids, cerotic acid, unsaturated acids, alcohols, hydrocarbons.[43,48]	Mucilaginous, emollient, suppurative inflammations, abscesses, ulcers.
N.A.	*Viola tricolor* L.	Saponins, mucilage, violin, salicylic compounds, tannins.[100,114]	Diuretic, diaphoretic, tonic, anti-inflammatory, blood-purifying properties.
China	*Vitex chinensis* Miller *V. jeguaod* L.	Essential oils, beta-caryophyllene, caryophyllene oxide.[33]	Antitussive, antiasthmatic, antibacterial.
N.A.	*Vitex labrusca* L. *V. agnus-castus* L.	Flavonoids, iridoids, agnuside, aucubin, cineol, casticin, viticine.[99,182,309]	Treatment of mastopathy, premenstrual syndrome, and luteae insufficiency. Regulates hormones, progesterogenic, increases breast-milk production.
China	*Wisteria sinensis* (Sims) Sweet	Toxic glycoside, toxic resin.[60] This herb is toxic.	Diuretic.
N.A.	*Wisteria floribunda* (Willd.) DC *W. brachybotrys* Sieb. et Zucc.	Isoflavonoids, triterpenoid saponins, dehydrosoyasaponin, triterpenoids.[330,331,332]	Antitumor; treats gastric cancer.

Table 3 Chinese and North American Medicinal Herbs Belonging to the Same Genus and Different Species: Major Constituents and Therapeutic Values (continued)

Source	Scientific Name	Major Constituents	Therapeutic Values*
China	*Zanthoxylum ailanthoides* Sieb. et Zucc.	Essential oils, methyl n-nonylketone, isopimpinellin, dictamine, skimmianine, magnoflorine, laurifoline.[64,94]	Treats chills, influenza, sunstroke, indigestion.
	Zanthoxylum bungeanum Maxim.	Essential oils, limonene, cumic alcohol, linalool, myrcene, benzene tert-butyl, sabinene, rerpinenol, piperitone, beta-gurjunene, alpha-piene, geraniol, estragole, cadinene, clovene.[53]	Anthelmintic, aromatic, astringent, carminative, emmenagogue, stimulant, sudorific.
	Zanthoxylum nitidum (Roxb.) DC	Nitidine, oxynitidine, vitexin, 6-ethoxy-chelerythrin, diosmin, oxynitidine, oxychelerythrine, skimmianine, N-desmethylchelerythrine.[33,50,53]	Analgesic, anodyne, analgesic, antitumor against leukemia, carminative, detoxicant; increases blood flow.
	Zanthoxylum schinifolium Sieb. et Zucc.	Estragol, citronellol, phellandrene, xanthoxylene, skimmianine, magnoflorine, xanthoplanine, dictamine, bergapten, berberine, esculetin dimethyl ether.[33,48,53]	Treats ascaris, relieves abdominal pain caused by ascariar obstruction.
N.A.	*Zanthoxylum americanum* Mill.	Chelerythrine, herclavin, asarinin, neoherculin, tannins, resins.[99]	Circulatory stimulant, increases sweating.

* This information should not be used for the diagnosis, treatment, or prevention of diseases in humans. The information contained herein is in no way intended to be a guide to medical practice or a recommendation that herbs be used for medicinal purposes. The information is presented here mainly for educational purposes and should not be used to promotes the sale of any product or replace the service of a physician.

References

1. Shibata, S. 1979. The chemistry of Chinese drugs. Am. J. Chin. Med. 7: 103–141.
2. Kariyone, T., Takahashi, M., and Takalshi, K. 1956. Studies on glycosides. VII. Components of Paeonia suffruticosa Andl. Yakugaku Zasshi 76: 917–921.
3. Kanada, M. et al. 1969. Chem. Pharm. Bull. (Tokyo) 17: 665.
4. Takido, M., Takahashi, S., Masuda, K., and Yasukawa, K. 1977. Torosachrysone, a new tetrahydroznthracene derivative from the seeds of Cassia torosa. Lloydia 40: 191.
5. Hartwell, J. L. and Shear, M. J. 1947. Chemotherapy of cancer. Classes of compounds under investigation and active components of podophyllin. Cancer Res. 7: 716–717.
6. Fujita, M., Itokawa, H., and Sasuda, Y. 1973. Studies on the components of Magnolia obovata Thunb. II. On the components of the methanol extract of the bark. Yakugaku Zasshi 93: 422–428.
7. Brekhman, I. I. and Dardymov, I. V. 1969. Pharmacological investigation of glycosides from ginseng and Eleutherococcus. Lloydia 32: 46–51.
8. Kochetkov, N. K., Khorlin, A., Chizhov, O. S., and Scheichenko, V. I. 1962. Deoxyschizandrin—structure and total synthesis. Tetrahedron Lett. 9: 361–363.
9. Shinoda, J. 1929. Über die Bestandteile des Arctium lapps L. (II. Mitteilung). Yakugaku Zasshi 49: 1165–1169.
10. Kawasaki, T., Yamauchi, T., and Itakura, N. 1963. Saponins of timo (Anemarrhenae rhizoma). I. Yakugaku Zasshi 83: 892–896.
11. Aritomi, M. and Kawasaki, T. 1968. A mangiferin monomethyl ether from Mangifera indica L. Chem. Pharm. Bull. 16: 760–761.
12. Shibata, S., Harads, M., and Murakami, T. 1959. Studies on the constituents of Japanese and Chinese crude drugs. II. Antispasmodic action of the constituents of pueraria root. Yakugaku Zasshi 79: 863–865.
13. Trunzler, van G. and Schuler, E. 1962. Vergicidende Studien über Wirkungen eines Crataegus Extraktes, von digitoxin, digoxin und G–strophanthin am isolierten Warmbluterherzen. Arzenetn. Forsch. 2: 198–202.
14. Kaneda, M., Iitaka, Y., and Shibata, M. 1972. Chemical studies on the oriental plant drugs. XXXIII. Tetrahedron 28: 4309–4317.
15. Aimi, N., Inaba, M., Watanabe, M., and Shibata, S. 1969. Tetrahedron 25: 1825.
16. Kitagawa, I., Nishimura, T., Kobayashi, A., and Yosoka, I. 1971. On the constituents of rhizome of Rehmannia glutinosa. Yakagaku Zasshi 91: 593–596.
17. Korte, F. 1954. Über neue glydosidische Pflanzeninhaltsstoffe V. Mitteil: Zur Konstitution des gentiopikrins und des Hexahydrogentiogenin–methylathers. Chem. Ber. 87: 780–783.
18. Inouye, H., Saitoh, S., Taguchi, H., and Endo, T. 1969. Zwei nhue: tridoidglucoside aus Gardenia jasminoides gardenosid und geniposid. Tetrahedron Lett. 28: 2347–2350.
19. Hikino, H., Hikino, Y., and Yosioka, I. 1964. Studies on the constituents of Atractylodes. IX. Structure and autoxidation of Atractylon. Chem. Pharm. Bull. 12: 755–760.
20. Shibata, S., Mihashi, S., and Tanaka, O. 1967. The occurrence of (–)fimarane-type diterpene in Aralia cordata Thunb. Tetrahedron Lett. 51: 5241–5243.

21. Kubota, T. and Hinoh, H. 1968. The constitution of saponins isolated from Bupleurum falcatum L. Tetrahedron Lett. 3: 303–306.
22. Kubota, T. and Hinoh, H. 1968. Triterpenoids from Bupleurum falcatum L. III. Isolation of genuine sapogenins saikogenins E, F and G. Tetrahedron 24: 675–686.
23. Kubota, T., Kitani, H., and Hinoh, H. 1969. The structure of platycogenic acids A, B, and C, further triterpenoid constituents of Platycodon grandiflorum A. De Candolle. Chem. Commun. 1969: 1313–1315.
24. Sakuma, S., Sugtura, N., Tanemoto, H., Amakawa, H., and Shuji, J. 1975. Abstr. 95th Annu. Meet. Pharm. Soc. Japan 11: 247.
25. Fujita, M., Itokawa, H., and Kumekawa, Y. 1974. The study on the constituents of Clematis and Akebia spp. I. Distribution of triterpenes and other components. Yakugaka Zasshi 94: 189–191.
26. Yahara, S., Kami, R., and Tanaka, O. 1977. New dammarane type saponins of leaves of Panax japonicus C. A. Meyer. (1) Chikusetsusaponins-L_5, L_9, and L_{10}. Chem. Pharm. Bull. (Tokyo) 23: 2041–2047.
27. Inoue, O., Takeda, T., and Ogihara, Y. 1977. Abstr. 97th Annu. Meet. Pharm. Soc. Japan 11: 220.
28. Saito, H., Lee, Y. M., Takagi, K., and Kondo, N. 1977. Chem. Pharm. Bull. (Tokyo) 25: 1017.
29. Kawanishi, S., Mitaral, Y., and Shoji, J. 1972. Studies on Wujiapi and some related crude drugs. I. Morphological studies on Bei-Wujiapi and some related drugs. Yakugaku Zasshi 89: 972–978.
30. Chen, K. K. 1929. Relationship between the pharmacological action and the chemical constitution and configuration of the optical isomers of ephedrine and related compounds. J. Pharmacol. 36: 363–400.
31. Yamasaki, K., Sankswa, U., and Shibata, S. 1969. Biosynthesis of ephedrine in Ephedra. Tetrahedron Lett. 47: 4099–4102.
32. Takagi S., Akiyama, T., Kinoshita, T., Sankawa, U., and Slubata, S. 1979. Minor basic constituents of evodia fruits. Shoyakugaka Zasshi 33: 30–34.
33. Huang, K. C. 1999. The Pharmacology of Chinese Herbs. 2nd ed. CRC Press. New York. 512 pp.
34. Liao, D. F., Lu, N., Lei, L. S., Yu., L., and Chen, J. X. 1995. Effects of gypenosides on mouse splenic lymphocyte transformation and DNA polymerase II activity in vitro. Acta Pharmacol. Sin. 16: 322–324.
35. Chen, K. K. and Henderson, F. G. 1965. Digitalis-like substances of antiaris. J. Pharmacol. Exp. Ther. 150: 53–56.
36. Zhang, B. H., Wang, N. S., Li, X. J., Kong, X. J., and Cai, Y. L. 1990. Anti-arrhythmic effects of matrine. Acta Pharmacol. Sin. 11: 253–257.
37. Chen, X. J., Fan, H. Y., Yao, Y. F., Zhang, J. X., and Gu, W. X. 1987. Relation of the antihypertensive effect and central alpha-adrenoceptor of rhomotoxin. Acta Pharmacol. Sin. 8: 247–250.
38. Kondo, H. and Tomita, M. 1936. Arch. Pharm. 274: 65–82.
39. Zhang, J. T. 1996. Studies on traditional Chinese drugs. Chin. Med. J. 109: 54–58.
40. Wu, P. C. 1983. Pharmacology of Zhong Cai Yao. Peoples' Health Publisher. Beijing. pp. 90–91.
41. Lei, S. W. and Lin, Z. B. 1991. J. Beijing Med. Univ. 23: 329–333 (English abstract only).
42. Itoigawa, M., Takeya, K., and Furukawa, H. 1994. Cardiotonic flavonoids from citrus plants. Biol. Pharmacol. Bull. (Japan) 78: 1519–1521.
43. Ju, H. S. 1991. Chin. Circ. J. 6: 227–229.

44. Hans, R. H. and Park, M. H. 1986. In Steiner (Ed.). Folk Medicine. American Chemical Society. Washington, D.C. p. 205–215.
45. Woo, W. S. 1985. In Advanced Chinese Medicinal Material Research. H. M. Chang et al. (Eds.). World Scientific Publishing. Singapore. pp. 129–146.
46. Zhon, S. G. 1931. Chin. J. Physiol. 5: 131–140.
47. Chow, K. and Kramer, I. 1990. All the Tea in China. China Books and Periodicals. San Francisco, CA.
48. Zhu, Y.-C. 1989. Plantae Medicinales China Roreali-Orientalis. Heilongjiang Sci. Technology Publ. House. 1300 pp.
49. Keys, J. D. 1976. Chinese Herbs—Their Chemistry and Pharmacodynamics. Charles E. Tuttle. Vermont. 388 pp.
50. Duke, J. A. and Ayensu, E. S. 1985. Medicinal Plants of China. Vol. 1 and 2. Reference Publ. Inc. Michigan. 705 p.
51. Bliss, B. 1973. Chinese Medicinal Herbs. Georgetown Press. San Francisco. 467 pp.
52. Perry, L. M. 1980. Medicinal Plants of East and Southeast Asia. MIT Press. Cambridge, MA. 620 pp.
53. Chen, F. C. 1997. Active Ingredients and Identification in Common Chinese Herbs. People Health Publ. Co. China. 872 pp.
54. Chiu, N. and Chang, K. 1995. The Illustrated Medicinal Plants in Taiwan. Vol. 1. SMC Publ. Inc. Taiwan. 283 pp.
55. Chiu, N. and Chang, K. 1995. The Illustrated Medicinal Plants in Taiwan. Vol. 2. SMC Publ. Inc. Taiwan. 315 pp.
56. Chiu, N. and Chang, K. 1995. The Illustrated Medicinal Plants in Taiwan. Vol. 3. SMC Publ. Inc. Taiwan. 312 pp.
57. Chiu, N. and Chang, K. 1995. The Illustrated Medicinal Plants in Taiwan. Vol. 4. SMC Publ. Inc. Taiwan. 319 pp.
58. Chiu, N. and Chang, K. 1995. The Illustrated Medicinal Plants in Taiwan. Vol. 5. SMC Publ. Inc. Taiwan. 289 pp.
59. Cordell, G. A. 1984. Studies in the Thymelacaceae. I. NMR spectral assignments of daphnoretin. J. Nat. Prod. 47: 84–88.
60. Perry, L. M. and Metzger, J. 1980. Medicinal plants of East and Southeast Asia: Attributed Properties and Uses. MIT Press. London. 620 p.
61. Crevost, C. and Petelot, A. 1929. Catalogue des produits de l'Indochine. Plantes Med. 32: 1–35.
62. Petelot, A. 1952. Les plantes medicinales du Cambodge, du Laos et du Vietnam. I. Arch. Recherches Agron. Cambodge, Laos, Vietnam 14: 1–408.
63. Arthur, H. R. and Cheung, H. T. 1960. A phytochemical survey of the Hong Kong medicinal plants. J. Pharm. Pharmacol. 12: 567–570.
64. Chiu, C. 1955. A New Manual of Chinese Materia Medica. Shanghai, China. 385 pp.
65. Kariyone, T. and Kimura, Y. 1949. Japanese-Chinese Medicinal Plants, Their Constituents and Medicinal Uses. Takeda Chemical Company. 2nd ed. Tokyo. 519 pp.
66. Gimlette, J. D. 1929. Malay Poisons and Charm Cures. 3rd ed. New York. 301 pp.
67. Steinmetz, E. F. 1954. Unsigned note: Soya bean, source of new edible antibiotic. Acta Phytother. 1: 15.
68. Ros, J. 1955. Traite des plantes medicinales chinoises. Encyc. Biol. 47: 1–500.
69. Chung, Y. C. 1959. New Chinese Materia Medica. Vol. 1. Roots. Beijing, China. 564 pp.
70. Johnston, J. C., Wetch, R. C., and Hunter, G. L. K. 1980. Volatile constituents of litchi (Litchi chinensis Sonn.). J. Agric. Food Chem. 28: 859–861.
71. Chung, K. T. 1959. A Chinese Native Medicinal Flora for Farmers. Beijing, China. 281 pp.

72. Kariyone, T. and Kimura, Y. 1949. Japanese-Chinese Medicinal Plants, Their Constituents and Medicinal Uses. 2nd ed. Tokyo, Japan. 519 pp.

73. Douglas, B. and Kiang, A. K. 1957. A phytochemical survey of the flora of Malaya. I. Alkaloids. Malayan Pharm. J. 6: 138–154.

74. Arata, Y. and Chashi, T. 1957. Studies on the constituents of Rhizoma nupharis. XII. On its third alkaloid (nupharmine). J. Pharm. Soc. Jpn. 77: 792.

75. Arata, Y. and Matsuda, H. 1961. Constituents of Rhizoma nupharis. XVII. Isolation of β-sitosterol, palmitic acid, and oleic acid. Kanazawa Daigaku Yakugakubu Kenyu Nempo 10: 35–39.

76. Arthur, H. and Cheung, T. 1960. A phytochemical survey of the Hong Kong medicinal plants. J. Pharm. Pharmacol. 12: 567–570.

77. Akhtardzhiev, K. and D. Koley. 1961. Composition of mucins. II. Mucins from Malya sylyestris flowers. Pharm. Zentralh. 100: 14–16.

78. Sreenis, C. G. S. van. 1958. Magic plants of the Dayak. Sarawsk Mus. J. n. s. 11: 430–436.

79. Youngken, H. W. (Ed.) 1936. A Textbook of Pharmacognosy. 4th ed. New York. 1063 p.

80. Kleipool, R. J. C. and Wibaut, J. P. 1959. Mimosine (leucaenine) 5th communication. Rec. Trav. Chim. Pays-Bas 69: 37–44.

81. Arthur, H. R. 1954. A phytochemical survey of some plants of North Borneo. J. Pharm. Pharmacol. 6: 66–72.

82. Chaudhry, G. R., Sharma, V. N., and Siddiqui, S. 1951. Chemical constituents of Nardostachys jatamanshi. I. Isolation of a crystalline acid and an essential oil. J. Sci. Ind. Res. 10B: 48.

83. Kiang, A. K., Douglas, B., and Morsingh, F. 1961. A phytochemical survey of Malaya. II. Alkaloids. J. Pharm. Pharmacol. 13: 98–104.

84. Li, C. S. 1996. Appraisal of Chinese Medicine. Shanghai Scientific and Technology Publishing Co., Shanghai, China. 699 pp.

85. Chiu, C. 1955. A New Manual of Chinese Materia Medica. Shanghai Scientific and Technology Publishing Co. China. 385 p.

86. Petelot, A. 1952. Les plantes medicinales du Cambodge, du Laos et du Vietnam. I. Arch. Rech. Agron. Pastor. Vietnam 18: 1–284.

87. Zheng, J. H. 1983. Commonly Used Chinese Herbs in the Treatment of Carcinoma. Department of Pharmacognosy, University of Chicago, Chicago, IL.

88. Roi, J. 1955. Traite des plantes medicinales chinoises. Encyc. Biol. 47: 1–500.

89. Arthur, R. 1954. A phytochemical survey of some plants of North Borneo. J. Pharm. Pharmacol. 6: 66–72.

90. Plugge, P. C. 1891. Das alkaloid von Sophora tomentosa L. Arch. Pharm. 229: 561–565.

91. Zhu, J. S., Halpern, G. M., and Jones, K. 1998. The scientific rediscovery of an ancient Chinese herbal medicine: Part I. Cordyceps sinensis. J. Alternative Complementary Med. 4: 289–305.

92. Zhu, J. S., Halpern, G. M., and Jones, K. 1998. The scientific rediscovery of an ancient Chinese herbal medicine: II. Cordyceps sinensis. J. Alternative Complementary Med. 4: 429–457.

93. Gorter, K. 1919. Sur la distribution de la lycorine dans la famille des Amaryllidacees. Bull. Jard. Bot. Buitenzorg III 1: 352–358.

94. Tomita, M. and H. Ishii. 1958. Studies on the alkaloids of rutaceous plants. IV. Alkaloids of Xanthoxylum ailanthoides Sieb. & Zucc. [Fagara ailanthoides (Sieb. & Zucc.) Engl.]. J. Pharm. Soc. Jpn. 78: 1441–1443.

95. Li, T. S. C. 2000. Medicinal Plants—Culture, Utilization and Phytopharmacology. Technomic Publishing Co., Inc. Lancaster, PA. 517 pp.

96. Crevost, C. and Petelot, A. 1929. Catalogue des produits de l'Indochine. Plantes Med. 37: 267–311.

97. Valenzuela, P., Santos, A. C., and Concha, J. A. 1954. The chemistry and pharmacy of drugs of pharmacopoeial value from Philippine medicinal plants. Proc. 8th Pacific Sci. Congr. 4A: 170–181.

98. Balandrin, M. F. and Klocke, J. A. 1988. Medicinal, aromatic, and industrial materials from plants. In Bajaj, Y. P. S. (Ed.). Biotechnology in Agriculture and Forestry 4. Medicinal and Aromatic Plants I. Springer-Verlag. New York. p. 3–36.

99. Chevallier, A. 1996. The Encyclopedia of Medicinal Plants. Dorling Kindersley Ltd. London. 336 pp.

100. Duke, J. A. 1985. CRC Handbook of Medicinal Herbs. CRC Press. Boca Raton, FL. 677 pp.

101. Guedon, D., Abbe, P., and Lamaison, J. L. 1993. Leaf and flower head flavonoids of Achillea millefolium L. subspecies. Biochem. Syst. Ecol. 21: 607–611.

102. Rogers, R. D. 1997. Sundew, Moonwort, Medicinal Plants of the Prairies. Vol. 1 and 2. Edmonton, Alberta. 282 p.

103. Small, E. and Catling, P. M. (Eds.) 1999. Canadian Medicinal Crops. NRC Research Press. Ottawa. 240 pp.

104. Shoyama, Y., Nishioka, I., and Hatano, K. 1991. Aconitum spp. (monkshood): somatic embryogenesis, plant regeneration, and the production of aconitine and other alkaloids. In Bajaj, Y. P. S. (Ed.). Biotechnology in Agriculture and Forestry 15. Medicinal and Aromatic plants III. Springer-Verlag. New York. pp. 58–72.

105. Bown, D. 1987. Acorus calamus L: A Species with a History. Aroideana (International Aroid Society, South Miami, FL) 10: 11–14.

106. Ducrey, B., Wolfender, J. L., Marston, A., and Hostenmann, K. 1955. Analysis of flavonol glycosides of thirteen Epilobium species (Onagraceae) by LC-UV and thermospray LC-MS. Phytochemistry 38: 129–137.

107. Small, E. (Ed.). 1997. Culinary Herbs. NRC Research Press. Ottawa. 710 pp.

108. Cavallini, A., Natali, L., and Castorena Sanchez, I. 1991. Aloe barbadensis Mill. (= A. vera L.). In Bajaj, Y. P. S. (Ed.). Biotechnology in Agriculture and Forestry 15. Medicinal and Aromatic Plants III. Springer-Verlag. New York. pp. 95–106.

109. Natali, A. C. and I. C. Sanchez. 1991. Aloe barbadensis Mill. (= A. vera L.). In Bajaj, Y. P. S. (Ed.). Biotechnology in Agriculture and Forestry 15. Medicinal and Aromatic Plants III. Springer-Verlag. New York. pp. 95–106.

110. Collin, H. A. and Isaac, S. 1991. Apium graveolens L. (celery): in vitro culture and the production of flavors. In Bajaj, Y. P. S. (Ed.). Biotechnology in Agriculture and Forestry 15. Medicinal and Aromatic Plants III. Springer-Verlag. New York. pp. 73–94.

111. Kato, M. 1989. Camellia sinensis L. (tea): in vitro regeneration. In Bajaj, Y. P. S. (Ed.). Biotechnology in Agriculture and Forestry 7. Medicinal and Aromatic Plants II. Springer-Verlag. New York. pp. 82–98.

112. Furuya, T. and Yoshikawa, T. 1991. Carthamus tinctorius L. (safflower): Production of vitamin E in cell culture. In Bajaj, Y. P. S. (Ed.). Biotechnology in Agriculture and Forestry 15. Medicinal and Aromatic Plants III. Springer-Verlag. New York. pp. 142–155.

113. Furmanowa, M., Sowinska, D., and Pietrosiuk, A. 1991. Carum carvi L. (caraway): in vitro culture, embryogenesis, and the production of aromatic compounds. In Bajaj, Y. P. S. (Ed.). Biotechnology in Agriculture and Forestry 15. Medicinal and Aromatic Plants III. Springer-Verlag. New York. pp. 177–192.

114. Willuhn, G. 1998. Arnica flowers: pharmacology, toxicology and analysis of the sesuiterpene lactonesn, their main active substance. In Lawsib, L. D. and Bauer, R. (Eds.). Phytomedicines of Europe, Chemistry and Biological Activity. American Chemical Society. Washington, D.C. pp. 118–132.

115. Jayatilake, G. S. and MacLeod, A. J. 1987. Volatile constituents of Centella asiatica. In Martens, M., Dalen, G. A., and Russwurm, H., Jr. (Eds.). Flavour Science and Technology, John Wiley & Sons. New York. pp. 79–82.

116. Solet, J. M., Simon-Ramiasa, A., Cosson, L., and Guignard, J. L. 1998. Centella asiatica (L.) urban (pennywort): Cell culture, production of terpenoids, and biotransformation capacity. In Bajaj, Y. P. S. (Ed.). Biotechnology in Agriculture and Forestry 41. Medicinal and Aromatic Plants X. Springer-Verlag. New York. pp. 81–96.

117. Cordeiro, M. C., Pats, M. S., and Brodelius, P. E. 1998. Cynara cardunculus subsp. flavescens (cardoon): in vitro culture, and the production of cyprosins-milk-clotting enzymes. In Bajaj, Y. P. S. (Ed.). Biotechnology in Agriculture and Forestry 41. Medicinal and Aromatic Plants X. Springer-Verlag. New York. pp. 132–153.

118. Bremness L. 1994. Herbs. Dorling Kindersley Ltd. London. 304 pp.

119. Sakurai, N. and Nagai, M. 1996. Chemical constituents of original plants of Cimicifuga rhizoma in Chinese medicine. Yakugaku Zasshi 116: 850–865.

120. Struck, D., Tegtmeier, M., and Harnischfeger, G. 1997. Flavones in extract of Cimicifaga racemosa. Planta Med. 63: 289.

121. Mansell, R. L. and McIntosh, C. A. 1991. Citrus spp.: in vitro culture and the production of naringin and limonin. In Bajaj, Y. P. S. (Ed.). Biotechnology in Agriculture and Forestry 15. Medicinal and Aromatic Plants III. Springer-Verlag. New York. pp. 193–210.

122. Khodzhimatov, K., Fakhrutdinov, S. F., Aprasidi, G. S., Kuchni, N. P., and Karimov, K. 1987. Codonopsis clematidea Schreuk—a valuable medicinal plant. Referativnyi Zh. 10: 55, 790.

123. Ishimaru, K., Tanaka, N., Kamiya, T., Sato, T., and Shimomura, K. 1998. Cornus kousa (dogwood): in vitro culture, and the production of tannins and other phenolic compounds. In Bajaj, Y. P. S. (Ed.). Biotechnology in Agriculture and Forestry 41. Medicinal and Aromatic Plants X. Springer-Verlag. New York. p. 113–131.

124. Strenath, H. L. and Jagadishchandra, K. S. 1991. Cymbopogon Spreng. (aromatic grasses): in vitro culture, regeneration, and the production of essential oils. In Bajaj, Y. P. S. (Ed.). Biotechnology in Agriculture and Forestry 15. Medicinal and Aromatic Plants III. Springer-Verlag. New York. pp. 211–236.

125. Tominaga, T. and Dubourdieu, D. 1997. Identification of 4-mercapto-4-methylpentan-2-one from the box tree (Buxus sempervirens L.) and broom (Sarothamnus scoparius (L.) Koch.). Flav. Fragrance J. 12: 373–376.

126. Furmanowa, M. and Guzewska, J. 1989. Dioscorea: in vitro culture and the micropropagation of diosgenin-containing species. In Bajaj, Y. P. S. (Ed.). Biotechnology in Agriculture and Forestry 7. Medicinal and Aromatic Plants II. Springer-Verlag. New York. pp. 162–184.

127. Halstead, B. W. and Hood, L. L. 1942. Eleutherococcus senticosus, Siberian Ginseng: An Introduction to the Concept of Adaptogenic Medicine. Oriental Healing Arts Institute. Long Beach, CA. 94 pp.

128. Li, T. S. C. 2001. Siberian Ginseng. HortTechnology 11: 1–7.

129. O'Dowd, N. A., McCauley, P. G., Wilson, G., Parnell, J. A. N., Kavanagh, T. A. K., and McConnell, D. J. 1998. Ephedra species: in vitro culture, micropropagation, and the production ephedrine and other alkaloids. In Bajaj, Y. P. S. (Ed.). Biotechnology in Agriculture and Forestry 41. Medicinal and Aromatic Plants X. Springer-Verlag. New York; pp. 154–193.

130. Duke, J. A. and duCellier, J. L. 1993. CRC Handbook of Alternative Cash Crops. CRC Press. London. 536 pp.

131. Zheng, G. Q., Kenney, P. M., and Lam, L. K. T. 1992. Sesquiterpenes from clove (Eugenia caryophyllata) as potential anticarcinogenic agents. J. Nat. Prod. 55: 999–1003.

132. Kameoka, H., Miyazawa, M., and Haze, K. 1975. 3-Ethyl-7-hydroxyphthalide from Forsythia japonica. Phytochemistry 14: 1676–1677.

133. Ahlemeyer, B. and Krieglstein, J. 1998. Neuroprotective effects of Ginkgo biloba extract. In Lawsib, L. D. and Bauer, R. (Eds.). Phytomedicines of Europe, Chemistry and Biological Activity. American Chemical Society. Washington, D.C. pp. 210–220.

134. Arenz, A., Klein, M., Fiehe, K., Gross, J., Drewke, C., Hemscheidt, T., and Leistner, E. 1996. Occurrence of neurotoxic 4'-O-methylpyridoxine in Ginkgo biloba leaves, Ginkgo medications and Japanese ginkgo food. Planta Med. 62: 548–551.

135. Xing, S., Huangpu, G., Zhang, Y., Hou, J., Sun, X., Han, F., and Yang, J. 1997. Analysis of the nutritional components of the seeds of promising ginkgo cultivars. J. Fruit Sci. 14: 39–41.

136. Rigney, U., Kimber, S., and Hindmarch, I. 1999. The effects of acute doses of standardized Ginkgo biloba extract on memory and psychomotor performance in volunteers. Phytother. Res. 13: 408–415.

137. Jacobs, B. P. and Browner, W. S. 2000. Ginkgo biloba: A living fossil. Am. J. Med. 108: 276–281.

138. Caballero, R., Haj-Ayed, M., Galvez, J. F., Hernaiz, P. J., and Ayed, M. H. 1995. Yield components and chemical composition of some annual legumes and oat under continental Mediterranean conditions. Agric. Mediterranea 125: 222–230.

139. Li, T. S. C. and Schroeder, W. R. 1996. Sea buckthorn (Hippophae rhamnoides L.): A multipurpose plant. HortTechnology 6: 370–380.

140. Li, T. S. C. and Wang, L. C. H. 1998. Physiological components and health effects of ginseng, echinacea and sea buckthorn. In Mazza, G. (Ed.). Functional Foods, Biochemical and Processing Aspects. Technomic Publishing Co., Inc. Lancaster, PA. 460 pp.

141. Oomah, D., Stephanie, L., and Godfrey, D. V. 1999. Properties of sea buckthorn (Hippophae rhamnoides L.) and ginseng (Panax quinquefolium L.) seed oils. Proc. Canadian Inst. Food Sci. Technology Annual Conf. Kelowna, BC. 55 p.

142. Heale, J. B., Legg, T., and Connell, S. 1989. Humulus lupulus L.: in vitro culture: attempted production of bittering components and novel disease resistance. In Bajaj, Y. P. S. (Ed.). Biotechnology in Agriculture and Forestry 7. Medicinal and Aromatic Plants II. Springer-Verlag. New York. pp. 264–285.

143. Stevens, R. 1967. The chemistry of hop constituents. Chem. Rev. 67: 19–71.

144. Tomlinson, B., Chan, T. Y. K., Chan, J. C. N., Critchley, J. A. J. H., and Bat, P. P. H. 2000. Toxicity of complementary therapies: Eastern perspective. J. Chin. Pharmacol. 40: 451–458.

145. List, P. H. and L. Horhammer. 1969. Hoger's Handbuch der Pharmaceutischen Praxis. Vol. 2–6. Springer-Verlag. Berlin.

146. Morton, J. F. 1981. Atlas of Medicinal Plants of Middle America. Charles C Thomas. Springfield, IL.

147. Bisset, N. G. 1994. Herbal Drugs and Phytopharmaceuticals. CRC Press. London. 566 pp.

148. Syrchina, A. I., Pechurina, N. N., Vereshchagin, A. L., Gorshkov, A. G., Tsapalova, I. E., and Semenov, A. A. 1993. A chemical investigation of Matteuccia struthiopteris. Chem. Nat. Compd. 29: 535–536.

149. Bastida, J., Bergonon, S., Viladomat, F., and Codina, C. 1994. Alkaloids from Narcissus primigenius. Planta Med. 60: 95–96.

150. Bracco, V. 1973. Determination of polycyclic aromatic hydrocarbons: Technique and application to coffee oil. Riv. Ital. Sostanze Grasse 50: 166–176.

151. Kreh, M., Matusch, R., and Witte, L. 1995. Acetylated alkaloids from Narcissus pseudonarcissus. Phytochemistry 40: 1303–1306.

152. Mulherjee, K. D. and Kiewitt, I. 1987. Formation of gamma linolenic acid in the higher plant evening primrose (Oenothera biennis L.). J. Agric. Food Chem. 35: 1009–1012.

153. Yamamoto, H. 1988. Paeonia spp.: in vitro culture and the production of paeoniflorin. In Bajaj, Y. P. S. (Ed.). Biotechnology in Agriculture and Forestry 4: Medicinal and Aromatic Plants I. Springer-Verlag. New York. p. 464–483.

154. Bunney, S. 1992. The Illustrated Encyclopedia of Herbs, Their Medicinal and Culinary Uses. Chancellor Press. London. 320 pp.

155. Li, T. S. C. 1995. Asian and American ginseng, a review. HortTechnology 5: 27–34.

156. Sticher, O. 1998. Biochemical, pharmaceutical, and medical perspectives of ginseng. In Lawsib, L. D. and Bauer, R. (Eds.). Phytomedicines of Europe, Chemistry and Biological Activity. American Chemical Society. Washington, D.C. pp. 221–240.

157. Gray, A. I., Bhandari, P., and Waterman, P. G. 1988. New protolimonoids from the fruits of Phelloderdron chinense. Phytochemistry 27: 1805–1808.

158. Chuda, Y., Ono, H., Ohnishi-Kameyama, M., Matsumoto, K., Nagata, T., and Kikuchi, Y. 1999. Mumefural, citric acid derivative improving blood fluidity from fruit-juice concentrate of Japanese apricot (Prunus mume Sieb. et Zucc.). J. Agric. Food Chem. 47: 828–831.

159. Fang, T. T. and Huang, J. H. 1999. Extraction, fractionation and identification of bitter oligopeptides and amino acids in mei fruit (Prunus mume Sieb. et Zucc.). J. Beijing Forestry Univ. 21: 61–71.

160. Short, K. C. and Roberts, A. V. 1991. Rosa spp. (roses): in vitro culture, micropropagation, and the production of secondary products. In Bajaj, Y. P. S. (Ed.). Biotechnology in Agriculture and Forestry 15. Medicinal and Aromatic Plants III. Springer-Verlag. New York. pp. 376–397.

161. McGimsey, J. 1993. Sage, Salvia officianalis. www.crop.cri.nz/broadshe/sage.htm.

162. Henry, M. 1989. Saponaria officinalis L.: in vitro culture and the production of triterpenoidal saponins. In Bajaj, Y. P. S. (Ed.). Biotechnology in Agriculture and Forestry 7. Medicinal and Aromatic Plants II. Springer-Verlag. New York. pp. 431–442.

163. Yamamoto, H. 1991. Scutellaria baicalensis Georgi: in vitro culture and the production of flavonoids. In Bajaj, Y. P. S. (Ed.). Biotechnology in Agriculture and Forestry 15. Medicinal and Aromatic Plants III. Springer-Verlag. New York. pp. 398–418.

164. Brown, M. S. and Molyneux, R. J. 1996. Effects of water and mineral nutrient deficiencies on pyrrolizidine alkaloid content of Senecio vulgaris flowers. J. Sci. Food Agric. 70: 209–211.

165. Chung, C. H., Yee, Y. J., Kim, D. H., Kim, H. K., and Chung, D. S. 1995. Changes of lipid, protein, RNA and fatty acid composition in developing seasame (Sesamum indicum L.) seeds. Plant Sci. Limerick 109: 237–243.

166. Ogasawara, T., Chiba, K., and Tada, M. 1998. Sesamum indicum L. (sesame): in vitro culture and the production of naphthoquinone and other secondary metabolites. In Bajaj, Y. P. S. (Ed.). Biotechnology in Agriculture and Forestry 41. Medicinal and Aromatic Plants X. Springer-Verlag. New York. pp. 366–393.

167. Omer, E. A., Refaat, A. M., Ahmed, S. S., Kamel, A., and Hammouda, F. M. 1993. Effect of spacing and fertilization on the yield and active constituents of milk thistle, Silybum marianum. J. Herbs Spices Med. Plants 1: 17–23.

168. Sonnenbichler, J., Sonnenbichler, I., and Scalera, F. 1998. Influence of the flavonolignan silibinin of milk thistle on hepatocytes and kidney cells. In Lawsib, L. D. and Bauer, R.(Eds.). Phytomedicines of Europe, Chemistry and Biological Activity. American Chemical Society. Washington, D.C. pp. 263–277.

169. Szentimihalyi, K., Then, M., Illes, V., Perneczky, S., Sandor, Z., Lakatos, B., and Vinkler, P. 1998. Phytochemical examination of oils obtained from the fruit of milk thistle (Silybum marianum L. Gaertner) by suercritical fluid extraction. Z. Naturforsch. Sec. C Biosci. 53: 9–10, 779–784.

170. Chapple, C. C. S. and Ellis, B. E. 1991. Syringa vulgaris L. (common lilac): in vitro culture and the occurrence and biosynthesis of phenylpropanoid glycosides. In Bajaj, Y. P. S. (Ed.). Biotechnology in Agriculture and Forestry 15. Medicinal and Aromatic Plants III. Springer-Verlag. New York. pp. 478–497.

171. Lawrence, B. M. 1985. Essential oils of the Tugetes genus. Perf. Flav. 10: 73–82.

172. Kussi, T., Hardh, K., and Kanon, H. 1984. Experiments on the cultivation of dandelion for salads use. II. The nutritive value and intrinsic quality of dandelion leaves. J. Agric. Sci. Finland 56: 23–31.

173. Kuusi, T., Pyysalo, H., and Autio, K. 1985. The bitterness properties of dandelion. II. Chemical investigations. Lebensm. Wiss. Technol. (Zurich) 18: 349–359.

174. Li, J. X., Shi, Q., Xiong, Q. B., Prasain, J. K., and Texuka, Y. 1998. Tribulusamide A and B, new hepatoprotective lignanamides from the fruits of Tribulus terrestris: Indications of cytoprotective activity in murine hepatocyte culture. Planta Med. 64: 628–631.

175. Kaushalya, G., Thakral, K. K., Arora, S. K., Chowdhary, M. L., and Gupta, K. 1996. Structural carbohydrate and mineral contents of fenugreek seeds. Indian Cocoa, Arecanut Spices J. 20: 120–124.

176. Baguena, C. L. 1942. Notes on some longicorns harmful to cultivated trees, including cacao and coffee, in the Spanish territories in the Gulf of Guinea Publ. Direcc. Agric. Territ. Exp. Golfi. Guinea 6: 39–91.

177. Cunio, L. 1994. Vaccinium myrtillus. Aust. J. Med. Herbalism 5: 81–85.

178. Nazarko, L. 1995. Infection control. The therapeutic uses of cranberry juice. Nurs. Stand. 9: 33–35.

179. Ofek, I., Goldhar, J., and Sharon, N. 1996. Anti-Escherichia coli adhesin activity of cranberry and blueberry juices. Adv. Exp. Med. Biol. 408: 179–183.

180. Schmidt, D. R. and Sobota, A. E. 1988. An examination of the anti-adherence activity of cranberry juice on urinary and nonurinary bacterial isolates. Microbios 55: 173–181.

181. Gabius, H. J. and Gabius, S. 1998. Phytotherapeutic immunomodulation as a treatment modality in oncology: lessons from research with mistletoe. In Lawsib, L. D. and Bauer, R. (Eds.). Phytomedicines of Europe, Chemistry and Biological Activity. American Chemical Society. Washington, D.C. pp. 278–286.

182. Ikan, R., Weinstein, V., Milner, Y., Bravdo, B., Shoseyov, O., Segal, D., Altman, A., Chet, I., Palevitch, D., and Putievsky, E. 1993. Natural glycosides as potential odorants and flavorants. Int. Symp. Medicinal Aromatic Plants, Tiberias on the Sea of Galilee, Israel 22–25 March, 1993. Acta Hort. 344: 17–28.

183. Bordia, A., Verma, S. K., and Srivastava, K. C. 1997. Effect of ginger (Zingiber officinale Rosc.) and fenugreek (Trigonella foenumgraecum L.) on blood lipids, blood sugar and platelet aggregation in patients with coronary artery diseases. Prostag. Leukot. Essent. Fatty Acids 56: 379–384.

184. Sakamura, F. and Suga, T. 1989. Zingiber officinale Roscoe (ginger): in vitro propagation and the production of volatile constituents. In Bajaj, Y. P. S. (Ed.). Biotechnology in Agriculture and Forestry 7. Medicinal and Aromatic Plants II. Springer-Verlag. New York. pp. 524–538.

185. Marles, R. J., Clavelle, C., Monteleone, L., Tays, N., and Burns, D. 1999. Aboriginal Plant Use in Canada's Northwest Boreal Forest. UBC Press. Vancouver, Canada. 368 pp.

186. Hetherington, M. and Steck, W. 1997. Natural chemicals from northern prairie plants: The phytochemical constituents of one thousand North American species. Fytokem Products. Saskatoon, SK.

187. Burkill, H. M. 1985. The Useful Plants of West Tropical Africa. Vol. 1. Families A-D. Royal Botanic Gardens. Kew, U.K.

188. Duke, J. A. and duCellier, J. L. 1993. CRC Handbook of Alternative Cash Crops. CRC Press. London. 536 pp.

189. Youngken, H. W. 1950. A Textbook of Pharmacognosy. 6th ed. Blakiston Publ. Co. New York. 1063 pp.

190. Kuhnlein, H. V. and Turner, N. J. 1991. Traditional Plant Foods of Canadian Indigenous Peoples. Nutrition, Botany and Use. Gordon and Breach Sci. Publ. Philadelphia, PA.

191. Small, E. (Ed.). 1997. Culinary Herbs. NRC Research Press. Ottawa. 710 pp.

192. Mathes, H. W. D., Luu, B., and Ourisson, G. 1980. Cytotoxic components of Zingiber zerumbet, Curcuma zedoria and C. domestica. Phytochemistry 19: 2643–2650.

193. Grover, G. S. and Rao, J. T. 1987. Antimicrobial properties of the essential oils of Cymbopogon martinii and Zingiber zerumbet. Pafai J. 9: 4, 19–22.

194. Bandara, B. M. R., Fernando, I. H. S., Hewage, C. M., Karunaratne, V., Adikaram, N. K. B., and Wijesundara, D. S. A. 1989. Antifungal activity of some medicinal plants of Sri Lanka. J. Nat. Sci. Counc. Sri Lanka 17: 1, 1–13.

195. Lechat-Vahirua, I., Francois, P., Menut, C., Lamaty, G., and Bessiere, J. M. 1993. Aromatic plants of French Polynesia. I. Constituents of the essential oils of rhizomes of three Zingiberaceae: Zingiber zerumbet Smith, Hedychium coronarium Koenig and Etlingera cevuga Smith. J. Essential Oil Res. 5: 1, 55–59.

196. Akihisa, T., Ghosh, P., Thakur, S., Oshikiri, S., Tamura, T., and Matsumoto, T. 1988. 24 beta-methylcholesta-5, 22E,25-trien-3beta-ol, and 24alpha-ethyl-5alpha-cholest-22E-en-3beta-ol from Clerodendrum fragrans. Phytochemistry 27: 241–244.

197. Komai, K. and Tang, C. S. 1989. Chemical constituents and inhibitory activities of essential oils from Cyperus brevifolius and C. kyllingia. J. Chem. Ecol. 15: 2171–2176.

198. Hellion-Ibarrola, M. C., Ibarrola, D. A., Montalbetti, Y., Villalba, D., Heinichen, O., and Ferro, E. A. 1999. Acute toxicity and general pharmacological effect on central nervous system of the crude rhizome extract of kyllinga brevifolia Rottb. J. Ethnopharmacol. 66: 271–276.

199. Ohmiya, S., Otomasu, H., Haginiwa, J., and Murakoshi, J. 1978. (+)-5,17-Dehydromatrine N-oxide, a new alkaloid in Euchresta japonica. Phytochemistry 17: 2021–2022.

200. Ohmiya, S., Otomasu, H., Haginiwa, J., and Murakoshi, J. 1979. (–)-12-Cytisineacetic acid, a new lupin alkaloid in Euchresta japonica. Phytochemistry 18: 649–650.

201. Shirataki, Y., Manaka, A., Yokoe, I., and Komatsu, M. 1982. Two prenylflavanones from Euchresta japonica. Phytochemistry 21: 2959–2963.

202. Lu, C., Yang, J., Wang, P., Lin, C., Lu, C. M., Yang, J. J., Wang, P. Y., and Lin, C. C. 2000. A new acylated flavonol glycoside and antioxidant effects of Hedyotis diffusa. Planta Med. 66: 374–377.
203. Nishihama, Y., Masuda, K., Yamaki, M., Takagi, S., and Sakina, K. 1981. Three new iridoid glucosides from Hedyotis diffusa. Planta Med. 43: 28–33.
204. Ho, T. I., Chen, G. P., Lin, Y. C., Lin, Y. M., and Chen, F. C. 1986. An anthraquinone from Hedyotis diffusa. Phytochemistry 25: 1988–1989.
205. Wu, H. M., Tao, X. L., Chen, Q., and Lao, X. F. 1991. Iridoids from Hedyotis diffusa. J. Nat. Prod. 54: 254–256.
206. Wong, K. C. and Tan, G. L. 1995. Composition of the essential oil of Hedyotis diffusa Willd. J. Essential Oil Res. 7: 537–539.
207. Ojewole, J. A. O. 1980. Studies on the pharmacology of tetramethylpyrazine from the stem of Jatropha podagrica. Planta Med. 39: 238.
208. Odebiyi, O. O. 1985. Steroids and flavonoids from Jatropha podagrica stem bark. Fitoterapia 56: 302–303.
209. Saito, K., Nishijima, M., Ohno, N., Yadomae, T., and Miyazaki, T. 1992. Structure and antitumour activity of the less-branched derivatives of an alkali-soluble glucan isolated from Omphalia lapidescens. Chem. Pharm. Bull. 40: 261–263.
210. Kuhnlein, H. V. and Soueida, R. 1992. Use and nutrient composition of traditional Baffin Inuit foods. J. Food Composition Anal. 5: 112–126.
211. Dogra, J. V. V. and Sinha, S. K. P. 1979. Observation of the age related changes in the level of alpha-ketoglutaric acid in leaves of Phyllanthus simplex (Retz.). Comp. Physiol. Ecol. 4: 35–37.
212. Negi, R. S. and Fakhir, T. M. 1988. Simplexine (14-hydroxy-4-methoxy-13, 14-dihydronorsecurinine): An alkaloid from Phyllanthus simplex. Phytochemistry 27: 3027–3028.
213. Komine, H., Takahashi, T., and Ayabe, S. 1996. Properties and partial purification of squalene synthase from culture of dandelion. Phytochemistry 42: 405–409.
214. Kim, C. J., Lim, J. S., and Cho, S. K. 1996. Anti-diabetic agents from medicinal plants inhibitory activity of Schizonepeta tenuifolia spikes on the diabttogenesis by streptozotocin in mice. Arch. Pharmacol. Res. Seoul 19: 441–446.
215. Tao, J., Tu, P., Xu, W., and Chen, D. 1999. Studies on chemical constituents and pharmacological effects of the stem of Cynomorium songaricum Rupr. Zhongguo Zhongyao Zaxhi 24: 292–294.
216. Ibrahim, N. A., El-Gengaihi, S. E., El-hamidi, A., Bashandy, S., and Svoboda, K. P. 1995. Chemical and biological evaluation of Tamarindus indica L. growing in Sudan. Int. Symp. Medicinal Aromatic Plants XXIV Int. Hort. Congress, Kyoto, Japan. 21–27 August 1994. Acta Hort. 390: 51–57.
217. Syrchina, A. I., Pechurina, N. N., Vereshchagin, A. L., Gorshkov, A. G., Tsapalova, I. E., and Semenov, A. A. 1993. A chemical investigation of Matteuccia struthiopteris. Chem. Nat. Compd. 29: 535–536.
218. Seth, R., Dorjee, S., and Morisco, P. 1993. Chemical constituents of Pedicularis longiflora var. tubiformis. Fitoterapia 64: 375–376.
219. Meyre, S. C., Mora, T. C., Biavatti, M. W., Santos, A. R. S., Dal, M. J., Yunes, R. A., and Cechinel, F. V. 1998. Preliminary phytochemical and pharmacological studies of Aleurites moluccana leaves (L.) Willd. Phytomedicine 5: 109–113.
220. Qiao, B. L., Wang, C. D., Mi, C. F., Li, F. X., Shi, H. L., and Takashima, J. 1997. Separation and identification of the ilungianin A and B from the root of Pimpinella thellungiana Wolff. Acta Pharm. Sin. 32: 56–58.
221. Mel'nikova, N. and Smirnov, V. 1972. Iodine contents of herbage species. Korma 5: 44.

222. Puffe, D. and Zerr, W. 1989. The content of various secondary plant products in wide-spread weeds in grassland with particular regard to higher lying areas. Eichhof-Berichte No. A 11, 127 pp.

223. Scehovic, J. 1988. Secondary metabolites in some grassland plants. Schweiz. Landwirtsch. Forsch. 27: 153–165.

224. Ahn, J. S., Kwon, Y. S., Ahn, S., and Kim, C. M. 1999. Anti-inflammatory constituents of Polygonum bistorta. Korean J. Pharmacognosy 30: 345–349.

225. Amoroso, V. B. 1988. Studies on medicinal ferns of the family Polypodiaceae. Philipp. J. Sci. 117: 1–15.

226. Wang, H. F., Li, X. D., Chen, Y. M., Yuan, L. B., and Foye, W. O. 1991. Radiation-protective and platelet aggregation inhibitory effects of five traditional Chinese drugs and acetylsali-cylic acid following high-dose gamma-irradiation. J. Ethnopharmacol. 34: 2–3, 215–219.

227. Kuo, Y. H., Li, S. Y., Wu, M. D., Huang, R. L., and Kuo, L. 1999. A new anti-HBeAg lig-nan, kadsumarin A, from Kadsura matsudai and Schizandra arisanensis. Chem. Pharm. Bull. 47: 1047–1048.

228. Longland, A. C., Theodorou, M. K., Sanderson, R., Lister, S. J., Powell, C. J., and Morris, P. 1995. Non-starch polysaccharide composition and in vitro fermentability of tropical forage legumes varying in phenolic content. Anim. Feed Sci. Technol. 55: 3–4, 161–177.

229. Wu, R. S., Huang, S. C., Chen, M. T., and Jong, T. T. 1989. Structure, synthesis and cytotoxicity of sphenone-A, a phenanthrene-1,4-quinone from Sphenomeris biflora. Phytochemistry 28: 1280–1281.

230. Jia, T., Li, J., Xie, S., and Zhang, J. 1996. Comparative research on the constituents of the volatile oil in the rhizome of Cibotium barometz (L.) J. Sm. and its processed prod-ucts. Zhongguo Zhongyao Zazhi 21: 216–218.

231. Ahn, K., Hahm, M., Park, E., and Lee, H. 1998. Corosolic acid isolated from the fruit of Crataegus pinnatifida var. psilosa is a protein kinase C inhibitor as well as a cytotoxic agent. Planta Med. 64: 468–470.

232. Akihisa, T., Matsubara, Y., Ghosh, P., Thakur, S., Shimizu, N., Tamura, T., and Matsumoto, T. 1988. The 24alpha- and 24beta-epimers of 24-ethylcholesta-5,22-dien-3beta-ol in two Clerodendrum species. Phytochemistry 27: 4, 1169–1172.

233. Ho, N. K. 1996. Traditional Chinese medicine and treatment of neonatal jaundice. Singapore Med. J. 37: 645–651.

234. Li, B. Q., Fu, T., Yan, Y. D., Baylor, N. W., Ruscetti, F., and Kung, H. F. 1993. Inhibition of HIV infection by baicalin: a flavonoid compound purified from Chinese herbal medi-cine. Cell. Mol. Biol. Res. 39: 119–124.

235. Fang, S. D., Xu, R. S., and Gao, Y. S. 1981. Some recent advances in the chemical stud-ies of Chinese herbal medicine. Am. J. Bot. 68: 300–303.

236. Li, W. K., Chan, C. L., and Lueng, H. W. 2000. Liquid chromatography atmospheric pressure chemical ionization mass spectrometry as a tool for the characterization of anthraquinone derivative from Chinese herbal medicine. J. Pharm. Pharmacol. 52: 723–729.

237. Hamasaki, N., Ishill, E., Tominaga, K., Texuka, Y., Nagaoka, T., Kadota, S., Kuroki, T., and Yano, I. 2000. Highly selective antibacterial activity of novel alkyl quinolone alkaloids from a Chinese herbal medicine, Gosyuyu (Wu-Chu-Yu), against Helicobacter pylori in vitro. Microbiol. Immunol. 44: 9–15.

238. Wu, T. N., Yang, K. C., Wang, C. M., Lai, J. S., and Ko, K. N. 1996. Lead poisoning caused by contaminated cordycep, a Chinese herbal medicine: two case reports. Sci. Total Environ. 182: 1–3, 193–195.

239. Numata, M., Yamamoto, A., Moribayashi, A., and Yamada, H. 1994. Antitumor components isolated from the Chinese herbal medicine Coix lachryma-jobi. Planta Med. 60: 356–359.

240. Ji, S. G., Chai, Y. F., Zhang, G. Q., Wu, Y. T., and Yin, X. P. 1999. Determination of tetraphydropalmatine in Chinese traditional medicine by nonaqueous capillary electrophoresis. Electrophoresis 20: 1904–1906.

241. Asano, K., Matsuishi, J., Yu, Y., Kasahara, T., and Hisamitsu, T. 1998. Suppressive effects of Triptergium wilfordii Hook f. a traditional Chinese medicine, on collagen arthritis in mice. Immunopharmacology 39: 117–126.

242. Izumi, S., Ohno, N., Kawakita, T., Nomoto, K., and Yadomae, T. 1997. Wide range of molecular weight distribution of mitogenic substance(s) in the hot water extract of a Chinese herbal medicine, Bupleurum chinense. Biol. Pharm. Bull. 20: 759–764.

243. Lin, G., Ho, Y. P., Li, P., and Li, X. G. 1995. Puqiedinone, a novel 5-alpha-cevanine alkaloid from the bulbs of Fritillaria puqiensis, and antitussive traditional Chinese medicine. J. Nat. Prod. Lloydia 58: 1662–1667.

244. Shi, D., He, S., Jiang, Y., and Qian, J. 1995. The effects of traditional Chinese medicine Cistanche species on the immune function and lipid peroxidation. Acta Acad. Med. Shanghai 22: 307–308.

245. Lin, H. W., Han, G. Y., and Liao, S. X. 1994. Studies on the active constituents of the Chinese traditional medicine Polygonatum odoratum (Mill.) Druce. Yaoxue Xuebao 29: 215–222.

246. Wei, M. J., Luo, X., Wang, X., and Zhu, J. S. 1991. Study of chemical pattern recognition as applied to quality assessment of the traditional Chinese medicine "Wei Ling Xian." Acta Pharm. Sin. 26: 772–776.

247. Rao, G. X., Sun, H. D., Lin, Z. W., and Hu, R. Y. 1991. Studies on the chemical constituents of the traditional Chinese medicine "Yun Qian-Hu" (Peucedanum rubricaule Shan et Shch.). Acta Pharm. Sin. 26: 30–36.

248. Guo, P., Li, Z., Hong, Z., Liu, S., and Wu, T. 1991. Determination of berberine hydrochloride in traditional Chinese medicine containing Coptis chinensis Franch by reversed phase high performance liquid chromatography. J. West China Univ. Med. Sci. 22: 90–92.

249. Kasuya, S., Goto, C., Koga, K., Ohtomo, H., Kagei, N., and Honda, G. 1990. Lethal efficacy of leaf extract from Perilla frutescens (traditional Chinese medicine) or perillaldehyde on Anisakis larvae in vitro. Jpn. J. Parasitol. 39: 220–225.

250. Baba, K., Yoneda, Y., Kozawa, M., Fujita, E., Wang, N. H., and Yuan, C. Q. 1989. Studies on Chinese traditional medicine "Fang-Feng". II. Comparison of several Fang-Feng by coumarins, chromones, and polyacetylenes. Shoyakugaku Xasshi 43: 216–221.

251. Kuang, H. X., Kasai, R., Ohtani, K., Liu, Z. S., Yuan, C. S., and Tanaka, O. 1989. Chemical constituents of pericarps of Rosa davurica Pall., a traditional Chinese medicine. Chem. Pharm. Bull. (Tokyo) 37: 2232–2333.

252. Ishida, H., Umino, T., Tsuji, K., and Kosuge, T. 1989. Studies on the antihemostatic substances in herbs classified as hemostatics in traditional Chinese medicine. I. On the antihemostatic principles in Sophora japonica. L. Chem. Pharm. Bull. (Tokyo) 37: 1616–1618.

253. Mizutani, K., Yuda, M., Tanaka, O., Saruwatari, Y. I., Fuwa, T., Jia, M. R., Ling, Y. K., and Fu, X. F. 1988. Chemical studies on Chinese traditional medicine, Dangshen. I. Isolation of (Z)-3- and (E)-2-hexenyl-beta-D-glucosides. Chem. Pharm. Bull. (Tokyo) 36: 2689–2690.

254. Chu, C. and Yang, S. 1986. Thin-layer chromatographic differentiation of Cinnamomum cassia in Chinese traditional medicine. Bull. Chin. Materia Med. 11: 609–611.

255. Ji, X. H., Li, Y., Liu, H. W., Yan, Y. N., and Li, J. S. 2000. Determination of alkaloids in rhizoma of some Mahonia plants hy HPCE. Acta Pharm. Sin. 35: 220–223.

256. Rogelj, B., Popovic, T., Ritonja, A., Strukelj, B., and Brzin, J. 1998. Chelidocystatin, a novel phytocystatin from Chelidonium majus. Phytochemistry 49: 1645–1649.

257. Yoshino, M., Ito, M., Okajima, H., Haneda, M., and Murakami, K. 1997. Role of baicalein compounds as antioxidant in the traditional herbal medicine. Biomed. Res. 18: 349–352.

258. Lin, Y. C., Jin, T., Wu, X. Y., Huang, Z. Q., Huang, J. S., Fan, J. S., and Chan, W. L. 1997. A novel bisesquiterpenoid, biatractylolide, from the Chinese herbal plant Atractylodes macrocephala. J. Nat. Prod. 60: 27–28.

259. Sakurai, M. H., Matsumoto, T., Kiyohara, H., and Yamada, H. 1996. Detection and tissue distribution of anti-ulcer pectic polysaccharides from Bupleurum falcatum by polyclonal antibody. Planta Med. 62: 341–346.

260. Kinoshita, K., Kawai, T., Imaizume, T., Akita, Y., Koyama, K., and Takahashi, K. 1996. Anti-emetic principles of Inula linariaefolia flowers and Forsythia suspensa fruits. Phytomedicine 3: 51–58.

261. Wong, R. N. S., Mak, N. K., Choi, W. T., and Law, P. T. W. 1995. Increased accumulation of trichosanthin in Trichosanthes kirilowii induced by microorganisms. J. Exp. Bot. 46: 284, 355–358.

262. Oyama, T., Isono, T., Suzuki, Y., and Hayakawa, Y. 1994. Anti-nociceptive effects of Aconiti tuber and its alkaloids. Am. J. Chin. Med. 22: 175–182.

263. Kasai, S., Watanabe, S., Kawabata, J., Tahara, S., and Mizutani, J. 1992. Antimicrobial catechin derivatives of Agrimonia pilosa. Phytochemistry 31: 787–789.

264. Kang, R., Helms, R., Stout, M. J., Jaber, H., Chen, Z. Q., and Nakatsu, T. 1992. Antimicrobial activity of the volatile constituents of Perilla frutescenes and its synergistic effects with polygodial. J. Agric. Food Chem. 40: 2328–2330.

265. Ishiguro, K., Yamaki, M., Kashihara, M., Takagi, S., and Isoli, K. 1990. A chromene from Hypericum japonicum. Phytochemistry 29: 1010–1011.

266. Yamade, H., Ra, K. S., Kiyohara, H., Cyong, J. C., Yang, H. C., and Otsuka, Y. 1988. Characterization of anti-complementary neutral polysaccharides from the roots of Bupleurum falcatum. Phytochemistry 27: 3163–3168.

267. Han, J. and Luan, D. H. K. 1984. Sow abortion caused by feeding Cynanchum auriculatum. Anim. Husb. Vet. Med. Humu Yu Shouyi 16: 266.

268. Tong, C. C. 1995. Nutraceutical properties of the mushroom ganoderma. Tunas Buletin Maklumat Pertanian Malaysis 15: 1–6.

269. Wei, Z., Pan, J., and Li, Y. 1992. Artemisinin G: a sesquiterpene from Artemisia Annua. Planta Med. (Germany) 58: 3090.

270. Jiang, Z., Chen, S., and Zhou, J. 1988. Study on chemical constituent of Aconitum austroyunnanense (I). Acta Bot. Yunnanica (China) 10: 317–323.

271. Farnsworth, N. R. 1999. NAPRALERT: natural products alert database. Program for collaborative research in the Pharmaceutical Science Department of Medicinal Chem. Pharmacognosy, College Pharmacy, University of Illinois, Chicago.

272. Zhang, S. and Cheng, K. 1989. Angelica sinensis (Oliv.) Diels: in vitro culture, regeneration, and the production of medicinal compounds. In Bajaj, Y. P. S. (Ed.). Biotechnology in Agriculture and Forestry 7. Medicinal and Aromatic Plants II. Springer-Verlag. New York. pp. 1–22.

273. Misik, V., Bezakova, L., Malekova, L., and Kostalova, D. 1995. Lipoxygenase inhibition and antioxidant properties of protoberberine and aporphine alkaloids isolated from Mahonia aquifolium. Planta Med. 61: 372–373.

274. Muller, K. and Ziereis, K. 1994. The antipsoriatic Mahonia aquifolium and its active constituents. I. Pro- and antioxidant properties and inhibition of 5-lipoxygenase. Planta Med. 60: 421–424.

275. Mallet, J. F., Cerrati, C., Ucciani, E., Gamisans, J., and Gruber, M. 1994. Antioxidant activity of plant leaves in relation to their alpha-tocopherol content. Food Chem. 49: 61–65.

276. Woldemariam, T. Z., Betz, J. M., and Houghton, P. J. 1997. Analysis of aporphine and quinolizidine alkaloids from Caulophyllum thalictroides by densitometry and HPLC. J. Pharm. Biomed. Anal. 15: 839–843.

277. Heptinstall, S. and Awang, D. V. C. 1998. Feverfew: A review of its history, its biological and medicinal properties, and the status of commercial preparations of the herb. In Lawsib, L. D. and Bauer, R. (Eds.). Phytomedicines of Europe, Chemistry and Biological Activity. American Chemical Society. Washington, D.C. pp. 158–175.

278. Shahat, A. A., Ismail, S. I., Hammouda, F. N., and Azzam, S. A. 1998. Anti-HIV activity of flavonoids and proanthocyanidins from Crataegus sinalica. Phytomedicine 5: 133–136.

279. Sticher, O. and Meier, B. 1998. Hawthorn (Crataegus): biological activity and new strategies for quality control. In: Lawsib, L. D. and Bauer, R. (Eds.). Phytomedicines of Europe, Chemistry and Biological Activity. American Chemical Society. Washington, D.C. pp. 241–262.

280. Henry, M., Edy, A. M., Desmarest, P., and Du Manuir, J. 1991. Glycyrrhiza glabra L. (licorice): cell culture, regeneration, and the production of glycyrrhizin. In Bajaj, Y. P. S. (Ed.). Biotechnology in Agriculture and Forestry 15. Medicinal and Aromatic Plants III. Springer-Verlag. New York. pp. 270–282.

281. Ihantola, V. A., Summanen, J., and H. Kankaanranta 1997. Anti-inflammatory activity of extracts from leaves of Phyllanthus emblica. Planta Med. 63: 518–524.

282. Brinda, S., Smitt, U. W., George, V., and Pushpangadan, P. 1998. Angiotensin converting enzyme (ACE) inhibitors from Jasminum azoricum and Jasminum grandiflorum. Planta Med. 64: 246–250.

283. Jonard, R. 1989. Jasminum spp. (jasmine): micropropagation and the production of essential oils. In Bajaj, Y. P. S. (Ed.). Biotechnology in Agriculture and Forestry 7. Medicinal and Aromatic Plants II. Springer-Verlag. New York. pp. 315–331.

284. Small, E. 1999. New crops for Canadian agriculture. In Janick, J. (Ed.). Perspectives on New Crops and New Uses. ASHS Press. Alexandra, VA. pp.15–52.

285. Muller, D., Carnat, A., and Lamaison, J. L. 1991. Fucus: Comparative study of Fucus vesiculosus L., Fucus serratus L. and Ascophyllum nodosum Le Jolis. Plantes Med. Phytother. 25: 194–201.

286. Ortega-Calvo, J. J., Mazuelos, C., Hermosin, B., and Saiz-Jimenez, C. 1993. Chemical composition of Spirulina and eukaryotic algae food products marketed in Spain. J. Appl. Phycol. 5: 425–435.

287. Teas, J. 1973. The dietary intake of Laminaria, a brown seaweed, and breast cancer prevention. Nutr. Cancer 4: 217–222.

288. Walkiw, O. and Douglas, D. E. 1975. Health food supplements prepared from kelp—A source of elevated urinary arsenic. Clin. Toxicol. 8: 325–331.

289. Murray, M. T. 1995. The Healing Power of Herbs. Prima Publishing. Rocklin, CA. 410 pp.

290. Turner, N. J. 1982. Traditional use of devil's-club (Oplopanax horridus; Araliaceae) by native peoples in western North America. J. Ethnobiol. 2: 17–38.

291. Moes, A. 1966. A parallel study of the chemical composition of Polygala senega and of "Securidaca longepedunculata" Fres. var. parvifolia, a Congolese polygalacea. J. Phar. Belg. 21: 347–362.

292. Takeda, O., Azuma, S., Mizukami, H., Ikenaga, T., and Ohashi, H. 1986. Cultivation of Polygala senega var. latifolia. II. Effect of soil moisture content on the growth and senegin content. Shoyakugaku Zasshi 40: 434–437.

293. Zapesochnaya, G. G. and Kurkin, V. A. 1983. The flavonoids of the rhizomes of Rhodiola rosea. II. A flavonolignan and glycosides of herbacetin. Chem. Nat. Comp. 19: 21–29.

294. Stevens, J. F., Elema, E. T., Bolck, A., and Hart, H. 1996. Flavonoid variation in Eurasian Sedum and Sempervivum. Phytochemistry 41: 503–512.

295. Whiterup, K. M., Look, S. A., Stasko, M. W., Ghiorzi, T. J., Muschik, G. M., and Cragg, G. M. 1990. Taxus spp. needles contain amounts of taxol comparable to the bark of Taxus brevifolia: analysis and isolation. J. Nat. Prod. 53: 1249–1255.

296. Wickremesinhe, E. R. M. and Arteca, R. N. 1998. Taxus species (yew): in vitro culture, and the production of taxol and other secondary metabolites. In Bajaj, Y. P. S. (Ed.). Biotechnology in Agriculture and Forestry 41. Medicinal and Aromatic Plants X. Springer-Verlag. New York. p. 415–442.

297. Keplinger, K., Laus, G., Wurm, M., Dierich, M. P., and Teppner, H. 1999. Uncaria tomentosa (Willd.) DC: ethnomedicinal use and new pharmacological, toxicological and botanical results. J. Ethnopharmacol. 64: 23–34.

298. Obregon Vilches, L. E. 1995. Cat's Claw. 3rd ed. Institute de Fitoterapia Americano. Lima, Peru. 169 pp.

299. Wurm, M., Kacani, L., Laus, G., Keplinger, K., and Dierich, M. P. 1998. Pentacyclic oxindole alkaloids from Uncaria tomentosa induce human endothelial cells to release a lymphocyte proliferation regulating factor. Planta Med. 64: 701–704.

300. Rogiers, S. Y. and Knowles, N. R. 1997. Physical and chemical changes during growth, maturation, and ripening of saskatoon. Can. J. Bot. 75: 1215–1225.

301. Fraisse, D., Carnat, A., and Lamaison, J. L. 1996. Polyphenolic composition of the leaf of bilberry. Am. Pharm. Fr. 54: 280–283.

302. Rogers, R. D. (Ed.) 1999. Sundew, Moonwort, Medicinal Plants of the Prairies. Vol. 3. R. D. Roger. Edmonton, Alberta. 181 pp.

303. Rogers, R. D. 2000. Sundew, Moonwort, Medicinal Plants of the Prairies. Vol. 4. R. D. Roger. Edmonton, Alberta. 237 p.

304. Awang, D. 1997. Saw palmetto, African prune and stinging nettle for benign prostatic hyperplasic (BPH). Can. Pharm. J. November 1997. pp. 37–40, 43–44, 60.

305. Baresova, H. 1988. Centaurium erythraea Rafn: Micropropagation and the production of secoiridoid glucosides. In Bajaj, Y. P. S. (Ed.). Biotechnology in Agriculture and Forestry 4. Medicinal and Aromatic Plants I. Springer-Verlag. New York. pp. 350–366.

306. Fuentes-Granados, R. G., Widrlechner, M. P., and Wilson, L. A. 1998. An overview of Agastache research. J. Herbs Spices Med. Plants 6: 69–97.

307. Shishoo, C. J., Shah, S. A., Rathod, I. S., and Patel, S. G. 1997. Determination of vitamin C content of Phyllanthus emblica and Chyavanprash. Indian J. Pharm. Sci. 59: 268–270.

308. Van den Berg, A. J. J. and Labadie, R. P. 1988. Rhamnus spp.: in vitro production of anthraquinones, anthrones, and dianthrones. In Bajaj, Y. P. S. (Ed.). Biotechnology in Agriculture and Forestry 4. Medicinal and Aromatic Plants I. Springer-Verlag. New York. pp. 513–528.

309. Winterhoff, H. 1998. Vitex agnus-castus (chaste tree): Pharmacological and clinical data. In: Lawsib, L. D. and Bauer, R. (Eds.) Phytomedicines of Europe, Chemistry and Biological Activity. American Chemical Society. Washington, D.C. pp. 299–307.

310. Zhao, X., Wang, J., Yang, M., and Pan, X. 1996. Study on the chemical constituents of the essential oil of berries of Physalis pubescene. J. NorthEast Forestry Univ. 24: 94–98.

311. Arenz, A., Klein, M., Fiehe, K., Gross, J., Drewke, C., Hemscheidt, T., and Leistner, E. 1996. Occurrence of neurotoxic 4'-O-methylpyridoxine in Ginkgo biloba leaves, Ginkgo medications and Japanese ginkgo food. Planta Med. 62: 548–551.

312. Henry, M., Edy, A. M., Desmarest, P., and du Manuir, J. 1991. Glycyrrhiza glabra L. (licorice): Cell culture, regeneration, and the production of glycyrrhizin. In Bajaj, Y. P. S. (Ed.). Biotechnology in Agriculture and Forestry 15. Medicinal and Aromatic Plants III. Springer-Verlag. New York. pp. 270–282.

313. Keplinger, K., Laus, G., Wurm, M., Dierich, M. P., and Teppner, H. 1999. Uncaria tomentosa (Willd.) DC. ethnomedicinal use and new pharmacological, toxicological and botanical results. J. Ethnopharmacol. 64: 23–34.

314. Srivastava, K. C. and Malhotra, N. 1991. Acetyl eugenol, a component of oil of cloves (Syzygium aromaticum L.) inhibits aggregation and alters arachidonic acid metabolism in human blood platelets. Prostag. Leukot. Essent. Fatty Acid 42: 73–81.

315. Yokoyama, M. and Inomata, S. 1998. Catharanthus roseus (periwinkle): in vitro culture, and high-level production of arbutin by biotrranformation. In Bajaj, Y. P. S. (Ed.). Biotechnology in Agriculture and Forestry 41. Medicinal and Aromatic Plants X. Springer-Verlag. New York. pp. 67–80.

316. Fujimoto, Y., Soemartono, A., and Sumatra, M. 1988. Sesquiterpene lactones from Blumea balsamifera. Phytochemistry 27: 1109–1111.

317. Lin, Y. C., Long, K. G., and Dcng, Y. J. 1988. Studies on the chemical constituents of the Chinese medicinal plant Blumea balsamifera. Acta Scientiarum Naturalium Univ. Sunyatseni 2: 77–80.

318. Fukuyama, Y., Kiriyama, Y., and Kodama, M. 1993. Concise synthesis of belamcandaquinones A and B by palladium (0) catalyzed cross-coupling reaction of bromoquinone with arylboronic acids. Tetrahedron Lett. 34: 7637–7638.

319. Ma, L., Song, Z. W., and Wu, F. 1996. Determination of five isoflavones in Belamcanda chinensis by RP-HPLC. Yaoxue Xuebao 31: 945–949.

320. Takahashi, K., Hoshino, Y., Suzuki, S., Hano, Y., and Nomura, T. 2000. Iridals from Iris tectorum and Belamcanda chinensis. Phytochemistry 53: 925–929.

321. Beerhues, L. and Berger, U. 1994. Xanthones in cell suspension cultures of two Centaurium species. Phytochemistry (Oxford) 35: 1227–1231.

322. Beerhues, L. 1996. Benzophenone synthase from cultured cells of Centaurium erythraea. FEBS Lett. 383: 264–266.

323. Nikolova, D. B. and Handjieva, N. 1996. Quantitative determination of swertiamarin and gentiopicroside in Centaurium erythrea and C. turcicum by densitometry. Phytochem. Anal. 7: 140–142.

324. Peumans, W. J., Winter, H. C., Berner, V., and Van-Leuven, L. 1997. Isolation of a novel plant lectin with an unusual specificity from Calystegia sepium. Glycoconjugate J. 14:259–265.

325. Goldmann, A., Message, B., Tepfer, D., Molyneux, R. J., Duclos, O., Boyer, F. D., and Elbein, A. D. Biological activities of the nortropane alkaloid, calystegine B2, and analogs: structure-function relationships. J. Nat. Prod. 59: 1137–1142.

326. Kapadia, G. J., Shuykla, Y. N., and Basak, S. P. 1978. Melovinone, an open chain analogue of melochinone from Melochia tomentosa. Phytochemistry 17: 1444–1445.

327. Kapadia, G. J. and Shukla, Y. N. 1993. Ju899Melosatin D: a new isatin alkaloid from Melochia tomentosa roots. Planta Med. 59: 568–569.

328. Harvala, E., Exner, J., and Becher, H. 1984. Flavonoids of Loranthus europaeus. J. Nat. Prod. 47: 1054–1055.

329. Berg, T., Damtoft, S., Jensen, S. R., Nielsen, B. J., and Rickelt, L. F. 1985. Iridoid glucosides from Pedicularis. Phytochemistry 24: 491–493.

330. Konoshima, T., Takasaki, M., Kozuka, M., Tokuda, H., Nishino, H., Matsuda, E., and Nagai, M. 1997. Anti-tumor promoting activities of isoflavonoids from Wistaria brachybotrys. Biol. Pharm. Bull. 20: 865–868.

331. Konoshima, T., Kozuka, M., Haruna, M., and Ito, K.1991. Constituents of leguminous plants. XIII. New triterpenoid saponins from Wisteria brachybotrys. J. Nat. Prod. 54: 830–836.

332. Konoshima, T., Kozuka, M., Haruna, M., Ito, K., and Kimura, T. 1989. Studies on the constituents of leguminous plants. XI. The structures of new triterpenoids from Wisteria brachybotrys Sieb. et Zucc. Chem. Pharm. Bull. 37: 1550–1553.

333. Perdue, R. E., Jr. 1977. Thalictrum dasycarpum: source of the antitumor agent thalicarpine. Lloydia 40: 607.

334. http://www.aabhealth.com/herbsdefined.htm

335. http://dbhs.wvusdk12ca.us/chem.history

336. http://www.botanical.com

337. Ishikura, N. and Sato, S. 1977. Isolation of new kaempferol glycosides from the leaves of Euonymus. Bot. Mag. Tokyo 90: 83–87.

338. Yamada, K., Shizuri, Y., and Hirata, Y. 1978. Isolation and structures of new alkaloid alatamine and an insecticidal alkaloid wilfordine. Tetrahedron 34: 1915–1920.

339. Ishiwata, H., Shizuri, Y., and Yamada, K.1983. Three sesquiterpenee alkaloids from Euonymus alatus forma striatus. Phytochemistry 22: 2839–2841.

340. Hikino, H., Tamada, M., and Yen, K. Y. 1978. Mallorepine, eyano-gamma-pyridone from Mallotus repandus. Planta Med. 33: 385–388.

341. Saijo, R., Nonaka, G., and Nishioka, I. 1989. Tannins and related compounds LXXXVII. Isolation and characterization of four new hydrolyzable tannins from the leaves of Mallotus repandus. Chem. Pharm. Bull. 37: 2624–2630.

342. Kohli, D. V. 1988. A new saponin, stigmasta-5,24(28)-diene-3beta-O-beta-D-galactopyranoside, from the seeds of Sesbania sesban. Fitoterapia 59: 478–479.

343. El-Sayed, N. H. 1991. A rare kaempferol trisaccharide anti-tumor promoter from Sesbania sesban. Pharmazie 46: 679–680.

344. Marston, A., Hostettmann, K., Harborne, J. B., and Tomas-Barberan, F. A. 1991. Plant saponins: chemistry and molluscicidal action. Ecological chemistry biochemistry plant terpenoids. Proc. Phytochem. Soc. Europe 31. pp. 264–286.

345. Bown, D. 1995. Encyclopedia of Herbs and Their Uses. The Reader's Digest Assn. (Canada) Ltd. Westmount, QC. 424 pp.

346. http://www.botanical.com/botanical/mgmh/s/sancom13.html.

347. Willard, T. 1992. Edible and Medicinal Plants of the Rocky Mountains and Neighbouring Territories. Wild Rose College of Natural Healing, Ltd. Calgary, AB. 278 pp.

348. Bremness, L. 1994. Herbs. Dorling Kindersley Ltd. London., 304 pp.

349. Hau, D. M., Chen, K. T., Wang, M., Lin, I., and Chen, W. C. 1996. Protective effects of Gynostemma pentaphyllum in gamma-irradiated mice. Am. J. Chin. Med. 24: 83–92.

350. Mei, K. F., Shen, X., Ye, Y. M., and Fei, H. M. 1993. Transformation of Gynostemma pentaphyllum by Agrobacterium rhizogenes and saponin production in hairy root cultures. Acta Bot. Sin. 95: 626–631.

351. Akihisa, T., Shimizu, N., Ghosh, P., Thakur, S., Rosenstein, F. U., Tamura, T., and Matsumoto, T. 1987. Sterols of the Cucurbitaceae. Phytochemistry 26: 1693–1706.

352. Zhang, J. S., Tian, Z., and Lou, Z. C. 1989. Quality evaluation of twelve species of Chinese ephedra (Ma Huang). Acta Pharm. Sin. 24: 865–871.

353. Nishiyama, N., Saito, H., and Chu, P. J. 1995. Beneficial effects of biota, a traditional Chinese herbal medicine, on learning impairment induced by basal forebrain lesion in mice. Biol. Pharm. Bull. 18: 1513–1517.

354. Mi, H. M., Li, C. G., Su, Z. W., Wang, N. P., Zhao, J. X., and Jiang, Y. G. 1987. Studies on chemical constituents and antifungal activities of essential oil from Oplopanax elatus Nakai. Acta Pharm. Sin. 22: 549–552.

355. Takino, Y., Koshioka, M., Shlokawa, M., Ishil, Y., Maruyama, S., Higashino, M., and Hayashi, T. 1979. Quantitative determination of glycyrrhizic acid liquorice roots and extracts by TLC-densitometry. Hippokrates Verlag GmbH 36: 74–78.

356. Killacky, J., Ross, M. S. F., and Turner, T. D. 1976. The determination of beta-glycyrrhetinic acid in liquorice by high pressure liquid chromatography. Planta Medica 30: 310–316.

357. Larry, D. 1973. Gas-chromatographic determination of β-asarome, a component of oil of calamus, in flavors and beverages. J. Assn. Off. Anal. Chem. 56: 1281–1283.

358. Pichon, P. 1971. The anthocyanin pigments of Epilobium rosmarinifolium Haenke (Onagraceae). Planta Med. 5: 115–117.

359. Luning, B. 1964. Studies of species Orchidaceae alkaloids. I. Screening of species for alkaloids. Acta Chem. Scand. 18: 1507–1516.

360. Bognar, R. 1965. Steroid alkaloid glycosides. VIII. Review of specific investigations on the occurrence of steroid alkaloid glycosides in genus Solanum. Pharmazie 20: 40–42.

361. Sohreiber, K. 1963. On the appearance of glycoside alkaloid in tuber-bearing species of the genus Solanum L. solanum alkaloids. XXVII. Kulturpflanze 11: 422–450.

362. Duke, J. A. 2000. Returning to Our Medicinal Roots. Mother Earth News. December–January 2000. pp. 23–24, 26–33.

363. Yarnell, E. 2000. The botanical roots of pharmaceutical discovery. Altern. Complementary Ther. June 2000, pp. 125–128.

364. Fang, S. D., Xu, R. S., and Gao, Y. S. 1980. Some recent advances in the chemical studies of Chinese herbal medicine. Am. J. Bot. 68: 300–303.

365. Perharic, L., Shaw, D., Leon, C., Murray, V. S. G., and De-Smet, P. 1995. Possible association of liver damage with the use of Chinese herbal medicine for skin disease. Vet. Hum. Toxicol. 37: 562–566.

366. Zhu, D. Y., Bai, D. L., and Tang, X. C. 1996. Recent studies on traditional Chinese medicinal plants. Drug Dev. Res. 39: 147–157.

367. Widrlechner, M. and Foster, S. 1991. Chinese medicinal plants in the U.S. National Plant Germplasm System. Herb Spice Med. Plant Dig. 9(4): 1–5.

368. Borchers, A. T., Hackman, R. M., Keen, C. L., Stern, J. S., and Eric Gershwin, M. 1997. Complementary medicine: a review of immunomodulatory effects of Chinese herbal medicines. Am. J. Clin. Nutr. 66: 1303–1312.

369. Zhu, D. Y., Bai, D. L., and Tang, X. C. 1997. Recent studies on traditional Chinese medicinal plants. Drug Dev. Res. 39: 147–157.

370. Nortier, J. L., Martinez, M. M., Schmeiser, H. H., Arlt, V. M., Bieler, C. A., Petein, M., Depierreux, M. F., De Pauw, L., Abramowicz, D., Vereerstraeten, P., and Vanherweghem, J. L. 2000. Urothelial carcinoma associated with the use of a Chinese herb (Aristolochia fangchi). N. Engl. J. Med. 342: 1686–1692.

371. Brekhman, I. I. and Grinevitch, M. A. 1981. Oriental medicine: a computerized study of complex recipes and their components: analysis of recipes intended to cure certain disease. Am. J. Chin. Med. 9(1): 34–38.

372. Shibata, S. 1979. The chemistry of Chinese drugs. Am. J. Chin. Med. 7(2): 103–141.

373. Cheng, J. T. 2000. Review: drug therapy in Chinese traditional medicine. J. Clin. Pharmacol. 40: 445–450.

374. Chen, J. K. 2000. Nephropathy associated with the use of Aristolchia. HerbalGram 48: 44–45.

375. Tomlinson, B., Chan, T. Y. K., Chan, J. C. N., Critchley, J. A. J. H., and Bat, P. P. H. Toxicity of complementary therapies: An eastern perspective. J. Clin. Pharmacol. 40: 451–456.

376. Lambert, J., Srivastava, J., and Vietmeyer, N. 1997. Medicinal Plants, Rescuing a Global Heritage. World Bank Technical Paper No. 355. 61 pp.

377. Tang, J. L., Shan, S. Y., and Ernst, E. 1999. Review of randomized controlled trials of traditional Chinese medicine (letter). Br. Med. J. 319: 160–161.

378. Koo, J. and Arain, S. 1998. Traditional Chinese medicine for the treatment of dermatologic disorders. Arch. Dermatol. 134: 1388–1393.

379. Fruehauf, H. 1999. Chinese medicine in crisis. J. Chin. Med. 61: 6–14.

380. Lahans, T. 2000. Integrated medicine: the interface between oriental and conventional medicine. J. Naturopathic Med. 7: 28–33.

381. Ninomiya, H., Mitsuma, T., Takara, M., Yokozawa, T., Terasawa, K., and Okuda, H. 1998. Effects of the oriental medical prescription Wen-Pi-Tang in patients receiving dialysis. Phytomedicine 5: 245–252.

382. Bensoussan, A., Tally, N. J., Hing, M., Menzies, R., Guo, A., and Ngu, M. 1998. Treatment of irritable bowel syndrome with a Chinese herbal medicine. J.A.M.A. 280: 1585–1589.

383. Monmaney, T. 1998. A dose of caution. The Los Angeles Times. September 1, 1998.

384. Eisenberg, D. M., Kessler, R. C., Foster, C., Norlock, F. E., Calkins, D. R., and Delbanco, T. L. 1993. Unconventional medicine in the United States: prevalence, costs, and patterns of use. N. Engl. J. Med. 328: 246–252.

385. Marwick, C. 1995. Growing use of medicinal botanicals forces assessment by drug regulators. J.A.M.A. 273: 607–609.

386. Frankel, E. N., Kanner, J., German, J. B., Parks, E., and Kinsella, J. E. 1993. Inhibition of oxidation of human low-density lipoprotein by phenolic substances in red wine. Lancet 341: 454–457.

387. Ramarathnam, N., Osawa, T., Ochi, H., and Kawakishi, S. 1995. The contribution of plant food antioxidants for human health. Trend. Food Sci. Technol. 6: 75–82.

388. Farnsworth, N. R. 1990. Bioactive Compounds from Plants. John Wiley & Sons. Chichester, U.K. pp. 2–14.

389. Reid, D. P. 1986. Chinese Herbal Medicine. Shambhala Publications. Boston, MA. 174 p.

390. Tao, S. and P. E. Lipsky. 2000. The Chinese anti-immunosuppressive herbal remedy Tripterygium wilfordii Hook F. Complement. Altern. Ther. Rheum. Dis. II 26(1): 29–50.

391. Anonymous. 1992. Rediscovering wormwood; qinghaosu for malaria. Lancet. March 14, 1992. pp. 649–651.

392. Vanherwegheum, J. 1993. Rapidly progressive interstitial renal fibrosis in young women: Association with slimming regimen including Chinese herbs. Lancet 341: 387–391.

393. E. O. Espinoza. 1995. Arsenic and mercury in traditional Chinese herbal balls. N. Engl. J. Med. 333: 803–804.

394. Hoffman, S. L. 1996. Artemether in severe malaria—still too many deaths. N. Engl. J. Med. 335: 124–125.

395. Day, M. 1996. Malaria falls to herbal remedy. New Scientist. July 13, 1996, p. 4.

396. Ernst, E. 1998. Harmless herbs? A review of the recent literature. Am. J. Med. 104: 170–178.

397. Bensoussan, A. 1998. Treatment of irritable bowel syndrome with Chinese herbal medicine. J.A.M.A. 280: 1585–1589.

398. Vimala, S. 1999. Anti-tumour promoter activity in Malaysian ginger rhizobia used in traditional medicine. Br. J. Cancer 80: 110–116.

399. Miller, L. G. 1998. Herbal medicinals. Arch. Intern. Med. 158: 2200–2011.

400. Havel, R. J. 1999. Dietary supplement or drug? The case of chloestin. Am. J. Clin. Nutr. 69: 175–176.

401. Bok, J. W., Lermer, L., Chilton, J., Klingeman, H. G., and Towers, G. H. N. 1999. Antitumor sterols from the mycelia of Cordyceps sinensis. Phytochemistry 51: 891–898.

402. Yong-Lu, L., Yin, L., Yang, J. W., and Liu, C. X. 1997. Studies on pharmacological activities of cultivated Cordyceps sinensis. Phytother. Res. 11: 237–239.

403. Kim, S. H., Kim, Y. S., Lee, C. K., and Han, S. S. 1999. In vitro chemopreventive effects of plant polysaccharides (Aloe barbadensis, Lentinus edodes, Ganoderma lucidum and Coriolus versicolor). Carcinogenesis 220: 1637–1640.

404. Eo, S. K. 1999. Antiviral activities of various water and methanol soluble substances isolated from Ganoderma lucidum. J. Ethnopharmacol. 68: 129–136.

405. Conney, A. H., Lu, Y. P., Lou, Y. R., Xie, J. G., and Huang, M. T. 1999. Inhibitory effect of green and black tea on tumor growth. Proc. Soc. Exp. Biol. Med. 220: 229–233.

406. Chung, F. L. 1999. The prevention of lung cancer induced by a tobacco-specific carcinogens in rodents by green and black tea. Proc. Soc. Exp. Biol. Med. 220: 244–248.

407. Zhu, M., Chang, Q., Wong, L. K., Chong, F. S., and Li, R. C. 1999. Triterpene antioxidants from Ganoderma lucidum. Phytotherapy 13: 529–531.

408. Lee-Huang, S., Chen, H. C., Huang, P. L., Bourinbaiai, A., Huang, H. I., and Kung, H. 1995. Anti-HIV and anti-tumor activities of recombinant MAP30 from bitter melon. Gene 161: 151–156.

409. Yang, G. Y., Liao, J., Kim, K., Yurkow, E. J., and Yang, C. S. 1998. Inhibition of growth and induction of apoptosis in human cancer cell lines by tea polyphenols. Carcinogenesis 19: 611–616.

410. Awe, S. O., Olajide, O. A., and Makinde, J. M. 1998. Inhibitory effects of isoflavones in roots of Astragalus membranaceus Bunge. (Astragali radix) on lipid peroxidation by reactive oxygen species. Phytother. Res. 12: 59–61.

411. Zheng, R. L., Kang, J. H., Chen, F. Y., Wang, P. F., Ren, J. G., and Liu, Q. L. 1997. Antioxidative components isolated from the roots of Astragalus membranaceus Bunge. (Astragali radix). Phytother. Res. 11: 603–605.

412. Zhu, J. S., Halpern, G. M., and Jones, K. 1998. The scientific rediscovery of an ancient Chinese herbal medicine: Cordyceps sinensis. I. J. Altern. Complement. Med. 4: 289–303.

413. Chen, Y. J., Shiao, M. S., Lee, S. S., and Wang, S. Y. 1997. Effect of Cordyceps sinensis on the proliferation and differentiation of human leukemic U937 cells. Life Sci. 60: 2349–2359.

414. Nakamura, K., Yamaguchi, Y., Kagota, S., Kwon, Y. M., Shinozuka, K., and Kunitomo, M. 1999. Inhibitory effect of Cordyceps sinensis on spontaneous liver metastasis of Lewis lung carcinoma and B16 melanoma cells in syngeneic mice. Jpn. J. Pharmacol. 79: 335–341.

415. Ng, C. R. A. 1997. Polysaccharopeptide from Coriolus versicolor has potential for use against human immunodeficiency virus type I infection. Life Sci. 60: 383–387.

416. Kobayashi, H., Matsunaga, K., and Oguchi, Y. 1995. Antimetastatic effects of PSK (krestin), a protein bound polysaccharide obtained from basidiomycetes: an overview. Cancer Epidemiol. Biomarkers Prev. 4: 275–281.

417. Ng, T. B. 1998. A review of research on the protein-bound polysaccharide (polysaccharopeptide, PSP) from the mushroom Coriolus versicolor (Basidiomycetes: Polyporaceae). Gen. Pharmacol. 30: 1–4.

418. Collins, R. A. 1997. Polysaccharopeptide from Coriolus versicolor has potential for use against human immunodeficiency virus type I infection. Life Sci. 60: 383–387.

419. Naito, T., Niitsu, K., Ikeya, Y., Okada, M., and Mitsuhashi, H. 1992. A phthalide and 2-farnesyl-6-methyl benzoquinone from Ligusticum chuangxiong. Phytochemistry 31: 1787–1789.

420. Naito, T., Katsuhara, T., Niitsu, K., Ikeya, Y., Okada, M., and Mitsuhashi, H. 1992. Two phthalides from Ligusticum chuangxiong. Phytochemistry 31: 639–642.

421. Gao, Z. and Meng, F. H. 1993. Effect of Fagopyrum cymosum root on clonal formation of four human tumor cells. Chung Kuo Chung Yao Tsa Chih 18: 498–500.

422. Lee, J. S., Cho, Y. S., Park, E. J., Oh, K. J., Lee, W. K., and Ahn, J. S. 1998. Phospholipase c gamma I inhibitory principles from the sarcotestas of Ginkgo biloba. J. Nat. Prod. 61: 867–871.

423. Bourinbaiar, A. S. and Lee-Huang, S. 1995. Potentiation of anti-HIV activity of anti-inflammatory drugs, dexamethasone and indomethacin, by MAP30, the antiviral agent from bitter melon. Biochem. Biophys. Res. Commun. 208: 779–785.

424. Hirazumi, A., Furusawa, E., Chou, S. C., and Hokama, Y. 1994. Anticancer activity of Morinda citrifolia (noni) on intraperitoneally implanted Lewis lung carcinoma in syngeneic mice. Proc. West. Pharmacol. Soc. 37: 145–146.

425. Li, S. H. and Chu, Y. 1999. Anti-inflammatory effects of total saponins of Panax notoginseng. Chung Kuo Yao Li Hsuch Pao 20: 551–554.

426. Ye, W. C., Ji, N. N., Zhao, S. X., Liu, J. H., Ye, T., McKervey, M. A., and Stevenson, P. 1996. Triterpenoids from Pulsatilla chinensis. Phytochemistry 42: 799–802.

427. Chen, B. X., Ding, Y. S., and Chen, L. G. 1994. Experimental study on the processed drug of castor seeds in the therapy of pulmonary carcinoma. Chung Kuo Chung Yao Tsa Chih 19: 726–727, 762.

428. Liu, Y., Yang, C. F., Lee, B. L., Shen, H. M., Ang, S. G., and Ong, C. N. 1999. Effect of Salvia miltiorrhiza on aflatoxin B1-induced oxidative stress in cultured rat hepatocytes. Free Radic. Res. 31: 559–568.

429. Wu, Y. J., Hong, C. Y., Lin, S. J., Wu, P., and Shiao, M. S. 1998. Increase of vitamin E content in LDL and reduction of atherosclerosis in cholesterol-fed rabbits by a water-soluble antioxidant-rich fraction of Salvia miltiorrhiza. Arterioscler. Thromb. Vasc. Biol. 18: 481–486.

430. Goldberg, V. E., Ryzhakov, V. M., Matiash, M. G., Stepovaia, E. A., and Boldyshev, D. A. 1997. Dry extract of Scutellaria baicalensis as a hemo-stimulant in anti-neoplastic chemotherapy in patients with lung cancer. Eksp. Klin. Farmakol. 60: 28–30.

431. Xu, J. Y., Yang, J., and Li, L. Z. 1992. Antitumor effect of Tripterygium wilfordii. Chung Kuo Chung Hsi I Chieh Ho Tsa Chih 12: 161–164.

432. Shinichi, I., Kaxunobu, S., Naomasa, Y., Ryouji, Y., Seiji, W., Keisuke, Y., and Taketoshi, K. 2000. Antitumor effects of Scutellariae radix and its components baicalein, baicalin, and wogonin on bladder cancer cell lines. Urology 55: 951–955.

433. Shan, B. E., Kazuya, Z., Tsutomu, S., Yasuhiro, Y., and Yamashita, U. 2000. Chinese medicinal herb, Acanthopanax gracilistylus, extract induces cell cycle arrest of human tumor cells in vitro. Jpn. J. Cancer Res. 91: 383–389.

434. Zheng, C., Feng, G., and Liang, H. 1998. Bletilla striata as a vascular embolizing agent in interventional treatment of primary hepatic carcinoma. Chin. Med. J. English ed. 111: 1060–1063.

435. Shan, B. E., Yoshida, Y., Sugiura, T., and Yamashita, U. 1999. Stimulating activity of Chinese medicinal herbs on human lymphocytes in vitro. Int. J. Immunopharmacol. 21: 149–159.

436. Kurashige, S., Akuzawa, Y., and Endo, F. 1999. Effects of Astragali radix extract on carcinogenesis, cytokin production, and cytotoxicity in mice treated with a carcinogen, N-butyl-N'-butanolnitrosoamine. Cancer Invest. 17: 30–35.

437. Li, B. S., Chen, X. H., Ren, W. L., Lu, H. Q., and Yao, Y. L. 1998. Antitumor activity of KangLaiTe injection. Zhongguo Yiyao Gongye Zazhi 29: 456–458.

438. Zheng, S., Yang, H., Zhang, S., Wang, X., Yu, L., Lu, J., and Li, J. 1997. Initial study on naturally occurring products from traditional Chinese herbs and vegetables for chemo-prevention. J. Cell. Biochem. Suppl. 27: 106–112.

439. Ha, M., Li, Z., and He, A. 1997. A laboratory study on Astragalus membranaceus mistura in the prophylaxis and treatment of myelosuppression caused by cancer chemotherapy. J. China Medical Univ. 26: 449–452.

440. Chen, C. Q., Lu, T. X., and Min, H. Q. 1996. Prospective study of radiosensitizing activity of three Chinese herbal agents in combination in the treatment of nasopharyngeal carcinoma. Zhongguo Zhongliu Linchuang 23: 483–485.

441. Kok, L. D. S., Wong, C. K., Leung, K. N., Tsang, S. F., Fung, K. P., and Choy, Y. M. 1995. Activation of the anti-tumor effector cells by radix Bupleuri. Immunopharmacology 30: 79–87.

442. Horikawa, K., Mohri, T., Tanaka, Y., and Tokiwa, H. 1994. Moderate inhibition of mutagenicity and carcinogenicity of benzo(a) pyrene, 1,6-dinitropyrene and 3,9-dinitrogluoranthene by Chinese medicinal herbs. Mutagensis 9: 523–526.

443. Zhang, C. P., Lu, X. Y., Ma, P. C., Chen, Y., Zhang, Y. G., and Yan, Z. 1993. Studies on diterpenoids from leaves of Tripterygium wilfordii. Acta Pharm. Sin. 28: 110–115.

444. Wang, J. Z., Tsumura, H., Shimura, K., and Ito, H. 1992. Antitumor activity of polysaccharide from a Chinese medicinal herb, Acanthopanax giraldii Harms. Cancer Lett. 65: 79–84.

445. Rittenhouse, J. R., Lui, P. D., and Lau, B. H. S. 1991. Chinese medicinal herbs reverse macrophage suppression induced by urological tumors. J. Urol. 146: 486–490.

446. Ma, Y. P. and Et, A. L. 1989. Prediction of responsiveness of human lung cancer xenografts to extracts of Fagopyrum cymosum (Trev) Meisn by SRC assay. Chin. J. Clin. Oncol. 16: 309–312.

447. Hu, Q., Jia, B., Gao, T., Zheng, E., Gao, Y., and Huo, L. 1989. A study on the anti-cancer effect of Ningxia wolfberry. J. Traditional Chin. Med. 9: 117–124.

448. Wang, D. C. and Et, A. L. 1989. Influence of Astragalus membranaceus (AM) polysaccharide F-8 on immunologic function of human periphery blood lymphocyte. Shonghua Zhongliu Zazhi 11: 180–183.
449. Shi, G. Z. and Et, A. L. 1992. Blockage of Glycyrrhiza uralensis and Chelidonium majus in MNNG induced cancer and mutagenesis. Chin. J. Prevent. Med. 26: 165–187.
450. Chauhan, N. S. 1999. Medicinal and Aromatic Plants of Himachal Pradesh. Indus Publishing Co. New Delhi, India. 652 p.
451. Tomoda, M., Gonda, R., Shimizu, N., and Ohara, N. 1994. Characterization of an acidic polysaccharide having immunological activities from the tuber of Alisma orientale. Biol. Pharm. Bull. 17: 572–576.
452. Yisgujawa, M., Yamaguch, S., Chatani, N., Nishino, Y., Matsuoka, T., and Yamahara, J. 1994. Crude drugs from aquatic plants. III. Quantitative analysis of triterpene constituents in Alismatis rhizoma by means of high performance liquid chromatography on the chemical change of the constituents during Alismatis rhizoma processing. Yakugaku Zasshi 114: 241–247.
453. Iwashina, T. and Kitajima, J. 2000. Chalcone and flavonol glycosides from Asaurm canadense. Phytochemistry 55: 971–974.
454. Krogh, A. 1971. The content of trans-aconitic acid in Asarum europaeum L. determined by means of a chromatogram spectrophotometer. Acta Chem. Scand. 25: 1495–1496.
455. Cui, B., Sakai, Y., Takeshita, T., Kinjo, J., and Nohara, T. 1992. Four new oleanene derviatives from the seeds of Astragalus complanatus. Chem. Pharm. Bull. 40: 136–138.
456. Koo, H. N., Heong, H. J., Choi, J. Y., Choi, S. D., and Choi, T. J. 2000. Inhibition of tumor necrosis factor-alpha-induced apoptosis by Asparagus cochinchinensis in Hep G2 cells. J. Ethnopharmacol. 73: 137–143.
457. Knight, V., Koshkina-Nadezhda, V., Clifford, W. J., Giovanella, B. C., and Gilbert, B. E. 1999. Anticancer effect of 9-nitrocamptothecin liposome aerosol on human cancer xenografts in nude mice. Cancer Chemother. Pharmacol. 44: 177–186.
458. Zhang, R., Li, Y., Cai, Q., Liu, T., She, H., and Chambless, B. 1998. Preclinical pharmacology of the natural product anticancer agent 10-hydroxycamptothecin, an inhibitor of topoisomerase I. Cancer Chemother. Pharmacol. 41: 257–267.
459. Chiou, W. F., Chang, P. C., Chou, C. J., and Chen. C. F. 2000. Protein constituent contributes to the hypotensive and vasorelaxant activities of Cordyceps sinensis. Life Sci. 66: 1369–1376.
460. Surh, Y. J. 1999. Molecular mechanisms of chemopreventive effects of selected dietary and medicinal phenolic substances. Mutat. Res. Fundam. Mol. Mech. Mutagen. 428: 1–2, 305-327.
461. Li, S. S., Deng, J. Z., and Zhao, S. X. 1999. Steroids from tubers of Dioscorea bulbifera L. J. Plant Resour. Environ. 8: 2, 61–62.
462. Komori, T. 1997. Glycosides from Dioscorea bulbifera. Toxicon Oxford 35: 1531–1535.
463. Matsuda, H., Kageura, T., Toguchida, I., Murakami, T., Kishi, A., and Yoshikawa, M. 1999. Effects of sesquiterpenes and triterpenes from the rhizome of Alisma orientale on nitric oxide production in lipopolysaccharide-activated macrophages: absolute stereostructures of alismaketones-B 23-acetate and -C 23 acetate. Bioorg. Med. Chem. Lett. 9: 3081–3086.
464. Shimizu, N., Ohtsu, S., Tomoda, M., Gonda, R., and Ohara, N. 1994. A glucan with immunological activities from the tuber of Alisma orientale. Biol. Pharm. Bull. 17: 1666–1668.

465. Gracza, L. 1981. In vitro study of the expectorant effect of phenylpropane derivatives of hazelwort. XII. Active substances from Asarum europaeum L. Planta Med. 42: 155–159.

466. Wierzchowska-Renke, K., Tokarz, H., and Skorkowska, M. 1970. Studies of Asarum europaeum L. II. Preliminary studies of the content of volatile oil and L-ascorbic acid in Herba Asari cum radicibus obtained at Gdansk coastal region. Acta Pol. Phar. 27: 63–69.

467. Li, Z. X., Huang, C. G., Cai, Y. J., Chen, X. M., Wang, F., and Chen, Y. Z. 2000. The chemical structure and antioxidative activity of polysaccharide from Asparagus cochinensis. Yaoxue Xuebao 35: 358–362.

468. Tsui, W. Y. and Brown, G. D. 1996. (+)-Nyasol from Asparagus cochinensis. Phytochemistry 43: 1413–1415.

469. Jaeger, E., Jaeger, D., Orth, J., and Knuth, A. 2000. Irinotecan in second-line therapy of metastatic colorectal cancer. Onkologie 23: 15–17.

470. Wong, M. P., Chiang, T. C., and Chang, H. M. 1983. Chemical studies on dangshen, the root of Codonopsis pilosula. Planta Med. 49: 60.

471. Liu, T., Liang, W., and Tu, G. 1988. Perlolyrine, a beta-carboline alkaloid from Codonopsis pilosula. Planta Med. 54: 472–473.

472. Dong, Y. F., Pan, Z. H., Zhuang, T. D., Liu, X. T., and Feng, X. 2000. Chemical analysis of fatty oil and polysaccharides in seeds from the genus Coix plants in China. J. Plant Resour. Environ. 9: 57–58.

473. Tian, R. H., Ding, Y., Nohara, T., Takia, K., and Takiguchi, Y. 1997. Study on fatty constituents in Coicis semen. Nat. Med. 51: 177–185.

474. Nagao, T., Otsuka, H., Kohda, H., Sato, T., and Yamasaki, K. 1985. Benzoxazinones from Coix lachryma jobi var. ma yuen. Phytochemistry 24: 2959–2962.

475. Otsuka, H., Takeuchi, M., Inoshiri, S., Sato, T., and Tamasaki, K. 1989. Phenolic compounds from Coix lachryma-jobi var. ma-yuen. Phytochemistry 28: 883–886.

476. Phillipson, J. D. 2000. Phytochemistry and medicinal plants. Phytochemistry 56: 217–243.

477. Kitamura, Y., Ohata, M., Ikenaga, T., and Watanabe, M. 1999. Different responses between anthocyanin-producing and non-producing cell cultures of Glehnia littoralis to stress. Proc. 9th Int. Congr., Internatinal Association of Plant Tissue Culture and Biotechnology, Jerusalem, Israel, 14–19 June, 1998. pp. 503–506.

478. Nakano, Y., Matsunaga, H., Saita, T., and Mori, M. 1998. Antiproliferative constituents in Umbelliferae plants II. Screening for polyacetylenes in some Umbelliferae plants, and isolation of panaxynol and falcarindiol from the root of Heracelum moellendorffii. Biol. Pharm. Bull. 21: 257–261.

479. McCutcheon, A. R., Stokes, R. W., Thorson, L. M., Ellis, S. M., and Towers, G. H. N. 1997. Anti-mycobacterial screening of British Columbian medicinal plants. Int. J. Pharmacog. 35: 77–83.

480. Satoh, A., Narita, Y., Endo, N., and Nishimura, H. 1996. Potent allelochemical falcalindiol from Glehnia littoralis F. Schm. Biosci. Biotechnol. Biochem. 60: 152–153.

481. Cheng, D. L., Shao, Y., Yang, L., and Zhan, P. E. 1994. Triterpene glycosides in the leaves of Acanthopanax giraldii. Acta Bot. Sin. 36: 75–79.

482. Chang, Q., Chen, D., Si, J., Zhu, Z., and Wang, X. 1993. Studies on chemical constituents of Acanthopanax giraldii var. hispidus Hoo. China J. Chin. Materia Med. 18: 162–164.

483. Matsuura, H., Saxena, G., Farmer, S. W., Hancock, R. E. W., and Towers, G. H. N. 1996. Antibacterial and antifungal polyine compounds from Glehnia littoralis spp. leiocarpa. Planta Med. 62: 256–259.

484. Fang, X. D., You, M., Ying, W. B., Sun, Z. M., and Shen, P. Z. 1986. The immuno-suppressive activities of polysaccharides from Glehnia littoralis (Bei Sha Shen). Acta Pharm. Sin. 21: 931–934.

485. Tang, X., Ma, Y., and Li, P. 1995. Separation and identification of the anti-inflammatory diterpene from the root cortices of Acanthopanax gracilistylus W. W. Smith. Zhongguo Zhongyao Zazhi 20: 231.

486. Okuyama, T., Takata, M., Nishino, H., Nishino, A., Takayasu, J., and Iwashima, A. 1990. Studies on the antitumor-promoting activity of naturally occurring substances. II. Inhibition of tumor-promoter-enhanced phospholipid metabolism by umbelliferous materials. Chem. Pharm. Bull. 38: 1084–1086.

487. Kaegi, E. 1998. Unconventional therapies for cancer. I. Essiac. Can. Med. Assn. J. 158: 897–902.

488. Ku, H. Y. and Luo, S. R. 1995. Separation and determination of five lignans in the seeds of Arctium lappaol Linne by RP-HPLC. Acta Pharm. Sin. 30: 41–45.

489. Ichihara, A., Oda, K., Numata, Y., and Sakamura., S. 1976. Lappaol A and B, novel lignans from Arctium lappa L. Tetrahedron Lett. 44: 3961–3964.

490. Peng, J. P., Yao, X. S., Kobayashi, H., and Ma, C. Y. 1995. Novel furostanol glycosides from Allium macrostemon. Planta Med. 61: 58–61.

491. Bai, L., Kato, T., Inoue, K., Yamaki, M., and Takagi, S. 1993. Stilbenoids from Bletilla striata. Phytochemistry 33: 1481–1483.

492. Yamaki, M., Bai, L., Kato, T., and Tomita, K. I. 1993. Blespirol, a phenanthrene with a spirolactone ring from Bletilla striata. Phytochemistry 33: 1497–1498.

493. Yamaki, M., Kato, T., Bai, L., Inoue, K., and Takagi, S. 1993. Phenanthrene glucosides from Bletilla striata. Phytochemistry 34: 535–537.

494. Yamaki, M., Kato, T., Bai, L., Inoue, K., and Takagi, S. 1992. Bisphenanthrene ethers from Bletilla striata. Phytochemistry 31: 3985–3987.

495. Bai, L., Kato, T., Inoue, K., Yamaki, M., and Takagi, S. 1991. Blestrianol A, B, and C, bisphenanthrenes from Bletilla striata. Phytochemistry 30: 2733–2735.

496. Yen, G. C., Chen, H. W., and Duh, P. D. 1998. Extraction and identification of an antioxidative component from jue ming zi (Cassia tora L.). J. Agric. Food Chem. 46: 820–824.

497. Kim, D. J., Ahn, B. W., Han, B. S., and Tsuda, H. 1997. Potential preventive effects of Chelidonium majis L. (Papaveraceae) herb extract on glandular stomach tumor development in rats treated with N-methyl-N'-nitro-N nitrosoguanidine (MNNG), and hypertonic sodium chloride. Cancer Lett. 112: 203–208.

498. Goda, Y., Sakai, S., Nakamura, T., Kondo, K., Akiyama, H., and Toyoda, M. 1998. Determination of digitoxigenin glycosides in "Moroheiya" (Corchorus olitorium) and its products by HPLC. J. Food Hyg. Soc. Jpn. 39: 415–420.

499. Goda, Y., Sakai, S., Nakamura, T., Kondo, K., Akiyama, H., and Toyoda, M. 1998. Identification and analyses of main cardiac glycosides in Corchorus olitorius seeds and their acute oral toxicity to mice. J. Food Hyg. Soc. Jpn. 39: 256–265.

500. Yoshikawa, M., Murakami, T., Shimada, H., and Yoshizumi, S. 1998. Medicinal food-stuffs. XIV. On the bioactive constituents of moroheiya. II: New fatty acids, corchorifatty acids A, B, C, D, E, and F, from the leaves of Corchorus olitorius L. Chem. Pharm. Bull. 46: 1008–1014.

501. Hall, I. H., Lee, K. H., Williams, W. L., Kimura, T., and Hiryama, T. 1980. Antitumor agents XLI: Effects of eupaformosanin on nucleic acid, protein, and anerobic and aerobic glycolytic metabolism of Ehrlich ascites cells. J. Pharm. Sci. 69: 294–297.

502. Perez, C., Canal, J. R., Campillo, J. E., Romero, A., and Torres, M. D. 1999. Hypotriglyceridaemic activity of Ficus carica leaves in experimental hypertriglyceridaemic rats. Phytother. Res. 13: 188–191.

503. Li, S. L., Li, P., Lin, G., Zhou, G. H., and Que, N. N. 1999. Existence of 5alpha-cevenine isosteroidal alkaloids in bulbs of Fritillaria L. Acta Pharm. Sin. 34: 842–847.

504. Bratati, D. 1997. Alkaloids of Strychnos nux-vomica. J. Med. Aromatic Plant Sci. 19: 432–439.

505. Guo, D. L., Zeng, F. D., Hu, C. J., and Guo, D. L. 1997. Effects of dauricine, quinidine, and sotalol on action potential duration of papillary muscles in vitro. Acta Pharmacol. Sin. 18: 348–350.

506. Singh, S. B. and Thakur, R. S. 1980. New furostanol and spirostanol saponins from tubers of Paris polyphylla. Planta Med. 40: 301–303.

507. Kasimu, R., Basnet, P., Tezuka, Y., Kadota, S., and Namba, T. 1997. Danshenols A and B, new aldose reductase inhibitors from the root of Salvia miltiorrhiza Bunge. Chem. Pharm. Bull. 45: 564–566.

508. You, Y. 1997. Determination of fumaric acid in Sarcandra glabra (Thunb.) Nakai by HPLC. Zhongguo Zhongyao Zazhi 22: 554, 576–577.

509. Tsui, W. Y. and G. D. Brown. 1996. Cycloeudesmanolides from Sarcandra glabra. Phytochem. Oxford 43: 819–821.

510 Perry, L. M. 1980. Medicinal Plants of East and Southeast Asia. MIT Press, Cambridge, MA. 620 pp.

511. Bliss, B. 1973. Chinese Medicinal Herbs. Georgetown Press. San Francisco. 467 pp.

512. NAS. 1975. Herbal Pharmacology in the People's Republic of China. A trip report of the American Herbal Pharmacology Delegation. National Academy of Science. Washington, DC. 269 pp.

513. Thomas S. C. Li. 2002. Chinese and Related North American Herbs Phytopharmacology and Therapeutic Values. CRC Press, Boca Raton, FL. 598 pp.

514. List, P. H. and Horhammer, L. 1979. Hagers Handbuch der Pharmazeutischen Praxix. Vols. 2–6. Springer-Verlag, Berlin.

515. Jiangsu New Medical College. Eds. 1979. Dictionary of Chinese Traditional Medicine. 3 vols. 2444 pp. Shanghai Science and Technology Publishing Co. Shanghai.

516. Duke, J. A. and Ayensu,E. S. 1985. Medicinal Plants of China. Vol. 1 and 2. Reference Publ. Inc. Algonac, Michigan. 705 pp.

517. Hachiro, H. 1983. Treatment of chronic hepatitis with posterity formulas. Bull. Oriental Healing Arts. Inst. USA 8(3):8–17.

518. Perdue, R. E. Jr. and Hartwell, J. L. Eds. 1976. Plants and cancer. Proc. 16th Annual Meeting Soc. Econ. Bot. Cancer Treatment Reports 60(8): 973–1215.

519. Huang, K. C. 1979. The Pharmacology of Chinese Herbs. 2nd ed. CRC Press. New York. 512 pp.

520. Duke, J. A, and Wain, K. K. 1981. Medicinal Plants of the World. Computer index with more than 85,000 entries. 3 vols. 1654 pp.

521. C.S.I.R. (Council of Scientific and Industrial Research). 1948–1976. The Wealth of India. New Delhi. 11 vols.

522 Zhu, Y. C. 1989. Plantal Medicinales China Roreali-Orientalis. Heilongjiang, Sci. Technology Publishing. House. 1300 pp.

523. Tomlinson, B., Chan, T. Y. K., Chan, J. C. N., Critchley, J. H., and Bat, P.P.H. 2000. Toxicity of complementary therapies: Eastern perspective. J. Chin Pharmacol. 40: 451–458.

524. Keys, J. D. 1976. Chinese Herbs: Their Botany, Chemistry and Pharmacodynamics. Charles. E. Tuttle Co., Tokyo. 388 pp.

525. Leung, A. Y. 1980. Encyclopedia of Common Natural Ingredients Used in Food Drugs and Cosmetics. John Wiley & Sons, New York. 409 pp.

526. Hsu, H. Y. 1980. *How to Treat Yourself with Chinese Herbs.* Oriental Healing Arts Institute, LA, 295 pp.

527. Monachino, J. 1956. Chinese herbal medicine—recent studies. Econ. Bot 10(1): 42–48.

528 Brekhman, I. I. and Grinevitch, M. A. 1981. Oriental Medicine: A computerized study of complex recipes and their components: Analysis of recipes intended to cure certain diseases. Am. J. Chin. Med. 9(1): 34–38.

529 Chen, J. Y. P. 1973. Chinese health foods and herbs and herb tonics. Am. J. Chin. Med. 1(2): 225–247.

530. Yeh, S. D. J. 1973. Anti-cancer Chinese herbal medicines. Am. J. Chin. Med. 1(2): 271–274.

531. Zhang, J. 1980. Recent achievements of the Institute of Material Medica on studies of natural products. pp. 15–54 in *Proceedings US–China Pharmacology Symposium.* October 29–31, 1979. National Academy of Science. Washington. 345 pp.

532. Duke, J. A. 1984b. Encyclopedia of Economic Plants. Unpublished data files on economic plants. Germplasm Resources Laboratory. USDA. Beltsville, MD.

533. Baba K., Yoneda, Y., Kozawa, N., Fujita, E., Wang, N. H., and Yuan, C. Q. 1989. Studies on Chinese traditional medicine "Fang-Feng." II. Comparison of several Fang-Feng by coumarins, chromones, and polyacetylenes. Shoyakugaku Zassh 43: 216–221.

534. Chauhan, N. S. 1999. Medicinal and Aromatic plants of Himachal Pradesh. Indus Publishing Co., New Delhi, India. 652 pp.

535 NIH. 1974. A Barefoot Doctor's Manual. John E. Fogarty International Center. National Institutes of Health. Washington, D.C. DHEW Publication. No. (NIH) 75–695.

536. Lewis, W. H. and Elvin-Lewis, M. P. F. 1977. Medical Botany. John Wiley & Sons, New York. 515 pp.

537. Duke, J. A. 1983. An Herb a Day. Typescript: 365 herbs which have been found to contain pharmacologically active compounds.

538. Leung, A. Y. 1980. Encyclopedia of Common Natural Ingredients Used in Food, Drugs and Cosmetics. John Wiley & Sons, New York. 409 pp.

539. Bever, B. O. and Zahnd. G. R. 1979. Plants with oral hypoglycaemic action. Quart J. Crude Drug. Res. 17(3&4): 139–196.

540. Pettit, G. R. 1977. Biosynthetic Products for Cancer Chemotherapy. Vol. 1. Plenum Press, New York. 215 pp.

541. Duke, J. A. 1984. Handbook of Medicinal Plants. CRC Press. Boca Raton, FL.

542 Hartwell, J. L. 1967–1971. Plants used against cancer: A survey. Lloydia 30–34.

543. Burkill, J. H. 1966. A Dictionary of Economic Products of the Malay Peninsula. Art Printing works. Kuala Lumpur. 2 vols.

544. But, P. P. H., Hu, S. Y., and King, Y. C. 1980. Vascular plants used in Chinese medicine. Fitoterapia 51(5): 245–264.

545. Shibata, S., Mihashi, S., and Tanaka, O. 1967. The occurrence of (-) fimarane-type diterpene in Aralia cordata Thumb. Tetrahedron Lett. 51, 5241–5243.

546. Ben-Hur, E. and Fulder, S. 1981. Effect of panax ginseng saponins and *Eleutherococcus senticosus* on survival of cultured mammalian cells after ionizing radiation. Am. J. Chin. Med. 9(1): 48–56.

547. Kawanishi, S., Mitaral, Y., and Shoji, J. 1972. Studies on wujiapi and some related crude drugs. An morphological studies on bei-wujiapi and some related drugs. Yakugaku Zasshi 89: 972–78.

548. Chandler, R. F., Hooper, S. N., and Harvey, M. J. 1982. Ethnobotany and phytochemistry of yarrow, *Achillea millefolium*, Compositae. Econ. Bot. 36(2): 203–223.

549. Puffie, D. and Zerr, W. 1989. The content of various secondary plant products in widespread weeds in grassland with particular regard to higher lying areas. Eichhof-Berichie No. A. 11, 127 pp.

550. Kong, Y. C., Hu, S. Y., Lau, F. K., Che, C. T., Yeung, H. W. Cheung, S., and Hwang, J. C. C. 1976. Potential anti-fertility plants from Chinese medicine. Am. J. Chin. Med. 4(2): 105–128.

551. Wat, C. K., Johns, T., and Towers, G. H. N. 1980. Phototoxic and antibiotic activities of plants of plants of the asteraceae used in folk medicine. J. Ethnopharm. 2: 279–290.

552. Arthur, H. R. and Cheung, H. T. 1960. A phytochemical survey of the Hong Kong medicinal plants. J. Pharmacol. 12: 567–570.

553. Watt, J. M. and Breyer-Brandwijk, M. G. 1962. The Medicinal and Poisonous Plants of Southern and Eastern Africa. 2nd ed. E. & S. Livingston, Ltd. Edinburgh and London. 1457 pp.

554. Duke, J. A. and Atchley, A. A. 1984. Proximate Analysis Chapter. Submitted to CRC Press, Boca Raton, FL.

555. Duke, J. A. 1981. Handbook of Legumes of World Economic Importance. Plenum Press. New York. 345 pp.

556. Wat, C. K., Johns, T., and Towers, G. H. N. 1980. Phototoxic and antibiotic activities of plants of the Asteraceae used in folk medicine. J. Ethnopharm. 2: 279–290.

557. Li, C. P. 1974. Chinese Herbal Medicine. PHEW Publication (NIH) 75-732. National Institutes of Health. Washington, DC. 120 00.

558. Wong, M. 1976. La Medicine Chinoise par les plantes (Le Corps a vivre). Editions Tchou, Tchou 285 pp.

559. Chen, J. Y. P. 1973. Chinese health foods and herb tonics. Am. J. Chin. Med. 5(1): 31–37.

560. Zheng, J. H. 1983. Commonly used Chinese herbs in the treatment of carcinoma. Typescript of paper, June 10, 1983. Presented at University of Chicago, Department of Pharmacognosy.

561. Zhao, Y. X. 1981. Traditional Chinese medicine in pneumonia of children. Chin Med. J. 44(9): 601–606.

562. Toth, I., Bathory, M., Szendrei, K., Minker, E., and Blazso, G. 1981. Ecdysteroids in chenopodiaceae: Chenopodium album. Fitoterapia 52(2): 77–88.

563. Duke, J. A. 1977. Vegetarian vitachart. Quart. J. Crude Drug Res. 15: 45–66.

564. Hu, S. Y. 1961. The economic botany of the Paulownias. Econ. Bot. 15(1): 11–27.

565. Ishikura, N. 1981. Anthocyanins and flavones in leaves and seeds of perilla plant. Agric. Biol. Chem. 45(8): 1855–1860.

566. Farnsworth, N. R. 1980. The development of pharmacological and chemical research for application to traditional medicine in developing countries. J. Ethnopharm. 2: 173–181.

567. John D. Keys 1976. Chinese Herbs: Their Botany, Chemistry and Pharmacodynamics. Charles E. Tuttle Co. Inc. Rutland, Vermon & Tokyo, Japan. 388 pp.

568. Li, Thomas S. C. 2000. Medicinal Plants: Culture, Utilization and Phytopharmacology. Technomic Publishing Co. Inc. Lancaster, Pennsylvania. 517 pp.

569. Chevallier, A. 1990. The Encyclopedia of Medicinal Plants. Dorling Kindersley Ltd. London. 336 pp.

570. Duke, J. A. 1985. CRC Handbook of Medicinal Herbs. CRC Press. Boca Raton, FL. 677 pp.
571. Reid, D. P. 1993. Chinese Herbal Medicine. Shambhala, Boston. 174 pp.
572. Li, Thomas S. C. 2002. Chinese and Related North American Herbs: Phytopharmacology and Therapeutic Values. CRC Press, Boca Raton. FL. 598 pp.
573. Nakano, Y., Matsunaga, H., Saita, T., and Mori, M. 1998. Antiproliferative constituents in umbelliferae plants. II. Screening for polyacetylenes in some umbelliferae plants, and isolation of panaxynol and falcarindiol from the root of Heracleum moellendorffii. Biol Pharm. Bull. 21: 257–261.
574. Rogers, R. D. 1997. Sundew, Moonwort, Medicinal Plants of the Prairies. Vol. 1 and 2. Edmonton, Alberta, Canada. 2820.
575. Zhu, D. Y., Bai, D. L., and Tang, X. C. 1997. Recent studies on traditional Chinese medicinal plants. Drug Dev. Res. 39: 147–157.
576. Li, B. Q., Fu, T., Yan, Y. D., Baylor, N. Y., Ruscetti, F., and Kung, H. F. 1993. Inhibition of HIV infection by baicalin, a flavonoid compound purified from Chinese herbal medicine. Cell Mol. Biol. Res. 39: 119–124.
577. Tomlinson, B., Chan, T. Y. K., Chan, J. C. N., Critchley, J. H., and Bat, P. P. H. Toxicity of complementary therapies: An eastern perspective. J. Clin. Pharmacol. 40: 451–456.
578. Chen, F. C. 1997. Active Ingredients and Identification in Common Chinese Herbs. People Health Publ. Co. China. 872 pp.
579. Perry, L. M. and Metzger, J. 1980. Medicinal Plants of East and Southeast Asia: Attributed Properties and Uses. MIT Press, London. 620 pp.
580. Cai, X. H. et al. 2008. A cage-monoterpene indole alkaloid from *Alstonia scholaris*. Org Lett. 10(4): 577–580.
581. Cui, Y., Zhang, X. M. Chen, J. J. et al. 2007. Chemical constituents from root of *Actinidia chinensis*. Zhongguo Zhong Yao Za Zhi 2007 August 32(16): 1663–5.
582. Lee, M. K., Ha, N. R., Yang, H., Sung, S. H. et al. 2007. Antiproliferative activity of triterpenoids from Eclipta prostrata on hepatic stellate cells. Phytomedicine 2007 November 29.
583. Shi, J., Yamashita, T., Todo, A. et al. 2007. Repellent from traditional Chinese medicine. Periplocasepium Bunge. Z. Naturforsch 2007 November–December 62(11–12): 821–825.
584. Wang, R. R., Gu, Q., Wang, Y. H. et al., 2008. Anti-HIV-1 activities of compounds isolated from the medicinal plant Rhus Chinensis. J. Ethnopharmacol. 2008 February 9 (E-publishing ahead of printing).
585. Berestetskiy, A., Dmitriev, A., Mitina, G. et al. 2008. Nonenolides and cytochalasins with phytotoxic activity against cirsium arvense and sonchus arvensis: A structure-activity relationships study. Phytochemicstry 2008, 69(4): 953–960.
586. Ghosh, A. 2007. Anti-oxidative effect of a protein from Cajanus indicus L. against acetaminophen-induced hepato-nephro toxicity. J. Biochem. Mol. Biol. 2007 November 30, 40(6): 1039–1049.
587. Manna, P. and Sinha, M. 2007. A 43KD protein isolated from the herb Cajanus indicus L. attenuates sodium fluoride-induced hepatic and renal disorders in vivo. J. Biochem Mol. Biol 2007 May 31, 40(3): 382–395.
588. Cui, C. B., Tezuka, Y., Kikuchi, T. et al., 1992. Constituents of a fern Davallia mariesii Moore. Chem. Pharm. Bull (Tokyo) 1992 April 40(4): 889–898.
589. Zheng, Yu Chun. 1955. Poison plant from Taiwan Holiday Publishing Limited Co, Taipei, Taiwan. 238 pp.
590. Chung, K. T. 1959. A Chinese Native Medicinal Flora for Farmers. Beijing, China, 281 pp.

591. Zi, J., Liu, M., Gan, M., Lin, S. et al., 2008. Glycosidic constituents of the tubers of Gymnadenia conopsea. J. Nat. Prod. 2008 March 19. (E-published ahead of printing.)

592. Zhang, T. J. 1993. Herbal textural analysis on the Chinese drug xuancaogen. Xhonnnue Zhong Yao Za Zhi. 1993 September 18(9): 515–517, 573.

593. Wu, Y. T., Lin, L. C., Sung, J. S., and Tsai, T. H. 2006. Determination of acteoside in Cistanche deserticola and Boschiakia rossica and its pharmacokinetics in freely-moving rats using LC-MS/MS. J. Chromatogr. B Analyt Technol Biomed Life Sci. 2006 November 21; 844(1): 89–95.

594. Shyr, M. H., Tsai, T. H., and Lin, L. C. 2006. Rossicasins A, B and rosicaside F, three new phenylpropanoid glycosides from Boschniakia yossica. Chem. Pharm. Bull (Tokyo) 2006. February 54(2): 252–254.

595. Choi, E. M. and Kim, Y. H. 2008. A preliminary study of the effects of an extract of Ligularia fischeri leaves on type II collagen-induced arthritis in DBA/1J mice. Food Chem. Toxicol. 2008 January 46(1): 375–379.

596. Sagiura, S., Yamaza, Ki, K. 2005. Moth pollination of Metaplexis japonica pollinaria transfer on the tip of the proboscis. J. Plant Res. 2005 August 118(4): 257–262.

597. Yoshida, M. Liu, Y., Uchida, S. et al. 2008. Effects of cellulose crystallinity, hemicellulose, and lignin on theenzymatic hydrolysis of miscanthus sinensis to monosaccharides. Biosci. Biotechnol. Biochem. 2008 March 72(3): 805–810.

598. Wang, Y., Wang, O. F., Guo, Y. H., and Barrett, S. C. 2005. Reproductive consequences of interactions between clonal growth and sexual reproduction in Nymphoides peltata: A distylous aquatic plant. New Phytol. 2005. January 165(1): 329–335.

599. Kwon, D., Yoon, S. Carter, O. et al. 2006. Antioxidant and antigenotoxic activities of Angelica keiskei, Oenanthe javanica, and Brassica oleracea in the Salmonella mutagenicity assay and in HCT 116 human colon cancer cells. Journal Biofactors. Vol. 26, 231–244.

600. Lin, L. C., Wang, Y. H., Hou, Y. C. et al., 2006. The inhibitory effect of phenylpropanoid glycosides and iridoid glucosides on free radical production and B-2 integrin expression in human leucocytes. J. Pharm. Pharmacol 2006 January 58(1): 129–135.

601. Zhang, T., Li, Q., Li, K. et al. 2008. Antitumor effects of saponin extract from Patrinia villosa (Thunb.) Juss on mice bearing U14 cervical cancer. Phytother. Res. 2008 March 18. (E-publishing ahead of printing.)

602. Kobayashi, S., Asai, T., Fujimoto, Y., and Kohshima S. 2008. Anti-herbivore structures of Paulownia tomentosa: Morphology, Distribution. Chemical constituents and changes during shoot and leaf development. Journal Annals of Botnay 101(7): 1035–1047.

603. Guo, R., Pittler, M. H., and Ernest, E. 2007. Herbal medicines for the treatment of allergic rhinitis: A systematic review. Ann Allerayasthma Immunol. 2007 December 99(6): 483–495.

604. Park, I. K., Shin, S. C., Kim, C. S. et al. 2005. Larvicidal activity of lingans identified in phryma leptostachya var. asiatica roots against three mosquito species. J. Agric. Food Chem. 2005 February 23, 53(4): 969–972.

605. Yamasald, K., Nakano, M., Kawahata, T. et al. 1998. Anti-HIV-1 activity of herbs in labiatae boil. Pharm Bull. 1998 August 21(8): 829–833.

606. Tai, J. and Cheung, S. 2007. Anti-proliferative and antioxidant activities of saposhnikovia divaricata. Oncol Rep. 2007 July 18(1): 227–234.

607. Wei, S., Zhou, Q., and Mathews, S. 2008. A newly found cadmium accumulator—Taraxacum mongolicum. J. Hazard Mater. 2008 February 23 (E-publishing ahead of printing).

608. Wang, J., Cai, Y., and Wu, Y. 2008. Anti-inflammatory and analgesic activity of topical administration of siegesbeckia pubesceens. Pak. J. Pharm. Sci. 2008 April 21(2) 89–91.

609. Cho, W. I., Choi, J. B., Leek et al., 2008 Antimicrobial activity of torilin isolated from Torilis japonica fruit against Bacillus subtilis. J. Food Sci. 2008 March 73(2): M37-46.
610. Produced by Elena Mannes CBS TV network, June 23, 1981.
611. Other useful chemicals—workshop proceedings U.S. Congress, Office of Technology Assessment. OTA-BP-F-23. September 1981. Washington DC. 252 pp.
612. Bao, L. 1979. China's "miracle" fruit. Prog. Horticulture 11(1): 37–40.
613. Chas, C. Thomas, Bahamas to Yucatan, Springfield, IL. 142 pp.
614. Influenza and bronchitis. Am. J. Chin. Med. 3(2): 187–205.
615. Villareal, Ruben. L. Medicinal Plants. Vol. 1, University of the Philippines, Los Baños.
616. Sharma, M. M., Lal, G., and Jacob, D. 1976. Estrogenic and pregnancy interceptory effects of carrot Daucus carota seeds. Indian J. Exper. Biol. 14: 506–508.
617. Anon. 1979. NIH to Study Spiney Ginseng. Bio Science 29(5): 324.
618. Menthal, M. Eleutherococcus senticosus on survival of cultured mammalian cells after ionizing radiation. Am. J. Chin. Med. 9(1): 48–56.
619. Davis, J. M. Clinical efficacy (Abstract). Am. J. Chin. Med. 2(2): 225–226.
620. Neumeyer, J. L. and Guan, J. H. 1981. Pharmacy in the People's Republic of China. American Pharmacy NS22(1): 14–21. (Grad. school pharmacy. Northeastern University, Boston, MA 02115.)
621. Kirtikar, K. R. and Basu, B. D. 1975. Indian Medicinal Plants. 4 vols. 2nd edition, Jayyed Press. New Delhi.

Appendix 1

Chinese and Scientific Names of Chinese Medicinal Herbs

Chinese Name	Scientific Name
A. Er Tai Zi Zi Wan	Heteropappus altaicus (Willd) NOvopokr..
A Wei	Ferula assa-foetida L., F. bungeana Kitag.
Ai Di Cha	Ardisia japonica (Hornst.) Blume
Ai Lei	Centaurium meyeri (Bunge.) Druce
Ai Na Xian	Blumea balsumifera (L.) DC var. microcephala Kitumura, Styrax tonkinensis
Ai Ye	Artemisia argyi Leveille & Vaniot, A. vulgarts L., A. argyi Leveille & Vaniot f. eximia Pamp, A. scoparia Waldst. & Kitaib., A. keiskeana Miq., A. selengensis Turcz. ex Bess., A. lagocephala Fisch. ex Bess., A. argyi Leveille & Vaniot f. gracilis (Pamp.) Kitag., A. integrifolia L., A. sieversiana Ehrh. ex Willd., A. halodendron Turez. ex Bess., A. indica Willd., A. japonica Thunb., A. igniaria Max., A. lavandulaefolia DC, A. japonica Thunb. var. manshurica (Kom.) Kitag.
Ai Yu Zi	Ficus awkeotsang Makino
Ba Dou	Croton tiglium L., C. cascarilloides Raeushel
Ba Ji Tian	Morinda officinalis
Ba Jiao Feng	Alangium chinense (Lour.) Harms.
Ba Jiao Hui Xiang	Illicium verum Hook f., I. lanacedatum A. S. Smith
Ba Jiao Lian	Dysosma pleiantha (Hance) Woodson.
Ba Jiao Lian	Podophylium peltatum L., P. pleianthum Hance.
Ba Li Ma	Rhododendron molle (Blume) G. Don
Bai Bao Zi	Mallotus paniculatus (Lam.) Muell-Arg.
Bai Ben Dou	Dolichos lablab L.
Bai Bu	Stemona japonica (Bl.) Miq.

Bai chang	Acorus calamus L.
Bai Chen	Cynanchum japonicum Moore et Decne
Bai Dou Ku	Amomum cardamomum L.
Bai Guo	Artemisia frigida Willd.
Bai He	Lilium brownii F. E. Brown var. viridulum Baker, L. dauricum Ker-gawler, L. distichum Nakai ex Kamibayashi, L. concolor Salisb. var. buschianum (Ledeb.) Baker, L. pumilum DC, L. lancifolium Thunb., L. japonicum Thunb., L. concolor Salisb. var. partheneion (Sieb. & De Vries) Baker
Bai Hua	Betula mandshurica (Regel) Nakai,
	B. platyphylla Suk.
Bai Hua Teng	Plumbago zeylanica L.
Bai Ji	Cymbidium hyacinthinum Sm., C. striatum Sw., Bletilla hyacinthina R. Br., B. striata (Thunb.) Reichb., B. hyacinthina R. Br, Epidendrum striatum Thunb., E. tuberosum Lour.
Bai Jiang Cao	Patrina scabiosaefolia Fisch ex Link.
Bai Jie	Sinapis alba L.
Bai Jie Zi	Brassica alba (L.) Rabenh., B. juncea (L.) Czern. et Coss.
Bai Lian	Ampelopsis japonica (Thunb.) Mak.
Bai Lian Guo	Artemisia gmelini Weber ex Stechmann
Bai Mao	Imperata arundinaceae Cyrill., I. cylindrica Beauv.
Bai Qian	Cyanchum glaucescens Decaisne
Bai Qu Cai	Chelidonium album L., C. hybridum L., C. majus L., C. serotinum L.
Bai Rui Cao	Thesium chinense Turcz.
Bai Shao	Paeonia albiflora Pall.,

Chinese Name	Scientific Name
	P. japonica (Makino) Miyabe. et Takeda, *P. edulis* Salisb., *P. officinalis* L., *P. moutan* Sims., *P. lactiflora* Pall., *P. lactiflora* Pall. var. trichocarpa (Bunge.) Stern
Bai Tou Went	*Pulsatilla ambigua* Turcz. ex Pritz., *P. cernua* (Thunb.) Bercht. et Opiz., *P. chinensis* (Bunge.) Regel, *P. chinensis* (Bunge.) Regel var. kissii (Mandl) S. H. Li et Y. H. Huang
Bai Tu Own	*Anemone cernua* Thunb., A. pulsatilla, A. pulsatilla var. chinensis Bunge.
Bai Way	*Cynanchum atratum* Bunge.
Bai Xian Pi	*Dictamnus albus* L. subsp. dasycarpus (Turcz.) Winter, Dictamnus dasycarpus, *Fraxinella dictamnus* Moench
Bai Yin Shu	*Securinega virosa* (Roxb.) Pax & Hoffmann
Bai Ying	*Solanum lyratum* Thunb, *S. melongena* L.
Bai Yu Lan	*Michelia alba* DC
Bai Zhi	*Angelica amurensis* Schischk., A. dahurica (Fisch.) Benth. et Hook., A. anomala Lallem.
Bai Zhu	*Atractylodes chinensis* (Bunge.) Koidz., A. chinensis (Bunge.) Koidz. f. simplicifolia (Loes.) Y. C. Chu, A. chinensis (Bunge.) Koidz. f. quinqueloba, (Baranov et Skv.) Y. C. Chu, A. chinensis (Bunge.) Koidz. var. liaotungensis (Kitag.) Y. C. Chu, A. japonica Koidz. ex Kitam., A. koreana (Nakai) Kitam., A. macrocephala Koidz.
Bai Du Juan Hua	*Rhododendron mucronatum* G. Don
Bai Hua Shi Shi Cao	*Oldenlandia diffusa, Hedyotis diffusa* Willd
Bai Hua Yi Mu Cao	*Leonurus sibiricus* L. f. albiflorus (Nakai Kitag.) G. Wu
Bai Jiang	*Patrinia villosa* Thunb.
Bai Long Chuan Hua	*Clerodendrum paniculatum* L. var. albiflorum Hsieh.
Bai Qian Ceng	*Melaleuca leucadendra* L.

Bai Su Zi	Perilla frutescens (L.) Britt.
Bai Ye Diao Zhang	Lindera glauca (Sieb. et Zucc.) Blume
Bai Zhu Shu	Gaultheria leucocarpa f. var. cumingiana (Vidal) Sleumer.
Ban Bian Lian	Lobelia chinensis L.
Ban Jiu Jiu	Vernonia andersonii C. B. Carke
Ban Lan	Baphicanthus cusia (Nees.) Bremek.
Ban Lan Gen	Isatis chinensis (Thunb.) Nakai, I. chinensis (Thunb.) Nakai var. graminifolia H. C. Fu
Ban Lan Geng	Polygonum tinctorium Lour.
Ban Xia	Pinellia ternata (Thunb.) Breit, P. tuberifera Tenore
Ban Zi Lian	Scutellaria barbata
Bang Chui Hui	Neoalsomitra integrifolioia (Cogr.) Hutch
Bao Gai Cao	Lamium amplexicaule L.
Be Han Cao	Melilotus alba Medicus, M. suaveolens Ledeb.
Bei Mei Do Xing Cao	Lepidium virginicum L.
Bei Mu	Fritillaria thunbergii Miq., F. ankeunensis Chen et Yin, F. collicola Hance, F. roylei Hook, F. maximowiczii Freyn, F. ussuriensis Maxim., F. verticillata Willd.
Bei Pu Jiang	Buddleia formosana Hatushima
Bei Xian	Cyathula prostrate (L.) Blume
Bei Za Seng	Glehnia littoralis F. S. Schmidt et Miq.
Ben Sao	Mylabris phalerata
Bi Ba	Piper longum L.
Bi Cheng Qie	P. cubeba L.

Chinese Name	Scientific Name
Bi Li Go	*Ficus pumila* L.
Bi Ma Zi	*Ricinus communis* L.
Bi Qao Jiang	*Costus specious* (Koen.) Smith
Bian Xu	*Polygonum aviculare* L., *P. vivipara* (L.) S. F. Gray, *P. lapidosa* Kitag., *P. aviculare* L var. *vegetum* Ledeb., *P. manshuriensis* Komarov
Bing Lang	*Areca catechu* L., *A. hortonsis* Lour.
Bo Hoo	*Mentha arvensis* L., *M. dahurica* Fisch. ex Benth., *M. haplocalyx* Briq., *M. sachalinensis* (Briq.) Kudo, *M. sachalinensis* (Briq.) Kudo f. arguta (Kitag.) Y. C. Chu
Bo Lo Mi	*Artocarpus heterophyllus* Lamarck
Bo Lou Hui	*Macleaya cordata*
Bu Gu Zi	*Psoralea corylifolia* L.
Cai Fu	*Raphanus sativus* L.
Cam Dou	*Vicia faba* L.
Canada Pon	*Erigeron canadensis* L.
Cang Er	*Xanthium chinense* Mill., *X. strumarium* L. *X. sibiricum* Patr. ex. Widd., *X. japonicum* Widder, *X. mongolicum* Kitag.
Cang Zhu	*Atractylodes lancea*
Cao Bai Ching	*Ampelopsis aconitifolia*
Cao Cong	*Boschniakia rossica* Cham& Schlecht.
Cao Guo	*Amomum globosum* Lour.
Cao He Che	*Polygonum bistorta* L.

Cao Jue Ming	Celosia margariacea L., C. argentea L.
Cao Wu	Aconitum laciniatum Stapf., A. chinensis Paxt., A. vilmorinianum Kom., A. pariculigerum Nakei., A. kusnezoffii Reichenbach
Cao Yu Mei	Anemone rivularis Buch-Hamilton ex DC, A. rivularis Buch-Hamilton ex DC var. flore-minore Maxim.
Ce Bai Ye	Platycladus orientalis (L.) Franco, Thuja chinensis Hort., T. koraiensis Nakai, T. orientalis L.
Ce Bai Ye	Biota chinensis Hort., B. orientalis L.
Ce Yan	Alnus japonica (Thunb.) Steudel, A. japonica (Thunb.) Steudel var. koreana Callier
Cha	Camellia bohea Griff., C. viridis Link., C. theifera Griff., C. sinensis (L.) Kuntze, Thea assamica Mast., T. sinensis L., T. bohea L., T. cochinchinensis Lour., T. viridis Link., T. cantoniensis Lour., T. chinensis Sims.
Chai Hu	Bupleurum chinense DC, B. scorzoneraefolium Willd., B. falcatum L.
Chang Bai Rui Xian	Daphne koreana Nakai
Chang Chun Hua	Catharanthus roseus (L.) G. Don
Chang Chun Ton	Hedera rhombea (Miq.) Bean
Chang Pu	Acorus calamus L. var. angustatus Besser, A. tatarinowii L., A. gramineus Ait.
Chang Shan	Dichroa febrifuga Lour., D. cyanitis Miq., D. febrifuga Lour., D. latifolia Miq., Adamia cyanea Wall., A. versicolof Fortune
Chang Shu	Cinnamomum camphora (L.) J. S. Presl.
Che Chen Zi	Plantago asiatica L., P. major L., P. major L. var. asiatica DC, P. exaltata Horn., P. loureiri Roem. et Schult., P. depressa Willd.
Che Sang Zi	Dodonaea viscosa (L.) Jacquin
Che Ye Sha Seng	Adenophora triphylla (Thunb.) DC
Che Zhou Cao	Trifolium pratense L., T. repens L.

Chinese Name	Scientific Name
Chen Pi	*Citrus reticulata* Blanco
Chen Wei	*Rosa multiflora* Thunb.
Chen Xiang	*Aquilaria agallocha* Roxb., *A. sinensis* (Lour.) Gilg.
Chi Chang Ts'ao	Trigonotis peduncularis Trevir.
Chin Nin Ts'ao	Kyllingia brevifolia Bottb.
Chinese Ji	Cirsium chinense Gardn. et. Champ.
Ching Mian Hua	Pileostegia viburnoides Hooker f. Thomson
Chiu Chung Ko	*Bougainvillea brasiliensis* Raeusch, *B. glabra* Choisy var. sanderiana Hort.
Cho Chong Jiu	*Pyrethrum cinerariifolium* (L.) Trev.
Cho Mo	*Fagopyrum esculentum* Moench, *F. sagittatum* Gilib.
Chong Guo	*Artemisia finita* Kitag.
Chou Chie Cao	*Boenninghausenia albiflora* (Hook.) Meisn.
Chou Fu Yong	*Tagetes erecta* L.
Chou Lee	*Prunus padus* L.
Chou Mu Lee	*Clerodendrum fragrans* Ventenat
Chou Wu Tong	*Clerodendrum trichotomum* Thunb., *C. trichotomum* Thunb. var. ferrugineum Nakai
Chou Xing	*Chenopotium ambrosiodes* L.
Chu Gu Jiu	*Chrysanthemum cinerriaefolium* Visiont
Chu Kao Mu	Siegcsbeckia pubescens Makino
Chu Kui	*Althaea rosea* (L.) Cav.
Chu Ye	*Phyllostachys bambusoide* Sieb. et Zucc., *P. nigra* Munro. var. henonis Mak.

Chuan Duan Chang Cao	*Corydalis incisa* (Thunb.) Pars.
Chuan Jian	*Xanthoxylum piperitum* DC
Chuan Jin Pi	*Hibiscus rhombifolius* Cav.
Chuan Lian	*Melia japonica* G. Don, *M. toosendan* L.
Chuan Shan Long	*Dioscorea nipponica* Makino
Chuan Xiang	*Ligustium chuanziang* Hort.
Chuan Xin Lian	*Andrographis paniculata* (Burm. f.) Nees
Chui Pen Chao	*Sedum sarmentosum* Bunge.
Chui Zhi Shi Song	*Lycopus phlegmaria* L.
Chun Hsiang-Chun	*Cedrela sinensis* A. Juss.
Chun Pi	*Ailanthus altissima* (Mill.) Swingle
Chung Way	*Leonurus sibiricus* L.
Ci Gu	*Sagittania sagitifolia* L.
Ci Hi Li	*Tribulus terrestsis* L.
Ci Luo Shi	*Maytenus diversifolia* Hou.
Ci Lao Ya	*Aralia mandschurica* (Rupr & Maxim) Seem
Ci Qiu Shu Pi	*Kalopanzx septemlobus* (Thunb.) Koidz.
Ci Seng	*Oplopanax elatus* (Nakai) Nakai
Ci Wu Jia	*Eleatherocossus senticosus* (Rupr ex Maxim.) Maxim., *Acanthopanax senticosus* (Rupr. et Maxim.) Harms.
Cong	*Allium fistulusum* L.
Cong Lan	*Zephyranthes candida* Herbert

Chinese Name	Scientific Name
Cu Fei	(Taiwan) *Cephalotaxus wilsoniana* Hayata
Cui Que	*Delphinium grandiflorum* L.
Cylon Rou Gui	*Cinnamomum zeglanicum* Blume
Da Dou	*Glycine max* (L.) Merrill
Da Fei Yang Cao	*Euphorbia hirta* L.
Da Feng Zi	*Hydnocarpus anthelmintica* Pierre, *H. castaneus H. F.* & Th.
Da Hua Tian Qing	*Sesbinia grundiflora* (L.) Persoon
Da Ji	*Cirsium japonicum* DC, *Euphorbia lasiocaula* Boiss., *E. sampsoni* Hance, *E. coraroides* Thunb., *E. pallasii* Turcz., *E. lunulata* Bunge., *E. sieboldiana* Moore et Decne., *E. pekinensis* Rupr.
Da Ji Ru Zi Shu	*Euphorbia resinfera* Berger
Da Ma Ren	*Cannabis chinensis* Del., *C. sativa* L.
Dang qui	*Angelica sinensis* (Oliv.) Diels.
Da Qing	*Isatis indigotica* Fortune ex Lindley, *I. oblongata* DC, *Clerodendrum cyrtophyllum* Turcsaninow
Da Qing Ye	*Baphicanthus cusia* (Nees.) Bremek.
Da Ri Jian Cao	*Oenathera terythrosepala* Borbus
Da Suan	*Allium chinense* Max, *A. tuberosum* Roxb, *A. sativum* L., *A. uliginosum* G. Don, *A. odorum* L.
Da Wan Hua	*Calystegia hederacea* Willich ex Roxb., *C. japonica* Choisy iu Zoll.
Da Xian Mao	*Curculigo capitulata* (Lour.) O. Kuntze
Da Xian Ye Shu	*Lindera megaphylla* Hemsley
Da Yi Zhi Jian	*Lycoris aurea* Herb.
Da Ye An	*Eucalyptus robusta* Sm.

Dan Gui	Angelica polymorpha Max., A. sinensis (Oliv.) Diels
Dan Ye Mu Jing	Vitex rotundifolia L. f.
Dan Ye Xiz Zhu	Phyllanthus simplex Ketziu
Dan Zhu Ye	Lophatherum gracile Brongn.
Dang Gui	Levisticum officinale Koch.
Dao Dou	Canavalia gladiata (Jacq.) DC
De Jin	Parthenocissus tricuspidata Planchon.
De Qing Cao	Euphorbia humifusa Willd.
Deng Tai Ye	Alstonia scholaris (L.) R. Br.
Di Dan Tou	Elephantopus elatus L.
Di Ding	Guelden staedtia Maxim
Di Ding Zi Jing	Corydalis bungeana Turcz.
Di Er Cao	Hypericum ascyron L. var. longisttylum Maxim.
Di Gu Pi	Lycium chinense Miller
Di Huang	Rehmannia chinensis Fisch., R. giutinosa (Gaertn.) Libosch.
Di Jiao	Thymus amurensis Klokov, T. komarovii Sergievskaja, T. przewalskii (Kom.) Nakai, T. kitagawianus Tscherneva, T. quinquecostatus Celakovsky, T. disjunctus Klokov
Di Jin Cao	Euphorbia helioscopia L.
Di Yu	Poterium officinale Benth.
Ding Gong Teng	Erycibe henryi Prain
Ding Xian	Eugenia ulmoides Oliv., E. aromctica Baill., E. caryophyllata (L.) Thunb.
Dio Ue Nao Bu	Stemona tuberosa Lour.

Chinese Name	Scientific Name
Do Xing Cao	*Lepidium apetalum* Willd.
Don Gua	*Benincase cerifera* Savi., *B. hispida* Cogn.
Don Shin	*Juncus communis* Meyer
Don Sin Cao	*Juncus effusus* L.
Dong Chong Xia Chao	*Cordyceps sinensis*
Dong Kui Zi	*Malva chinensis* Mill., *M. verticillata* L., *M. pulchella* Berhn.
Dong Ling Cao	*Rabdosia rubescens*
Dong San Hu	*Solanum pseudo-capsicum* L.
Dong Seng	*Codonopsis pilosula* (Franch.) Nannfeldt, *C. tangshen* Oliv., *C. ussuriensis* (Rupr. et Maxim.) Hemsl.
Dong Yao	*Swertia pseudochinensis* Hara
Dou Kou	*Alpinia katsumadai*
Dou Tu Si	*Cuscuta australis* R. Brown
Du Huo	*Angelica gigas* Maxim.
Du Jing Sha	*Maesa japonica* (Thunb.) Moritzi
Du Zhong	*Eucommia ulmoides* D. Oliver
Duan Geng Ren Dong	*Lonicera apodonta* Ohwi
Duan Geng Wu Jia	*Acanthopanax sessiliflorus* (Rupr. et Maxim.) Seem
Duan Xue Liu	*Clinopodium chinense* (Benth.), C. polycephalum
Dui Ye Dou	*Cassia alata* L.
Duo Ti Hu	*Cynoglossum divaricatum* Stemphan
E. Chang Coo	*Malachium aquaticum* L.

E Zhu	*Curcuma pallida* Lour.
Er Cha	*Acacia catechu* Willd.
Fan Lu	*Stellaria media* (L.) Cyrillo
Fan Mu Pen	*Strychnos nux-vomica* L.
Fan Qie	*Lycopersicon esculentum* L.
Fan Shi Lui	*Psidium guajava* L.
Fan Sui Xian	*Amaranthus paniculatus* L.
Fan Tian Hua	*Urena procumbens* L.
Fan Xie Ye	*Cassia angustifolia* Vahl., Tinnevelly *senna* (Syn. *Cassia angustifolia*)
Fan Xing	*Tetragonia tetragonoides* (Pall.) O. Kuntz.
Fang Chi	*Menispermum dauricum* DC. f. pilosum (Schneider) Kitag.
Fang Feng	*Ledebouriella divaricata, Saposhnikovia divaricata (Turcz) Schisehk.*
Fang Ji	*Cocculus diversifolius* Miq., *C. thunbergii* DC
Fei Bai	*Allium macrostemon* Bunge., *A. tartaricum* Ait.
Fei Hin Cao	*Aletris formosuna* (Hayata) Sasaki, *A. spicata* Franch
Fei Ji Cao	*Eupatorium odoratum* L.
Fei Lan	*Zephyranthes carinata* Herbert
Fen Fang Ji	*S. tetrandraq* Moore
Fen Wei Ju	*Pueraria wallichiana* Agardh.
Feng Dou Cai	*Petasites japonicus* F. Schmidt.
Feng Huan Mu	*Delonix regia* (Boj.) Raf.
Feng Lee	*Ananas comosus* (L.) Merrill

Chinese Name	Scientific Name
Feng Lin Cao	*Campanula glomerata* L. f. canescens (Maxim.) Kitag., *C. punctata* Lam, *C. glomerata* L. var. dahurica Fisch. ex Ker-Gawler
Feng Lun Cai	*Clinopodium umbrosum* (Bleb.) C. Koch.
Feng Wei Cao	*Pteris cretica* L., *P. ensiformis* Burmann, *P. multifida* Poir, *P. vittata* L., *P. wallichiana* Agardh.
Feng Xian Hua	*Impatiens balsamina* L., *I. noli-tangere* L., *I. textori* Miq.
Feng Yang	*Pterocarya stenoptera* DC
Fo Jia Cao	*Sedum lineare* Thunb.
Fo Jia Cao (Taiwan)	*S. formosanum* N. E. Brown
Fon Xian Chi	*Liquidambar acerifolia* Max., *L. formosana* Hance, *L. maximowiczii* Miq.
Fong Chang	*Eclipta thermalis* Bunge., *E. alba* Hassk., *E. prostrata* (L.) L., *E. marginata* Boiss.
Fong Kui	*Peucedanum japonicum* Thunb., *P. praeruptorum* L.
Foo Shao Mai	*Triticum vulgare* Vill.
Fu Ling	*Poria cocos* (Polyporaceae)
Fu Pen Zi	*Rubus coreanus* Miq., *R. matsumuranus* Levelle & Vaniot var. concolor (Kom.) Kitag., *R. saxatilis* L., *R. crataegifolius* Bunge., *R. matsumuranus* Levelle & Vaniot
Fu Ping	*Spirodela polyrhiza* Schleid.
Fu Rong Yie	*Hibiscus mutabilis* L.
Fu Shou Cao	*Adonis chrysocyathus* Hook F. & T. Thoms., *A. brevistyla* Franch., *A. vernalis* L. Adonis amuresis Regel & Radde
Fu Wei Lan	*Sansevieria trifosciate* Prain

Fu Zi	*Aconitum balfouri* Stapf., *A. praeparata*, *A. jaluense* Kom. F. glabrescene (Nakai) Kitag., *A. carmichaelii* Debeaux, *A. volubile* Pall. ex Koelle var. oligotrichum Kitag., *A. napellus* L., *A. koreanum* R. Raymund, *A. deinorrhizum* Stapf., *Aconitum chasmanthum* Stapf., *A. fischeri* Reichb.
Fuag Ji (Japanese)	*Sinomenium acutum*, *S. diversifolium* Diels.
Ga Song Xiang	*Nardostachys jatamansi* DC
Gan Cao	*Glycyrrhiza uralensis* Fisch. ex DC, *G. pallidiflora* Maxim.
Gan Feng Cao	*Zephyranthes candida* Herb.
Gan Lan	*Canarium album* Raeusch., *C. sinense* Rumph., *C. album* Raeusch., *Pimela alba* Lour.
Gan Qi	*Rhus verniciflua* Stokes
Gan Su	*Iris buatatas* (L.) Lamarck.
Gang Ban Gui	*Polygonum perfoliatum*
Gao Ben	*Ligustium jeholense* (Nakai et Kitag.) Nakai et Kitag., *L. tenuissimum* (Nakai) Kitag., *L. sinense* Oliv., *L. pyrenacum* Couan.
Gao Liang Jiang	*Alpinia officinarum* Hance
Gao Mu	*Nothosmyrnium japonicum* Miq.
Gao Shan Liao	*Oxyria digyna* (L.) Hill
Ge Cong	*Allium victorialis* L. var. platyphyllum (Hult.) Makino
Ge Gen	*Pachyrhizus thunbergianus* Sieb. et Zucc., *Pueraria thunbergiana* Benth., *P. lobata* (Willd.) Ohwi.
Giang Huo	*Notopterygium incisium* Ting
Gong Chong	*Conioselinum univittatum* Turcz.
Gong Lao Mu	*Magnolia fortunei* (Lindl.) Fedde
Gong Xian Teng	*Mallotus repandus* (Willd.) Muell.

Chinese Name	Scientific Name
Gou Gi	*Lycium barbarum* L., *L. ovatum* Loisel., *L. turbinatum* Loisel., *L. megistocarpum* Dun., *L. trewianum* G. Don, *Poncirus trifoliata* Rafin
Gou Gu	*Mahonia japonica* DC
Gou Ji Guan Zhong	*Woodwardia japonica*
Gou Ma	*Abutilon theophrasti* Malv., *A. avicennae* Gaertn. Fruct. Sem.
Gou Mei	*Myrica rubra* (Lour.) Sieb. et Zucc.
Gou Min	*Gelsemium sempervirens* (L.) Ait.
Gou Shi Cao	*Senecio campestris* (Retz.) DC
Gou Teng	*Nauclea rhynchophylla* Miq., *N. sinensis* Oliv., *Uncaria hirsuta* Havil, *U. rhynchophylla* Miq.
Gua Di	*Cucumis melo* L.
Gua Lou	*Trichosanthes kirilowii* Maxim., *T. uniflora* Hao
Guan Ye Lean Qiao	*Hypericum perforatum* L.
Guan Zhong	*Dryopteris crassirhizoma* Nakai, *D. crassirhizoma* Nakai, *Aspidium falcatum* Sw.
Guang Feng Lun Cai	*Clinopodium gracile* (Benth.) O. Kuntze.
Gui Hua	*Osmanthus fragrans* (Thunb.) Lour.
Gui Zhi	*Cinnamomum cassia* Presl., *C. aromaticum* Nees.
Guo Gang Long	*Entada phaseoloides* (L.) Merrill.
Guo Ko Yi	*Erythroxylum coca* Lam.
Guo Tan Loan	*Adiantum flabellulatum* L.
Hai Dai	*Laminaria angusta* Kjellim., *L. japonica* Aresch., *L. longipedalis* Okam., *L. religiosa* Miyabe., *L. cichorioides* Miyabe.

Hai Jin Sha Teng	*Lygodium japonicum* Swartz.
Hai Tong Pi	*Erythrina indica* Lam., *E. variegata* L.
Hai Zao	*Sargassum pallidum*
Han Cai	*Rorippa indica* (L.) Hiern., *R. islandica* (Oeder) Borbas
Han Mai Bin Cao	*Silene jenisseensis* Willd., *S. jenisseensis* Willd. f. parviflora (Turcz.) Schischk., *S. jenisseensis* Willd. f. dasyphylla (Turcz.) Schischk., *S. jenisseensis* Willd. var. viscifera Y. C. Chu, *S. jenisseensis* Willd. f. setifolia (Turcz.) Schischk., *S. jenisseensis* Willd. var. oliganthella (Nakai ex Kitag.) Y. C. Chu
Han Xiao Hua	*Michelia figo* DC
Han Xiou Cao	*Mimosa pudica* L., *M. arborea* Thunb.
Han Xiou Cao (American)	*Mimosa invisa* Mart. et Colla
He Huan Pi	*Acacia nemu* Willd. (Syn. *Albizia julibrissin*)
He Que She	*Portulaca pilosa* L.
He Shi	*Carpesium abrotanoides* L., *C. athunbergianum Torilis japonica* (Houtt) DC.
He Shou Wu	*Polygonum multifolrum* Thunb
He Ye	*Nelumbium nelumbo* Druce.
He Zi	*Terminalia chebula* Retz.
Hei Nan Pu Tao	*Syzygium cuminii* (L.) Skeels
Hei Shuo	*Melasma arvense* (Benth) Handel-Maxxetti
Hi Lu	*Anagalis arvensis* L.
Hi Tong	*Pittosporum tobira* (Thunb.) Aiton
Hie Quin Cao	*Cibotium barometz* (L.) J. Smith

Chinese Name	Scientific Name
Hin Gu Cao	*Ajuga bracteosa* Wallich, *A. decumbens*
Ho	*Nelumbo nucifera* Gaertner, Fruct et Sem.
Hong Gin Tian	*Rhodiola elongata* (Ledeb.) Fisch. & Meyer
Hong Guan Yao	*Aster ageratoides*
Hong Je Dan Hua	*Plumeria rubra* L.
Hong Jua	*Carthamus tinctorius* L.
Hong Ma Feng Shu	*Jatropha gospiifolia* L. var. delgans Muel.
Hong Mao Dan	*Nephelium lappaceum*
Hong Mei Xiao	*Rubus parvifolius* L.
Hong Men Lan	*Orchis latifolia* L.
Hong Nan	*Machilus thunbergii* Sieb. et Zucc.
Hong Pi	*Styrax suberifolus* Hook. et Arnott.
Hong Si Xian	*Solanum biflorum* Loureiro
Hong Teng	*Sargentodoxa cuneata*
Hong Tu Cao	*Blumea lacera* (Burm. f.) DC
Hong Xian Ren Dong	*Lonicera hypoglauca* Miq.
Hou Po	*Magnolia hypoleuca* Diels., *M. officinalis* Rehd. et Wils.
Hsia Ku Ts'ao	*Prunella asiatica* Nakai.
Hsii Tuan	*Lamium album* L.
Hu Chang	*Polygonum cuspidatum* Siebold & Zucc.
Hu Gu Xiao	*Sambucus formosana* Nakai

Hu Hua Pi	*Albizia julibrissin* Duraz., *A. lebbeck* (L.) Bentham
Hu Huang Lain	*Picrorhiza kurroa* Royle.
Hu Ji Shang	*Viscum album* L. subsp. coloratum Kom., *V. album* L. subsp. coloratum Kom. f. rubroaurantiacum (Makino) Kitag., *V. coloratum,* Hu Jiao *Piper nigrum* L.
Hu Li Wei	*Uraria crinita* Desvaux
Hu Lo Po	*Fsuvus carota* L.
Hu Lu Cao	*Desmodium triquetrum* (L.) DC
Hu Lu Qi Cho	*Ligularia fischeri* Ledeb.
Hu Tao Ren	*Juglans mandshurica* Maxim., *J. regia* L.
Hu Tin Chi	*Elaeagnus umbellata* Thunb., *E. pungens* Thunb.
Hu Tin Chi (Tiawan)	*E. formosana* Nakai
Hua Jiao	*Zanthoxylum schinifolium* Sieb. et Zucc.
Hua Qian Jin Teng	*Stephania sinica*
Hua Shan Seng	*Physochlaina infundibularis*
Huai Hua	*Sophora japonica* L.
Huai Niu Teng	*Achyranthes bidentata* L.
Huan Hun Cao	*Senecio cannabifolius* Lessing
Huang Bai	*Phellodendron amurense* Rupr., *P. amurense* Rupr. f. molle (Nakai) Y. C. Chu, *P. chinensis* Schneid
Huang Bai Mu	*Magnolia japonica* (Thunb.) DC
Huang Gen Cao	*Rubia akane* Nakai
Huang Ging	*Polygonatum chinense* Kunth., *P. cirrhifolium* Royle., *P. macropodium* Turez. *P. sibiricum* Delar. ex Redoute, *P. stenophyllum* Maxim.

Chinese Name	Scientific Name
Huang Gua	*Cucumis sativus* L.
Huang Hua Jia Zhu Tao	*Thevetai peruviana* (Pers.) K. Schum.
Huang Hua Xuan Cao	*Hemerocallis flava* L.
Huang Lian	*Coptis japonica* Makino, *C. teeta* Wall., *C. chinensis* Franch.
Huang Lu	*Cotinus coggygria* Scop.
Huang Ma	*Corchorus capsularis* L.
Huang Qin (Taiwan)	*Scutellaria formosana* Brown
Huang Qin	*S. baicalensis* Georgi, *S. grandiflora* Adams, *S. lanceolaria* Miq., *S. macrantha* Fisch., *S. rivulararis* Benth., *S. viscidula* Bunge
Huang Shui Jia	*Solanum incanum* L.
Huang Teng	*Fibraurea recisa, Daemonorops margaritae* (Hance) Beccari
Huang Wu Tien	*Caesalpinia pulcherrima* Swartz
Huang Yang (Taiwan)	*Buxus microophylla* Sieb. et Zucc.
Huang Yao Zi	*Dioscorea bulbifera* L.
Huang Zhi	*Astragalus complanatus* R. Fr. ex Bunge., *A. melilotoides* Pallas, *A. mongholicus* Bunge., *A. reflexistipulus* Franch., *A. hoantchy* Franch., *A. membranaceus* (Fisch.) Bunge., *A. sinensis* L., *A. henyri* Oliv.
Hui Hsien	*Chenopodium album* L.
Hui Hui Suan	*Ranunculus chinensis* Bunge.
Hui Mao Dou	*Tephrosia purpurea* Persoon
Hui Qin	*Pimpinella thellungiana* Wolff, *P. thellungiana* Wolff var. tenuisecta Chu
Huo Ma Ren	*Cannabis chinensis* Del., *C. sativa* L.

Huo Qin Hua	*Haemanthus multiflorus* Mart. ex Willd.
Huo Tan Mo Cao	*Polygonum chinensis* L.
Huo Xiang	*Pogostemon cablin* Benth., *Agastache rugosa* (Fisch. & Mey.) O. Kuntze, *A. rugosa* (Fisch. & Mey.) O. Kuntze f. hypoleuca (Maxim.) Hara, *Lophanthus chinensis* Walp., *L. rugosus* Fisch. et Mey.
Huo Yu Jin	*Euphorbia antiquorum* L.
Huong Jing	*Vitex nequndo* L.
India Bian Teng	*Flagellaria indica* L.
Indian Ren Xian	*Acalypha indica* L.
Indian Tian Qing	*Sesbinia sesbin* (L.) Merrill.
Japan Jin Fen Ju	*Onychium japonicum* (Thunb.) Kunze.
Japan Liu Shan	*Cryptotaenia japonica* Hasskarl
Japan Mu Fang Ji	*Cocculus sarmentosus* DC
Japan Mu Gua	*Chaenomeles japonica* (Thunb.) Lind.
Japan Niu Teng	*Achyranthes japonica* (Miq.) Nakai
Japan Nu Zhen	*Ligustium japonicum* Thunb.
Japan She Gen Cao	*Ophiorrhiza japonica* Blume.
Japan Su	*Castanea mollissima* Blume., *C. crenuta* Sieb. et Zucc.
Je Koo Cai	*Calloglossa lepieurii*
Je She	*Morinda citrifolia* L.
Ji Dan Hua	*Plumeria rubra* L. var. acutifolia (Poir) Ball.
Ji Guan Hua	*Celosia argentea* var. cristata Bth., *C. cristata* L.
Ji Mu	*Loropetalum chinense* (R. Br.) D. Oliver

Chinese Name	Scientific Name
Ji Xue Cao	*Centella asiatica* (L.) Urb.
Ji Xue Teng	*Milletia reticulata* Bentham, *Spatholobus suberectus*
Jia Gou Ju	*Matteuccia struthiopteris* (L.) Todaro
Jia Mu	*Aralia chinensis* L., *A. cordata* Thunb. var. continentalis (Kitag.) Y. C. Chu, *A. elata* (Miq.) Seem., *A. elata* (Miq.) Seem. F. subinermis Y. C. Chu
Jia Zhu Tao	*Nerium indicum* Mill.
Jian Xui Fuan Hou	*Antiaris toxicaris* (Pers.) Lesch.
Jian Zi Mu	*Cornus macrophylla* Wallich
Jiang Sun	*Polygonatum odoratum* (Mill.) Druce var. pluriflorum (Miq.) Ohwi f. ovarifolium Y. C. Chu, *P. vulgare* Desf.
Jiang Zhen Xiang	*Acronychia pedunculata* (L.) Miquel.
Jiao Cao	*Valeriana alternifolia* Bunge, *V. amurensis* P. Smiru. ex. Kom., *V. subbipinnatifolia* A. Baranow., *V. fauriei* Briq., *V. alternifolia* Bunge. var. stolonifera Baranov & Skv., *V. fauriei* Briq. Var. dasycarpa Hara, *V. alternifolia* Bunge var. stolonifera Baranov & Skv. F. angustifolia (Kom.) Kitag.
Jiao Wei Xiang	*Hypecoum erectum* L.
Jie Cai	*Capsella bursa-pastoris* (L.) Medicus
Jie Geng	*Platycodon grandiflorum* DC, *P. grandiflorum* DC, *P. autumnalis* Decne., *P. sinensis* Lam.
Jie Geng	*Campanula gentianoides* Lam., *C. grandiflora* Jacq., *C. giauca* Thunb.
Jie Gu Mu	*Sambucus coreana* Kom. & Klob. Alisova, *S. sieboldiana* (Miq.) Blume ex Graebner var. miquelii (Nakai) Hara, *S. manshurica* Kitag., *S. latipinna* Nakai, *S. williamsii* Hance, *S. peninsularis* Kitag.

Jin Cai — *Viola acuminata* Ledeb., *V. patrini* DC ex Ging., *V. alisoviana* Kiss, *V. alisoviana* Kiss f. intermedia (Kitag.) Takenouchi, *V. verecunda* A. Gray, *V. mandshurica* W. Becker, *V. collina* Bess., *V. alisoviana* Kiss f. candida (Kitag.) Takenouchi, *V. dissecta* Ledeb., *V. dissecta* Ledeb. f. pubescens (Regel) Kitag., *V. prionantha* Bunga

Jin Cao — *Arthraxon hispidus* (Thunb.) Makino

Jin Chuang Xian Cao — *Ajuga pygmaea* A. Gray

Jin Deng Long — *Physalis alkekengi* L. var. francheti (Mast.) Makino

Jin Gan — *Fortunella crassifolia* Swingle

Jin Gi Er — *Caragana sinica* Lam.

Jin Gi Er — *C. microphylla* Lam.

Jin Gi Er — *C. intermedia* Kuang

Jin Gi Er — *C. franchetiana* Koma

Jin Gu Cao — *Lycopodium cernum* L.

Jin He Huan — *Acacoa farnesiana* Willd.

Jin Jia Dou — *Phaseolus angularis* (Willd.) W. F. Wight, *P. radiatus* L., *P. vulgaris* L., *P. lunatus* L.

Jin Jing Zi — *Rosa acicularis* Lindl, *R. koreana* Kom., *R. davurica* Pall., *R. amygdalifolia* Ser, *R. davurica* Pall. var. alpestris (Nakai) Kitag., *R. laevigata* Michx., *R. maximowicziana* Regel

Jin Ju — *Fortunella margarita* (Lour.) Swin.

Jin Moa — *Thlaspi arvense* L.

Jin Qian Cao — *Glechoma longituba* (Nakai) Kuprijan , *G. hederacea* L. var. grandis (A. Gray) Kudo

Jin Qian Chao — *Lysimachia barystachys* Bunge., *L. clethroides* Duby, *L. christinae, L. davurica* Ledeb., *L. davurica* Ledeb. f. latifolia Korsh, *L. salicaria* L. var. glabrum Ledeb.

Jin Que Hua — *Cytisus scoparius* (L.) Link.

Chinese Name	Scientific Name
Jin Si Tao	*Hypericum attenuatum* Choisy, *H. japonicum* Thunb., *H. ascyron* L.
Jin Tsan Jiu	*Calendula officinalis* L.
Jin Xian Diao Wu Gui	*Stephania cepharantha*
Jin Yang Huo	*Epimedium koreanum* Nakai, *E. macranthum* Moore et Decne., *E. brevicorum*
Jin Yin Hua	*Lonicera brachypoda* DC
Jing Jie	*Schizonepeta multifida* (L.) Briquet, *S. tenuifolia* (Benth.) Briquet
Jing Mian Cao	*Lemmaphyllum microphyllum* Presl.
Jing Tian	*Sedum erythrostichum* Miq., *S. kamtschaticum* Fisch., *S. verticillatum* L., *S. sarmentosum* Bunge.
Jing Tian San Qi	*S. aizoon* L.
Jiu Hong	*Citrus reticulata* Blanco var. chachiensis
Jiu Hua	*Chrysanthemum koraiense* Nakai, *C. sinense* Sabine., *C. morifolium* Ramat., *C. jucundum* Nakai & Kitag., *C. boreale* (Makino) Makino
Jiu Hua Teng	*Bauhinia championi* Bentham
Jiu Jie Cha	*Chloranthus glubra* (Thunb.) Nakia
Jiu Jie Mu	*Psychotria rubra* (Lour.) Poir.
Jiu Li Xiang	*Murraya paniculata* (L.) Jack
Jiu Pi	*Citrus deliciosa* Tenore, *C. nobilis* Lour.
Joe Koo Lan	*Gynostemma pentaphyllum* (Thunb.) Makino
Jou Tsung Jung	*Orobanche caerulescens* Stephan
Ju Shi Cai	*Sonchus arvensis* L., *S. oleraceus* L.
Juan Bai	*Selaginella tamarisina* (Beauv.) Spring

Jue	Pteridium aquihum (L.) Kuhu.
Jue Ming Zi	Cassia obtusifolia L., C. nomame (Siet.) Honda
Jun Zi Lian	Clivia miniata Lindley
Knog Que Cao	Tagetes patula L.
Ko Cho Mo	Fagopyrum tataricum (L.) Gaertn.
Kong Xin Lian Zi Cao	Alternanthera philoxeroides (Mart.) Griseb.
Koo Jing Cao	Eriocaulon sieboldianum Stend.
Korean Si Zhao Hua	Cornus walteri Wangerin
Korean Yan Hu Suo	Corydalis ternata (Nakai) Nakai
Ku Cai	Sonchus oleraceus L.
Ku Dong Zi	Sophora alopecurosides
Ku Gua	Momordica charantia L.
Ku Lian Chi	Melia azedarach L.
Ku Lian Pi	Melia azedarach L.
Ku Seng	Sophora flavescens Ait.
Ku Shu	Picrasma quassioides (D. Don) Benn., P. quassioides (D. Don) Benn. f. dasycarpa Kitag.
Ku Zhi	Physalis angulata L.
Kuan Dong Hua	Tussilago farfara L.
Kuei Chen Gao	Bidens bipinnata L., B. parviflora Willd.
Kui Shu Zi	Livistona chinensis (Jacq.) R. Br. ex Mart.
Kun Bu	Laminaria angusta Kjellim., L. longipedalis Okam., L. religiosa Miyabe., L. japonica Aresch., L. cichorioides Miyabe.

Chinese Name	Scientific Name
La Lian	*Polygonum hydropiperoides* Michx.
Lai Ye Sheng Ma	*Actaea asiatica* Hara
Lang Ba Cao	*Bidens tripartita* L.
Lao Huan Cao	*Geranium eriostemon* Fisch. ex DC f. hypoleucum (Nakai) Y. C. Chu, *G. dahuricum* DC, *G. eriostemon* Fisch. ex DC, *G. wilfordi* Maxim., *G. wlassowianum* Fisch. ex Link, *G. sibiricum* L., *G. eriostemon* Fisch. ex DC f. megalanthum (Nakai) Y. C. Chu
Lao Jium Xiu	*Usnea diffracta* Dill. ex Adans.
Lei Gong Teng	*Tripterygium wilfordii* Hook. f.
Lei Wan	*Omphalia lapidescens*
Li Chi	*Litchi chinensis* Sonn.
Li Chun Hua	*Papaver rhoeas* L.
Li Lu	*Veratrum dahuricum* (Turcz.) Loes f., *V. maackii* Regel, *V. nigrum* L.
Li Lu (Taiwan)	*V. formosanum* Loesener
Li Shu	*Quercus acutissima* Carr., *Q. dentata* Thunb., *Q. variabilis* Blume, *Q. mongolica* Fisch. ex Turcz., *Q. aliena* Blume var. acutiserrata Maxim. ex Wenzig, *Q. liaotungensis* Koidz.
Li Tou Cao	*Typhonium divaricatum* (L.) Decaisne
Li Zhi Cao	*Salvia plebeia* R. Brown
Lian	*Nelumbium nuciferum* Gaertner, *N. speclosum* Willd.
Lian Qiao	*Forsythia suspensa* (Thunb.) Vahl., *Syringa dilatata* Nakai, *S. oblata* Lindley, *S. reticulata* (Blume) Hara var. mandshurica (Maxim.) Hara, *S. oblata* Lindley var. alba Hort. ex Rehd., *S. vulgaris* L., *S. suspensa* Thunb.(Syn. *Forsythia suspensa*)
Lian Sheng	*Asclepias curassavica* L.

Lian Zi Xin	*Nelumbium nelumbo* Druce.
Liang Shi	*Hyoscyamus bohemicus* F. W. Schmidt, *H. niger* L.
Liao Ge Wang	*Wikestroemia indica*
Lie Xiang Du Juan	*Rhododendron anthopogon* D. Don
Ling	*Trapa bispinosa* Roxburgh, *Trapa bicornis* Osbeck, *Trapa manshurica* Flerov.
Ling Bi Long	*Psychotria serpens* L.
Ling Lan	*Convallaria keiskei* Miq.
Ling Nan Huai	*Sophora tomatosa* L.
Ling Zhi	*Ganoderma lucidum* (Polyporaceae)
Liu Chun Yu	*Linaria vulgaris* Miller subsp. sinensis (Bebeaux) Hong
Liu Lan	*Chamaenerion angustifolium* (L.) Scop. f. pubescens (Hausskn.) Kitag., *C. angustifolium* (L.) Scop.
Liu Shen Cao	*Spilanthes acmella* L. var. oleracea Clarke
Liu Xing Zi	*Vaccaria pyramidata* Medic.
Liu Ye	*Salix babylonica* L.
	S. matsudana Koidz., *S. microstachya* Turcz. ex Trautv.
Liu Ye Cai	*Epilobium amurense* Haussku., *E. palustre* L., *E. hirsutum* L.
Lo Han Song	*Podocarpus macrophyllus* (Thunb.) D. Don
Lo Huang Zi	*Tamarindus indicus* L.
Lo Mo	*Metaplexis japonica* (Thunb.)
Lo Sheng Kui	*Hibiscus sabdariffa* L.
Loan Mao Cao	*Agrimonia eupatoria* L., *A. viscidula* Eunge., *A. pilosa* Ledeb., *A. pilosa* Ledeb. var. simplex T. Shimizu,
	A. pilosa Ledeb. var. japonica (Miq.) Nakai, *A. pilosa* Ledeb. var. viscidula (Bunge.) Kom.

Chinese Name	Scientific Name
Loan Now Xiang	*Dryobalanops aromatica* Gaertn., *D. camphora* Colebr.
Long Dan	*Gentiana squarrosa* Ledeb., *G. manshurica* Kitag., *G. algida* Pall., *G. scabra* Bunge., *G. barbata* Froel., *G. olivieri* DC, *G. triflora* Pall.
Long Kui	*Solanum nigrum* L.
Lou Lu	*Echinops grijsii* Hance, *E. gmelini* Ledeb., *E. dahuricus* Fisch., *E. sphacrocephalus* Miq.
Lour Lu	*Rhaponticum uniflorum* Ludl.
Lu Cao	*Humulus scandens* (Lour.) Merr.
Lu Er Jin	*Laggera alata* (D. Don) Schultz-Bip ex Oliver
Lu Gen	*Phragmites communis* Trin.
Lu Teng	*Milletia taiwaniana* (Matsum.) Hayata
Lu Wen	*Aloe barbadensis* Miller var. chinensis Berger, *A. vera* L.
Lu Xian	*Amaranthus lividus* L., *A. viridis* L.
Lu Xian Cao	*Pyrola decorata, P. incarnata* Fisch. ex DC, *P. rotundifolia* L., *P. japonica* Klenze ex Alefeld, *P. renifolia* Maxim.
Lu Yao	*Smilacina japonica* A. Gray
Lu Zhu	*Arundo donax* L.
Luan Hua	*Koelreuteria paniculata* Laxm.
Lun Ye Sha Seng	*Adenophora verticillata* Fisch.
Lung Kuei	*Solidago pacifica* Juzepczuk.
Luo Bu Ma	*Apocynum venetum*
Luo Fu Mu	*Rauvolfia verticilata* (Lour.) Baill.

Luo Han Guo	*Momordica grosvenori* Swingle
Luo Hua	*Arachis hypogaea* L.
Luo Hua Zi Zhu	*Callicarpa nudiflora* Hook & Arn.
Luo Shi	*Trachelospermum jasminoides* Lam., *Parechites adnascens* Hance, *P. thunbergii* A. Gray
Luo Ti Cao	*Caltha palustris* L. var. *membranacea* Turcz., *C. palustris* L. var. *sibirica* Regel
Luo Xing Fu	*Astilbe longicarpa* (Hay.) Hayata
Ma An Teng	*Iris pes-caprae* (L.) Sweet subsp. *brasiliensis* (L.) Oostst.
Ma Bian Cao	*Verbena officinalis* L., *V. oxysepalum* Turcz.
Ma Bo	*Lasiosphaera nipponica* Reichardt
Ma Chi Xian	*Portulaca oleracea* L.
Ma Dou Ling	*Aristolochia contorta* Bunge., *A. recurvilabra* Hance, *A. kaempferi* Willd., *A. longa* Thunb.
Ma Dou Ling (Taiwan)	*A. shimadai* Hayata
Ma Huang	*Ephedra distachya* L., *E. intermedia* Schrenk ex Mey., *E. sinica* Stapf, *E. monosperma* Gmel. ex. Mey., *E. equisetina* Bunge.
Ma Lan Zi	*Iris pallasii*
Ma Qian Zi	*Strychnos pierriana*
Ma Wei Lian	*Thalictrum ichangense* Lecoyer ex Oliver
Ma Xian Gao	*Pedicularis resupinata* L., *P. resupinata* L. f. *ramosa* Kom., *P. resupinata* L. f. *pubescens* Kom.
Mai Dong	*Ophioglossum japonicus* (Thunb.) Ker-Gawl.
Mai Liang Cai	*Elephantopus molis* H. B. K.
Mai Meng Dong	*Ophiopogon gracilus* Kunth., *O. longifolius* Decne., *O. spicatus* Ker-Gawl., *Draceana graminifolia* L., *Liriope graminifolia* Bak., *L. spicata* Lour., *L. platyphylla* Wang & Tang

Chinese Name	Scientific Name
Mai Ya	*Hordeum vulgare* L.
Man Jiang Hong	*Azolla imbricata* (Roxb.) Nakai
Man Shan Hong	*Rhododendron dahuricum* DC
Man Ti Xian	*Alternanthera sessilis* (L.) R. Brown
Man Tu Luo	*Datura tatula* L., *D. stramonium* L., *D. metel* L., *D. alba* Nees., *D. fastuosa* L. var. alba Clark, *D. innoxia* Mill.
Mang Guo	Mangifera indica L.
Mang Ji	*Dicranopteris linearis* (Burm. f.) Under.
Mang Jing	Miscanthus sinensis Andress.
Mao Di Huang	*Digitalia sanguinalis* (L.) Scop. var. ciliaris (Retz.) Parl., *D. purpurea* L., *D. sanguinalis* (L.) Scop.
Mao Dong Qing	*Ilex pubescens* Hook & Am.
Mao Dong San Hu	*Solanum capsicastrum* Link.
Mao Gao Cai	*Drosera burmunni* Vahl., *D. anglica* Hudson, *D. rotundifolia* L.
Mao Guo Yan Ming Cao	*Rabdosia lasiocarpus* (Hayata) Hara
Mao Liang	*Ranunculus japonicus* Thunb., *R. sarmentosa* Adams
Mao Xian	*Hierochloe odorata* (L.) Beauv.
Mao Zhua Chao	*Ranunculus ternatus* Thunb.
Mar Dong	*Ophiopogon japonicus* Wall.
Mei Deng Mu	*Maytenus serrata* (Hochst. ex A. Rich) Wilcz.
Mei Gui Hua	*Rosa rugosa* Thunb.
Mei Hua Cao	*Parnassia palustris* L.

Mei Li Cao	*Chimaphila umbellata* (L.) W. Barton
Mei She Chao	*Calloglossa lepieurii*
Mi Hou Tao	*Actinidia arguta* (Sieb. et Zucc.) Planch ex Miq., *A. japonica* Nakai, *A. kolomikta* (Maxim. ex. Rupr.) Maxim., *A. polygama* (Sieb. et Zucc.) Planch. ex Maxim. Chang, *A. chinensis*
Mi Meng Hua	*Buddleia madagascariensis* Hance, *B. officinalis* Maxim.
Mian Bao Shu	*Artocarpus altilis* (Park.) Fosberg.
Mian Hua Gen	*Gossypium herbaceum* L.
Mian Ma Guan Zhong	*Dryopteris laeta* (Kom.) Christ.
Mian Zi Soo	*Gossypium herbaceum* L.
Min Dong Seng	*Changium smyrnioides* Wolff.
Mo Han Lian	*Eclipta erecta* L.
Mo Ja Chao	*Equisetum hyemale* L., *E. ramosissimum* Desf., *E. arvense* L.
Mo Li Hua	*Jasminum samba* (L.) Aiton
Mo Yao	*Commiphora myrrha* Engler
Mo Yue	*Amorphophallus rivieri* Durieu
Moo Tune	*Akebia quinata* (Hoytt.) Decne.
Mu Dan Pi	*Paeonia obovata* Maxim., *P. suffruticosa* Andr., *P. veitchii* Lynch.
Mu Dou	*Cajanus cajan* L.
Mu Er Cao	*Gynura bicolor* DC
Mu Fang Ji	*Cocculus trilobus* (Thunb.) DC, *C. laurifolius* DC
Mu Gui	*Osmanthus fragrans* (Thunb.) Lour
Mu Jin	*Hibiscus chinensis* DC, *H. syriacus* L. *H. trionum* L.

Chinese Name	Scientific Name
Mu Jing	*Vitex chinensis* Miller, *V. jeguaod*
Mu Jing Chi	*Vitex trifolia* L. var. simplicifolia Cham.
Mu Lan	*Magnolia liliflora* Desr.
Mu Tian Liao	Actinidia polygama (Sieb. & Zucc)
Mu Tong	*Clematis heracleifolia* DC var. davidiana (Decaisue ex Verlot) O. Kuntze, *C. heracleifolia* DC, *C. armandii* Franch.
Mu Tou Hui	*Aristolochia manshuriensis* Kom.
Mu Xiang	*Patrinia heterophylla*
Mu Xu	Aucklandia costus Falc, Saussurea lappa Clarke
Mu Yu Ma	*Medicago falcata* L., *M. polymorpha* L., *M. lupulina* L., *M. sativa* L., *M. ruthenica* (L.) Ledeb.
Mu Zei	*Boehmeria densiflora* Hooker et Arnott
Mu Ziang	*Eauisetum palustre* L.
	Saussurea japonica (Thunb.) DC, *S. japonica* (Thunb.) DC var. maritima Kitag., *S. japonica* (Thunb.) DC f. alata (Chen) Kitag., *S. lapa* Clarke
Na Yang Shan	*Araucaria cunninghamii* Aitonex Sweet
Nan Gua Zi	*Cucurbita moschata* Duch. var. melonaeformis (Carr.) Makino, *C. pepo* L.
Nan He Chi	*Daucus carota* L. subsp. sative Hoffm.
Nan Tian Zhu	*Nandina domestica* Thunb.
Nan Wu Wei Zi	*Kadsura japonica* (L.) Dunal
Nan Zhu	*Lyonia ovalifolia* (Wall.) Drude.
Nao Yang Hua	*Rhododendron molle* (Bl) G. Don

Nei Don Zi	*Lindera obtusiloba* Blume f. villosa (Blume) Kitag., *L. akoensis* Hayata
Ning Meng Sian Mao	*Cymbopogon eitratus* (DC) Stapf.
Niu Bang Chi	*Arctium lappa* L., *Lappa communis* Coss et Germ., *L. edulis* Sieb., *L. minor* DC, *L. major* Gaerth.
Niu Fang Feng	*Heracleum dissectum* Ledeb.
Niu She Tou	*Sonchus arvensis* L.
Niu Xin Qie Zi	*Cerbera manghas* L.
Niu Zi Qie	*Solanum indicum* L.
Nou Me	*Oryza sativa* L.
Nu Zhen Zi	*Ligustium lucidum* Mill
Pa Jiao Lian	*Dysosma pleiantha* (Hance) Woodson
Pai chiang	*Phryma leptostachya* L.
Pai Lan (Taiwan)	*Eupatorium formosanum* L.
Pai Qian Chao	*Desmodium pulehellum*
Pai Qian Shu	*Phyllodium pulchellum* (L.) Desvaux.
Pan Chan Teng	*Cassytha filliformis* L.
Pei Lan	*Eupatorium odoratum* L.
Peng Lai Teng	*Pericamylus formosanus* Diels
Peng Wo Mao	*Curcuma phaeocoulis* Val.
Peng Zi Cao	*Galium verum* L. var. leiocarpum Ledeb., *G. verum* L. var. trachycarpum DC
Pi Jiang	*Alpinia kumatake* Mak.
Pi Pa Yie	*Eriobotrya japonica* Linkdl.
Pin Di Mu	*Ardisia japonica* (Hornst.) Blume

Chinese Name	Scientific Name
Pin Peng Cao (Japan)	*Nuphar japonicum* DC, *N. pumilum* (Timm) DC
Ping	Marsilea quadrifolia L.
Ping Er Xiao Cao	Ophioglossum thermale Kom
Po Po Na	*Veronica sibirica* L., *V. sibirica* L. f. glabra (Nakai) Kitag., *V. undulata* Wallich
Po Shu	*Celtis bungeana* Blume, *C. sinensis* Pers.
Po Ts'ai	Spinacia oleracea L.
Po Yen	*Rhus semialata* Murr.
P'u Kung Ying	Taraxacum mongolicum Hand-Mazz Monogr Tarax.
Pu Gong Ying	*Taraxacum mongolicum* Hand. Mazz., *T. sinicum* Kitag.
Pu Gong Ying (Taiwan)	Taraxacum formosanum Kitamura
Pu Gong Ying (Western)	*Taraxacum officinale* G. H. Weber ex Wigg.
Pu Huang	*Typha angustata* Bory et Chaub., *T. latifolia* L., *T. angustifolia* L., *T. orientalis* Presl., *T. minima* Hoppe, *T. davidiana* (Kronfeld) Hand. Mazz., *T. przegalskii* Skv.
Qan Qi	Rhus vernicipiua Strokes
Qi Zhou Yi Zhi Hao	Equisetum arvense L., Equisetum Canadensis L.
Qian Cao	*Rubia chinensis* Regel & Maack, *R. cordifolia* Thunb., *R. cordifolia* L. f. pratensis (Maxim.) Kitag., *R. mungista* Roxb., *R. sylvatica* (Maxim.) Nakai
Qian Hu	*Angelica decursiva* (Miq.) Franch. et Savat., *A. pubescens* Maxim., *A. grosserrata* Maxim.
Qian Hu	Peucedanum decursivum Max.
Qian Hu (Taiwan)	Peucedanum formosanum Hayata
Qian Jin Teng	*Stephania japonica* (Thunb.) Miers.

Qian Jin Zi	*Euphorbia kansui*
Qian Li Guang	*Senecio argunensis* Turcz., *S. scandens* Buch-Ham ex D. Don, *S. nemorensis* L.
Qian Li Guang (European)	*S. vulgaris* L.
Qian Qu Cai	*Lythrum salicaria* L.
Qian Ri Hong	*Gomphrena globosa* L.
Qian Shi	*Euryale ferox* Salisb.
Qiang Cao	*Parietaria micrantha* Ledeb.
Qin Cai	*Apium graveolens* L.
Qin Jiu	*Justicia gendarussa* L., *J. procumbens* L., *Gentiana dahurica* Fisch., *G. macrophylla* Pall., *G. lutea* L.
Qing Feng Teng	*Sinomenium acutum* var. cinereum
Qing Guo	*Artemisia annua* L., *A. apiacea* Hance ex Walpers
Qing Mu Xiang	*Aristolochia debilis* Sieb. et. Zucc.
Qing Ping	*Lemna minor* L., *L. perpusilla* Torrey
Qing Ye Dan	*Swertia diluta* (Turcz.) Benth. et Hook. f., *S. mileensis* L.
Qiong Zhi	*Gelidium amansii*
Qiu Jiang	*Zingiber zerumbet* Smith
Qu Mai	*Dianthus superbus, D. barbatus* L. var. asiaticus Nakai, *D. oreadum* Hance
Quan Li	*Senecio scandens* Buch-hami
Quan Yuan Guan Zhong	*Cytomium falcatum* (L. f.) Presel.
Quao Ye Ging Lan	*Dracocephalum integrifolium* L.
Quian Niu Zi	*Pharbitis diversifolia* Lindl., *P. triloba* Miq., *P. nil, P. hederacea* Choisy

Chinese Name	Scientific Name
Quian Niu	*Ipomoea barbata* Both., *I. caerulea* Koeh., *I. hederacea* Jacq., *I. triloba* Thunb
Quin Pi	*Fraxinus obovata* Blume, *F. rhynchophylla* Hance, *F. chinensis* Roxb., *F. floribunda* Bunge., *F. ornus* L. var. *bungeana* Hance, *F. bungeana* DC
Ren Dong	*Lonicera chinensis* Wats., *L. japonica* Thunb., *L. flexuosa* Thunb., *L. maackii* (Rupr.) Maxim., *L. japonica* var. chinensis Bak., *L. confusa* Miq.
Ren Seng	*Panax ginseng* C. A. Meyer
Ri Jian Ca Sha Jiang Cao	*Oenathera odorata* Jacq.
Ron Yen Raw	*Nephelium longana* Camb.
Rong Cai	*Iris aqyatuca* Forskal
Rong Shu	*Ficus inicrocarpa* L.
Rou Dau Kou	*Myristica fragrans* Houtt.
Ru Xiang	*Pistacia lentiscus* L.
Rui Ye Ren Dong	*Lonicera acuminata* Wallich
San Bai Cao	*Saururus chinensis* (Lour.) Baillon
San Dian Jin Cao	*Desmodium triforum* (L.) DC
San Hai Ton	*Tripterygium hypoglaucum* (Levl.) Hutch.
San Hu Ci Tong	*Erythrina corallodendron* L.
San Hu Shu	*Datura suaveolens* Humb. & Bonpl. ex Willd.
San Hu You Tong	*Jatropha podagrica* Hooker
San Jian Shan	Cephalotaxus fortunei

San Long Zhi	*Scopolia tangutica* Max.
San Qi	*Gynura japonica* Mak., *G. pinnatifida* Vanniot, *G. segetum* Merr.
San Se Xian	*Panax zingiberensis* C. Y. Wu & K. M. Feng
San Ya Ko	*Amaranthus tricolor* L.
San Ye Wu Jia	*Evodia lepta* (Spreng.) Merrill., *E. triphylla* DC
Sang Gen Bai Pi	*Acanthopanax trifoliatus* (L.) Merr.
Sang Zhi	*Morus alba* L., *M. constantinopolitana* Poir., *M. indica* L.
Se Gua	*M. alba* L, *M. constantinopolitana* Poir., *M. indica* L.
	Luffa faetida Sieb. et Zucc., *L. aegyptiaca* Mill, *L. petola* Ser., *L. cylindrica* Roem., *Momordica cylindrica* L.
Sha Cao	*Cyperus difformis* L., *C. glomeratus* L., *C. iria* L.
Sha Cha Hua	*Camellia japonica* L.
Sha Gen Cai	*Lobelia pyramidalis* Wallich., *L. sessilifolia* Lambert
Sha Gui Hua	*Maesa perlarius* (Lour.) Merrill.
Sha Gui Hua (Taiwan)	*Maesa tenera* Mez.
Sha Hong Fan Cao	*Blumea riparia* (Blume) DC var. megacephala Randeria
Sha Ji	*Hippophae rhamnoides* L.
Sha Jiang Cao	*Oxalis corriculaza* L., *O. corymbosa* DC
Sha Ma	*Corchorus olitorius* L.
Sha Ren	*Hedychium coronarium* Koen., *Amomum tsao-ko*, *A. villosum*

Chinese Name	Scientific Name
Sha Seng	*Adenophora coronopifolia* Fisch., *A. pereskiaefolia* (Fisch.) G. Don, *A. paniculata* Nannf., *A. tetraphylla*, *A. remotiflora* (Sieb. et Zucc.) Miq., *A. polymorpha* Ledeb., *A. stenanthina* (Ledeb.) Kitag.
Sha Yuan Zi	*Astragalus chinensis* L. fil.
Sha Zhu Yu	*Macrocarpium officinalis* (Sieb. et Zucc.) Nakai
Shan	*Adamia chinensis* Gard. et Champ.
Shan Ci Ko	*Iphigenia indica* (Syn. *Tulipa edulis*), *Tulipa edulis* Bak.
Shan Cong Zi	*Litsea cubeba*
Shan Dou Gen	*Menispermum dauricum* L., *Euchresta japonicum* Benth., *Sophora subprostrata*
Shan Ge	*Pueraria montana* (Lour.) Merrill., *P. thunbergiana* Benth.
Shan Guo	*Artemisia brachyloba* Franch.
Shan Hua Jiao	*Zanthoxylum bungeanum* Maxim.
Shan Jian Chai	*Cacalia hastate* L.
Shan Jiang	*Alpinia speciosa* K. Schum.
Shan Jiu	*Sapium discolor* Mueller-Arg.
Shan Liu Jiu	*Hieracium umbellatum* L.
Shan Ma Zi	*Ribes mandshurica* (Maxim.) Kom., *R. mandshurica* (Maxim.) Kom. f. subglabrum (Kom.) Kitag.
Shan Na	*Kaempferia galanga* L.
Shan Pu Tao	*Vitis amurensis* Rupr., *V. vinifera* L.
Shan Teng Zi	*Viburnum sargenti* Koehne f. glabrum Kom., *V. sargenti* Koehne var. puberulum (Kom.) Kitag., *V. sargenti* Koehne f. intermedium (Kom.) Kitag.
Shan Tou	*Cajanus indicus* L.

Shan Wo Ju	*Lactuca raddeana* Maxim., *L. indica* L.
Shan Yan Cao	*Solanum verbascifolium* L.
Shan Yao	*Dioscorea opposita* Thunb.
Shan Ye Man Shi Song	*Lycopodium annotinum* L.
Shan Ye Shi Song	*Lycopus fargesii* Herter
Shan Zha	*Crataegus pinnatifida* Bunge., *C. maximowiczii* Schneid., *C. dahurica* Koehne ex Schneid., *C. chlorusarca* Maxim., *C. pentagyna* Waldst. et Kit., *C. sanguinea* Pall., *C. cuneata* Sieb. et Zucc.
Shan Zhi	*Gardenia angusta* (L.) Merrill.
Shan Zhi Ma	*Oenathera biennis* L.
Shan Zhu Yu	*Cornus officinalis* Sieb. et Zucc.
Shang Han Cao	*Vernonia cinerea* (L.) Less.
Shang Lu	*Phytolacca acinosa* Roxb., *P. kaempferi* A. Gray, *P. octandra* Bge., *P. pekinensis* Hance, *P. japonica* Makino, *P. americana* L.
Shao Ci Wu Jia	*Acanthopanax senticosus* (Rupr. et Maxim.) var. subinermis (Regel) Kitag.
Shao Lan	*Cypripedium macranthum* Swartz f. albiflorum Y. C. Chu, *C. macranthum* Swartz, *C. guttatum* Swartz
She Cheung Zi	*Cnidium monnieri* (L.) Cusson
She Gan	*Belamcanda chinensis* (L.) DC, *B. panctata* Moench., *Iris dichotoma* Pallas
She Ma	*Humulus lupulus* L.
She Mei	*Fragaria indica* Andr., *Duchesnea indica* (Andr.) Focke.
Shen Jin Cao	*Lycopodium clavatum* L. var. nipponicum Nakai, *L. obscurum* L., *L. selago* L., *L. serratum* Thunb.
Shen Liu	*Tamarix juniperina* Bunge.
Sheng Hong Yu	*Ageratum conyzoides* L., *A. houstonianum* Mill

Chinese Name	Scientific Name
Sheng Hong Ji	Agerodum conzoides L.
Sheng Jiang	Zingiber officinale Roscoe
Sheng Ma	Cimicifuga foetida L., C. dahurica (Turcz.) Maxim., C. heracleifolia Kom., C. racemosa (L.) Nutt., C. ussuriensis Oettingen
Sheng Teng	Calamus margaritae Hance
Shi Cao	Achillea alpina L., A. millefolium L.
Shi Da Chuan	Oldenlandia chrysotricha L.
Shi Diao Lan	Lysionotus pauciflorus G. Don
Shi Dou	Dendrobium nobile Lindl, Epidendrum monile Thunb.
Shi Hong Hua	Crocus sativus L.
Shi Ji Qing	Ilex chinensis Sims
Shi Jian Chuan	Salvia chinensis, S. pogonocalyx Hance, S. przewalskii
Shi Jiun Zi	Quisqualis grandiflora Miq., Q. pubescens Burm., Q. longifolia Presl., Q. indica L., Q. sinensis Lindl., Q. loureiri G. Don
Shi Juan Bai	Selaginella involvens (Sw.) Sprengel
Shi Li	Aleurites moluceanu (L.) Willd.
Shi Liu Pi	Punica granatum L.
Shi Long Nei	Ranunculus sceleratus L.
Shi Luo Zi	Anethum graveoleus L.
Shi Mu Zi	Urtica laetevirens Maxim.
Shi Nan Ye	Photinia serrulata Lindl.

Shi Shan	*Hyperzia serrata* (Thunb.) Trev.
Shi Shang Bai	*Selaginella doederleinii* Heironyus
Shi Sheng Yu	*Bistorta lapidosa* Kitag, *Polygonum lapidosum* Kitag.
Shi Song	*Lycopus lucidus* Turcz., *L. ramosissimus* (Makino) Makino var. japonicus (Matsum et Kudo) Kitam., *L. parviflorus* Maxim., *L. lucidus* Turcz. f. hirtus (Regel) Kitag., *L. maackianus* (Maxim.) Makino
Shi Suan	*Lycoris radiata* (L'Her.) Herb., *L. longituba* Y. Han et Fan, *L. aura* (L'Her.) Herb.
Shi Suan Hua	*Hippeastrum hybridum* Hortorum
Shi Wei	*Pyrrosia lingua* (Thunb.) Farwell, *P. sheareri* (Baker) Ching, *P. petiolosa*
Shi Wu Tou	*Centipeda minima* (L.) A. Braun. et Ascherison
Shi Xiiang Rou	*Mosla chinensis* Maxim
Shi Zhu Yu	*Zanthoxylum ailanthoides* Sieb. et Zucc.
Shi Zi	*Diospyros lotus* L., *D. kaki* L., *D. roxburgii* Carr., *D. chinensis* Blume, *D. costata* Carr.
Shih nan	*Rhododendron miscranthum* Turcz.
Shin Chu	*Dianthus chinensis* L.
Shin Ts' as	*Carex kobomug* Ohwi.
Shong Jie Fong	*Sarcandra glabra*
Shou Jao Sun	*Gymnadenia conopsea* (L.) R. Brown.
Shu Gi	*Ardisia sieboldii* Miq.
Shu Li	*Rhamnus davurica* Pall, *R. parvifolia* Bunge., *R. davurica* Pall. var. nipponica Makino
Shu Liang	*Dioscorea cirrhosa*, *D. japonica* Thunb., *D. hispida* Dennst.
Shu Long	*Pyrrosia adnascens* (Sw.) Ching

Chinese Name	Scientific Name
Shu Qu Cao	*Gnaphalium affine* L., *G. multiceps* Wall., *G. confusum* DC, *G. luteo-album* L. var. multiceps Hook, *G. arenarium* Thunb., *G. ramigerum* DC, *G. javanum* DC, *G. uliginosum* L., *G. tranzschelii* Kirpicznikov
Shu Shu	*Manihot esculenta* Crantz.
Shu Yu	*Dioscorea batatus* Decaisue
Shuang Mian Ci	*Zanthoxylum nitidum* (Roxb.) DC
Shui Cai	*Menyanthes trifoliata* L.
Shui Ding Xiang	*Ludwigia* prostrate Roxy
Shui Fei Ji	*Silybum marianum* (L.) Gaertn.
Shui Gui Jiao	*Hymenocallis speciosa* Salisbury
Shui Hong Cao	*Polygonum orientale* L.
Shui Huang Pi	*Pongamia pinnata* (L.) Pierre ex Merrill.
Shui Jin	*Oenathera javanica* (Bl) DC
Shui Ku Shi	*Veronica anagallis-aquatica* L., *V. anagallis-aquatica* L. f. pumila Kitag.
Shui Lian	*Nymphaea tetragona* Georgi, *N. tetragona* Georgi var. crasifolia (Hand. Mazz.) Y. C. Chu
Shui Man Chin	*Veronica linariaefolia* Pall. ex Link, *V. linariaefolia* Pall. ex Link subsp. dilatata (Nakai et Kitag.) Hong
Shui Qin	*Oenanthe javanica* Blume
Shui Shai	*Narcissus tazetta* L. var. Chinensis Roem.
Shui Shai Gen	*Narcissus tazetta* L. var. Chinensis Roem.
Shui Shai Gen	*Polyanthus narcissus*
Shui Su	*Stachys chinensis* Bunge. ex Benth., *S. japonica* Miq., *S. baicalensis* Fisch. ex Benth., *S. baicalensis* Fisch. ex Benth. var. angustifolia Honda

Shui Tuan Hua	*Adina ratemosa* (Sieb. et Zucc.) Miquel
Shui Xian Cao	*Hedyotis corymbosa* (L.) Lamarck.
Shui Yang Mei	*Geum aleppicum* Jacquin f. glabricaule (Juzepczuk) Kitag., *G. aleppicum* Jacquin
Shui Yang Mei Gen	*Adina rubella* Hance
Shui Yu	*Sorbus alnifolia* (Sieb. & Zucc.) K. Koch, *S. amurensis* Koehne, *S. alnifolia* (Sieb. & Zucc.) K. Koch var. lobulata Rehd, *S. pohuashanensis* (Hance) Hedl. var. manshuriensis (Kitag.) Y. C. Chu.
Shun	*Cunninghamia lanceolata* (Lamb.) Hooker
Si Gua	*Citrullus edulis* Spach., *C. anguria* Duch., *C. lanatus* Matsumura & Nakai, *C. vulgaris* Schrad.
Si Yang Bai Hua Cai	*Cleome spinosa* Jacquin
Si Yang Seng	*Panax quinquefolium* L.
Si Ye Huang Yang	*Buxus harlandii* Hance
Si Ye Lian	*Chloranthus oldhami* Solms.
Si Ye Lu	*Galium bungei* Stead.
Si Zhao Hua	*Cornus kousa* Hance, *C. alba* L.
Siang Si Zi	*Abrus precatorius* L.
Sien Feng Cao	*Bidens pilosa* L. var. minor (Blume) Sherff.
Song Ji Shang	*Loranthus parasiticus, L. yadoriki* Sieb. et Zucc.
Song Lo	*Usnea longissima* Acharius
Song Ta	*Pinus bungeana* Zucc. ex Endl., *P. koraiensis* Sieb. et Zucc., *P. tabulaeformis* Carr., *P. densiflora* Sieb. et Zucc., *P. sylvestris* L. var. mongolica Litv., *P. sylvestris* L. var. sylvestriformis (Takenouchi) Cheng et C. D. Chu
Song Ye Mo	*Portulaca grandiflora* Hooker

Chinese Name	Scientific Name
Su Cao	*Polygala japonica* Houtt., *P. sibirica* L., *P. tatarinowii* Regel.
Su Mu	*Caesalpinia sappan* L.
Su Yang	*Cynomorium songarium* L., *C. coccineum* L.
Suan Cheng	*Citrus aurantium* (Christm.) Swingle var. amara
Suan Mo	*Rumex acetosa* L., *R. marschallianus* Rehb., *R. acetosella* L., *R. aquaticus* L., *R. gmelini* Turcz., *R. longifolius* DC, *R. maritimus* L., *R. stenophyllus* Ledeb. var. ussuriensis (A. Los.) Kitag., *R. amurensis* Fr. Schm., *R. thyrsiflorus* Fingerh.
Suan Zao	*Ziziphus jujuba* Mill., *Z. spinosa*
Suan Zao Ren	*Z. jujuba* Mill.
Sui Me Chai	*Desmodium microphyllum* (Thunb.) DC
Sui Mi Jie	*Cardamine lyrata* Bunge, *C. leucantha* (Tausch.) O. E. Schulz.
Sun Cha	*Clerodendrum spicatus* (Thunb.) C. Y. Wu
Sung	*Pinus madshurica* Rapr.
Suo Lou Zi	*Aesculus chinensis* L., *A. hippocastanum* L.
Tai Huang	*Rheum officinale* Baill., *R. undulatum* L., *R. palmatum* L.
Tai Zi Shen	*Pseudostellaria heterophylla* (Miq.)
Tan Seng	*Salvia miltiorhiza* Bunge
Tan Xian	*Santalum album* L., *S. myrtifolium* Roxb, *S. verum* L.
Tang Jie	*Erysimum cheiranthoides* L., *E. amurense* Kitag. var. bungei (Kitag.) Kitag.

Tang Song Cao	*Thalictrum aquilegifolium* L. var. sibiricum Regel & Tiling, *T. fauriel* Hayata, *T. glandulissimum, T. petaloideum* L., *T. baicalense* Turcz., *T. petaloideum* L. var. supradecompositum (Nakai) Kitag., *T. squarrosum* Steph. ex Willd., *T. simplex* L. var. brevipes Hara, *T. simplex* L. var. affine (Ledeb.) Regel, *T. simplex* L., *T. thunbergii* DC, *T. baicalense* Turcz. f. levicarpum Tamura
Tang Song Cao (Taiwan)	*Thalictrum foetidum* L.
Ten Min Qing	*Carpesium abrotanoides* L., *C. athunbergianum* Sieb. et Zucc.
Teng Hu Tin Chi	*Elaeagnus glabra* Thunb.
Tian Bao Cao	*Hypericum sumpsonii* Hance
Tian Cai	*Artemisia lactiflora* Wallich
Tian Hua Fen	*Trichosanthes kirilowii* Maxim.
Tian Ja Cai	*Elephantopus scaber* L.
Tian Jiu	*Stevia rebaudiana* (Bertoni) Hemsl.
Tian Kui Zi	*Semiaquilegia adoxoides* (DC) Mak.
Tian Ma	*Gastrodia elata* Blume f. pallens (Kitaig.) Tuyama, *G. elata* Blume
Tian Men Dong	*Asparagus cochinenesis* (Lour.) Merr., *A. falcatus* Benth, *A. insularis* Hance, *A. lucidus* Lindl., *A. officinalis* L.
Tian Ming Jing	*Carpesium abrotanoides* L.
Tian Nan Xing	*Arisaema amurense* Maxim., *A. peninsulae* Y. C. Chu et D. C. Wu, *A. peninsulae* Nakai, *A. heterophyllum* Blume, *A. erubescens* (Wall.) Schott., *A. consanguineum, A. amurense* Maxim. f. purpureum (Nakai) Kitag., *A. amurense* Maxim. f. serratum (Nakai) Kitag., *A. thunbergii* Blume, *A. amurense* Maxim. f. violaceum (Engler) Kitag.
Tian Peng Cao	*Stellaria alsine* Grimm var. undulata (Thunb.) Ohwi

Chinese Name	Scientific Name
Tian Qi	*Panax notoginseng* (Burk) F. H. Chen
Tian Qing	*Sesbinia javanica* (L.) Persoon
Tian Xuan Hua	*Convolvulus arvensis* L.
Tian We Cao	*Spilanthes acmella* (L.) Murray
Tian Zhu Cao	*Scoparia dulcis* L.
Tie Dao Mu	*Cassia siamea* Lamark
Tie Shu	*Cycas revoluta* Thunb.
Tie Xian Cai	*Acacoa australis*
Tie Xian Cao	*Cynodon dactylon* (L.) Persoon
Tie Xian Jiu	*Adiantum boreale* Presl., *A. pedatum* L., *A. capillus-junonis* Rupr.
Tie Xian Lian	*Clematis intriicata* Bunge., *C. mandshurica* Rupr.
Ting Li Zi	*Draba nemorosa* L.
Tong Guan Teng	*Marsdenia tenacissima*
Tong Pi	*Paulownia tomefosa* Thunb.
Tou Gu Cao	*Impatiens balsamina* L., *I. noli-tangere* L., *I. textori* Miq.
Tou Ren	*Prunus persica* (L.) Batsch.
Tu Chiian	*Rhododendron aurvum* Georgi
Tu Er Cao	*Blumea hieracifolia* (D. Don) DC
Tu Fang Ji	*Paracyclea insularis* Kudo et Yamamoto
Tu Fang Ji (Taiwan)	*Paracyclea ochiaiana* Kudo et Yamamoto

Tu Gu Ling	*Smilax china* L., *S. riparia* DC subsp. ussuriensi (Regel) Kitag, *S. sieboldii* Miq., *S. nipponica* Miq. subsp. manshurica Kitag.
Tu Hung Hua	*Callicarpa formosana* Rolfe.
Tu Mu Xiang	Inula helenium L.
Tu Niu Teng	*Achyranthes asperia* L. var. indica L.
Tu Ren Shen	*Taliunm triangulare* Willd.
Tu Si Zi	*Cuscuta chinensis* Lam., *C. lupuliformis* Krocker, *C. japonica* Choisy. *C. europaea* L.
Tu Soon	*Juniperus rigida* Sieb. et Zucc., *J. rigida* Sieb. et Zucc. f. modesta (Nakai) Y. C. Chu
Tu Tai Huang	*Rumex patientia* L. var. callosus Fr. Schm.
Tu Wei Cao	*Uraria lagopodiodes* (L.) Dexvaux
Tu Xian	*Ledum palustre* L. subsp. decumbens (Aiton) Hulten
Tzu Su	Perilla frutescens (L.) Britt.
Wa Song	Orostachys fimbriatus Turcz.
Wan Shou Jiu	*Tagetes patula* L.
Wang Bu Liu Xing	*Saponaria officinalis* L., *S. vaccaria* L., *Vaccaria segetalis* (Neck.) Garcke
Wang Jiang Nan	*Cassia tora* L., *C. occidentalis* L.
Wei Ling Cai	*Potentilla bifurca* L. var. canescens Bong. et Mey., *P. bifurca* L. var. glabrata Lehm., *P. kleiniana* Wight & Arnott var. robusta (Franch. & Savat.) Kitag., *P. fragariodes* L. var. major Maxim., *P. discolor* Bunge., *P. fragariodes* L., *P. freyaiana* Bormmuller, *P. chinensis* Seringe
Wei Ling Xian	*Clematis florida* Thunb., *C. sinensis* Lour, *C. hexapetala* Pall. f. longiloba (Freyn) S. H. Li et Y. H. Huang, *C. minor* Lour., *C. terniflora* DC, *C. chinensis* Retz., *C. hexapetala* Pall.
Wei Mao	*Evonymus alatus* Regel

Chinese Name	Scientific Name
	E. alatus (Thunb.) Sieb., *E. maackii* Rupr., *E. bungeanus* Maxim., *E. alatus* Regel, *E. alatus* (Thunb.) Sieb. var. apterus Regel, *E. thunbergianus* Blume, *E. subtriflorus* Blume, *Celastrus alatus* Thunb., *C. striatus* Thunb.
Wei Sui Xian	*Amaranthus caudatus* L.
Wei Yan Xian	*Caulophyllum robustum* Maxim.
Wen Jing	Equisetum arvense L.
Wo Seng	*Anthriscus aemula* (Woron.) Schischk., *A. aemula* (Woron.) Schischk. f. hirtifructa (Ohwi) Kitag.
Wo Zu	*Curcuma zedoaria* (Christ.) Roscoe
Won Nian Qing	*Rhodea japonica*
Wu An	*Coriandrum sativum* L.
Wu Bei Zi	*Rhus chinensis* Mill., *R. cotinus* L., *R. javanica* L., *R. osbeckii* Decne.
Wu Fan Shu	*Vaccinium bracteatum* Thunb., *V. vitis-idaea* L.
Wu Gan	*Iris lactea* Pall. subsp. chinensis (Fisch.) Kitag.
Wu Gong	*Scolopendrium subspinipes*, *Phyllitis scolopendrium* (L.) Newm., *P. scolopendrium* (L.) Newm.
Wu Hua Go	*Ficus carica* L.
Wu Huan Shu	*Sapindus mukorossi* Gaertner
Wu Jia Pi	*Acanthopanax gracilistylus*, *A. spinosum* Miq., Eleutherococcus senticosus Maxim.
Wu Jiu	*Sapium sebiferum* (L.) Roxb.
Wu Ju	*Sphenomeris chusana* (L.) Copel.
Wu Ma	Sessamum indicum L.
Wu Mai	*Prunus mume* (Sieb.) Sieb. et Zucc.

Wu Ru Ba	*Trigonella foenum-graecum* L.
Wu Song Ju	*Pueraria vittata* L.
Wu Tao	*Aconitum balfouri* Stapf., *A. koreanum* R. Raymund, *A. volubile* Pall. ex Koelle var. oligotrichum Kitag., *A. carmichaeli* Debeaux, *A. praeparata*, *A. jaluense* Kom. F. glabrescene (Nakai) Kitag., *A. fischeri* Reichb., *A. deimorrhizum* Stapf., *A. chasmanthum* Stapf., *A. napellus* L.
Wu Tong	*Firmiana simplex* (L.) W. F. Wight
Wu Wei Zi	*Schisandra chinensis* (Turcz.) Baill.
Wu Wei Zi (Taiwan)	*S. arisanensis* Hayata
Wu Yao	*Daphnidium myrrha* Sieb. et Zucc., *D. strychnifolius* Sieb. et Zucc., *Lindera strychnifolia* Vill.
Wu Zhao Jin Long	*Ipomoea cairica* (L.) Sweet
Wu Zhu Yu	*Evodia rutaecarpa* (Juss.) Berth
Xi Sheng Teng	*Cissampelos pareira* L.
Xi Shin	*Hepatica asiatica* Nakai
Xi Shu	*Camptotheca acuminata* Decne.
Xi Xian	*Soegesbecloa orientallis* L.
Xi Xin	*Asarum canaderse* L., *A. terotropoides* Fr. Schmidt var. mandshuricum (Maxim.) Kitag., *A. sieboldii* Miq., *A. heterotropoides* Fr. Schmidt var. seouleuse (Nakai) Kitag.
Xi Ye Sha Seng	*Wahlenbergia marginata* (Thunb.) A. DC
Xi Ye Zhu Chi Cao	*Phyllanthus virgatus* Forster
Xia Ku Chao	*Prunella vulgaris* L.
Xia Tian Wu	*Corydalis decurbens* (Thunb.) Pers.
Xian	*Armeniaca ansu* (Maxim.) Kostina, *A. mandshurica* (Maxim.) Skvortzov, *A. sibirica* (L.) Lam., *A. vulgaris* Lam.

Chinese Name	Scientific Name
Xian He Cao	*Agrimonia eupatoria* L., *A. pilosa* Ledeb., *A. pilosa* Ledeb. var. *japonica* (Miq.) Nakai, *A. pilosa* Ledeb. var. *viscidula* (Bunge.) Kom., *A. viscidula* Bunge.
Xian Huang Lian	*Jeffersonia dubia* (Maxim) Benth et Hook f.
Xian Mao	*Curculigo stans* Labill., *C. ensifolia* R. Br., *C. malabarica* Labill., *C. orchiodes* gaertn., *Cymbopogon nardus* Rendle
Xian Xia Hua	*Vernonia patula* (Ait.) Merr.
Xian Ye Chiu	*Filifolium sibiricum* (L.) Kitam
Xian Ye Shu	*Lindera communis* Hemsley
Xiang Fu	*Cyperus rotundus* L.
Xiang Jia Pi	*Periploca sepium* Bunge.
Xiang Jiao	*Musa paradisiaca* L. var. *sapientum* O. Ktze.
Xiang Ri Kui Zi	*Helianthus annuus* L.
Xiang Si Shu	*Acacia confusa* Merrili
Xiang Ru	*Elshoitzia cihate* (hunb) Hylander
Xiang Tian Huang	*Cleome viscosa* L.
Xiang Xu	*Elsholtzia souliei* Lev., *E. feddei* Lev., *E. cristata* Willd., *E. argyi* Lev., *Hyssopus ocymifolius* Lam., *Perilla frutescens* (L.) Britt., *P. polystachya* D. Don (Syn. *Elsholtzia cristata*), *P. ocymoides* L. var. *crispa* Benth., *P. ocymoides* L.
Xiang Pi Mu	*Alstonia scholaris* (L.) R. Br.
Xiao Bo	*Berberis amurensis* Rupr.
Xiao Fei Yang Cao	*Euphorbia thymifolia* L.
Xiao Hui Xiang	*Foeniculum vulgare* Mill., *F. officinale* All.

Xiao Ji	*Cirsium japonicum* DC, *C. setosum* (Willd.) Bieb, *C. littorale* Max., *C. albescens* Kitamura, *C. segetum* Bunge., *C. vlassovianum* Fisch. ex DC, *C. maakii* Max., *C. brevicaule* A. Grey, *Carduus acaulis* Thunb., *C. japonicus* Franch., *C. crispus* L., *Cephalanoplos segetum*
Xiao Shan Ju	*Glycosmis cochinchinensis* Pierre
Xiao Ye Yang Jiao Teng	*Morinda parvifolia* Bartling
Xiao Yeh	*Berberis amurensis* Rupr., *B. sibirica* Pall., *B. poiretii* Schneid, *B. soulieana* C. K. Schneid
Xiao Ying Qie	*Solanum aculeatissimum* Jacquin
Xin Ye Chiu	*Filifolium sibiricum* (L.) Kitam
Xin Yi	*Magnolia biloba* Cheng, *M. discolor* Vent.. *M. denudata* Desr., *M. purpurea* Curt.
Xing Cai	*Nymphoides peltate.* S. G. Gmelin
Xing Ren	*Prunus armeniaca* L.
Xiu Qiu	*Hydrangea macrophylla* (Thunb.) Seringe
Xiu Xian Jiu	*Spiraea salicifolia* L., *S. salicifolia* L. var. grosseserrata Liou & Liou fil., *S. salicifolia* L. var. oligodonta Yu
Xu Chang Qing	*Cynanchum paniculatum* L.
Xu Duan	*Dipsacus asper* Wall.
Xu Sui Zi	*Euphorbia lathyrus* L., *E. lucorum* Rupr.
Xuan Caogen	*Hemerocallis minor* Miller
Xuan Fu Hua	*Inula britannica* L., *I. japonica* Thunb., *I. salsoloides* (Turcz.) Ostenfeld, *I. linariaefolia* Turcz., *I. linariaefolia* Turcz. f. simplex Kom.
Xuan Mu Gua	*Chaenomeles speciosa* (Sweet) Nakai, *C. sinensis* Koch., *Cydonia sinensis* Thou.
Xuan Seng	*Scrophularia buergeriana* Miq., *S. puergeriana* Miq., *S. kakudensis* Franch var. latisepala (Kitag.) Kitag., *S. oldhami* Oliv., *S. ningpoensis* Hemsl.

Chinese Name	Scientific Name
Xue Jian Chou	*Aquilegia buergeriana* Sieb. et Zucc. f. pallidiflora (Nakai) Kitab., *A. parviflora* Ledeb., *A. buergeriana* Sieb. et Zucc. var. oxysepala (Trautv. Et Mey.) Kitam.
Xue Jie	*Daemonorops draco* Blume.
Xue Shang Yi Zhi Hao	*Aconitum barbatum* Persoon
Ya Dan Zi	*Brucea javanica* (L.) Merrill, *B. sumatrana* Roxb.
Ya Er Qin	*Cryptotaenia canadensis* (L.) DC
Ya Ma	*Linum stelleroides* Planch., *L. usitatissimum* L., *Commelina communis* L.
Yan Cao	*Nicotiana tabacum* L.
Yan Fu Zi	*Rhus chinensis* Mill
Yan Hu Suo	*Corydalis yanhusuo* W. T. Wang ex Z. Y. Su et C. Y. Wu, *C. turtschaninovii* Besser Bess. f. yanhusa, *C. incisa* (Thunb.) Pers., *C. repens* Mandl. et Muehld. var. watnabei (Kitag.) Y. C. Chu, *C. ambigua* Cham. et Schlecht. var. amurensis Maxim.
Yan Jie Cao	*Ophiopogon japonicus* Wall.
Yan Lai Hong	*Amaranthus tricolor* L.
Yang Gan Jiu	*Matricaria chamomilla* L.
Yang Guo Nau	*Strophanthus divaricatus* (Lour.) Hook & Arn.
Yang Lu	*Codonopsis lanceolata* (Sieb. et Zucc.) Trautv.
Yang Lu Kui	*Anredera cordifolia* (Tenore) Van Steen
Yang Shi Cao	*Achillea millefolium* L.
Yang Ti Cao	*Emilia sonchifolia* (L.) D.C.
Yang Ti Gen	*Rumex crispus* L., *R. japonicus*

Yang Yu Lan	*Magnolia grandiflora* L.
Yang Zhi Zu	*Rhododendron sinensis* Sw., *Azalea japonica* A. Gray, *A. pontica* var. *sinensis* Lindl., *A. mollis* Blume
Yao Jiu Hua	*Chrysanthemum procumbens* Lour., *C. indicum* L., *C. lavandulaefolium*, *C. tripartium* Sw.
Ye Bai He	*Crotalaria sessiliflora* L.
Ye Da Dou	*Glycine soja* Sieb. & Zucc.
Ye Dou Gen	*Menispermum dauricum* L.
Ye Gan Zi	*Phyllanthus emblica* L.
Ye Guan Men	*Lespedeza cuneata* G. Don
Ye He Hua	*Magnolia coco* (Lour.) DC
Ye Lu Kui	*Melochia corchonfolia* L.
Ye Mu Gua	*Stauntonia hexaphylla* Dence
Ye Wo	*Carum carvi* L.
Ye Wu Tong	*Mallotus japonicus* (Thunb.) Muell.
Ye Yen Me	*Avena fatua* L.
Yeh Chin Ts'ai	*Cicuta virosa* L.
Yeh Wan Tou	*Lathyrus pratensis* L.
Yen Lin Cao	*Trillium camschatcense* Ker-Gawler
Yen Xing	*Ginkgo biloba* L.
Yeu Je Hua	*Rosa chinensis* Jacq., *R. indica* Lindl.
Yi Dian Hong	*Nervilia purpurea* (Hayata) Schltr.
Yi Mu Cao	*Leonurus heterophyllus, L. japonicus* Houttuyn., *L. macranthus* Maxim., *L. mongolicus* V. Kreczet. et Kupr., *L. pseudo-macranthus* Kitag.

Chinese Name	Scientific Name
Yi Nian Pon	*Erigeron annuus* (L.) Persoon
Yi Wu	*Elaeagnus oldhamii* Maixmowicz
Yi Ye Chan	*Securinega suffruticosa* (Pall.) Rehd.
Yi Yi	*Coix chinensis* Tod, *C. agrestis* Lour., *C. lachryma* L., *C. lachryma-jobi* L. var. ma-yuen (Roman) Stapf
Yi Zhi	*Alpinia oxyphylla* Miq.
Yi Zhi Huang Hua	*Solidago dahurica* (Kitag.) Kitag., *S. virgaurea* L., *S. pacifica* Juzepczuk.
Yi Zhi Huang Hua (N. Am.)	*S. canadensis* L.
Yi Zhi Jian	*Ophioglossum vulgatum* L.
Yi Zhi Zi	*Elettaria cardamomum* Maton.
Yie Huang Hua	*Patrina scabiosaefolia* Fisch. ex Link.
Yie Mian Hua	*Anemone vitifolia* (Buch. Ham.) Nakai
Yie Pu Tao Teng	*Ampelopsis brevipedunculata* (Maxim.) Trautv.
Yie Xiz Zhu	*Phyllanthus urinaria* L.
Yin Bai Yang	*Populus alba* L., *P. davidiana* Dode, *P. tomentosa* Carr.
Yin Cao	*Primula sieboldii* E. Morren var. patens (Turcz.) Kitag., *P. asiatica* Nakai f. albiflora (Koidz.) Kitag., *P. asiatica* Nakai f. lilacina (Nakai) Kitag., *P. vulgaris* L., *P. asiatica* Nakai
Yin Chen	*Artemisia capillaris* Thunb.
Yin Dijue	*Botrychium strictum* Underw.
Yin He Huan	*Leucaena leucocephala* (Lam.) De Wit
Yin Lian Hua	*Anemone raddeana* Regel

Ying Chun Hua	*Jasminum mesnyi* Hance, *J. nudiflorum* Lindley
Ying Su	*Papaver amurense* (N. Busch) N. Busch ex Tolmatchev., *P. nudicaule* L., *P. radicatum* Rottb. var. pseudoradicatum (Kitag.) Kitag.
You Tong	*Aleurites fordii* Hemsl.
Yu	*Potamogeton perfoliatus* L.
Yu Bai	*Lycopus obscurum* L.
Yu Bai Pi	*Ulmus campestris* L.
Yu Dei Mei	*Hoya carnosa* (L. F.) R. Brown
Yu Jin	*Curcuma aromatica*, *C. longa* L.
Yu Jin Xian	*Tulipa gesneriana* L.
Yu Lee Ren	*Prunus domestica* L., *P. glandulosa* Thunb., *P. japonic* Thunb.
Yu Ma	*Urtica angustifolia* Fisch. ex Hornem., *U. urens* L., *U. cannabina* L. f. angustiloba Chu, *U. lobata* L., *U. tenacissima* Roxb., *U. utillis* Hort., *U. cannabina* L., *Boehmeria nivea* Gaudich.
Yu Ma Gen	*Boehmeria tenacissima* Gaudick
Yu Mei Ku	*Papaver somniferum* L.
Yu Mi Xu	*Zea mays* L.
Yu Shan Dou	*Lupinus luteus* L.
Yu Shan Shi Song	*Lycopus veitchii* Christ.
Yu Xing Cao	*Houttynia cordata* Thunb.
Yu Ye Jin Hua	*Mussaenda parviflora* Miq.
Yuan Bai	*Sabina chinensis* (L.) Antoine
Yuan Cao Gen	*Hemerocallis fulva* L.

Chinese Name	Scientific Name
Yuan Hua	*Daphne fortunei* Lindl., *D. genkwa* Sieb. et Zucc.
Yuan Jin Gan	*Fortunella japonica* (Thunb.) Swin.
Yuan Xi Huang San	*Allamanda cathatica* L.
Yuan Zhi	*Polygala tenuifolia* Willd.
Yue Tao	*Alpinia japonica* Miq.
Yun Shi	*Caesalpinia decapetula* (Roth.) Alston
Yun Xian Cao	*Cymbopogon goeringii* (Steud.) A. Camus, *C. distans* (Nees ex Steud.) J. F. Watson
Zang Hui Xiang	*Carum carvi* L.
Zao Ci	*Gleditschia sinensis* Lam, *G. xylocarpa* Hance, *G. horrida* Willd.
Zao Zhui	*Arenaria juncea* Bieb., *A. juncea* Bieb. var. abbreviata Kitag., *A. juncea* Bieb. var. glabra Regel, *A. serpyllifolia* L.
Ze Lan	*Arethusa japonica* A. Gr.
Ze Qi	*Euphorbia esula* L.
Ze Xie	*Alisma cordifolia* Thunb., *A. plantago* L., *A. plantago-aquatica* L., *A. orientalis* (Sam.) Juzep.
Zhang Shu	*Atractylis chinensis* DC, *A. ovata* Thunb., *A. lyrata* Sieb. et Zucc., *A. lancea* Thunb.
Zhe Gu Cai	*Calloglossa lepieurii* (Mont.) J. Ag.
Zhen	*Corylus mandshurica* Maxim. ex Rupr. f. brevituba (Kom.) Kitag., *C. heterophylla* Fisch. ex Besser, *C. mandshurica* Maxim. ex Rupr.
Zhi	*Gardenia maruba* Sieb., *G. florida* L., *G. grandiflora* Sieb. et Zucc., *G. jasminoides* Ellis, *G. pictorum* Hassk., *G. radicans* Thunb.
Zhi Bei Zi	*Hovenia dulcis* Thunb.
Zhi Jia Hua	*Lawsonia inermis* L.

Zhi Jin Niu	*Ardisia quinquegona* (Blume) Nakai
Zhi Mu	*Anemarrhena asphodeloides* Bunge.
Zhi Wen	*Aster tataricus* L.
Zhi Xie Mu Pi	*Holarrhena antidysenterica* Wall.
Zhow Sho	*Paris polyphylla* Smith, *P. quadrifolia* L.
Zhu Cao	*Lithospermum erythrorhizon* Sieb. et Zucc.
Zhu Chun Hua	*Salvia coccinea* L.
Zhu Je Seng	*Panax japonicum* C. A. Meyer
Zhu Jin	*Hibiscus rosa-sinensis* L.
Zhu Ling	*Polyporus umbeliatus*
Zhu Long Cao	*Nepenthes rafsiana* Masilus
Zhu Mao Chao	*Salsola collina*
Zhu Shan	*Taxus cuspidata* Sieb. et Zucc.
Zhu Shi Tou	*Crotalaria mucronata* Desv.
Zhu Wei	*Campsis adrepens* Lour., *C. grandiflora* (Thunb.) Loiseleur, *C. chinensis* Voss.
Zhu Wei	*Bignonia grandiflora* Thunb.
Zhu Ye Lan	*Arundina gramirifolia* (D. Don) Hochrentiner
Zhu Yin Yin	*Galium spurium* L.
Zhu Zi Cao	*Phyllanthus niruri* Li, *P. reticulatus* Poiret
Zi Bai Pi	*Catalpa ovata* G. Don
Zi Bei Cao	*Emilia sonchifolia* (L.) DC
Zi Cao	*Arnebia euchroma*

Chinese Name	Scientific Name
Zi Duan	*Tilia amurensis* Rupr., *T. mongolica* Maxim., *T. mandshurica* Rupr. & Maxim.
Zi Jin Pi	*Tripterygium hypoglaucum*
Zi Kee Guan Zhong	*Osmunda japonica* L.
Zi Lan	*Eupatorium chinense* L. var. simplicifolium (Malcino) Kitam., *E. lindleyanum* DC
Zi Su	*Perilla arguta* Benth.
Zi Teng	*Wisteria sinensis* (Sims) Sweet
Zi Wei Hua	*Bignonia grandiflora* Thunb., *B. chinensis* Lam., *Campsis chinensis* Voss.
Zi Yu	*Sanguisorba officinalis* L., *S. grandiflora* (Maxim.) Makino, *S. officinalis* L. var. longa Kitag., *S. officinalis* L. var. longa Kitag. f. dilutiflora Kitag., *S. parviflora* (Maxim.) Takeda, *S. officinalis* L., *S. officinalis* L. f. latifoliata (Liou et C. Y. Li) Y. C. Chu, *S. x tenuifolia* Fisch. ex Link
Zi Zhu Cao	*Callicarpa macrophylla* L.
Ziang Jia Pi	*Periploca sepium* Berge.
Zong Lu	*Trachycarpus wagnerianus* Beccari
Zong Lu Zi	*Trachycarpus excelsa* WendZong Shi *Lappa communis* Coss et Germ., *L. minor* DC, *L. major* Gaerth., *L. edulis* Sieb.
Zu Si Ma	*Daphne giraldii* Nitsche, *D. gurakaluu* Nitsche, *D. retusa* Hemsl, *D. tangutica* Maxim.
Zuo Yie He Cao	*Cotyledon malaacophylla* Pall., *C. fimbriatum* Turcz.

Appendix 2

Major Chemical Components and Their Sources in Chinese Medicinal Herbs

Component	Source
(+)-5,17-dehydromatrine N-oxide	*Euchresta japonicum*
(−)-12-cytisineacetic acid	*Euchresta japonicum*
1,8-cineol	*Glechoma hederacea, G. longituba, Hedychium coronarium, Lysimachia barystachys, L. christinae, L. clethroides, L. davurica, Thymus vulgaris*
α-amyrin	*Ageratum conyxoldes* L.
α-amyrin/acetate	*Alstonia scholaris* (L.) R. Br.
α-naginatene	*Elshoitzia ciliatai* Thunb.
α-phellandrene	*Daucus carota* L.
α-pinene	*Oenanthe javanica* (BI) DC, *Perilla frutescens* L.
α-spinasterol	*Spinacia oleracea* L.
α-terthienyl	*Tabetes erecia* L.
α-tocopherol	*Spinacia oleracea* L.
β-amyrin	*Eupatorium odoratum* L.
β-carotene	*Amaranthus tricolor* L., *Helianthus annuus* L.
	Marsilea quadrifolia L. *Sonchus oleraceus* L.
β-caryophyllene	*Ageratum conyzoldes* L.
β-ecdysone	*Chenopodium album* L.
β-naginatene	*Elshoitzia ciliatai* Thunb.
β-phellandrene	*Pinus madshurica* Rupr.
β-sitosterol	*Alstonia scholaris* (L.) R. Br., *Trachelospermum jasminoides* (Lindl) Lem, *Trapa manshurica* Flerov.

β-vicianosyl-3-quercetin	*Nymphoides peltata* (S. G. Gmelin)
1β, 10β-epoxyfuranoer emophilane	*Ligularia fischeri* (Ledeb)
1-acetyl-4-isopropylidenecyclopentene	*Cinnamomum camphora*
1 (10) eremophilen-11-ol.	*Ligularia fischeri* (Ledeb.)
2-hydroxymethyl prop-2-enoate	*Ligularia fischeri* (Ledeb)
10-deacetylbaccatin	*Taxus cuspidata, T. chinensis, T. yunnanensis*
10-hydroxycamptothecin	*Camptotheca acuminata*
12-benzoxydaphnetoxin	*Daphne fortunei, D. genkwa*
12-di-dehydroandrog-rapholide	*Andrographis paniculata*
2'-deoxyadenosine	*Cordyceps sinensis*
2-hydroxphenylacetic acid	*Astilbe longicarpa, A. chinensis*
2-methyl-1,2,3,4-tetrahydro-β-carboline	*Elaeagnus pungens, E. umbellata*
2-o-caffeoylarbutin	*Vaccirium bracteatum, V. vitis-idaea*
2α-hydroxyursolic acid	*Actinidia chinensis* Planch.
22-dihydrostigmast-4-en-3,6-dione	*Trapa manshurica* Flerov.
22E-dehydroclerosterol	*Clerodendrum fragrans*
22-ergostate traen-3-one	*Trapa manshurica* Flerov.
2,4-dichloro-6-aminopyridine	*Picrasma quassioides*
2,4,4',t-tetrahydroxybenzophenone	*Morus alba, M. constantinopolitana, M. indica*

Component	Source
24alpha-epimer stigmasterol	*Clerodendrum paniculatum*
24alpha-ethyl-5alpha-cholest	*Clerodendrum fragrans*
24beta-epimer poriferasterol	*Clerodendrum paniculatum*
24beta-methylcholesta	*Clerodendrum fragrans*
25-D-spirosta-3,5-diene	*Dioscorea nipponica*
2,6-dimethoxy-p-benzo-quinone	*Picrasma quassioides*
2,6-nonadienol	*Cucumis sativus*
3≤,4≤-O-diacetylafzelin	*Zingiber zerumbet*
3'-angeloyloxy-4'-isovaleroyloxy	*Angelica decursiva*
3-(4-hydroxyphenyl)-2 (E)-propenoate	*Costus specious, Curcuma zedoaria, C. aromatica, C. kwangsiensis*
3-butyl phthalide	*Ligularia japonica* (Thunb.) Less
3-chloroplumbagin	*Plumbago zeylanica*
3-hydroxy-30-horoleana-12,18- dien-29-oate	*Anredera cordifolia*
3-indolylmethylgluco-sinolate	*Clerodendrum cyrtophyllum*
3-methoxypyridine	*Equisetum arvense, E. hyemale, E. ramosissimum*
3-O-β-glucosylplatycodigenin	*Platycodon autumnalis, P. grandiflorum, P. sinensis*
3-O-demethylhernandifoline	*Stephania hernendifolia*
3-O-methylquercetin	*Ophioglossum vulgatum*

3-O-methylquer-cetin-T-O-diglucoside -4'-0-glucoside	*Ophioglossum thermale* Kom
3β- hydroxyeremophile mophilenolide, furanoeremophilane, petasalbin	*Ligularia fischeri* (Ledeb.)
3-oxykojie acid	*Maytenus diversifolia, M. confertiflorus*
3-p-coumarylglycoside-5-glucoside	*Perilla frutescens, P. ocymoides, P. polystachya, P. arguta*
3,4-dihydroxyacetophenone	*Ilex pubescens*
3,4-dihydroxycinnamic acid	*Ananas comosus*
3,4-dihydroxyphenethyl alcohol	*Syringa dilatata, S. oblata, S. reticulata, S. suspensa, S. vulgaris*
3,6-dime thoxy-quercetagetin	*Filifolium sibiricum* (L.) Kitam
4-dementhyl-hasubanonine	*Stephania hernendifolia*
4-epiisocembrol	*Pinus madshurica* Rupr.
4-hydroxycinnamic acid	*Ananas comosus, Melilotus alba, M. suaveolens, M. indica*
4-methoxy salicyl aldehyde	*Acanthopanax gracilistylus, A. spinosum. Periploca sepium Bunge*
4-quinazolone	*Dichroa cyanitis, D. febrifuga, D. latifolia*
4,5-dimethoxycanthin-6-one	*Picrasma quassioides*
5 alpha-ergost-8(-14)-en-3B-ol	*Ophiorrhiza japonica, O. mungos*
5 alpha-ergost-en-3 beta-ol	*Ophiorrhiza japonica, O. mungos*
5-β-cholanic acid	*Abrus precatorius*
5-fluorouracil	*Rabdosia lasiocarpus*

Component	Source
5-guaizulene	*Murraya paniculata*
5-hydroxytryptamine	*Ranunculus sceleratus, Urtica angustifolia, U. cannabina, U. cannabina, U. lobata, U. tenacissima, U. urens, U. utillis*
5-methyl kaempferol	*Rhododendron mucronatum*
5-methyl myricetin	*Rhododendron mucronatum*
5-stigmautena-3β,7d-diol	*Ananas comosus*
6,9,12-octadecatrienoic acid	*Oenothera biennis, O. odorata*
6-C-galactopyranosyl-isoscutellarein	*Silene jenisseensis*
6-ethoxy-chelerythrin	*Zanthoxylum nitidum*
6-hydroxymethy-luumazin	*Spinacia oleracea* L.
6-isoinosine	*Acanthopanax giraldii*
6-methylocodine	*Papaver somniferum*
6-O-β-sophoruside	*Bougainvillea brasiliensis, B. glabra*
6-O-acetyl-arbutin	*Vaccinium bracteatum, V. vitis-idaea*
6-O-rhamnosyl cophoroside	*Bougainvillea brasiliensis, B. glabra*
6,6'-dimethoxylgossypol	*Gossypium herbaceum*
6,8-di-C-galactopyranosylapigenin	*Silene jenisseensis*
7-β-glucopyranoside	*Veronica sibirica, V. undulata*
7-caffeyl-glucosides	*Perilla frutescens, P. ocymoides, P. polystachya, P. arguta*
7-hydroxylathyrol	*Euphorbia lathyrus, E. lucorum, E. resinfera, E. thymifolia*

7-methoxy-2,2-dimethylchromene	*Ageratum conyzoides, A. houstonianum*
7-methoxy-baicalein	*Scutellaria baicalensis, S. grandiflora, S. lanceolaria, S. macrantha, S. rivularis, S. viscidula*
7-methoxynorwogonin	*Scutellaria baicalensis, S. grandiflora, S. lanceolaria, S. macrantha, S. rivularis, S. viscidula*
7-O-methyl-morroniside	*Cornus officinalis*
8-(O-methyl-p-coumaroyl)-harpagide	*Scrophularia buergeriana, S. kakudensis, S. ningpoensis, S. oldhami, S. puergeriana*
Abamagenin	*Sansevieria trifosciate*
Abeotaxanes	*Taxus cuspidata, T. chinensis, T. yunnanensis*
Abscisic acid	*Pileostegia viburnoides*
Acacetin-7-glucoside	*Chrysanthemum jucundum, C. koraiense, C. morifolium, C. sinense*
Acacetin-7-glucurono-(1,2)-glucuronide	*Clerodendrum trichotomum, C. spicatus*
Acacetin-7-β-D-glucurono-B-D-glucuronide	*Chrysanthemum jucundum, C. koraiense, C. morifolium, C. sinense, Cirsium albescens, C. brevicale, C. littorale, C. maakii, C. segetum, C. setosum, C. vlassovianum*
Acacetin	*Clerodendrum trichotomum* Thunb.
Acalyphine	*Acalypha australis, A. farnesiana, A. indica*
Acanthosides	*Acanthopanax sessiliflorus*
Acetadehyde	*Carum carvi L.Prunus persica*
Acetic acid	*Ajuga bracteosa, Jasminum samba, Luffa aegyptiaca, L. cylindrica, L. faetida, L. petola, Michelia alba, M. figo, Prunus persica*
Acetone	*Coriandrum sativum*

Component	Source
Acetovanillone	*Gossypium herbaceum*
Acetycophalotaxine	*Cephalotaxus wilsoniana*
Acetyl lupeol	*Plumeria rubra*
Acetyl oleanolic acid	*Syzygium cuminii*
Acetylcholine	*Diospyros chinensis, D. costata, D. khaki, D. lotus, D. roxburgii, Viscum album*
Acetylcorynoline	*Corydalis incisa, C. bungeana*
Acetyleugenol	*Eugenia aromatica, E. caryophyllata, E. ulmoides*
Acetylinic compound	*Carum carvi* L.
Acetylsalicylic acid	*Ligusticum chuanziang,* Paeonia albiflora, P. edulis, P. japonica, P. lactiflora, P. moutan, P. officinalis, Salvia militiorhiza. Sargentodoxa cuneata
Acetylshikonin	*Arnebia euchroma, Lithospermum erythrorhizon, L. officinalis*
Achilleine	*Achillea alpina, A. millefolium*
Achillin	*Achillea alpina, A. millefolium*
Acidic resin	*Wikestroemia indica*
Aconine	*Aconitum laciniatum, A. kusnezoffii, A. chinense, A. vilmorinianum, A. pariculigerum*
Aconitic acid	*Arthraxon hispidus, Avena fatua, Equisetum palustre* L.
Aconitine	*Aconitum laciniatum, A. kusnezoffii, A. chinense, A. vilmorinianum, A. pariculigerum, A. barbatum, A. austroyunnanense, A. balfouri, A. carmichaelii, A. chasmanthum, A. deinorrhizum, A. fischeri, A. jaluense, A. koreanum, A. napellus, A. praeparata, A. volubile*
Acoric acid	*Acorus calamus* var. angustatus, *A. gramineus, A. tatarinowii*
Acornes	*Acorus calamus* L.

Actinidine	*Actinidia arguta, A. chinensis, A. japonica, A. kolomikta, A. polygama, Boschniaka rossica* Cham & Schlecht
Actronycine	*Acronychia pedunculata, A. laurifolia*
Acutumidine	*Cocculus diversifolius, C. thunbergii*
Acutumine	*Cocculus diversifolius, C. thunbergii, Menispermum dauricum, Sinomenium acutum*
Acutuminine	*Menispermum dauricum*
Acyclic diterpene glycosides	*Lycium chinense*
Acyl flavonol di-gycoside	*Hedyotis diffusa*
Adenine	*Angelica polymorpha, A. sinensis, Artemisia brachyloba, Chrysanthemum cinerriaefolium, C. jucundum, C. koraiense, C. morifolium, C. sinense, Perilla frutescens* L. *Pueraria montana, P. thunbergiana, Pyrethrum cinerariifolium, P. sinense, Solanum lyratum, S. melongena, Tetragonia tetragonoides*
Adenosine	*Cordyceps sinensis, Ganoderma lucidum, Verbena officinalis, V. oxysepalum*
Adiantone	*Adiantum boreale, A. capillus-junonis, A. pedatum, A. flabellulatum*
Adipedatol	*Adiantum boreale, A. capillus-junonis, A. pedatum, A. flabellulatum*
Adonilide	*Adonis amurensis* Regel & Radde *Adonis chrysocyathus, A. brevistyla, A. vernalis*
Adynerin	*Nerum indicum*
Aegicerin	*Primula sieboldii, P. asiatica, P. vulgaris*
Aescilom	*Euphorbia lathyrus, E. lucorum, E. resinfera, E. thymifolia*
Aescine	*Aesculus indica*
Aesculetin	*Azolla imbricata, Viburnum sargenti*
Aesculin	*Fraxinus bungeana, F. chinensis, F. floribunda, F. obovata, F. ornus, F. rhynchophylla*
Aesculine	*Aesculus indica*

Component	Source
Aflatoxin B	*Gossypium herbaceum*
Aflatoxins	*Coriandrum sativum*
Afzelin	*Ailanthus altissima*, *Dicranopteris linearis*
Agarol	*Aquilaria agallocha, A. sinensis*
Agaropectin	*Gelidium amansii*
Agarose	*Gelidium amansii*
Agarospirol	*Aquilaria agallocha, A. sinensis*
Agathin	*Sesbinia grandiflora*
Agathodienediol	*Pinus madshurica* Rupr.
Ageniadin	*Plumeria rubra* L.
Agerato-chromene	*Ageratum conyzoides, A. houstonianum*
Agglutinins	*Viscum album, V. coloratum*
Aglucones	*Siegesbeckia orientallis* L.
Aglycone	*Corchorus capsularis, C. olitorius, Glycyrrhiza pallidiflora, G. uralensis, Picrorhiza kurroa*
Agnuside	*Vitex nequndo, V. trifolia, V. rotundifolia*
Agoniadin	*Plumeria rubra*
Agrimols	*Agrimonia eupatoria, A. pilosa, A. viscidula*
Agrimonine	*Agrimonia eupatoria, A. pilosa, A. viscidula*
Agrimonolide	*Agrimonia eupatoria, A. pilosa, A. viscidula*
Agrimophol	*Agrimonia eupatoria, A. pilosa, A. viscidula*
Ailanthone	*Ailanthus altissima*

Ajugasterone	*Ajuga bracteosa, A. decumbens, A. pygmaea, Alangium lamarckii*
Alangicine	*Alangium lamarckii*
Alangimarckine	*Alangium lamarckii*
Alanine	*Fagropyrum esculentum, Litchi chinensis, Taraxacum mongolicum, T. sinicum*
Alantolactone	*Inula helenium* L.
Alatamine	*Euonymus alatus, E. bungeanus, E. maackii*
Albaspidin	*Dryopteris crassirhizoma, D. laeta, D. filix-mas*
Albiflorin	*Paeonia albiflora, P. edulis, P. japonica, P. lactiflora, P. moutan, P. officinalis*
Albigenic acid	*Codonopsis lanceolata*
Albumin	*Aesculus indica, Hyoscyamus bohemicus*
Alcohol derivatives	*Ledebouriella divaricata*
Alcohols	*Viola acuminata, V. alisoviana, V. collina, V. dissecta, V. mandshurica, V. patrini, V. prionantha, V. verecunda*
Aldehyde	*Hedychium coronarium, Perilla frutescens, P. ocymoides, P. polystachya, P. arguta, Plumeria rubra*
Algin	*Laminaria angusta, L. cichorioides, L. japonica, L. longipedalis, L. religiosa, Sargassum pallidum*
Alginic acid	*Sargassum pallidum*
Alisarin	*Galium bungei, G. spurium, G. verum*
Alisol A	*Alisma cordifolia, A. orientalis, A. plantago-aquatica, A. plantago*
Alisol B	*Alisma cordifolia, A. orientalis, A. plantago-aquatica, A. plantago*
Alisol monoacetate	*Alisma cordifolia, A. orientalis, A. plantago-aquatica, A. plantago*

Component	Source
Alizarin	*Morinda citrifolia, M. officinalis, Rubia akane, R. chinensis, R. cordifolia, R. cordifolia, R. mungista, R. sylvatica*
Alizarin-1-methyl ether	*Morinda parvifolia*
Alkaloid lamine	*Dipsacus asper*
Alkaloids	*Achillea alpina, A. millefolium, Arisaema amurense, A. consanguineum, A. erubescens, A. heterophyllum, A. peninsulae, A. thunbergii, Caesalpinia pulcherrima Capsella bursa-pastoris, Caragana sinica, C. microphylla, C. intermedia, C. franchetiana, Catharanthus roseus (L.) G. Don Centipeda minima, Cephalanoplos segetum, Delonix regia, Dodonaea viscosa, Eclipta alba, E. marginata, E. prostrata, E. thermalis, Emilia sonchifolia, Erythrina corallodendron, E. indica, E. variegata, Flagellaria indica, Hyoscyamus bohemicus, H. niger; Maesa perlarius, Magnolia hypoleuca, M. coco, M. fortunei, M. officinalis, M. japonica, Mallotus japonicus, Ophiorrhiza japonica, O. mungos, Orchis latifolia, Psychotria rubra, P. serpens, Selaginella involvens, S. doederieninii, S. campestris, Solanum incanum, Spiraea salicifolia, S. chinensis, S. baicalensis, S. japonica, Tripterygium hypoglaucum, Tulipa edulis, T. gesneriana, Urtica angustifolia, U. cannabina, U. lobata, U. tenacissima, U. urens, U. utillis, Veratrum formosanum, Vernonia andersonii; V. cinerea, V. patula, Viscum album, V. coloratum, Wahlenbergia marginata, Zephyranthes carinata*
Alkamin-B	*Arnebia euchroma*
Alkannan	*Lithospermum erythrorhizon, L. officinalis*
Alkyl methyl quinolone alkaloids	*Evodia rutaecarpa*
Allamandin	*Allamanda cathatica*
Allantoin	*Dioscorea batatus, D. opposita*

Allelopathic essential oils	*Cyperus brevifolius, C. difformis, C. glomeratus, C. iria*
Allicin	*Allium chinense, A. odorum, A. sativum, A. tuberosum, A. liginosum*
Allistatin	*Allium chinense, A. odorum, A. sativum, A. tuberosum, A. uliginosum*
Allocryptopine	*Chelidonium album, C. hybridum, C. majus, C. serotinum*
Allomatatabiol	*Actinidia arguta, A. chinensis, A. japonica, A. kolomikta, A. polygama*
Allosecurinine	*Securinega suffruticosa*
Allyl-, benzyl- and propenyl- isothiocyanate	*Descurainia Sophia* (L.) Schur.
Allyl-disulpide	*Descurania Sophia* Webb ex Prantl.
Allyl isothiocyanate	*Brassica alba, B. juncea, Descurania Sophia* Webb ex Proantl.
Allyl sinapic oil	*Draba nemorosa*
Allyl-1-propenyl disulfide	*Allium victorialis*
Aloe-emodin	*Aloe barbadensis, A. vera, Cassia alata, C. angustifolia, Rhamnus davurica, R. parvifolia, Rheum officinale, R. palmatum, R. tanguticum, R. undulatum, R. koreanum*
Aloins	*Aloe barbadensis, A. vera*
Alpha-agarofuran	*Aquilaria agallocha, A. sinensis*
Alpha-allocryptopine	*Corydalis ambigua, C. repens, C. turtschaninovii, C. yanhusuo, C. ternata, Macleaya cordata, Thalictrum aquilegifolium, T. baicalense, T. fauriel, T. petaloideum, T. simplex, T. squarrosum, T. thunbergii*
Alpha-amyrenol	*Spilarthes acmella*
Alpha-amyrin	*Aleurites fordii, Alnus japonica, Cirsium chinense, C. japonicum, Pedicularis resupinata*
Alpha-amyrin palmitate	*Sambucus formosana*
Alpha-antiarin	*Antiaris toxicaris*

Component	Source
Alpha-antioside	*Antiaris toxicaris*
Alpha-bergamotene	*Perilla frutescens, P. ocymoides, P. polystachya, P. arguta*
Alpha-camphorene	*Commiphora myrrha*
Alpha-carotene-5,6-epoxide	*Cuscuta australis*
Alpha-caryophylline	*Eugenia aromatica, E. caryophyllata, E. ulmoides*
Alpha-crocetin	*Gardenia angusta, G. jasminoides*
Alpha-cyperene	*Cyperus rotundus*
Alpha-cyperol	*Cyperus rotundus*
Alpha-cyperone	*Cyperus brevifolius, C. difformis, C. glomeratus, C. iria*
Alpha-dichroine	*Adamia chinensis, A. cyanea, A. versicolof*
Alpha-elaeo stearic	*Aleurites fordii*
Alpha-euphol	*Euphorbia kansui*
Alpha-euphorbol	*Euphorbia antiquorum, E. kansui*
Alpha-fenchene	*Valeriana alternifolia, V. amurensis, V. fauriei, V. subbipinnatifolia*
Alpha-globuline	*Phaseolus angularis, P. lunatus, P. radiatus, P. vulgaris*
Alpha-humulene	*Cyperus brevifolius, C. difformis, C. glomeratus, C. iria, Zingiber zerumbet*
Alpha-hydrojuglone-4-β-D-glucoside	*Juglans mandshurica, J. regia*
Alpha-ionone	*Lawsonia inermis*
Alpha-kainic acid	*Calloglossa lepieurii*
Alpha-ketoglutaric acid	*Phyllanthus simplex*

Alpha-leucodelphinidin	*Phyllanthus emblica*
Alpha-lupanine	*Caulophyllum robustum*
Alpha-methyl ether	*Morinda citrifolia, M. officinalis*
Alpha-obscurine	*Lycopodium annotinum, L. cernum, L. compianatum*
Alpha-onocerin	*Lycopodium clavatum, L. obscurum, L. selago, L. serratum*
Alpha-paristyphnin	*Paris polyphylla, P. quadrifolia*
Alpha-phellandrene	*Daucus carota, Elettaria cardamomum, Ledum palustre, Thymus vulgaris*
Alpha-phenylethylisothiocyante	*Rorippa indica, R. islandica, R. montana*
Alpha-phenylpropyl cinnamyl cinnamate	*Styrax tonkinensis, S. benzoin*
Alpha-pinene	*Aconitum deinorrtuzum, Chrysanthemum boreale, C. indicum, C. lavandulaefolium, C. procumbens, C. tripartium, Daucus carota, Glechoma hederacea, G. longituba, Juniperus rigida, Ledum palustre, Oenothera javanica, Zanthoxylum bungeanum*
Alpha-santalene	*Santalum album, S. myrtifolium, S. verum*
Alpha-santalol	*Santalum album, S. myrtifolium, S. verum*
Alpha-santenone	*Santalum album, S. myrtifolium, S. verum*
Alpha-spinasterol	*Codonopsis lanceolata, Menyanthes trifoliata*
Alpha-taralin	*Aralia chinensis, A. cordata, A. elata*
Alpha-taraxerol	*Euphorbia antiquorum*
Alpha-terpinene	*Oenothera javanica*
Alpha-terpineol	*Cymbopogon citratus, Glechoma hederacea, G. longituba, Thyrus amurensis, T. disjunctus, T. kitagawianus, T. komarovii, T. przewalskii, T. quinquecostatus, T. vulgaris*
Alpha-terpinyl acetate	*Thymus vulgaris*

Component	Source
Alpha-terthienyl	*Tagetes erecta*
Alpha-tertiary methanol	*Eclipta erecta*
Alpha-trevilline	*Erythroxylum coca*
Alpha-typhasterol	*Typha angustata, T. angustifolia, T. davidiana, T. latifolia, T. minima, T. orientalis, T. przeqalskii*
Althaeine	*Althaea rose*
Aluminum	*Portulaca pilosa*
Aluminum oxide	*Phyllostachys bambusoide, P. nigra*
Amaranthin	*Gomphrena globosa*
Amarbelin	*Cuscuta chinensis, C. europaea, C. japonica, C. lupuliformis*
Amarolide	*Ailanthus altissima*
Ambroide	*Chenopotium ambrosiodes*
Amellin	*Scopalia dulcis*
Amentoflavone	*Cycas revoluta, Selaginella tamarisina, Thuja koraiensis, T. orientalis, T. chinensis*
Amino acids	*Acacia confusa. Ampelopsis aconitifolia, A. brevipedunculata, A. japonica, A. bodinieri, A. contonensis, A. humulifolia, Arachis hypogaea, Codonopsis pilosula, C. tangshen, C. ussuriensis, Elaeagnus glabra, Evodia lepta, E. triphylla, Ganoderma lucidum, Laggera alata, Laminaria angusta, L. cichorioides, L. japonica, L. longipedalis, L. religiosa, Litchi chinensis, Mallotus paniculatus, Nepenthes raffsiana, Nymphaea tetragona, Oryza sativa, Phyllanthus virgatus, Pinellia ternata, P. tuberifera, Polygonum perfoliatum, P. tinctorium, Saururus chinensis, Urena procumbens*
Aminoadipic acid	*Avena fatua*

Aminol	*Cimicifuga dahurica, C. foetida, C. heracleifolia, C. racemosa, C. ussuriensis*
Amorphous dracoalban	*Daemonorops Draco*
Amorphous dracoresene	*Daemonorops Draco*
Amritoside	*Psidium guajava*
Amurine	*Papaver amurense, P. nudicaule, P. radicatum*
Amuroine	*Papaver amurense, P. nudicaule, P. radicatum*
Amuroline	*Papaver amurense, P. nudicaule, P. radicatum*
Amygdalin	*Armeniaca ansu, A. mandsharica, A. sibirica, A. vulgaris, Eriobotrya japonica, Prunus mume, P. domestica, P. glandulosa, P. japonica, P. padus, P. armeniaca, Pyrrosia adnascens*
Amylase	*Hordeum vulgare*
Amylodextrins	*Myristica fragrans*
Amylose	*Aesculus chinensis, A. hippocastanum, Trapa bispinosa*
Amyrenol	*Sedum formosanum*
Amyrenone	*Sedum formosanum*
Amyrin	*Ficus carica*
Anacardic acid	*Ginkgo biloba*
Anagalligenone	*Anagalis arvensis*
Anagalline	*Anagalis arvensis*
Anagyrine	*Sophora subprostrata*
Ananasic acid	*Ananas comosus*
Ancubin	*Veronica sibirica, Veronica undulata*
Andelin	*Angelica decursiva.*

Component	Source
Andrographan	*Andrographis paniculata* (Burm.f.) Nees
Andrographolide	*Andrographis paniculata*
Andromedotoxin	*Chimaphila umbellata, Lyonia ovalifolia, Rhododendron sinensis, R. mucronatum* Turcz
Anemonin	*Caltha palustris, Clematis chinensis, C. florida, C. hexapetala, C. minor, C. sinensis, C. terniflora, Pulsatilla ambigua, P. cernua, P. chinensis, Ranunculus chinensis, R. sceleratus, R. japonicus, R. sarmentosa*
Anemonol	*Clematis chinensis, C. florida, C. hexapetala, C. minor, C. sinensis, C. terniflora*
Anethol	*Foeniculum officinale, F. vulgare, Illicium verum*
Anethole	*Agastache rugosa, A. rugosa* f. *hypoleuca, Juniperus rigida, Magnolia biloba, M. denudata, M. discolor, M. liliflora, M. purpurea*
Angelic acid	*Angelica pubescens, A. sinensis* (Oliv) Diels. *A. grosserrata, Blumea lacera, Matricaria chamomilla*
Angelicin	*Psoralea corylifolia*
Angelicotoxin	*Angelica pubescens, A. sinensis* (Oliv) Diels.
Angelol	*Angelica pubescens*
Angenomalin	*Angelica amurensis, A. anomala, A. dahurica*
Anhydroderrid	*Milletia reticulata, M. taiwaniana*
Anisaldehyde	*Agastache rugosa, A. rugosa* f. *hypoleuca, Foeniculum officinale, F. vulgare, Illicium verum*
Anisatin	*Illicium lanacedatum*
Anisic acid	*Eupatorium odoratum*
Anisic ketone	*Illicium verum*
Anisodamine	*Datura suaveolens, Scopolia tangutica*

Anisodine	*Datura suaveolens, Scopolia tangutica*
Ankorine isotubulosine	*Alangium lamarckii*
Anneparine	*Nelumbium nuciferum, N. speciosum*
Anodyne	*Ligularia japonica* (Thunb.) Less
Anomalin	*Angelica amurensis, A. anomala, A. dahurica, Peucedanum formosanum*
Anonaine	*Nelumbium nuciferum, N. speciosum*
Anromedotoxin	*Rhododendron dauricum*
Anthelmic acid	*Matricaria chamomilla*
Antheraxanthin	*Taraxacum mongolicum, T. sinicum*
Anthocyanin	*Cynomorium coccineum, C. songarium*
Anthocyanidines	*Achillea alpina, A. millefolium, Anthriscus aemula, A. sylvestris, Heracleum dissectum, H. lanatum, Polygonum bistorta, Sanguisorba officinalis, S. grandiflora, S. parviflora, S. x tenuifolia. Taraxacum officinale*
Anthocyanins	*Glehnia littoralis, Perilla frutescens, P. ocymoides, P. polystachya, P. arguta*
Anthranil acid	*Jasminum samba*
Anthranol	*Rumex patientia*
Anthraquinoids	*Nepenthes raffsiana*
Anthraquinone derivative	*Cassia angustifolia*
Anthraquinones	*Achillea alpina, A. millefolium, Anthriscus aemula, A. sylvestris, Cassia nomame, C. obtusifolia, C. tora, C. occidentalis, C. torosa, Hedyotis diffusa, Heracleum dissectum, H. lanatum, Polygonum bistorta, Rheum officinale, R. palmatum, R. tanguticum, R. undulatum, R. koreanum, Sanguisorba officinalis, S. grandiflora, S. officinalis, S. parviflora, S. x tenuifolia, Taraxacum officinale*

Component	Source
Anthraxin	*Arthruxon hispidus*
Anthricin	*Anthriscus aemula, A. sylvestris*
Anti-HIV protein MAP 30	*Momordica charantia*
Anti-proliferative activity of triterpenoids	*Eclipta prostrate* L.
Apigenin	*Achillea millefolium* L. *Cassia occidentalis, C. torosa, Clinopodium chinense, C. polycephalum, C. gracile, C. umbrosum, Codonopsis lanceolata, C. mucronata, Daphne fortunei, D. genkwa, Jatropha podagrica, Juncus effusus, Perilla frutescens, P. ocymoides, P. polystachya, P. arguta, Selaginella tamarisina, Thymus amurensis, T. disjunctus, T. kitagawianus, T. komarovii, T. przewalskii, T. quinquecostatus*
Apigenin-7-β-D-glucoside	*Agrimonia eupatoria, A. pilosa, A. viscidula*
Apigenin-7-diglucuronide	*Clerodendrum trichotomum, C. spicatus*
Apigenin-7-O-glucoside	*Spirodela polyrhiza*
Apigenin-8-C-glucoside	*Spirodela polyrhiza*
Apiin	*Apium graveolens*
Appendicitis	*Senecio scandens* Buch-Ham
Apyrocatechol	*Citrus reticulata*
Arabinan polymer	*Bupleurum chinense, B. falcatum, B. scorzoneraefolium*
Arabinon	*Gleditsia horrida, G. sinensis, G. xylocarpa*
Arabinose	*Aster tataricus, Camellia japonica, Juncus communis, Tamarindus indicus, T. officinale*
Arabinose ester	*Psidium guajava*
Arachic	*Acanthopanax gracilistylus, A. spinosum*

Arachidic acid	*Jatropha gospiifolia, J. curcas Sinapis alba*
Arachine	*Arachis hypogaea*
Aragome	*Asparagus cochinenesis, A. falcatus, A. insularis, A. lucidus, A. officinalis*
Araligenin	*Aralia chinensis, A. cordata, A. elata*
Araloside A, B, C	*Aralia chinensis, A. cordata, A. elata, A. mandschurica* (Rupr & Maxim) Seem
Arasaponins	*Panax zingiberensis*
Arborinol	*Imperata arundinaceae, I. cylindrica*
Arborinone	*Imperata arundinaceae, I. cylindrica*
Arbutin	*Chimaphila umbellata, Pyrola decorata, P. japonica, P. incarnata, P. renifolia, P. rotundifolia, Pyrrosia adnascens, Sedum aizoon, Senecio cannabifolius, Vaccinium bracteatum, V. vitis-idaea, Veronica sibirica. V. undulata*
Arbutin ericolin	*Ledum palustre*
Archangelicin	*Cnidium monnieri*
Arctigenin	*Arctium lappa, Lappa communis, L. edulis, L. major; L. miror*
Arctin	*Arctium lappa, Lappa communis, L. edulis, L. major; L. miror; Trachelospermum jasminoides* (Lindl) Lem
Arecholidine	*Areca catechu, A. hortonsis*
Arecholine	*Areca catechu, A. hortonsis*
Arecoline	*Areca catechu* L.
Aresentic acid	*Juncus effusus*
Argamolic acid	*Psidium guajava*
Arginine	*Cajanus cajan* L. Cucumis sativus, Dioscorea opposita, D. batatus, Dolichos lablab, Litchi chinensis, Perilla frutescens L.

Component	Source
Arginine glucoside	*Solanum lyratum, S. melongen*
Aricine	*Rauvolfia verticilata*
Aristolochic acid	*Aristolochia shimadai, A. debilis, Clematis armandii, C. heracleifolia*
Aristolochic acid A	*Aristolochia contorta, A. kaempferi, A. longa, A. recurvilabra*
Aristolochic acid D	*Aristolochia contorta, A. kaempferi, A. longa, A. recurvilabra*
Aristolone	*Aristolochia debilis*
Aristoloside	*Aristolochia contorta, A. kaempferi, A. longa, A. recurvilabra*
Arjunolic acid	*Elaeagnus oldhumii*
Arnidiol	*Calendula officinalis*
Aromadendrene	*Cinnamomum camphora*
Aromadendrin	*Thuja koraiensis, T. orientalis, T. chinensis*
Aromadendrine	*Menyanthes trifoliata*
Aromatic acids	*Eupatorium odoratum*
Aromatic bitter principle (Anthemic acid)	*Matricarin chamonilla* L.
Arrenin	*Anagalis arvensis*
Artemisia alcohol	*Artemisia argyi, A. halodendron, A. igniaria, A. indica, A. integrifolia, A. japonica, A. keiskeana, A. lagocephala, A. lavandulaefolia, A. scoparia, A. selengensis, A. ieversiana, A. vulgarts*
Artemisine	*Artemisia brachyloba*
Artemisinin	*Artemisia annua, A. apiacea*

Artesunate	*Artemisia annua, A. apiacea*
Articulatin	*Equisetum arvense, E. hyemale, E. ramosissimum*
Arundoin	*Imperata arundinaceae, I. cylindrica, Lophatherum gracile*
Asarensinotannol	*Ferula assa-foetida, F. bungeana*
Asariline	*Asarum canadense, A. heterotropoides, A. sieboldii*
Asarone	*Daucus carota*
Ascaridol	*Chenopotium ambrosiodes, Ledum palustre*
Asclepiadin	*Asclepias curassavica* L.
Asclepin	*Asclepias curassavica* L.
Asclepogenin	*Asclepias curassavica* L.
Ascorbic acid	*Achyranthes asperia, Amaranthus tricolor, A. lividus, A. blitum, A. viridis, Benincase cerifera, B. hispida, Boehmeria densiflora, Canarium album, C. sinense. Castanea crenuta, C. mollissima, Corylus heterophylla, C. mandshurica, Eleusine indica* (L.) Gaertner, *Helianthus annuus* L. *Marsilea quadrifolia* L. *Petasites japonicus, Rosa multiflora, Sonchus oleraceus* L. *Syzygium aromaticum, Zea mays*
Ash	*Urtica angustifolia, U. cannabina, U. lobata, U. tenacissima, U urens, U. utillis*
Asiaticoside	*Centella asiatica*
Asoryl-ketone	*Asarum canadense, A. heterotropoides, A. sieboldii*
Asparagic acid	*Litchi chinensis*
Asparagine	*Arundo donax, A. phragmites, Arnebia euchroma Hemerocallis fulva* L. Humulus scandens Phragmites communis, Pueraria montana, P. thunbergiana, Sagittardia sagitifolia, Taraxacum mongolicuma, T. sinicum
Asparaginic acid	*Avena fatua*

Component	Source
Asperuloside	*Oldenlandia diffusa*
Aspidinol	*Pteridium aquilinum* (L.) Kuhn.
Astilbin	*Astilbe longicarpa, A. chinensis*
Astragalin	*Astragalus chinensis, A. complanatus, A. henyri, A. hoantchy, A. membranaceus, A. melilotoides, A. mongholicus, A. reflexistipulus, A. sinensis, Cyrtomium falcatum, Dysosma pleiantha, Equisetum arvense, E. hyemale, E. ramosissimum, Gymnadenia conopsea* (L.) R. Brown Matteuccia struthiopteris, *Paeonia obovata, P. suffruticosa, P. veitchii,* Solidago *dahurica, S. pacifica, S. virgaurea, Tribulus terrestsis*
Astragalosides	*Astragalus complanatus, A. henyri, A. hoantchy, A. membranaceus, A. melilotoides, A. mongholicus, A. reflexistipulus, A. sinensis*
Atractylodine	*Atractylis chinensis, A. lancea, A. lyrata, A. ovata*
Atractylol	*Atractylis chinensis, A. lancea, A. lyrata, A. ovata*
Atractylone	*Atractylodes lancea, A. chinensis, A. japonica, A. koreana, A. lancea, A. lyrata, A. macrocephala, A. ovata*
Atropine	*Datura suaveolens*
Aucubin	*Melasma arvense, Plantago asiatica, P. depressa, P. exaltata, P. loureiri, P. major, Veronica anagallis-aquatica, Vitex trifolia, V. rotundifolia, V. nequndo*
Aurantio-obtusin rubrofusarin	*Cassia nomame, C. obtusifolia, C. tora*
Aurapten	*Dictamnus albus, D. dasycarpus*
Auroxanthin	*Physalis alkekengi*
Austroinulin	*Stevia rebaudiana*
Avenasterol	*Avena fatua, Linum stelleroides, L. usitatissimum*

Avicularin	*Chimaphila umbellata, Loranthus parasiticus, L. yadoriki, Persicaria amphibia, Polygonum aviculare, P. bistorta, P. lapidosa, P. manshuriensis, P. vivipara, Psidium guajava, Saururus chinensis, Vaccinium bracteatum, V. vitis-idaea*
Awobanin	*Commelina communis*
Azadarachtin	*Melia japonica, M. toosendan, M. azedarach*
Azaleatin	*Rhododendron mucronatum*
Azelaic acid	*Ailanthus altissima, Lycopodium clavatum, L. obscurum, L. selago, L. serratum*
Azulene	*Artemisia gmelini, Ferula assafoetida L. Matricaria chamomilla, Melaleuca leucadendra, Piper cubeba*
Baccatin	*Taxus cuspidata, T. chinensis, T. yunnanensis*
Bacteriostatic	
Baicalein	*Scutellaria baicalensis, S. grandiflora, S. lanceolaria, S. macrantha, S. rivulararis, S. viscidula*
Baicalin	*Scutellaria baicalensis, S. grandiflora, S. lanceolaria, S. macrantha, S. rivulararis, S. viscidula, S. formosana*
Balsam	*Liquidambar acerifolia, L. formosana, L. maximowiczii*
Balsamic acid	*Styrax tonkinensis, S. benzoin*
Barbaloin	*Aloe barbadensis, A. vera*
Barbatic acid	*Usnea diffracta., U. longissima*
Barium	*Juncus effusus*
Bassorine	
Bauerenol	*Acronychia pedunculata, A. laurifolia*
Bavachinin	*Psoralea corylifolia*

Component	Source
Behenic acid	*Brassica alba, B. juncea, Pongamia pinnata*
Bellidifolin	*Swertia pseudochinensis*
Bensaldehyde	*Rosa rugosa, Cinnamomum zeglanicum, Melaleuca leucadendra, Prunus persica*
Benzene tert-butyl	*Zanthoxylum bungeanum*
Benzoic acid	*Arisaema amurense, A. consanguineum, A. erubescens, A. heterophyllum, A. peninsulae, A. peninsulae, A. thunbergii, Daemonorops draco, Jasminum samba, Paeonia albiflora, P. edulis, P. japonica, P. lactiflora, P. moutan, P. officinalis, Phyllostachys bambusoide, P. nigra*
Benzolacetic ester	*Daemonorops margaritae*
Benzoquinone	*Ligusticum chuangxiong*
Benzyl benzoate	*Dianthus chinensis* L.
Benzoyl isothiocyanate	*Descurania Sophia* Webb ex Prantl.
Benzoyl paeoniflorin	*Paeonia albiflora, P. edulis, P. japonica, P. lactiflora, P. moutan, P. officinalis*
Benzoyl salicin	*Populus alba, P. davidiana, P. tomentosa*
Benzoyl ecgonine	*Erythroxylum coca*
Benzoyl alcohol	*Prunus persica*
Benzoyl isothiocyanate	*Brassica alba, B. juncea*
Benzoyl acetone	*Aquilegia buergeriana, A. parviflora*
Benzoyl ramanone	*Metaplexis japonica* Thunb.
Benzoyl bezoic acid	*Conyza Canadensis, Rhododendron dauricum* L.
Berbamine	*Berberis amurensis, B. poiretii, B. sibirica, B. soulieana, Stephania cepharantha, Thalictrum aquilegifolium, T. baicalense, T. fauriel, T. petaloideum, T. simplex, T. squarrosum, T. thunbergii*

Berberine	*Berberis amurensis, B. poiretii, B. sibirica, B. soulieana, Caltha palustris, Coptis chinensis, C. japonica, C. teeta, Chenopotium ambrosiodes, Jeffersonia dubia (Maxim) Benth et Hook f. Mahonia japonica, Nandina domestica, Papaver somniferum, Phellodendron amurense, P. chinensis, Scutellaria formosana, Thalictrum aquilegifolium, T. baicalense, T. fauriel, T. foetidum, T. glandulissimum, T. ichangense, T. petaloideum, T. simplex, T. squarrosum, T. thunbergii, Xanthoxylum piperitum, Zanthoxylum schinifolium*
Bergapten	*Anethum graveoleus, Argelica amurensis, A. gigas* Maxim. *A. anomala, A. dahurica, A. pubescens, A. sinensis* (Oliv) Diels *Dictamnus albus, D. dasycarpus, Heracleum lanatum, H. moellendorffii* Hance *Poncirus trifoliata, Zanthoxylum schinifolium*
Bergaptin	*Ficus carica*
Bergenin	*Ardisia quinquegona, A. sieboldii, A. longicarpa, A. chinensis, Cuscuta chinensis, C. europaea, C. japonica, C. lupuliformis, Mallotus repandus*
Bergenine glucoside	*Ardisia japonica*
Beta-agarofuran	*Aquilaria agallocha, A. sinensis*
Beta-amirine	*Jatropha podagrica*
Beta-amyrenol	*Spilanthes acmella*
Beta-amyrin	*Aleurites fordii, Chimaphila umbellata, Cirsium chinense, C. japonicum, Cuscuta chinensis, C. europaea, C. japonica, C. lupuliformis, Eclipta erecta, Eupatorium odoratum, Euphorbia antiquorum, Firmiana simplex, Pedicularis resupinata, Tamarindus indicus, Taraxacum mongolicum, T. sinicum, Viburnum sargenti, Viscum album, V. coloratum*
Beta-amyrin acetate	*Artocarpus altilis, Firmiana simplex*

Component	Source
Beta-asarone	Acorus calamus, Achyranthes asperia, Amaranthus tricolor, Boehmeria densiflora, Canarium album, C. sinense, Castanea crenata, C. mollissima, Corylus heterophylla, C. mandshurica, Petasites japonicus, Zea mays
Beta-carotenoid	Crocus sativus, Hippophae rhamnoides
Beta-caryophyllene	Ageratum conyzoides, A. houstonianum, Artemisia argyi, A. halodendron, A. igniaria, A. indica, A. integrifolia, A. japonica, A. keiskeana, A. lagocephala, A. lavandulaefolia, A. scoparia, A. selengensis, A. sieversiana, A. vulgarts, Murraya paniculata, Perilla frutescens, P. ocymoides, P. polystachya, P. arguta, Vitex chinensis, V. jeguaod
Beta-caryophylline	Eugenia aromatica, E. caryophyllata, E. ulmoides
Beta-cyperene	Cyperus rotundus
Beta-cyperol	Cyperus rotundus
Beta-D-glucosyloxy	Melilotus alba, M. suaveolens, M. indica
Beta-dichroine	Adamia chinensis, A. cyanea, A. versicolof
Beta-dihydropseudoionone	Cymbopogon citratus
Beta-dimethylacrylate	Arnebia euchroma
Beta-dimethylacrytoylshikonin	Arnebia euchroma
Beta-elemene	Juniperus rigida, Panax ginseng
Beta-eudesmol	Atractylis chinensis, A. lancea, A. lyrata, A. ovata, Magnolia hypoleuca, M. officinalis, M. japonica
Beta-globulin	Phaseolus angularis, P. lunatus, P. radiatus, P. vulgaris
Beta-glycyrrhetinic acid	Glycyrrhiza pallidiflora, G. uralensis
Beta-guaienen	Artemisia lactiflora

Beta-gurjunene   Zanthoxylum bungeanum

Beta-ionone   Lawsonia inermis

Beta-methylaesculetin   Convolvulus arvensis

Beta-OH-isovalerylshikonin   Arnebia euchroma

Beta-p glucophyranoside   Bauhinia championi

Beta-phellandrene   Osmanthus fragrans

Beta-phenethyl alcohol   Litchi chinensis

Beta-pinene   Agastache rugosa, A. rugosa f. hypoleuca, Asarum canadense, A. heterotropoides, A. sieboldii, Glechoma hederacea, G. longituba, Hedychium coronarium, Ledum palustre, Lindera glauca, Oenothera javanica

Beta-santalene   Santalum album, S. myrtifolium, S. verum

Beta-santalol   Santalum album, S. myrtifolium, S. verum

Beta-selinene   Cyperus brevifolius, C. difformis, C. glomeratus, C. iria

Beta-sitosterol   Acanthopanax giraldii, A. gracilistylus, A. spinosum, Adenophora triphylla, A. verticillata, Adina rubella, A. ratemosa, Ailanthus altissima, Ajuga bracteosa, Aletris formosuna, A. spicata, Aristolochia contorta, A. kaempferi, A. longa, A. recurvilabra, Arnebia euchroma, Avena fatua, Bauhinia championi, Cirsium chinense, C. japonicum, Cuscuta chinensis, C. europaea, C. japonica, C. lupuliformis, Cynodon dactylon, Cynomorium coccineum, C. songarium, Cyperus rotundus, Eugenia aromatica, E. caryophyllata, E. ulmoides, Firmiana simplex, Glehnia littoralis, Ipomoea cairica, Jatropha podagrica, Lespedeza cuneata, Matteuccia struthiopteris, Oldenlandia diffusa, Ophioglossum japonicus, Ophiopogon japonicus, Ophiorrhiza japonica, O. mungos, Papaver somniferum, Petasites japonicus, Polygonatum chinense, P. cirrhifolium, P. macropodium, P. officinale, P. sibiricum,

Component	Source
Beta-sitosterol	*P. stenophyllum, P. odoratum, P. vulgare, Prunus padus, Rauvolfia verticilata, Rubus coreanus, R. crataegifolius, R. matsumuranus, R. saxatilis, Schizonepeta multifida, S. tenuifolia, Scopalia dulcis, Scutellaria baicalensis, S. grandiflora, S. lanceolaria, S. macrantha, S. rivulararis, S. viscidula, Solanum incanum, Syzygium cuminii, Tamarindus indicus, Taraxacum mongolicum, T. sinicum, Trapa bispinosa, Viburnum sargenti*
Beta-sitosterol glucoside	*Eleutherocossus senticosus*
Beta-sitosterolm	*Aleurites fordii*
Beta-sitosteryl palmitate	*Adenophora triphylla, A. verticillata*
Beta-solamargine	*Solanum aculeatissimum*
Beta-sotpsterols	*Dioscorea bulbifera*
Beta-sterol	*Ginkgo biloba*
Beta-taralin	*Aralia chinensis, A. cordata, A. elata*
Beta-trevilline	*Erythroxylum coca*
Betacyamines	*Gomphrena globosa*
Betacyanin	*Portulaca grandiflora*
Betaine	*Amaranthus caudatus, A. paniculatus, Astragalus complanatus, A. henyri, A. hoantchy, A. membranaceus, A. melilotoides, A. mongholicus, A. reflexistipulus, A. sinensis, Chenopodium album L. Firmiana simplex, Lycium chinense, L. barbarum, L. megistocarpum, L. ovatum, L. trewianum, L. turbinatum, Salsola collina*
Betanidin	*Bougainvillea brasiliensis, B. glabra, Portulaca grandiflora*
Betanin	*Portulaca grandiflora*
Betonicine	*Achillea alpina, A. millefolium*

Betulafolienetetraol	*Betula mandshurica, B. platyphylla*
Betulafolienetriol	*Betula mandshurica, B. platyphylla*
Betulic acid	*Ziziphus jujuba, Z. spinosa*
Betulin	*Betula mandshurica, B. platyphylla, Euphorbia lathyrus, E. lucorum, E. resinfera, E. thymifolia, Platycodon autumnalis, P. grandiflorum, P. sinensis, Vicia faba, Ziziphus jujuba, Z. spinosa*
Betulinic acid	*Adina rubella, A. ratemosa, Alnus japonica, Diospyros chinensis, D. costata, D. khaki, D. lotus, D. roxburgii, Melaleuca leucadendra, Menyanthes trifoliata, Pedicularis resupinata, Syzygium cuminii, Ziziphus jujuba, Z. spinosa*
Betuloside	*Betula mandshurica, B. platyphylla*
Bianthraquinonyl	*Rheum officinale, R. palmatum, R. tanguticum, R. undulatum, R. koreanum*
Biatractylolide	*Atractylodes chinensis, A. japonica, A. koreana, A. lancen, A. macrocephala, A. ovata*
Biflorine	*Oldenlandia chrysotricha, C. corymbosa*
Biflorone	*Oldenlandia chrysotricha, C. corymbosa*
Bigelovin	*Forsythia suspensa, Inula britannica, I. japonica, I. linariaefolia, I. salsoloides*
Bilobal	*Ginkgo biloba*
Bilobetin	*Ginkgo biloba*
Biotin	*Angelica polymorpha, A. sinensis, Arachis hypogaea, Polyporus umbellatus*
Bis (2-ethyl butyl) phthalate	*Oenothera javanica*
Bisabolene	*Daucus carota, Murraya paniculata, Valeriana alternifolia, V. amurensis, V. fauriei, V. subbipinnatifolia*
Bisesquiterpenoid	*Atractylodes chinensis, A. japonica, A. koreana, A. lancen, A. macrocephala, A. ovata*
Bitter glycoside	*Centaurium meyeri, Siegesbeckia orientalis L.*

Component	Source
Bitter principle	*Elephantopus scaber*
Bocconine	*Macleaya cordata*
Bocconoline	*Macleaya cordata*
Bogoroside	*Antiaris toxicaris*
Bonducin	*Caesalpinia decapetula*
Borneol	*Artemisia argyi, A. gmelini, A. halodendron, A. igniaria, A. indica, A. integrifolia, A. japonica, A. keiskeana, A. lagocephala, A. lavandulaefolia, A. scoparia, A. selengensis, A. sieversiana, A. vulgarts, Blumea balsumifera (L.) DC, Chrysanthemum boreale, C. indicum, C. lavandulaefolium, C. procumbens, C. tripartium, Coriandrum sativum, Cunninghamia lanceolata, Curcuma pallida, C. phaeocoulis, Dryobalanops aromatica, D. camphora, Hedyotis corymbosa, Juniperus rigida, Kaempferia galanga, Thymus amurensis, T. disjunctus, T. kitagawianus, T. komarovii, T. przewalskii, T. quinquecostatus, T. vulgaris, Valeriana alternifolia, V. amurensis, V. fauriei, V. subbipinnatifolia*
Borneol acetate	*Amomum cardamomum, A. globosum, A. tsao-ko, A. villosum, A. xanthloides*
Bornol	*Chrysanthemum jucundum, C. koraiense, C. morifolium, C. sinense*
Bornyl acetate	*Eupatorium chinense, E. lindleyanum, E. japonicum, Hedyotis corymbosa, Thymus vulgaris*
Bornyl isovalerate	*Valeriana alternifolia, V. amurensis, V. fauriei, V. subbipinnatifolia*
Bornylautate	*Lindera glauca*
Boschniakine	*Cistanche deserticola, Boschniaka rossica* Cham & Schlocht
Boschiniakinic acid	*Boschniaka rossica* Cham & Schlocht
Boschnialactone	*Cistanche deserticola*
Bourbonene	*Luffa aegyptiaca, L. cylindrica, L. faetida, L. petola*

Brahmissoside	Centella asiatica
Brahmoside	Centella asiatica
Brasilin	Caesalpinia sappan
Bromelin	Ananas comosus
Bruceines	Brucea javanica, B. sumatrana
Bruceolide	Brucea javanica, B. sumatrana
Brucine	Strychnos nux-vomica, S. pierriana
Brusatol	Brucea javanica, B. sumatrana
Buddleoglycoside	Buddleia formosana, B. madaguscariensis, B. officinalis
Bufotenine	Desmodium pulehellum, Phyllodium pulchellum
Bulbocapnine	Corydalis decumbens
Bullatines	Aconitum barbatum, A. austroyunnanense
Bupleuran	Bupleurum chinense, B. falcatum, B. scorzoneraefolium
Bursic acid	Capsella bursa-pastoris
Buteine	Bidens tripartita
Butin	Bidens tripartita
Butyl phthalate	Oenanthe javanica (Bl) DC
Butylidenephalide	Angelica polymorpha, A. sinensis
Butyric acid	Ajuga bracteosa, Euphorbia coraroides, E. lasiocaula, E. lunulata, E. pallasii, E. pekinensis, E. sampsoni, E. sieboldiana, E. esula, E. helioscopia, Melia japonica, M. toosendan, M. azedarach, Pueraria montana, P. thunbergiana, Ulmus campestris, U. macrocarpa, U. pumila

Component	Source
Buxanmine E	*Buxus harlandii, B. microphylla*
Buxpiine	*Buxus microphylla*
Buxpiine K	*Buxus harlandii*
Buxtauine	*Buxus microphylla*
Byak-angelicin	*Angelica pubescens, A. amurensis, A. anomala, A. dahurica*
Byak-angelicol	*Angelica amurensis, A. anomala, A. dahurica, A. pubescens, A. sinensis* (Oliv) Diels
c-3-epi-wilsonine	*Cephalotaxus wilsoniana*
Cadinene	*Alpinia officinarum, Menyanthes trifoliata, Piper cubeba, Podocarpus macrophyllus, Solidago canadensis, Valeriana alternifolia, V. amurensis, V. fauriei, V. subbipinnatifolia, Zanthoxylum bungeanum*
Caffeic acid	*Ajuga bracteosa, Azolla imbricata, Johnston Buglossoides arvense L.. Cirsium albescens, C. brevicaule, C. littorale, C. maakii, C. segetum, C. setosum, C. vlassovianum, Convolvulus arvensis, Crataegus cuneata, C. chlorusarca, C. dahurica, C. maximowiczii, C. pentagyna, C. pinnatifida, C. sanguinea, Cucumis sativus, Elaeagnus pungens, E. umbellata, Fagopyrum esculentum, F. sagittatum, Impatiens balsamina, I. noli-tangere, I. textori, Inula britannica, I. japonica, I. linariaefolia, I. salsoloides, Matteuccia struthiopteris, Melilotus alba, M. suaveolens, M. indica, Polygonum aviculare, P. lapidosa, P. manshuriensis, P. vivipara, P. bistorta, Prunella vulgaris, Solidago dahurica, S. pacifica, S. virgaurea, Valeriana alternifolia, V. amurensis, V.fauriei, V. subbipinnatifolia, Viburnum sargenti, Xanthium chinense, X. japonicum, X. mongolicum, X. sibiricum, X. strumarium*
Caffeine	*Thea assamica, T. bohea, T. cantoniensis, T. chinensis, T. cochinchinensis, T. sinensis, T. viridis*

Cajuputol	*Melaleuca leucadendra*
Calamene	*Actinidia chinensis* Planch *Aconitum deinorrtuzum, Agastache rugosa, A. rugosa* f. *hypoleuca*
Calamenol	*Aconitum deinorrtuzum*
Calamenone	*Aconitum deinorrtuzum*
Calcium	*Curculigo orohiodes* Gaertn. *Duchesnea indica,* Laminaria *angusta, L. cichorioides, L. japonica, L. longipedalis, L. religiosa, Oxalis corriculaza, O. corymbosa, Phyllostachys bambusoide, P. nigra, Portulaca pilosa*
Calcium malate	*Euphorbia coraroides, E. lasiocaula, E. lunulata, E. pallasii, E. pekinensis, E. sampsoni, E. sieboldiana*
Calcium oxalate	*Acanthopanax gracilistylus, A. spinosum, Achyranthes japonica, Curculigo capitulata, C. ensifolia, C. malabarica, C. orchiodes, C. stams, Euphorbia coraroides, E. lasiocaula, E. lunulata, E. pallasii, E. pekinensis, E. sampsoni, E. sieboldiana, Sesamum indicum*
Calechin	*Rheum officinale, R. palmatum, R. tanguticum, R. undulatum, R. koreanum*
Calenduline	*Calendula officinalis*
Calotropin	*Asclepias curassavice* L.
Camelliagenins	*Camellia japonica*
Camellin	*Camellia japonica*
Campesterol	*Aleurites fordii,Alstonia scholaris* (L.) *R. Br. Arundinaria graminifolia, Cuscuta chinensis, C. europaea, C. japonica, C. lupuliformis, Dioscorea bulbifera, Lindera obtusiloba, Linum stelleroides, L. usitatissimum, Matteuccia struthiopteris, Rehmannia chinensis, R. glutinosa, Rubus coreanus, R. crataegifolius, R. matsumuranus, R. saxatilis, Schizonepeta multifida, S. tenuifolia, Syzygium aromaticum, Tamarindus indicus*

Component	Source
Camphene	*Anethum graveoleus, Aucklandia costus, A. lappa, Cunninghamia lanceolata, Curcuma longa, C. pallida, C. phaeocoulis, Daucus carota, Dryobalanops aromatica, D. camphora, Elettaria cardamomum, Hedychium coronarium, Ledum palustre, Lindera glauca, Liquidambar acerifolia, L. formosana, L. maximowiczii, Piper cubeba, Podocarpus macrophyllus, Poncirus trifoliata, Thymus vulgaris, Valeriana alternifolia, V. amurensis, V. fauriei, V. subbipinnatifolia, Vitex negundo, V. trifolia, V. rotundifolia, Zingiber officinale*
Campherene	*Myristica fragrans*
Camphol	*Euphorbia humifusa, E. hirta*
Camphor	*Artemisia gmelini, Blumea balsumifera, B. lacera, Chrysanthemum jucundum, C. koraiense, C. morifolium, C. sinense, Curcuma longa, C. pallida, C. phaeocoulis, Kaempferia galanga*
Camphore	*Artemisia argyi, A. halodendron, A. igniaria, A. indica, A. integrifolia, A. japonica, A. keiskeana, A. lagocephala, A. lavandulaefolia, A. scoparia, A. selengensis, A. sieversiana, A. vulgarts, Chrysanthemum boreale, C. indicum, C. lavandulaefolium, C. procumbens, C. tripartium*
Camphorm citral	*Elettaria cardamomum*
Camptothecin	*Camptotheca acuminata*
Canaline	*Medicago falcata, M. lupulina, M. polymorpha, M. ruthenica, M. sativa*
Canavalia gibberellin I	*Canavalia gladiata, C. ensiformis*
Canavalia gibberellin II	*Canavalia gladiata, C. ensiformis*
Canavaline	*Canavalia gladiata, C. ensiformis*
Canavanine	*Astragalus complanatus, A. henyri, A. hoantchy, A. membranaceus, A. melilotoides, A. mongholicus, A. reflexistipulus, A. sinensis, A. chinensis, Canavalia gladiata, C. ensiformis*

Candicine	*Phellodendron amurense, P. chinensis*
Canescein	*Erysimum amurense, E. cheiranthoides*
Caniferyl aldehyde	*Santalum album, S. myrtifolium, S. verum*
Cannabidiol	*Cannabis chinensis, C. sativa*
Cannabinol	*Cannabis chinensis, C. sativa*
Caoutchoue	*Artocarpus heterophyllus*
Capadiene	*Cyperus rotundus*
Capillanol	*Artemisia capillaris*
Capillarin	*Artemisia capillaris*
Capillene	*Artemisia capillaris*
Capillin	*Artemisia capillaris*
Capillon	*Artemisia capillaris*
Capric acid	*Citrullus anguria, C. edulis, C. lanatus, C. vulgaris, Lindera obtusiloba, Ulmus campestris, U. macrocarpa, U. pumila*
Capronic acid	*Aquilegia vulgaris*
Caprylic	*Cymbopogon citratus*
Capsularin	*Corchorus capsularis, C. olitorius*
Carbohydrate. fiber	*Arachis hypogaea, Bauhinia championi, B. variegata, B. variegata. indicum,Emilia sonchifolia (L.) DC. Erigeron Canadensis L. Triticum vulgare, Urtica angustifolia, U. cannabina, U. canrabina, U. lobata, U. tenacissima, U. urens, U. utillis, Zea mays*
Cardiac glucoside	*Thalictrum foetidum*
Cardioactive glycosides	*Periploca sepium Bunge*

Component	Source
Cardoside	*Gardenia florida, G. grandiflora, G. maruba, G. pictorum, G. radicans*
Cardiotonic constituent	*Cynanchum bunge* Decaisne.
Carene	*Juniperus rigida, Murraya paniculata, Piper cubeba*
Carenolide	*Periploca sepium*
Carnaubic acid	*Chenopodium album* L.
Cariandrol	*Coriandrum sativum* L.
Carosine	*Catharanthus roseus*
Carotenes	*Achillea alpina, A. millefolium, Aleurites moluceanu, Alpinia katsumadai, A. globosum, A. kumatake, Anthriscus aemula, A. sylvestris, Blumea lacera, Dasiphora fruticosa (Nestler) Kom. Daucus carota, L. Eriobotrya japonica, Calendula officinalis, Gnaphalium affine, G. arenarium, G. confusum, G. javanum, G. luteo-album, G. multiceps, G. ramigerum, G. tranzschelii, G. uliginosum, Heracleum dissectum, H. lanatum, Lycium barbarum, L. megistocarpum, L. ovatum, L. trewianum, L. turbinatum, Polygonum bistorta, Sanguisorba officinalis, S. grandiflora, S. parviflora, S. x tenuifolia, Spinacia oleracea L. Taraxacum officinale*
Carotenoids	*Cuscuta australis, Neoalsomitra integrifoliola, Spiraea salicifolia*
Carotol	*Daucus carota*
Caprylic acid	*Citrullus anguria, C. edulis, C. lanatus, C. vulgaris*
Cartharmin	*Carthamus tinctorius*
Carvacrol	*Thymus amurensis, T. disjunctus, T. kitagawianus, T. komarovii, T. przewalskii, T. quinquecostatus, T. vulgaris*
Carveol	*Carum carvi* L.

Carvone — *Andthum graveolens L. Carum carvi L. Chrysanthemum boreale, C. indicum, C. lavandulaefolium, C. procumbens, C. tripartium, L. aegyptiaca, L. cylindrica, L. faetida, L. petola*

Caryophyllene — *Agastache rugosa, A. rugosa f. hypoleuca, Cinnamomum zeglanicum, Daucus carota, Juniperus rigida, Lindera glauca, Luffa aegyptiaca, L. cylindrica, L. faetida, L. petola, Piper nigrum, Thuja koraiensis, T. orientalis, T. chinensis*

Caryophyllene oxide — *Vitex chinensis, V. jeguaod*

Cassiollin — *Cassia occidentalis, Cassia torosa*

Cassyfiline — *Cassytha filliformis*

Cassythidine — *Cassytha filliformis*

Cassythine — *Cassytha filliformis*

Casticin — *Vitex trifolia, V. rotundifolia, V. trifolia, V. rotundifolia, V. nequndo*

Castor oil — *Ricinus communis*

Catalpol — *Rehmannia chinensis, R. glutinosa.*

Catalpalactone — *Catalpa ovata* G. Don.

Catalposide — *Catalpa ovata* G. Don.

Catechin — *Cynomorium coccineum, C. songarium, Elaeagnus pungens, E. umbellata, Rosa acicularis, R. amygdalifolia, R. davurica, R. koreana, R. laevigata, R. maximowicziana, R. multiflora*

Catechin derivatives — *Agrimonia eupatoria, A. pilosa, A. viscidula*

Catecholamines — *Portulaca oleracea*

Catechutanic acid — *Acacia cutechu*

Catharanthine — *Catharanthus rosous* (L.) G. Don

Cathartic acid — *Picrorhiza kurroa*

Component	Source
Caudoside	*Strophanthus divaricatus*
Caudostroside	*Strophanthus divaricatus*
Cauloside	*Caulophyllum robustum*
Celastrol	*Tripterygium wilfordii*
Cellulose	*Quercus acutissima, Q. aliena, Q. dentata, Q. liaotungensis, Q. mongolica, Q. variabilis*
Celosiaol	*Celosia argentea, C. cristata*
Cembrene	*Commiphora myrrha*
Centaur X	*Conyza canadensis*
Cephaeline	*Alangium lamarckii*
Cephalomannine	*Taxus cuspidata, T. chinensis, T. yunnanensis*
Cephalotaxine	*Cephalotaxus fortunei, C. qinensis, C. oliveri, C. wilsoniana*
Cephalotaxinone	*Cephalotaxus fortunei, C. qinensis, C. oliveri, C. wilsoniana*
Cepharamine	*Stephania cepharantha*
Cepharanoline	*Stephania cepharantha*
Cepharanthine	*Stephania cepharantha*
Cerberin	*Cerbera manghas* L.
Cereberigenin	*Cerbera manghas* L.
Cerberose	*Cerbera manghas* L.
Cerotic acid	*Ajuga bracteosa, Artocarpus heterophyllus, Plumeria rubra, Viola acuminata, V. alisoviana, V. collina, V. dissecta, V. mandshurica, V. patrini, V. prionantha, V. verecunda*
Cerotinic acid	*Ficus carica, Plumeria rubra*

Cerylalcohol	*Buglossoides arvense* (L.) *Johnston Calendula officinalis, Chamaenerion angustifolium, Eupatorium odoratum L.Lactuca raddeana, L. indica, L. sativa*
Cerylic alcohol	*Solidago pacifica Juzepczuk. Taraxacum officinale*
Cetylalcohol	*Cedrela sinensis* A. Juss.
Cevadine	*Caltha palustris*
Chalcone glucose	*Asarum europaeum, Glycyrrhiza pallidiflora, G. uralensis*
Chanerol	*Chamaenerion angustifolium*
Chatinine	*Valeriana alternifolia, V. amurensis, V. fauriei, V. subbipinnatifolia*
Chaulmoogric acid	*Hydnocarpus anthelmintica, H. castaneus*
Chavicine	*Piper nigrum*
Chavicol	*Eugenia aromatica, E. caryophyllata, E. ulmoides*
Chebulic acid	*Terminalia chebula*
Chebulinic acid	*Phyllanthus emblica, Terminalia chebula*
Chelerythrine	*Chelidonium album, C. hybridum, C. majus, C. serotinum, Hypecoum erectum* L. *Macleaya cordata*
Chelidocystatin	*Chelidonium album, C. hybridum, C. majus, C. serotinum*
Chelidonine	*Chelidonium album, C. hybridum, C. majus, C. serotinum*
Chelilutine	*Macleaya cordata*
Chelirubine	*Hypecoum erectum* L. *Macleaya cordata*
Chikusetsa saponin II	*Panax japonicum*
Chikusetsa saponin IV	*Panax japonicum*
Chimaphilin	*Chimaphila umbellata, Pyrola decorata, P. japonica, P. incarnata, P. renifolia, P. rotundifolia*

Component	Source
Chinic acid	*Chimaphila umbellata*
Chiretta	*Centaurium meyeri*
Chlogogenic acid	*Melilotus alba, M. suaveolens, M. indica*
Chlorogen acid	*Sambucus coreana, S. latipinna, S. manshurica, S. peninsularis, S. sieboldiana, S. williamsii*
Chlorogenic acid	*Ajuga bracteosa, Campanula glomerata, C. punctata, Cirsium albescens, C. brevicaule, C. littorale, C. maakii, C. segetum, C. setosum, C. vlassovianum, Crataegus cuneata, C. chlorusarca, C. dahurica, C. maximowiczii, C. pentagyna, C. pinnatifida, C. sanguinea, Cucumis sativus, Elaeagnus pungens, E. umbellata, Inula britannica, I. japonica, I. linariaefolia, I. salsoloides, Lonicera acuminata, L. apodonta, L. brachypoda, L. japonica, L. confusa, L. hypoglauca, L. chinensis, L. flexuosa, L. maackii, Lythrum salicaria, Matteuccia struthiopteris, Orobanche caerulescens Stephan. Polygonum aviculare, P. lapidosa, P. manshuriensis, P. vivipara, Senecio argunensis, S. nemorensis, S. scandens, Sesamum indicum, Solidago dahurica, S. pacifica, S. virgaurea, Urtica angustifolia, U. cannabina, U. lobata, U. tenacissima, U. urens, U. utillis, Viburnum sargenti*
Chlorogenin	*Gardenia angusta, G. jasminoides*
Chlorophenolic acid	*Euphorbia hirta*
Chloroquine	*Artemisia annua, A. apiacea*
Cholestanol	*Rubus coreanus, R. crataegifolius, R. matsumuranus, R. saxatilis*
Cholesterol	*Cuscuta chinensis, C. europaea, C. japonica, C. lupuliformis, Linum stelleroides, L. usitatissimum, Pedicularis resupinata, Spinacia oleracea L.*

Choline — *Achillea millefolium L. Caltha palustris, Cannabis chinensis, C. sativa, Capsella bursa-pastoris, Cephalanoplos segetum, Chrysanthemum jucundum, Chrysanthemum jucundum, C. koraiense, C. morifolium, C. sinense, C. cinerriaefolium, Dictamnus albus, D. dasycarpus, Dioscorea opposita, Diospyros chinensis, D. costata, D. khaki, D. lotus, D. roxburgii, Firmiana simplex, Glycine max, G. soja, Humulus scandens, Hyoscyanus bohemicus, Menyanthes trifoliata, Pinellia ternata, P. tuberifera, Pyrethrum cinerariifolium, P. sinense, Sesamum indicum, Solanum lyratum, S. melongena, Taraxacum officinale, T. formosanum, Tetragonia tetragonoides, Viscum album*

Choline asparaginer — *Humulus lupulus*

Chromone — *Carum carvi, Ligusticum jeholense, L. pyrenacum, L. sinense, L. tenuissimum*

Chrysanthemaxanthin — *Chrysanthemum boreale, C. indicum, C. lavandulaefolium, C. procumbens, C. tripartium, Senecio argunensis, S. nemorensis, S. scandens*

Chrysanthemin — *Chrysanthemum jucundum, C. koraiense, C. morifolium, C. sinense*

Chrysanthinin — *Chrysanthemum boreale, C. indicum, C. lavandulaefolium, C. procumbens, C. tripartium*

Chrysarobin — *Cassia siamea*

Chryso-obtusin — *Cassia nomame, C. obtusifolia, C. tora*

Chrysophanein — *Rumex acetosa, R. acetosella, R, amurensis, R. aquaticus, R. gmelini, R. longifolius, R. maritimus, R. marschallianus, R. stenophyllus, R. thyrsiflorus, R. crispus, R. japonicus*

Chrysophanic acid — *Cassia alata, C. siamea, Duchesnea indica, Polygonum multifolrum, P. chinensis*

Component	Source
Chrysophanol	*Cassia nomame, C. obtusifolia, C. tora, Hemerocallis minor Miller, H. fulva L. Polygonum perfoliatum, P. tinctorium, P. multifolrum, P. chinensis, Rhamnus davurica, R. parvifolia, Rumex acetosa, R. acetosella, R. amurensis, R. aquaticus, R. gmelini, R. longifolius, R. maritimus, R. marschallianus, R. patientia, R. stenophyllus, R. thyrsiflorus, R. officinale, R. palmatum, R. tanguticum, R. undulatum, R. koreanum*
Cicerose	*Isatis chinensis, I. chinensis, I. tinctoria*
Dihydrocarveol	*Carum carvi* L.
Cimifugenol	*Cimicifuga dahurica, C. foetida, C. heracleifolia, C. racemosa, C. ussuriensis*
Cimigenol	*Cimicifuga dahurica, C. foetida, C. heracleifolia, C. racemosa, C. ussuriensis*
Cimitin	*Cimicifuga dahurica, C. foetida, C. heracleifolia, C. racemosa, C. ussuriensis*
Cincholic acid	*Adina rubella, A. ratemosa*
Cincole	*Alpinia oxyphylla*
Cineol acid	*Alpinia officinarum, Artemisia argyi, A. halodendron, A. igniaria, A. indica, A. intergrifolia, A. japonica, A. keiskeana, A. lagocephala, A. lavandulaefolia, A. scoparia, A. selengensis, A. sieversiana, A. vulgarts, Chrysanthemum boreale, C. indicum, C. lavandulaefolium, C. procumbens, C. tripartium, Eucalyptus robusta, Kaempferia galanga, Magnolia biloba, M. denudata, M. discolor, M. lilifloria, M. puruea, Piper cubeba, Vitex nequndo*
Cineole	*Alpinia japonica, A. katsumadai, A. globosum, A. kumatake, Artemisia brachyloba, A. gmelini, Blumea balsumifera, B. lacera, B. balsumifera, Cinnamomum camphora, Cunninghamia lanceolata, Curcuma pallida, C. phaeocoulis, Elettaria cardamomum, Lindera glauca, Liquidambar acerifolia, L. formosana, L. maximowiczii, L. acerifolia, Melaleuca leucadendra, Menyanthes trifoliata*

Cinnamic acid	*Cinnamomum aromaticum, C. cassia, Daemonorops draco, Lycium chinense*
Cinnamic aldehyde	*Cinnamomum aromaticum, C. cassia, C. aromaticum, C. cassia, C. zeglanicum*
Cinnamyl acetate	*Cinnamomum aromaticum, C. cassia*
Cinnamylococaine	*Erythroxylum coca*
Cissamine	*Cissampelos pareira*
Cissampareine	*Cissampelos pareira*
Citral	*Alpinia katsumadai, A. globosum, A. kumatake, Blumea lacera, Citrus deliciosa, C. nobilis, C. reticulata, Cymbopogon citratus, Daucus carota, Litsea cubeba, Magnolia biloba, M. denudata, M. discolor, M. liliflora, M. purpurea, Rosa rugosa, Xanthoxylum piperitum, Thymus vulgaris, Zingiber officinale*
Citrene	*Cunninghamia lanceolata*
Citric acid	*Capsella bursa-pastoris, Chaenomeles japonica, C. sinensis, C. speciosa, Crataegus cuneata, C. chlorusarca, C. dahurica, C. maximowiczii, C. pentagyna, C. pinnatifida, C. sanguinea, Cydonia sinensis, Drosera anglica, D. burmunni, D. rotundifolia, Eriobotrya japonica, Lactuca raddeana, L. indica, L. sativa, Litchi chinensis, Oxalis corriculaza, O. corymbosa, Prunus domestica, P. glandulosa, P. japonica, P. mume P. persica, Ribes mandshurica, Viburnum sargenti*
Citrifoliol	*Poncirus trifoliata*
Citrogellol	*Cymbopogon citratus*
Citrol	*Magnolia biloba, M. denudata, M. discolor, M. liliflora, M. purpurea, Rosa rugosa*
Citronellal	*Cymbopogon citratus, Elettaria cardamomum*
Citronellic	*Cymbopogon citratus*

Component	Source
Citronellol	*Daucus carota, Elettaria cardamomum, Juniperus rigida, Plumeria rubra, Vitex nequndo, Xanthoxylum piperitum, Zanthoxylum schinifolium*
Citrulline	*Citrullus anguria, C. edulis, C. lanatus, C. vulgaris, Medicago falcata, M. lupulina, M. polymorpha, M. ruthenica, M. sativa*
Clavatine	*Lycopodium clavatum, L. obscurum, L. selago, L. serratum, L. annotinum, L. cernum, L. compianatum*
Clavatoxine	*Lycopodium clavatum, L. obscurum, L. selago, L. serratum*
Clavoloninine	*Lycopodium clavatum, L. obscurum, L. selago, L. serratum*
Clematoside A	*Clematis intriicata, C. mandshurica*
Cleomin	*Cleome spinosa, C. gynandra, C. viscosa*
Cleroden drin A,	*Clerodendrum trichotomum Thunb. C. spicatus*
Clerodolone	*Clerodendrum trichotomum, C. spicatus*
Clerodone	*Clerodendrum trichotomum Thunb.*
Clerosterol	*Clerodendrum fragrans. C. trichotomum Thunb.*
Clividine	*Clivia miniata*
Clovene	*Zanthaxylum bungeanum*
Cnidiadin	*Cnidium monnieri*
Cnidilide	*Ligusticum chuanziang*
Cnidimine	*Cnidium monnieri*
Cocculolidine	*Cocculus laurifolius, C. sarmentosus, C. trilobus*
Colchicine	*Hemerocallis fulva L.*

Coclobine	*Cocculus laurifolius, C. sarmentosus, C. trilobus*
Codeine	*Papaver somniferum*
Coessidine	*Holarrhenia antidy-senterica* Wall
Coixenolide	*Coix agrestis, C. chinensis, C. lachryma*
Coixol	*Coix agrestis, C. chinensis, C. lachryma*
Colchicine	*Iphigenia indica, Lilium brownii, L. concolor, L. dauricum, L. distichum, L. japonicum, L. lancifolium, L. pumilum,Tulipa edulis, T. gesneriana*
Colchicine amide	*Iphigenia indica*
Columbamine	*Coptis chinensis, C. japonica, C. teeta, Corydalis ambigua, C. repens, C. turtschaninovii, C. yanhusuo, C. ternata*
Columbianetin	*Cnidium monnieri*
Comarinic acid-β-glucoside	*Hierochloe odorata*
Commelin	*Commelina communis*
Complanatine	*Lycopodium annotinum, L. cernum, L. compianatum*
Conamine	*Holarrhenia antidy-senterica* Wall
Concuressine	*Holarrhenia antidy-senterica* Wall
Condensed tannin	*Leucaena leucocephala*
Condurangin	*Hoya carnosa*
Conessimine	*Holarrhenia antidy-senterica* Wall
Confertifolin	*Persicaria hydropiper*
Coniferyl alcohol	*Blumea lacera*
Coniferyl cinnamate	*Styrax tonkinensis, S. benzoin*

Component	Source
Conkurchine	*Holarrhenia antidy-senterica* Wall
Convallamarin	*Convallaria keiskei, Polygonatum chinense, P. cirrhifolium, P. macropodium, P. officinale, P. sibiricum, P. stenophyllum, P. odoratum, P. vulgare*
Convallarin	*Polygonatum chinense, P. cirrhifolium, P. macropodium, P. officinale, P. sibiricum, P. stenophyllum, P. odoratum, P. vulgare*
Convallatoxin	*Adonis chrysocyathus, A. brevistyla, A. vernalis, Antiaris toxicaris, Convallaria keiskei*
Convallatoxol	*Convallaria keiskei*
Convalloside	*Convallaria keiskei*
Copaene	*Artemisia lactiflora*
Coptisine	*Chelidonium album, C. hybridum, C. majus, C. serotinum, Coptis chinensis, C. japonica, C. teeta, Corydalis ambigua, C. repens, C. turtschaninovii, C. yanhusuo, C. ternata, C. incisa, C. bungeana, Hypecoum erectum L. Macleaya cordata, Papaver amurense, P. nudicaule, P. radicatum, T. aquilegifolium, T. baicalense, T. fauriel, T. petaloideum, T. simplex, T. squarrosum, T. thunbergii*
Corchorin	*Corchorus capsularis, C. olitorius*
Corchoritin	*Corchorus capsularis, C. olitorius*
Corchoroside A	*Adonis chrysocyathus, A. brevistyla, A. amurensis Regel & Radde. A. vernalis, Corchorus olitorius, Erysimum amurense, E. cheiranthoides*
Corchotoxin	*Corchorus capsularis, C. olitorius*
Cordycepic acid	*Veronica linariaefolia*
Cordycepin	*Cordyceps sinensis*
Coreximine	*Corydalis incisa, C. bungeana*

Coriandrol — *Coriandrum sativum*

Corilagin — *Sepium sebiferum, S. discolor*

Cornigerine — *Iphigenia indica*

Cornin — *Cornus officinalis, Macrocarpium officinalis*

Coroglaucigenin — *Asclepias curassavica* L.

Corosolic acid — *Crataegus cuneata, C. chlorosarca, C. dahurica, C. maximowiczii, C. pentagyna, C. pinnatifida, C. sanguinea*

Corticosteroids — *Costus specious*

Corycavine — *Corydalis incisa, C. bungeana*

Corydalamine — *Corydalis ambigua, C. repens, C. turtschaninovii, C. yanhusuo, C. ternata*

Corydalis — *Corydalis ambigua, C. repens, C. turtschaninovii, C. yanhusuo, C. ternata*

Corydamine — *Corydalis incisa, C. bungeana*

Coryfolin — *Psoralea corylifolia*

Corylifolinin — *Psoralea corylifolia*

Corynantheine — *Uncaria hirsuta, U. rhynchophylla*

Corynoline — *Corydalis incisa, C. bungeana*

Corynoloxin — *Corydalis incisa, C. bungeana*

Corynoxeine — *Uncaria hirsuta, U. rhynchophylla*

Cosin — *Astragalus complanatus, A. henryi, A. hoantchy, A. membranaceus, A. melilotoides, A. mongholicus, A. reflexistipulus, A. sinensis*

Cosmosiin — *Agrimonia eupatoria, A. pilosa, A. viscidula, Chrysanthemum jucundum, C. koraiense, C. morifolium, C. sinense*

Component	Source
Costene	*Aucklandia costus, A. lappa*
Costol	*Aucklandia costus, A. lappa*
Costulactone	*Aucklandia costus, A. lappa*
Costunolide	*Eupatorium formosanum*
Coumarinic acid	*Melilotus alba, M. suaveolens, M. indica*
Coumarins	*Achillea alpina, A. millefolium, Ageratum conyzoides, A. houstonianum, Alternanthera philoxeroides, A. sessilis, Angelica pubescens, A. gigas Maxim A. sinensis (Oliv) Diels Anthriscus aemula, A. sylvestris, Artemisia lactiflora, Astragalus complanatus, A. henyri, A. hoantchy, A. membranaceus, A. melilotoides, A. mongholicus, A. reflexistipulus, A. sinensis, Bidens tripartita, Coriandrum sativum, Carum carvi, C. aromaticum, C. cassia, Heracleum dissectum, H. lanatum, Hierochloe odorata, Ligusticum jeholense, L. pyrenacum, L. sinense, L. tenuissimum, Peucedanum formosanum, Polygonum bistorta, Rhododendron mucronatum, Sanguisorba officinalis, S. grandiflora, S. parviflora, S. x tenuifolia, S. chinensis, S. baicalensis, S. japonica, Taraxacum officinale*
Coumaroylquinic acid	*Campanula glomerata, C. punctata*
Coumesterol	*Medicago falcata, M. lupulina, M. polymorpha, M. ruthenica, M. sativa*
Couvallatoxin	*Adonis amurensis* Regel & Radde
Crataegol acid	*Syzygium aromaticum*
Crataegolic acid	*Chamaenerion angustifolium, Crataegus cuneata, C. chlorusarca, C. dahurica, C. maximowiczii, C. pentagyna, C. pinnatifida, C. sanguinea, Psidium guajava*
Crocetin	*Crocus sativus*
Crocetin di-glucose ester	*Crocus sativus*

Crocetin geniobiose glucose ester	*Crocus sativus*
Crotin	*Crocus sativus, Croton cascarilloides, C. tiglium*
Croton resin	*Croton cascarilloides, C. tiglium*
Crotonic acid	*Croton cascarilloides, C. tiglium*
Crotonoside	*Croton cascarilloides, C. tiglium*
Crustecdysone	*Pteridium aquilinum* (L.) Kuhu
Crybulbine	*Corydalis ambigua, C. repens, C. turtschaninovii, C. yanhusuo, C. ternata*
Cryptone	*Piper nigrum*
Cryptopine	*Thalictrum aquilegifolium, T. baicalense, T. fauriel, T. petaloideum, T. simplex, T. squarrosum, T. thunbergii*
Cryptotaenen	*Cryptotaenia japonica, C. canadensis*
Cryptotanshinone	*Salvia miltiorhiza*
Cryptoxan-thin-epoxide	*Hippophae rhamnoides, Physalis alkekengi, Taraxacum mongolicum, T. sinicum*
Crytomeridiol	*Magnolia hypoleuca, M. officinalis, M. japonica*
Crytoxanthin	*Eriobotrya japonica*
Cubebin	*Piper cubeba*
Cucurbitacin B	*Cucurtis melo, Neoalsomitra integrifoliola*
Cucurbitacin E	*Cucurtis melo*
Cucurbitacins	*Anagalis arvensis, Citrullus anguria, C. edulis, C. lanatus, C. vulgaris, Cucumis sativus*
Cucurbitine	*Cucurbita moschata, C. pepo*
Cuercetin	*Fagopyrum esculentum, F. sagittatum*
Cumaldehyde	*Cinnamomum camphora*

Component	Source
Cumaric acid	*Melilotus alba, M. suaveolens, M. indica*
Cumarin	*Eupatorium chinense, E. lindleyanum, E. japonicum*
Cumic alcohol	*Zanthoxylum bungeanum*
Cuminic aldehyde	*Cinnamomum zeglanicum*
Cumulene	*Conyza canadensis.*
Curcasin	*Jatropha gospiifolia, J. curcas*
Curcin	*Jatropha gospiifolia, J. curcas*
Curcolone	*Curcuma longa, C. zedoaria, C. aromatica, C. kwangsiensis*
Curcumin	*Curcuma longa, C. zedoaria, C. aromatica, C. kwangsiensis, C. pallida, C. phaeocoulis*
Curcuminoids diferuloylmethane	*Zingiber zerumbet*
Curcumol	*Curcuma longa*
Curdione	*Curcuma zedoaria, C. aromatica, C. kwangsiensis, C. longa*
Curmarin	*Curcuma longa*
Curzenene	*Curcuma zedoaria, C. aromatica, C. kwangsiensis, C. longa*
Curzerenone	*Curcuma zedoaria, C. aromatica, C. kwangsiensis*
Cuscutalin	*Cuscuta chinensis, C. europaea, C. japonica, C. lupuliformis*
Cuscutin	*Cuscuta chinensis, C. europaea, C. japonica, C. lupuliformis*
Custeodysine	*Osmunda japonica*
Cyandidin-3-sophoroside	*Hibiscus rosa-sinensis, H. rhombifolius*
Cyanic acid	*Nandina domestica*

Cyanidin	*Fagopyrum esculentum, F. sagittatum, Parthenocissus tricuspidata, Perilla frutescens, P. ocymoides, P. polystachya, P. arguta, Prunella vulgaris, Tagetes patula L.*
Cyanidin-3-gentiobioside	*Solidago dahurica, S. pacifica, S. virgaurea*
Cyanidin-3-glucoside	*Solidago dahurica, S. pacifica, S. virgaurea*
Cyanidin-3-monogalactoside	*Lythrum salicaria*
Cyanidin-3-rutinoside	*Campsis adrepens, C. chinensis, C. grandiflora*
Cyanidin diglycoside	*Tagetes patula*
Cyanidin rhamno-glucoside	*Syzygium cuminii*
Cyanin	*Rosa rugosa*
Cyanogenic glucoside	*Ageratum conyzoides, A. houstonianum, Flagellaria indica*
Cyasterone	*Trillium camschatcense*
Cyclanoline	*Aristolochia debilis, Stephania japonica, S. tetrandraq, S. cepharantha*
Cycloartemol	*Abrus precatorius, Euphorbia antiquorum*
Cycloastrangenol	*Astragalus complanatus, A. henyri, A. hoantchy, A. membranaceus, A. melilotoides, A. mongholicus, A. reflexistipulus, A. sinensis*
Cycloencalenol	*Melia japonica, M. toosendan, M. azedarach*
Cyclomonerviol	*Nerviia purpurea*
Cyclomulberrochromene	*Morus alba, M. constantinopolitana, M. indica*
Cyclonervilol	*Nerviia purpurea*
Cyclonerviol	*Nerviia purpurea*
Cycloprotobuxamines	*Buxus microophylla*
Cycloprotobuxines	*Buxus harlandii, B. microophylla*

Component	Source
Cycloshikonin	*Lithospermum erythrorhizon, L. officinalis*
Cyclovivobuxine C	*Buxus microophylla*
Cyclovivobuxine D	*Buxus harlandii, B. microophylla*
Cylindrin	*Imperata arundinaceae, I. cylindrica, Lophatherum gracile*
Cymarigenin	*Adonis amurensis* Regel & Radde.
Cymarin	*Adonis amurensis* Regel & Radde. *Apocynum venetum*
Cymarol	*Adonis chrysocyathus, A. brevistyla, A. amurensis* Regel & Radde. *A. vernalis*
Cymarose	*Trachelospermum jasminoides* (Lindl) Lem
Cymbopogonol	*Cymbopogon citrates*
Cymene	*Agastache rugosa, A. rugosa* f. hypoleuca, *Coriandrum sativum, Myristica fragrans*
Cynanchin	*Cynanchum bunge* Decaisne *Cynanchum atratum, C. auriculatum. Cynanchum glaucescens* Decaisne
Cynanchocerin	*Cynanchum atratum, C. auriculatum, Cynanchum glaucescens* Decaisne
Cynanchol	*Cynanchum glaucescens* Decaisne *Cynanchum atratum, C. auriculatum*
Cynarin	*Senecio argunensis, S. nemorensis, S. scandens*
Cyperoone	*Cyperus rotundus*
Cyrtomin	*Cyrtomium falcatum*
Cysteic acid	*Taraxacum mongolicum, T. sinicum*
Cysteine	*Taraxacum mongolicum, T. sinicum*
Cystine	*Cajanus cajan* L. *Taraxacum mongolicum, T. sinicum*
Cytisine	*Sophora tomatosa, S. flavescens, S. alopecurosides*

Cytochrome C	*Ricinus communis*
Cytokinin	*Ginkgo biloba*
d-abscisin	*Dioscorea batatus*
d-apiose	*Lemmaphyllum microphyllum*
d-asarinin	*Paulownia tometosa* Thunb.
d-backuchiol	*Psoralea corylifolia*
d-borneol	*Amomum cardamomum, A. globosum, A. tsao-ko, A. villosum, A. xanthloides, Zingiber officinale*
d-camphor	*Achillea alpina, A. millefolium, A. cardamomum, A. globosum, A. tsao-ko, A. villosum, A. xanthloides, Aconitum deinorrtuzum, Chenopotium ambrosiodes, Cinnamomum camphora, Prunella vulgaris*
d-carvone	*Anethum graveoleus, Carum carvi*
d-catechin	*Acacia catechu, C. cuneata, C. chlorusarca, C. dahurica, C. maximowiczii, C. pentagyna, C. pinnatifida, C. sanguinea, Potentilla bifurca, P. chinensis, P. discolor, P. fragariodes, P. freyaiana, P. kleiniana*
d-catechol	*Camellia japonica, Polygonum aviculare, Vaccinium bracteatum, V. vitis-idaea*
d-corydaline	*Corydalis ambigua, C. repens, C. turtschaninovii, C. yanhusuo, C. ternata*
d-fenchone	*Foeniculum officinale, F. vulgare, Prunella vulgaris*
d-fructose	*Sagittardia sagittifolia*
d-galactose	*Sagittardia sagittifolia, Sesbinia javanica*
d-galacturonic acid	*Malva chinensis, M. pulchella, M. verticillata, M. sylvestris, Plantago asiatica, P. depressa, P. exaltata, P. loureiri, P. major*
d-gallocatechol	*Vaccinium bracteatum, V. vitis-idaea*

Component	Source
d-glucaric acid	Ginkgo biloba
d-glucose	Bupleurum chinense, B. falcatum, B. scorzoneraefolium, Selaginella involvens, S. doederieninii, Solanum incanum
d-guereitol	Cissampelos pareira
d-isochondrodendrine	Cissampelos pareira
d-limonene	Agastache rugosa, A. rugosa f. hypoleuca, Carum carvi, Elettaria cardamomum, Schizonepeta multifida, S. tenuifolia, Tagetes erecta
d-lupaine	Lupinus luteus
d-mannitol	Ailanthus altissima, Cordyceps sinensis
d-mamnose	Sesbinia javanica
d-matrine	Sophora flavescens, S. alopecurosides
d-menthone	Schizonepeta multifida, S. tenuifolia
d-N-methylpseudoephedrine	Ephedra distachya, E. equisetina, E. intermedia, E. monosperma, E. sinica
d-oxymatrine	Sophora flavescens, S. alopecurosides
d-pinene	Agastache rugosa, A. rugosa f. hypoleuca
d-pseudoephedrinem	Ephedra distachya, E. equisetina, E. intermedia, E. monosperma, E. sinica
d-raffinose	Sagittardia sagittifolia
d-sesamine	Acanthopanax giraldii, Paulownia tometosa Thunb.
d-sophoranol	Sophora flavescens, S. alopecurosides
d-stachyose	Sagittardia sagittifolia
d-terpineol	Valeriana alternifolia, V. amurensis, V. fauriei, V. subbipinnatifolia

d-tetrahydropalmatine	*Corydalis decumbens*
d-tetrandrine	*Stephania tetrandraq*
d-verbascose	*Sagittardia sagittifolia*
d-xylose	*Plantago asiatica, P. depressa, P. exaltata, P. loureiri., P. major*
Daechu alkaloids	*Ziziphus jujuba, Z. spinosa*
Daidzein	*Glycine max, G. soja, Medicago falcata, M. lupulina, M. polymorpha, M. ruthenica, M. sativa, Pueraria lobata, P. pseudo-hirsuta, Sophora subprostrata, Trifolium pratense, T. repens*
Daidzin	*Pueraria lobata, P. pseudo-hirsuta*
Dambonitol	*Nerium indicum* Mill, *Trachelospermum jasminoides* (Lindl) Lem
Daphnetin	*Daphne giraldii, D. gurakduu, D. retusa, D. tangutica, D. koreana, Euphorbia lathyrus, E. lucorum, E. resinfera, E. thymifolia, Wikestroemia indica*
Daphnoretin	*Boenninghausenia albiflora*
Darutoside	*Siegesbeckia orientallis* L.
Daturodiol	*Datura alba, D. fastuosa, D. innoxia, D. metel, D. stramonium, D. tatula*
Daturolone	*Datura alba, D. fastuosa, D. innoxia, D. metel, D. stramonium, D. tatula*
Daucine	*Daucus carota*
Daucol	*Daucus carota*
Daucosterin	*Acanthopanax sessiliflorus*
Daucosterol	*Acanthopanax sessiliflorus, Adenophora triphylla, A. verticillata, Cynomorium coccineum, C. songarium, Daucus carota*
Dauricine	*Menispermum dauricum*
Dauricinoline	*Menispermum dauricum*

Component	Source
Daurinoline	*Menispermum dauricum*
Deacetylfawcetine	*Lycopodium clavatum, L. obscurum, L. selago, L. serratum*
Deacetylo-leandrin	*Nerium indicum* Mill
Deacylcynanchogenin	*Cynanchum paniculatum*
Deacylmetaplexigenin	*Cynanchum paniculatum, Metaplexis japonica* Thunb.
Deaspidin	*Dryopteris laeta, D. filix-mas*
Debilic acid	*Aristolochia debilis*
Decalactone	*Prunus persica*
Decanal	*Coriandrum sativum, Cymbopogon citratus*
Decanol	*Coriandrum sativum*
Decanoylacetaldehyde	*Houttynia cordata*
Decuroside	*Angelica decursiva*
Decursidin	*Angelica decursiva, Peucedanum japonicum, P. praeruptorum, P. rubricaule*
Decursind	*Angelica gigas* Maxim, Regel & Radde *Angelica decursiva*
Decussatin	*Swertia pseudochinensis*
Decylic aldehyde	*Coriandrum sativum*
Degueline	*Tephrosia purpurea*
Dehydroandrographolide	*Andrographis paniculata*
Dehydrocorydaline	*Corydalis ambigua, C. repens, C. turtschaninovii, C. yanhusuo, C. ternata*
Dehydrocorydalmine	*Corydalis ambigua, C. repens, C. turtschaninovii, C. yanhusuo, C. ternata*
Dehydroevodiamine	*Evodia rutaecarpa*

Dehydromatricaria	*Erigeron canadensis, E. annuus*
Dehydromatricaria ester	*Conyza Canadensis* L.
Dehydrosilybin	*Silybum marianum*
Delphinidin	*Medicago falcata, M. lupulina, M. polymorpha, M. ruthenica, M. sativa, Prunella vulgaris*
Delphinidin-3-monoglucoside	*Solanum lyratum, S. melongena*
Delphinidin-3,5-diglucoside	*Aquilegia vulgaris, Trigonotis peduncularis* Trevir
Delta-3-carene	*Vitex nequndo*
Demethyl cephalotaxine	*Cephalotaxus wilsoniana*
Demethyl-coclaurine	*Nelumbium nelumbo*
Demethyl-tubulosine	*Alangium lamarckii*
Demethyl cephaeline	*Alangium lamarckii*
Demethyl cephalotaxine	*Cephalotaxus fortunei, C. qinensis, C. oliveri*
Demethyl nobiletin	*Heteropappus altaicus* (Willd.) Novopokr.
Demethyl psychotrin	*Alangium lamarckii*
Demethyl wedolactone	*Eclipta erecta*
Dencichine	*Panax notoginseng*
Dendrobine	*Dendrobium nobile*
Deoxyandrograppholide	*Andrographis paniculata*
Deoxyelephantopin	*Elephantopus molis*
Deoxypodophyllotoxin	*Anthriscus aemula, A. sylvestris, Dysosma pleiantha*
Deoxysantalin	*Santalum album, S. myrtifolium, S. verum*
Deoxyschizandrin	*Schisandra chinensis*

Component	Source
Deoxytubulosine	*Alangium lamarckii*
Dephenyl methane-2-carboxylic acid	*Conyza canadensis*
des-O-methyl-licariine	*Epimedium brevicorum, E. koreanum, E. macranthum*
Desacetylasperuloside	*Oldenlandia diffusa*
Desacetylmatricarin	*Achillea alpina, A. millefolium*
Desoxypodophyllotoxin	*Podophylium peltatum, P. pleianthum*
Destrose	*Campsis adrepens, C. chinensis, C. grandiflora, Hordeum vulgare*
di-p-coumaroylmethane	*Zingiber zerumbet*
Diacetate	*Araucaria cunninghamii*
Diacetyl-atractylodiol	*Atractylis chinensis, A. lancea, A. lyrata, A. ovata*
Diadzin-4,7-diglucoside	*Pueraria lobata, P. pseudo-hirsuta*
Dially disulfide	*Allium victorialis*
Diallyl sulfide	*Allium chinense, A. odorum, A. sativum, A. tuberosum, A. uliginosum*
Dianthrone glucoside	*Cassia angustifolia*
Dianthronic heteroside	*Cassia occidentalis, C. torosa*
Dianthus saponin	*Dianthus barbatus, D. superbus, D. oreadum*
Dibasic acids	*Rhus verniciflua* Strokes
Dibenzoylgagaimol	*Metaplexis japonica* Thunb.
Dibilone	*Aristolochia debilis*
Dicaffeoxylquinic acid	*Xanthium chinense, X. japonicum, X. mongolicum, X. sibiricum, X. strumarium*

Dichrins	*Dichroa cyanitis, D. febrifuga, D. latifolia*
Dichroidine	*Dichroa cyanitis, D. febrifuga, D. latifolia*
Dichroines	*Dichroa cyanitis, D. febrifuga, D. latifolia*
Dicoumarol	*Medicago falcata, M. lupulina, M. polymorpha, M. ruthenica, M. sativa*
Dictamine	*Dictamnus albus, D. dasycarpus, Zanthoxylum ailanthoides, Z. schinifolium*
Dicumarol	*Melilotus alba, M. suaveolens, M. indica*
Diethyl phthalate	*Oenothera javanica (Bl) DC*
Diffractaic acid	*Usnea diffracta, U. longissima*
Dihydrono petalactol	*Actinidia polygama* Sieb& Zucc.
Digeneaside	*Calloglossa lepieurii*
Digicirin	*Digitalia purpurea, D. sanguinalis*
Digicoside	*Digitalia purpurea, D. sanguinalis*
Digifolein	*Digitalia purpurea, D. sanguinalis*
Digipurin	*Digitalia purpurea, D. sanguinalis*
Digitonid	*Cerbera manghas* L.
Digitonin	*Digitalia purpurea, D. sanguinalis*
Digitoxigenin	*Corchorus olitorius, Digitalia purpurea, D. sanguinalis. Nerium indicum* Mill.
Digitoxin	*Digitalia purpurea, D. sanguinalis*
Diglycoside	*Acanthopanax giraldii, Tagetes patula* L. *Yagetes patula* L.
Dihydro-N-methyl-isopelletierine	*Sedum sarmentosum*
Dihydro-artemisinin	*Artemisia annua, A. apiacea*

Component	Source
Dihydro-bigelovin	*Forsythia suspensa, Inula britannica, I. japonica, I. linariaefolia, I. linariaefolia, I. salsoloides*
Dihydro-carveol	*Anethum graveolens L. Piper nigrum*
Dihydrocarvone	*Anethum graveoleus*
Dihydro-conessine	*Holarrhenia antidy-senterica Wall*
Dihydro-cyclonervilol	*Nervilia purpurea*
Dihydro-foliamenthin	*Menyanthes trifoliata*
Dihydro-harman	*Elaeagnus pungens, E. umbellata*
Dihydroisopelletierine	*Sedum sarmentosum*
Dihydrokaempterol	*Morus alba, M. constantinopolitana, M. indica*
Dihydrolycorine	*Lycoris radiata, L. longituba, L. aura*
Dihydromorin	*Morus alba, M. constantinopolitana, M. indica*
Dihydronepetalactol	*Actinidia arguta, A. chinensis, A. japonica, A. kolomikta, A. polygama*
Dihydronuciferine	*Nelumbium nuciferum, N. speclosum*
Dihydroquercetin	*Rhododendron mucronatum*
Dihydrosecurinine	*Securinega suffruticosa, S. virosa*
Dihydroshikonin	*Lithospermum erythrorhizon, L. officinalis*
Dihydrostigmast	*Trapa bispinosa*
Dihydroterpene	*Pittosporum tobira*
Dihydroxy methyl anthraquinone	*Morinda citrifolia, M. officinalis*
Dihydroxycoumarin	*Bidens tripartita*

Dihydroxyglutamic acid	*Cornus alba, C. kousa, C. macrophylla*
Dillapiole	*Anethum graveoleus*
Dimethoxyallylbenzene	*Nothosmyrnium japonicum*
Dimethy ether	*Blumea balsumifera*
Dimethyl thymohydroquinone	*Eupatorium chinense, E. lindleyanum, E. japonicum*
Diogenin	*Costus specious*
Diol	*Curcuma zedoaria, C. aromatica, C. kwangsiensis, C. longa*
Diosbulbin	*Dioscorea bulbifera*
Diosbulbines	*Dioscorea bulbifera*
Dioscin	*Dioscorea nipponica*
Dioscorecin	*Dioscorea bulbifera*
Dioscoretoxin	*Dioscorea bulbifera*
Diosgenin	*Aletris formosana, A. spicata, Arnebia euchroma, Dioscorea bulbifera, D. opposita, D. nipponica, D. batatus, Solanum indicum, Trillium camschatcense*
Diosgenin glycoside	*Paris polyphylla, Paris quadrifolia*
Diosmetin-7-glucoside	*Chrysanthemum jucundum, C. koraiense, C. morifolium, C. sinense. Clintonia udensis Trautv et Mey*
Diosmin	*Zanthoxylum nitidum*
Dioxybenzoic acid	*Althaea rosea*
Dioxyflavonol	*Alpinia officinarum*
Dipalmiin	*Typhonium giganteum*

Component	Source
Dipentene	*Anethum graveolens, Coriandrum sativum, Cymbopogon citratus, Dryopteris crassirhizoma, Erigeron canadensis, E. annuus, Liquidambar acerifolia, L. formosana, L. maximowiczii, Melaleuca leucadendra, Myristica fragrans, Piper cubeba, Pyrrosia lingua, P. petiolosa, P. sheareri, Pinus madshurica* Rupr. *Vitex nequndo*
Diploptene	*Dryopteris laeta, D.filix-mas*
Disinomenine	*Cocculus diversifolius, C. thunbergii, Menispermum dauricum, Sinomenium acutum*
Diterpenes	*Acanthapanax gracilistylus, Ginkgo biloba*
Diterpenoids	*Aralia chinensis, A. cordata, A. elata, Tripterygium wilfordi*
Divaricoside	*Strophanthus divaricatus*
dl-anabasine	*Alangium lamarckii, A. chinense*
dl-beheerine	*Cissampelos pareira*
dl-borneol	*Elettaria cardamomum*
dl-catechol	*Machilus thunbergii*
dl-curine	*Cissampelos pareira*
dl-methylisopelletierine	*Sedum sarmentosum*
dl-N-noramepavine	*Machilus thunbergii*
dl-tetrahydropalmatine	*Corydalis ambigua, C. repens, C. turtschaninovii, C. yanhusuo, C. ternata*
dl-tetrandrine	*Stephania hernendifolia*
Dodecen-4-oic acid	*Lindera obtusiloba*
Domesticine	*Nandina domestica*
Donoxime	*Desmodium pulehellum*

Dopamine	*Portulaca oleracea*
Dotriacontanol	*Elephantopus molis*
Doumimidine	*Gelsemium sempervirens, G. elegans*
Dracoalban	*Daemonorops margaritae*
Dracoresene	*Daemonorops margaritae*
Dracoresinotannol	*Daemonorops margaritae*
Dracorhodin	*Draceana graminifolia*
Dracorubin	*Draceana graminifolia*
Dronin A.	*Clerodendrum trichotomum* Thunb.
Dryocrassin	*Dryopteris crassirhizoma, D. laeta, D. filix-mas*
Dulcilone	*Scopalia dulcis*
Dulciol	*Scopalia dulcis*
Dulcite	*Euonymus alatus, E. bungeanus, E. maackii*
Dulcitol	*Maytenus diversifolia, M. confertiflorus*
Dydimin	*Clinopodium chinense, C. polycephalum, C. gracile, C. umbrosum*
Ebelin lactone	*Hovenia dulcis*
Eburicoic acid	*Poria cocos*
Ecdysone	*Osmunda japonica*
Ecdysones cyasterone	*Ajuga bracteosa, A. decumbens, A. pygmaea*
Ecdystecide	*Chenopodium album* L.

Component	Source
Ecdysterone	*Achyranthes bidentata, A. japonica, Ajuga bracteosa, A. decumbens, A. pygmaea, Cyathula prostrate, Matteuccia struthiopteris. Osmunda japonica, Podocarpus macrophyllus, Trillium camschatcense*
Ecgonine	*Erythroxylum coca*
Echinocoside	*Cistanche deserticola*
Echinocystic acid	*Codonopsis lanceolata*
Echinopanacene	*Oplopanax elatus*
Echinopanacol	*Oplopanax elatus*
Echinopsine	*Echinops dahuricus, E. gmelini, E. grijsii, E. sphacrocephalus, E. latifollus*
Echitamidine	*Alstonia scholaris*
Echitamine	*Alstonia scholaris*
Ecliptine	*Eclipta alba, E. marginata, E. prostrata, E. thermalis, E. erecta*
Eicosanoic acid	*Tamarindus indicus*
Eicosenic acid	*Brassica alba, B. juncea*
Eikosanol	*Linum stelleroides, L. usitatissimum*
Eissampeline	*Chenopotium ambrosiodes*
Elaeocarpid	*Elephantopus elatus, E. grandiflorus*
Elcosanedicarboxylic acid	*Rhus verniciflua* strokes
Elegic acid	*Punica granatam*
Elemene	*Agastache rugosa, A. rugosa* f. *hypoleuca, Artemisia lactiflora*
Elemicin	*Cymbopogon citrates, Perilla frutescens* L.

Elephantin — *Elephantopus molis*

Elephantopin — *Elephantopus molis*

Eleutherosides — *Eleutherocossus senticosus*

Ellagic acid — *Dasiphora fruticosa* (Nestler) Kom. *Eugenia aromatica, E. caryophyllata, E. ulmoides, Lythrum salicaria, Syzygium cuminii, Terminalia chebula*

Elsholtzia ketone — *Elsholtzia argyi, E. cristata, E. splendens, E. feddei, E. souliei*

Elsholtzianic acid — *Elsholtzia argyi, E. cristata, E. splendens, E. feddei, E. souliei*

Elskoliziaketon — *Elshoitzia ciliatai* Thunb.

Emetine — *Alangium lamarckii*

Emilsin-like enzyme — *Flagellaria indica*

Emmenagogue — *Ligularia japonica* (Thunb.) Less

Emodin — *Cassia noname, C. obtusifolia, C. tora, Duchesnea indica, Polygonum multifolrum, P. chinensis, P. perfoliatum, P. tinctorium, P. cuspidatum, Rhamnus davurica, R. davurica, R. parvifolia, Rheum officinale, R. palmatum, R. tanguticum, R. undulatum, R. koreanum, Rumex acetosa, R. acetosella, R. amurensis, R. aquaticus, R. gmelini, R. longifolius, R. maritimus, R. marschallianus, R. patientia, R. stenophyllus, R. thyrsiflorus*

Emodin methyl ester — *Polygonum multifolrum, P. chinensis*

Emodin-monomethylether — *Rumex patientia*

Entageric acid — *Entada phaseoloides*

Ephedrine — *Ephedra distachya, E. equisetina, E. intermedia, E. monosperma, E. sinica*

Epialisol A — *Alisma cordifolia, A. orientalis, A. plantago-aquatica, A. plantago*

Epibrassicasterol — *Nervilia purpurea*

Component	Source
Epicatechin	*Acacia catechu, Davallia mariesii Moore ex Baker Elaeagnus pungens, E. umbellata, Rosa acicularis, R. amygdalifolia, R. davurica, R. davurica, R. koreana, R. laevigata, R. maximowicziana*
Epicatechin gallate	*Rosa acicularis, R. amygdalifolia, R. davurica, R. koreana, R. laevigata, R. maximowicziana*
Epicephalotaxin	*Cephalotaxus wilsoniana*
Epicephalotaxine	*Cephalotaxus fortunei, C. qinensis, C. oliveri*
Epifriedelanol	*Clerodendrum trichotomum Thunb.Syzygium cuminii*
Epifriedelin	*Clerodendrum trichotomum, C. spicatus. Clerodendrum trichotomum Thunb.*
Epifriedelinol	*Elephantopus molis, E. alatus, E. bungeanus, E. maackii*
Epigallocatechin	*Elaeagnus glabra, R. acicularis, R. amygdalifolia, R. davurica, R. koreana, R. laevigata, R. maximowicziana*
Epiliguloxide	*Ligularia fischeri (Ledeb.)*
Epimedoside A	*Epimedium brevicorum, E. koreanum, E. macranthum*
Epistephanine	*Stephania japonica*
Epiwilsonine	*Cephalotaxus fortunei, C. qinensis, C. oliveri*
Epoxyquaine	*Cyperus rotundus*
Equisetonin	*Equisetum arvense, E. hyemale, E. ramosissimum*
Equisetrin	*Equisetum arvense, E. hyemale, E. ramosissimum*
Equistic acid	*Equisetum arvense L.*
Eremophilene	*Valeriana alternifolia, V. amurensis, V. fauriei, V. subbipinnatifolia*
Ergostatetraen	*Trapa bispinosa*

Ergosterol peroxide	*Ananas comosus*
Ergosterol	*Ganoderma lucidum, Lactuca raddeana, L. indica, L. sativa, Nervilia purpurea, Polyporus umbellatus*
Ericolin	*Chimaphila umbellata*
Erigeron	*Erigeron canadensis, E. annuus*
Eriodictyol	*Filifolium sibiricum* (L.) Kitam
Erisin	*Populus alba, P. davidiana, P. tomentosa*
Erpinene	*Valeriana alternifolia, V. amurensis, V. fauriei, V. subbipinnatifolia*
Erucic acid	*Avena fatua, Brassica alba, B. juncea, Cardamine leucantha, C. lyrata,* Descurainia Sophia (L.) Schur. *Erysimum amurense, E. cheiranthoides, Sinapis alba*
Erycbelline	*Erycibe henryi, E. aenea*
Erychroside	*Corchorus olitorius, Erysimum amurense, E. cheiranthoides*
Erysimoside	*Erysimum amurense, E. cheiranthoides*
Erysimosol	*Erysimum amurense, E. cheiranthoides*
Erysimotoxin	*Erysimum amurense, E. cheiranthoides*
Erythriside	*Erysimum amurense, E. cheiranthoides*
Escigenin	*Aesculus chinensis, A. hippocastanum*
Esculetin	*Euphorbia lathyrus, E. lucorum, E. resinfera, E. thymifolia*
Esculetin dimethyl ether	*Zanthoxylum schinifolium*
Esgoside	*Momordica grosvenori*

Component	Source
Essential oils	*Achillea alpina, A. millefolium, Agastache rugosa, A. rugosa f. hypoleuca, Alpinia japonica, A. officinarum, Anethum graveoleus, Ardisia japonica, Arethusa japonica, Artemisia brachyloba, A. gmelini, Asarum canadense, A. heterotropoides, A. sieboldii, Asparagus cochinensis, A. falcatus, A. insularis, A. lucidus, A. officinalis, Aspidium falcatum, A. gmelini, Atractylis chinensis, A. lancea, A. lyrata, A. ovata, Atractylodes lancea, Bidens bipinnata, B. parviflora, Biota chinensis, B. orientalis, Bletilla hyacinthina, B. striata, Blumea balsumifera, Caesalpinia sappan, Carduus acaulis, C. crispus, C. japonicus, Carpesium abrotanoides, C. athunbergianum, Carum carvi, Celtis bungeana, C. sinensis, Centipeda minima, Chloranthus glabra, C. oldhnami, Chrysanthemum cinerriaefolium, Cirsium albescens, C. brevicaule, C. littorale, C. maakii, C. segetum, C. setosum, C. vlassovianum, Commiphora myrrha, Conioselinum univittatum, Conyza canadensis, Cunninghamia lanceolata, Cymbidium hyacinthinum, C. striatum, Cymbopogon distans, C. goeringii, C. nardus, Cyperus rotundus, Dianthus barbatus, D. superbus, D. oreadum, Dipsacus asper, Dracocephalum integrifolium, Eclipta erecta, Elsholtzia argyi, E. cristata, E. splendens, E. feddei, E. souliei, Erigeron canadensis, E. annuus, Eriobotrya japonica, Eucalyptus robusta, Fortunella crassifolia, F. japonica, F. margarita, Gnaphalium affine, G. arenarium, G. confusum, G. javanum, G. luteo-album, G. multiceps, G. ramigerum, G. tranzschelii, G. uliginosum, Hedyotis diffusa, Hippophae rhamnoides, Houttynia cordata, Laggera alata, Lappa communis, L. edulis, L. major, L. minor, Ledebouriella divaricata, Leonurus sibiricus, Lindera megaphylla, Lophanthus chinensis, L. rugosus, Lysimachia barystachys, L. christinae, L. clethroides, L. davurica, Nardostachys jatamansi, Nothosmyrnium japonicum, Notopterygium incisium, Oplopanax elatus,*

Essential oils (continued)	*Patrinia scabiosaefolia, P. heterophylla, Pinus bungeana, P. densiflora, P. koraiensis, P. sylvestris, P. tabulaeformis, Plumeria rubra, Pogostemon cablin, Pyrethrum cinerariifolium, P. sinense, Rabdosia rubescens Rhaponticum uniflorum, R. mucronatum, R. anthopogon, Rosa chinensis, R. indica, R. rugosa, Salvia plebeia, Sarcandra glabra, Schizonepeta multifida, S. tenuifolia, Scrophularia buergeriana, S. kakudensis, S. ningpoensis, S. oldhami, S. puergeriana, Silene jeniseensis, Sorbus alnifolia, S. amurensis, S. pohuashanensis, Taraxacum officinale, Thlaspi arvense, Tilia amurensis, T. mandshurica, T. mongolica, Tussilago farfara, Verbena officinalis, V. oxysepalum, Viburnum sargenti, Vitex nequndo, V. chinensis, V. jeguaod, Xanthoxylum piperitum, Zanthoxylum bungeanum, Z. ailanthoides, Zingiber officinale, Z. zerumbet*
Estragol	*Angelica decursiva, Zanthoxylum schinifolium*
Estragole	*Magnolia biloba, M. denudata, M. discolor, M. liliflora, M. purpurea, Zanthoxylum bungeanum*
Estrogenic and thiophene activity	*Eclipta prostrate L.*
Ethanol	*Prunus persica*
Ether oils	*Achillea alpina, A. millefolium, Anthriscus aemula, A. sylvestris, Heracleum dissectum, H. lanatum, Polygonum bistorta, Sanguisorba officinalis, S. grandiflora, S. parviflora, S. x tenuifolia, Taraxacum officinale*
Ethoxychelerythrine	*Macleaya cordata*
Ethoxysanguinarine	*Macleaya cordata*
Ethyl alcohol	*Kaempferia galanga*
Ethyl beta-fructopyranoside	*Rosa cicularis, R. amygdalifolia, R. davurica, R. koreana, R. laevigata, R. maximowicziana*
Ethyl ester	*Anredera cordifolia*

Component	Source
Etoposide	*Dysosma pleiantha*
Eucalyptol	*Artemisia argyi, A. halodendron, A. igniaria, A. indica, A. integrifolia, A. japonica, A. keiskeana, A. lagocephala, A. lavandulaefolia, A. scoparia, A. selengensis, A. sieversiana, A. vulgarts, Melaleuca leucadendra*
Eucalyptole	*Cinnamomum camphora*
Euchrestaflavanones	*Euchresta japonicum*
Eudesnol	*Atractylodes chinensis, A. japonica, A. koreana, A. lancen, A. macrocephala, A. ovata*
Eugenitin	*Eugenia aromatica, E. caryophyllata, E. ulmoides*
Eugenol	*Alpinia officinarum, Cinnamomum zeglanicum, C. aromaticum, C. cassia, Dianthus barbatus, Dianthus chinensis L. D. superbus, D. oreadum, Geum aleppicum, Magnolia biloba, M. denudata, M. discolor, M. liliflora, M. purpurea, Myristica fragrans, Rosa rugosa, R. rugosa, Vitex nequndo*
Eugianin	*Syzygium cuminii*
Eupafolin	*Salvia plebeia*
Eupaformonin	*Eupatorium formosanum*
Eupaformosanin	*Eupatorium formosanum*
Euparin	*Eupatorium chinense, E. lindlleyanum, E. japonicum*
Eupatol	*Eupatorium odoratum*
Eupatolide	
Euphol	*Euphorbia antiquorum, E. lathyrus, E. lucorum, E. resinfera, E. thymifolia*
Euphorbetin	*Euphorbia lathyrus, E. lucorum, E. resinfera, E. thymifolia*

Euphorbias	*Euphorbia coraroides, E. lasiocaula, E. lunulata, E. pallasii, E. pekinensis, E. sampsoni, E. sieboldiana*
Euphorbiasteroid	*Euphorbia lathyrus, E. lucorum, E. resinfera, E. thymifolia*
Euphorbine	*Euphorbia esula, E. helioscopia*
Euphorbol	*Euphorbia lathyrus, E. lucorum, E. resinfera, E. thymifolia*
Euphorbon	*Euphorbia humifusa, E. hirta, E. coraroides, E. lasiocaula, E. lunulata, E. pallasii, E. pekinensis, E. sampsoni, E. sieboldiana*
Evocarpine	*Evodia rutaecarpa*
Evodiamine	*Evodia rutaecarpa*
Evodol	*Evodia rutaecarpa*
Evomonoside	*Descurania Sophia* Webb ex Prantl.
Eycinnuik	*Eucommia ulmoides*
Fabric	*Urtica angustifolia, U. cannabina, U. lobata, U. tenacissima, U. urens, U. utillis*
Fagomine	*Fagropyrum esculentum*
Fagopyrin	*Fagropyrum esculentum*
Falcarindiol	*Glehnia hittoralis, Notoptergium incisium*
Falcarindone	*Carum carvi* L.
Falvins	*Alternanthera philoxeroides, A. sessilis*
Fangchinoline	*Stephania hernendifolia, S. tetrandraq*
Faradiol	*Tussilago farfara*
Farnesal	*Cymbopogon citratus*
Farnesene	*Agastache rugosa, A. rugosa* f. *hypoleuca*

Component	Source
Farnesiferol A, B, C.	*Ferula assa-foetida, F. bungeana*
Farnesol	*Cymbopogon citratus, Plumeria rubra*
Farreol	*Rhododendron mucronatum, R. dauricum*
Fat	*Ficus awkeotsang, Gnaphalium affine, G. arenarium, G. confusum, G. javanum, G. luteo-album, G. multiceps, G. ramigerum, G. tranzschelii, G. uliginosum, Triticum vulgare, Urtica angustifolia, U. cannabina, U. lobata, U. tenacissima, U. urens, U. utillis*
Fatty acids	*Aesculus indica, Cassia alata, Cornus walteri, Euonymus alatus, E. bungeanus, E. maackii, Geum aleppicum, Hippophae rhamnoides, Lindera glauca, L. communis, Linum stelleroides, L. usitatissimum, Magnolia grandiflora, Parthenocissus tricuspidata, Phaseolus angularis, P. lunatus, P. radiatus, P. vulgaris, Prunus padus, P. domestica, P. glandulosa, P. japonica, P. armeniaca, Pueraria lobata, P. pseudo-hirsuta, Sorbus alnifolia, S. amurensis, S. pohuashanensis, Thlaspi arvense, Viscum album*
Fatty oil	*Lappa communis, L. edulis, L. major, L. minor, Leonurus heterophyllus, L. japonicus, L. macranthus, L. mongolicus, L. pseudo-macranthus, Lygodium japonicum, Terminalia chebula*
Fawcetimine	*Lycopodium clavatum, L. obscurum, L. selago, L. serratum*
Fawcetine	*Lycopodium clavatum, L. obscurum, L. selago, L. serratum*
Febrifugin	*Hydrangea macrophylla*
Fenchone	*Blumea lacera, Thuja koraiensis, T. orientalis, T. chinensis*
Feriol	*Rhododendron dauricum*
Fernadiene	*Adiantum boreale, A. capillus-junonis, A. pedatum, A. flabellulatum*
Fernene	*Adiantum boreale, A. capillus-junonis, A. pedatum, A. flabellulatum, Dryopteris crassirhizoma*

Fernenol	*Imperata arundinaceae, I. cylindrica*
Ferulic acid	*Angelica polymorpha, A. sinensis, Catalpa ovata G. Don, Chenopodium album L. Cimicifuga dahurica, C. foetida, C. heracleifolia, C. racemosa, C. ussuriensis, Dasiphora fruticosa (Nestler) Kom. Ferula assa-foetida, F. bungeana, Impatiens balsamina, I. noli-tangere, I. textori, Lycopodium clavatum, L. obscurum, L. selago, L. serratum, Matteuccia struthiopteris*
Feruloyl-p-coumaroylmethane	*Zingiber zerumbet*
Fetidine	*Thalictrum aquilegifolium, T. baicalense, T. foetidum, T. fauriel, T. petaloideum, T. simplex, T. squarrosum, T. thunbergii*
Fibralactone	*Fibraurea recisa*
Fibramine	*Fibraurea recisa*
Fibraminine	*Fibraurea recisa*
Ficusin	*Ficus carica*
Filicenal	*Adiantum boreale, A. capillus-junonis, A. pedatum, A. flabellulatum*
Filicene	*Adiantum boreale, A. capillus-junonis, A. pedatum, A. flabellulatum*
Filicic acid	*Aspidium falcatum, Dryopteris crassirhizoma, D. laeta, D. filix-mas, Pteridium aquilinum (L.) Kuhn. Pteris cretica, P. ensiformis, P. multifida, P. vittata, P. wallichiana*
Filicin	*Dryopteris laeta, D. filix-mas, Matteuccia struthiopteris*
Filifolin	*Filifolium sibiricum (L.) Kitam*
Filmarone	*Dryopteris crassirhizoma, D. laeta, D. filix-mas*
Finitin	*Artemisia finita, A. frigida*
Fisetine	*Cotinus coggygria, Gleditsia horrida, G. sinensis, G. xylocarpa, Pistacia lentiscus, Rhus verniciflua Strokes*
Flavaspidic acids	*Dryopteris laeta, D. filix-mas*

Component	Source
Flavaspidin	*Dryopteris crassirhizoma*
Flavocommelitin	*Commelina communis*
Flavon glucoside	*Ajuga bracteosa, A. decumbens, A. pygmaea*
Flavone	*Callicarpa macrophylla, Citrus aurantium, Dracocephalum integrifolium, Fagopyrum esculentum, F. tataricum, Geum aleppicum, Glycyrrhiza pallidiflora, G uralensis, Ilex pubescens, Inula britannica, I. japonica, I. linariaefolia, I. linariaefolia, I. salsoloides, Iris aqyatuca, I. buatatas, I. dichotoma, Lindera akoensis, Loropetalum chinense, Lysionotus pauciflorus, Vitex trifolia, V. rotundifolia*
Flavonoid derivatives	*Astragalus complanatus, A. henyri, A. hoantchy, A. membranaceus, A. melilotoides, A. mongholicus, A. reflexistipulus, A. sinensis*
Flavonoid glycoside	*Aesculus indica, Asarum europaeum, Laggera alata, Thalictrum foetidum, Urena procumbens*
Flavonoids	*Achillea alpina, A. millefolium, Ampelopsis aconitifolia, A. brevipedunculata, A. japonica, A. bodinieri, A. contonensis, A. humulifolia, Artemisia lactiflora, Bauhinia championi, B. variegata, Bidens bipinnata, B. parviflora, Chloranthus glubra, C. oldhnami, Crataegus cuneata, C. chlorusarca, C. dahurica, C. maximowiczii, C. pentagyna, C. pinnatifida, C. sanguinea, Cypripedium guttatum, C. macranthum, C. pubescens, Cyrtomium falcatum, Elaeagnus glabra, Gynura bicolor, Hippophae rhamnoides, Jatropha podagrica, Lemmaphyllum microphyllum, Lespedeza cuneata, Nepenthes raffsiana, Oplopanax elatus, Panax notoginseng, Podophylum peltatum, P. pleianthum, Polygonum cuspidatum, P. perfoliatum, P. tinctorium, Rhaponticum uniflorum, Rhodiola elongata, Rhododendron dauricum, Rosa acicularis, R. amygdalifolia, R. davurica, R. koreana, R. laevigata,*

Flavonoids (continued)	*R. maximowicziana, Rubus parvifolius, R. coreanus, R. crataegifolius, R. matsumuranus, R. saxatilis, Salvia plebeia, Scrophularia buergeriana, S. kakudensis, S. ningpoensis, S. oldhami, S. puergeriana, Solidago dahurica, S. pacifica, S. virgaurea, Sorbus alnifolia, S. amurensis, S. pohuashanensis, Spiraea salicifolia, Taraxacum officinale, Thalictrum aquilegifolium, T. baicalense, T. fauriel, T. petaloideum, T. simplex, T. squarrosum, T. thunbergii, Thesium chinense, Tilia amurensis, T. mandshurica, T. mongolica, Veronica linariaefolia*
Flavoxanthin	*Calendula officinalis*
Flavoyadorinin	*Viscum album, V. coloratum*
Fluggein	*Securirega virosa*
Foliamenthin	*Menyanthes trifoliata*
Folinic acid	*Angelica polymorpha, A. sinensis*
Formic acid	*Jasminum samba, Pyrrosia adnascens*
Formonetin	*Medicago falcata, M. lupulina, M. polymorpha, M. ruthenica, M. sativa*
Formononetin	*Astragalus complanatus, A. henryi, A. hoantchy, A. membranaceus, A. melilotoides, A. mongolicus, A. reflexistipulus, A. sinensis. Trifolium pratense, T. repens*
Fragarine	*Dictamnus albus, D. dasycarpus*
Framine	*Phyllodium pulchellum*
Fraxin	*Clerodendrum trichotomum thumb.Fraxinus bungeana, F. chinensis, F. floribunda, F. obovata, F. ornus, F. rhynchophylla*
Fraxinella	*Melia azedarach, M. japonica, M. toosendan*
Fraxinellone	*Dictamnus albus, D. dasycarpus*
Friedelaun-3-ol	*Euphorbia antiquorum*

Component	Source
Friedelin	*Clerodendrum trichotomum, C. spicatus, Codonopsis pilosula, C. tangshen, C. ussuriensis, Euonymus alatus, E. bungeanus, E. maackii, Hemerocallis fulva* L. *Lophatherum gracile, Spatholobus suberectus, Syzygium cuminii*
Fritillarine	*Fritillaria anheunensis, F. collicola, F. maximowiczii, F. roylei, F. thunbergii, F. ussuriensis, F. verticillata*
Fritilline	*Fritillaria anheunensis, F. collicola, F. maximowiczii, F. roylei, F. thunbergii, F. ussuriensis, F. verticillata*
Fructose	*Cucumis sativus, Ficus awkeotsang, Plumbago zeylanica, Pseudostellaria heterophylla, Sedum lineare*
Fudujusone	Adonis amurensis Regel & Radde
Fulvoplumierin	*Plumeria rubra*
Fumaric acid	*Sarcandra glabra, Vicia faba*
Fungal lysozyme	*Ganoderma lucidum*
Fungiridal	*Acorus calamus* L., *Buglossoides arvense* (L.) Johnston
Furane	*Elsholtzia argyi, E. cristata, E. splendens, E. feddei, E. souliei*
Furanocoumarin	*Glehnia hittoralis*
Furanodiene	*Curcuma aromatica, C. kwangsiensis, C. longa C. zedoaria*
Furanodienone	*Curcuma aromatica, C. kwangsiensis, C. longa, C. zedoaria*
Furanoses quiterpenes	*Ligularia japonica* (Thunb.) Less
Furostanol saponins	*Allium macrostemon, Corchorus olitorius*
Furylisobutyl ketone	*Elsholtzia argyi, E. cristata, E. splendens, E. feddei, E. souliei*
Furylmethyl ketone	*Elsholtzia argyi, E. cristata, E. splendens, E. feddei, E. souliei*

Component	Sources
Furylpropyl ketone	*Elsholtzia argyi, E. cristata, E. splendens, E. feddei, E. souliei*
Fustin	*Cotinus coggygria, Gleditsia horrida, G. sinensis, G. xylocarpa, Pistacia lentiscus Rhus verniciflua* Strokes
Gagaminin	Metaplexis japonica Thunb.
Galactan	*Luffa aegyptiaca, L. cylindrica, L. faetida, L. petola, Quercus acutissima, Q. aliena, Q. dentata, Q. liaotungensis, Q. mongolica, Q. variabilis, Trachycarpus wagnerianus, T. fortunei*
Galactitol	*Cassytha filliformis*
Galactomannan	*Cassia occidentalis, C. torosa*
Galactose	*Alpinia katsumadai, A. globosum, A. kumatake Cucumis sativus, Fortunella crassifolia, F. japonica, F. margarita, Rhus verniciflua* Strokes *Tamarindus indicus*
Galangin	
Galangol	*Alpinia katsumadai, A. globosum, A. kumatake, A. officinarum*
Galanthamine	*Hippeastrum hybridum,* Lycoris aurea Herb. *L. radiata, L. longituba, L. aura*
Gallic acid	*Acalypha australis, Caesalpinia pulcherrima, C. sappan, Cornus officinalis,* Cotinus coggygria Scop. *Erigeron canadensis, E. annuus, Eucalyptus robusta, Euphorbia humifusa, E. hirta, Geranium dahuricum, G. eriostemon, G. sibiricum, G. wlassowianum, G. wilfordi, Lawsonia inermis, Macrocarpium officinalis, Melaleuca leucadendra, Pistacia lentiscus, Polygonum bistorta, P. zeylanica Punica granatam, Rheum officinale, R. palmatum, R. tanguticum, R. undulatum,* Mill. *R. koreanum, Rhus chinensis, R. cotinus, R. javanica, R. osbeckii*
Gallocatechin	*Rosa acicularis, R. amygdalifolia, R. davurica, R. koreana, R. laevigata, R. maximowicziana*
Gallotannic acid	*Punica granatam, Rhus chinensis, R. cotinus, R. javanica, R. osbeckii*

Component	Source
Galuteolin	*Equisetum arvense, E. hyemale, E. ramosissimum*
Gamatin	*Pongamia pinnata*
Gambir-fluorescein	*Acacia catechu*
Gambirine	*Acacia catechu*
Gamma-aminobutyric acid	*Astragalus complanatus, A. henyri, A. hoantchy, A. membranaceus, A. melilotoides, A. mongholicus, A. reflexistipulus, A. sinensis*
Gamma-dichroline	*Adamia chinensis, A. cyanea, A. versicolof*
Gamma-fernene	*Adiantum boreale, A. capillus-junonis, A. pedatum, A. flabellulatum*
Gamma-sitosterol	*Oldenlandia chrysotricha, O. corymbosa*
Gamma-terpinene	*Poncirus trifoliata*
Gardenin	*Gardenia angusta, G. jasminoides*
Gardenoside	*Gardenia florida, G. grandiflora, G. maruba, G. pictorum, G. radicans, Gentiana algida, G. barbata, G. manshurica, G. olivieri, G. scabra, G. squarrosa, G. triflora*
Gastrodin	*Gastrodia elata*
Gedunin	*Melia japonica, M. toosendan, M. azedarach*
Gein	*Geum aleppicum*
Gelatin	*Bletilla hyacinthina, B. striata*
Gelsemidine	*Gelsemium sempervirens, G. elegans*
Gelsemine	*Gelsemium sempervirens, G. elegans*
Geniposide	*Gentiana algida, G. barbata, G. manshurica, G. olivieri, G. scabra, G. squarrosa, G. triflora*
Genisteine	*Cytisus scoparius, Glycine max, G. soja, Sophora japonica, Trifolium pratense, T. repens*

Genkwadaphnin	*Daphne fortunei, D. genkwa*
Genkwanin	*Artemisia gmelini, Daphne fortunei, D. genkwa*
Gentialutine	*Menyanthes trifoliata*
Gentianidine	*Gentiana dahurica, G. lutea, G. macrophylla*
Gentianine	*Gentiana algida, G. barbata, G. dahurica, G. lutea, G. macrophylla, G. manshurica, G. olivieri, G. scabra, G. squarrosa, G. triflora, Halenia corniculata L. Menyanthes trifoliata*
Gentianol	*Gentiana dahurica, G. lutea, G. macrophylla*
Gentiatibetin	*Menyanthes trifoliata*
Gentiatibetine	*Menyanthes trifoliata*
Gentiopicrin	*Gentiana algida, G. barbata, G. manshurica, G. olivieri, G. scabra, G. squarrosa, G. triflora*
Gentisic acid	*Impatiens balsamina, I. noli-tangere, I. textori*
Geoside	*Geum aleppicum*
Geranic	*Cymbopogon citratus*
Geraniol	*Chenopotium ambrosiodes, Citrus reticulata, Coriandrum sativum, Cymbopogon citratus, Daucus carota, Elettaria cardamomum, Myristica fragrans, Plumeria rubra, Rosa rugosa, Thymus vulgaris, Vitex negurdo, Xanthoxylum piperitum, Zanthaxylum bungeanum*
Geranylgeraniol	*Linum stelleroides, L. usitatissimum*
Gerariol	*Murraya paniculata*
Germacrene	*Kadsura japonica, Rhododendron dauricum, R. mucronatum*
Germacron	*Rhododendron micranthum* Turcz.
Germacrene-D-apigenin-7-glucuronide	*Eregeron annuus* L.
Germine	*Veratrum dahuricum, V. maackii, V. nigrum*

Component	Source
Gibberelin A_{21}	*Canavalia gladiata, C. ensiformis*
Gibberelin A_{22}	*Canavalia gladiata, C. ensiformis*
Gibberellin	*Ginkgo biloba, Pharbitis diversifolia, P. hederacea, P. nil, P. triloba*
Gingerol	*Zingiber officinale*
Ginkgetin	*Ginkgo biloba*
Ginkgol	*Ginkgo biloba*
Ginkgolic acid	*Ginkgo biloba*
Ginkgolides	*Ginkgo biloba*
Ginnol	*Ginkgo biloba*
Ginsenoside R_o	*Panax japonicum*
Ginsenosides	*Panax quinquefolium, P. ginseng, P. notoginseng*
Gitaloxigenin	*Digitalia purpurea, D. sanguinalis*
Gitaloxin	*Digitalia purpurea, D. sanguinalis*
Gitanin	*Digitalia purpurea, D. sanguinalis*
Gitoxigenin-Karabin	*Digitalia purpurea, D. sanguinalis, Nerium indicum Mill.*
Gitoxin	*Digitalia purpurea, D. sanguinalis*
Glabralactone	*Angelica pubescens, A. gigas Maxim, A. sinensis (Oliv) Diels*
Glaucine	*Thalictrum aquilegifolium, T. baicalense, T. fauriel, T. petaloideum, T. simplex, T. squarrosum, T. thunbergii*
Gleditsin	*Gleditsia horrida, G. sinensis, G. xylocarpa*
Globulin	*Arachis hypogaea*

Glucan	*Kalopanax septemlobus* (Thunb) Koidz.,*Omphalia lapidescens*
Glucides	*Fortunella crassifolia, F. japonica, F. margarita.*
Glucobrassicin	*Clerodendrum cyrtophyllum, Isatis indigotica, I. oblongata*
Glucoevatromonoside	*Corchorus olitorius*
Glucofragulin	*Polygonum cuspidatum*
Glucokinin	*Dolichos lablab*
Glucominol	*Allium chinense, A. odorum, A. sativum, A. tuberosum, A. uliginosum*
Gluconasturtin	*Rorippa indica, R. islandica, R. montana*
Glucononitol	*Vitex negundo*
Glucopyranosides	*Codonopsis pilosula, C. tangshen, C. ussuriensis*
Glucose	*Ficus awkeotsang, Lycopus lucidus, L. maackianus, L. parviflorus, L. ramosissimus, L. fargesii, L. veitchii, Nephelium longana, N. lappaceum, Plumbago zeylanica, Sedum lineare, Sagittardia sagittifolia, Tamarindus indicus*
Glucoside	*Ampelopsis aconitifolia, A. brevipedunculata, A. japonica, A. bodinieri, A. contonensis, A. humulifolia, Arachis hypogaea, Caragana sinica, C. microphylla, C. intermedia, C. franchetiana, Dodonaea viscosa, Sarcandra glabra*
Glucoside asiaticoside	*Centella ascatica*
Glucoside morindin	*Morinda citrifolia, M. officinalis*
Glucovanillin	*Avena fatua*
Glucuronic acid	*Aralia mandschurica* (Rupr & Maxim) Seed
Glutamic acid	*Litchi chinensis, Pueraria montana, P. thunbergiana*
Glutamine	*Ficus carica, Dioscorea opposita*
Glutin-5-en-3-ol	*Alnus japonica*

Component	Source
Glutinol	*Imperata arundinaceae, I. cylindrica*
Glycine	*Taraxacum mongolicum, T. sinicum*
Glycoalkaloids	*Sesbinia sesbin*
Glycogen	*Cymbidium hyacinthinum, C. striatum*
Glycolic acid	*Asparagus cochinenesis, A. falcatus, A. insularis, A. lucidus, A. officinalis, Physalis alkekengi*
Glycoside alkaloids	*Solanum biflorum*
Glycoside nodakenin	*Peucedanum decursivum, Peucedanum terebinthaceum* Fisch
Glycoside pharbitin	*Pharbitis diversifolia, P. hederacea, P. nil, P. triloba*
Glycoside rhaphantin	*Polygonum multifolrum, P. chinensis*
Glycosides G. and K.	*Arundo donax, A. phragmites, Caesalpinia decapetula, Carduus acaulis, C. crispus, C. japonicus, Centipeda minima, Corchorus capsularis, C. olitorius, Gardenia angusta, G. jasminoides, Periploca sepium* Bunge *Phragmittes communis, Wisteria sinensis*
Glycosides clerodendrin	*Clerodendrum trichotomum, C. spicatus*
Glycosides tribuloside	*Tribulus terrestsis*
Glycosine	*Glycosmis cochinchinensis, G. pentaphylla*
Glycosmine	*Glycosmis cochinchinensis, G. pentaphylla*
Glycosminine	*Glycosmis cochinchinensis, G. pentaphylla*
Glycyrrhiza	*Glycyrrhiza pallidiflora, G. uralensis*
Glycyrrhizic acid	*Glycyrrhiza pallidiflora, G. uralensis*
Glycyrrhizin	*Arachis hypogaea*
Glypenosides	*Gynostemma pentaphyllum*

Gobosterin	*Lappa communis, L. edulis, L. major, L. minor*
Gomphrenin	*Gomphrena globosa*
Gorlic acid	*Hydnocarpus anthelmintica, H. castaneus*
Gossypetin	*Rhododendron anthopogon, R. mucronatum*
Gossypitrin	*Equisetum arvense, E. hyemale, E. ramosissimum*
Gossypol	*Gossypium herbaceum*
Graveobiosides	*Apium graveolens*
Grayanotoxin	*Rhododendron mucronatum* Turcz
Guaiacol	*Ficus carica*
Guaiaxulene	*Ficus carica*
Guajaverin	*Psidium guajava*
Guercitrin	*Plumbago zeylanica*
Guggulsterol	*Commiphora myrrha*
Guijaverin	*Celosia argentea, C. cristata, C. margariacea*
Gum	*Curcuma pallida, C. phaeocoulis, Delonix regia, Ficus awkeotsang*
Gurjuncene	*Menyanthes trifoliata*
Guvacine	*Areca catechu, A. hortonsis*
Guvacoline	*Areca catechu, A. hortonsis*
Gypsogenin	*Vaccaria segetalis, V. pyramidata*
Haemanthidiene	*Haemanthus multiflorus, Zephyranthes candida*
Haementhamine	*Haemanthus multiflorus*
Haemolytic sapogenin	*Sansevieria trifosciate*

Component	Source
Hallucinogenic	*Acorus calamus* L.
Harman	*Elaeagnus pungens, E. umbellata, Hippophae rhamnoides*
Harmane	*Tribulus terrestsis*
Harmine	*Tribulus terrestsis*
Harmol	*Hippophae rhamnoides*
Harpagoside	*Scrophularia buergeriana, S. kakudensis, S. ningpoensis, S. oldhami, S. puergeriana*
Harringtonine	*Cephalotaxus fortunei, C. qinensis, C. oliveri, C. wilsoniana, Stephania japonica*
Hastanecine	*Cacalia hastate* L. *Cajanus indicus* L.
Hayatidine	*Cissampelos pareira*
Hayatine	*Cissampelos pareira*
Hayatinine	*Cissampelos pareira*
Hectalactone	*Prunus persica*
Hedera acid	*Hedera rhombea, H. helix*
Hederagenin	*Aristolochia contorta, A. kaempferi, A. longa, A. recurvilabra, Caulophyllum robustum*
Hederasaponin B	*Anemone raddeana, A. rivularis, A. vitifolia*
Hederin	*Hedera rhombea, H. helix*
Helenin	*Tagetes patula*
Heliscopiol	*Euphorbia esula, E. helioscopia*
Hellebrin	*Caltha palustris*
Helminthosporin	*Cassia occidentalis, C. torosa*
Helveticoside	*Corchorus olitorius*

Helveticosol	*Erysimum amurense, E. cheiranthoides*
Hemerocallin	*Hemerocallis minor Miller, H. fulva L. Inula helenium L.*
Hemigossypol	*Gossypium herbaceum*
Hemin	*Hippophae rhamnoides*
Hentriacontane	*Carpesium abrotanoides L. Plantago asiatica, P. depressa, P. exaltata, P. loureiri, P. major*
Heptacosane	*Alnus japonica, Dicranopteris linearis, Hemerocallis minor Miller*
Heptane	*Pittosporum tobira*
Herbacetrin	*Equisetum arvense, E. hyemale, E. ramosissimum*
Herberinecorysamine	*Macleaya cordata*
Hernandezine	*Thalictrum aquilegifolium, T. baicalense, T. fauriel, T. petaloideum, T. simplex, T. squarrosum, T. thunbergii*
Hernandine	*Stephania hernendifolia*
Hernandoline	*Stephania hernendifolia*
Hernandolinol	*Stephania hernendifolia*
Herniarin	*Carum carvi*
Hesperidin	*Citrus deliciosa, C. nobilis, C. reticulata, Clinopodium chinense, C. polycephalum, C. gracile, C. umbrosum, Schizonepeta multifida, S. tenuifolia*
Heutriacontane	*Firmiana simplex*
Hexacosanol	*Scopalia dulcis*
Hexadeceonic acid	*Myristica fragrans*
Hexahydromatricaria	*Erigeron canadensis, E. annuus*
Hexalactone	*Prunus persica*

Component	Source
Hexanoic acid	*Prunus persica*
Hexanol	*Agastache rugosa, A. rugosa* f. hypoleuca, *Prunus persica*
Hexenal	*Pteridium aquilinum* (L.) Kuhu.
Hexylenaldehyde	*Ulmus campestris, U. macrocarpa, U. pumila*
Hibifolin	*Melochia corchonfolia*
Pinene	*Ferula assafoetica* L.
Hinesol	*Atractylodes chinensis, A. japonica, A. koreana, A. lancea, A. lyrata, A. ovata, A. macrocephala*
Hinokiflavone	*Cycas revoluta, Podocarpus macrophyllus, Selaginella tamarisina, Thuja koraiensis, T. orientalis, T. chinensis*
Hirsuteine	*Uncaria hirsuta, U. rhynchophylla*
Hirsutine	*Uncaria hirsuta, U. rhynchophylla*
Hirsutrin	*Gossypium herbaceum*
Hispidulin	*Salvia plebeia*
Histamine	*Viscum album*
Histidin	*Cajanus cajan* L.
Homoarbutin	*Chimaphila umbellata, Pyrola decorata, P. japonica, P. incarnata, P. renifolia, P. rotundifolia*
Homoaromoline	*Stephania cepharantha*
Homoflavoyadorinin	*Viscum album, V. coloratum*
Homolycorine	*Lycoris radiata, L. longituba, L. aurea*
Homoorentin	*Swertia pseudochinensis*

Homoorientin   *Fagopyrum esculentum, F. sagittatum*

Homoplantagin   *Plantago asiatica, P. depressa, P. exaltata, P. loureiri, P. major*

Homoplantaginin   *Salvia plebeia*

Homoserine   *Astragalus chinensis, A. complanatus, A. henyri, A. hoantchy, A. membranaceus, A. melilotoides, A. mongholicus, A. reflexistipulus, A. sinensis*

Homostephanoline   *Stephania japonica*

Homotrilobine   *Cocculus laurifolius, C. sarmentosus, C. trilobus*

Hopadiene   *Adiantum boreale, A. capillus-junonis, A. pedatum, A. flabellulatum*

Hormoharringtonine   *Cephalotaxus wilsoniana*

Houttuynium   *Houttynia cordata*

Hovenosides   *Hovenia dulcis*

Hoyin   *Hoya carnosa*

Humulene   *Eugenia aromatica, E. caryophyllata, E. ulmoides, Juniperus rigida, Humulus lupulus, H. scandens*

Huperzine A   *Hyperzia serrata*

Hydnocarpus oil   *Hydnocarpus anthelmintica, H. castaneus*

Hydrangeic acid   *Hydrangea macrophylla*

Hydrangenol   *Hydrangea macrophylla*

Hydroagarofuran   *Aquilaria agallocha, A. sinensis*

Hydrocarbons   *Viola acuminata, V. alisoviana, V. collina, V. dissecta, V. mandshurica, V. patrini, V. prionantha, V. verecunda*

Hydrocinnamic acid   *Aquilaria agallocha, A. sinensis*

Component	Source
Hydrocinnamic aldehyde	*Cinnamomum zeglanicum*
Hydrocotylene	*Vitex nequndo*
Hydrocyanic acid	*Achillea alpina, A. millefolium, Anthriscus aemula, A. sylvestris, Armeniaca ansu, A. mandsharica, A. sibirica, A. vulgaris, Chaenomeles japonica, C. sinensis, C. speciosa, Cydonia sinensis, Heracleum dissectum, H. lanatum, Manihot esculenta, Photinia serrulata, Polygonum bistorta, Sanguisorba officinalis, S. grandiflora, S. parviflora, S. x tenuifolia, Taraxacum officinale*
Hydrogen cyanide	*Ophiorrhiza japonica, O. mungos*
Hydroquinone	*Ilex pubescens, Preridium aquilinum (L.) Kuhn.*
Hydroxybenzoic acids	*Achillea alpina, A. millefolium, Anthriscus aemula, A. sylvestris, Heracleum dissectum, H. lanatum, Polygonum bistorta, Sanguisorba officinalis, S. orba grandiflora, S. officinalis, S. parviflora, S. x tenuifolia, Solidago dahurica, S. pacifica, S. virgaurea, Taraxacum officinale, Vitex nequndo*
Hydroxycinnamic acid	*Melilotus alba, M. suaveolens, M. indica, Taraxacum officinale*
Hydroxyeephalotaxine	*Cephalotaxus wilsoniana*
Hydroxyevodiamine	*Evodia rutaecarpa*
Hydroxygenkwanin	*Daphne fortunei, D. genkwa, Wikestroemia indica*
Hydroxyleamptothecine	*Camptotheca acuminata*
Hydroxypeucedanin	*Angelica decursiva*
Hylander	*Elshoitzia ciliatai* Thunb.
Hynocarpic acid	*Hydnocarpus anthelmintica, H. castaneus*
Hyoscine	*Cyperus rotundus, Hyoscyamus bohemicus*

Hyoscyamine	*Datura alba, D. fastuosa, D. innoxia, D. metel, D. stramonium, D. tatula, Scopolia tangutica, Hyoscyamus bohemicus, Physochlaina infundibularis*
Hyoscypierin	*Hyoscyamus bohemicus*
Hypaconitine	*Aconitum laciniatum, A. kusnezoffii, A. chinense, A. vilmorinianum, A. pariculigerum, A. balfouri, A. carmichaelii, A. chasmanthum, A. deinorrhizum, A. fischeri, Aconitum jaluense, A. koreanum, A. napellus, A. praeparata, A. volubile*
Hypaphorine	*Abrus precatorius*
Hypericin	*Hypericum chinensis, H. perforatum H. triquetrifolium*
Hyperin	*Chimaphila umbellata, Dysosma pleiantha (Hance) Woodson Dysosma pleiantha, Hypericum triquetrifolium, H. chinensis, Parnassia palustris, Persicaria hydropiper, Prunus padus, Rhododendron dauricum L. Rumex acetosa, R. acetosella, R. arnurensis, R. aquaticus, R. gmelini, R. longifolius, R. maritimus, R. marschallianus, R. sterophyllus, R. thyrsiflorus, Saururus chinensis, Tussilago farfara, Vaccinium bracteatum, V. vitis-idaea*
Hyperoside	*Campanula glomerata, C. punctata, Celosia argentea, C. cristata, C. margariacea, Crataegus cuneata, C. chlorusarca, C. dahurica, C. maximowiczii, C. pentagyna, C. pinnatifida, C. sanguinea, Hibiscus mutabilis, Persicaria amphibia, Prunella vulgaris, Vaccinium bracteatum, V. vitis-idaea*
Hypoepistephanine	*Stephania japonica*
Hypoglycemic	*Catharanthus roseus* (L.) G. Don
Hypophyllanthin	*Phyllanthus urinaria, P. niruri, P. reticulatus*
Hypotensive	*Acorus calamus* L.
Hystonin	*Physalis alkekengi, P. angulata*
Icariine	*Epimedium brevicorum, E. koreanum, E. macranthum*

Component	Source
Icarlin	*Epimedium brevicorum, E. koreanum, E. macranthum*
Ifflaionic acid	*Scopalia dulcis*
Ilungianins	*Pimpinella thellungiana*
Imidazolylothylamine	*Solanum lyratum, S. melongena*
Imperatorin	*Angelica amurensis, A. anomala, A. dahurica, A. decursiva, A. gigas Maxim. Poncirus trifoliata*
Indican	*Clerodendrum cyrtophyllum, Glehnia littoralis, Isatis indigotica, I. oblongata, Polygonum perfoliatum, P. tinctorium*
Indigo	*Baphicanthus cusia, Isatis indigotica, I. oblongata*
Indirubin	*Baphicanthus cusia, Clerodendrum cyrtophyllum*
Indo-brown	*Baphicanthus cusia*
Indo-yellow	*Baphicanthus cusia*
Indoid compounds	*Gentiana algida, G. barbata, G. manshurica, G. olivieri, G. scabra, G. squarrosa, G. triflora*
Indoxyl-5-ketogluconate	*Isatis chinensis, I. tinctoria*
Ineole	*Luffa aegyptiaca, L. cylindrica, L. faetida, L. petola*
Ingigo	*Clerodendrum cyrtophyllum*
Inulin	*Carpesium abrotanoides, C. athunbergianum Inula Britannica L.*
Inokosterone	*Achyranthes bidentata, A. japonica, Woodwardia japonica*
Inositol	*Lonicera acuminata, L. apodonta, L. brachypoda, L. japonica, L. confusa, L. hypoglauca, L. chinensis, L.flexuosa, L. maackii, Sonchus arvensis, S. oleraceus*
Insecticidal	*Acorus calamus L.*

Insulanoline	*Paracyclea insularis*
Insularine	*Paracyclea ochiaiana, P. insularis, Stephania japonica*
Insulin	*Dolichos lablab*
Inteolin-7-β-neohesperidoside	*Veronica sibirica, V. undulata*
Inulicin	*Inula britannica, I. japonica, I. linariaefolia, I. salsoloides*
Inulin	*Adenophora triphylla, A. verticillata, Ajuga bracteosa, Calendula officinalis, Campanula gentianoides, C. grandiflora, Carpesium abrotanoides L. Codonopsis pilosula, C. tangshen, C. ussuriensis, Cirsium chinense, C. japonicum, Lappa communis, L. edulis, L. major, L. minor, Senecio vulgaris, Taraxacum formosanum, T. officinale*
Inusterols	*Inula britannica, I. japonica, I. linariaefolia. I. salsoloides*
Invertase	*Hordeum vulgare, Plumbago zeylanica*
Invertin	*Menyanthes trifoliata*
Iocyanate	*Descurania Sophia* Webb ex Prantl.
Iodine	*Dioscorea bulbifera, Laminaria angusta, L. cichorioides, L. japonica, L. longipedalis, L. religiosa, Polygonum bistorta, Salix babylonica, S. matsudana, S. microstachya, Trifolium pratense, T. repens*
Ipolamiide	*Lamiu mamplexi-caule* L.
Iridin	*Iris aqxatuca, I. buatatas, I. dichotoma, I. lactea*
Iridoid glucosides	*Hedyotis diffusa, Scrophularia buergeriana, S. kakudensis, S. ningpoensis, S. oldhami, S. puergeriana*
Iridoidglycosides	*Vitex negundo*
Iridomyrmecin	*Actinidia arguta, A. chinensis, A. japonica, A. kolomikta, A. polygama*
Irigenin	*Iris lactea*

Component	Source
Irisflorentin	*Iris lactea*
Irisquinone	*Iris pallasii*
Iron	*Actinidia chinensis* Planch *Portulaca pilosa, Sargassum pallidum*
Iron oxide	*Phyllostachys bambusoide, P. nigra*
Isatan	*Isatis chinensis, I. tinctoria*
Isatan B	*Clerodendrum cyrtophyllum, Isatis indigotica, I. oblongata*
Iso-adiantone	*Adiantum boreale, A. capillus-junonis, A. pedatum, A. flabellulatum*
Iso-butyliso	*Elshoitzia ciliatai* Thunb.
Iso-cerberin	*Cerbera manghas* L.
Iso-chondrodendrine	*Paracyclea insularis*
Iso-corynoxeine	*Uncaria hirsuta, U. rhynchophylla*
Iso-cucurbitacin B	*Neoalsomitra integrifoliola*
Iso-darutigenals B & C	*Siegesbeckia orientallis* L.
Iso-myristicin	*Anethum graveolens* L.
Iso-quercitrine	*Lamium album* L.
Iso-rhynchophylline	*Uncaria hirsuta, U. rhynchophylla*
Iso-amaranthin	*Gomphrena globosa*
Iso-amyl	*Matricaria chamomilla*
Iso-anthricin	*Anthriscus aemula, A. sylvestris*
Iso-bavachin	*Psoralea corylifolia*
Iso-bergaten	*Heracleum moellendorffii* Hance

Iso-beturudin	*Bougainvillea brasiliensis, B.glabra*
Iso-butyl	*Matricaria chamomilla*
Iso-chaksine	*Cassia alata,*
Iso-leucine	*Cajanus cajan* L.
Iso-chlorogenic acid	*Sorbus alnifolia, S. amurensis, S. ohuashanensis, Viburnum sargenti*
Iso-chondrodendrine	*Stephania hernendifolia*
Iso-corynoline	*Corydalis incisa, C. bungeana*
Iso-corypalmine	*Papaver somniferum*
Iso-cryptomerin	*Selaginella tamarisina*
Iso-cryptotanshinone	*Salvia miltiorhiza*
Iso-dexyelephantopin	*Elephantopus molis*
Iso-dydrone-petalactol	*Actinidia polygama* Sieb & Zucc.
Iso-egomaketone	*Perilla frutescens* L.
Iso-eugenitin	*Eugenia aromatica, E. caryophyllata, E. ulmoides*
Iso-eugenol	*Myristica fragrans*
Iso-fernene	*Adiantum boreale, A. capillus-junonis, A. pedatum, A. flabellulatum*
Iso-ferulic acid	*Catalpa ovata* G. Don *Cimicifuga dahurica, C. foetida, C. heracleifolia, C. racemosa, C. ussuriensis*
Iso-flavone derivatives	*Glycine max, G. soja*
Iso-flavones	*Pueraria lobata, P. pseudo-hirsuta*
Iso-furanogermacrene	*Lindera strychnifolia*
Iso-ginketine	*Ginkgo biloba*

Component	Source
Iso-harringtonine	*Cephalotaxus wilsoniana*
Iso-homoarbutin	*Chimaphila umbellata, Pyrola decorata, P. japonica, P. incarnata, P. renifolia, P. rotundifolia*
Iso-humulone	*Humulus lupulus*
Iso-hydrocarveol	*Carum carvi* L.
Iso-imperatorin	*Notopterygium incisium*
Iso-hyperine	*Rhododendron dauricum* L.
Iso-indigo	*Baphicanthus cusia, Clerodendrum cyrtophyllum*
Iso-leucin	*Avena fatua*
Iso-leucine	*Oryza sativa*
Iso-liensinine	*Nelumbium nelumbo*
Iso-linderalactone	*Lindera strychnifolia*
Iso-linderoxide	*Lindera strychnifolia*
Iso-lineolone	*Adonis amurensis* Regel & Radde
Iso-liquiritigenen	*Glycyrrhiza pallidiflora, G. uralensis*
Iso-liquiritigenin	*Astragalus complanatus, A. henyri, A. hoantchy, A. membranaceus, A. melilotoides, A. mongholicus, A. reflexistipulus, A. sinensis*
Iso-liquiritin	*Glycyrrhiza pallidiflora, G. uralensis*
Iso-lobelamine	*Lobelia chinensis, L. pyramidalis, L. sessilifolia*
Iso-maculosindine	*Dictamnus albus, D. dasycarpus*
Iso-mangiferin	*Anemarrhena asphodeloides, Pyrrosia lingua, P. petiolosa, P. sheareri*
Iso-menthone	*Glechoma hederacea, G. longituba*

Iso-mesityl oxide	Cryptotaenia japonica, C. canadensis
Iso-myristicin	Anethum graveoleus
Iso-neomatatabiol	Actinidia arguta, A. chinensis, A. japonica, A. kolomikta, A. polygama
Iso-orientin	Lathyrus pratensis L. Polygonum orientale, Trigonella foenum-graecum, Vitex negundo, V. trifolia, V. rotundifolia
Iso-patrinene	Patrina scabiosaefolia, P. villosa (Thunb.)
Iso-paulownin	Paulownia tometosa Thunb
Iso-pelletierine	Punica granatum
Iso-pentenic acid	Ligularia fischeri (Ledeb)
Iso-pimpinellin	Zanthoxylum ailanthoides
Iso-pinocamphone	Glechoma hederacea, G. longituba
Iso-propylidere	Orostachys fimbriatus Turcz.
Iso-quercetin	Campanula glomerata, C. punctata, Hypericum attenuatum, H. ascyron, H. japonicum, H. perforatum, H. sumpsonii
Iso-quercitrin	Apocynum venetum, Celosic argentea, C. cristata, C. margariacea, Cucumis sativus, Cyrtomium falcatum, Equisetum arvense, E. hyemale, E. ramosissimum, Gymnadenia conopsea (L.) R. Brown Hibiscus mutabilis, Houttynia cordata, Inula britannica, I. japonica, I. linariaefolia, I. salsoloides, Loropetalum chinense, Melaleuca leucadendra, Pteridium aquilinum (L.) Kuhu Saururus chinensis
Iso-quinoline	Chenopotium ambrosiodes
Iso-ramanone	Adonis chrysocyathus, A. brevistyla, A. vernalis, Metaplexis japonica Thunb.
Iso-rhamnetin	Caltha palustris, Campanula glomerata, C. punctata, Ginkgo biloba, Hippophae rhamnoides, Sophora japonica

Component	Source
Iso-rhamnetin-3-mono-beta-D-glucoside	*Hippophae rhamnoides*
Iso-rhynchophylline	*Nauclea rhynchophylla, N. sinensis*
Iso-ricinoleic acid	*Ricinus communis*
Iso-orientin	*Lathyrus pratensis* L.
Iso-sakuranetin	*Eupatorium odoratum*
Iso-sinomenine	*Cocculus diversifolius, C. thunbergii, Sinomenium acutum*
Iso-steroidal alkaloids	*Fritillaria* species
Iso-tadeonal	*Persicaria hydropiper, Polygonum hydropiper*
Iso-talatizidine	*Aconitum barbatum, A. austroyunnanense, Stemona japonica, S. tuberosa*
Iso-tetrandrine	*Stephania cepharantha, Thalictrum aquilegifolium, T. baicalense, T. fauriel, T. petaloideum, T. simplex, T. squarrosum, T. thunbergii*
Iso-thalidenzine	*Thalictrum aquilegifolium, T. baicalense, T. fauriel, T. petaloideum, T. simplex, T. squarrosum, T. thunbergii*
Iso-thamnetin	*Typha angustata, T. angustifolia, T. davidiana, T. latifolia, T. minima, T. orientalis, T. przegalskii*
Iso-thiocyanates	*Lepidium apetalum, L. virginicum*
Iso-thujene	*Ledum palustre*
Iso-trilobine	*Cocculus laurifolius, C. sarmentosus, C. trilobus*
Iso-quercitrin	*Equisetum arvense* L. *Lathyrus pratensis* L. *Nymphoides peltata* (S. G. Gmelin)
Iso-valeraldehyde	*Santalum album, S. myrtifolium, S. verum*

Iso-valeric acid — *Artemisia gmelini, Cymbopogon citratus, Humulus lupulus, Valeriana alternifolia, V. amurensis, V. fauriei, V. subbipinnatifolia*

Iso-vanihyperzine A — *Hyperzia serrata*

Iso-vitexin — *Jatropha podagrica, Polygonum orientale, Swertia pseudochinensis*

Iso-xanthanol — *Xanthium chinense, X. japonicum, X. mongolicum, X. sibiricum, X. strumarium*

Izalpinin — *Alpinia japonica*

Jaligonic acid — *Phytolacca acinosa, P. americana, P. japonica, P. kaempferi, P. octandra, P. pekinensis*

Jambolin — *Syzygium cuminii*

Jasmiflorin — *Jasminum mesnyi, J. nudiflorum*

Jasmipierin — *Jasminum mesnyi, J. nudiflorum*

Jatamansic acid — *Nardostachys jatamansi*

Jatrorhizine — *Berberis amurensis, B. poiretii, B. sibirica, B. soulieana, Coptis chinensis, C. japonica, C. teeta. Fibraurea recisa, Mahonia japonica, Thalictrum aquilegifolium, T. baicalense, T. fauriei, T. foetidum, T. glandulissimum, T. ichangense, T. petaloideum, T. simplex, T. squarrosum, T. thunbergii*

Jervine — *Hemerocallis flava, Veratrum dahuricum, V. formosanum, V. maackii, V. nigrum*

Juglandic acid — *Juncus effusus*

Juglanin — *Juglans mandshurica, J. regia*

Juglonone — *Juncus effusus*

Jugone — *Juglans mandshurica, J. regia*

Jujuboside A — *Ziziphus jujuba, Z. spinosa*

Jujuboside B — *Ziziphus jujuba, Z. spinosa*

k-strophanthin-β — *Adonis amurensis Regel & Radde. Apocynum venetum*

Component	Source
Kadsurarin A	*Kadsura japonica*
Kadsuric acid	*Kadsura japonica*
Kadsurin	*Kadsura japonica*
Kaempferin	*Alpinia katsumadai, A. globosum, A. kumatake, Cassia angustifolia*
Kaempferitrin	*Desmodium microphyllum, Geranium dahuricum, G. eriostemon, G. sibiricum, G. wlassowianum, G. wilfordi*
Kaempferol	*Aster ageratoides, Astragalus complanatus, A. henyri, A. hoantchy, A. membranaceus, A. melilotoides, A. mongholicus, A. reflexistipulus, A. sinensis, Calystegia hederacea, C. japonica, Campanula glomerata, C. punctata, Chimaphila umbellata, Convolvulus arvensis, Coriandrum sativum, Cornus alba, C. kousa, C. macrophylla, Eugenia aromatica,Dysosma pleiantha (Hance) Woodson, E. caryophyllata, E. ulmoides, Heracleum moellendorffii Hance Hypericum perforatum, Isatis chinensis, I. tinctoria, Lathyrus pratensis L. Parnassia palustris, Pedicularis resupinata, Persicaria amphibia, Plumeria rubra, Pongamia pinnata, Rhamnus davurica, R. parvifolia, Vicia faba*
Kaempferol-3-β-glycoside	*Gymnadenia conopsea* (L.) R. Brown, *Lamium album* L.
Kaempferol glucosides	*Euonymus alatus, E. bungeanus, E. maackii*
Kaempferol trisaccharide	*Sesbinia sesbin*
Kaempferol-3-galactoside daempferol	*Bauhinia championi, B. variegata*
Kaempferol-3-glucoside	*Rosa multiflora, Viburnum sargenti*
Kaempferol-3-glucosylgalactoside	*Ophiopogon japonicus*
Kaempferol-3-rhamnoglucoside	*Calystegia hederacea, C. japonica, Ginkgo biloba*

Kaempferol-3-robinobioside	*Phaseolus angularis, P. lunatus, P. radiatus, P. vulgaris*
Kaempferol-3,7-diglucoside	*Equisetum arvense, E. hyemale, E. ramosissimum*
Kaempferol-7-β-glycoside	*Gymnadenia conopsea* (L.) R. Brown
Kaempferol-7-shamnoside	*Chenopotium ambrosiodes*
Kaempferol-rhamno glucoside	*Solidago dahurica, S. pacifica, S. virgaurea*
Kaempferol-rhamnoside	*Onychium japonicum, Pueraria lobata, P. pseudo-hirsuta, Solidago virgaurea* L.
Kansuinine	*Euphorbia kansui*
Kanugin	*Pongamia pinnata*
Kalopanax sapenin A, B.	*Kalopanax septemlobus* (Thunb) Koidz.
Kalosaponin	*Kalopanax septemlobus* (Thunb.) Koidz.
Kalotoxin	*Kalopanax septemlobus* (Thunb.) Koidz.
Karabin	*Nerum indicum*
Karanjin	*Pongamia pinnata*
Kaurene	*Podocarpus macrophyllus*
Kaurene derivatives	*Aralia chinensis, A. cordata, A. elata*
Ketone	*Hedychium coronarium, Plumeria rubra*
Khellol	*Cimicifuga dahurica, C. foetida, C. heracleifolia, C. racemosa, C. ussuriensis*
Kiganen	*Cryptotaenia japonica, C. canadensis*
Kiganol	*Cryptotaenia japonica, C. canadensis*
Kino-tannic acid	*Pterocarya stenoptera*
Klob-Alisova	*Dasiphora fruticosa* (Nestler) Kom.
Kobusone	*Cyperus rotundus*

Component	Source
Koelreuteria A, B.	*Koelreuteria paniculata* Laxm.
Konokiol	*Magnolia hypoleuca, M. officinalis, M. japonica*
Korepimedoside A	*Epimedium brevicorum, E. koreanum, E. macranthum*
Korepimedoside B	*Epimedium brevicorum, E. koreanum, E. macranthum*
Koumine	*Gelsemium sempervirens, G. elegans*
Kouminicine	*Gelsemium sempervirens, G. elegans*
Kouminine	*Gelsemium sempervirens, G. elegans*
Kukoamines	*Lycium chinense*
Kulinone	*Melia japonica, M. toosendan, M. azedarach*
Kumatakinin	*Alpinia japonica*
Kuraridin	*Sophora flavescens, S. alopecurosides*
Kurrin	*Picrorhiza kurroa*
Kutkin	*Picrorhiza kurroa*
l-(d)-isoleucine betaine	*Cannabis chinensis, C. sativa*
l-abrine	*Abrus precatorius*
l-anagyrine	*Sophora flavescens, S. alopecurosides*
l-arabinose	*Bupleurum chinense, B. falcatum, B. scorzoneraefolium, Malva chinensis, M. pulchella, M. verticillata, M. sylvestris, Plantago asiatica, P. depressa, P. exaltata, P. loureiri, P. major*
l-baptifoline	*Sophora flavescens, S. alopecurosides*
l-beta-santonin	*Artemisia finita, A. frigida*
l-borneol	*Liquidambar acerifolia, L. formosana, L. maximowiczii*

l-cadinene	*Murraya paniculata*
l-camphen	*Cnidium monnieri*
l-camphor	*Hedyotis corymbosa*
l-caryophyllene	*Valeriana alternifolia, V. amurensis, V. fauriei, V. subbipinnatifolia*
l-citronellol	*Rosa rugosa*
l-cocaine	*Erythroxylum coca*
l-curcamene	*Curcuma longa*
l-ephedrin	*Pinellia ternata, P. tuberifera*
l-ephedrine	*Ephedra distachya, E. equisetina, E. intermedia, E. monosperma, E. sinica*
l-epicatechin	*Crataegus cuneata, C. chlorosarca, C. dahurica, C. maximowiczii, C. pentagyna, C. pinnatifida, C. sanguinea*
l-epicatechol	*Camellia japonica, Vaccinium bracteatum, V. vitis-idaea*
Leptostachyd acetate	*Phryma leptostachya* L.
l-hexacosene	*Acanthopanax giraldii*
l-hydroxy-2,3,4,7-tetramethoxy xanthone	*Halenia corniculata* L.
l-hyoscyamine	*Datura suaveolens*
l-limonene	*Chenopotium ambrosiodes, Melaleuca leucadendra, Perilla frutescens* L. *Valeriana alternifolia, V. amurensis, V. fauriei, V. subbipinnatifolia*
l-linalool	*Cinnamomum zeglanicum, Rosa rugosa, Tagetes erecta*
l-marmesin	*Peucedanum terebinthaceum* Fisch
l-menthone	*Glechoma hederacea, G. longituba, Lysimachia barystachys, L. christinae, L. clethroides, L. davurica*

Component	Source
l-methylcytisine	*Sophora flavescens, S. alopecuroides*
l-methylephedrine	*Ephedra distachya, E. equisetina, E. intermedia, E. monosperma, E. sinica*
l-norephedrine	*Ephedra distachya, E. equisetina, E. intermedia, E. monosperma, E. sinica*
l-p-menthene	*Rosa rugosa*
l-perilla	*Perilla frutescens, P. ocymoides, P. polystachya, P. arguta*
l-perilla-aiuehyde	*Perilla frutescens L.*
l-phellandrine	*Cinnamomum zeglanicum*
l-pimara-8,15-dien-19-oic acid	*Aralia chinensis, A. cordata, A. elata*
l-pinene	*Cnidium monnieri, Lysimachia barystachys, L. christinae, L. clethroides, L. davurica*
l-pinocamphone	*Glechoma hederacea, G. longituba, Lysimachia barystachys, L. christinae, L. clethroides, L. davurica*
l-propenylsulforic acid	*Allium victorialis*
l-pulegone	*Glechoma hederacea, G. longituba*
l-rhamnose	*Malva chinensis, M. pulchella, M. verticillata, M. sylvestris, Plantago asiatica, P. depressa, P. exaltata, P. loureiri, P. major*
l-sesamen	*Eleutherocossus senticosus*
l-sesamin	*Acanthopanax sessiliflorus*
l-stepharine	*Menispermum dauricum*
l-tetrahydropalmatine	*Stephania sinica*
Labdadien	*Araucaria cunninghamii*
Labenzyme	*Cirsium chinense, C. japonicum*

Laccerol	Baphicanthus cusia
Lacerol	Clerodendrum cyrtophyllum
Lacnophyllum	Erigeron caradensis, E. annuus
Lactiflorenol	Artemisia lactiflora
Lactone	Cleome spinosa, C. gynandra, C. viscosa, Eupatorium chinense, E. lindleyanum, E. japonicum, Rhaponticum uniflorum
Lactones-xanthatin	Vitex nequndo
Lactucerol	Sonchus arvensis, S. oleraceus
Lambertianic acid	Pinus madshurica Rupr.
Lamber-tianic methylate, longifolene	Pinus madshurica Rupr.
Laminarin	Laminaria angusta, L. cichorioides, L. japonica, L. longipedalis, L. religiosa
Laminine	Laminaria angusta, L. cichorioides, L. japonica, L. longipedalis, L. religiosa
Lamiol	Lamiu mamplexi-caule L.
Lapase	Aquilegia vulgaris
Lamioside	Lamiu mamplexi-caule L.
Lappaol	Arctium lappa
Lappatin	Lappa communis, L. edulis, L. major, L. minor
Lappine	Lappa communis, L. edulis, L. major, L. minor
Larreagenin	Anredera cordifolia
Latex	Ficus pumila, F. inicrocarpa
Lathyrol diacetate benzoate	Euphorbia lathyrus, E. lucorum, E. resinfera, E. thymifolia

Component	Source
Lathyrol diacetate nicotinate	*Euphorbia lathyrus, E. lucorum, E. resinfera, E. thymifolia*
Laudanine	*Papaver somniferum*
Lauric acid	*Ajuga bracteosa, Citrullus anguria, C. edulis, C. lanatus, C. vulgaris, Lindera obtusiloba, Melia japonica, M. toosendan, M. azedarach, Myristica fragrans, Sepium sebiferum, S. discolor, Taraxacum mongolicum, T. sinicum*
Laurifoline	*Zanthoxylum ailanthoides*
Laurotetanine	*Cassytha filliformis, Litsea cubeba*
Lavoxanthin	*Senecio argunensis, S. nemorensis, S. scandens*
Lawsone	*Impatiens balsamina, I. noli-tangere, I. textori, Lawsonia inermis*
Laxogenin	*Smilax china, S. nipponica, S. sieboldii, S. riparia*
Lecithin	*Polygonum multifolrum, P. chinensis, Sesamum indicum*
Lecithine	*Cardamine leucantha, C. lyrata*
Lenrosine	*Catharanthus roseus*
Lenrosivine	*Catharanthus roseus*
Leonaridine	*Leonurus heterophyllus, L. japonicus, L. macranthus, L. mongolicus, L. pseudo-macranthus*
Leonurine	*Leonurus heterophyllus, L. japonicus, L. macranthus, L. mongolicus, L. pseudo-macranthus, L. sibiricus*
Leonurinine	*Leonurus heterophyllus, L. japonicus, L. macranthus, L. mongolicus, L. pseudo-macranthus*
Leucaenine	*Leucaena leucocephala*
Leucanol	*Leucaena leucocephala*

Leucine	*Avena fatua, Dioscorea opposita, Linum stelleroides, L. usitatissimum, Litchi chinensis, Oryza sativa*
Leucoanthocyanins	*Camellia japonica, Fagopyrum esculentum, F. sagittatum, Persicaria hydropiper, Plumbago zeylanica, Prunus persica*
Leucocyanidin	*Pileostegia viburnoides*
Leucocyanidol	*Euphorbia hirta*
Leucylphenylalanine anhydride	*Ligusticum chuanziang*
Leurosine sulphate	*Catharanthus roseus* (L.) G. Don
Levidulinase	*Amorphophallus rivieri*
Leviduline	*Amorphophallus rivieri*
Levulose	*Eriobotrya japonica*
Lichenin	*Usnea diffracta, U. longissima*
Liderane	*Lindera strychnifolia*
Liensinine	*Nelumbium nelumbo*
Lignin	*Arctium lappa, Boehmeria densiflora, Melaleuca leucadendra, Quercus acutissima, Q. aliena, Q. dentata, Q. liaotungensis, Q. mongolica, Q. variabilis*
Lignoceric acid	*Machilus thunbergii, Sinapis alba*
Liguloxidol acetate	*Ligularia fischeri* (Ledeb)
Ligustilide	*Angelica polymorpha, A. sinensis, Ligusticum chuanziang*
Limonene	*Anethum graveoleus, Artemisia lactiflora, Blumea balsumifera, Chrysanthemum boreale, C. indicum, C. lavandulaefolium, C. procumbens, C. tripartium, Citrus deliciosa, C. nobilis, Conyza canadensis, Coriandrum sativum, Cunninghamia lanceolata, Erigeron canadensis, E. annuus, Glechoma hederacea, G. longituba, Juniperus rigida, Ledum palustre, Lindera glauca,*

Component	Source
Limonene (continued)	*Luffa aegyptiaca, L. cylindrica, L. faetida, L. petola, Lysimachia barystachys, L. christinae, L. clethroides, L. davurica, Medicago falcata, M. lupulina, M. polymorpha, M. ruthenica, M. sativa, Notopterygium incisium, Oenothera javanica, Perilla frutescens, P. ocymoides, P. polystachya, P. arguta, Pinus bungeana, P. densiflora, P. koraiensis, P. sylvestris, P. tabulaeformis, Tagetes patula, Thymus vulgaris, Xanthoxylum piperitum, Zanthoxylum bungeanum*
Limonin	*Dictamnus albus, D. dasycarpus, Evodia rutaecarpa, Poncirus trifoliata*
Linalol	*Elettaria cardamomum*
Linalool	*Agastache rugosa, A. rugosa f. hypoleuca, Amomum cardamomum, A. globosum, A. tsao-ko, A. villosum, A. xanthoides, Artemisia argyi, A. halodendron, A. igniaria, A. indica, A. integrifolia, A. japonica, A. keiskeana, A. lagocephala, A. lavandulaefolia, A. scoparia, A. selengensis, A. sieversiana, A. vulgarts, Citrus reticulata, Conyza canadensis, Coriandrum sativum, Cymbopogon citratus, Eupatorium chinense, E. lindleyanum, E. japonicum, Glechoma hederacea, G. longituba, Hedyotis corymbosa, H. diffusa, Litsea cubeba, Luffa aegyptiaca, L. cylindrica, L. faetida, L. petola, Medicago falcata, M. lupulina, M. polymorpha, M. ruthenica, M. sativa, Michelia alba, M. figo, Myristica fragrans, Perilla frutescens, P. ocymoides, P. polystachya, P. arguta, Plumeria rubra, Tagetes patula, Thymus vulgaris, Zanthoxylum bungeanum, Zingiber officinale*
Linalyl acetate	*Tagetes patula, Thymus vulgaris*
Linamarase	*Fagopyrum esculentum, F. sagittatum*
Linamarin	*Linum stelleroides, L. usitatissimum*
Linaracrine	*Lindera akoensis*
Linarezine	*Lindera akoensis*

Linarin	*Lindera akoensis*
Linderalactone	*Lindera strychnifolia*
Linderene	*Lindera strychnifolia*
Linderic acid	*Lindera obtusiloba*
Linderol	*Lindera obtusiloba*
Linderoxide	*Lindera strychnifolia*
Lindestrene	*Lindera strychnifolia*
Lindestreolide	*Lindera strychnifolia*
Lineolone	*Adonis amurensis* Regel & Radde
Linoleic acid	*Acanthopanax gracilistylus, A. spinosum, Angelica grosserrata, Ajuga bracteosa, Aquilegia vulgaris, Benincase cerifera, B. hispida, Cardamine leucantha, C. lyrata, Cibotium barometz, Citrullus anguria, C. edulis, C. lanatus, C. vulgaris, Coix agrestis, C. chinensis, C. lachryma, Cucumis sativus, Descurainia Sophia* (L.) Schur. *Elettaria cardamomum, Jatropha gospifolia, J. curcas, Lindera obtusiloba, Myristica fragrans, Sinapis alba, Tamarindus indicus, Sesamum indicum*
Linolein	*Typhonium giganteum*
Linolenic acid	*Cardamine leucantha, C. lyrata, Cornus walteri, Descuraina Sophia* (L.) Schur. *Oenothera terythrosepala*
Linoleyl acetate	*Conyza canadensis.*
Linolic acid	*Corchorus capsularis, C. olitorius*
Linositol	*Solidago pacifica* Juzepczuk
Lipase	*Ulmus campestris, U. macrocarpa, U. pumila*
Liquiritigenin	*Glycyrrhiza pallidiflora, G. uralensis*

Component	Source
Liquiritin	*Glycyrrhiza pallidiflora, G. uralensis*
Liquloxidol	*Ligularia fischeri* (Ledeb)
Liriodenine	*Magnolia hypoleuca, M. officinalis, M. japonica Nelumbium nuciferum, N. speclosum*
Lithospermin	*Lithospermum erythrorhizon, L. officinalis*
Llysine	*Taraxacum mongolicum, T. sinicum*
Lobelanidine	*Lobelia chinensis, L. pyramidalis, L. sessilifolia*
Lobelanine	*Lobelia chinensis, L. pyramidalis, L. sessilifolia*
Lobeline	*Lobelia chinensis, L. pyramidalis, L. sessilifolia*
Lochnerine	*Catharanthus roseus* (L.) G. Don
Loganin	*Cornus officinalis, C. walteri, Lonicera acuminata, L. apodonta, L. brachypoda, L. chinensis, L. confusa, L. flexuosa, L. hypoglauca, L. japonica, L. maackii, Menyanthes trifoliate, Patrinia villosa* (Thunb.)
Loliolide	*Maytenus diversifolia, M. confertiflorus*
Longiceroside	*Cornus officinalis*
Lonicerin	*Lonicera acuminata, L. apodonta, L. brachypoda, L. chinensis, L. confusa, L. flexuosa, L. hypoglauca, L. japonica, L. maackii, Menyanthes trifoliata*
Lotaustralin	*Linum stelleroides, L. usitatissimum*
Lotusine	*Nelumbium nelumbo*
Lucernol	*Medicago falcata, M. lupulina, M. polymorpha, M. ruthenica, M. sativa*
Lumicaerulic acid	*Coptis chinensis, C. japonica, C. teeta*
Lunularic acid	*Pileostegia viburnoides*

Lupenone	*Adenophora triphylla, A. verticillata, Alnus japonica, Firmiana simplex*
Lupeol	*Elephantopus molis, Eupatorium odoratum, Ficus carica, Plumeria rubra, Prunus padus, Viscum album, V. coloratum*
Lupeol acetate	*Artocarpus altilis, Elephantopus molis*
Lupeol palmitate	*Jatropha podagrica*
Lupeose	*Isatis chinensis, I. tinctoria*
Lupin alkaloid	*Euchresta japonicum*
Lupinidin	*Lupinus luteus*
Lupinine	*Lupinus luteus*
Lupulin	*Humulus lupulus*
Lupulone	*Humulus lupulus, H. scandens*
Lutein	*Cuscuta australis, Potamogeton perfoliatus L. Taraxacum mongolicum, T. sinicum*
Luteioic acid	*Psidium guajava*
Luteolin	*Ajuga bracteosa, A. decumbens, A. pygmaea, Anthriscus aemula, A. sylvestris, Bidens tripartita, Botrychium strictum Underw. Codonopsis lanceolata, Humulus scandens, Lonicera acuminata, L. apodonta, L. brachypoda, L. chinensis, L. confusa, L. flexuosa, L. hypoglauca, L. japonica, L. maackii; Perilla frutescens, P. ocymoides, P. polystachya, P. arguta, Physalis alkekengi, Veronica sibirica. V. undulata*
Luteolin-7-glucoside	*Arthruxon hispidus, Daucus carota, Juncus effusus, Persicaria amphibia, Thymus amurensis, T. disjunctus, T. kitagawianus, T. komarovii, T. przewalskii, T. quinquecostatus, Vitex trifolia, V. rotundifolia*
Luteolin-7-rhamnoglucoside	*Lonicera acuminata, L. apodonta, L. brachypoda, L. chinensis, L. confusa, L. flexuosa, L. hypoglauca, L. japonica, L. maackii*

Component	Source
Luteolin-7-β-D-glucoside	*Agrimonia eupatoria, A. pilosa, A. viscidula*
Luteolin-7-β-D-glucopyranoside	*Lemmaphyllum microphyllum*
Luteolin-monoarabinoside	*Arthruxon hispidus, A. hispidus*
Luteoline	*Arthruxon hispidus*
Luteolinidin	*Juncus effusus*
Luteolinidin 5-glucoside	*Azolla imbricata*
Lycoclavanin	*Lycopodium clavatum, L. obscurum, L. selago, L. serratum*
Lycoclavanol	*Lycopodium clavatum, L. obscurum, L. selago, L. serratum*
Lycodoline	*Lycopodium clavatum, L. obscurum, L. selago, L. serratum*
Lycopene	*Calendula officinalis, Crocus sativus, Daucus carota, Hippophae rhamnoides*
Lycopodine	*Lycopodium annotinum, L. cernum, L. compianatum*
Lycopose	*Lycopus fargesii, L. lucidus, L. maackianus, L. parviflorus, L. ramosissimus, L. veitchii*
Lycoramine	*Hippeastrum hybridum, Lycoris radiata, L. longituba, L. aura*
Lycorenine	*Lycoris radiata, L. longituba, L. aura*
Lycoricidine	*Lycoris radiata, L. longituba, L. aurea*
Lycoricidinol	*Lycoris radiata, L. longituba, L. aura*
Lycorin	*Hymenocallis speciosa*
Lycorine	*Clivia miniata, Hippeastrum hybridum, Lycoris radiata, L. longituba, L. aura, Narcissus tazetta, Zephyranthes candida* Herb.
Lyoniols	*Lyonia ovalifolia*
Lysine	*Cajanus cajan* L. *Dolichos lablab, Litchi chinensis, Oryza sativa*

Lysopine   *Parthenocissus tricuspidata*

Maclurin   *Morus alba, M. constantinopolitana, M. indica*

Macrephyllic acid   *Podocarpus macrophyllus*

Macrophylline   *Senecio argunensis, S. nemorensis, S. scandens*

Madecassoside   *Centella asiatica*

Maesaguinone   *Maesa japonica, M. tenera*

Magnesium   *Portulaca pilosa, P. oleracea*

Magnocurarine   *Magnolia biloba, M. denudata, M. discolor, M. grandiflora, M. hypoleuca, M. japonica M. liliflora, M. officinalis, M. purpurea*

Magnoflorine   *Aristolochia debilis, A. contorta, A. kaempferi, A. longa, A. recurvilabra, Caulophyllum robustum, Cocculus laurifolius, C. sarmentosus, C. trilobus, Epimedium brevicorum, E. macranthum, E. macranthum, Magnolia hypoleuca, M. japonica, M. officinalis, Menispermum dauricum, Papaver somniferum, Sinomenium acutum, Thalictrum foetidum, Zanthoxylum ailanthoides, Z. schinifolium*

Magnolol   *Magnolia hypoleuca, M. officinalis, M. japonica*

Mairin   *Eugenia aromatica, E. caryophyllata, E. ulmoides*

Makisterones   *Podocarpus macrophyllus*

Makulor   *Commiphora myrrha*

Malic acid   *Chaenomeles japonica, C. sinensis, C. speciosa, Coriandrum sativum, Cornus officinalis, Cydonia sinensis, Drosera anglica, D. burmanni, D. rotundifolia, Eriobotrya japonica, Lactuca raddeana, L. indica, L. sativa, Macrocarpium officinalis, Matricaria chamomilla, Oxalis corriculaza, O. corymbosa, Prunus mume, P. persica, Ribes mandshurica, Viburnum sargenti, Vitis amurensis, V. vinifera*

Component	Source
Mallorepine	*Mallotus repandus*
Mallotinin	*Mallotus repandus*
Maltase	*Fagopyrum esculentum, F. sagittatum, Solanum indicum*
Maltose	*Hordeum vulgare*
Maluidin glucoside	*Syzygium cuminii*
Malvidin	*Lythrum salicaria, Medicago falcata, M. lupulina, M. polymorpha, M. ruthenica, M. sativa*
Malvin	*Lythrum salicaria*
Mandelonitrile	*Prunus armeniaca*
Mangasese	*Portulaca pilosa*
Mangiferin	*Anemarrhena asphodeloides*
Mannan	*Dioscorea batatus*
Manneotetrose	*Isatis chinensis, I. tinctoria*
Mannit	*Gardenia angusta, G. jasminoides*
Mannitol	*Scopolia dulcis, Solidago pacifica Juzepczuk Sonchus arvensi L. s, S. oleraceus, Thesium chinense, Veronica sibirica, V. undulata*
Mannosan	*Luffa aegyptiaca, L. cylindrica, L. faetida, L. petola*
Mannosan	*Trachycarpus wagnerianus, T. fortunei*
Mannose	*Amorphophallus rivieri, Cucumis sativus, Jasminum mesnyi, J. nudiflorum*
Margaric acid	*Sepium sebiferum, S. discolor*
Markogenin	*Anemarrhena asphodeloides*
Matabic acid	*Actinidia polygama* Sieb & Zucc.

Marmesin	*Angelica amurensis, A. anomala, A. dahurica*
Marsdeoreophisides	*Marsdenia tenacissima*
Martaicaria ester	*Conyza canadensis.*
Maslinic acid	*Chamaenerion angustifolium, Crataegus cuneata, C. chlorusarca, C. dahurica, C. maximowiczii, C. pentagyna, C. pinnatifida, C. sanguinea, Purica granatam,*
Masperuloside	*Morinda citrifolia, M. officinalis*
Masticinic acid	*Pistacia lentiscus*
Masticonic acid	*Pistacia lentiscus*
Masticoresene	*Pistacia lentiscus*
Matai-resinol	*Arctium lappa*
Matairesinoside	*Trachelospermum jasminoides*
Matatabic acid	*Actinidia arguta, A. chinensis, A. japonica, A. kolomikta, A. polygama*
Matatabiether	*Actinidia arguta, A. chinensis, A. japonica, A. kolomikta, A. polygama*
Matatabistic acid	*Actinidia arguta, A. chinensis, A. japonica, A. kolomikta, A. polygama*
Matricaria	*Erigeron canadensis, E. annuus*
Matrine	*Sophora subprostrata*
Matteucinol	*Paulownia tometosa* Thunb.
Maytanacine	*Maytenus serrata, M. hookeri*
Maytanbutine	*Maytenus serrata, M. hookeri*
Maytanprine	*Maytenus serrata, M. hookeri*
Maytansine	*Maytenus diversifolia, M. confertiflorus, M. hookeri, M. serrata*
Maytansinol	*Maytenus serrata, M. hookeri*

Component	Source
Maytanvaline	*Maytenus serrata, M. hookeri*
Meconine	*Papaver somniferum*
Medicagemic acid	*Medicago falcata, M. lupulina, M. polymorpha, M. ruthenica, M. sativa*
Melaleucin	*Melaleuca leucadendra*
Melialactone	*Melia azedarach, M. japonica, M. toosendan*
Melianodiol	*Melia japonica, M. toosendan, M. azedarach*
Melianol	*Melia japonica, M. toosendan, M. azedarach*
Melianotriol	*Melia azedarach. M. japonica, M. toosendan*
Meliatin	*Menyanthes trifoliata*
Melibiase	*Solanum indicum*
Melilotic acid	*Melilotus alba, M. suaveolens, M. indica*
Melilotoside	*Melilotus alba, M. suaveolens, M. indica*
Melocorin	*Melochia corchonfolia*
Melotoxin	*Cucumis melo*
Menisnine	*Cissampelos pareira*
Menisperine	*Menispermum dauricum*
Menispermine	*Menispermum dauricum*
Menthene	*Rosa rugosa*
Menthiafolin	*Menyanthes trifoliata*
Menthol	*Glechoma hederacea, G. longituba, Luffa aegyptiaca, L. cylindrica, L. faetida, L. petola, Mentha arvensis, M. dahurica, M. haplocalyx, M. sachalinensis*

Menthone	*Luffa aegyptiaca, L. cylindrica, L. faetida, L. petola, Mentha arvensis, M. dahurica, M. haplocalyx, M. sachalinensis*
Menthyl acetate	*Mentha arvensis, M. dahurica, M. haplocalyx, M. sachalinensis*
Menyanthin	*Menyanthes trifoliata*
Meoglucobrassicin	*Isatis indigotica, I. oblongata*
Mesaconitine	*Aconitum balfouri, A. carmichaelii, A. chasmanthum, A. chinense, A. deinorrhizum, A. fischeri, A. jaluense, A. koreanum, A. kusnezoffii, A. laciniatum, A. napellus, A. pariculigerum, A. praeparata, A. vilmorinianum, A. volubile*
Mesityl oxide	*Cryptotaenia japonica, C. canadensis*
Mesoinositol	*Clerodendrum trichotomum, C. spicatus, Viscum album, V. coloratum*
Metaphanine	*Stephania japonica*
Metaploxigenin	*Marsdenia tenacissima, M. japonica* Thunb.
Methanethiol	*Asparagus cochinenesis, A. falcatus, A. insularis, A. lucidus, A. officinalis*
Methanolic	*Morinda parvifolia*
Methioine	*Cajanus cajan* L. *Oryza sativa*
Methoxyl-camptothecine	*Camptotheca acuminata*
Methoxylhemigosipol	*Gossypium herbaceum*
Methyl-3-O-beta-glucopyranosyl-gallate	*Rosa acicularis, R. amygdalifolia, R. davurica, R. davurica, R. koreana, R. laevigata, R. maximowicziana*
Methyl acetyl-isocupressate	*Araucaria cunninghamii*
Methyl allyltrisulfide	*Allium victorialis*
Methyl allyldisulfide	*Allium victorialis*
Methyl amentoflavone	*Araucaria cunninghamii*

Component	Source
Methyl anthranilate	*Citrus deliciosa, C. nobilis*
Methyl-bellidifolin	*Swertia pseudochinensis*
Methyl caffeate	*Campanula glomerata, C. punctata*
Methyl cinnamate	*Alpinia officinarum*
Methyl communate	*Araucaria cunninghamii*
Methyl-corypalline	*Nelumbium nelumbo*
Methyl eugenol	*Michelia alba, M. figo*
Methyl isobutyl ketone	*Cryptotaenia japonica, C. canadensis*
Methyl isocupressate	*Araucaria cunninghamii*
Methyl-1-propenyl disulfide	*Allium victorialis*
Methyl-laurate	*Osmanthus fragrans*
Methyl-n-amyl ketone	*Cinnamomum zeglanicum*
Methyl n-nonyl ketone	*Zanthoxylum ailanthoides*
Methyl nigakinone	*Picrasma quassioides*
Methyl palmitate	*Codonopsis pilosula, C. tangshen, C. ussuriensis*
Methyl-pelletierine	*Punica granatum*
Methyl-salicylate	*Gaultheria leucocarpa*
Methyl-swertianin	*Swertia pseudochinensis*
Methyl-acetic acid	*Erigeron canadensis, E. annuus*
Methy-lanthranilate	*Citrus reticulata, Murraya paniculata*
Methylchavicol	*Agastache rugosa, A. rugosa* f. *hypoleuca, Foeniculum officinale, F. vulgare*

Methyl-cytisine	*Caulophyllum robustum, Sophora subprostrata*
Methylene-bishydroxy-coumarin	*Medicago falcata, M. lupulina, M. polymorpha, M. ruthenica, M. sativa*
Methy-ephedrine	*Ephedra distachya, E. equisetina, E. intermedia, E. monosperma, E. sinica*
Methylethylacetic ester	*Michelia alba, M. figo*
Methylheptenol	*Cymbopogon citratus*
Methylheptenone	*Cymbopogon citratus, Zingiber officinale*
Methylisopelletierine	*Punica granatum*
Methylkulonate	*Melia japonica, M. toosendan, M. azedarach*
Methylcaconitine	*Delphinium grandiflorum*
Methylmyristate	*Osmanthus fragrans*
Methylpentosans	*Abutilon theophrasti, A. avicennae*
Methylpalmitate	*Osmanthus fragrans*
Methylpentose	*Abutilon theophrasti, A. avicennae*
Methylsalicylate	*Dianthus chinensis* L.
Methyl-vanillin	*Gymnadenia conopsea* (L.) R. Brown
Mi-hem erocallin	*Hemerocallis minor* Miller
Michelabine	*Cocculus diversifolius, C. thunbergii, Michelia alba, M. figo*
Minerals	*Actinia chinensis* Planch
Michelenolide	*Eupatorium formosanum*
Miltirone	*Salvia miltiorhiza*
Mineral elements	*Oxyria digyna*
Minerals	*Artocarpus heterophyllus*

Component	Source
Miniatine	*Clivia miniata*
Minosine	*Mimosa invisa, M. pudica*
Mitraphylline	*Acacia catechu*
Molephantin	*Elephantopus molis*
Monocrotalines	*Crotalaria sessiliflora*
Monomeric tertiary indol alkaloids	*Strychnos nux-vomica*
Monoterpene	*Achillea alpina, A. millefolium, Anethum graveolens L. Anthriscus aemula, A. sylvestris, Heracleum dissectum, H. lanatum, Polygonum bistorta, Sanguisorba officinalis, S. grandiflora, S. parviflora, S. x tenuifolia, Taraxacum officinale*
Monotropin	*Pyrola decorata, P. japonica, P. incarnata, P. renifolia, P. rotundifolia*
Morin	*Morus alba, M. constantinopolitana, M. indica*
Morindadiol	*Morinda citrifolia, M. officinalis*
Morindaparvin-A	*Morinda parvifolia*
Morolic acid	*Adina rubella, A. ratemosa*
Morphine	*Papaver somniferum*
Morroniside	*Cornus officinalis, Patrinia villosa* Thunb.
Motephantinin	*Elephantopus molis*
Mucic acid	*Phyllanthus emblica*

Mucilage	*Ailanthus altissima, Ajuga bracteosa, Cymbidium hyacinthinum, C. striatum, Draceana graminifolia, Hyoscyamus bohemicus, Liriope graminifolia, L. platyphylla, L. spicata, Pericamylus formosanus, Polygonatum chinense, P. cirrhifolium, P. macropodium, P. officinale, P. sibiricum, P. stenophyllum, P. odoratum, P. vulgare, Plumbago zeylanica, Tamarindus indicus*
Mucronatine	*Crotalaria mucronata*
Mucronatinine	*Crotalaria mucronata*
Mucus	*Dioscorea cirrhosa, D. hispida, D. japonica*
Mukorosside	*Sapindus mukorossi*
Mulberrin	*Morus alba, M. constantinopolitana, M. indica*
Mulberrochromene	*Morus alba, M. constantinopolitana, M. indica*
Multiflorin	*Rosa multiflora*
Munjistin	*Rubia chinensis, R. cordifolia, R. mungista, R. sylvatica*
Muramine	*Papaver amurense, P. nudicaule, P. radicatum*
Muricatin A	*Ipomoea cairica*
Musaenoide	*Melasma arvense*
Muscarine	*Cannabis chinensis, C. sativa*
Muslinic acid	*Elaeagnus oldhumii*
Mustard oil	*Brassica alba, B. juncea*
Mutaxanthin	*Physalis alkekengi*

Component	Source
Myrcene	*Artemisia lactiflora, Commiphora myrrha, Daucus carota, Elettaria cardamomum, Juniperus rigida, Ledum palustre, Medicago falcata, M. lupulina, M. polymorpha, M. ruthenica, M. sativa, Oenothera javanica, Oinus madshurica* Rupr. *Poncirus trifoliata, Thymus vulgaris, Valeriana alternifolia, V. amurensis, V. fauriei, V. subbipinnatifolia, Zanthoxylum bungeanum*
Myricetin	*Cotinus coggygria, Myrica rubra, Rhododendron dauricum, Syzygium cuminii, Thuja koraiensis, T. orientalis, T. chinensis*
Myricetol	*Cotinus coggygria* Scop.
Myricitrin	*Cotinus coggygria*
Myricyl	*Spilanthes acmella*
Myricyl alcohol	*Cassia angustifolia*
Myriogynine	*Centipeda minima*
Myristic acid	*Blumea balsamifera, Citrullus anguria, C. edulis, C. lanatus, C. vulgaris, Coix agrestis, C. chinensis, C. lachryma, Jatropha gospiifolia, J. curcas, Lindera obtusiloba, Myristica fragrans, Sesamum indicum, Taraxacum mongolicum, T. sinicum, Viscum album, V. coloratum*
Myristicin	*Myristica fragrans*
Myrcene	*Oenanthe javanica* (Bl) DC.
Myrocin	*Brassica alba, B. juncea, Thlaspi arvense*
Myrosinase	*Cardamine leucantha, C. lyrata, Sinapis alba, Thlaspi arvense*
Myrtenol	*Valeriana alternifolia, V. amurensis, V. fauriei, V. subbipinnatifolia*
N-desmethylchelerythrine	*Zanthoxylum nitidum*
N,N-dimethyltryptamine	*Phyllodium pulchellum*

N,N-dimethyltryptamine oxide	*Phyllodium pulchellum*
n-butyl allophanate	*Codonopsis pilosula, C. tangshen, C. ussuriensis*
n-butyl-2-ethyl butylphthalata	*Oenanthe javanica* (Blume) DC
n-butyl-z-ethyl butyl phthalate	*Oenothera javanica* (Blume) DC
n-caprylaldehyde	*Oplopanax elatus*
n-formyl-N-deacetylcolchine	*Iphigenia indica*
n-hentriacontane	*Aleurites fordii*
n-hexacosane	*Jatropha podagrica*
n-methyl anthranilic acid	*Evodia rutaecarpa*
n-methyl-2-(β-OH-propyl) piperidine	*Sedum sarmentosum*
n-methyl-isopelletierine	*Sedum sarmentosum*
n-methylanthranflamide	*Evodia rutaecarpa*
n-methylcoclaurine	*Nelumbium nuciferum, N. speclosum*
n-methylisococlaurine	*Nelumbium nuciferum, N. speclosum*
n-methylmorpholine	*Cassia occidentalis, C. torosa*
n-methyltyramine	*Citrus aurantium*
n-n-dimethyl-5-methoxytryptamine	*Evodia rutaecarpa*
n-nonyl aldehyde	*Tagetes erecta*
n-norarmepavine	*Machilus thunbergii*
n-phenylethyl alcohol	*Rosa rugosa*

Component	Source
n-triacontane	*Buglossoides arvense* (L.) Johnston
Naginataketone	Elshoitzia ciliatai Thunb.
Nandazurine	*Nandina domestica*
Nandinine	*Nandina domestica*
Naphthaquinone	*Plumbago zeylanica*
Naphthopyrones	*Cassia nomame, C. obtusifolia, C. tora*
Narcitine	*Narcissus tazetta*
Narcotic alkaloid	*Pericamylus formosanus*
Narcotine	*Papaver somniferum*
Naringenin-4′-O-pyranogluoside	*Cynomorium coccineum, C. songarium*
Naringin	*Citrus reticulata*
Nasunin	*Solanum lyratum, S. melongena*
Naucleoside	*Adina rubella, A. ratemosa*
Neferine	*Nelumbium nelumbo*
Negundoside	*Vitex nequndo*
Neo-allicin	*Allium chinense, A. odorum, A. sativum, A. tuberosum, A. uliginosum*
Neo-lignans	*Magnolia hypoleuca, M. japonica, M. officinalis*
Neo-nepetalactone	*Actinidia arguta, A. chinensis, A. japonica, A. kolomikta, A. polygama*
Neoandrographolide	*Andrographis paniculata*
Neoanisatin	*Illicium lanacedatum*
Neoboschnialactone	*Cistanche deserticola*

Neocarthamin	*Carthamus tinctorius*
Neochlorogenic acid	*Elaeagnus pungens, E. umbellata*
Neocnidilide	*Ligusticum chuanziang*
Neocryuptomerin	*Podocarpus macrophyllus*
Neogitogenin	*Anemarrhena asphodeloides*
Neoglucobrassicin	*Clerodendrum cyrtophyllum*
Neohesperidin	*Poncirus trifoliata*
Neohespiridin	*Citrus reticulata*
Neolinarin	*Lindera akoensis*
Neolinderalactone	*Lindera strychnifolia*
Neomatabiol	*Actinidia arguta, A. chinensis, A. japonica, A. kolomikta, A. polygama*
Neotigogenin	*Smilax china, S. nipponica, S. sieboldii, S. riparia*
Neoxanthin	*Potamogeton perfoliatus L.Taraxacum mongolicum, T. sinicum*
Nepodin	*Rumex acetosa, R. acetosella, R. amurensis, R. aquaticus, R. crispus, R. gmelini, R. japonicus, R. longifolius, R. maritimus, R. marschallianus, R. stenophyllus, R. thyrsiflorus*
Neriantin	*Nerium indicum* Mill.
Nerinine	*Zephyranthes candida*
Nerioderin	*Nerium indicum Mill*
Neriodin	*Nerium indicum Mill*
Neriodorin	*Nerium indicum Mill*
Neriororin	*Nerium indicum Mill*
Nerol	*Cymbopogon citratus, Osmanthus fragrans, Rosa rugosa*

Component	Source
Nerolidiol	*Melaleuca leucadendra*
Nerolidol	*Amomum cardamomum, A. globosum, A. tsao-ko, A. villosum, A. xanthloides, Hedyotis corymbosa*
Nervisterol	*Nervilia purpurea*
Nevadersin	*Lysionotus pauciflorus*
Niacin	*Achyranthes asperia, Amaranthus tricolor, Arachis hypogaea, Benincase cerifera, B. hispida, Boehmeria densiflora, Canarium album, C. sinense, Castanea crenuta, C. mollissima, Corylus heterophylla, C. mandshurica, Eleusine indica* (L.) Gaertner *Glycine max, G. soja, Helianthus annuus* L. *Hibiscus rosa-sinensis, H. rhombifolius, Marsilea quadrifolia* L. *Petasites japonicus, Sonchus oleraceus* L. *Syzygium aromaticum, Zea mays*
Nicotelline	*Nicotiana tabacum*
Nicotimine	*Nicotiana tabacum*
Nicotine	*Eclipta alba, E. marginata, E. prostrata, E. thermalis, E. erecta, Equisetum palustre* L. *Lycopodium clavatum, L. obscurum, L. selago, L. serratum, Nicotiana tabacum*
Nicotinic acid	*Angelica polymorpha, A. sinensis, Celosia argentea, C. cristata, Lycium barbarum, L. megistocarpum, L. ovatum, L. trewianum, L. turbinatum, Lycoperiscon esculentum, Solanum nigrum*
Nigakihemiacetal A	*Picrasma quassioides*
Nigakilactone A	*Picrasma quassioides*
Nigakinone	*Picrasma quassioides*
Nilgirine	*Crotalaria mucronata, Desmodium pulehellum*
Nimbin	*Melia japonica, M. toosendan, M. azedarach*

Nimbolins	*Melia japonica, M. toosendan, M. azedarach*
Niranthin	*Phyllanthus urinaria, P. niruri, P. reticulatus*
Nirtetralin	*Phyllanthus urinaria, P. niruri, P. reticulatus*
Nishindaside	*Vitex negundo*
Nishindine	*Vitex negundo*
Nitidine	*Zanthoxylum nitidum*
Nitroacronycine	*Acronychia pedunculata, A. laurifolia*
Nitryl-glycoside	*Aquilegia vulgaris*
Nobiletin	*Citrus reticulata*
Nocoteine	*Nicotiana tabacum*
Nodakenetin	*Angelica decursiva, A. gigas Maxim. Peucedanum japonicum, P. praeruptorum, P. rubricaule, Peucedanum terebinthaceum* Fisch
Nodakenin	*Angelica decursiva, A. gigas Maxim. Peucedanum japonicum, P. praeruptorum, P. rubricaule*
Nonacosan-10-ol	*Dicranopteris linearis*
Nonacosan-10-one	*Dicranopteris linearis*
Nonacosane	*Chenopodium album* L. *Dicranopteris linearis, Prunus padus, Rosa rugosa*
Nonalactone	*Prunus persica*
Nonanal	*Coriandrum sativum*
Nonyl aldehyde	*Rosa rugosa, Zingiber officinale*
Nonylic aldehyde	*Cinnamomum zeglanicum*
nor-rubrofusarin	*Cassia nomame, C. obtusifolia, C. tora*
Nordamnacanthal	*Morinda citrifolia, M. officinalis*

Component	Source
Nordracorubin	*Draceana graminifolia*
Norepinephrine	*Musa paradisiaca, Portulaca oleracea*
Noreugenin	*Adina rubella, A. ratemosa*
Noricariin	*Epimedium brevicorum, E. koreanum, E. macranthum*
Norkurarinone	*Sophora flavescens, S. alopecurosides*
Normenisarine	*Cocculus laurifolius, C. sarmentosus, C. trilobus*
Nornuciferine	*Nelumbo nucifera* Gaetn.
Norpseudoephedrine	*Ephedra distachya, E. equisetina, E. intermedia, E. monosperma, E. sinica*
Norsecurinine	*Securinega viro*
Nortracheloside	*Trachelospermum jasminoides*
Nothosmyrnol	*Ligusticum jeholense, L. pyrenacum, L. sinense, L. tenuissimum, Nothosmyrnium japonicum*
Notoptero	*Notopterygium incisium*
Novacine	*Strychnos pierriana*
Nuciferine	*Nelumbium nuciferum, N. speclosum, Papaver amurense, P. nudicaule, P. radicatum*
Nudicaulin	*Papaver amurense, P. nudicaule, P. radicatum*
Nupharamine	*Nuphar japonicum, N. pumilum*
Nuzhenide	*Ligustrum lucidum, L. japonicum*
o-acetylcolumbianetin	*Cnidium monnieri*
o-cumaric acid	*Eupatorium chinense, E. lindleyanum, E. japonicum*
o-isovaleryl columbianetin	*Cnidium monnieri*
o-nornuciferine	*Nelumbium nuciferum, N. speclosum*

Obacunone	*Phellodendron amurense, P. chinensis*
Obakinone	*Dictamnus albus, D. dasycarpus*
Obtusifolin	*Cassia nomame, C. obtusifolia, C. tora*
Obtusin	*Cassia nomame, C. obtusifolia, C. tora*
Ocimene	*Tagetes patula*
Octacosane	*Ficus carica*
Octacosanol	*Firmiana simplex*
Octadecatetraenoic acid	*Stellaria media*
Octalactone	*Prunus persica*
Octanol	*Agastache rugosa, A. rugosa* f. *hypoleuca*
Octopinic acid	*Parthenocissus tricuspidata*
Odine	*Sargassum pallidum*
Odoilin	*Cerbera manghas* L.
Odoratin	*Eupatorium odoratum*
Okinalein	*Pulsatilla ambigua, P. cernua, P. chinensis*
Okinalin	*Pulsatilla ambigua, P. cernua, P. chinensis*
Olcanolic acid	*Chenopodium album* L.
Oldenlandoside	*Oldenlandia diffusa*
Oleanolic acid	*Cyperus rotundus*
Oleandrin	*Nerium indicum*
Oleandrose	*Nerium indicum*
Oleanene derivatives	*Asparagus cochinensis*

Component	Source
Oleanic acid	*Chenopodium album* L.
Oleanolic acid	*Achyranthes japonica, Anemone raddeana, A. rivularis, A. vitifolia, Aralia chinensis, A. cordata, A. elata, A. mandschurica (Rupr & Maxim) Seem. Aristolochia contorta, A. kaempferi, A. longa, A. recurvilabra, Calendula officinalis, Clematis intricata, C. mandshurica, Codonopsis lanceolata, Eugenia aromatica, E. caryophyllata, E. ulmoides, Ligustrum lucidum, L. japonicum, Melaleuca leucadendra, Oldenlandia chrysotricha,*
Oleanolic acid	*O. corymbosa, Panax ginseng, Prunella vulgaris, Swertia diluta, S. mileensis, Viscum album, V. coloratum*
Oleic acid	*Aleurites fordii, Angelica grosserrata, Aquilegia vulgaris, Ajuga bracteosa, Brucea javanica, B. sumatrana, Cardamine leucantha, C. lyrata, Citrullus anguria, C. edulis, C. lanatus, C. vulgaris, Coix agrestis, C. chinensis, C. lachryma, Corchorus capsularis, C. olitorius, Cucumis sativus, Descurainia Sophia (L.) Schur. Elettaria cardamomum, Hedera rhombea, H. helix, Jatropha gospiifolia, J. curcas, Lindera obtusiloba, Myristica fragrans, Nuphar japonicum, N. pumilum, Tamarindus indicus*
Olein	*Cedrela sinensis* A. Juss. *Ricinus communis*
Olein acid	*Sesamum indicum*
Oleoresin	*Dryopteris laeta, D. filix-mas*
Oleyl alcohol	*Chenopodium album* L.
Oligosaccharides	*Aesculus chinensis, A. hippocastanum, Typha angustata, T. angustifolia, T. davidiana, T. latifolia, T. minima, T. orientalis, T. przeqalskii*
Olitoriside	*Corchorus olitorius*
Onjisaponin A	*Polygala tenuifolia*

Onjisaponin B	*Polygala tenuifolia*
Ononitol	*Medicago falcata, M. lupulina, M. polymorpha, M. ruthenica, M. sativa*
Ophelic acid	*Centaurium meyeri*
Ophiopogenins	*Ophiopogon japonicus*
Oplopanaxosides	*Oplopanax elatus*
Organic acids	*Ganoderma lucidum, Ledebouriella divaricata, Lysionotus pauciflorus, Polygonum perfoliatum, P. tinctorium, Ribes mandshurica, Sansevieria trifosciate, Trichosanthes kirilowii, T. uniflora*
Oridonin	*Rabdosia lasiocarpus, R. rubescens*
Orientin	*Fagopyrum esculentum, F. sagittatum, Lathyrus pratensis L. Linum stelleroides, L. usitatissimum, Lythrum salicaria, Persicaria orientalis, Polygonum orientale, Vitex negundo, V. rotundifolia, V. trifolia*
Orientin-7-0-glucoside	*Uraria crinita, U. lagopodiodes*
Orientoside	*Persicaria orientalis*
Orobanchin	*Orobanche caerulescens* Stephan.
Orthomethylcoumaric aldehyde	*Cinnamomum aromaticum, C. cassia*
Osalic acid	*Plumbago zeylanica*
Osmane	*Osmanthus fragrans*
Osthenol-7-o-β-gentiobioside	*Glehnia littoralis*
Osthol	*Angelica pubescens, A. gigas* Maxim *A. sinensis* (Oliv) Diels *Murraya paniculata*
Oxalate	*Curculigo orchiodes* Gaertn. *Oxalis corriculaza, O. corymbosa*

Component	Source
Oxalic acids	*Achillea alpina, A. millefolium, Anthriscus aemula, A. sylvestris, Chenopodium album* L. *Coriandrum sativum, Heracleum dissectum, H. lanatum, Juncus effusus, Lactuca raddeana, L. indica, L. sativa, Polygonum bistorta, Sanguisorba officinalis, S. grandiflora, S. parviflora, S. x tenuifolia, Taraxacum officinale, Vitis amurensis, V. vinifera*
Oxoushinsunine	*Michelia alba, M. figo*
Oxycanthine	*Berberis amurensis, B. poiretii, B. sibirica, B. soulieana, Thalictrum aquilegifolium, T. baicalense, T. fauriel, T. petaloideum, T. simplex, T. squarrosum, T. thunbergii*
Oxychelerythrine	*Zanthoxylum nitidum*
Oxylysin	*Avena fatua*
Oxymatrine	*Sophora subprostrata*
Oxymethyl anthraquinone	*Cassia alata, C. siamea*
Oxynitidine	*Zanthoxylum niti*
Oxypaeoniflorin	*Paeonia albiflora, P. edulis, P. japonica, P. lactiflora, P. moutan, P. officinalis*
Oxypeucedanine	*Angelica amurensis, A. anomala, A. dahurica*
Oxypurpureine	*Thalictrum aquilegifolium, T. baicalense, T. fauriel, T. petaloideum, T. simplex, T. squarrosum, T. thunbergii*
Oxyristic acid	*Phytolacca acinosa, P. americana, P. japonica, P. kaempferi, P. octandra, P. pekinensis*
Oxysanguinarine	*Macleaya cordata*
p-coumaric acid	*Catalpa ovata* G. Don *Impatiens balsamina, I. noli-tangere, I. textori, Matteuccia struthiopteris, Plumbago zeylanica*

p- cymene	*Chenopotium ambrosiodes, Cinnamomum zeglanicum, Daucus carota, Elettaria cardamomum, Glechoma hederacea, G. longituba, Juniperus rigida, Pinus madshurica* Rupr. *Thymus amurensis, T. disjunctus, T. kitagawianus, T. komarovii, T. przewalskii, T. quinquecostatus, T. vulgaris*
p-hydroxyacetophenone	*Senecio cannabifolius*
p-hydroxybenzoic acid	*Matteuccia struthiopteris, Rhododendron mucronatum*
p-hydroxycinnamic acid	*Catalpa ovata* G. Don
p-hydroxylbenzoyl	*Catalpa ovata* G. Don
p-lumicolchicine	*Iphigenia indica*
P-methoxybenzylacetone	*Aquilegia buergeriana, A. parviflora*
p-methoxycinnamaldehyde	*Agastache rugosa, A. rugosa* f. hypoleuca
p-methoxylcinnamic acid	*Scrophularia buergeriana, S. kakudensis, S. ningpoensis, S. oldhami, S. puergeriana*
p-terpinene	*Thymus amurensis, T. disjunctus, T. kitagawianus, T. komarovii, T. przewalskii, T. quinquecostatus*
p-tyrosol	*Rhodiola elongata*
p-vinylguaiacol	*Hedyotis diffusa*
p-vinylphenol	*Hedyotis diffusa*
Pachymarose	*Poria cocos*
Pachymic acid	*Poria cocos*
Paeonidin	*Rosa rugosa*
Paeoniflorin	*Paeonia albiflora, P. edulis, P. japonica, P. lactiflora, P. moutan, P. officinalis*
Paeonin	*Cynanchum paniculatum, Paeonia obovata, P. suffruticosa, P. veitchii, Viburnum sargenti*
Paeonol	*Cynanchum paniculatum, Paeonia obovata, P. suffruticosa, P. veitchii*

Component	Source
Paeonoside	*Paeonia obovata, P. suffruticosa, P. veitchii*
Palamatine	*Berberis amurensis, B. poiretii, B. sibirica, B. soulieana*
Palderoside	*Oldenlandia diffusa*
Pallidine	*Corydalis incisa, C. bungeana*
Palmaline	*Coptis chinensis, C. japonica, C. teeta*
Palmatine	*Calystegia hederacea, C. japonica, Fibraurea recisa, Phellodendron amurense, P. chinensis, Thalictrum aquilegifolium, T. baicalense, T. fauriel, T. glandulissimum, T. ichangense, T. petaloideum, T. simplex, T. squarrosum, T. thunbergii*
Palmitic acid	*Acanthopanax gracilistylus, A. spinosum, Aquilegia vulgaris, Ajuga bracteosa, Aleurites fordii, Angelica grosserrata, Benincase cerifera, B. hispida, Blumea balsumifera, Cibotium barometz, Citrullus anguria, C. edulis, C. lanatus, C. vulgaris, Coix agrestis, C. chinensis, C. lachryma, Corchorus capsularis, C. olitorius, Cucumis sativus, Cynomorium coccineum, C. songarium, Descurainia Sophia (L.) Schur. Elettaria cardamomum, Jatropha gospiifolia, J. curcas, Matteuccia struthiopteris, Melia japonica, M. toosendan, M. azedarach, Nuphar japonicum, N. pumilum, Sepium sebiferum, S. discolor, Sesamum indicum, Sonchus arvensis L. S. oleraceus, Tamarindus indicus, Taraxacum mongolicum, T. sinicum*
Palmitine	*Cedrela sinensis A. Juss. Thalictrum foetidum*
Palustrine	*Equisetum arvense, E. hyemale, E. ramosissimum*
Palustridine	*Equisetum palustre L.*
Panaxadiol	*Gynostemma pentaphyllum, Panax notoginseng, P. zingiberensis*
Panaxynol	*Panax ginseng*
Paniculatincoumurrayin	*Murraya paniculata*

Pantothenic acid — *Glycine max, G. soja*

Papain — *Ficus carica*

Papaverine — *Papaver somniferum*

Paraaspidin — *Dryopteris laeta, D. filix-mas*

Parasorbic acid — *Sorbus alnifolia, S. amurensis, S. pohuashanensis*

Parietin — *Polygonum multifolrum, P. chinensis*

Parthenolide — *Eupatorium formosanum*

Patchoulenone — *Cyperus rotundus*

Patrinene — *Patrinia villosa* (Thunb.)

Patrinoside — *Patrina scabiosaefolia*

Patuletin — *Spinacia oleracea* L. *Tagetes patula*

Patulitrin — *Tagetes patula*

Paulownin — *Paulownia tometosa* Thunb.

Paulownioside — *Paulownia tometosa* Thunb.

Pectic acid — *Centella ascatica, Kalopanax septemlobus* (Thunb)

Pectic compound — *Lactuca raddeana, L. indica, L. sativa*

Pectins — *Ajuga bracteosa, Myristica fragrans, Plumeria rubra, Taraxacum formosanum, T. indicus*

Pectolinarigenin — *Lindera akoensis*

Pectolinarin — *Cirsium chinense, C. japonicum, Lindera akoensis*

Peganine — *Lindera akoensis*

Peimidine — *Fritillaria anheunensis, F. collicola, F. maximowiczii, F. roylei, F. thunbergii, F. ussuriensis, F. verticillata*

Component	Source
Peimilidine	*Fritillaria anheunensis, F. collicola, F. maximowiczii, F. roylei, F. thunbergii, F. ussuriensis, F. verticillata*
Peimine	*Fritillaria anheunensis, F. collicola, F. maximowiczii, F. roylei, F. thunbergii, F. ussuriensis, F. verticillata*
Peiminine	*Fritillaria anheunensis, F. collicola, F. maximowiczii, F. roylei, F. thunbergii, F. ussuriensis, F. verticillata*
Peimisine	*Fritillaria anheunensis, F. collicola, F. maximowiczii, F. roylei, F. thunbergii, F. ussuriensis, F. verticillata*
Peiniphine	*Fritillaria anheunensis, F. collicola, F. maximowiczii, F. roylei, F. thunbergii, F. ussuriensis, F. verticillata*
Pelargonidin-3-rhamnosylglucoside	*Chloranthus glubra, C. oldhnami*
Pelargonin	*Paeonia obovata, P. suffruticosa, P. veitchii*
Pelletierine	*Punica granatum*
Peltatin	*Dysosma pleiantha*
Pencordin	*Peucedanum japonicum, P. praeruptorum, P. rubricaule*
Penicillin	*Senecio scandens* Buch-Ham
Penta-m-digalloyl-β-glucose	*Rhus chinensis* Mill
Penta-o-galloyl-β-d-glucose	*Chamaenerion angustifolium*
Pentanoic acid	*Prunus persica*
Pentosan	*Quercus acutissima, Q. aliena, Q. dentata, Q. liaotungensis, Q. mongolica, Q. variabilis, Sesamum indicum, Sesbinia javanica*

Pentosane	*Abutilon theophrasti, A. avicennae, Fortunella crassifolia, F. japonica, F. margarita*
Pentose	*Abutilon theophrasti, A. avicennae*
Pepsin	*Ficus carica*
Peptides	*Lycium chinense*
Peraksine	*Rauvolfia verticilata*
Pergularin	*Adonis chrysocyathus, A. brevistyla, A. vernalis, Metaplexis japonica* Thunb.
Pericalline	*Catharanthus roseus*
Perilladehyde	*Perilla frutescens, P. ocymoides, P. polystachya, P. arguta*
Perillaketon	*Perilla frutescens* L.
Perillanin	*Perilla frutescens* L.
Peripalloside	*Antiaris toxicaris*
Periplocin	*Periploca sepium* Bunge
Periplocymarin	*Periploca sepium* Bunge
Perividine	*Catharanthus roseus*
Perivine	*Catharanthus roseus*
Perlolyrine	*Ligusticum chuanziang*
Persicarin	*Oenothera javanica, Persicaria hydropiper, Polygonum hydropiper*
Peruvosides	*Thevetia peruviana*
Pesticidal for maggots	*Rhododendron molle* (Bl) G. Don
Petroselenic acid	*Glehnia littoralis*
Petroselic acid	*Cryptotaenia japonica, C. canadensis*
Petroselidinic acid	*Glehnia littoralis*

Component	Source
Petroselinic acid	*Daucus carota, Oenothera javanica*
Petunidin	*Medicago falcata, M. lupulina, M. polymorpha, M. ruthenica, M. sativa*
Petunidin glucoside	*Syzygium cuminii*
Peuformosin	*Peucedanum formosanum*
Phantomolin	*Elephantopus molis*
Pharbilic acid	*Pharbitis diversifolia, P. hederacea, P. nil, P. triloba*
Phasin	*Euphorbia esula, E. helioscopia*
Phasine	*Euphorbia esula, E. helioscopia*
Phellandrene	*Anethum graveolens* L. *Aucklandia costus, A. lappa, Cinnamomum aromaticum, C. cassia, Coriandrum sativum, Saussurea japonica, S. lappa, Valeriana alternifolia, V. amurensis, V. fauriei, V. subbipinnatifolia, Xanthoxylum piperitum, Zanthoxylum schinifolium, Zingiber officinale*
Phellandrine	*Cunninghamia lanceolata*
Phellodendrine	*Phellodendron amurense, P. chinensis*
Phellopterin	*Angelica amurensis, A. anomala, A. dahurica*
Phenanthrene-1,4-quinone	*Sphenomeris chusana*
Phene	*Saussurea japonica, S. lappa*
Phenethylamine	*Cornus alba, C. kousa, C. macrophylla*
Phenolic acid	*Ranunculus ternatus, Rhododendron mucronatum*
Phenolic compounds	*Leucaena leucocephala*
Phenolic derivatives	*Vitex negundo*

Phenols — *Cypripedium guttatum, C. macranthum, C. pubescens, Hedychium coronarium, Laggera alata, Nepenthes rafsiana, Urena procumbens*

Phenyl ethyl alcohol — *Dianthus chinensis* L. *Eriobotrya japonica*

Phenylacetic acid — *Rosa rugosa*

Phenylalanine — *Cajanus cajan* L. *Oryza sativa*

Phenylethyl alcohol — *Plumeria rubra*

Phenylpropane derivatives — *Acorus calamus* L.

Phenylpropyl alcohol — *Cinnamomum aromaticum, C. cassia*

Phenytheptatriyne — *Bidens pilosa*

Phetidine — *Thalictrum aquilegifolium, T. baicalense, T. fauriel, T. petaloideum, T. simplex, T. squarrosum, T. thunbergii*

Phillyrin — *Forsythia suspensa, Syringa dilatata, S. oblata, S. reticulata, S. suspensa, S. vulgaris*

Phlobaphene — *Ulmus campestris, U. macrocarpa, U. pumila*

Phorbol — *Croton cascarilloides, C. tiglium*

Phosphatase — *Sinapis alba*

Phosphates — *Portulaca pilosa*

Phosphatides — *Fagopyrum esculentum, F. sagittatum*

Phosphatidyl-ethanolamine — *Tetragonia tetragonoides*

Phosphatidyl-inositol — *Tetragonia tetragonoides*

Phosphatidyl-serine — *Tetragonia tetragonoides*

Phosphatidylcholine — *Tetragonia tetragonoides*

Phosphorus — *Actinidia chinensis* Planch

Component	Source
Phototoxic	*Matricarin chamonilla* L.
Phrymarolin-I, II	*Phryma leptotachya* L.
Phthalide	*Ligusticum chuangxiong*
Phthatate	*Oenanthe javanica* (Bl) DC.
Phyllanthin	*Phyllanthus urinaria, P. niruri, P. reticulatus*
Phyllanthine	*Phyllanthus urinaria, P. niruri, P. reticulatus*
Phyllantidine	*Phyllanthus urinaria, P. niruri, P. reticulatus, Securinega suffruticosa*
Phylteralin	*Phyllanthus urinaria, P. niruri, P. reticulatus*
Physalein	*Lycium barbarum, L. megistocarpum, L. ovatum, L. trewianum, L. turbinatum, Physalis alkekengi*
Physalin A	*Physalis alkekengi*
Physalin B	*Physalis alkekengi*
Physalin C	*Physalis alkekengi*
Physanols	*Physalis alkekengi*
Physcim-1-gluco-rhamnoside	*Phyllodium pulchellum*
Physcion	*Cassia nomame, C. obtusifolia, C. tora, Rheum officinale, R. palmatum, R. tanguticum, R. undulatum, R. koreanum, Rumex acetosa, R. acetosella, R. amurensis, R. aquaticus, R. gmelini, R. longifolius, R. maritimus, R. marschallianus, R. patientia, R. stenophyllus, R. thyrsiflorus*
Physoxanthin	*Physalis alkekengi*
Phytic acid	*Dioscorea batatus*

Phytin   *Sesamum indicum*

Phytoestrogens   *Trifolium pratense, T. repens*

Phytofluere   *Daucus carota*

Phytolaccatoxin   *Phytolacca acinosa, P. americana, P. japonica, P. kaempferi, P. octandra, P. pekinensis*

Phytolacine   *Phytolacca acinosa, P. americana, P. japonica, P. kaempferi, P. octandra, P. pekinensis*

Phytosterindigitonid   *Hoya carnosa*

Phytosterines   *Achillea alpina, A. millefolium, Anthriscus aemula, A. sylvestris, Heracleum dissectum, H. lanatum, Lindera akoensis, Polygonum bistorta, Sanguisorba grandiflora, S. officinalis, S. parviflora, S. x tenuifolia, Taraxacum officinale*

Phytosterol   *Aleurites fordii, Carum carvi, Duchesnea indica, Elettaria cardamomum, Gnaphalium affine, G. arenarium, G. confusum, G. javanum, G. luteo-album, G. multiceps, G. ramigerum, G. tranzschelii, G. uliginosum, Panax quinquefolium, Syzygium aromaticum, Ulmus campestris, U. macrocarpa, U. pumila*

Phytotoxin   *Jatropha gospiifolia, J. curcas*

Picein   *Clerodendrum trichotomum* Thunb.

Picralinal   *Alstonia scholaris*

Picrasmin   *Picrasma quassioides*

Picrinine   *Alstonia scholaris*

Picrorhizin   *Picrorhiza kurroa*

Pienen acid   *Vitex negundo*

Pimaradene   *Aralia chinensis, A. cordata A. elata*

Pinacene   *Pinus madshurica* Rupr.

Component	Source
Pinene	*Alpinia officinarum, Anethum graveolens* L. *Artemisia brachyloba, Cinnamomum camphora, C. zeglanicum, Cunninghamia lanceolata, Cyperus rotundus, Elettaria cardamomum, Elsholtzia argyi, E. cristata, E. splendens, E. feddei, E. souliei, Luffa aegyptiaca, L. cylindrica, L. faetida, L. petola, Melaleuca leucadendra, Myristica fragrans, Piper cubeba, Podocarpus macrophyllus, Thuja koraiensis, T. orientalis, T. chinensis, T. vulgaris, Valeriana alternifolia, V. amurensis, V. fauriei, V. subbipinnatifolia, Vitex trifolia, V. rotundifolia*
Pinicolic acid	*Poria cocos*
Pinipicrin	*Biota chinensis, B. orientalis*
Pinitol	*Lespedeza cuneata, Pinus bungeana, P. densiflora, P. koraiensis, P. sylvestris, P. tabulaeformis*
Pinnatin	*Pongamia pinnata*
Pinocarveol	*Cinnamomum camphora*
Pinoresinol-di-β-D-glycoside	*Eucommia ulmoides*
Piperamine	*Piper nigrum*
Piperine	*Piper longum, P. nigrum*
Piperitone	*Cymbopogon distans, C. goeringii, C. nardus, Gymnadenia conopsea* (L.) R. Brown *Zanthoxylum bungeanum*
Piperonal	*Piper nigrum*
Plantagin	*Plantago asiatica, P. depressa, P. exaltata, P. loureiri, P. major*
Plantasan	*Plantago asiatica, P. depressa, P. exaltata, P. loureiri, P. major*
Plantenolic acid	*Plantago asiatica, P. depressa, P. exaltata, P. loureiri, P. major*
Plastoquinone	*Persicaria orientalis*
Plastoquinone-9	*Polygonum orientale*

Platycodigenic acids	*Platycodon autumnalis, P. grandiflorum, P. sinensis*
Platycodigenin	*Campanula gentianoides, C. grandiflora, Platycodon autumnalis, P. grandiflorum, P. sinensis*
Platycodonin	*Platycodon autumnalis, P. grandiflorum, P. sinensis*
Platycodosides	*Platycodon autumnalis, P. grandiflorum, P. sinensis*
Platyconin	*Platycodon autumnalis, P. grandiflorum, P. sinensis*
Pleridine	*Pteridium aquilinum* (L.) Kuhn.
Plumbagin	*Plumbago zeylanica*
Plumieric acid	*Plumeria rubra*
Plumieride	*Plumeria rubra*
Podocarpene	*Podocarpus macrophyllus*
Podocarpusflavones	*Podocarpus macrophyllus*
Podophyllotoxin	*Dysosma pleiantha, Podophyllium peltatum, P. pleianthum*
Podototarin	*Podocarpus macrophyllus*
Polyacetylene	*Bidens pilosa, Carum carvi, Glehnia hittoralis, Ligusticum jeholense, L. pyrenacum, L. sinense, L. tenuissimum*
Polydatin	*Polygonum cuspidatum*
Polygalacic acid	*Platycodon autumnalis, P. grandiflorum, P. sinensis, Solidago dahurica, S. pacifica, S. virgaurea*
Polygodiol	*Persicaria hydropiper*
Polygonin	*Polygonum cuspidatum*
Polygonone	*Persicaria hydropiper*
Polyine	*Glehnia littoralis*

Component	Source
Polyphenols	*Hippophae rhamnoides, Paulownia tometosa Thunb.Thea assamica*
Polysaccharides	*Acanthopanax giraldii, Achyranthes bidentata, Alisma orientalis, Allium chinense, A. odorum, A. sativum, A. tuberosum, A. uliginosum, Angelica polymorpha, A. sinensis, Astragalus complanatus, A. henyri, A. hoantchy, A. membranaceus, A. melilotoides, A. mongholicus, A. reflexistipulus, A. sinensis, Coriolus versicolor, Epimedium brevicorum, E. koreanum, E. macranthum, Ganoderma lucidum, Glehnia littoralis, Lycium chinense, Ophiopogon japonicus, Rehmannia chinensis, R. glutinosa, Rhus chinensis, R. cotinus, R. javanica, R. osbeckii, Trichosanthes kirilowii, T. uniflora*
Polysaccharuperptide	*Coriolus versicolor*
Polythienyls	*Tagetes patula*
Ponasterone	*Podocarpus macrophyllus, Pteridium aquilinum (L.) Kuhu*
Ponasterone A	*Matteuccia struthiopteris, Osmunda japonica*
Poncirin	*Citrus reticulata, Poncirus trifoliata*
Pongapin	*Pongamia pinnata*
Ponicidine	*Rabdosia rubescens*
Populin	*Populus alba, P. davidiana, P. tomentosa*
Populnin	*Equisetum arvense, E. hyemale, E. ramosissimum*
Portulal	*Portulaca grandiflora*
Potasium	*Actinidia chinensis Planch, Cacalia hastate L. Cajanus indicus L. Laminaria angusta, L. cichorioides, L. japonica, L. longipedalis, L. religiosa, Sargassum pallidum*
Potassium hydroxide	*Phyllostachys bambusoide, P. nigra*
Potassium malate	*Hovenia dulcis*

Potassium myronate	*Brassica alba, B. juncea*
Potassium nitrate	*Cynoglossum divaricatum, Hovenia dulcis*
Potassium oxide	*Desmodium triforum, D. triquetrum*
Potassium salts	*Portulaca oleracea*
Potassium sodium	*Portulaca pilosa*
Precatorine	*Abrus precatorius*
Pregnenes	*Periploca sepium*
Preskinnianine	*Dictamnus albus, D. dasycarpus*
Primulagenin A	*Primula sieboldii, P. asiatica, P. vulgaris*
Procurcumenol	*Curcuma zedoaria, C. aromatica, C. kwangsiensis, C. longa*
Procyanidin	*Davallia mariesii* Moore ex Baker
Proesapanin A	*Caesalpinia sappan*
Proline	*Litchi chinensis*
Prometaphanine	*Stephania japonica*
Pronuciferine	*Nelumbium nuciferum, N. speclosum*
Propeimin	*Fritillaria anheunensis, F. collicola, F. maximowiczii, F. roylei, F. thunbergii, F. ussuriensis, F. verticillata*
Propionic acid	*Ajuga bracteosa*
Prosapogenin	*Platycodon autumnalis, P. grandiflorum, P. sinensis*
Protease	*Fagopyrum esculentum, F. sagittatum, Plumbago zeylanica*

Component	Source
Proteins	Abutilon theophrasti, A. avicennae, Achyranthes asperia, Aleurites moluceanu, Amaranthus lividus, A, blitum, A, viridis, Arachis hypogaea, Arundo donax, A. phragmites, Artocarpus heterophyllus, Bauhinia championi, B. variegata, Boehmeria densiflora, Campsis adrepens, C. chinensis, C. grandiflora, Castanea crenuta, C. mollissima, Coix agrestis, C. chinensis, C. lachryma, Cordyceps sinensis, Dioscorea batatus, Eleusine indica (L.) Gaertne. rErigeron Canadensis L. Euryale ferox, Ficus awkeotsang, Glycine max, G. soja, Lemmaphyllum microphyllum, Hibiscus rosa-sinensis, H. rhombifolius, Lilium brownii, L. concolor; L. dauricum, L. distichum, L. japonicum, L. lancifolium, L. pumilum, Lycopersicon esculentum, Oxyria digyna, Phragmites communis, Polygonum perfoliatum, P. tinctorium, Polyporus umbellatus, Quercus acutissima, Q. aliena, Q. dentata, Q. liaotungensis, Q. mongolica, Q. variabilis, Triticum vulgare, Trapa bispinosa, Urtica angustifolia, U. cannabina, U. cannabina, U. lobata, U. tenacissima, U. urens, U. utillis
Protein (TAP29)	Trichosanthes kirilowii, T. uniflora
Proteinase	Ganoderma lucidum, Hordeum vulgare
Proto-isoerubosides	Allium chinense, A. odorum, A. sativum, A. tuberosum, A. uliginosum
Protoanemonin	Anemone cernua, A. pulsatilla, Caltha palustris, Pulsatilla ambigua, P. cernua, P. chinensis, Ranunculus chinensis, R. japonicus, R. sarmentosa, R. sceleratus
Protocatechinic acid	Pterocarya stenoptera, Rhododendron mucronatum Turcz
Protocatechuic acid	Cirsium albescens, C. brevicaule, C. littorale, C. maakii, C. segetum, C. setosum, C. vlassovianum, Ilex chinensis, Matteuccia struthiopteris, Parietaria micrantha, Polygonum bistorta, Rhododendron mucronatum
Protocatechuic aldehyde	Ilex chinensis
Protoescigenine	Aesculus chinensis, A. hippocastanum

Protohypericin	*Hypericum perforatum*
Protopine	*Chelidonium album, C. hybridum, C. majus, C. serotinum, Corydalis ambigua, C. bungeana, C. decumbens, C. incisa, C. repens, C. ternata, C. turtschaninovii, C. yanhusuo, Hypecoum erectum L. Macleaya cordata, Thalictrum aquilegifolium, T. baicalense, T. fauriel, T. petaloideum, T. simplex, T. squarrosum, T. thunbergii*
Protoprimulagenin A	*Primula sieboldii, P. asiatica, P. vulgaris*
Protostemonine	*Stemona japonica, S. tuberosa*
Protostephanine	*Stephania japonica*
Protoveratrine	*Hemerocallis flava, Veratrum formosanum*
Prudomenin	*Prunus mume*
Prunasin	*Prunus armeniaca*
Prunelin	*Prunella asiatica* NaKa
Pseudoaconitine	*Aconitum deinorrtuzum*
Pseudoanisatin	*Illicium lanacedatum*
Pseudobrucine	*Strychnos pierriana*
Pseudohypericin	*Hypericum triquetrifolium, H. chinensis*
Pseudojervine	*Hemerocallis flava, Veratrum dahuricum, V. maackii, V. nigrum*
Pseudolycorine	*Lycoris radiata, L. longituba, L. auera*
Pseudomorphine	*Papaver somniferum*
Pseudopelletierine	*Punica granatum*
Pseudopurpurin	*Rubia chinensis, R. cordifolia, R. mungista, R. sylvatica*
Pseudostrychnine	*Strychnos pierriana*

Component	Source
Psoralen	*Dictamnus albus, D. dasycarpus, Ficus carica, Glehnia littoralis, Psoralea corylifolia*
Psoralidin	*Psoralea corylifolia*
Psychotrine	*Alangium lamarckii*
Psyllostearyl alcohol	*Guelden staedtia* Maxim
pterosterone	*Matteuccia struthiopteris*
Puerarin	*Pueraria lobata, P. pseudo-hirsuta*
Puqiedinone	*Fritillaria anheunensis, F. collicola, F. maximowiczii, F. roylei, F. thunbergii, F. ussuriensis, F. verticillata*
Purpureal glycosides	*Digitalia purpurea, D. sanguinalis*
Purpurin	*Galium bungei, G. spurium, G. verum, Rubia akane, R. chinensis, R. cordifolia, R. mungista, R. sylvatica*
Purulent	*Senecio scandens* Bush-Ham,
Putrescine	*Citrus reticulata, Panax ginseng*
Pyrocaledol	*Salix babylonica, S. matsudana, S. microstachya*
Pyrocatechic tannin	*Blumea balsumifera*
Pyrocatechine acid	*Pterocarya stenoptera*
Pyrogallol tannin	*Ranunculus sceleratus*
Pyrrolidine	*Daucus carota*
Pyrryl--methyl ketone	*Valeriana alternifolia, V. amurensis, V. fauriei, V. subbipinnatifolia*
Pyromeconic acid	*Eregeron annuus* L.
Qianhucocumarin	*Peucedanum japonicum, P. praeruptorum, P. rubricaule*

Qluconic acid	*Nelumbo nucifera* Gaetn.
Quassin	*Cedrela sinensis.* A. Juss. *Picrasma quassioides*
Querbrachitol	*Viscum album, V. coloratum*
Quercetagetin	*Tagetes patula*
Quercetagetrin	*Tagetes patula*
Quercetin	*Apocynum venetum, Aster ageratoides, A. tataricus Astilbe longicarpa, A. chinensis, Campanula glomerata, C. punctata, Cassia alata, Castanea crenata, C. mollissima, Convolvulus arvensis, Coriandrum sativum, Crataegus cuneata, C. chlorusarca, C. dahurica, C. maximowiczii, C. pentagyna, C. pinnatifida, C. sanguinea, Dysosma pleiantha (Hance) Woodson Eregeron annuus L. Euonymus alatus, E. bungeanus, E. maackii, Heracleum moellendorffi Hance Geranium dahuricum, G. eriostemon, G. sibiricum,*
Quercetin	*G. wlassowianum, G. wilfordi, Hibiscus mutabilis, Hypericum attenuatum, H. ascyron, H. japonicum, H. perforatum, H. sumpsonii, Inula britannica, I. japonica, I. linariaefolia, I. salsoloides, Isatis chinensis, I. tinctoria, Jatropha podagrica, Loranthus parasiticus, L. yadoriki, Machilus thunbergii, Persicaria amphibia, P. hydropiper, Pileostegia viburnoides, Pistacia lentiscus, Plumeria rubra, R.ododendron anthopogon, R. dauricum, Thuja koraiensis, T. orientalis, T. chinensis, Viscum album*
Quercetin glucoside	*Tephrosia purpurea*
Quercetin-3-β-glycoside-7-β-glycoside-7-β-glycoside	*Gymnadenia conopsea* (L.) R. Brown
Quercetin-3-galacto-xylo-glucoside	*Prunus padus*

Component	Source
Quercetin-3-galactoside	*Rumex acetosa, R. acetosella, R. amurensis, R. aquaticus, R. gmelini, R. longifolius, R. maritimus, R. marschallianus, R. stenophyllus, R. thyrsiflorus*
Quercetin-4-glucoside	*Hibiscus mutabilis*
Quercetin-monomethylether	*Tamarix juniperina*
Quercetol	*Rosa multiflora*
Quercimeritrin	*Hibiscus mutabilis, Melaleuca leucadendra, Persicaria amphibia, P. hydropiper, Polygonum bistorta, P. hydropiper*
Quercitin	*Celosia argentea, C. cristata, C. margariacea, Ficus carica, Hippophae rhamnoides, Inula Britannica* L. Solidago virgaurea L.
Quercitol	*Cornus alba, C. kousa, C. macrophylla, Euphorbia hirta, Fagopyrum esculentum, F. sagittatum, Viscum album, V. coloratum*
Quercitrin	*Biota chinensis, B. orientalis, Dicranopteris linearis, Euphorbia hirta, Ginkgo biloba, Houttynia cordata, Hypericum attenuatum, H. ascyron, H. japonicum, H. perforatum, H. sumpsonii, Loropetalum chinense, Persicaria hydropiper, Saururus chinensis, Solidago canadensis*
Quercitol	*Cotinus coggygria* Scop.
Quereetin	*Astragalus complanatus, A. henyri, A. hoantchy, A. membranaceus, A. melilotoides, A. mongholicus, A. reflexistipulus, A. sinensis, Ginkgo biloba, Lathyrus pratensis* L.
Quinic acid	*Ledum palustre*
Quinochalone	*Solidago dahurica, S. pacifica, S. virgaurea*
Quinochalone	*Carthamus tinctorius*
Quinonic substance	*Maesa perlarius*

Quinonoid	*Lithospermum erythrorhizon, L. officinalis*
Quinoric acid	*Adina rubella, A. ratemosa*
Quinquenosides	*Panax ginseng*
Quisqualic acid	*Quisqualis grandiflora, Q. indica, Q. longifolia, Q. loureiri, Q. pubescens, Q. sinensis*
r-cadinene	*Juniperus rigida*
r-glutamyl-valyl-glutamic acid	*Juncus effusus*
r-linolenic acid	*Stellaria media*
r-terpinene	*Pinus madshurica* Rupr.
Racemic acid	*Vitis amurensis, V. vinifera*
Raddanoside	*Anemone raddeana, A. rivularis, A. vitifolia*
Raddeanin A	*Anemone raddeana, A. rivularis, A. vitifolia*
Raffinose	*Lycopus lucidus, L. maackianus, L. parviflorus, L. ramosissimus, L. fargesii, L. veitchii*
Ramalic acid	*Usnea diffracta, U. longissima*
Ranunculin	*Anemone raddeana, A. rivularis, A. vitifolia, Pulsatilla ambigua, P. cernua, P. chinensis, Ranunculus chinensis, R. sceleratus*
Raphanin	*Raphanus sativus*
Rebaudiosides	*Stevia rebaudiana*
Rebixanthin	*Calendula officinalis*
Rehmannin	*Rehmannia chinensis, R. glutinosa*
Reliculine	*Corydalis incisa, C. bungeana*
Renifolin	*Chimaphila umbellata*
Repandusinic acids	*Mallotus repandus*

Component	Source
Repandusinin	*Mallotus repandus*
Rerpinenol	*Zanthoxylum bungeanum*
Reserpine	*Chenopotium ambrosiodes, Rauvolfia verticilata*
Resin	*Artocarpus heterophyllus, Caesalpinia pulcherrima, Callicarpa macrophylla, Centella ascatica, Commiphora myrrha, Cornus officinalis, Curculigo capitulata, C. ensifolia, C. malabarica, C. orchiodes Gaertn, C. stams, Curcuma pallida, C. phaeocoulis, Daemonorops draco, Dodonaea viscosa, Euonymus alatus, E. bungeanus, E. maackii, Ficus awkeotsang, Gnaphalium affine, G. arenarium, G. confusum, G. javanum, G. luteo-album, G. multiceps, G. ramigerum, G. tranzschelii, G. uliginosum, Humulus lupulus, Lappa communis, L. edulis, L. major, L. minor, Lemmaphyllum microphyllum, Lycopus lucidus, L. maackianus, L. parviflorus, L. ramosissimus, L. fargesii, L. veitchii, Macrocarpium officinalis,*
Resin (contineued)	*Mallotus japonicus, Myristica fragrans, Ophiorrhiza japonica, O. mungos, Rhus chinensis, R. cotinus, R. javanica, R. osbeckii, Smilax china, S. nipponica, S. riparia, S. sieboldii, Trichosanthes kirilowii, T. uniflora, Wisteria sinensis*
Resin albaspidin	*Dryopteris laeta, D.filix-mas*
Resinous oil urushiol	*Rhus verniciflua*
Reticuline	*Machilus thunbergii*
Retinol	*Oxyria digyna*
Retroresine	*Crotalaria mucronata, Thalictrum aquilegifolium, T. baicalense, T. fauriel, T. petaloideum, T. simplex, T. squarrosum, T. thunbergii*
Rhamnazin	*Persicaria hydropiper, Polygonum hydropiper*
Rhamnetin	*Eugenia aromatica, E. caryophyllata, E. ulmoides*

Rhamnocitrin	*Alpinia japonica, Astragalus complanatus, A. henyri, A. hoantchy, A. membranaceus, A. melilotoides, A. mongholicus, A. reflexistipulus, A. sinensis*
Rhamnodiastase	*Rhamnus davurica, R. parvifolia*
Rhamnose	*Camellia japonica, Euphorbia hirta, Fagopyrum esculentum, F. sagittatum, Ficus carica*
Rhein	*Cassia angustifolia, Hemerocallis minor Miller Polygonum multifolrum, P. chinensis, Rheum officinale, R. palmatum, R. tanguticum, R. undulatum, R. koreanum*
Rhein aurantioobtusin	*Cassia nomame, C. obtusifolia, C. tora*
Rhein chrysarobin	*Cassia alata*
Rhein monoglucoside	*Cassia angustifolia*
Rhodexins	*Rhodea japonica*
Rhodioloside	*Rhodiola elongata*
Rhodotoxin	*Rhododendron dauricum*
Rhoeadine	*Papaver rhoeaes, P. somniferum*
Rhoeagenine	*Papaver rhoeaes*
Rhomotoxin	*Rhododendron molle*
Rhymohydroquinone	*Eupatorium chinense, E. lindleyanum, E. japonicum*
Rhynchophylline	*Nauclea rhynchophylla, N. sinensis, Uncaria hirsuta, U. rhynchophylla*

Component	Source
Riboflavin	*Achyranthes asperia, Alpinia katsumadai, A. globosum, A. kumatake, Amaranthus tricolor, A. lividus, A. blitum, A. viridis, Arachis hypogaea, Benincase cerifera, B. hispida, Boehmeria densiflora, Canarium album, C. sinense, Castanea crenuta, C. mollissima, Corylus heterophylla, C. mandshurica, Eleusine indica (L.) Gaertner Glycine max, G. soja, Helianthus annuus L. Hibiscus rosa- sinensis, H. rhombifolius, Lycopersicon esculentum, Marsilea quadrifolia L. Petasites japonicus, Solanum nigrum,Sonchus oleraceus L. Zea mays*
Ricinine	*Ricinus communis*
Ricinolein	*Ricinus communis*
Robinin	*Phaseolus angularis, P. lunatus, P. radiatus, P. vulgaris, Pueraria lobata, P. pseudo-hirsuta, Rauvolfia verticilata*
Roemerine	*Nelumbium nuciferum, N. speclosum*
Rorifamide	*Rorippa indica, R. islandica, R. montana*
Rorifone	*Rorippa indica, R. islandica, R. montana*
Rosenoxide	*Rosa rugosa*
Rotenone	*Milletia reticulata, M. taiwaniana, Tephrosia purpurea*
Rotundone	*Cyperus rotundus*
Rotunol	*Cyperus rotundus*
Rovidine	*Catharanthus roseus*
Roxburghine D	*Acacia catechu*
Rrechts–lupinine	*Lupinus luteus*
Rubescensine B	*Rabdosia rubescens*

Rubescensins	*Rabdosia lasiocarpus*
Rubiadin-I-methyl ether	*Morinda citrifolia, M. officinalis*
Rubichloric acid	*Morinda citrifolia, M. officinalis*
Rubierythrinic acid	*Rubia chinensis, R. cordifolia, R. mungista, R. sylvatica*
Rubijervine	*Veratrum dahuricum, V. maackii, V. nigrum*
Rubricauloside	*Peucedanum japonicum, P. praeruptorum, P. rubricaule*
Rubrierythrinic acid	*Galium bungei, G. spurium, G. verum, Rubia akane*
Rutaecarpine	*Evodia rutaecarpa*
Rutin	*Abutilon theophrasti, A. avicennae, Achillea millefolium L.Cassia alata, Cirsium albescens, C. brevicaule, C. littorale, C. maakii, C. segetum, C. setosum, C. vlassovianum, Cucumis sativus, Coriandrum sativum, Fagopyrum esculentum, F. sagittatum, F. tataricum, Ficus carica, Firmiana simplex, Forsythia suspensa, Heracleum moellendorffii Hance Ginkgo biloba, Hibiscus mutabilis, Hydrangea macrophylla, Jatropha podagrica, Nymphoides peltata (S.G. Gmelin) Prunella vulgaris, Sophora japonica, Tephrosia purpurea, Tussilago farfara*
s-guaiazulene	*Artemisia lactiflora*
Sabinene	*Elettaria cardamomum, Ledum palustre, Pinus madshurica Rupr. Piper cubeba, Zanthoxylum bungeanum*
Saccharase	*Solanum indicum*
Saccharides	*Cordyceps sinensis*
Saccharose	*Trachycarpus wagnerianus, T. fortunei*
Safflomin A	*Carthamus tinctorius*
Safflower yellow	*Carthamus tinctorius*

Component	Source
Safrole	*Asarum canadense, A. heterotropoides, A. sieboldii, Illicium verum, Magnolia biloba, M. denudata, M. discolor, M. liliflora, M. purpurea, Myristica fragrans*
Saikosaponins	*Bupleurum chinense, B. falcatum, B. scorzoneraefolium*
Salicarin	*Lythrum salicaria*
Salicifoline	*Magnolia biloba, M. denudata, M. discolor, M. grandiflora, M. liliflora, M. purpurea, Michelia alba, M. figo*
Salicin	*Populus alba, P. davidiana, P. tomentosa*
Salicinase	*Populus alba, P. davidiana, P. tomentosa*
Salicortin	*Populus alba, P. davidiana, P. tomentosa*
Salicylate	*Dianthus chinensis* L.
Salicylic acid	*Calendula officinalis, Gaultheria leucocarpa, Pterocarya stenoptera, Scopalia dulcis, Siegesbeckia orientalis* L.
Salidroside	*Vaccinium bracteatum, V. vitis-idaea*
Saligenin glucoside	*Salix babylonica, S. matsudana, S. microstachya*
Salireposide	*Populus alba, P. davidiana, P. tomentosa*
Salsolidine	*Salsola collina*
Salsoline	*Salsola collina*
Saluianin	*Salvia coccinea*
Salvigenin	*Eupatorium odoratum*
Salviol	*Salvia miltiorhiza*
Sanguinarine	*Macleaya cordata, Papaver somniferum*

Sanguisorbins	*Sanguisorba grandiflora, S. officinalis, S. parviflora, S. x tenuifolia*
Sanjoinines	*Ziziphus jujuba, Z. spinosa*
Sanquinarine	*Hypecoum erectum* L.
Sanshol	*Xanthoxylum piperitum*
Santalic acid	*Santalum album, S. myrtifolium, S. verum*
Santalin	*Santalum album, S. myrtifolium, S. verum*
Santalone	*Santalum album, S. myrtifolium, S. verum*
Santamarine	*Eupatorium formosanum*
Santene	*Santalum album, S. myrtifolium, S. verum*
Sapnons	*Gynura japonica, G. pinnatifida, G. segetum, Syringa dilatata, S. oblata, S. reticulata, S. suspensa, S. vulgaris*
Sapogenins	*Bupleurum chinense, B. falcatum, B. scorzoneraefolium*
Saponaretin	*Fagopyrum esculentum, F. sagittatum, Hibiscus sabdariffa, Thalictrum ichangense, T. glandulissimum, Trigonella foenum-graecum*
Saponarin	*Hibiscus chinensis, H. syriacus, H. trionum, H. rhombifolius, Saponaria officinalis, S. vaccaria*
Saponartin-4'-0-glucoside	*Uraria crinita, U. lagopodiodes*

Component

Saponins

Source

Acacia nemu, Achyranthes aspera, A. japonica, Adenophora coronopifolia, A. paniculata, A. pereskiaefolia, A. polymorpha, A. remotiflora, A. stenanthina, A. tetraphylla, Adina rubella, A. ratemosa, Aleurites fordii, Alternanthera philoxeroides, A. sessilis, Anemone cernua, A. pulsatilla, Arenaria juncea, A. serpyllifolia, Arisaema amurense, A. consanguineum, A. erubescens, A. heterophyllum, A. peninsulae, A. thunbergii, Aster tataricus, Astragalus complanatus, A. henyri, A. hoantchy, A. membranaceus, A. melilotoides, A. mongholicus, A. reflexistipulus, A. sinensis, Caesalpinia decapetula, C. sappan, Caltha palustris, Campanula gentianoides, C. grandiflora, Centipeda minima, Cephalanoplos segetum, Chenopotium ambrosiodes, Clematis chinensis, C. florida, C. hexapetala, C. minor, C. sinensis, C. terniflora, Crataegus chlorusarca, C. dahurica, C. cuneata, C. maximowiczii, C. pentagyna, C. pinnatifida, C. sanguinea, Delonix regia, Dictamnus albus, D. dasycarpus, Dioscorea bulbifera, Eclipta erecta, Elephantopus elatus, E. grandiflorus, Eleutherocccus senticosus Maxim, Euphorbia esula, E. helioscopia, Gentiana algida, G. barbata, G. manshurica, G. olivieri, G. scabra, G. squarrosa, G. triflora, Gleditsia horrida, G. sinensis, G. xylocarpa, Gomphrena globosa, Hibiscus sabdariffa, Lonicera acuminata, L. apodonta, L. brachypoda, L. japonica, L. confusa, L. hypoglauca, L. chinensis, L. flexuosa, L. maackii, Loranthus parasiticus, L. yadoriki, Luffa aegyptiaca, L. cylindrica, L. faetida, L. petola, Marsdenia tenacissima, Nephelium longana, N. lappaceum, Panax japonicum, P. notogineng, P. zingiberensis, Phytolacca acinosa, P. americana, P. japonica, P. kaempferi, P. octandra, P. pekinensis, Polygala japonica, P. sibirica, P. tatarinowii, Pulsatilla ambigua, P. cernua, P. chinensis, Rhododendron anthopogon, Salix babylonica, S. matsudana, S. microstachy, Sesbinia sesbin, Sapindus mukorossi, Smilax china, S. nipponica, S. sieboldii, S. riparia, Solanum nigrum, Thalictrum foetidum, Trichosanthes kirilowii, T. uniflora, Trigonella foenum-graecum, Tussilago farfara, Vernonia andersonii, V. cinerea, V. patula, Xanthoxylum piperitum, Ziziphus jujuba, Z. spinosa

Saponin akebin	*Clematis armandii, C. heraclei*
Saponin alpha-methylester	*Achyranthes aspera*
Saponin beta-methylester	*Achyranthes aspera*
Sarcostin	*Cynanchum paniculatum, Marsdenia tenacissima*
Sarmentosin	*Sedum erythrostichum, S. kamtschaticum, S. verticillatum*
Sarmentoslin	*Sedum sarmentosum*
Sarmutoside	*Strophanthus divaricatus*
Sarcostin	*Metaplexis japonica* Thunb.
Sarolactone	*Hypericum attenuatum, H. ascyron, H. japonicum, H. perforatum, H. sumpsonii*
Sarothamine	*Cytisus scoparius*
Sarracine	*Senecio argunensis, S. nemorensis, S. scandens*
Sarsasapogenin	*Anemarrhena asphodeloides, Arnebia euchroma*
Sativol	*Medicago falcata, M. lupulina, M. polymorpha, M. ruthenica, M. sativa*
Saturated acids	*Viola acuminata, V. alisoviana, V. collina, V. dissecta, V. mandshurica, V. patrini, V. prionantha, V. verecunda*
Saussurine	*Aucklandia costus, A. lappa, Saussurea japonica, S. lappa*
Savinin	*Acanthopanax sessiliflorus*
Schisantherins	*Schisandra arisanensis, S. sphenanthera*
Schizandrer	*Schisandra chinensis*
Schizandrin	*Schisandra chinensis*
Schizandrol	*Schisandra chinensis*
Sciadopitysin	*Podocarpus macrophyllus*

Component	Source
Scopanol	*Scopalia dulcis*
Scoparin	*Cytisus scoparius*
Scoparon	*Artemisia capillaris*
Scoplin	*Physochlaina infundibularis*
Scopolamine	*Datura alba, D. fastuosa, D. innoxia, D. metel, D. stramonium, D. suaveolens, D. tatula, Physochlaina infundibularis, Scopolia tangutica*
Scopoletin	*Adonis amurensis* Regel & Radde *Angelica amurensis, A. anomala, A. dahurica, Artemisia gmelini, Bidens tripartita, Caltha palustris, Carum carvi, Coriandrum sativum, Erycibe henryi, E. aenea, Ilex pubescens, Impatiens balsamina, I. noli-tangere, I. textori, Melilotus alba, M. suaveolens, M. indica, Nerum indicum, Physochlaina infundibularis, Rhododendron dauricum* L. *Viburnum sargenti*
Scopoline	*Erycibe henryi, E. aenea, Nerum indicum*
Scopolomine	*Hyoscyamus bohemicus*
Scoulerine	*Corydalis incisa, C. bungeana*
Scrophularin	*Scrophularia buergeriana, S. kakudensis, S. ningpoensis, S. oldhami, S. puergeriana*
Scutellarein heteroside	*Thymus amurensis, T. disjunctus, T. kitagawianus, T. komarovii, T. przewalskii, T. quinquecostatus*
Scutellarin	*Salvia chinensis, S. pogonocalyx, S. przewalskii*
Sebiferic acid	*Sepium sebiferum, S. discolor*
Secalose	*Avena fatua*
Secologanin	*Menyanthes trifoliata*
Securinine	*Securinega suffruticosa*

Securinol	*Securinega suffruticosa*
Securitinine	*Securinega suffruticosa*
Sedocaulin	*Sedum aizoon*
Sedocitrin	*Sedum aizoon*
Sedoflorin	*Sedum aizoon*
Sedoheptulosan	*Orostachys fimbriatus* Turcz. *Sedum lineare, S. erythrostichum, S. kamtschaticum, S. verticillatum*
Sedoheptulose	*Sedum aizoon*
Selinene	*Valeriana alternifolia, V. amurensis, V. fauriei, V. subbipinnatifolia*
Sempervirine	*Gelsemium elegans, G. sempervirens*
Senecionine	*Senecio vulgaris*
Sennosides	*Cassia angustifolia, Rheum officinale, R. palmatum, R. tanguticum, R. undulatum, R. koreanum*
Sequiterpine	*Eupatorium odoratum* L.
Seratonin	*Ranunculus sceleratus*
Serine	*Litchi chinensis, Taraxacum mongolicum, T. sinicum*
Serotonin	*Hippophae rhamnoides, Musa paradisiaca*
Serpentine	*Rauvolfia verticilata*
Serratenediol	*Lycopodium annotinum, L. cernum, L. compianatum*
Sesamin	*Sesamum indicum*
Sesamol	*Sesamum indicum*
Sesquijasmine	*Jasminum samba*

Component	Source
Sesquilignans	*Arctium lappa*
Sesquiterpenes	*Paatrinia villosa* (Thunb.)
Sesquiterpene Ketones	*Acorus calamus* L. *Alpinia officinarum, Artemisia annua, A. apiacea, Curcuma longa, Cyperus rotundus, Dryobalanops aromatica, D. camphora, Ginkgo biloba, Hedychium coronarium, Jasminum samba, Melaleuca leucadendra, Nardostachys jatamansi, Xanthoxylum piperitum*
Sesquiterpene alcohol	*Blumea balsumifera, Melaleuca leucadendra*
Sesquiterpene alkaloids	*Euonymus alatus, E. bungeanus, E. maackii*
Sesquiterpene glucosides	*Achillea alpina, A. millefolium, Anthriscus aemula, A. sylvestris, Heracleum dissectum, H. lanatum, Polygonum bistorta, Sanguisorba officinalis, S. grandiflora, S. parviflora, S. x tenuifolia, Taraxacum officinale*
Sesquiterpine lactones	*Eupatorium formosanum*
Shanzhiside	*Gardenia florida, G. grandiflora, G. maruba, G. pictorum, G. radicans*
Shibuol	*Diospyros chinensis, D. costata, D. khaki, D. lotus, D. roxburgii*
Shikimic acid	*Ginkgo biloba, Illicium lanacedatum*
Shikonin	*Arnebia euchroma, Lithospermum erythrorhizon, L. officinalis*
Shionon	*Aster tataricus*
Shisonin	*Solanum lyratum, S. melongena*
Shobakunine	*Mahonia japonica*
Shyobunones	*Acorus calamus* L.
Silica	*Phyllostachys bambusoide, P. nigra*
Silicic acid	*Desmodium triforum, D. triquetrum,* Equisetum arvense L. *Plumbago zeylanica*

Silybin	*Silybum marianum*
Silybinomer	*Silybum marianum*
Silyckristin	*Silybum marianum*
Silydiamin	*Silybum marianum*
Silymarin	*Silybum marianum*
Simiarenol	*Imperata arundinaceae, I. cylindrica*
Simplexine	*Phyllanthus simplex*
Sinactine	*Cocculus diversifolius, C. thunbergii, Sinomenium acutum*
Sinalbine	*Sinapis alba*
Sinapic acid	*Brassica alba, B. juncea, Impatiens balsamina, I. noli-tangere, I. textori*
Sinapine	*Brassica alba, B. juncea*
Sinapyl aldehyde	*Santalum album, S. myrtifolium, S. verum*
Sinigrin	*Brassica alba, B. juncea, Thlaspi arvense*
Sinigroside	*Cardamine leucantha, C. lyrata*
Sinoacutine	*Cocculus diversifolius, C. thunbergii, Sinomenium acutum*
Sinocecatine	*Corydalis incisa, C. bungeana*
Sinodiosgenin	*Dioscorea opposita*
Sinomenine	*Cocculus diversifolius, C. thunbergi, Menispermum dauricum, Sinomenium acutum*
Sinoside	*Strophanthus divaricatus*
Sinostroside	*Strophanthus divaricatus*
Sioimperatorin	*Angelica decursiva.*
Sioquercitrin	*Vaccinium bracteatum, V. vitis-idaea*

Component	Source
Siosakuranetin	*Clinopodium chinense, C. polycephalum, C. gracile, C. umbrosum*
Sitosterol	*Arundinaria graminifolia, Buglossoides arvense* (L.) Johnston Centella ascatica, *Chenopodium album* L. *Elaeagnus oldhumii, Ficus carica, Nuphar japonicum,* N. pumilum, *Punica granatam, Syzygium aromaticum, Ulmus campestris, U. macrocarpa, U. pumila*
Sitosteryl glucopyranosid	*Elaeagnus oldhumii*
Sitosteryl-o-β-d-glucoside	*Spilanthes acmella*
Skimmianine	*Dictamnus albus, D. dasycarpus, Glycosmis cochinchinensis, G. pentaphylla, Zanthoxylum nitidum, Z. schinifolium, Z. ailanthoides*
Skullcapflavones	*Scutellaria baicalensis, S. grandiflora, S. lanceolaria, S. macrantha, S. rivulararis, S. viscidula*
Slliptinone	*Plumbago zeylanica*
Sloeemodin	*Rumex patientia*
Smilacin	*Smilax china, S. nipponica, S. sieboldii, S. riparia*
Solamargine	*Solanum incanum*
Solanidine	*Solanum indicum*
Solanigrines	*Solanum nigrum*
Solanine	*Solanum indicum, S. lyratum, S. melongena*
Solanocapsin	*Solanum capsicastrum*
Solanocapsine	*Solanum pseudo-capsicum*
Solasodine	*Solanum indicum, S. lyratum, S. melongena*
Solasonine	*Solanum aculeatissimum, S. verbascifolium*
Solasurine	*Solanum aculeatissimum*

Somalin	*Adonis amurensis* Regel & Radde
Sophorabioside	*Sophora japonica*
Sophoradin	*Sophora subprostrata*
Sophoradiol	*Sophora japonica*
Sophoraflavonoloside	*Sophora japonica*
Sophoranochromene	*Sophora subprostrata*
Sophoranone	*Sophora subprostrata*
Sophoricoside	*Sophora japonica*
Soranjudiol	*Morinda citrifolia, M. officinalis*
Sorbose	*Rhus verniciflua* Strokes
Sotelsuflavone	*Cycas revoluta, Selaginella tamarisina*
Soyasapogenol B, E.	*Gueldenstaedtia* Maxim
Sparteine	*Chelidonium album, C. hybridum, C. majus, C. serotinum, Cytisus scoparius*
Spathulenol	*Artemisia lactiflora*
Spemine	*Panax ginseng*
Spermindine	*Panax ginseng*
Sphenone A	*Sphenomeris chusana*
Spilanthol	*Spilanthes acmella*
Spinacetin	*Spinacia oleracea* L.
Spinasterol	*Codonopsis pilosula, C. tangshen, C. ussuriensis, Platycodon autumnalis, P. grandiflorum, P. sinensis*
Spirostanol saponins	*Paris polyphylla*

Component	Source
Spongesterol	*Angelica decursiva*
Springic acid	*Catalpa ovata* G. Don
Squalene	*Abrus precatorius, Taraxacum officinale*
Stachydrine	*Chrysanthemum cinerriaefolium C. jucundum, C. koraiense, C. morifolium, C. sinense*
Stachydrine chloride	*Achillea millefolium* L. *Stachys chinensis, S. baicalensis, S. japonica*
Stachyose	*Isatis chinensis, I. chinensis, I. tinctoria, Lycopus lucidus, L. maackianus, L. parviflorus, L. ramosissimus, L. fargesii, L. veitchii, Prunella asiatica* NaKa
Starch	*Pteris cretica, P. ensiformis, P. multifida, P. vittata, P. wallichiana, Pseudostellaria heterophylla, Tulipa edulis, T. gesneriana*
Stauntonin	*Stauntonia hexaphylla*
Stearic acid	*Aleurites fordii, Angelica grosserrata, Ajuga bracteosa, Benincase cerifera, B. hispida, Citrullus anguria, C. edulis, C. lanatus, C. vulgaris, Coix agrestis, C. chinensis, C. lachryma, Corchorus capsularis, C. olitorius, Cucumis sativus, Descurainia Sophia* (L.) Schur. *Jatropha gospiifolia, J. curcas, Melia azedarach, M. toosendan, M. japonica, Myristica fragrans, Sonchus arvensis, S. oleraceus, Taraxacum mongolicum, T. sinicum*
Stearin acid	*Cedrela sinensis* A. Juss.*Ricinus communis, Sesamum indicum*
Stemonidine	*Stemona japonica, S. tuberosa*
Stemonine	*Stemona japonica, S. tuberosa*
Stephanine	*Stephania japonica*
Stephanoline	*Stephania japonica*
Stepharine	*Menispermum dauricum, Stephania sinica*
Stepharotine	*Stephania sinica*

Stephisoferuline	*Stephania hernandifolia*
Stepholidine	*Menispermum dauricum*
Stepinonine	*Stephania japonica*
Steponine	*Stephania japonica*
Stereoisomer	*Arctium lappa*
Steroid alkaloid glycosides	*Solanum biflorum*
Steroid glycosides	*Periploca sepium, Sesbinia sesbin*
Steroid saponins	*Allium chinense, A. odorum, A. sativum, A. tuberosum, A. uliginosum, Anemarrhena asphodeloides, Arnebia euchroma*
Steroidal saponin POD-II	*Polygonatum chinense, P. cirrhifolium, P. macropodium, P. odoratum, P. officinale, P. sibiricum, P. stenophyllum, P. vulgare*
Sterol	*Cordyceps sinensis, Cucurbita moschata, C. pepo, Cucumis melo, Cypripedium guttatum, C. macranthum, C. pubescens, C. sativus, Fibraurea recisa, Luffa aegyptiaca, L. cylindrica, L. faetida, L. petola, Momordica charantia*
Steviolbioside	*Stevia rebaudiana*
Stevioside	*Stevia rebaudiana*
Stigmast-7-enol	*Menyanthes trifoliata*
Stigmasterol	*Adina rubella, A. ratemosa, Aletris formosuna, A. spicata, Aleurites fordii Arundinaria graminifolia, Bauhinia championi, Cirsium chinense, C. japonicum, Codonopsis lanceolata, C. pilosula, C. tangshen, C. ussuriensis, Cuscuta chinensis, C. europaea, C. japonica, C. lupuliformis, Dioscorea bulbifera, Eclipta erecta, Glehnia littoralis, Linum stelleroides, L. usitatissimum, Matteuccia struthiopteris, Nervilia purpurea, Oldenlandia chrysotricha,*

Component	Source
Stigmasterol	*O. corymbosa, O. diffusa, Ophiopogon japonicus, Papaver somniferum, Rubus coreanus, R. crataegifolius, R. matsumuranus, R. saxatilis, Schizonepeta multifida, S. tenuifolia, Spilanthes acmella, Syzygium aromaticum*
Streptomycin	*Senecio scandens* Bush-Ham
Strophalloside	*Antiaris toxicaris*
Strophantidin	*Apocynum venetum, Corchorus olitorius*
Strospeside	*Digitalia purpurea, D. sanguinalis*
Strumaroside	*Xanthium chinense, X. japonicum, X. mongolicum, X. sibiricum, X. strumarium*
Strychnine	*Strychnos nux-vomica, S. pierriana*
Stylopine	*Chelidonium album, C. hybridum, C. majus, C. serotinum*
Styracin	*Styrax tonkinensis, S. benzoin*
Succinacid	*Cynomorium coccineum, C. songarium*
Succinic acid	*Angelica polymorpha, A. sinensis, Geranium dahuricum, G. eriostemon, G. sibiricum, G. wlassowianum, G. wilfordi, Maytenus diversifolia, M. confertiflorus, Prunus mume, Sarcandra glabra, Typhonium giganteum*
Sucrose	*Codonopsis pilosula, C. tangshen, C. ussuriensis, Eriobotrya japonica, Nephelium longana, N. lappaceum*
Sulfuretin	*Cotinus coggygria* Scop
Sumaresinolic acid	*Styrax tonkinensis, S. benzoin*
Swertiamarin	*Swertia pseudochinensis*
Swertifrancheside	*Swertia pseudochinensis*
Swertisin	*Swertia pseudochinensis*

Sworoside

Cornus officinalis

Synephrine

Citrus aurantium

Syringareinol

Eleutherocossus senticosus

Syringaresinol

Acanthopanax giraldii, A. sessiliflorus

Syringen

Paulownia tomentose Thunb.

Syringic acid

Ailanthus altissima, Maytenus diversifolia, M. confertiflorus, Rhododendron mucronatum

Syringic aldehyde

Santalum album, S. myrtifolium, S. verum

Syringin

Jasminum mesnyi, J. nudiflorum, Lonicera acuminata, L. apodonta, L. brachypoda, L. japonica, L. confusa, L. hypoglauca, L. chinensis, L. flexuosa, L. maackii, Paulownia tometosa Thunb. *Syringa dilatata, S. oblata, S. reticulata, S. suspensa, S. vulgaris*

Tabetone

Tagetes erecia L.

Tadeonal

Persicaria hydropiper, Polygonum hydropiper

Taeniafuge

Pteridium aquilinum (L.) Kuhn.

Tagetone

Tagetes erecta, T. patula

Talatisamine

Aconitum austroyunnarense, A. balfouri, A. barbatum, A. carmichaelii, A. chasmanthum, A. chinense, A. deinorrhizum, A. fischeri, A. jaluense, A. koreanum, A. kusnezoffii, A. lacinatum, A. napellus, A. pariculigerum, A. praeparata, A. vilmorinianum, A. volubile

Talictrine

Thalictrum ichangense, T. glandulissimum

Tangeratin

Citrus aurantium

Tannates

Ophiorrhiza japonica, O. mungos

Tannic acid

Acalypha australis, Acanthopanax gracilistylus, A. spinosum, Caesalpinia sappan, Coriandrum sativum, Cornus officinalis, Eclipta erecta, Erigeron canadensis, E. annuus, Hedera rhombea, H. helix, Hieracium umbellatum, Ilex chinensis, Macrocarpium officinalis,

Component	Source
Tannic acid	*Rabdosia rubescens, Rheum officinale, R. palmatum, R. tanguticum, R. undulatum, R. koreanum, Thea assamica, T. bohea, T. cantoniensis, T. chinensis, T. cochinchinensis, T. sinensis, T. viridis, Vitex negundo*
Tannins	*Acacia cutechu, A. nemu, Agrimonia eupatoria, A. pilosa, A. viscidula, Aleurites fordii, Alternanthera philoxeroides, A. sessilis, Arethusa japonica, Artemisia brachyloba, Aspidium falcatum, Callicarpa macrophylla, Caesalpinia pulcherrima, Cedrela sinensis A. Juss. Centella ascatica, Cleome spinosa, C. gynandra, C. viscosa, Curculigo capitulata, C. ensifolia, C. malabarica, C. orchiodes, C. stams, Curculigo orchiodes Gaertn. Desmodium triforum, D. triquetrum, Dioscorea bulbifera, D. cirrhosa, D. hispida, D. japonica, Dodonaea viscosa Geranium dahuricum, G. eriostemon, G. sibiricum, G. wlassowianum, G. wilfordi, Jasminum mesnyi, J. nudiflorum, Hippophae rhamnoides, Lappa communis, L. edulis, L. major, L. minor, Lonicera acuminata, L. apodonta, L. brachypoda, L. chinensis, L. confusa, L. flexuosa, L. hypoglauca, L. japonica, L. maackii, Lythrum salicaria, Mallotus japonicus, Matricaria chamomilla, Nephelium longana, N. lappaceum, Morinda citrifolia, M. officinalis, Photinia serrulata, Polygonum aviculare, P. lapidosa, P. manshuriensis, P. vivipara, Populus alba, P. davidiana, P. tomentosa, Portulaca pilosa, Prunus persica, Pteris cretica, P. ensiformis, P. multifida, P. vittata, P. wallichiana, Pteridium aquilinum (L.) Kuhn Pyrrosia adnascens Ranunculus ternatus, Rhus semialata, Scopalia dulcis, Smilax china, S. nipponica, S. sieboldii, S. riparia, Tamarindus indicus, Terminalia chebula, T. wagnerianus, Trachycarpus fortunei, Tussilago farfara, Vaccinium bracteatum, V. vitis-idaea, Verbena officinalis, V. oxysepalum*
Tanshinol	*Salvia miltiorhiza*
Tanshinone	*Salvia miltiorhiza*

Taraxacerin	*Taraxacum mongolicum, T. sinicum*
Taraxacin	*Taraxacum mongolicum, T. sinicum*
Taraxanthin	*Cuscuta australis, Tussilago farfara*
Taraxasterol	*Inula britannica, I. japonica, I. linariaefolia, I. salsoloides, Sonchus arvensis, S. oleraceus, Taraxacum formosanum, T. mongolicum, T. sinicum*
Taraxasteryl acetate	*Forsythia suspensa, Inula britannica, I. japonica, I. linariaefolia, I. salsoloides*
Taraxasteryl palmitate	*Forsythia suspensa, Inula britannica, I. japonica, I. linariaefolia, I. salsoloides*
Taraxerol	*Acanthopanax trifoliatus, Alnus japonica, Codonopsis pilosula, C. tangshen, C. ussuriensis, Euphorbia hirta, Taraxacum mongolicum, T. sinicum*
Taraxerone	*Adenophora triphylla, A. verticillata, Euphorbia hirta, Spatholobus suberectus*
Taraxeryl acetate	*Codonopsis pilosula, C. tangshen, C. ussuriensis*
Taraxol	*Taraxacum mongolicum, T. sinicum*
Taraxasterol acetate	*Cirsium chinense, C. japonicum, Sonchus arvensis* L.
Tartaric acid	*Cacalia hastate* L. *Chaenomeles japonica, C. sinensis, C. specicsa, Cornus officinalis, Cydonia sinensis, Eriobotrya japonica, Macrocarpium officinalis, Nephelium longana, N. lappaceum, Oxalis corriculaza, O. corymbosa, Prunus mume, Pyrrosia adnascens, Sonchus arvensis, S. oleraceus, Vitis amurensis, V. vinifera*
Tartrate	*Cajanus indicus* L.
Taspine	*Caulophyllum robustum*
Tataxasterol	*Taraxacurn mongolicum* Hand-Mazz
Taurine	*Gelidium amansii*
Taxettin	*Zephyranthes candida*
Taxifolin	*Pistacia lentiscus*

Component	Source
Taxinine E	*Taxus cuspidata, T. chinensis, T. yunnanensis*
Taxol	*Taxus cuspidata, T. chinensis, T. yunnanensis*
Tazettine	*Hippeastrum hybridum, Lycoris radiata, L. longituba, L. aura, Narcissus tazetta, Zephyranthes Candida* Herb.
Tectoridin	*Belamcanda chinensis, B. panctata, Iris aqyatuca, I. buatatas, I. dichotoma*
Tenuidine	*Polygala japonica, P. sibirica, P. tatarinowii*
Tenuifolin	*Polygala japonica, P. sibirica, P. tatarinowii*
Tephrosin	*Tephrosia purpurea*
Teresantalic	*Santalum album, S. myrtifolium, S. verum*
Teresantalol	*Santalum album, S. myrtifolium, S. verum*
Terpene	*Amomum cardamomum, A. globosum, A. tsao-ko, A. illosum, A. xanthloides, Aquilegia buergeriana, A. parviflora, Artemisia brachyloba, Cyperus brevifolius, C. difformis, C. glomeratus, C. iria, C. rotundus, Elsholtzia argyi, E. cristata, E. splendens, E. feddei, E. souliei, Liquidambar acerifolia, L. formosana, L. maximowiczii, Rabdosia lasiocarpus*
Terpeneol	*Erigeron canadensis, E. annuus*
Terpenylacetate	*Vitex trifolia, V. rotundifolia*
Terpinen-4-ol	*Thymus vulgaris*
Terpinene	*Coriandrum sativum, Elettaria cardamomum, Juniperus rigida*
Terpinenol-4	*Artemisia argyi, A. halodendron, A. igniaria, A. indica, A. integrifolia, A. japonica, A. keiskeana, A. lagocephala, A. lavandulaefolia, A. scoparia, A. selengensis, A. sieversiana, A. vulgarts*

Terpineol	*Cunninghamia lanceolata, Dryobalanops aromatica, D. camphora, Elettaria cardamomum, Myristica fragrans, Piper cubeba*
Terpinol	*Melaleuca leucadendra*
Terpinolene	*Coriandrum sativum, Cryptotaenia japonica, C. canadensis, Oenothera javanica, Valeriana alternifolia, V. amurensis, V. fauriei, V. subbipinnatifolia*
Teteracylic acid	*Rosa acicularis, R. amygdalifolia, R. davurica, R. davurica, R. koreana, R. laevigata, R. maximowicziana*
Tetra-hydrocannobinol	*Cannabis chinensis, C. sativa*
Tetraacetylbrazilin	*Caesalpinia sappan*
Tetradecen-4-oic acid	*Lindera obtusiloba*
Tetragonin	*Tetragonia tetragonoides*
Tetrahydroalstonine	*Catharanthus roseus* (L.) G. Don
Tetrahydrocolumbamine	*Corydalis ambigua, C. repens, C. turtschaninovii, C. yanhusuo, C. ternata*
Tetrahydrocoptisine	*Corydalis ambigua, C. repens, C. turtschaninovii, C. yanhusuo, C. ternata*
Tetrahydroharman	*Elaeagnus pungens, E. umbellata*
Tetrahydroxy flavone	*Xanthium chinense, X. japonicum, X. mongolicum, X. sibiricum, X. strumarium*
Tetramethylpyrazine	*Ligusticum chuanziang*
Tetramethylpyrazinesteroids	*Jatropha podagrica*
Tetrandrine	*Menispermum dauricum*
Tetulinic acid	*Scopolia dulcis*
Thalfoetidine	*Thalictrum aquilegifolium, T. baicalense, T. fauriel, T. foetidumT. petaloideum, T. simplex, T. squarrosum, T. thunbergii*
Thalicarpine	*T. halictrum, T. ichangense, T. glandulissimum*

Component	Source
Thalidasine	*T. halictrum, T. ichangense, T. glandulissimum*
Thalidezine.BP	*Thalictrum aquilegifolium, T. baicalense, T. fauriel, T. petaloideum, T. simplex, T. squarrosum, T. thunbergii*
Thalphinine	*Thalictrum aquilegifolium, T. baicalense, T. fauriel, T. foetidum T. petaloideum, T. simplex, T. squarrosum, T. thunbergii*
Thalpine	*Thalictrum aquilegifolium, T. baicalense, T. fauriel, T. foetidum, T. petaloideum, T. simplex, T. squarrosum, T. thunbergii*
Theasaponin	*Camellia japonica*
Thebaine	*Papaver somniferum*
Thelic simidine	*Thalictrum aquilegifolium, T. baicalense, T. fauriel, T. petaloideum, T. simplex, T. squarrosum, T. thunbergii*
Theobromine	*Thea assamica, T. bohea, T. cantoniensis, T. chinensis, T. cochinchinensis, T. sinensis, T. viridis*
Theophylline	*Thea assamica, T. bohea, T. cantoniensis, T. chinensis, T. cochinchinensis, T. sinensis, T. viridis*
Theronine	*Litchi chinensis*
Theveside	*Thevetai peruviana*
Thevetin A	*Thevetai peruviana*
Thevetin B	*Thevetai peruviana*
Theviridoside	*Thevetai peruviana*

Thiamine	Achyranthes asperia, Aleurites moluceanu, Alpinia katsumadai, A. globosum, A. kumatake, Amaranthus lividus, A. blitum, A. tricolor, A. viridis, Amaranthus tricolor L. Arachis hypogaea, Benincase cerifera, B. hispida, Boehmeria densiflora, Canarium album, C. sinense, Castanea crenata, C. mollissima, Corylus heterophylla, C. mandshurica, Eleusine indica (L.) Gaertner Glycine max, G. soja, Helianthus annuus L. Hibiscus rosa-sinensis, H. rhombifolius, Lycopersicon esculentum, Marsilea quadrifolia L. Petasites japonicus, Sonchus oleraceus L. Zea mays
Threonin	Avena fatua
Threonine	Oryza sativa
Thujone	Biota chinensis, B. orientalis Elettaria cardamomum, Thuja koraiensis, T. orientalis, T. chinensis
Thymine	Equisetum palustre L.
Thymol	Eucalyptus robusta, Thymus amurensis, T. disjunctus, T. kitagawianus, T. komarovii, T. przewalskii, T. quinquecostatus
Tienmulilmine	Veratrum dahuricum, V. maackii, V. nigrum
Tienmulilminine	Veratrum dahuricum, V. maackii, V. nigrum
Tiglic acid	Angelica pubescens, Ajuga bracteosa, Matricaria chamomilla
Tigloidine	Physalis alkekengi
Tigogenin	Costus specious, Smilax china, S. nipponica, S. sieboldii, S. riparia
Tirucallol	Euphorbia kansui
Tithymalin	Euphorbia esula, E. helioscopia
Tocopherol	Calendula officinalis
Tohogenol	Lycopodium annotinum, L. cernum, L. complanatum

Component	Source
Tomentogenin	*Cedrela sinensis A. Juss Cynanchum paniculatum.*
Toosendanin	*Melia japonica, M. toosendan, M. azedarach*
Toralacton	*Cassia nomame, C. obtusifolia, C. tora*
Torosachrysone	*Cassia occidentalis, C. torosa*
Totarol	*Podocarpus macrophyllus*
Tracheloside	*Trachelospermum jasminoides*
trans-β-farnesene	*Artemisia lactiflora*
trans-aconitic acid	*Actaea asiatica*
trans-beta-ocimene	*Cryptotaenia japonica, C. canadensis*
trans-caryophyllene	*Artemisia lactiflora*
Trehalose	*Selaginella involvens, S. doederieninii, S. tamarisina*
Tremulacin	*Populus alba, P. davidiana, P. tomentosa*
Triboline	*Cocculus laurifolius, C. sarmentosus, C. trilobus*
Trachitin	*Kalopanax septemlobus (Thunb.) Koidz.*
Trichosanthin	*Trichosanthes kirilowii, T. uniflora*
Tricin	*Medicago falcata, M. lupulina, M. polymorpha, M. ruthenica, M. sativa*
Tricycloekasantal	*Santalum album, S. myrtifolium, S. verum*
Trifolin	*Campanula glomerata, C. punctata, Lathyrus pratensis L. Melochia corchorfolia*
Trifolioside	*Menyanthes trifoliata*
Trifolirhizin	*Sophora alopecurosides, S. flavescens*
Triglyceride	*Livistona chinensis*

Trigonelline	*Abrus precatorius, Achillea millefolium* L. *Avena fatua, Cannabis chinensis, C. sativa, Dictamnus albus, D. dasycarpus, Quisqualis grandiflora, Q. indica, Q. longifolia, Q. loureiri, Q. pubescens, Q. sinensis, Solanum lyratum, S. melongena, Tetragonia tetragonoides, Trigonella foenum-graecum*
Trihydric alcohol	*Eupatorium odoratum* L.
Trihydroxytriptolide	*Tripterygium wilfordi*
Trillarin	*Trillium camschatcense*
Trillin	*Dioscorea nipponica, Trillium camschatcense*
Trilobamine	*Cocculus laurifolius, C. sarmentosus, C. triiobus*
Tripchilorolide	*Tripterygium wilfordii*
Tripeptide	*Juncus effusus*
Triptein	*Tripterygium wilfordii*
Triptolide	*Tripterygium hypoglaucum, T. wilfordii*
Tripdiolide	*Tripterygium wilfordi*
Tripdioltonide	*Tripterygium wilfordi*
Triptolidenol	*Tripterygium wilfordii*
Triptonide	*Tripterygium wilfordii*
Triptophenolide	*Tripterygium wilfordii*
Triterpene acid	*Eupatorium odoratum* L. *Rosa acicularis, R. amygdalifolia, R. davurica, R. koreana, R. laevigata, R. maximowicziana*
Triterpene glycosides	*Acanthopanax giraldii, Sesbinia sesbin*
Triterpenes	*Alisma orientalis, Artocarpus altilis, Sedum formosanum*
Triterpenoid saponins	*Bupleurum chinense, B. falcatum, B. scorzoneraefolium, Glycyrrhiza pallidiflora, G. uralensis*

Component	Source
Triterpenoids	*Clematis armandii, C. heracleifolia, Ganoderma lucidum, Mussaenda parviflora, Panax ginseng, Pulsatilla chinensis, Vernonia andersonii, V. cinerea, V. patula*
Tryptanthrin	*Baphicanthus cusia, Clerodendrum cyrtophyllum*
Tryptophane	*Dolichos lablab, Oryza sativa*
Tsudzuic acid	*Lindera obtusiloba*
Tubulosine	*Alangium lamarckii*
Tuduranine	*Cocculus diversifolius, C. thunbergii, Sinomenium acutum, Stephania sinica*
Tumulosic acid	*Poria cocos*
Turmerone	*Curcuma aromatica, C. kwangsiensis, C. zedoaria*
Tymol	*Thymus vulgaris*
Tyramine	*Viscum album*
Tyrosine	*Dioscorea opposita, Dolichos lablab, Ficus carica, Oryza sativa, Typhonium giganteum*
Ucarvone	*Asarum canadense, A. heterotropoides, A. sieboldii*
Ugenol	*Eugenia aromatica, E. caryophyllata, E. ulmoides*
Umbelliferone	*Adonis amurensis* Regel & Radde *Anethum graveolens* L. *Angelica decursiva, A. gigas* Maxim *A. pubescens, Artemisia gmelini, Bidens tripartita, Caltha palustris, Carum carvi, Coriandrum sativum, Daucus carota,* Ferula assafoetida L. *Heracleum moellendorffii* Hance *Melilotus alba, M. suaveolens, M. indica, Peucedanum japonicum, P. praeruptorum, P. rubricaule*
Umbelliprenin	*Anethum graveoleus, Angelica decursiva*
Unsaturated acids	*Viola acuminata, V. alisoviana, V. collina, V. dissecta, V. mandshurica, V. patrini, V. prionantha, V. verecunda*

Uracil	*Angelica polymorpha, A. sinensis, Typhonium giganteum*
Urbenine	*Coptis chinensis, C. japonica, C. teeta*
Urea	*Castanea crenuta, C. mollissima, Portulaca pilosa*
Urease	*Canavalia gladiata, C. ensiformis, Ficus carica, Fagopyrum esculentum, F. sagittatum*
Uronic acid	*Abutilon avicennae, A. theophrasti, Tamarindus indicus*
Ursolic aicd	*Anredera cordifolia, Chimaphila umbellata, Clinopodium chinense, C. polycephalum, C. gracile, C. umbrosum, Crataegus cuneata, C. chlorusarca, C. dahurica, C. maximowiczii, C. pentagyna, C. pinnatifida, C. sanguinea, Cynomorium coccineum, C. songarium, Dasiphora fruticosa* (Nestler) Kom., *Ilex chinensis, I. pubescens, Ligustrum lucidum, L. japonicum, Melaleuca leucadendra, Nerum indicum, Oldenlandia chrysotricha, O. corymbosa, O. diffusa,* Paulownia tometosa Thunb., *Plantago asiatica, P. depressa, P. exaltata, P. loureiri, P. major, Prunella vulgaris, Prunella asiatica* NaKa, *Punica granatam, Rubus coreanus, R. crataegifolius, R. matsumuranus, R. saxatilis, Solanum incanum, Thymus amurensis, T. disjunctus, T. kitagawianus, T. komarovii, T. prezewalskii, T. quinquecostatus*
Ursone	*Vaccinium bracteatum, V. vitis-idaea*
Urushiol	*Rhus verniciflua* Strokes.
Usaramine	*Crotalaria mucronata*
Ushinsunine	*Michelia alba, M. figo*
Usigtoercin	*Hypericum perforatum*
Usnic acid	*Usnea diffracta, U. longissima*
Utendin	*Metaplexis japonica* Thunb.
Uvaol	*Osmanthus fragrans*

Component	Source
Uzarigenin	*Asclepias curassavica* L.
Vaccarin	*Vaccaria segetalis, V. pyramidata*
Vaccaroside	*Vaccaria segetalis, V. pyramidata*
Valeraldehyde	*Melaleuca leucadendra*
Valerenone	*Valeriana alternifolia, V. amurensis, V. fauriei, V. subbipinnatifolia*
Valerianic acid	*Melia japonica, M. toosendan, M. azedarach*
Valerianol	*Valeriana alternifolia, V. amurensis, V. fauriei, V. subbipinnatifolia*
Valeric acid	*Luffa aegyptiaca, L. cylindrica, L. faetida, L. petola*
Valine	*Linum stelleroides, L. usitatissimum, Litchi chinensis, Oryza sativa, Typhonium giganteum*
Vallarine	*Centella ascatica*
Vanillic acid	*Ailanthus altissima, Catalpa ovata* G. Don *Chenopodium album* L. *Lycopodium clavatum, L. obscurum, L. selago, L. serratum, Matteuccia struthiopteris, Picrorhiza kurroa, Rhododendron mucronatum.* Turcz
Vanillin	*Catalpa ovata* G. Don *Ferula assa-foetida, F. bungeana, Gastrodia elata, Styrax tonkinensis, S. benzoin*
Vanillyl alcohol	*Gastrodia elata*
Vellosimine	*Rauvolfia verticilata*
Venoterpine	*Camptotheca acuminata*
Veratramine	*Veratrum formosanum*
Veratrine alkaloids	*Rhododendron sinensis*
Verbenalin	*Verbena officinalis, V. oxysepalum*

Verbenalol	*Verbena officinalis, V. oxysepalum*
Veronicastroside	*Veronica sibirica,. V. undulata*
Vertiaflavone	*Thevetai peruviana*
Verticine	*Fritillaria anheunensis, F. collicola, F. maximowiczii, F. roylei, F. thunbergii, F. ussuriensis, F. verticillata*
Verticinine	*Fritillaria anheunensis, F. collicola, F. maximowiczii, F. roylei, F. thunbergii, F. ussuriensis, F. verticillata*
Villoside	Patrinia villosa (Thunb.)
Vilmorrianines	*Aconitum barbatum, A.. austroyunnanense*
Vinblastine	*Catharanthus roseus*
Vincristine	*Catharanthus roseus*
Vindoline	*Catharanthus roseus* (L.) G. Don
Vindolinine	*Catharanthus roseus*
Vinrosidine	*Catharanthus roseus*
Violaxanthin	*Calendula officinalis, Potamogeton perfoliatus L. Rumex acetosa, R. acetosella, R. amurensis, R. aquaticus, R. gmelini, R. longifolius, R. maritimus, R. marschallianus, R. stenophyllus, R. thyrsiflorus, Taraxacum mongolicum, T. sinicum*
Viroallosecurinine	*Securinega virosa*
Virosecurinin	*Securinega virosa*
Virosine	*Securinega virosa*

Component	Source
Vitamin A	*Acanthopanax gracilistylus, A. spinosum, Amaranthus tricolor L. Capsella bursa-pastoris, Citrus deliciosa, C. nobilis, Hemerocallis fulva L., Gastrodia elata, Leonurus heterophyllus, L. japonicus, L. macranthus, L. mongolicus, L. pseudo-macranthus, Litchi chinensis, Luffa aegyptiaca, L. cylindrica, L. faetida, L. petola, Lycopersicon esculentum,*
Vitamin A	*Nephelium longana, N. lappaceum, Phaseolus angularis, P. lunatus, P. radiatus, P. vulgaris, Portulaca oleracea, Sesamum indicum, Taraxacum officinale, Triticum vulgare*
Vitamin B	*Actinidia chinensis Planch Actinidia arguta, A. chinensis, A. japonica, A. kolomikta, A. polygama, Citrus deliciosa, C. nobilis, Hemerocallis fulva L Hordeum vulgare, Litchi chinensis, Luffa aegyptiaca, L. cylindrica, L. faetida, L. petola, Nephelium longana, N. lappaceum, Phaseolus angularis, P. lunatus, P. radiatus, P. vulgaris, Portulaca oleracea, Sagittardia sagittifolia, Sesamum indicum, Tamarindus indicus, T. officinale,Triticum vulgare*
Vitamin B$_1$	*Cannabis chinensis, C. sativa, Gnaphalium affine, G. arenarium, G. confusum, G. javanum, G. luteo-album, G. multiceps, G. ramigerum, G. tranzschelii, G. uliginosum*
Vitamin B$_2$	*Cannabis chinensis, C. sativa, Phaseolus angularis, P. lunatus, P. radiatus, P. vulgaris*
Vitamin B$_{12}$	*Angelica polymorpha, A. sinensis*
Vitamin C	*Actinidia arguta, A. chinensi Planchs, A. japonica, A. kolomikta, A. polygama, Agrimonia eupatoria, A. pilosa, A. viscidula, Amaranthus tricolor L. Blumea lacera, Chaenomeles japonica, C. sinensis, C. speciosa, Citrus deliciosa, C. nobilis, Cydonia sinensis, Cypripedium guttatum, C. macranthum, C. pubescens, Dasiphora fruticosa (Nestler) Kom. Euphorbia coraroides, E. lasiocaula, E. lunulata, E. pallasii, E. pekinensis, E. sampsoni, E. sieboldiana, Fortunella crassifolia, F. japonica, F. margarita, Hemerocallis fulva L Hieracium umbellatum, Hippophae rhamnoides, Hordeum vulgare, Litchi chinensis,*

	Luffa aegyptiaca, L. cylindrica, L. faetida, L. petola, Lycium barbarum, L. megistocarpum, L. ovatum, L. trewianum, L. turbinatum, Oxalis corriculaza, O. corymbosa, Phyllanthus emblica, P. virgatus, Rumex acetosa, R. acetosella, R. amurensis, R. aquaticus, R. gmelini, R. longifolius, R. maritimus, R. marschallianus, R. stenophyllus, R. thyrsiflorus, Solanum nigrum, Spiraea salicifolia, Taraxacum officinale, Viscum album, V. coloratum
Vitamin E	*Angelica polymorpha, A. sinensis, Hippophae rhamnoides, Lactuca raddeana, L. indica, L. sativa, Polygonum aviculare, P. lapidosa, P. manshuriensis, P. vivipara, Triticum vulgare, Viscum album, V. coloratum*
Vitamin G	*Triticum vulgare*
Vitamin K	*Agrimonia eupatoria, A. pilosa, A. viscidula*
Vitamins	*Amaranthus lividus, A. blitum, A. viridis, Ananas comosus, Juncus effusus, Lemmaphyllum microphyllum, Rosa acicularis, R. amygdalifolia, R. davurica, R. koreana, R. laevigata, R. maximowicziana*
Vitex	*Trigonella foenum-graecum*
Vitexicarpin	*Vitex trifolia, V. rotundifolia*
Vitexin	*Crotalaria mucronata, Fagopyrum esculentum, F. sagittatum, Hibiscus sabdariffa, Jatropha podagrica, Lythrum salicaria, Persicaria orientalis, Polygonum orientale, Rumex acetosa, R. acetosella, R. amurensis, R. aquaticus, R. gmelini, R. longifolius, R. maritimus, R. marschallianus, R. stenophyllus, R. thyrsiflorus, Uraria crinita, U. lagopodiodes, Zanthoxylum nitidum*
Vitexin cycloartenol	*Linum stelleroides, L. usitatissimum*
Vitexin-7-glucoside	*Trigonella foenum-graecum*
Vitexin-7-0-glucoside	*Uraria crinita, U. lagopodiodes*

Component	Source
Vitrexin-4-O-xyloside	*Crotalaria mucronata*
Vitricine	*Vitex trifolia, V. rotundifolia*
Volatile carbonyl	*Ranunculus ternatus*
Volatile oil	*Caesalpinia decapetula, Chenopotium ambrosiodes, Cleome spinosa, C. gynandra, C. viscosa, Curcuma pallida, C. phaeocoulis, Duchesnea indica, Eupatorium chinense, E. lindleyanum, E. japonicum, Gardenia angusta., G. jasminoides, Matnolia grandiflora, Matricaria chamomilla, Piper longum, Thymus amurensis, T. disjunctus, T. kitagawianus, T. komarovii, T. przewalskii, T. quinquecostatus*
Volatile phenols	*Ranunculus ternatus*
Vomicine	*Strychnos pierriana*
Vomifliol	*Ilex pubescens*
Wedolactone	*Eclipta erecta*
Wikstroemin	*Wikestroemia indica*
Wilfordine	*Euonymus alatus, E. bungeanus, E. maackii, Tripterygium wilfordii*
Wilsonine	*Cephalotaxus fortunei, C. qinensis, C. oliveri, C. wilsoniana*
Wognoside	*Scutellaria baicalensis, S. grandiflora, S. lanceolaria, S. macrantha, S. rivulararis, S. viscidula*
Wogonin	*Scutellaria baicalensis, S. grandiflora, S. lanceolaria, S. macrantha, S. rivulararis, S. viscidula*
Woodorien	*Woodwardia japonica*
Woodwardic acid	*Woodwardia japonica*
Worenine	*Coptis chinensis, C. japonica, C. teeta*

Xamthophyll	Spinacia oleracea L.
Xanthine	Thea assamica, T. bohea, T. cantoniensis, T. chinensis, T. cochinchinensis, T. sinensis, T. viridis
Xanthanol	Xanthium chinense, X. japonicum, X. mongolicum, X. sibiricum, X. strumarium
Xanthinin	Xanthium chinense, X. japonicum, X. mongolicum, X. sibiricum, X. strumarium
Xanthoagathin	Sesbinia grandiflora
Xanthophyllepoxyl	Avena fatua, Caltha palustris
Xanthoplanine	Zanthoxylum schinifolium
Xanthorin	Cassia occidentalis, C. torosa
Xanthotoxine	Angelica amurensis, A. anomala, A. dahurica, Heracleum lanatum,
Xanthoxylene	Zanthoxylum schinifolium
Xanthoxylin	Sepium sebiferum, S. discolor
Xanthoxylinin	Xanthoxylum piperitum
Xanthumin	Xanthium chinense, X. japonicum, X. mongolicum, X. sibiricum, X. strumarium
Xylopurarin	Pueraria lobata, P. pseudo-hirsuta
Xylose	Juncus communis, Luffa aegyptiaca, L. cylindrica, L. faetida, L. petola, Tamarindus indicus
Y-sitosterol	Ajuga bracteosa
Yatanoside	Brucea javanica, B. sumatrana
Yejuhualactone	Chrysanthemum boreale, C. indicum, C. lavandulaefolium, C. procumbens, C. tripartium
Ylangene	Eugenia aromatica, E. caryophyllata, E. ulmoides
Yuanhuacine	Daphne fortunei, D. genkwa
Yuanhuafine	Daphne fortunei, D. genkwa

Component	Source
Yuanhuatine	*Daphne fortunei, D. genkwa*
Yunnanxana	*Taxus cuspidata, T. chinensis, T. yunnanensis*
Z-guggulsterol	*Commiphora myrrha*
Zanthaline	*Papaver somniferum*
Zeaxanthin	*Lycium barbarum, L. megistocarpum, L. ovatum, L. trewianum, L. turbinatum, Physalis alkekengi, Taraxacum mongolicum, T. sinicum*
Zederone	*Curcuma aromatica, C. kwangsiensis, C. longa, C. zedoaria*
Zedoarin	*Curcuma pallida, C. phaeocoulis*
Zerumbone	*Curcuma zedoaria, C. aromatica, C. kwangsiensis, Zingiber zerumbet*
Zerumbone epoxide	*Zingiber zerumbet*
Zi Yu glucoside I	*Sanguisorba officinalis, S. grandiflora, S. parviflora, S. x tenuifolia*
Zi Yu glucoside II	*Sanguisorba officinalis, S. grandiflora, S. parviflora, S. x tenuifolia*
Zingiberene	*Alpinia oxyphylla, A. speciosa, Curcuma aromatica, C. kwangsiensis, C. pallida, C. phaeocoulis, C. zedoaria, Thymus amurensis, T. disjunctus, T. kitagawianus, T. komarovii, T. przewalskii, T. quinquecostatus, Zingiber officinale*
Zingiberol	*Alpinia oxyphylla, A. speciosa, Zingiber officinale*
Zygadenine	*Veratrum dahuricum, V. maackii, V. nigrum*

Appendix 3

Major Chemical Components and Their Sources in Related North American Medicinal Herbs

Component	Source
1, 8-dihydroxy-anthracene derivatives	*Aloe barbadensis, A. vera*
2-(6'-cinnamoyl) glucosido-	*Silene ocaulis, S. virginica*
2-β-glucuronosyl	*Glycyrrhiza glabra*
2-3,4-dihydroxyphenyl-ethanol	*Jasaminum grandiflorum, J. officinale*
2-methoxy-1,4-naphthoquinone	*Impatiens balsamina*
2-vinyl-4H-1,3-dithin	*Allium sativum, A. fistulosum, A. tuberosum*
22-dihydrospinasterol	*Silene ocaulis, S. virginica*
2,5-dimethoxypara-quinone	*Phragmites australis*
3-0-glucoside	*Platycladus occidentalis*
3-n-pentadecylcatechol	*Rhus radicans, R. glabra, R. toxicodendron*
3-O-β-D-glucuronide	*Chamaenerion angustifolium, Epilobium angustifolium*
4'-0-methylpyridoxine	*Ginkgo biloba*
4-hydroxybenzaldehyde	*Phragmites australis*
4-hydroxycinnamic acid	*Phragmites australis*
Aabrin	*Abrus precatorius*
Ascorbic acid	*Rubus chamaemorus*
Abrin	*Abrus precatorius*
Abrotamine	*Artemisia annua*
Absinthol	*Artemisia absinthium*
Acalyphine	*Acalypha indica*

Acetopenone glucoside	*Veronica officinalis*
Acetyl harpagide	*Scrophularia ningpoensis*
Acetylated alkaloids	*Narcissus tazetta*
Acetylcholine	*Capsella bursa-pastoris, Thlaspi arvense, Viscum album*
Acetylenes	*Bidens tripartita, B. connata*
Acetylenic compounds	*Panax quinquefolium, P. ginseng*
Acetylinic	*Pimpinella anisum*
Achilleine	*Achillea millefolium*
Acidic polysaccharides	*Commiphora myrrha, C. molmol*
Aconitine	*Aconitum napellus, A. carmichaelii*
Acoric acid	*Acorus calamus, A. gramineus*
Acrylic acid	*Ananas comosus*
Actein	*Cimicifuga racemosa, C. foetida*
Actinidine	*Actinidia polygama*
Acutomidine	*Menispermum canadense*
Acutumine	*Menispermum canadense*
Adiantone	*Adiantum capillus-junonis*
Adonitoxin	*Adonis vernalis*
Aescin	*Aesculus hippocastanum*
Afzelechin	*Platycladus occidentalis*
Agaropectin	*Gelidium cartilagineum*
Agarose	*Gelidium cartilagineum*

Component	Source
Aglycone	*Hedera helix*
Agnuside	*Vitex labrusca, V. agnus-castus*
Ailanthone	*Ailanthus altissima, A. glandulosa*
Ajmaline	*Rauvolfia serpentina*
Ajoene	*Allium sativum, A. fistulosum, A. tuberosum*
Albiflorin	*Paeonia albiflora, P. lactiflora*
Albumin	*Papaver somniferum*
Albuminoides	*Dioscorea opposita*
Aldehyde antioxine	*Perilla frutescens*
Aldehydes	*Catharanthus roseus*
Alkaloid lamine	*Dipsacus asper*
Alkaloids	*Ailanthus altissima, A. glandulosa, Alstonia scholaris, Catharanthus roseus, Desmodium gangeticum, Ephedra distachya, Fritillaria verticillata, Justicia adhatoda, Lobelia siphilitica, Medicago sativa, Nicotiana tabacum, Pedicularis palustris, P. canadensis, Physalis alkekengi, P. franchetti, P. pubescene, Picrasma excelsa, Swertia chirata, Valeriana officinalis, Veratrum viride, Verbena officinalis, Viscum album*
Alkanes	*Aspidium filix-mis, Dryopteris filix-mas, D. filix-mas, Euphorbia hirta, Galium verum*
Alkenyl	*Codonopsis pilosula, C. tangshen*
Alkenyl glycoside	*Codonopsis pilosula, C. tangshen*
Allantoin	*Phaseolus vulgaris*
Alliin	*Allium sativum, A. fistulosum, A. tuberosum*
Allocryptopine	*Chelidonium majus*

Allomatatabiol	*Actinidia polygama*
Alnulin	*Alnus crispus, A. glutinosa*
Aloe-emodin	*Cassia angustifolia, C. senna*
Aloeresins	*Aloe barbadensis, A. vera*
Aloesin glycone	*Aloe barbadensis, A. vera*
Aloesone	*Aloe barbadensis, A. vera*
Aloin isobarbaloin	*Aloe barbadensis, A. vera*
Alpha-acid	*Humulus lupulus*
Alpha-bisabolol	*Matricaria chamomilla*
Alpha-masticoresin	*Pistacia lentiscus*
Alpha-phytosterol	*Iris versicolor, Iris pseudacorus*
Alpha-pinene	*Alpinia galanga, Coriandrum sativum, Juniperus rigida, Perilla frutescens, Pinus sylvestris, P. albicaulis, P. contorta, P. mugo, P. palustris, P. strobus, Pistacia lentiscus*
Alpha-spinasterol	*Impatiens pallida, I. capensis*
Alpha-terpineol	*Melaleuca leucadendra*
Alpha-terthienylmethanol	*Eclipta alba, E. prostrata*
Alpha-thujone	*Biota orientalis*
Alpha-tocopherol	*Buxus sempervirens*
Alzarin	*Rubia tinctorum*
Amarogentin	*Swertia chirata*
Amine choline	*Capsella bursa-pastoris, Thlaspi arvense*
Amines	*Crataegus laevigata, C. monongyna, C. oxyacantha*

Component	Source
Amino acids	*Lycopersicon esculentum*
Aminobutyric acid	*Lycopersicon esculentum*
Amygdalin	*Crataegus laevigata, C. monongyna, C. oxyacantha, Cydonia oblonga, Prunus armeniaca, P. armericana*
Anagalline	*Anagalis arvensis*
Anagyrine	*Caulophyllum thalietroides*
Andole alkaloids	*Abrus precatorius*
Anemoni	*Ranunculus ficaria*
Anemonin	*Anemone pulsatilla, A. hepatica, A. patens, A. pulsatilla, Pulsatilla chinensis, Ranunculus occidentalis*
Anemonol	*Pulsatilla chinensis*
Anerhole	*Agastache anethrodora, A. foeniculum, Dictamnus albus, Foeniculum vulgare, Illicium verum, Pimpinella anisum*
Angelicide	*Angelica archangelica*
Anisaldehyde	*Agastache anethrodora, A. foeniculum*
Anthaquinone glycosides	*Aloe barbadensis, A. vera*
Anthemidin	*Matricaria chamomilla*
Anthocyanidin	*Paeonia officinalis*
Anthocyanin	*Abrus precatorius, Hibiscus sabdariffa, Malva sylvestris, M. rotundifolia, Morus alba, Perilla frutescens, Pinus albicaulis, P. contorta, P. mugo, P. palustris, P. strobus, Rubus idaeus, Vitis vinifera*

Anthocyanosides	*Ribes lacustre, R. nigrum, Vaccinium vitis-idaea, V. myrtilloides, V. myrtillus, V. oreophilum, V. macrocarpon*
Anthraquinone	*Cassia angustifolia, C. senna*
Anthraquinone derivatives	*Rubia tinctorum*
Anthraquinone glycosides	*Rhamnus catharticus, R. frangula, R. purshianus*
Anthraquinone compounds	*Rheum officinale, R. palmatum, R. tanguticum*
Anthraquinones	*Galium verum, G. aparine, Lobelia pulmonaria, Polygonum hydropiper, P. bistorta, Rumex crispus, R. obtusifolia, R. acetosella, R. aquaticus, Terminalia chebula*
Antiprotease	*Ribes nigrum*
Apigenin	*Jatropha gospiifolia, Thymus vulgaris, T. capitatus, T. citriodorus, T. praecox, T. pulegiodes, T. serpyllum, T. vulgaris*
Apiin	*Apium graveolens, Cryptotaenia japonica*
Apiole	*Cryptotaenia japonica*
Aplotaxene	*Saussurea lappa*
Apocynein	*Apocynum androsaemifolium*
Apocynin	*Apocynum androsaemifolium*
Arachidic acid	*Thevetai peruviana*
Arbutin	*Pyrola rotundifolia, Vaccinium myrtilloides, V. myrtillus, V. oreophilum*
Arctic acid	*Arctium lappa*
Arctiin	*Areca catechu, Arctium lappa*
Arctiol	*Arctium lappa*
Arecaidine	*Areca catechu*

Component	Source
Arecaine	*Areca catechu*
Arecolidine	*Areca catechu*
Arecoline	*Areca catechu*
Aretylcholine	*Urtica urens*
Arginine	*Citrullus vulgaris, Lemna minor, Phaseolus vulgaris, Raphanus sativus*
Arisllochic acids	*Asarum canadense, Aristolochia clematitis, A. serpentaria*
Artemisinin	*Artemisia annua*
Artocapin	*Morus alba*
Asarinin	*Zanthoxylum americanum*
Asarone	*Acorus calamus, A. gramineus*
Ascaridole	*Chenopodium ambrosiodes*
Ascorbic acid	*Artemisia vulgars, Camellia sinensis, Prunus mume*
Ash	*Oxyria digyna*
Asiaticoside	*Centella asiatica*
Asparagine	*Abutilon indicum, Althaea officinalis, Asparagus officinalis, Astragalus membranaceus, A. americana, Euonymus atropurpureus, Phragmites australis*
Asparagosides	*Asparagus officinalis*
Asparamide	*Phragmites australis*
Asperuloside	*Galium verum, Rubia tinctorum*
Astragalin	*Matteuccia struthiopteris, Paeonia lactiflora*
Astragalosides	*Astragalus membranaceus*

Athocyanin	*Hibiscus rosa-sinensis*
Atractylenolide II	*Atractylodes macrocephala*
Atractylenolide III	*Atractylodes macrocephala*
Atractylol	*Atractylodes macrocephala*
Atropine alkaloids	*Solanum tuberosum*
Aucubin	*Plantago major, Prunella vulgaris, Scrophularia ningpoensis, Vitex labrusca, V. agnus-castus*
Aucuboside	*Veronica officinalis*
Azulene	*Matricaria chamomilla*
Baicalein	*Scutellaria baicalensis, S. macrantha, S. lateriflora*
Baicalin	*Scutellaria baicalensis, S. macrantha, S. lateriflora*
Baldrianic acid	*Sambucus racemosa*
Balsaminasterol	*Impatiens balsamina*
Balsaminones	*Impatiens balsamina*
Barbaloin	*Aloe barbadensis, A. vera*
Bavachin	*Psoralea corylifolia*
Belamcandaquinones A	*Belamcanda chinensis*
Belamcandaquinones B	*Belamcanda chinensis*
Benzaldehyde	*Styrax benzoin*
Benzoic acid	*Arisaema consanguineum, Citrus aurantium, Paeonia lactiflora, P. lactiflora, P. albiflora, P. suffruticosa, Rubus chamaemorus, Scutellaria baicalensis, S. macrantha, S. lateriflora, Styrax benzoin, S. benzoin*
Benzol-aconitine	*Aconitum carmichaelii, A. napellus*

Component	Source
Benzophenone	*Centaurium erythraea*
Benzoquinene	*Lysimachia vulgaris*
Benzoylecgonine	*Erythroxylum coca*
Benzyl benzoate	*Dianthus caryophyllus*
Berbamine	*Berberis vulgaria*
Berberine	*Berberis vulgaria, B. aquifolium, Chelidonium majus, Coptis chinensis, C. groenlandica, C. trifolia, Mahonia aquifolium, Phellodendron amurense, P. chinensis*
Berberubine	*Berberis vulgaria*
Bergapten	*Apium graveolens*
Bergegin	*Cuscuta chinensis, C. epithymum*
Beta-acid	*Humulus lupulus*
Beta-amyrin	*Anemone pulsatilla*
Beta-carotene	*Camellia sinensis, Crocus sativus, Prunus mume*
Beta-D-glucoside	*Melochia tomentosa*
Beta-D-glucopyranosil	*Campanula rotundifolia, C. palustris*
Beta-elemone	*Hedera helix*
Beta-ergosterol	*Impatiens pallida, I. capensis*
Beta-masticoresin	*Pistacia lentiscus*
Beta-methyl-adipic acid	*Menispermum palmatum*
Beta-pachyman	*Poria cocos*
Beta-pachymanase	*Poria cocos*

Beta-pinene	*Juniperus rigida, Melaleuca leucadendra, Pinus sylvestris, P. albicaulis, P. contorta, P. mugo, P. palustris, P. strobus*
Beta-sitosterol	*Alnus crispus, A. glutinosa, Anemone pulsatilla, Cassia angustifolia, C. senna, Corylus avellana, C. cornuta, C. rostrata, C. americana, Lycium barbarum, L. chinense, L. pallidum, Matteuccia struthiopteris, Melochia tomentosa, Paeonia lactiflora, Rehmannia glutinosa, Scutellaria baicalensis, S. macrantha, S. lateriflora, Smilax aristolochifolia, S. china, Ulmus rubra, U. procera, Urtica urens*
Beta-thujone	*Biota orientalis*
Betaine	*Lycium barbarum, L. chinense, L. pallidum, Stachys officinalis*
Betonicine	*Stachys officinalis*
Betulinic acid	*Prunella vulgaris*
Beyerene	*Biota orientalis*
Bilobalide	*Ginkgo biloba*
Bilobetin	*Ginkgo biloba*
Bioflavonoids	*Citrus aurantium, Fagopyrum esculentum, F. tutricum, F. esculentum*
Biotin	*Lycoperiscon esculentum*
bis-norargemonine	*Thalictrum dasycarpum, T. occidentale*
Bishomophinolenic acid	*Pinus albicaulis, P. contorta, P. mugo, P. palustris, P. strobus*
Bonducin	*Caesalpinia ascendens, C. bonducella, C. sylvatica*
Borneol	*Asarum canadense, Biota orientalis, Chrysanthemum vulgare, Elettaria cardamomum, Juniperus communis, J. sabina, J. horizontalis, Kaempferia galanga, Salvia officinalis*
Borneol acetate	*Biota orientalis, Pinus albicaulis, P. contorta, P. mugo, P. palustris, P. strobus*
Bornyl esters	*Valeriana officinalis*

Component	Source
Bornyl isovalerate	*Cnidium monnieri*
Brefeldin A	*Angelica archangelica*
Bromelain	*Ananas comosus*
Bulnesene	*Pogostemon cablin*
Bupleurumol	*Bupleurum falcatum*
Bursine	*Capsella bursa-pastoris, Thlaspi arvense*
Butylphthalide	*Angelica polymorpha*
Cadinene	*Angelica polymorpha*
Caffeic acid	*Aconitum napellus, A.carmichaelii, Campanula rotundifolia, C. palustris, Digitalia purpurea, Eucalyptus citriodora, E. globulus, Glycine max, Hieracium pilosella, Matteuccia struthiopteris, Prunella vulgaris, Tilia cordata, T. europaea, Trifolium pratense, Viscum album*
Caffeic derivatives	*Lycopus virginicus*
Caffeine	*Camellia sinensis, I. aquifolium, I. paraguensis*
Calcium	*Chaenomeles speciosa, Portulaca oleracea, Rubia tinctorum*
Calcium oxalate	*Rheum officinale, R. palmatum, R. tanguticum*
Calcyosin	*Astragalus membranaceus, A. americana*
Calystegins	*Calystegia sepium*
Campesterol	*Matteuccia struthiopteris, Ulmus rubra, U. procera*
Camphene	*Cinnamomum cassia, Cnidium monnieri, Kaempferia galanga, Mentha arvensis, M. haplocalyx*
Camphesterol	*Scutellaria baicalensis, S. macrantha, S. lateriflora*

Camphor	*Achillea millefolium, Biota orientalis, Blumea balsamifera, Chrysanthemum parthenium, C. vulgare, Cinnamomum camphora, C. cassia, Dryobalanops aromatica, Elettaria cardamomum, Salvia officinalis*
Caoutchouc	*Apocynum androsaemifolium, Thevetai peruviana*
Capronic acid	*Aquilegia vulgaris*
Carbohydrate	*Oxyria digyna, Phragmites communis, Glycine max, Solanum aculeatissimum, S. melongena*
Carboxylic acid	*Lobelia inflata*
Cardenolides	*Convallaria majalis, C. sepium*
Cardiac glycosides	*Adonis vernalis, Convallaria majalis, C. sepium, Euonymus atropurpureus, Strophanthus gratus, S. kombe*
Cardienolides	*Euonymus atropurpureus*
Carene	*Kaempferia galanga*
Cariaester	*Solidago canadensis, S. virgaurea*
Carotene	*Avena sativa, Daucus carota, Lycium barbarum, L. pallidum, Lycopersicon esculentum, Plantago major*
Carotenoids	*Hippophae rhamnoides, Calendula officinalis, Citrus aurantium, Ginkgo biloba, Viscum album*
Carthamone	*Carthamus tinctorius*
Carvacrol	*Angelica polymorpha, Thymus vulgaris, T. capitatus, T. citriodorus, T. praecox, T. pulegiodes, T. serpyllum*
Carvone	*Anethum graveoleus, Carum carvi, Elettaria cardamomum, Mentha spicata, M. x piperita, Peucedanum graveolens*
Caryophyllen	*Phytolacca americana*

Component	Source
Caryophyllene	*Elettaria cardamomum, Piper nigrum, P. longum*
Casticin	*Vitex labrusca, V. agnus-castus*
Catalpine	*Bignonia catalpa*
Catalpol	*Scutellaria lateriflora, S. baicalensis, S. macrantha*
Catechins	*Crataegus laevigata, C. monongyna, C. oxyacantha, Leonurus cardiaca, Platycladus occidentalis, Potentilla erecoa, P. tormentilla, Rheum officinale, R. palmatum, R. tanguticum, Uncaria gambir*
Caulophylline	*Caulophyllum thalietroides*
Caulosaponin	*Caulophyllum thalietroides*
Chaconine	*Solanum nigrum*
Chalcones flavonoids	*Glycyrrhiza uralensis*
Chamazulene	*Achillea millefolium*
Charantin	*Momordica charantia*
Chebulic acid	*Terminalia chebula*
Chelerythrine	*Zanthoxylum americanum*
Chelidonic acid	*Veratrum viride*
Chelidonine	*Chelidonium majus*
Chimpahilin	*Pyrola rotundifolia*
Chlorogenic acid	*Digitalia purpurea, Hypericum perforatum, Matteuccia struthiopteris, Plantago major, Strychnos nux-vomica*
Chlorogenic derivatives	*Lycopus virginicus*

Cholesterol — *Ulmus rubra, U. procera*

Choline — *Cannabis sativa, Digitalia purpurea, Euphorbia hirta, Glycine max, Leonurus cardiaca, Potentilla anserina, Stachys officinalis, Taraxacum officinale, Trigonella foenum-graecum, Urtica urens, Viscum album*

Chrysophanic acid — *Aloe barbadensis, A. vera, Polygonum hydropiper, P. bistorta*

Chrysophanol — *Rhamnus catharticus, R. frangula, R. purshianus, Rumex acetosella, R. aquaticus, R. crispus*

Cimicifugin — *Cimicifuga racemosa*

Cimicifugoside — *Cimicifuga foetida*

Cineole — *Achillea millefolium, Alpinia galanga, Artemisia vulgaris, Crocus sativus, Eucalyptus citriodora, E. globulus, Juniperus rigida, Melaleuca leucadendra, Mentha spicata, M. x piperita, Salvia officinalis, Vitex labrusca, V. agnus-castus*

Cinerins — *Chrysanthemum cinerriaefolium, Pyrethrum cinerariifolium*

Cinnamaldehyde — *Cinnamomum zeglanicum, C. cassia*

Cinnamic acid — *Citrus aurantium, Liquidambar orientalis, L. styraciflua, Lycium barbarum, L. pallidum, Populus alba, Rheum officinale, R. palmatum, R. tanguticum, Styrax benzoin*

Cinnamon — *Blumea balsumifera*

Cinnamyl cinnamate — *Liquidambar orientalis, L. styraciflua*

Cinnamylcoaine — *Erythroxylum coca*

Cissampeline — *Cissampelos pareira*

Citral — *Cymbopogon citratus, C. nardus, C. martinii, C. winterianus, Perilla frutescens*

Citric acid — *Aesculus hippocastanum, Ananas comosus, Fragaria vesca, Hibiscus rosa-sinensis, H. sabdariffa, Lonicera caerulea, L. caprifolium, Paris quadrifolia, Prunus mume, Ribes nigrum, Rosa rugosa, R. acicularis, R. canina, R. damascena, R. gallica*

Component	Source
Citronellal	*Cymbopogon citratus, C. nardus, C. martinii, C. winterianus*
Citronellol	*Rosa canina, R. damascena, R. gallica*
Citrullin	*Citrullus vulgaris*
Clerodendrin acacetin	*Clerodendrum trichotomum*
Cnicin	*Carduus benedita, Geum urbanum*
Cocain	*Erythroxylum coca*
Codeine	*Papaver somniferum, P. rhoeaes, P. bracteatum*
Columbamine	*Berberis aquifolium, Mahonia aquifolium*
Colxol	*Phragmites australis*
Condurangogenins	*Marsdenia condurango*
Coniferaldehyde	*Phragmites australis*
Convallatoxol	*Convallaria majalis, C. sepium*
Convalloside	*Convallaria majalis, C. sepium*
Convallotoxin	*Convallaria majalis, C. sepium*
Convolvulin	*Convolvulus jajapa, Ipomoea purga*
Coptisine	*Coptis chinensis, C. groenlandica, C. trifolia*
Cornerin	*Nerium oleander*
Cornic acid	*Cornus canadensis*
Cornine	*Cornus canadensis*
Cortenerin	*Nerium oleander*
Corydaline	*Corydalis yanhusuo, C. solida*

Corydalis   *Corydalis yanhusuo, C. solida*

Corynoxeine   *Uncaria gambir*

Corypalline   *Thalictrum dasycarpum, Thalictrum occidentale*

Coumaric acid   *Eucalyptus citriodora, E. globulus, Populus alba*

Coumarin derivatives   *Aesculus hippocastanum, Tagetes minuta, T. lucida*

Coumarins   *Angelica archangelica, Agrimonia eupatoria, Anethum graveoleus, Anthriscus cerefolium, Apium graveolens, Artemisia dracunculus, Aster tataricus, Cinnamomum zeglanicum, Citrus aurantium, C. aurantium, Crataegus laevigata, C. monongyna, C. oxyacantha, Cryptotaenia japonica, Datura innoxia, D. metel, D. stramonium, Eleutherocossus senticosus, Ferula assa-foetida, Fraxinus ornus, F. americana, F. excelsior, Gelsemium sempervirens, Hieracium pilosella, Hierochloe odorata, Lactuca serriola, Lawsonia inermis, Ledum palustre, Matricaria chamomilla, Medicago sativa, Melilotus alba, M. arvensis, M. officinalis, Menyanthes trifoliata, Peucedanum graveolens Picrasma excelsa, Pimpinella anisum, Stellaria media, Trigonella foenum-graecum, Viburnum opulus, V. prunifolium*

Coumestrol   *Glycine max*

Crataegus acid   *Crataegus laevigata, C. monongyna, C. oxyacantha*

Creosol   *Pimpinella anisum*

Cresols   *Menispermum palmatum*

Crocine glycosides   *Crocus sativus*

Croton oil   *Croton tiglium*

Crude fiber   *Phragmites communis*

Cubebin   *Piper cubeba*

Cucurbitacins   *Anagalis arvensis, Cucurbita maxima*

Component	Source
Curcumin	*Curcuma longa, Santalum album*
Curcuminoids	*Curcuma aromatica*
Cutins	*Quercus robur*
Cyanide	*Prunus mume*
Cyanidin	*Jatropha gospiifolia*
Cyanogenic glycosides	*Acalypha indica, Cydonia oblonga, Hydrangea arborescens, Manihot esculenta, Prunus domestica, P. armeniaca, P. armericana, Sambucus nigra, S. canadensis, S. racemosa*
Cymarin	*Apocynum androsaemifolium*
Cymene	*Juniperus communis, J. sabina, J. sabina, J. horizontalis, Mentha spicata, M. x piperita*
Cynaroside	*Matricaria chamomilla*
Cypripedin	*Cypripedium calceolus, C. pariflorum*
d-borneol	*Dryobalanops aromatica*
d-camphene	*Myristica fragrans*
d-pseudoephedrine	*Ephedra distachya, E. sinica, E. nevadensis*
d-Usnic acid	*Lobelia pulmonaria*
Daidzein	*Glycine max, Pueraria thunbergiana*
Daidzin	*Pueraria lobata*
Dammaranedienol	*Inula helenium*
Daphnetoxin	*Daphne genkwa, D. mezereum*
Dasycarponin	*Thalictrum dasycarpum, T. occidentale*
Dauricine	*Menispermum canadense*

Daurinoline	*Menispermum canadense*
Deanolic acid	*Prunella vulgaris*
Deguelin	*Tephrosia virginiana*
Dehydrofukinone	*Arctium lappa, Tephrosia virginiana*
Dehydrosoyasaponin	*Wisteria floribunda, W. brachybotrys*
Delphinidin	*Anemone pulsatilla*
Delphinidin-3,5-diglucoside	*Aquilegia vulgaris*
Delta-limonene	*Pinus sylvestris*
Delta-linalool	*Asarum canadense*
Dendrolasin	*Santalum album*
Diadzein	*Pueraria lobata*
Diallyl disulfide	*Allium sativum, A. fistulosum, A. tuberosum*
Diallyl trisulfide	*Allium sativum, A. fistulosum, A. tuberosum*
Dianthrone glucosides	*Cassia angustifolia, C. senna*
Diastase	*Rorippa nasturtium-aquaticum*
Dicoumarol	*Melilotus alba, M. arvensis*
Dictamnin	*Dictamnus albus*
Digitoxin	*Digitalia purpurea*
Digoxin	*Digitalia purpurea*
Dihydro-β-agarofuran	*Santalum album*
Dihydrolycopodine	*Lycopodium clavatum, L. obscurum*
Dihydronepetalactol	*Actinidia polygama*

Component	Source
Dilactone	*Poterium officinale, Sanguisorba officinalis*
Dimeric indole alkaloids	*Catharanthus roseus*
Diosgenin	*Aletris farinosa, Dioscorea opposita, Solanum nigrum*
Dipentene	*Cinnamomum cassia*
Disulphides	*Ferula assa-foetida*
Diterpene acids	*Aralia catechu, A. nudicaulis, A. racemosa*
Diterpene jatrophone	*Jatropha gospiifolia*
Diterpenes	*Aster tataricus, Eupatorium perfoliatum, Juniperus rigida*
Dodium	*Chaenomeles speciosa*
Ecdysterones	*Silene ocaulis, S. virginica*
Ecgonine	*Erythroxylum coca*
Ecliptine	*Eclipta alba, E. prostrata*
Eleutherosides	*Eleutherocossus senticosus*
Elixen	*Hedera helix*
Ellagic acids	*Lycopus virginicus*
Ellagitannins	*Potentilla erecoa, P. anserina, P. tormentilla, Punica granatum*
Emetic	*Jatropha gospiifolia*
Emodin	*Aloe barbadensis, A. vera, Rhamnus catharticus, R. frangula, R. purshianus, Rheum officinale, R. palmatum, R. tanguticum, Rhumex acetosella, R. aquaticus, R. crispus*
Emulsin	*Crataegus laevigata, C. monongyna, C. oxyacantha*
Enzymes	*Drosera rotundifolia, Ficus carica*

Ephedrine — *Ephedra distachya, E. sinica*

Epiafzalechin — *Platycladus occidentalis*

Epicatechin — *Platycladus occidentalis*

Epigallocatechin — *Platycladus occidentalis*

Equisitine — *Equisetum arvense*

Esculetin — *Campanula rotundifolia, C. palustris*

Essential oils — *Betula lenta, B. pendala, B. verrucosa, Corylus avellana, C. cornuta, C. rostrata, C. americana, Curcuma aromatica, Dipsacus asper, Heracleum maximum, H. lanatum, H. sphondylium, Hippophae rhamnoides, Jasaminum grandiflorum, J. officinale, Ligusticum scoticum, Ligustrum lucidum, L. vulgare, Prunus persica, Rorippa nasturtium-aquaticum, Tagetes patula, Valeriana officinalis, Veronica officinalis, Zea mays*

Estragol — *Dictamnus albus*

Estragole — *Artemisia dracunculus*

Ethyl cinnamate — *Kaempferia galanga*

Ethyl-p-methoxycinnamate — *Kaempferia galanga*

Eucalptole — *Elettaria cardamomum*

Eucalyptol — *Eucalyptus citriodora, E. globulus*

Eugenal — *Menispermum palmatum*

Eugenol — *Cinnamomum zeglanicum, C. camphora, Dianthus caryophyllus, Eugenia caryophyllata, Geum aleppicum, G. urbanum, Syzygium aromaticum, S. aromaticum*

Eupafolin polysaccarides — *Eupatorium perfoliatum*

Euphorbone — *Euphorbia lathyrus*

Falvonoids — *Chimaphila umbellata, Eriobotrya japonica*

Component	Source
Fat	*Oxyria digyna*
Fatty acids	*Glycine max, Hippophae rhamnoides, Lobelia chinensis, Zea mays*
Fatty oil	*Pimpinella anisum*
Fenchone	*Biota orientalis, Foeniculum vulgare*
Ferment	*Rorippa nasturtium-aquaticum*
Ferric oxide	*Corylus avellana, C. cornuta, C. rostrata, C. americana*
Ferulic acid	*Angelica archangelica, Campanula rotundifolia, C. palustris, Matteaccia struthiopteris, Phragmittes australis*
Flavonols	*Ailanthus altissima*
Fiber	*Glycine max, Plantago asiatica*
Filicin	*Dryopteris filix-mas*
Fix oils	*Apocynum androsaemifolium, Caesalpinia ascendens, C. bonducella, C. sylvatica, Cydonia oblonga, Cyperus esculentus, C. brevifolius, Euphorbia lathyrus, Momordica charantia, Trillium erectum*
Flavanol glycosides	*Rhodiola rosea, Sedum acre*
Flavanones	*Populus alba*
Flavone glycoside	*Alnus crispus, A. glutinosa, Perilla frutescens*
Flavones	*Chamaenerion angustifolium, Cytisus scoparius, Ephedra sinica, Platycladus occidentalis, Populus alba, Swertia chirata*
Flavonlignans	*Carduus marianus*

Flavonoids

Acacia catechu, A. hippocastanum, Agrimonia eupatoria, Althaea officinalis, Anethum graveoleus, Anthriscus cerefolium, Artemisia dracunculus, A. absinthium, Asparagus officinalis, Aster tataricus, Bidens tripartita, B. connata, Biota orientalis, Blumea balsamifera, Bupleurum falcatum, Calendula officinalis, Capsella bursa-pastoris, Carum carvi, Conyza canadensis, Cuscuta chinensis, C. epithymum, Datura innoxia, D. metel, D. stramonium, Drosera rotundifolia, Epilobium parviflorum, Equisetum hyemale, Erigeron canadensis, Eupatorium perfoliatum, Euphorbia hirta, Ficus carica, Fraxinus ornus, F. americana, F. excelsior, F. ornus, Galium verum, Ginkgo biloba, Glechoma hederacea, Glycyrrhiza glabra, Gossypium herbaceum, Hieracium pilosella, Hydrangea arborescens, Hypericum perforatum, Hippophae rhamroides, Inula britannica, I. japonica, Lactuca serriola, Lawsonia inermis, Ledum palustre, Loranthus europaeus, Lycopodium clavatum, L. obscurum, L. annotinum, Lysimachia vulgaris, Matricaria chamomilla, Melia azedarach, Melilotus alba, M. officinalis, M. arvensis. Morus alba, Myrica cerifera, M. penxylvanica, Oenothera biennis, Parietaria judaica, Peucedanum graveolens, Physalis alkekengi, P. franchetti, P. pubescene, Pimpinella anisum, Polygonatum odoratum, P. multiflorum, P. biflorum, Polygonum aviculare, P. viviparum, P. multifolrum, Populus alba, P. balsamifera, P. cardicans, Potentilla arserina, Primula vulgaris, P. veris, Rubus idaeus, Salix alba, S. discolor, Sambucus nigra, S. canadensis, Scrophularia ningpoensis, Stellaria media, Thlaspi arvense, Thuja occidentalis, Tilia cordata, T. europaea, Trifolium incarnatum, Veronica officinalis, Viscum album, Vitex labrusca, V. agnus-castus, Vitis vinifera, Ziziphus jujuba

Flavonoid glycosides

Convallaria majalis, C. sepium, Crataegus laevigata, C. monongyna, C. oxyacantha, Malva sylvestris, M. rotundifolia, Menyanthes trifoliata, Pyrola rotundifolia, Spiraea ulmaria, Silybum marianum

Flavonols

Ailanthus glandulosa, Populus alba

Component	Source
Foetidin	*Ferula assa-foetida*
Folic acid	*Lycopersicon esculentum, Sesamum indicum*
Folinerin	*Nerium oleander*
Formononetin	*Astragalus membranaceus, A. americana*
Forsythin	*Forsythia suspensa*
Frangulin A	*Rhamnus catharticus, R. frangula, R. purshianus*
Frangulin B	*Rhamnus catharticus, R. frangula, R. purshianus*
Fumaric acid	*Salvia officinalis*
Furanoid	*Artemisia tridentata*
Galactan	*Luffa aegyptiaca, L. cylindrica*
Galactose	*Pinus albicaulis, P. contorta, P. mugo, P. palustris, P. strobus*
Galactoside-specific lectin	*Viscum album*
Galangin	*Alpinia galanga*
Galangol	*Alpinia galanga*
Gallic acid	*Betula lenta, B. pendala, B. verrucosa, Cornus florida, Cypripedium calceolus, C. pariflorum, Eucalyptus citriodora, E. globulus, Rheum officinale, R. palmatum, R. tanguticum, Rubus idaeus, Tagetes minuta, T. lucida, Tussilago farfara*
Gallic acid derivatives	*Epilobium parviflorum*
Gallocatechin	*Platycladus occidentalis*
Gallotannin	*Paeonia lactiflora*
Gardenin crocin	*Gardenia angusta*

Gaultherin	*Polygala vulgaris*
Gelsedine	*Gelsemium sempervirens*
Gelsemine	*Gelsemium sempervirens*
Geniposide	*Gardenia angusta*
Gentianindine	*Gentiana lutea, G. macrophylla, G. scabra*
Gentianine	*Gentiana lutea, G. macrophylla, G. scabra*
Gentiopicroside	*Centaurium erythraea*
Gentisic acid	*Eucalyptus citriodora, E. globulus*
Geraniol	*Chenopodium ambrosiodes, Rosa acicularis, R. rugosa, R. canina, R. damascena, R. gallica*
Germacrene B	*Hedera helix*
Gingerol	*Zingiber officinale*
Ginkgetin	*Ginkgo biloba*
Ginkgocide A	*Ginkgo biloba*
Ginkgocide B	*Ginkgo biloba*
Ginkgocide C	*Ginkgo biloba*
Ginkgocide J	*Ginkgo biloba*
Ginkgocide M	*Ginkgo biloba*
Ginsenosides	*Panax quinquefolium, P. ginseng*
Givacoline	*Areca catechu*
Glabridin	*Glycyrrhiza glabra*
Glechomine	*Glechoma hederacea*
Globulin	*Lycopersicon esculentum*

Component	Source
Globuline	*Glycine max*
Glucans	*Sambucus racemosa*
Glucofrangulin A	*Rhamnus catharticus, R. frangula, R. purshianus*
Glucofrangulin B	*Rhamnus catharticus, R. frangula, R. purshianus*
Gluconasturin	*Rorippa nasturtium-aquaticum*
Glucoquinone	*Urtica urens*
Glucose	*Ficus carica, Tagetes minuta, T. lucida*
Glucoside apocynamarin	*Apocynum androsaemifolium*
Glucosinolates	*Raphanus sativus*
Glucuronic acid	*Glycyrrhiza glabra*
Glutamic acid	*Lycopersicon esculentum*
Glutamine	*Heracleum maximum, H. lanatum, H. sphondylium*
Gluten	*Taraxacum officinale*
Glycans	*Eleutherocossus senticosus*
Glycine	*Glycine max, Lycopersicon esculentum*
Glycorrhizin	*Glycyrrhiza glabra*
Glycosides	*Aralia catechu, A. nudicaulis, A. racemosa, Paeonia officinalis, Phragmites australis, Salix alba, S. discolor*
Glycyrrhetinic acid	*Glycyrrhiza glabra*
Glycyrrhizin	*Abrus precatorius*
Gossypetin	*Hibiscus rosa-sinensis, H. sabdariffa*

Gossypin-3-sulfate	*Malva sylvestris, M. rotundifolia*
Gossypol	*Gossypium herbaceum*
Gramme	*Hordeum vulgare*
Guaridine	*Benincase hispida*
Gum	*Commiphora myrrha, C. molmol, C. myrrha, Eugenia caryophyllata, Ferula assa-foetida, Poterium officinale, Sanguisorba officinalis, Syzygium aromaticum, S. aromaticum, Tagetes minuta, T. lucida, Taraxacum officinale*
Guvacine	*Areca catechu*
Harpagide	*Ajuga reptans*
Harpagoside	*Scrophularia ningpoensis*
Havonoids	*Cryptotaenia japonica*
Hederacoside B	*Hedera helix*
Hederacoside C	*Hedera helix*
Hederin	*Hedera helix*
Helenalin	*Inula helenium*
Heliotropin	*Spiraea ulmaria*
Heneicosanic acid	*Areca catechu*
Heraclein	*Heracleum maximum, H. lanatum, H. sphondylium*
Herclavin	*Zanthoxylum americanum*
Herniarin	*Matricaria chamomilla*
Hesperidin	*Hyssopus officinalis*
Hexadecenoic acid	*Myristica fragrans*

Component	Source
Hibiscus acid	*Hibiscus sabdariffa, H. rosa-sinensis, Vaccinium vitis-idaea, V. macrocarpon*
Hirsutine	*Uncaria gambir*
Histamine	*Capsella bursa-pastoris, Jatropha gospiifolia, Thlaspi arvense, Urtica urens*
Histidine	*Raphanus sativus*
Hordenine	*Hordeum vulgare*
Hosenkosides	*Impatiens balsamina*
Humulene	*Elettaria cardamomum, Humulus lupulus*
Hydrangein	*Hydrangea arborescens*
Hydrociannic acid	*Acalypha indica, Aleurites moluceanu*
Hydrocotyline	*Centella asiatica*
Hydrocoumarin	*Melilotus officinalis*
Hydrocyanic acid	*Aquilegia flavescens, Prunus armeniaca, P. armericana*
Hydrojuglone	*Juglans regia*
Hydroquinones	*Chimaphila umbellata, Viburnum opulus, V. prunifolium*
Hydroxybenzoic acid	*Eucalyptus citriodora, E. globulus*
Hydroxycinnamic acid	*Cuscuta chinensis, C. epithymum*
Hydroxycoumarin	*Melilotus officinalis*
Hydroxyphenylethanol glycosides	*Syringa suspensa, S. vulgaris*
Hyoscine	*Datura innoxia, D. metel, D. stramonium, Hyoscyamus niger*
Hyoscyamine	*Datura innoxia, D. metel, D. stramonium, Hyoscyamus niger*
Hypaconitine	*Aconitum napellus, A. carmichaelii*

Hypericin	*Hypericum perforatum*
Hyperoside	*Betula lenta, B. pendala, B. verrucosa, Hypericum perforatum*
Idoflavones	*Glycine max*
Indole alkaloids	*Rauvolfia serpentina, Strychnos nux-vomica*
Inositol	*Cannabis sativa*
Inositol	*Juniperus communis, J. sabina, J. horizontalis, Lonicera caerulea, L. caprifolium, Phaseolus vulgaris*
Insulin-like peptide	*Momordica charantia*
Insulins	*Vaccinium myrtilloides, V. myrtillus, V. oreophilum*
Inulin	*Arctium lappa, Artemisia vulgarts, Dipsacus fullonum, Inula helenium, Solidago virgaurea, S. canadensis, Taraxacum officinale, Tussilago farfara*
Iodine	*Allium sativum, A. fistulosum, A. tuberosum, Artemisia dracunculus, Hedera helix, Laminaria digitata, L. saccharine, L. longicruris*
Iridals	*Belamcanda chinensis*
Iridin	*Belamcanda chinensis*
Iridoid glycosides	*Cornus officinalis, Menyanthes trifoliata, Pedicularis palustris, P. canadensis*
Iridoid valepotriates	*Galium aparine*
Iridoids	*Gelsemium sempervirens, Rubia tinctorum, Swertia chirata, Vaccinium myrtilloides, V. myrtillus, V. oreophilum, Vitex labrusca, V. agnus-castus*
Iridomyrmecin	*Actinidia polygama*
Irigenin	*Belamcanda chinensis*
Irisflorentin	*Belamcanda chinensis*
Iron	*Chaenomeles speciosa*

Component	Source
Iron manganese	*Lemna minor*
Isobetanine	*Phytolacca americana*
Isoborneol	*Cnidium monnieri*
Isocaproic acid	*Ananas comosus*
Isoferulic acid	*Cimicifuga foetida, C. racemosa*
Isoflavones	*Belamcanda chinensis, Cimicifuga racemosa, C. foetida, Medicago sativa*
Isoflavonoids	*Glycyrrhiza uralensis, Pueraria lobata, P. thunbergiana, Wisteria floribunda, W. brachybotrys*
Isofraxin	*Eleutherocossus senticosus*
Isoginkgetin	*Ginkgo biloba*
Isoguvacine	*Areca catechu*
Isomenthone	*Mentha spicata, M. x piperita*
Isoneomatatabiol	*Actinidia polygama*
Isophthalic acid	*Ilex aquifolium, I. paraguensis, Iris versicolor, I. pseudacorus*
Isophytosterol	*Jatropha gospiifolia*
Isoprebetanine	*Phytolacca americana*
Isopsorlin	*Psoralea corylifolia*
Isopulegone	*Mentha pulegium*
Isoquercitin	*Adiantum capillus-junonis*
Isoquercitrin	*Jasaminum grandiflorum, J. officinale*
Isoquiniline	*Coptis chinensis*

Isoquiniline alkaloids	*Coptis groenlandica, C. trifolia, Chelidonium majus, Phellodendron amurense, P. chinensis*
Isorhamnetin	*Typha angustifolia, T. latifolia*
Isorhyncophylline	*Uncaria gambir*
Isosafrole	*Angelica polymorpha*
Isovaltrate	*Valeriana officinalis*
Isovitexin	*Jatropha gospiifolia*
Jacoline	*Senecio vulgaris, S. aureus*
Juglandin	*Juglans regia*
Juglone	*Juglans regia*
Juniperin	*Juniperus communis, J. sabina, J. horizontalis*
Jutrophine	*Jatropha gospiifolia*
Kaempferol	*Aesculus hippocastanum, Loranthus europaeus*
Kaempferol derivatives	*Impatiens balsamina*
Koenigin	*Murraya koenigii*
Kumatakenin	*Astragalus membranaceus, A. americana*
l-asparagine	*Paris quadrifolia*
l-citronellol	*Rosa acicularis, R. rugosa*
l-ephedrine	*Ephedra distachya, E. sinica, E. nevadensis*
l-homostarchydrine	*Medicago sativa*
l-laudanidine	*Thalictrum dasycarpum, T. occidentale*
l-limonene	*Perilla frutescens*
l-phyllandrene	*Piper longum, P. nigrum*

Component	Source
l-zingiberene	*Zingiber officinale*
Laburnine	*Caulophyllum thalietroides*
Lactone protoanemonin	*Pulsatilla vulgaris*
Lactones	*Achillea millefolium, A. archangelica, Atractylodes macrocephala, Hierochloe odorata, Pulsatilla chinensis*
Lactones atractylenolide II	*Atractylodes macrocephala*
Lactones atractylenolide III	*Atractylodes macrocephala*
Lactrile	*Prunus armeniaca*
Lactucerin	*Lactuca serriola*
Lactucopicrin	*Lactuca serriola*
Laetrile	*Prunus mume, P. armericana*
Lanatoside	*Digitalia purpurea*
Lapase	*Aquilegia vulgaris*
Lauric acid	*Areca catechu, Menispermum palmatum, Myristica fragrans*
Lawsone	*Impatiens pallida, I. capensis, Lawsonia inermis, Ledum palustre*
Lecithin	*Glycine max, Polygonum bistorta, P. hydropiper*
Lectins	*Calystegia sepium, Narcissus tazetta, Phytolacca acinosa, Ricinus communis*
Leonticine	*Corydalis yanhusuo, C. solida*
Leonuride	*Leonurus cardiaca*
Leonurin	*Leonurus cardiaca*
Leucine	*Phaseolus vulgaris*

Levulin — *Taraxacum officinale*

Lignans — *Artemisia absinthium, Biota orientalis, Carduus benedita, Carthamus tinctorius, Cinnamomum camphora, Eleutherocossus senticosus, Schisandra chinensis, Sessamum indicum, Syringa suspensa, S. vulgaris, Viscum album,*

Ligustilide — *Angelica archangelica*

Ligustrin — *Syringa suspensa, S. vulgaris*

Lilacin — *Syringa suspensa, S. vulgaris*

Limonene — *Anethum graveoleus, Angelica archangelica, Apium graveolens, Carum carvi, Cinnamomum cassia, Conyza canadensis, Erigeron canadensis, Hyssopus officinalis, Juniperus communis, J. sabina, J. sabina, J. horizontalis, Menispermum palmatum, Mentha arvensis, M. haplocalyx, M. spicata, M. x piperita, Schizonepeta tenuifolia, Tagetes erecta, T. patula*

Limonic acid — *Lonicera caerulea, L. caprifolium*

Linalool — *Acorus calamus, A. gramineus, Alpinia galanga, Conyza canadensis, Coriandrum sativum, Erigeron canadensis, Tagetes patula, T. erecta*

Linalyl acetate — *Tagetes patula*

Linamarin — *Linum usitatissimum*

Linarin — *Linaria vulgaris*

Linoleic acid — *Aleurites moluceanu, Allium sativum, A. fistulosum, A. tuberosum, Angelica polymorpha, Apium graveolens, Aquilegia vulgaris, Chrysanthemum cinerriaefolium, Cucumis sativus, Cucurbita maxima, Linum usitatissimum, Myristica fragrans, Oenothera biennis, Plantago asiatica, P. psyllium, Pyrethrum cinerariifolium, Sesamum indicum, Solanum nigrum, Trigonella foenum-graecum, Vitis vinifera*

Component	Source
Linolenic acid	*Aleurites moluceanu, Linum usitatissimum, Oenothera biennis, Stellaria media, Trigonella foenum-graecum*
Linseed oil	*Linum usitatissimum*
Lipids	*Daucus carota*
Lithospermic acid	*Lithospermum erythrorhizon, L. officinale*
Lobelanidine	*Lobelia inflata*
Lobelidiol	*Lobelia inflata*
Lobeline	*Lobelia inflata*
Loganin	*Strychnos nux-vomica*
Lupulone	*Humulus lupulus*
Lutein-7-primveroside	*Campanula rotundifolia, C. palustris*
Luteolin	*Agrimonia eupatoria, Matricaria chamomilla, Mentha spicata, M. x piperita, Thymus vulgaris*
Luteolin-7-0-beta-D-glucopyranosil	*Campanula rotundifolia, C. palustris*
Lutolin	*Thymus capitatus, T. citriodorus, T. praecox, T. pulegiodes, T. serpyllum, T. vulgaris*
Lycopodine	*Lycopodium clavatum, L. obscurum, L. annotinum*
Lysine	*Lemna minor*
Madasiatic acids	*Centella asiatica*
Madecassic	*Centella asiatica*
Magnesium	*Chaenomeles speciosa*
Magnocurarine	*Magnolia liliflora, M. officinalis*

Magnoflorine	*Berberis aquifolium, Caulophyllum thalietroides, Mahonia aquifolium, Menispermum canadense*
Malic acid	*Ananas comosus, Cornus florida, Fragaria vesca, Hibiscus sabdariffa, H. rosa-sinensis, Lonicera caerulea, L. caprifolium, Prunus mume, Rosa acicularis, R. rugosa, R. canina, R. damascena, R. gallica, Salvia officinalis, Tussilago farfara, Vitis vinifera*
Mallol	*Pinus albicaulis, P. contorta, P. mugo, P. palustris, P. strobus*
Malonic acid	*Aconitum carmichaelii, A. napellus*
Malvin	*Malva sylvestris, M. rotundifolia*
Mannitol	*Rehmannia glutinosa*
Margaric acid	*Areca catechu*
Massoilactone	*Hierochloe odorata*
Mastic acid	*Pistacia lentiscus*
Masticin	*Pistacia lentiscus*
Matatabic acid	*Actinidia polygama*
Matatabiether	*Actinidia polygama*
Matatabistic acid	*Actinidia polygama*
Maysin	*Zea mays*
Meconic acid	*Papaver somniferum*
Meliacins	*Melia azedarach*
Melosatin D	*Melochia tomentosa*
Melovinone	*Melochia tomentosa*
Menthol	*Achillea millefolium, Mentha arvensis, M. haplocalyx, M. pulegium, M. spicata, M. x piperita*

Component	Source
Menthone	*Mentha arvensis, M. haplocalyx, M. spicata, M. x piperita, Schizonepeta tenuifolia*
Menthyl acetate	*Mentha arvensis, M. haplocalyx*
Mesaconitine	*Aconitum carmichaelii, A. napellus*
Mesoinositol	*Clerodendrum trichotomum*
Methlxanthines	*Camellia sinensis*
Methyl chavicol	*Artemisia dracunculus*
Methyl salicylate	*Chimaphila umbellata, Dianthus caryophyllus, Erythroxylum coca, Gaultheria procumbens*
Methyl chavicol	*Illicium verum*
Methyl salicylate	*Betula lenta, B. pendala, B. verrucosa, Polygala senega*
Methyl-n-propyl ketone	*Ananas comosus*
Methylamine	*Acorus calamus, A. gramineus*
Methylchavicol	*Agastache anethrodora, A. foeniculum*
Methylcytisine	*Caulophyllum thalietroides, Cimicifuga racemosa*
Mezerein	*Daphne genkwa, D. mezereum*
Mineral elements	*Oxyria digyna*
Minerals	*Cardamine pratensis, Glycine max, Phragmites communis, Smilax china, S. aristolochiifolia, Solanum tuberosum*
Monoterpene glycosides	*Vitis vinifera*
Monoterpenes	*Aster tataricus*
Monoterpenoid glycosides	*Paeonia suffruticosa, P. lactiflora, P. albiflora*
Morindin	*Morinda didyma, M. fistulosa, M. punctata*

Mormordicine	*Momordica charantia*
Mormordin	*Momordica charantia*
Morphine	*Papaver somniferum, P. rhoeas, P. bracteatum*
Mucilage	*Abutilon indicum, Acacia catechu, Acorus calamus, A. gramineus, Adiantum capillus-junonis, Althaea officinalis, Arctium lappa, Calendula officinalis, Cassia angustifolia, C. senna, Castanea sativa, Celtis australis, Chamaenerion angustifolium, Chrysanthemum parthenium, Cinnamomum zeglanicum, Citrus aurantium, Cydonia oblonga, Daphne genkwa, D. mezereum, Epilobium angustifolium, Gelidium cartilagineum, Glycyrrhiza glabra, Hibiscus rosa-sinensis, H. sabdariffa, Inula helenium, Linaria vulgaris, Linum usitatissimum, Lobelia chinensis, Malva sylvestris, M. rotundifolia, Orchis mascula, Papaver somniferum, Phytolacca acinosa, Plantago major, P. asiatica, P. psyllium, Polygala vulgaris, P. multifolrum, Polygonum, P. aviculare, P. viviparum, Portulaca oleracea, Thuja occidentalis, T. cordata, T. europaea, Tussilago farfara, Ulmus rubra, U. procera, Verbena officinalis, Viola tricolor, Zea mays, Ziziphus jujuba.*
Mustard oil	*Sinapis alba*
Mustard-oil glycosides	*Brassica alba, B. juncea*
Myoinositol	*Paeonia lactiflora*
Myrcene	*Chenopodium ambrosioides, Mentha spicata, M. x piperita, Juniperus rigida*
Myricetin	*Platycladus occidentalis*
Myricyl alcohol	*Iris versicolor, I. pseudacorus*
Myristic acid	*Apium graveolens, Areca catechu, Hibiscus rosa-sinensis, H. sabdariffa, L. clavatum, L. obscurum, Menispermum palmatum*
Myristicin	*Cryptotaenia japonica, Myristica fragrans*

Component	Source
Myristoleic acid	*Apium graveolens*
Myrtocyan	*Vaccinium myrtilloides, V. myrtillus, V. oreophilum*
N'-desmethyldauricine	*Menispermum canadense*
n-butyldenephthalide	*Angelica archangelica*
n-coumaric	*Campanula rotundifolia, C. palustris*
n-coumaric acid	*Matricaria chamomilla*
n-dodecanol	*Angelica polymorpha*
n-methyl anabasine	*Sedum acre*
n-pentadecane	*Kaempferia galanga*
n-tetradecanol	*Angelica polymorpha*
n-*trans*-coumaroyltyramine	*Tribulus terrestsis*
n-*trans*-feruloyltyramine	*Tribulus terrestsis*
Napelline	*Aconitum napellus, A. carmichaelii*
Naphthaquinones	*Drosera rotundifolia, Lawsonia inermis, Ledum palustre*
Naphthalene glycosides	*Cassia senna, C. angustifolia*
Narcotine	*Papaver somniferum*
Neo-nepetalactone	*Actinidia polygama*
Neoherculin	*Zanthoxylum americanum*
Neoline	*Aconitum napellus, A. carmichaelii*
Neomatabiol	*Actinidia polygama*
Neothujic acid	*Platycladus occidentalis*

Nepodin	*Rumex crispus*
Neriin	*Nerium oleander*
Niacin	*Camellia sinensis, Rorippa nasturtium-aquaticum*
Nicotine	*Equisetum arvense, E. hyemale, Erythroxylum coca, Nicotiana tabacum*
Nicotinic acid	*Daucus carota, Lycopersicon esculentum, Trigonella foenum-graecum, Uncaria gambir*
Nigerine (N,N-dimethyltryptamine)	*Mimosa pudica, M. hostilis*
Nitryl-glycoside	*Aquilegia vulgaris*
Nonadecanoid	*Areca catechu*
Norargemonine	*Thalictrum dasycarpum, T. occidentale*
Nupharine	*Nymphaea alba*
Nymphaeine	*Nymphaea alba*
Ocimene	*Tagetes patula*
Octacosanol	*Melochia tomentosa*
Octadecatetraenic acid	*Stellaria media*
Oleandomycin	*Nerium oleander*
Oleandrin	*Nerium oleander*
Oleic acid	*Aleurites moluceanu, Apium graveolens, Aquilegia vulgaris, Centella asiatica, Cucumis sativus, C. maxima, Linum usitatissimum, Myristica fragrans, Plantago asiatica, P. psyllium, Sesamum indicum, Trigonella foenum-graecum, Vitis vinifera*
Oleo-resins	*Aspidium filix-mis, Dryopteris filix-mas, Catharanthus roseus, Matteuccia struthiopteris*
Oleostearic acid	*Aleurites moluceanu*
Oligopeptides	*Prunus mume*

Component	Source
Organic acids	*Rubus coreanus, R. fruiticosus, Sorbus aucuparia*
Oripavine	*Papaver rhoeaes, P. bracteatum*
Oxalates	*Rumex crispus, R. acetosella, R. aquaticus, R. obtusifolia*
Oxalic acid	*Salvia officinalis*
Oxyberberine	*Berberis aquifolium, Mahonia aquifolium*
Oxylenzoic acid	*Bignonia catalpa*
p-coumaric acid	*Matteuccia struthiopteris, Trifolium pratense, T. pratense*
p-hydroxybenzoic	*Matteuccia struthiopteris*
p-methoxystyrene	*Kaempferia galanga*
Pachymic acid	*Poria cocos*
Paenoiflorin	*Paeonia albiflora*
Paeonine	*Paeonia officinalis*
Paeonol	*Paeonia lactiflora*
Paliloleic acid	*Apium graveolens*
Palmitic acid	*Acorus calamus, A. gramineus, Angelica polymorpha, Apium graveolens, Aquilegia vulgaris, Areca catechu, Chrysanthemum cinerriaefolium, Cucumis sativus, Hibiscus sabdariffa, H. rosa-sinensis, Matteaccia struthiopteris, Menispermum palmatum, Paeonia lactiflora, Plantago asiatica, P. psyllium, Pyrethrum cinerariifolium, Solanum nigrum, Thevetai peruviana, Trigonella foenum-graecum,Vitis vinifera*
Panaxosides	*Panax quinquefolium, P. ginseng*
Pantothenic acid	*Lycopersicon esculentum*
Papaverine	*Papaver somniferum*

Paradin	*Paris quadrifolia*
Paridol	*Paris quadrifolia*
Parinaric acid	*Impatiens balsamina*
Paristyphnine	*Paris quadrifolia*
Pectin	*Althaea officinalis, Catharanthus roseus, Cydonia oblonga, Eleutherocossus senticosus, Fragaria vesca, Paris quadrifolia, Rhamnus catharticus, R. frangula, R. purshianus, Ribes nigrum, Rosa acicularis, R. rugosa, R. canina, R. damascena, R. gallica, Rubus idaeus, Sorbus aucuparia, Tagetes minuta, T. lucida, Crataegus laevigata, C. monongyna, C. oxyacantha*
Peimine	*Fritillaria verticillata*
Pelargonidin glycosides	*Anemone pulsatilla*
Pelletierene alkaloids	*Punica granatum*
Pentacosane	*Typha angustifolia, T. latifolia*
Pentagallotannin	*Paeonia lactiflora*
Pentagalloyl glucoside	*Paeonia lactiflora, P. albiflora*
Pentane	*Artemisia tridentata*
Pentoses	*Ulmus rubra, U. procera*
Peregrinine	*Paeonia officinalis*
Perillanin chloride	*Perilla frutescens*
Perlolyrin	*Codonopsis pilosula, C. tar-gshen*
Petroselaidic acid	*Apium graveolens*
Petroselinic acid	*Apium graveolens*
Phanol	*Rumex obtusifolia*

Component	Source
Phelandrine	*Artemisia dracunculus*
Phenol	*Menispermum palmatum, Rumex acetosella, R. aquaticus, Sesamum indicum*
Phenolic acid	*Artemisia absinthium, Equisetum hyemale, E. hirta, Inula britannica, I. japonica, Lycopus virginicus, Menyanthes trifoliata, Oenothera biennis, Polygala senega, Salix alba, S. discolor, Sambucus nigra, S. canadensis, Scrophularia ningpoensis*
Phenolic glycosides	*Geum aleppicum, G. urbanum, Populus balsamifera, P. candicans, Trifolium pratense*
Phenolic flavonols	*Rhamnus catharticus, R. frangula, R. purshianus*
Phenols	*Laminaria digitata, L. saccharine, L. longicruris, Myrica cerifera, M. penxylvanica, Primula vulgaris, P. veris*
Phenyl ethyl	*Rorippa nasturtium-aquaticum*
Phenyl-propanoid glycosides	*Pedicularis palustris, P. canadensis*
Phenylprepyl cinnamate	*Liquidambar orientalis, L. styraciflua*
Phenylpropanoids	*Aster tataricus, Eleutherocossus senticosus*
Phlobaphenes	*Alnus crispus, A. glutinosa, Potentilla erecoa, P. tormentilla*
Phthalides	*Cryptotaenia japonica, Ligusticum scoticum, L. lucidum, L. vulgare*
Phyllandrene	*Angelica archangelica, Cinnamomum cassia*
Physalin	*Lycium barbarum, L. pallidum, Physalis alkekengi*
Physcion	*Rhamnus catharticus, R. frangula, R. purshianus, Rumex acetosella, R. aquaticus, R. obtusifolia*
Phytic acid	*Glycine max*
Phytin	*Daucus carota*

Phytoene	*Crocus sativus*
Phytofluene	*Crocus sativus*
Phytosterols	*Calendula officinalis, Cannabis sativa, Marsdenia condurango, Physalis franchetti, P. pubescene, Rehmannia glutinosa, Schisandra chinensis, Smilax china, S. aristolochiifolia, Tilia cordata, T. europaea, Typha angustifolia, T. latifolia*
Pigments	*Tussilago farfara*
Pinecamphene	*Hyssopus officinalis*
Pinene	*Asarum canadense, Cinnamomum cassia, Cnidium monnieri, Crocus sativus, Cryptotaenia japonica, Elettaria cardamomum, Hyssopus officinalis, Juniperus communis, J. sabina, J. sabina, J. horizontalis, Mentha spicata, M. x piperita, Menispermum palmatum*
Piperidine	*Piper cubeba*
Piperine	*Piper nigrum, P. longum*
Plastoquinones	*Castanea sative*
Podophyllotoxin type lignans	*Platycladus occidentalis*
Polyacetylenes	*Aster tataricus, Carduus benedita*
Polygalic acid	*Solidago canadensis, S. virgaurea*
Polygalitol	*Polygala senega*
Polynes	*Oplopanax horridus*
Polypeptides	*Capsella bursa-pastoris, Thlaspi arvense*
Polyphenolic acids	*Galium aparine*
Polyphenols	*Camellia sinensis, Humulus lupulus, Lycopodium annotinum, L. clavatum, L. obscurum, Polygonum multifolrum, P. aviculare, P. viviparum*

Component	Source
Polysaccharides	*Agrimonia eupatoria, Carthamus tinctorius, Carum carvi, Codonopsis pilosula, C. tangshen, Eleutherococcus senticosus, Gelidium cartilagineum, Laminaria digitata, L. saccharine, L. longicruris, Lobelia pulmonaria, Panax ginseng, P. quinquefolium, Phragmites australis, Prunus mume, Urtica urens*
Populin	*Populus tremuloides*
Porphyrins	*Medicago sativa*
Potassium	*Chaenomeles speciosa*
Proanthocyanidins	*Crataegus laevigata, C. monongyna, C. oxyacantha*
Procyanidins	*Crataegus laevigata, C. monongyna, C. oxyacantha, Platycladus occidentalis*
Progestoron	*Dioscorea opposita*
Protein	*Aleurites moluceanu, Avena sativa, Coriandrum sativum, Glycine max, Oxyria digyna, Perilla frutescens, Phragmites communis, P. australis, Pimpinella anisum, Piper longum, P. nigrum, Sesamum indicum, Solanum aculeatissimum, S. melongena, Trigonella foenum-graecum, Viscum album*
Protoalkaloids	*Ephedra sinica*
Protoalnulin	*Alnus crispus, A. glutinosa*
Protoanemonin	*Clematis vitalba, C. virginiana, Pulsatilla chinensis*
Protoanemonoid compound	*Actaea rubra, A. alba*
Protoberberine alkaloids	*Berberis aquifolium, Mahonia aquifolium*
Protocatechetic acid	*Bignonia catalpa*
Protocatechuic	*Matteuccia struthiopteris*
Protopine	*Corydalis yanhusuo, C. solida*

Provitamin A	*Hippophae rhamnoides, Taraxacum officinale*
Prussic acid	*Prunus armeniaca*
Pseudocuramine	*Nerium oleander*
Pseudoephedrine	*Ephedra nevadensis*
Pseudohypericin	*Hypericum perforatum*
Psoralen	*Heracleum maximum, H. lanatum, H. sphondylium*
Psoraline	*Psoralea corylifolia*
Psyllic acid	*Lycium barbarum, L. pallidum*
Puerarin	*Pueraria thunbergiana, P. lobata*
Pulegone	*Menispermum palmatum, Mentha pulegium*
Pulin	*Taraxacum officinale*
Pulsatoside	*Pulsatilla chinensis*
Purgative oil	*Jatropha gospiifolia*
Purine	*Camellia sinensis*
Purpurea-glycosides A	*Digitalia purpurea*
Purpurea-glycosides B	*Digitalia purpurea*
Purpurin	*Rubia tinctorum*
Pyrethrins	*Chrysanthemum cinerriaefolium, Pyrethrum cinerariifolium*
Pyrogallol	*Leonurus cardiaca*
Pyrrolizidine	*Senecio vulgaris, S. aureus*
Pyrrolizidine alkaloids	*Tussiiago farfara*
Quassin	*Ailanthus altissima, A. glandulosa, Picrasma excelsa*

Component	Source
Quassinoids	*Ailanthus altissima, A. glandulosa, Picrasma excelsa*
Querbrachitol	*Acalypha indica*
Quercetin	*Aesculus hippocastanum, Cornus canadensis, Crataegus laevigata, C. monongyna, C. oxyacantha, Hippophae rhamnoides, Hypericum perforatum, Impatiens balsamina, Loranthus europaeus, Urtica urens*
Ranunculin	*Anemone pulsatilla*
Raphanol	*Rorippa nasturtium-aquaticum*
Raphanolide	*Rorippa nasturtium-aquaticum*
Rehmannin	*Rehmannia glutinosa*
Rescinnamine	*Rauvolfia serpentina*
Reserpine	*Alstonia scholaris, Rauvolfia serpentina*
Resins	*Acacia catechu, Acalypha indica, Actaea rubra, A. alba, Aesculus hippocastanum, Aletris farinosa, Alnus crispus, A. glutinosa, Anemone pulsatilla, Arctium lappa, Arenaria rubra, Artemisia vulgarts, Aspidium filix-mis, Berberis vulgaria, Calendula officinalis, Cimicifuga foetida, C. racemosa, Commiphora myrrha, C. molmol, Convolvulus jajapa, Cornus florida, Curcuma longa, Dryopteris filix-mas, Eleutherocossus senticosus, Eriobotrya japonica, Euphorbia lathyrus, Ferula assa-foetida, Glechoma hederacea, Humulus lupulus, Inula helenium, Jatropha gospiifolia, Juniperus rigida, J. communis, J. sabina, J. horizontalis, Lycopodium clavatum, L. obscurum, Melilotus alba, M. arvensis, M. officinalis, Myrica cerifera, M. penxylvanica, Nymphaea alba, Papaver somniferum, Phytolacca acinosa, Pinus albicaulis, P. contorta, P. mugo, P. palustris, P. strobus, Piper cubeba, Rubia tinctorum, Saponaria officinalis, Saussurea lappa, Senecio aureus, S. vulgaris Smilax china,*

Resins	*S. aristolochiifolia, Tagetes minuta, T. lucida, Taxus x media, T. brevifolia, Terminalia chebula, Thevetai peruviana, Trillium erectum, Viburnum opulus, V. prunifolium, Viscum album, Zanthoxylum americanum, Zea mays*
Retinol	*Oxyria digyna*
Rhamnetin-3-0-beta-D-galactoside	*Campanula rotundifolia, C. palustris*
Rhein	*Cassia angustifolia, C. senna, Rheum officinale, R. palmatum, R. tanguticum*
Rhein anthrones	*Rheum officinale, R. palmatum, R. tanguticum*
Rhodioloside	*Rhodiola rosea*
Rhyncophylline	*Uncaria gambir*
Riboflavin	*Lycopersicon esculentum*
Ricin	*Ricinus communis*
Ricinine	*Ricinus communis*
Ricinoleic acid	*Ricinus communis*
Rosagenin	*Nerium oleander*
Rotenone	*Tephrosia virginiana*
Ruberythric acid	*Rubia tinctorum*
Rutin	*Adiantum capillus-juronis, Artemisia dracunculus, Crataegus laevigata, C. monongyna, C. oxyacantha, Fagopyrum tataricum, F. esculentum, Hypericum perforatum, Nerium oleander, Sambucus racemosa*
Sabinen	*Biota orientalis*
Sabinene	*Elettaria cardamomum, Juniperus rigida*
Safranal	*Crocus sativus*
Safrole	*Angelica polymorpha, Cinnamomum camphora, Illicium verum, Myristica fragrans*

Component	Source
Saikosides	*Bupleurum falcatum*
Salicarin	*Lythrum salicaria*
Salicin	*Populus tremuloides, Salix alba, Salix discolor*
Salicortin	*Salix alba, S. discolor*
Salicylates	*Cimicifuga racemosa, Spiraea ulmaria*
Salicylic acid	*Cimicifuga foetida, Ilex aquifolium, I. paraguensis, Iris versicolor, I. pseudacorus, Lonicera japonica, L. caerulea, L. caprifolium, Rubus chamaemorus, Solidago virgaurea, S. canadensis, Trifolium pratense, T. incarnatum*
Salicylic compounds	*Viola tricolor*
Salvin	*Salvia officinalis*
Sanguisorbic acid	*Poterium officinale, Sanguisorba officinalis*
Santalols	*Santalum album*
Santhophylls	*Bidens tripartita*
Sapogenin	*Dioscorea opposita, Saponaria officinalis*
Saponins	*Acorus calamus, A. gramineus Aesculus hippocastanum, Anagalis arvensis, Anemone pulsatilla, Aster tataricus, A. tataricus, Avena sativa, Benincase hispida, Betula lenta, B. pendala, B. verrucosa, Calendula officinalis, Catharanthus roseus, Centella asiatica, Chenopodium ambrosiodes, Clematis vitalba, C. virginiana, Cornus officinalis, C. florida, Digitalia purpurea, Eclipta alba, E. prostrata, Entada phaseoloides, Ephedra sinica, Glechoma hederacea, Glycine max, Glycyrrhiza glabra, Hydrangea arborescens, Impatiens balsamina, Leonurus cardiaca, Lysimachia vulgaris, Polygonatum odoratum, P. multiflorum, P. biflorum, Prunella vulgaris, Ranunculus ficaria, Saponaria officinalis,*

Saponins	*Solidago virgaurea, S. canadensis, Thymus vulgaris, T. capitatus, T. citriodorus, T. praecox, T. pulegiodes, T. serpyllum, T. vulgaris, Tribulus terrestsis, Trillium erectum, Viola tricolor, Zea mays, Ziziphus jujuba*
Sarsapic acid	*Smilax china, S. aristolochiifolia*
Saussarine	*Saussurea lappa*
Scabioside	*Dipsacus fullonum*
Sciadopitysin	*Ginkgo biloba*
Scoparoside	*Cytisus scoparius*
Scopoletin	*Picrasma excelsa*
Scordinins	*Allium sativum, A. fistulosum, A. tuberosum*
Scutellarin	*Scutellaria baicalensis, S. macrantha, S. lateriflora*
Secoiridoid glucosides	*Centaurium erythraea*
Secoiridoids	*Erythrina centaurium*
Sedacrine	*Sedum acre*
Sedacryptine	*Sedum acre*
Sedinine	*Sedum acre*
Selenium	*Allium sativum, A. fistulosum, A. tuberosum*
Senecionine	*Senecio vulgaris, S. aureus*
Seneciphyline	*Senecio vulgaris, S. aureus*
Sennoside A	*Cassia angustifolia, C. senna*
Sennoside B	*Cassia angustifolia C. senna*
Serine	*Lycopersicon esculentum*

Component	Source
Serotonin	*Phragmites australis, Urtica urens*
Sesquiterpene	*Achillea millefolium, Acorus calamus, A. gramineus, Angelica polymorpha, Aster tataricus, Biota orientalis, Catharanthus roseus, Eugenia caryophyllata, Glechoma hederacea, Oplopanax horridus, Pyrola rotundifolia, Saussurea lappa, S. lappa, Syzygium aromaticum, S. aromaticum*
Sesquiterpenic alcohol	*Angelica polymorpha*
Sesquiterpene hydrocarbons	*Santalum album*
Sesquiterpene lactones	*Alpinia galanga, Blumea balsumifera, Carduus benedita, Chrysanthemum cinerriaefolium, C. parthenium, Eupatorium perfoliatum, Geum aleppicum, G. urbanum, Lactuca serriola Phellodendron amurense, P. chinensis, Pyrethrum, cinerariifolium*
Sesquiterpenes patchoulol	*Pogostemon cablin*
Sesquiterpenoid	*Ferula assa-foetida*
Shikimic acid	*Euphorbia hirta*
Shishonin	*Perilla frutescens*
Shogaols	*Zingiber officinale*
Silibinin	*Silybum marianum*
Silicates	*Equisetum arvense, E. hyemale*
Silicic acid	*Equisetum hyemale, E. arvense, Polygonum aviculare, P. viviparum, P. multifolrum, Trifolium pratense*
Silymarin	*Silybum marianum*
Silymarin polyacetylenes	*Carduus marianus*
Sitosterol	*Centella asiatica, Epilobium parviflorum, Solanum nigrum, Tribulus terrestsis*

Solamargine	*Solanum nigrum*
Solanine	*Solanum nigrum*
Solanocarpine	*Solanum xanthocarpum*
Solasodine	*Solanum nigrum, S. dulcamara*
Soldulcamaridine	*Solanum dulcamara*
Sorbitol	*Lonicera caerulea, L. caprifolium*
Sparteine	*Chelidonium majus, Cytisus scoparius*
Sphondin	*Heracleum maximum. H. lanatum, H. sphondylium*
Spinasterol	*Silene ocaulis, S. virginica*
Stachydrine	*Medicago sativa, Stachys officinalis*
Starch	*Smilax china, S. aristolochifolia*
Stearic acid	*Apium graveolens, Areca catechu, Cucumis sativus, Linum usitatissimum, Myristica fragrans, Solanum nigrum, Thevetai peruviana, Vitis vinifera*
Sterins	*Codonopsis pilosula, Codonopsis tangshen*
Steroidal	*Veratrum viride*
Steroidal alkaloids	*Buxus sempervirens, Solanum xanthocarpum, S. dulcamara*
Steroidal saponins	*Aletris farinosa, Caulophyllum thalietroides, Dioscorea opposita, D. batatus, Smilax china, S. aristolochifolia, Solanum dulcamara*
Steroids	*Humulus lupulus*

Component	Source
Sterols	*Achillea millefolium, Astragalus membranaceus, A. americana, Bidens tripartita, B. connata, Cimicifuga racemosa, Equisetum hyemale, Euonymus atropurpureus, Eupatorium perfoliatum, Lawsonia inermis, Ledum palustre, Linaria vulgaris, Menyanthes trifoliata, Phellodendron amurense, P. chinensis, Physalis alkekengi, Polygala senega, Pueraria thunbergiana, P. lobata, Sambucus nigra, S. canadensis, Saponaria officinalis, Tussilago farfara*
Sticinic acid	*Lobelia chinensis*
Stictic acid	*Lobelia chinensis*
Stigmast-4-3-one	*Urtica urens*
Stigmasterol	*Matteuccia struthiopteris, Melochia tomentosa, Rehmannia glutinosa, Scutellaria baicalensis, S. macrantha, S. lateriflora, Urtica urens*
Strychnine	*Strychnos nux-vomica*
Suberins	*Quercus robur*
Sugars	*Linaria vulgaris, Papaver somniferum, Phaseolus vulgaris, Ziziphus jujuba*
Sumaresinolic acid esters	*Styrax benzoin*
Swertiamarin	*Centaurium erythraea*
Syringaldehyde	*Phragmites australis*
Syringic acid	*Eucalyptus citriodora, E. globulus*
Tagetone	*Tagetes patula*
Tangshenoside	*Codonopsis pilosula, C. tangshen*
Tannic acid	*Cornus florida, Cypripedium calceolus, C. pariflorum, Glycyrrhiza glabra, Rumex acetosella, R. aquaticus, R. obtusifolia*

Tannin

Adiantum capillus-junonis, Aesculus hippocastanum, Aleurites moluceanu, Anemone pulsatilla, Artemisia vulgaris, Dodonaea viscosa, Hydrangea arborescens, Jatropha gospiifolia, Juglans regia, Lythrum salicaria, Rheum officinale, R. palmatum, R. tanguticum, Salvia officinalis, Trillium erectum

Tannins

Abutilon indicum, Acacia catechu, Acalypha indica, Achillea millefolium, Agrimonia eupatoria, Ailanthus altissima, A. glandulosa, Alnus crispus, A. glutinosa, Amaranthus hypochondriacus, Aragalis arvensis, Anemone hepatica, A. patens, A. pulsatilla, Aralia catechu, A. nudicaulis, A. racemosa, Arctium lappa, Areca catechu, Aristolochia clematitis, A. serpentaria, Artemisia absinthium, A. dracunculus, Berberis vulgaria, Betula lenta, B. pendala, B. verrucosa, Bidens tripartita, B. connata, Caesalpinia ascendens, C. bonducella, C. sylvatica, Castanea sative, Catharanthus roseus, Celtis australis, Chamaenerion angustifolium, Chimaphila umbellata, Chrysanthemum partherium, Cimicifuga foetida, Cinnamomum zeglanicum, Conyza canadensis, Cornus officinalis, C. canadensis, C. florida, Corylus avellana, C. cornuta, C. rostrata, C. americana, C. laevigata, C. monongyna, C. oxyacantha, Cydonia oblonga, C. oblonga, Daphne genkwa, D. mezereum, Datura innoxia, D. metel, D. stramonium, Ephedra sinica, Epilobium angustifolium, Erigeron canadensis, Eugenia caryophyllata, Euonymus atropurpureus, Fragaria vesca, Fraxinus ornus, Galium aparine, Gelsemium sempervirens, Geranium macrorrhizum, G. robertianum, G. maculatum, Geum aleppicum, G. urbanum, Glechoma hederacea, Gnaphalium uliginosum, Hedera helix, Hippophae rhamnoides, Humulus lupulus, Hyssopus officinalis, Lawsonia inermis, Ledum palustre, Linaria vulgaris, Lobelia chinensis, Lonicera japonica, L. caerulea, L. caprifolium, Lysimachia vulgaris, Malva sylvestris, M. rotundifolia, Matricaria chamomilla, Melia azedarach, Melilotus alba, M. arvensis, M. officinalis, Menyanthes trifoliata, Murraya koenigii, Myrica cerifera, M. penxylvanica,

Component	Source
Tannins	*Oenothera biennis, Paeonia officinalis, Parietaria judaica, Pinus albicaulis, P. contorta, P. mugo, P. palustris, P. strobus, Pistacia lentiscus, Plantago major, Polygonum aviculare, P. viviparum, P. multifolrum, Populus tremuloides, Potentilla erecoa, P. tormentilla, Poterium officinale, Primula vulgaris, P. veris, Prunella vulgaris, Prunus armericana, Pulsatilla vulgaris, Quercus robur, Ranunculus ficaria, Ribes nigrum, Rubus coreanus, R. fruiticosus, R. idaeus, Rumex crispus, Salix alba, S. discolor, Sambucus racemosa, Sanguisorba officinalis, Scutellaria baicalensis, S. macrantha, S. lateriflora, Solanum dulcamara, Solidago canadensis, S. virgaurea, Sorbus aucuparia, Spiraea ulmaria, Stachys officinalis, Syzygium aromaticum, Tagetes minuta, T. lucida, Taraxacum officinale, Terminalia chebula, Thuja occidentalis, Thymus vulgaris, T. capitatus, T. citriodorus, T. praecox, T. pulegiodes, T. serpyllum, Tilia cordata, T. europaea, Tribulus terrestsi, Trifolium pratense, Tussilago farfara, Ulmus rubra, U. procera, Vaccinium myrtilloides, V. myrtillus, V. oreophilum, Verbena officinalis, Veronica officinalis, Viburnum opulus, V. prunifolium, Viola tricolor, Vitis vinifera, Zanthoxylum americanum, Zea mays*
Taraxacin	*Taraxacum officinale*
Taraxasterol	*Inula britannica, I. japonica, Taraxacum officinale*
Taraxerol	*Alnus crispus, A. glutinosa, Taraxacum officinale, Tilia cordata, T. europaea*
Tartalic acid	*Cornus florida, Crataegus laevigata, C. monongyna, C. oxyacantha, Hibiscus rosa-sinensis, H. sabdariffa, Tussilago farfara, Vitis vinifera*
Taxol	*Taxus x media, T. brevifolia*
Tectoridin	*Belamcanda chinensis*
Tectorigenin	*Belamcanda chinensis*
Tephrosin	*Tephrosia virginiana*

Terpenes	*Conyza canadensis, Erigeron canadensis, Hyssopus officinalis, Saussurea lappa*
Terpenoids	*Adiantum capillus-junonis, Aster tataricus, Euphorbia hirta, Ligustrum lucidum, L. scoticum, L. vulgare, Melaleuca leucadendra, Mentha arvensis, M. haplocalyx, M. pulegium, Populus alba*
Terpinene	*Coriandrum sativum, Elettaria cardamomum, Juniperus communis, J. sabina, J. horizontalis, Mentha spicata, M. x piperita*
Terpineol	*Asarum canadense, Cinnamomum camphora, Conyza canadensis, Erigeron canadensis*
Terrestriamide	*Tribulus terrestsis*
Tetrahydro-cannabinols	*Cannabis sativa*
Tetrahydropalmatine	*Corydalis yanhusuo, C. solida*
Tetrandrine	*Menispermum canadense*
Thalicarpine	*Thalictrum dasycarpum, T. occidentale*
Thalidasine	*Thalictrum dasycarpum, T. occidentale*
Thalisopavine	*Thalictrum dasycarpum, T. occidentale*
Thamnolic	*Lobelia pulmonaria*
Thebaine	*Papaver rhoeaes, P. bractectum*
Theophylline	*Camellia sinensis*
Thiamine	*Aleurites moluceana, Artemisia vulgarts, Camellia sinensis, Cannabis sativa, Daucus carota, Hibiscus rosa-sinensis, H. sabdariffa, Lycoperiscon esculentum, Prunus mume, Zea mays*
Thiophenes	*Tagetes patula*
Thujone	*Artemisia vulgarts, Chrysanthemum vulgare, Salvia coccinea, S. officinalis, Thuja occidentalis*
Thujyl alcohol	*Artemisia absinthium*

Component	Source
Thymol	*Thymus vulgaris, T. capitatus, T. citriodorus, T. praecox, T. pulegiodes, T. serpyllum*
Tiger nut oil	*Cyperus rotundus*
Tigonenin	*Solanum nigrum*
Tiliadine	*Tilia cordata, T. europaea*
Timnins	*Angelica archangelica, Fraxinus americana, F. excelsior, F. ornus, Juniperus rigida, Nymphaea alba, Senecio vulgaris, S. aureus*
Tocopherol	*Glycine max, Rubus chamaemorus*
Toxicodendrol	*Rhus radicans, R. glabra, R. toxicodendron*
Trans-aconitic acid	*Actaea rubra, A. alba*
Trans-5, cis-9-octadecadienoic acid	*Thalictrum dasycarpum, T. occidentale*
Trans-5-hexadecenoic acid	*Thalictrum dasycarpum, T. occidentale*
Triacetonamine	*Acalypha indica*
Triacylglycerols	*Lythrum salicaria*
Triandrin	*Salix alba, S. discolor*
Tribulusamide B	*Tribulus terrestsis*
Tribulusamide A	*Tribulus terrestsis*
Trichosanic acid	*Trichosanthes kirilowii*
Tricin	*Phragmites australis*
Trigonelline	*Cannabis sativa, Trigonella foenum-graecum*
Trillin	*Trillium erectum*
Triterpene	*Chimaphila umbellata*

Triterpene glycosides	*Cimicifuga racemosa, C. foetida*
Triterpene saponins	*Glycyrrhiza uralensis*
Triterpene acid	*Liquidambar orientalis, L. styraciflua*
Triterpenenes	*Lycopodium annotinum*
Triterpenes	*Achillea millefolium, Anethum graveoleus, Aspidium filix-mas, Aster tataricus, Calendula officinalis, Dryopteris filix-mas, Inula britannica, I. japonica, Lycopodium clavatum, L. obscurum, Myrica cerifera, M. penxylvanica, Peucedanum graveolens, Sambucus nigra, S. canadensis*
Triterpenoid bitters	*Melia azedarach*
Triterpenoid saponins	*Achyranthes bidentata, Anemone hepatica, A. patens, A. pulsatilla, Arisaema consanguineum, Bupleurum falcatum, Codonopsis pilosula, C. tangshen, Eleutherocossus senticosus, Phytolacca acinosa, Polygala vulgaris, P. senega, Primula vulgaris, P. veris, Pulsatilla vulgaris, Stellaria media, Wisteria floribunda, W. brachybotrys*
Triterpenoids	*Centella asiatica, Crataegus laevigata, C. monongyna, C. oxyacantha, Ilex aquifolium, I. paraguensis, Iris versicolor, I. pseudacorus, Menyanthes trifoliata, Punica granatum, Wisteria floribunda, W. brachybotrys*
Tropane alkaloids	*Hyoscyamus niger*
Turmerone	*Curcuma longa*
Tyramine	*Capsella bursa-pastoris, Thlaspi arvense*
Tyrosine	*Phaseolus vulgaris*
Umbelliferone	*Hieracium pilosella, Matricaria chamomilla*
Uric acid	*Aesculus hippocastanum*
Ursolic acid	*Jasminum grandiflorum, J. officinale, Prunella vulgaris, Pyrola rotundifolia*

Component	Source
Urushiol	*Rhus radicans, R. glabra, R. toxicodendron*
Valepotriates	*Valeriana officinalis*
Valerianic acid	*Ananas comosus*
Valeric acid	*Angelica archangelica*
Valtrate	*Valeriana officinalis*
Vanillic acid	*Ananas comosus, Eucalyptus citriodora, E. globulus, Matteuccia struthiopteris, Phragmites australis, Spiraea ulmaria, Styrax benzoin, Tilia cordata, T. europaea*
Vellarin	*Centella asiatica*
Verbenalin	*Cornus officinalis, C. florida, Verbena officinalis*
Vervenin	*Verbena officinalis*
Viburnito	*Menispermum canadense*
Vinblastine	*Catharanthus roseus*
Violin	*Viola tricolor*
Viscin	*Viscum album*
Viscotoxin	*Viscum album*
Vitamin A	*Artemisia annua, Citrus aurantium, Physalis franchetti, Morus alba, P. pubescene, Pinus albicaulis, P. contorta, P. mugo, P. palustris, P. strobus, Polygonatum odoratum, P. multiflorum, P. biflorum, Raphanus sativus, Sambucus nigra, S. canadensis, Solanum aculeatissimum, S. melongena, S. tuberosum, Vaccinium vitis-idaea, V. macrocarpon*
Vitamin B	*Actinidia polygama, Citrus aurantium, Prunella vulgaris, Raphanus sativus, Sesamum indicum*
Vitamin B complex	*Avena sativa, Daucus carota, Hippophae rhamnoides, Rosa canina, R. damascena, R. gallica*

Vitamin B₁	Lycium barbarum, Morus alba, Picrasma excelsa, Ribes nigrum
Vitamin B₁₂	Lycium barbarum
Vitamin B₂	Morus alba, Solanum aculeatissimum, S. melongena, S. tuberosum
Vitamin C	Actinidia polygama, Cardamine pratensis, Citrus aurantium, Coriandrum sativum, Daucus carota, Fragaria vesca, Hippophae rhamnoides, Lepidium virginicum, Lycium barbarum, Morinda didyma, M. fistulosa, M. punctata, Morus alba, Physalis alkekengi, P. franchetti, P. pubescene, Pinus albicaulis, P. contorta, P. mugo, P. palustris, P. strobus, Prunella vulgaris, Raphanus sativus, Rhamnus catharticus, R. frangula, R. purshianus, Ribes nigrum, Rosa canina, R. damascena, R. gallica, Rubus coreanus, R. idaeus, R. chamaemorus, R. fruticosus, Sambucus nigra, S. canadensis, Schisandra chinensis, Sesamum indicum, Solanum aculeatissimum, S. melongena, S. tuberosum, Sorbus aucuparia, Stellaria media, Taraxacum officinale, Taraxacum officinaie, Vaccinium vitis-idaea, V. macrocarpon, Veronica officinalis
Vitamin E	Carthamus tinctorius, Hippophae rhamnoides, Ribes nigrum, Schisandra chinensis
Vitamin K	Prunella vulgaris, Solanum tuberosum
Vitamin P	Forsythia suspensa, Ribes nigrum
Vitamins	Rorippa nasturtium-aquaticum, Cucurbita maxima, Ficus carica, Glycine max, Medicago sativa, Rosa rugosa, R. acicularis, Ziziphus jujuba
Vitexin	Lythrum salicaria
Viticine	Vitex labrusca, V. agnus-castus

Component	Source
Volatile oil	*Acalypha indica, Aletris farinosa, Alpinia galanga, Anemone hepatica, A. patens, A. pulsutilla, Anthriscus cerefolium, Apocynum androsaemifolium, Aralia catechu, A. nudicaulis, A. racemosa, Aristolochia clematitis, A. serpentaria, Artemisia tridentata, Aspidium filix-mis, Bidens tripartita, B. connata, Carduus benedita, Cryptotaenia canadensis, Curcuma longa, Cymbopogon winterianus, C. citratus, C. martinii, Desmodium gangeticum, Drosera rotundifolia, Dryopteris filix-mas, Ephedra sinica, Eriobotrya japonica, Eupatorium perfoliatum, Fraxinus ornus, F. americana, F. excelsior, Gardenia angusta, Geum urbanum, Gnaphalium uliginosum, Inula britannica, I. japonica, Justicia adhatoda, Lonicera japonica, Lophanthus rugosus, Magnolia liliflora, M. officinalis, Marsdenia condurango, Melilotus alba, M. arvensis, Murraya koenigii, Nicotiana tabacum, Piper longum, P. cubeba, P. nigrum, Polygala vulgaris, Primula vulgaris, Primula veris, Pulsatilla vulgaris, Ranunculus ficaria, Rumex crispus, Saponaria officinalis, Senecio vulgaris, Syzygium aromaticum, Trillium erectum, Verbena officinalis, Zingiber officinale*
Wax	*Papaver somniferum, Thuja occidentalis*
Withanolides	*Datura innoxia, D. metel, D. stramonium*
Wogonin	*Scutellaria baicalensis, S. macrantha, S. lateriflora*
Worenine	*Coptis chinensis*
Xanthones	*Anethum graveoleus, Centaurium erythraea, Peucedanum graveolens, Swertia chirata*
Xanthophylls	*Bidens tripartita, B. connata*
Xylan	*Luffa aegyptiaca, L. cylindrica*
Xylose	*Cannabis sativa, Luffa aegyptiaca, L. cylindrica*
Yohimbine	*Rauvolfia serpentina*
Zingiberen	*Curcuma longa*

Index

1 (10) eremophilen-11-ol3β-hydroxyeremophile, 166

1,8-cineol, 73, 92, 146, 203, 273, 396

1,8-dihydroxy-anthracene derivatives, 179, 582

1-acetyl-4-isopropylidenecyclopentene, 188

1β, 10β-epoxyfuranoeremophilane, 166

2-3,4-dihydroxyphenyl-ethanol, 268, 582

2,4,4't-tetrahydroxybenzophenone, 98, 214, 397

2,4-dichloro-6-aminopyridine, 111, 281, 397

2,5-dimethoxypara-quinone, 280, 582

2,6-dimethoxy-p-benzo-quinone, 111, 281, 398

2,6-nonadienol, 50, 193, 398

2α-hydroxyursolic acid, 157, 397

2'-deoxyadenosine, 47, 397

2-hydroxphenylacetic acid, 25, 397

2-hydroxymethyl prop-2-enoate, 166, 397

2-o-caffeoylarbutin, 150, 235, 301, 397

2-vinyl-4H-1,3-dithin, 179, 582

3-0-glucoside, 282, 582

3,4-dihydroxyacetophenone, 79, 266, 399

3,4-dihydroxycinnamic acid, 15, 180, 399

3,4-dihydroxyphenethyl alcohol, 141, 231, 399

3-(4-hydroxyphenyl)-2 (E)-propenoate, 49, 398

3',4'-O-diacetylafzelin, 156, 398

3'-angeloyloxy-4'-isovaleroyloxy, 398

3-butyl phthalide, 398

3-chloroplumbagin, 113, 398

3-hydroxy-30-horoleana-12,18-dien-29-oate, 17

3-indolylmethylgluco-sinolate, 43, 398

3-methoxypyridine, 63, 199, 398

3-n-pentadecylcatechol, 288, 582

3-O-demethylhernandifoline, 139, 398

3-oxykojie acid, 95, 399

3-p-coumarylglycoside-5-glucoside, 217

4'-0-methylpyridoxine, 203, 582

4,5-dimethoxycanthin-6-one, 111, 281, 399

4-dementhyl-hasubanonine, 139, 399

4-epiisocembrol, 169, 399

4-hydroxybenzaldehyde, 280, 582

4-hydroxycinnamic acid, 15, 96, 180, 213, 274, 399, 582

4-methoxysalicyladehyde, 5

4-quinazolone, 57, 399

5-guaizulene, 99, 276, 400

5-hydroxytryptamine, 122, 149, 235, 285, 400

5-methyl kaempferol, 287, 400

5-methyl myricetin, 124, 287, 400

6,6'-dimethoxygossypol, 75, 204, 400

6,8-di-C-galactopyranosylapigenin, 134, 294, 400

6,9,12-octadecatrienoic acid, 101, 215, 400

6-C-galactopyranosyl-isoscutellarein, 134, 294

6-ethoxy-chelerythrin, 155, 304, 400

6-isoinosine, 5, 400

6-methylocodine, 106, 217, 279, 400

6-O-acetyl-arbutin, 150, 235, 301, 400

6-O-rhamnosyl cophoroside, 30, 400

7-caffeyl-glucosides, 169, 217, 400

7-hydroxylathyrol, 67, 200, 400

7-methoxy-2,2-dimethylchromene, 401

7-methoxy-baicalein, 229, 292, 401

7-methoxynorwogonin, 131, 229, 292, 401

7-O-methyl-morroniside, 48, 191, 255, 401

8-(O-methyl-p-coumaroyl)-harpagide, 131, 228, 401

10-deacetylbaccatin, 143, 298, 397

10-hydroxycamptothecin, 33, 328, 397

12-benzoxydaphnetoxin, 55, 195, 258, 397

12-di-dehydroandrog-rapholide, 158, 397

22-dihydrospinasterol, 294, 582

22-dihydrostigmast-4-en-3,6-dione, 397

22E-dehydroclerosterol, 43, 397

22-ergostate traen-3-one, 173, 397

24alpha-epimer stigmasterol, 398

24alpha-ethyl-5alpha-cholest, 398

24beta-epimer poriferasterol, 43, 398

24bets-methylcholesta, 398

25-D-spirosta-3,5-diene, 58, 398

A

A Wei, 68, 163, 338

Aabrin, 582

Abamagenin, 128, 401

Abeotaxanes, 143, 298, 401

Abrin, 176, 582

Abrotamine, 182, 582

Abrus precatorius, 4, 176, 379, 399, 447, 485, 496, 539, 560, 571, 582, 586, 606

Abscisic acid, 111, 401

Absinthol, 245, 582

Abutilon avicennae, 573

Abutilon indicum, 237, 588, 617, 633

Acacetin, 160, 190, 401, 596

Acacetin-7-glucoside, 401

Acacia catechu, 4, 176, 349, 449, 462, 474, 514, 548, 603, 617, 626, 633

Acacia confusa, 4, 386, 410

Acacia nemu, 4, 11, 353, 552

Acacoa australis, 382

Acalypha australis, 4, 401, 473, 563
Acalypha farnesiana, 4
Acalypha indica, 176, 357, 598, 608, 626, 633,
        636, 640
Acalyphine, 4, 176, 401, 582
Acanthopanax giraldii, 327, 329, 400, 423, 450,
        455, 497, 538, 563, 571
Acanthopanax gracilistylus, 5, 327, 330, 384,
        399, 414, 429, 503, 528, 563, 576
Acanthopanax senticosus, 5, 345, 375
Acanthopanax sessiliflorus, 5
Acanthosides, 5, 401
Acetadehyde, 119, 223, 401
Acetic acid, 81, 89, 97, 119, 210, 223, 268, 401
Acetone, 47, 161, 191, 401, 420
Acetopenone glucoside, 302, 583
Acetovanillone, 75, 204, 402
Acetycophalotaxine, 38, 402
Acetyl harpagide, 228, 583
Acetyl lupeol, 113, 402
Acetyl oleanolic acid, 141, 298, 402
Acetylated alkaloids, 215, 312, 583
Acetylcholine, 59, 152, 163, 185, 233, 235, 402,
        583
Acetylcorynoline, 48, 255, 402
Acetylenes, 183, 248, 583
Acetylenic compounds, 216–217, 583
Acetyleugenol, 65, 200, 402
Acetylsalicylic acid, 85, 104, 127, 129, 402
Acetylshikonin, 22, 88, 209, 402
Achillea alpina, 5, 376, 402, 406, 413, 424, 432,
        444, 449, 454, 464–465, 470, 484,
        514, 526, 535, 556
Achillea millefolium, 157, 176, 309, 333, 388, 414,
        437, 549, 560, 571, 583, 593–595, 612,
        615, 630, 632–633, 637
Achilleine, 5, 176, 402, 583
Achillin, 5, 176, 402
Achyranthes asperia, 6, 383, 417, 422, 520, 540,
        548, 569
Achyranthes bidentata, 6, 177, 355, 460, 486,
        538, 637
Achyranthes japonica, 6, 357, 429, 524
Acidic polysaccharides, 191, 254, 583
Acidic resin, 153, 402
Aconine, 7, 402
Aconitic acid, 24, 27, 162, 247, 402
Aconitine, 3, 6–7, 177, 309, 402, 583
Aconitum austroyunnanense, 563
Aconitum balfouri, 6, 351, 385, 511
Aconitum barbatum, 6, 388, 427, 492, 575
Aconitum carmichaelii, 177, 589, 615–616
Aconitum chasmanthum, 351
Aconitum jaluense, 485

Aconitum laciniatum, 7, 343, 402, 485
Aconitum napellus, 177, 583, 592, 608, 618
Acoric acid, 7, 157, 177, 402, 583
Acornes, 402
Acorus calamus, 7, 157, 177, 309, 339, 343, 402,
        422, 472, 480, 485–486, 533, 556, 583,
        588, 613, 616–617, 620, 628, 630
Acronychia pedunculata, 7, 358, 403, 419, 521
Acrylic acid, 180, 583
Actaea asiatica, 7, 237, 362, 570
Actaea rubra, 237, 624, 626, 636
Actein, 188, 583
Actinidia arguta, 7, 367, 403, 407, 456, 487, 491,
        509, 518–519, 576
Actinidia chinensis, 157, 334, 397, 429, 488, 533,
        538, 576
Actinidia polygama, 177, 368, 455, 489, 508, 583,
        585, 609–610, 615, 618, 638–639
Actinidine, 7, 159, 177, 403, 583
Actronycine, 7, 403
Acutumidine, 44, 403
Acutumine, 44, 96–97, 135, 275, 403, 583
Acutuminine, 97, 275, 403
Acyclic diterpene glycosides, 90, 210, 403
Adamia chinensis, 7, 374, 408, 422, 474
Adamia cyanea, 343
Adenine, 17, 23, 39–40, 120, 136, 144, 181, 188,
        224–225, 241, 252, 265, 295, 403
Adenophora coronopifolia, 8, 237, 374, 552
Adenophora stricta, 237
Adenophora triphylla, 8, 237, 343, 423–424, 451,
        487, 505, 565
Adenophora verticillata, 364
Adenosine, 45, 47, 71, 151, 235, 403
Adiantone, 8, 178, 403, 583
Adiantum boreale, 8, 382, 403, 468–469, 474,
        483, 488–489
Adiantum capillus-junonis, 583, 610, 617, 627,
        633, 635
Adina ratemosa, 379
Adina rubella, 8, 379, 423, 425, 438, 514, 518,
        522, 545, 552, 561
Adipedatol, 8, 178, 403
Adonilide, 9, 157, 178, 237, 403
Adonis amurensis, 403, 444, 448, 472, 490, 493,
        503, 554, 559, 572
Adonis brevistyla, 9
Adonis chrysocyathus, 237, 350, 403, 442, 448,
        491, 531
Adonis vernalis, 178, 238, 583, 593
Adonitoxin, 178, 238, 583
Adynerin, 101, 403
Aegicerin, 118, 222, 284, 403
Aescilom, 403

Aescin, 178, 583
Aescine, 9, 403
Aesculetin, 27, 151, 303, 403
Aesculine, 9, 403
Aesculus chinensis, 9, 380, 411, 463, 524, 540
Aesculus hippocastanum, 178, 583, 595, 597, 611,
    626, 628, 633, 637
Aesculus indica, 403, 405, 468, 470
Afzelechin, 282, 583
Afzelin, 10, 57, 179, 238, 404
Agarol, 18, 242, 404
Agaropectin, 71, 263, 404, 583
Agarose, 71, 263, 404, 583
Agarospirol, 18, 242, 404
Agastache anethrodora, 238, 586, 616
Agastache rugosa, 9, 89, 238, 357, 412, 423, 429,
    433, 448, 450, 460, 464, 467, 482,
    502, 512, 523, 527
Agathin, 133, 404
Agathodienediol, 404
Ageniadin, 404
Agerato-chromene, 9, 404
Ageratum conyzoides, 9, 375, 401, 404, 422, 444,
    447
Agglutinins, 152, 235, 404
Aglucones, 171, 404
Agnuside, 153, 303, 404, 584
Agoniadin, 113, 169, 404
Agrimols, 10, 178, 404
Agrimonia eupatoria, 10, 178, 363, 386, 404, 414,
    433, 443, 506, 564, 576–577, 597, 603,
    614, 624, 633
Agrimonia pilosa, 318
Agrimonine, 10, 178, 404
Agrimonolide, 10, 178, 404
Agrimophol, 10, 178, 404
Ai Di Cha, 20, 338
Ai Lei, 37, 338
Ai Na Xian, 29, 338
Ai Ye, 23, 338
Ai Yu Zi, 68, 338
Ailanthone, 10, 179, 238, 404, 584
Ailanthus altissima, 10, 179, 238, 345, 404, 410,
    419, 423, 450, 515, 563, 574, 584, 602,
    625–626, 633
Ailanthus glandulosa, 238, 603
Ajmaline, 286, 584
Ajoene, 179, 584
Ajuga bracteosa, 10, 238, 354, 401, 405, 423,
    427–428, 434, 436, 459–460, 470, 487,
    500, 503, 505, 515, 524, 528–529, 539,
    560, 569, 579
Ajuga pygmaea, 359
Ajuga reptans, 238, 607

Ajugasterone, 10, 238, 405
Akebia quinata, 10, 238, 367
Akebia trifoliata, 239
Alangicine, 11, 405
Alangimarckine, 11, 405
Alangium chinense, 11, 338
Alangium lamarckii, 11, 405, 413, 434, 453–454,
    458, 461, 542, 572
Alanine, 68, 88, 143, 172, 405
Alatamine, 65, 322, 405
Albaspidin, 60, 197, 405, 546
Albigenic acid, 45, 405
Albizia julibrissin, 4, 11, 353, 355
Albumin, 9, 79, 217, 405, 584
Albuminoides, 196, 584
Alcohol derivatives, 84, 405
Alcohols, 95, 152, 303, 405
Aldehyde, 40–41, 47, 75, 79, 107, 125, 129, 142,
    161, 172, 189, 191, 217, 226–227, 232,
    266, 289, 298, 399, 405, 431, 439, 446,
    452, 484, 517, 521, 525, 540, 557, 563,
    584
Aletris formosuna, 11, 239, 349, 423, 457, 561
Aleurites fordii, 11, 391, 407–408, 421, 424, 429,
    517, 524, 528, 535, 552, 560–561, 564
Aleurites moluceanu, 11, 179, 376, 432, 540, 569,
    608, 613–614, 619, 624, 633, 635
Algin, 83, 129, 269, 292, 405
Alginic acid, 129, 292, 405
Alisarin, 70, 202, 263, 405
Alisma cordifolia, 11, 392, 405, 461
Alisma orientalis, 538, 571
Alisol A, 11, 405
Alisol B, 11, 405
Alisol monoacetate, 11, 405
Alizarin, 98, 125–126, 276, 289, 406
Alizarin-l-*methyl ether,* 276, 406
Alkaloid lamine, 59, 259, 406, 584
Alkamin-B, 22, 406
Alkanes, 197, 200, 202, 245, 260, 584
Alkannan, 88, 209, 406
Alkenyl, 190, 584
Alkenyl glycoside, 190, 584
Alkyl methyl quinolone alkaloids, 67, 406
Allamanda cathatica, 12, 392, 406
Allamandin, 12, 406
Allantoin, 58, 196, 218, 406, 584
Allelopathic essential oils, 54, 258, 407
Allicin, 12, 179, 407
Alliin, 179, 584
Allistatin, 12, 179, 407
Allium chinense, 12, 346, 407, 454, 477, 518, 538,
    540, 561
Allium macrostemon, 330, 349, 472

Allium sativum, 179, 582, 584, 599, 609, 613, 629
Allocryptopine, 39, 187, 407, 584
Allomatatabiol, 7, 177, 407, 585
Allosecurinine, 131, 407
Allyl isothiocyanate, 30, 184, 407
Allyl sinapic oil, 59, 407
Alnulin, 239, 585
Alnus crispus, 239, 585, 591, 602, 622, 624, 626, 633–634
Alnus japonica, 12, 239, 343, 407, 425, 481, 505, 565
Aloe barbadensis, 12, 179, 309, 325, 364, 407, 419, 582, 585–586, 589, 595, 600
Aloe-emodin, 12, 35, 123, 179, 186, 225, 251, 286, 407, 585
Aloeresins, 585
Aloesin glycone, 179, 585
Aloesone, 179, 585
Aloin isobarbaloin, 179, 585
Aloins, 12, 179, 407
Alpha-acid, 585
Alpha-agarofuran, 18, 242, 407
Alpha-allocryptopine, 48, 92, 192, 255, 407
Alpha-amyrenol, 138, 407
Alpha-amyrin, 11–12, 41, 107, 128, 239, 279, 292, 407
Alpha-*amyrin palmitate,* 128, 292, 407
Alpha-antiarin, 18, 407
Alpha-antioside, 18, 408
Alpha-bergamotene, 107, 408
Alpha-bisabolol, 212, 585
Alpha-camphorene, 45, 191, 253, 408
Alpha-carotene-5,6-epoxide, 52, 257, 408
Alpha-caryophylline, 200, 408
Alpha-crocetin, 71, 202, 408
Alpha-cyperene, 54, 194, 258, 408
Alpha-cyperol, 54, 194, 258, 408
Alpha-cyperone, 54, 408
Alpha-dichroine, 7, 408
Alpha-*elaeo stearic,* 11, 179, 408
Alpha-euphol, 67, 408
Alpha-euphorbol, 66, 408
Alpha-fenchene, 150, 302, 408
Alpha-globuline, 109, 218, 408
Alpha-humulene, 54, 156, 408
Alpha-ionone, 84, 208, 408
Alpha-kainic acid, 32, 408
Alpha-ketoglutaric acid, 109, 315, 408
Alpha-leucodelphinidin, 409
Alpha-lupanine, 36, 409
Alpha-masticoresin, 220, 585
Alpha-*methyl ether,* 98, 276, 409
Alpha-obscurine, 90, 409
Alpha-onocerin, 91, 211, 409

Alpha-paristyphnin, 106, 217, 409
Alpha-phellandrene, 84, 197, 208, 233, 409
Alpha-phenylethylisothiocyante, 125, 288, 409
Alpha-*phenylpropyl cinnamyl cinnamate,* 141, 297, 409
Alpha-phytosterol, 267, 585
Alpha-pinene, 7, 39, 56, 83–84, 102, 191, 195, 203, 207–208, 217, 219, 240, 252, 268, 409, 585
Alpha-santalol, 129, 227, 409
Alpha-santenone, 129, 227, 409
Alpha-spinasterol, 45, 267, 409, 585
Alpha-taralin, 19, 243, 409
Alpha-taraxerol, 66, 409
Alpha-terpinene, 102, 409
Alpha-terpineol, 53, 73, 145–146, 194, 203, 213, 233, 257, 299, 409, 585
Alpha-*terpinyl acetate,* 146, 233, 409
Alpha-terthienyl, 142, 232, 298, 410
Alpha-terthienylmethanol, 197, 585
Alpha-*tertiary methanol,* 61, 410
Alpha-thujone, 184, 585
Alpha-tocopherol, 249, 319, 585
Alpha-trevilline, 199, 410
Alpha-typhasterol, 148, 234, 410
Alpinia galanga, 240, 585, 595, 604, 613, 630, 640
Alpinia japonica, 13, 239, 392, 438, 464, 493, 496, 547
Alpinia katsumadai, 13, 348, 432, 439, 473, 494, 548, 569
Alpinia kumatake, 369
Alpinia officinarum, 13, 239
Alpinia oxyphylla, 13, 239, 390, 438, 580
Alpinia speciosa, 13, 239, 374
Alstonia scholaris, 13, 157, 180, 334, 347, 386, 396, 429, 460, 535, 584, 626
Alternanthera philoxeroides, 13, 361, 444, 467, 552, 564
Alternanthera sessilis, 366
Althaea officinalis, 240
Althaea rosea, 13, 240, 344, 457
Althaeine, 13, 240, 410
Aluminum, 110, 117, 218, 410
Aluminum oxide, 110, 218, 410
Alzarin, 289, 585
Amaranthin, 75, 410
Amaranthus blitum, 240
Amaranthus caudatus, 13, 240, 384, 424
Amaranthus hypochondriacus, 240, 633
Amaranthus lividus, 14, 364, 540, 569, 577
Amaranthus paniculatus, 14, 349
Amaranthus tricolor, 14, 158, 240, 373, 388, 396, 417, 422, 520, 548, 569, 576

Amarbelin, 52, 194, 257, 410
Amarogentin, 297, 585
Amarolide, 10, 179, 238, 410
Ambroide, 39, 410
Amellin, 131, 410
Amine choline, 185, 233, 585
Amines, 256, 585
Amino acids, 4, 14–15, 19, 45, 61, 67, 71, 83, 88, 94, 100–101, 103, 110–111, 116, 130, 149, 190, 210, 269, 277, 312, 410, 586
Aminoadipic acid, 27, 247, 410
Aminobutyric acid, 210, 586
Aminol, 40, 188, 411
Amomum cardamomum, 14, 240, 339, 426, 449, 502, 520, 566
Amomum globosum, 342
Amomum tsao-ko, 373
Amorphophallus rivieri, 14, 367, 501, 508
Amorphous dracoalban, 55, 411
Amorphous dracoresene, 55, 411
Ampelopsis brevipedunculata, 14, 390
Ampelopsis japonica, 15, 339
Amritoside, 119, 411
Amurine, 105, 279, 411
Amuroine, 105, 279, 411
Amuroline, 105, 279, 411
Amygdalin, 22, 64, 118, 121, 199, 223, 244, 256–257, 411, 586
Amylase, 68, 78, 173, 200, 205, 262, 411
Amylodextrins, 99, 215, 411
Amylose, 9, 146, 178, 411
Amyrenol, 132, 293, 411
Amyrenone, 132, 293, 411
Amyrin, 69, 157, 163, 172, 201, 396, 411
Anacardic acid, 73, 203, 411
Anagalis arvensis, 15, 180, 353, 411, 416, 445, 586, 597, 628, 633
Anagalligenone, 15, 180, 411
Anagalline, 15, 180, 411, 586
Anagyrine, 137, 252, 411, 586
Ananas comosus, 15, 180, 349, 399–400, 411, 427, 463, 577, 583, 592, 595, 610, 615–616, 638
Ananasic acid, 15, 180, 411
Ancubin, 151, 302, 411
Andelin, 17, 241, 411
Andole alkaloids, 586
Andrographis paniculata, 15, 158, 345, 397, 412, 452–453, 518
Andrographolide, 15, 158, 412
Andrographon, 158
Andromedotoxin, 39, 91, 124, 170, 187, 287, 412
Anemarrhena asphodeloides, 15, 393, 490, 508, 519, 553, 561

Anemone cernua, 15, 241, 340, 540, 552
Anemone hepatica, 241, 633, 637, 640
Anemone pulsatilla, 180, 586, 590–591, 599, 621, 626, 628, 633
Anemone raddeana, 16, 390, 480, 524, 545
Anemone rivularis, 343
Anemone vitifolia, 390
Anemoni, 285, 586
Anemonin, 32, 43, 120, 122, 180, 224, 241, 249, 253, 284–285, 412, 586
Anemonol, 43, 224, 253, 412, 586
Anerhole, 238, 586
Anethol, 69, 80, 93, 201, 206, 211, 412
Anethole, 9, 83, 93, 195, 201, 206–207, 211, 238, 268, 281, 412
Anethum graveoleus, 16, 181, 376, 421, 430, 449, 456–458, 464, 491, 501, 572, 593, 597, 603, 613, 637, 640
Angelic acid, 17, 29, 94, 158, 212, 248, 412
Angelica amurensis, 16, 241, 340, 412–413, 421, 428, 486, 509, 526, 532, 554, 579
Angelica archangelica, 242, 586, 592, 597, 602, 613, 618, 622, 636, 638
Angelica decursiva, 17, 241, 370, 398, 411, 452, 465, 484, 521, 557, 560, 572
Angelica grosserrata, 17, 241, 503, 524, 528, 560
Angelica polymorpha, 17, 181, 347, 403, 425, 427, 469, 471, 501, 520, 538, 562, 573, 576–577, 592–593, 611, 613, 618, 620, 627, 630
Angelica pubescens, 17, 412, 428, 444, 476, 525, 569
Angelica sinensis, 158, 241, 318, 346
Angelicide, 242, 586
Angelicin, 119, 224, 412
Angelicotoxin, 17, 158, 412
Angelol, 17, 412
Angenomalin, 16, 241, 412
Anhydroderrid, 97, 412
Anisaldehyde, 9, 69, 80, 201, 206, 238, 412, 586
Anisatin, 80, 412
Anisic acid, 66, 412
Anisic ketone, 80, 206, 412
Anisodamine, 56, 131, 412
Anisodine, 56, 131, 413
Ankorine isotubulosine, 11, 413
Anneparine, 100, 277, 413
Anodyne, 5–6, 15, 17, 39, 63, 66, 73, 76, 83, 85, 95, 107, 111, 125, 155, 162–164, 180, 187, 199, 203, 241, 258, 265, 269–270, 281, 289, 304, 413
Anomalin, 16, 108, 241, 280, 413
Anonaine, 100, 277, 413
Anredera cordifolia, 17, 388, 398, 499, 573

Anromedotoxin, 123, 287, 413
Anthaquinone glycosides, 179, 586
Anthelmic acid, 94, 212, 413
Anthemidin, 212, 586
Antheraxanthin, 143, 172, 413
Anthocyanidin, 216, 586
Anthocyanidines, 5, 18, 76, 115, 128, 143, 221, 413
Anthocyanins, 107, 169, 176, 214, 217, 235, 290, 333, 413
Anthocyanosides, 235, 288, 301, 587
Anthranil acid, 81, 268, 413
Anthranol, 127, 413
Anthraquinoids, 100, 413
Anthraquinone compounds, 225, 286, 587
Anthraquinone derivatives, 289, 587
Anthraquinone glycosides, 286, 587
Anthraquinones, 5, 18, 35–36, 76, 115, 123, 128, 143, 202, 221, 225, 227, 232, 251, 263, 272, 286, 290, 320, 413, 587
Anthraxin, 24, 414
Anthricin, 18, 242, 414
Anthriscus aemula, 18, 242, 384, 413–414, 432, 444, 453, 465, 484, 488, 505, 514, 526, 535, 556
Anthriscus cerefolium, 242, 597, 603, 640
Anti-*HIV protein* MAP, 98, 214, 414
Antiproliferative activity, 334
Antiprotease, 288, 587
Apigenin, 36, 44–45, 50, 55, 82, 107, 133, 145, 157, 169, 195, 207, 217, 233, 251, 253, 258, 299–300, 414, 587
Apigenin-7-diglucuronide, 44, 190, 414
Apigenin-7-O-glucoside, 138, 414
Apigenin-8-C-glucoside, 414
Apiin, 18, 181, 193, 414, 587
Apiole, 193, 587
Apium graveolens, 18, 181, 309, 371, 414, 479, 587, 590, 597, 613, 617–621, 631
Aplotaxene, 171, 228, 292, 587
Apocynein, 242, 587
Apocynin, 242, 587
Apocynum androsaemifolium, 242, 587, 593, 598, 602, 606, 640
Apocynum venetum, 18, 242, 364, 448, 491, 543, 562
Appendicitis, 163, 166, 168, 171, 414
Apyrocatechol, 42, 414
Aquilaria agallocha, 18, 242, 344, 404, 407, 421, 483
Aquilaria flavescens, 242
Arabinan polymer, 31, 414
Arabinon, 73, 414
Arabinose, 25, 33, 82, 119, 142–143, 183, 232, 414

Arabinose ester, 119, 414
Arachic, 5, 414
Arachidic acid, 134, 229, 233, 415, 587
Arachine, 19, 415
Arachis hypogaea, 19, 365, 410, 415, 425, 431, 477–478, 520, 540, 548, 569
Aragome, 415
Aralia catechu, 243, 600, 606, 633, 640
Aralia chinensis, 19, 243, 358, 409, 415, 424, 458, 495, 498, 524, 535
Aralia cordata, 305, 332
Araligenin, 19, 243, 415
Aralosides, 19, 243
Araucaria cunninghamii, 19, 368, 454, 498, 511–512
Arborinol, 80, 415
Arborinone, 80, 415
Arbutin, 39, 84, 121, 132–133, 150–151, 187, 208, 225, 235, 293–294, 301–302, 321, 415, 587
Arbutin ericolin, 84, 208, 415
Archangelicin, 44, 190, 415
Arctic acid, 181, 587
Arctigenin, 20, 84, 181, 415
Arctiin, 173, 243, 587
Arctiol, 181, 587
Arctium lappa, 20, 84, 181, 330, 369, 415, 499, 501, 509, 556, 561, 587, 599, 609, 617, 626, 633
Ardisia japonica, 20, 338, 369, 421, 464
Ardisia quinquegona, 20, 393, 421
Ardisia sieboldii, 377
Areca catechu, 20, 158, 181, 342, 415, 479, 587–588, 605, 607, 610, 612, 615, 617, 619–620, 631, 633
Arecaidine, 181, 587
Arecaine, 181, 588
Arecholidine, 20, 181, 415
Arecholine, 20, 181, 415
Arecoline, 158, 181, 415, 588
Arenaria juncea, 20, 243, 392, 552
Arenaria rubra, 243, 626
Aresentic acid, 415
Arethusa japonica, 20, 392, 464, 564
Aretylcholine, 235, 588
Argamolic acid, 119, 415
Arginine, 50, 58–59, 88, 136, 159, 189, 193, 196, 208, 218, 225, 260, 295, 415–416, 588
Arginine glucoside, 136, 295, 416
Aricine, 122, 285, 416
Arisaema amurense, 21, 381, 406, 420, 552
Arisaema consanguineum, 182, 589, 637
Arislolochic acid, 182
Aristolochia clematitis, 244, 588, 633, 640

Aristolochia contorta, 21, 243, 365, 416, 423, 480, 524
Aristolochia debilis, 21, 243, 371, 416, 447, 452, 454, 507
Aristolochia manshuriensis, 21, 244, 368
Aristolochia shimadai, 21, 244, 416
Aristolone, 21, 243, 416
Aristoloside, 21, 243, 416
Arjunolic acid, 61, 416
Armeniaca ansu, 22, 244, 385, 411, 484
Arnebia euchroma, 22, 402, 406, 417, 422–423, 457, 553, 556, 561
Arnidiol, 32, 185, 416
Aromadendrene, 40, 188, 416
Aromadendrin, 145, 282, 299, 416
Aromadendrine, 97, 214, 416
Aromatic acids, 66, 416
Arrenin, 15, 180, 416
Artemisia absinthium, 245, 582, 613, 622, 633
Artemisia alcohol, 23, 182, 244, 416
Artemisia annua, 22, 182, 318, 371, 416–417, 436, 455, 556, 582, 588, 638
Artemisia apiacea, 244
Artemisia argyi, 23, 244, 338, 416, 422, 426, 430, 438, 466, 502, 566
Artemisia brachyloba, 23, 374, 403, 416, 438, 464, 536, 564, 566
Artemisia capillaris, 23, 245, 390, 431, 554
Artemisia dracunculus, 597, 601, 603, 609, 616, 622, 627
Artemisia frigida, 339
Artemisia gmelini, 24, 339, 419, 430, 475, 493, 554, 572
Artemisia lactiflora, 24
Artemisia tridentata, 604, 621, 640
Artemisia vulgarts, 182, 588, 595, 609, 626, 633, 635
Artemisine, 23, 416
Artemisinin, 22, 182, 244, 318, 416, 588
Artesunate, 22, 182, 244, 417
Arthraxon hispidus, 24, 359, 402
Articulain, 63, 162, 199
Artocapin, 214, 588
Artocarpus altilis, 24, 367, 421, 505, 571
Artocarpus heterophyllus, 24, 342, 431, 434, 540, 546
Arundinaria graminifolia, 24, 429, 558, 561
Arundo donax, 25, 364, 417, 478, 540
Arundoin, 80, 89, 417
Asarensinotannol, 68, 201, 417
Asariline, 25, 182, 417
Asarinin, 304, 588
Asarone, 56, 157, 177, 195, 417, 588

Asarum canadense, 25, 182, 385, 417, 423, 464, 550, 572, 588, 591, 599, 623, 635
Asarum europaeum, 328–329, 435, 470
Ascaridol, 39, 84, 187, 208, 417
Ascaridole, 187, 588
Asclepiadin, 159, 417
Asclepin, 159, 417
Asclepogenin, 159, 417
Ascorbic aci*d,* 6, 14, 28, 30, 34, 36, 49, 108, 125, 141, 155, 158, 163, 167, 171, 182–183, 185, 223, 231, 236, 240, 251, 256, 289–290, 417, 582, 588
Ash, 70, 104, 149, 215, 235, 417, 588
Asiaticoside, 37, 186, 417, 477, 588
Asoryl-ketone, 25, 182, 417
Asparagic acid, 88, 417
Asparagine, 25, 78, 127, 143, 182–183, 237, 240, 246, 262, 417, 588
Asparaginic acid, 27, 247, 417
Asparagosides, 182, 588
Asparagus cochinenesis, 25, 381, 415, 464, 478, 511
Asparagus officinalis, 182
Asparamide, 280, 588
Asperuloside, 102, 202, 289, 418, 588
Aspidinol, 170, 418
Aspidium falcatum, 25, 245, 352, 464, 469, 564
Aster ageratoides, 25, 245, 354, 494, 543
Aster tataricus, 25, 183, 246, 393, 414, 552, 556, 597, 600, 603, 616, 622–623, 628, 630, 635, 637
Astilbe longicarpa, 25, 365, 397, 418, 543
Astilbin, 25, 418
Astragalin, 26, 55, 60, 63, 95, 104, 136, 146, 161, 163, 171, 183, 199, 212, 216, 230, 233, 246, 418, 588
Astragalosides, 26, 183, 246, 418, 588
Astragalus americana, 246
Astragalus chinensis, 26, 246, 374, 418, 483
Astragalus complanatus, 26, 328, 356, 418, 424, 430, 443–444, 447, 470–471, 474, 490, 494, 538, 544, 547, 552
Astragalus membranaceus, 183, 325, 327–328, 588, 592, 604, 611, 632
Athocyanin, 589
Atractylenolide II, 183, 246, 589, 612
Atractylenolide III, 589, 612
Atractylis chinensis, 26, 392, 418, 422, 454, 464
Atractylodes chinensis, 26, 340, 425, 466, 482
Atractylodes lancea, 27, 246, 342, 418, 464
Atractylodes macrocephala, 183, 246, 318, 589, 612
Atractylodine, 26, 418
Atractylol, 26–27, 183, 246, 418, 589

Atractylone, 26, 183, 418
Atropine, 56, 296, 418, 589
Atropine alkaloids, 296, 589
Aucklandia costus, 27, 368, 430, 444, 532, 553
Aucubin, 65, 96, 113, 151, 153, 220, 223, 228,
    282, 302–303, 418, 589
Aucuboside, 302, 589
Aurantio-*obtusin rubrofusarin,* 35, 251, 418
Aurapten, 57, 195, 418
Auroxanthin, 110, 218, 281, 418
Austroinulin, 140, 418
Avena fatua, 27, 247, 389, 402, 418, 423, 463,
    477, 490, 501, 526, 554, 569, 571, 579
Avena sativa, 247, 593, 624, 628, 638
Avenasterol, 27, 87, 209, 247, 418
Avicularin, 39, 89, 107, 115, 119, 130, 150, 221,
    235, 272, 301, 419
Awobanin, 45, 419
Azadarachtin, 96, 213, 419
Azalea japonica, 27, 389
Azaleatin, 124, 287, 419
Azelaic acid, 10, 91, 179, 211, 238, 419
Azolla imbricata, 27, 366, 403, 428, 506
Azulene, 24, 94–95, 112, 212, 219, 245, 419, 589

B

Ba Dou, 50, 338
Ba Ji Tian, 338
Ba Jian Lian, 114
Ba Jiao Feng, 11, 338
Ba Jiao Hui Xiang, 80, 338
Ba Li Ma, 124, 338
Baccatin, 143, 298, 419
Bacteriostatic, 419
Bai Bao Zi, 94, 338
Bai Ben Dou, 59, 338
Bai Bu, 338
Bai Chen, 53, 339
Bai Dou Ku, 339
Bai Guo, 339
Bai He, 50, 86, 339, 389
Bai Hua, 28, 43, 76, 85, 102, 113, 339–340, 379
Bai Hua Shi Shi Cao, 102, 340
Bai Hua Teng, 113, 339
Bai Hua Yi Mu Cao, 85, 340
Bai Ji, 29, 53, 63, 339
Bai Jiang Cao, 107, 339
Bai Jie, 30, 134, 339
Bai Jie Zi, 30, 339
Bai Lian, 15, 24, 339
Bai Lian Guo, 24, 339
Bai Long Chuan Hua, 43, 340
Bai Mao, 80, 339

Bai Qian Ceng, 95, 340
Bai Qu Cai, 39, 339
Bai Rui Cao, 145, 339
Bai Shao, 104, 339
Bai Tou Went, 120, 340
Bai Tu Own, 15, 340
Bai Way, 53, 340
Bai Xian Pi, 57, 70, 340
Bai Ye Diao Zhang, 87, 341
Bai Yin Shu, 132, 340
Bai Ying, 136, 340
Bai Yu Lan, 97, 340
Bai Zhi, 16, 340
Bai Zhu, 26, 71, 340–341
Bai Zhu Shu, 71, 341
Baicalein, 131, 229, 292–293, 318, 327, 419, 589
Baicalin, 131, 229, 292–293, 316, 327, 334, 419,
    589
Baldrianic acid, 292, 589
Balsam, 80, 87, 271, 419
Balsamic acid, 141, 297, 419
Balsaminasterol, 206, 589
Balsaminones, 206, 589
Ban Bian Lian, 88, 341
Ban Jiu Jiu, 341
Ban Lan, 27, 81, 341
Ban Lan Gen, 81, 341
Ban Lan Geng, 341
Ban Xia, 111, 341
Ban Zi Lian, 341
Bang Chui Hui, 100, 341
Baphicanthus cusia, 27, 341, 346, 486, 490, 499,
    572
Barbaloin, 12, 179, 419, 589
Barbatic acid, 149, 419
Barium, 82, 419
Bassorine, 163, 419
Bauerenol, 7, 419
Bauhinia championi, 27, 360, 423, 431, 470, 494,
    540, 561
Bavachin, 224, 589
Bavachinin, 119, 224, 419
Be Han Cao, 96, 341
Behenic acid, 30, 116, 172, 184, 420
Bei Mei Do Xing Cao, 341
Bei Mu, 70, 341
Bei Pu Jiang, 341
Bei Xian, 52, 341
Bei Za Seng, 73, 341
Belamcanda chinensis, 28, 247, 321, 375, 566,
    589, 609–610, 634
Belamcanda panctata, 247
Belamcandaquinones A, 247, 321, 589
Belamcandaquinones B, 589

Bellidifolin, 141, 297, 420

Benincase cerifera, 28, 348, 417, 503, 520, 528, 548, 560, 569

Benincase hispida, 183, 607, 628

Bensaldehyde, 125, 226, 289, 420

*Benzene tert-*butyl, 154, 304, 420

Benzoic acid, 21, 55, 81, 104, 110, 182, 190, 216, 218, 229, 268, 290, 293, 297, 420, 589

Benzolacetic ester, 55, 420

Benzol-aconitine, 589

Benzophenone, 252, 321, 590

Benzoquinene, 273, 590

Benzoquinone, 85, 326, 420

Benzoyl ramanone, 420

Benzoyl salicin, 116, 221, 283, 420

Benzoylecgonine, 64, 199, 590

Benzyl alcohol, 119, 223

Benzyl benzoate, 161, 259, 420, 590

Benzyl isothiocyanate, 30, 184

Benzylacetone, 18–19, 242–243

Berbamine, 28, 139, 144, 247, 299, 420, 590

Berberine, 28, 32, 41, 46, 94, 99, 106, 109, 131, 144, 154–155, 159, 164, 187, 191, 217–218, 247, 249, 254, 273, 279, 292, 299, 304, 317, 421, 590

Berberis amurensis, 28, 159, 247, 386–387, 420–421, 493, 526, 528

Berberis aquifolium, 247, 596, 615, 620, 624

Berberis vulgaria, 247, 590, 626, 633

Berberubine, 247, 590

Bergapten, 16–17, 57, 76, 116, 155, 158, 163, 171, 181, 195, 241, 304, 421, 590

Bergaptin, 69, 201, 421

Bergegin, 194, 257, 590

Bergenin, 20, 25, 52, 94, 194, 257, 421

Bergenine glucoside, 20, 421

Beta-acid, 590

Beta-agarofuran, 18, 421

Beta-amirine, 82, 421

Beta-amyrenol, 138, 421

Beta-amyrin, 11, 24, 39, 41, 66, 69, 107, 142–143, 151–152, 180, 187, 194, 235, 257, 279, 303, 421, 590

Beta-amyrin acetate, 24, 421

Beta-asarone, 7, 422

Beta-carotene, 6, 14, 30, 34, 36, 49, 108, 155, 163, 185, 192, 223, 236, 240, 251, 256, 590

Beta-carotenoid, 77, 205, 422

Beta-caryophyllene, 99, 152, 276, 303, 422

Beta-caryophylline, 65, 200, 422

Beta-cyperene, 54, 194, 258, 422

Beta-D-glucopyranosil, 590

Beta-D-glucoside, 275, 590

Beta-D-glucosyloxy, 96, 213, 274, 422

Beta-dichroine, 7, 422

Beta-dihydropseudoionone, 194, 257, 422

Beta-dimethylacrylate, 422

Beta-dimethylacrytoylshikonin, 422

Beta-elemone, 265, 590

Beta-ergosterol, 267, 590

Beta-eudesmol, 93, 422

Beta-globulin, 109, 218, 422

Beta-glycyrrhetinic acid, 74, 422

Beta-guaienen, 24, 245, 422

Beta-gurjunene, 154, 304, 423

Betaine, 13–14, 26, 34, 69, 90, 127, 157, 160, 183, 185, 210, 240, 246, 273, 297, 424, 496, 591

Beta-ionone, 84, 208, 423

Beta-masticoresin, 220, 590

Beta-methyl-adipic acid, 275, 590

Beta-methylaesculetin, 46, 423

Betanidin, 30, 117, 424

Betanin, 117, 424

Beta-OH-isovalerylshikonin, 423

Beta-p glucophyranoside, 27, 423

Beta-pachyman, 222, 590

Beta-pachymanase, 222, 590

Beta-phellandrene, 103, 423

Beta-phenethyl alcohol, 88, 423

Beta-pinene, 9, 25, 73, 75, 84, 87, 102, 203, 207–208, 213, 219, 238, 271, 423, 591

Beta-santalene, 129, 227, 423

Beta-santalol, 129, 227, 423

Beta-selinene, 54, 258, 423

Beta-sitosterol, 5, 8, 10–11, 21, 25, 27, 52–54, 62, 65, 73, 81–82, 85, 95, 102–103, 106, 108, 115, 118, 126, 130–131, 135, 141–143, 146, 151, 179–180, 186, 194, 198, 200, 210, 212, 216–217, 226, 229–230, 235, 237–239, 243, 247, 251, 257, 267, 273, 275, 279, 285, 290, 292–293, 295, 298

Beta-sitosterol glucoside, 62, 198, 424

Beta-sitosteryl palmitate, 8, 237, 424

Beta-solamargine, 135, 230, 295, 424

Beta-sotpsterols, 58, 424

Beta-taralin, 19, 243, 424

Beta-thujone, 184, 591

Beta-trevilline, 64, 199, 424

Betonicine, 5, 176, 297, 424, 591

Betula lenta, 247, 601, 604, 609, 616, 628, 633

Betula mandshurica, 28, 247, 339, 425

Betulafolienetetraol, 28, 247, 425

Betulafolienetriol, 28, 247, 425

Betulic acid, 156, 236, 425

Betulin, 28, 67, 113, 151, 156, 200, 236, 247, 425

Betulinic acid, 8, 12, 59, 95, 97, 107, 141, 156,
        214, 223, 236, 239, 279, 298, 425, 591
Betuloside, 28, 247, 425
Beyerene, 184, 591
Bi Ba, 112, 341
Bi Cheng Qie, 112, 341
Bi Li Go, 342
Bi Ma Zi, 125, 342
Bi Qao Jiang, 49, 342
Bian Xu, 115, 342
Bianthraquinonyl, 123, 225, 286, 425
Biatractylolide, 26, 318, 425
Bidens bipinnata, 28, 248, 361, 464, 470
Bidens pilosa, 28, 379, 533, 537
Bidens tripartita, 28, 183, 248, 362, 427, 444,
        505, 554, 572, 583, 603, 628, 632–633,
        640
Bigelovin, 69, 80, 425
Bignonia catalpa, 248, 594, 620, 624
Bignonia chinensis, 248
Bignonia grandiflora, 28
Bilobal, 73, 203, 425
Bilobalide, 203, 591
Bilobetin, 73, 203, 425, 591
Bing Lang, 20, 158, 342
Biota chinensis, 29, 343, 464, 536, 544, 569
Biota orientalis, 145, 184, 585, 591, 593,
        602–603, 613, 627, 630
Biotin, 17, 19, 116, 181, 210, 241, 265, 425, 591
Bis (2-*ethyl butyl) phthalate,* 102, 425
Bisabolene, 56, 99, 150, 161, 195, 276, 302, 425
Bisesquiterpenoid, 26, 318, 425
Bishomophinolenic acid, 282, 591
Bis-norargemonine, 299, 591
Bistorta lapidosa, 29, 377
Bitter glycoside, 37, 252, 425
Bitter principle, 34, 62, 242, 250, 416, 426
Bletilla hyacinthina, 29, 53, 63, 339, 464, 474
Bletilla striata, 327, 330
Blumea balsumifera, 29, 184, 248, 338, 426, 430,
        438, 457, 464, 501, 516, 528, 542, 556,
        593, 595, 603, 630
Blumea hieraciifolia, 29, 248, 382
Blumea lacera, 29, 248, 354, 412, 432, 439, 441,
        468, 576
Bo Hoo, 97, 342
Bo Lo Mi, 24, 342
Bo Lou Hui, 92, 342
Bocconine, 92, 426
Bocconoline, 92, 426
Boehmeria densiflora, 30
Boehmeria tenacissima, 391
Bogoroside, 18, 426
Bonducin, 31, 249, 426, 591

Borneol, 14, 23–24, 29, 39, 47, 51, 59, 76, 83,
        145–146, 150, 159, 161, 182, 184, 191,
        197, 207, 233, 240, 244–245, 252–253,
        268–269, 282, 291, 299, 302, 426, 591
Borneol acetate, 14, 240, 282, 426, 591
Bornol, 40, 426
Bornyl acetate, 65, 76, 146, 184, 233, 426
Bornyl esters, 302, 591
Bornyl isovalerate, 150, 190, 302, 426, 592
Bornylautate, 87, 271, 426
Boschniakine, 41, 159, 426
Boschniakinic acid, 159
Boschnialactone, 41, 159, 426
Bougainvillea brasiliensis, 30, 344, 400, 424,
        489
Bourbonene, 89, 210, 426
Brahmissoside, 37, 186, 427
Brahmoside, 37, 186, 427
Brasilin, 31, 249, 427
Brassica alba, 30, 184, 339, 407, 420, 460, 463,
        515–516, 539, 557, 617
Brefeldin A, 242, 592
Bromelain, 180, 592
Bromelin, 15, 180, 427
Brucea javanica, 30, 388, 427, 524, 579
Bruceines, 30, 427
Bruceolide, 30, 427
Brucine, 140, 427
Brusatol, 30, 427
Bu Gu Zi, 119, 342
Buddleia formosana, 30, 341, 427
Buddleia madaguscariensis, 367
Buddleoglycoside, 30, 427
Bufotenine, 57, 110, 259, 427
Bulbocapnine, 48, 255, 427
Bullatines, 6, 427
Bulnesene, 220, 592
Bupleuran, 31, 427
Bupleurum chinense, 31, 317, 343, 414, 427, 450,
        496, 550–551, 571
Bupleurum falcatum, 184, 306, 318, 592, 603,
        628, 637
Bupleurumol, 184, 592
Bursic acid, 34, 185, 427
Bursine, 185, 233, 592
Buteine, 28, 183, 427
Butin, 28, 183, 427
Butylidenephalide, 17, 181, 241, 265, 427
Butylphthalide, 181, 592
Butyric acid, 66, 96, 120, 148, 224, 301, 427
Buxanmine E, 31, 249, 428
Buxpiine, 31, 249, 428
Buxpiine K, 31, 249, 428
Buxus harlandii, 31, 249, 379, 428, 447–448

Buxus microoophylla, 31, 249, 356, 428, 447–448
Buxus sempervirens, 249, 310, 585, 631
Byak-angelicin, 16–17, 158, 241, 428
Byak-angelicol, 16–17, 158, 241, 428

C

c-3-epi-wilsonine, 428
Cadinene, 13, 97, 112, 114, 136, 150, 154, 172,
        181, 214, 219, 230, 239, 302, 304, 428,
        592
Caesalpinia ascendens, 249, 591, 602, 633
Caesalpinia decapetula, 31, 249, 392, 426, 478,
        552, 578
Caesalpinia pulcherrima, 31, 356, 406, 473, 546,
        564
Caffeic acid, 27, 41, 46, 49–50, 61, 68, 80, 95–96,
        115, 118, 136, 150–151, 154, 164, 169,
        177, 193, 196, 200, 203, 206, 212–213,
        221, 223, 230, 234–235, 254, 256,
        261–262, 265, 267, 274, 300, 302–303,
        428, 592
Caffeic derivatives, 273, 592
Caffeine, 145, 185, 266, 428, 592
Cai Fu, 122, 342
Cajuputol, 95, 429
Calamene, 7, 9, 238, 429
Calamenol, 7, 429
Calamenone, 7, 429
Calamus margaritae, 31, 376
Calcium, 5–6, 51, 60, 66, 83, 104, 110, 117, 133,
        157, 161, 187, 218, 222, 225, 229, 262,
        269, 278, 289, 429, 592
Calcium malate, 66, 429
Calcium oxalate, 5–6, 51, 66, 133, 161, 225, 229,
        429, 592
Calcyosin, 183, 592
Calechin, 123, 225, 286, 429
Calendula officinalis, 32, 185
Calenduline, 32, 185, 429
Callicarpa formosana, 32, 383
Callicarpa macrophylla, 32, 394, 470, 546, 564
Calloglossa lepieurii, 32, 357, 367, 392, 408, 455
Calotropin, 159, 429
Caltha leptosepala, 249
Caltha palustris, 32, 249, 365, 412, 421, 435, 437,
        480, 491, 540, 552, 554, 572, 579
Calystegia hederacea, 32, 250, 346, 494, 528
Calystegia sepium, 250, 321, 592, 612
Calystegins, 250, 592
Cam Dou, 151, 342
Camellia bohea, 33, 343
Camellia japonica, 33, 373, 414, 429, 449, 497,
        501, 547, 568

Camellia sinensis, 185, 309, 588, 590, 592, 616,
        619, 623, 625, 635
Camelliagenins, 33, 429
Camellin, 33, 429
Campanula gentianoides, 33, 358, 487, 537, 552
Campanula glomerata, 33, 250, 350, 436, 444,
        485, 491, 494, 512, 543, 570
Campanula rotundifolia, 250, 590, 592,
        601–602, 614, 618, 627
Campesterol, 11, 24, 52, 58, 87, 95, 122, 126, 130,
        141–142, 157, 194, 209, 212, 225–226,
        231, 257, 271, 290, 301, 429, 592
Camphene, 16, 27, 51, 56, 59, 62, 75, 84, 87, 112,
        114, 116, 146, 150, 153, 155, 158,
        160 161, 171, 181, 189–190, 193, 195,
        197, 207–208, 213, 219, 233, 236, 271,
        302, 430, 592
Campherene, 99, 215, 430
Camphesterol, 229, 293, 592
Camphol, 67, 200, 430
Camphor, 24, 29, 39–40, 51, 59, 62, 83, 138, 159,
        176, 184, 188–189, 193, 197, 207, 245,
        248, 253, 291, 430, 593
Camphore, 23, 182, 244, 252, 430
Camphorm citral, 197, 430
Campsis adrepens, 33, 393, 447, 454, 540
Campsis chinensis, 28, 394
Camptotheca acuminata, 33, 385, 397, 430, 484,
        511, 574
Camptothecin, 430
Canada Pon, 342
Canaline, 95, 212, 430
Canarium album, 34, 111, 351, 417, 422, 520,
        548, 569
Canavalia gibberellin I, 430
Canavalia gibberellin II, 430
Canavalia gladiata, 34, 347, 430, 476, 573
Canavaline, 34, 430
Canavanine, 26, 34, 183, 246, 430
Candicine, 109, 218, 431
Canescein, 64, 261, 431
Cang Er, 154, 173, 342
Cang Zhu, 27, 342
Caniferyl aldehyde, 129, 227, 431
Cannabidiol, 34, 185, 431
Cannabinol, 34, 185, 431
Cannabis chinensis, 34, 346, 356, 431, 437, 496,
        515, 567, 571, 576
Cannabis sativa, 185, 595, 609, 623, 635–636,
        640
Cao Bai Ching, 14, 342
Cao Guo, 342
Cao He Che, 115, 342
Cao Jue Ming, 37, 343

Cao Wu, 7, 343
Cao Yu Mei, 16, 343
Capadiene, 54, 194, 258, 431
Capillanol, 23, 245, 431
Capillarin, 23, 245, 431
Capillene, 23, 245, 431
Capillin, 23, 245, 431
Capillon, 23, 245, 431
Capric acid, 42, 87, 148, 189, 271, 301, 431
Capronic acid, 243, 431, 593
Caprylic, 53, 194, 431
Capsella bursa-pastoris, 34, 358, 427, 439, 576,
    583, 585, 592, 603, 608, 623, 637
Capsularin, 47, 431
Caragana franchetiana, 34
Caragana sinica, 359, 406, 477
Carbohydrates, 27, 78, 203, 205, 230
Carboxylic acid, 272, 593
Cardamine leucantha, 34, 250, 463, 500, 503,
    516, 524, 557
Cardamine lyrata, 380
Cardamine pratensis, 250, 616, 639
Cardenolides, 254, 593
Cardiac glycosides, 238, 254, 262, 330, 593
Cardienolides, 262, 593
Cardioactive glycosides, 169, 431
Cardiotonic, 6, 10, 18, 49, 85, 117, 167, 169,
    177–178, 208, 222, 256, 277, 306, 432
Cardoside, 71, 432
Carduus acaulis, 34, 250, 387, 464, 478
Carduus benedita, 251, 596, 613, 623, 630, 640
Carduus marianus, 251, 630
Carene, 83, 99, 112, 207, 219, 268, 276, 432, 593
Carenolide, 107, 432
Carex kobomug Ohwi, 159, 377
Cariaester, 230, 593
Cariandrol, 161, 432
Carnaubic acid, 160, 432
Carosine, 36, 186, 432
Carotene, 5, 11, 13, 29, 74, 158, 167, 171–172,
    204, 210, 220, 239, 248, 273, 396, 593
Carotenoids, 52, 100, 138, 185
Carotol, 56, 195, 432
Carpesium abrotanoides, 35, 159, 353, 381, 464,
    481, 486–487
Carprylic acid, 42, 189, 432
Carthamone, 185, 593
Carthamus tinctorius, 35, 185, 309, 354, 432,
    519, 544, 549, 593, 613, 624, 639
Cartharmin, 35, 185, 432
Carum carvi, 35, 160, 186, 310, 389, 392,
    401–402, 432–433, 437–438, 444,
    449–450, 464, 467, 481, 490, 535, 537,
    554, 572, 593, 603, 613, 624

Carvacrol, 145–146, 167, 181, 233, 299–300, 432,
    593
Carveol, 160, 432
Carvone, 39, 89, 160, 181, 186, 197, 210, 252, 275,
    280, 433, 593
Caryophyllen, 219, 593
Caryophyllene, 9, 41, 56, 83, 87, 89, 112, 145,
    152, 157, 189, 195, 197, 207, 210, 219,
    238, 268, 271, 282, 299, 303, 396,
    433, 594
Caryophyllene oxide, 152, 303, 433
Cassia alata, 35, 251, 348, 407, 437, 468, 489,
    526, 543, 547, 549
Cassia angustifolia, 35, 146, 186, 251, 349, 413,
    454, 494, 516, 547, 555, 585, 587, 591,
    599, 617, 627, 629
Cassia nomame, 35, 251, 413, 418, 437–438, 461,
    518, 521, 523, 534, 547, 570
Cassia obtusifolia, 361
Cassia occidentalis, 36, 251, 414, 433, 454, 473,
    480, 517, 570, 579
Cassia senna, 251, 618
Cassia siamea, 36, 251, 382, 437
Cassia tora, 330, 383
Cassia torosa, 305, 433
Cassiollin, 36, 251, 433
Cassythidine, 36, 433
Cassythine, 36, 433
Castanea crenuta, 36, 251, 417, 422, 520, 540,
    543, 548, 569, 573
Castanea mollissima, 357
Castanea sative, 252, 617, 623, 633
Casticin, 153, 303, 433, 594
Castor oil, 125, 433
Catalpine, 248, 594
Catalposide, 160, 433
Catapol, 229
Catechin, 10, 54, 61, 125, 225–226, 282, 286,
    289, 301, 318, 433
Catechin derivatives, 10, 318, 433
Catecholamines, 117, 222, 433
Catechutanic acid, 4, 433
Catharanthine, 160, 433
Catharanthus roseus, 36, 160, 186, 321, 343,
    406, 432, 485, 500–501, 504, 531, 548,
    567, 575, 584, 600, 619, 621, 628, 630,
    633, 638
Cathartic acid, 111, 433
Caudoside, 140, 297, 434
Caudostroside, 140, 297, 434
Caulophylline, 252, 594
Caulophyllum robustum, 36, 252, 384, 409, 434,
    480, 507, 513, 565

Caulophyllum thalietroides, 252, 586, 594, 612, 615–616, 631
Caulosaponin, 252, 594
Cauloside, 36, 252, 434
Ce Bai Ye, 29, 113, 145, 343
Ce Yan, 12, 343
Celastrol, 147, 434
Celastrus alatus, 37, 384
Cellulose, 121, 285, 335, 434
Celosia argentea, 37, 357, 434, 479, 485, 491, 520, 544
Celosia margariacea, 343
Celosiaol, 37, 434
Celtis australis, 252, 617, 633
Celtis bungeana, 37, 252, 370, 464
Cembrene, 45, 191, 253, 434
Centaur X, 46, 191, 434
Centaurium erythraea, 252, 320–321, 590, 605, 629, 632, 640
Centaurium meyeri, 37, 252, 338, 425, 436, 525
Centella asiatica, 37, 186, 310, 358, 417, 427, 507, 588, 608, 614, 619, 628, 630, 637–638
Centipeda minima, 37, 377, 406, 464, 478, 516, 552
Cephaeline, 11, 434, 453
Cephalanoplos segetum, 37, 387, 406, 437, 552
Cephalomannine, 143, 298, 434
Cephalotaxine, 38, 434, 453
Cephalotaxinone, 38, 434
Cephalotaxus fortunei, 38, 434, 453, 462, 480, 578
Cephalotaxus wilsoniana, 38, 346, 402, 428, 453, 462, 483–484, 490
Cepharamine, 139, 434
Cepharanthine, 139, 434
Cerberin, 434
Cerberose, 434
Cereberigenin, 434
Cerotic acid, 10, 24, 113, 152, 303, 434
Cerotinic acid, 69, 113, 169, 201, 434
Ceryl alcohol, 83, 163, 269
Cerylalcohol, 32, 38, 171, 185, 187, 435
Cerylic alcohol, 143, 232, 435
Cevadine, 32, 249, 435
Cha, 1, 4, 20, 33, 39, 145, 175, 338, 343, 349, 360, 373, 380
Chaconine, 230, 594
Chaenomeles japonica, 38, 357, 439, 484, 507, 565, 576
Chaenomeles speciosa, 187, 387, 592, 600, 614, 624
Chai Hu, 31, 343
Chalcone glucose, 74, 204, 264, 435

Chamaenerion angustifolium, 38, 187, 363, 435, 444, 509, 530, 582, 602, 617, 633
Chamazulene, 176, 594
Chanerol, 38, 187, 435
Chang Bai Rui Xian, 343
Chang Chun Hua, 36, 160, 343
Chang Chun Ton, 75, 343
Chang Pu, 7, 343
Chang Shan, 7, 57, 343
Chang Shu, 40, 343
Changium smyrnioides, 38, 367
Charantin, 214, 594
Chatinine, 150, 302, 435
Chaulmoogric acid, 78, 265, 435
Chavicine, 112, 219, 435
Chavicol, 65, 200, 206, 245, 435, 616
Che Chen Zi, 113, 343
Che Sang Zi, 59, 343
Che Ye Sha Seng, 343
Che Zhou Cao, 147, 343
Chebulic acid, 143, 232, 435, 594
Chebulinic acid, 109, 143, 435
Chelerythrine, 39, 92, 164, 187, 304, 435, 594
Chelidocystatin, 39, 318, 435
Chelidonic acid, 302, 594
Chelidonine, 39, 187, 435, 594
Chelidonium album, 39, 339, 407, 435, 442, 541, 559, 562
Chelidonium majus, 187, 318, 328, 590, 594, 611, 631
Chelilutine, 92, 435
Chelirubine, 92, 164, 435
Chen Pi, 42, 344
Chen Wei, 125, 344
Chcn Xiang, 18, 344
Chenopotium ambrosiodes, 39, 344, 410, 417, 421, 449, 460, 475, 491, 495, 497, 527, 546, 552, 578
Chikusetsa saponin II, 105, 435
Chikusetsa saponin IV, 105, 435
Chimaphila umbellata, 39, 187, 367, 412, 415, 419, 421, 435–436, 463, 482, 485, 490, 494, 545, 573, 601, 608, 616, 633
Chimpahilin, 225, 594
Chinese Ji, 41, 344
Ching Mian Hua, 111, 344
Chinic acid, 39, 187, 436
Chiretta, 37, 252, 436
Chiu Chung Ko, 344
Chlogogenic acid, 96, 213, 274, 436
Chloranthus glubra, 39, 360, 464, 470, 530
Chlorogen acid, 128, 291, 436

Chlorogenic acid, 33, 41, 49–50, 61, 80, 89, 92,
        133, 136, 149, 151, 164, 168, 193, 196,
        206, 209, 211, 220–221, 229–231, 235,
        250, 256, 267, 272, 293, 303, 436, 594
Chlorogenic derivatives, 273, 594
Chlorogenin, 71, 202, 436
Chlorophenolic acid, 67, 200, 436
Chloroquine, 22, 182, 244, 436
Cholestanol, 126, 226, 290, 436
Cholesterol, 12, 52–53, 71, 87, 93, 101, 107, 123,
        147, 172, 179, 194, 209, 211, 215, 225,
        230, 234, 257, 279, 286, 301, 436, 595
Choline, 32, 34, 37, 39–40, 57–59, 69, 74, 78–79,
        97, 111, 120, 133, 136, 142–144, 152,
        157, 172, 185, 188, 195–196, 200, 203,
        214, 225, 229, 232–235, 249, 252, 270,
        284, 295, 297, 437, 585, 595
Choline asparaginer, 78, 437
Chou Chie Cao, 30, 344
Chou Fu Yong, 142, 344
Chou Lee, 118, 344
Chou Mu Lee, 43, 344
Chou Wu Tong, 44, 160, 344
Chou Xing, 39, 344
Chromone, 35, 85, 437
Chrysanthemaxanthin, 39, 133, 252, 293, 437
Chrysanthemin, 40, 437
Chrysanthemum boreale, 39, 252, 409, 426, 430,
        433, 437–438, 501, 579
Chrysanthemum cinerriaefolium, 39, 188, 344,
        403, 464, 560, 595, 613, 620, 625, 630
Chrysanthemum jucundum, 40, 401, 426, 430,
        437, 443, 457
Chrysanthemum koraiense, 360
Chrysanthemum parthenium, 253, 593, 617, 633
Chrysanthemum procumbens, 389
Chrysanthemum vulgare, 253, 591, 635
Chrysanthinin, 39, 252, 437
Chrysarobin, 35–36, 186, 251, 437, 547
Chryso-obtusin, 251, 437
Chrysophanein, 126–127, 227, 290, 437
Chrysophanic acid, 35–36, 60, 116, 179, 186, 221,
        251, 262, 437, 595
Chrysophanol, 35, 116, 123, 126–127, 164, 225,
        227, 251, 286, 290, 438, 595
Chu Chong Jiu, 120
Chu Gu Jiu, 39, 344
Chu Ye, 110, 344
Chuan Duan Chang Cao, 48, 345
Chuan Jian, 154, 345
Chuan Jin Pi, 76, 345
Chuan Lian, 96, 345
Chuan Shan Long, 58, 345
Chuan Xiang, 85, 345

Chuan Xin Lian, 15, 158, 345
Chui Pen Chao, 132, 345
Chun Pi, 10, 345
Chung Way, 85, 345
Ci Gu, 127, 345
Ci Hi Li, 146, 345
Ci Luo Shi, 95, 345
Ci Seng, 103, 345
Ci Wu Jia, 5, 62, 345, 375
Cibotium barometz, 40, 316, 353, 503, 528
Cicerose, 81, 207, 268, 438
Cicutol, 160
Cicutoxin, 160
Cimicifuga dahurica, 40, 411, 438, 469, 489, 495
Cimicifuga foetida, 188, 376, 595, 610, 626, 628,
        633
Cimicifuga racemosa, 188, 583, 595, 610, 616,
        628, 632, 637
Cimicifugin, 188, 595
Cimicifugoside, 188, 595
Cimifugenol, 40, 188, 438
Cimigenol, 40, 188, 438
Cimitin, 40, 188, 438
Cincholic acid, 8, 438
Cincole, 13, 239, 438
Cineol acid, 153, 438
Cineole, 13, 23–24, 29, 40, 51, 62, 87, 89, 95, 97,
        159, 176, 182, 184, 188, 192, 197, 207,
        210, 213–214, 239–240, 245, 248, 261,
        271, 275, 291, 438, 595
Cinerins, 188, 225, 595
Cinnamaldehyde, 189, 595
Cinnamic acid, 40, 55, 87, 90, 189–190, 210, 221,
        225, 271, 273, 286, 439, 595
Cinnamic aldehyde, 40–41, 189, 439
Cinnamomum aromaticum, 40, 439, 525,
        532–533
Cinnamomum camphora, 40, 188, 343, 397, 416,
        438, 449, 466, 536, 593, 613, 627, 635
Cinnamomum cassia, 189, 318, 352, 592, 600,
        613, 622–623
Cinnamomum zeglanicum, 41, 189, 346, 420,
        433, 446, 466, 484, 497–498, 512, 521,
        527, 595, 597, 601, 617, 633
Cinnamon, 40–41, 184, 595
Cinnamyl acetate, 40, 189, 439
Cinnamyl cinnamate, 141, 271, 297, 409, 595
Cinnamylococaine, 64, 199, 439
Cirsium albescens, 41, 401, 428, 436, 464, 540,
        549
Cirsium chinense, 41, 344, 407, 421, 423, 487,
        498, 529, 561, 565
Cirsium japonicum, 34, 346, 387
Cissamine, 41, 189, 439

Cissampeline, 41, 189, 595
Cissampelos pareira, 41, 189, 385, 439, 450, 458, 480, 510, 595
Cistanche deserticola, 41, 335, 426, 460
Citral, 13, 29, 42, 53, 56, 62, 88, 93, 125, 146, 154–155, 194–195, 197, 211, 217, 226, 233, 236, 239, 248, 257, 289, 430, 439, 595
Citrene, 51, 439
Citric acid, 34, 38, 49, 52, 59, 64, 83, 88, 104, 118–119, 124, 151, 178, 180, 185, 187, 196, 199, 217, 223, 226, 256–257, 263, 269, 272, 278, 288, 303, 312, 439, 595
Citrifoliol, 116, 439
Citrogellol, 53, 194, 257, 439
Citrol, 93, 125, 211, 226, 289, 439
Citronellal, 53, 62, 194, 197, 257, 439, 596
Citronellic, 53, 194, 257, 439
Citronellol, 56, 62, 83, 113, 153–155, 195, 197, 207, 268, 289, 304, 440, 596
Citrulline, 42, 95, 189, 212, 440
Citrullus anguria, 42, 431–432, 440, 445, 500, 503, 516, 524, 528, 560
Citrullus edulis, 379
Citrullus vulgaris, 189, 588, 596
Citrus aurantium, 42, 190, 380, 470, 517, 563, 589, 591, 593, 595, 597, 617, 638–639
Citrus deliciosa, 42, 360, 439, 481, 501, 512, 576
Citrus reticulata, 42, 344, 360, 414, 475, 502, 512, 518–519, 521, 538, 542
Clavatine, 90–91, 210–211, 440
Clavatoxine, 91, 211, 440
Clavoloninine, 91, 211, 440
Clematis armandii, 21, 42, 253, 416, 553, 572
Clematis chinensis, 43, 253, 412, 552
Clematis heracleifolia, 368
Clematis intriicata, 43, 253, 382, 440, 524
Clematis vitalba, 253, 624, 628
Clematoside A, 43, 253, 440
Cleome spinosa, 43, 379, 440, 499, 564, 578
Cleome viscosa, 386
Cleomin, 43, 440
Clerodendrin, 44, 160, 190, 478, 596
Clerodendrin acacetin, 190, 596
Clerodendrum cyrtophyllum, 43, 346, 398, 477, 486, 488, 490, 499, 519, 572
Clerodendrum fragrans, 43, 314, 344, 397–398, 440
Clerodendrum spicatus, 380
Clerodendrum trichotomum, 44, 160, 190, 344, 401, 414, 440, 459, 462, 471–472, 478, 511, 535, 596, 616
Clerodolone, 44, 160, 190, 440
Clerodone, 160, 440

Clerosterol, 43, 160, 440
Clinopodium acinos, 253
Clinopodium chinense, 44, 253, 348, 414, 459, 481, 558, 573
Clinopodium gracile, 352
Clinopodium umbrosum, 350
Clivia miniata, 44, 361, 440, 506, 514
Clividine, 44, 440
Clovene, 154, 304, 440
Cnicin, 251, 264, 596
Cnidiadin, 44, 190, 440
Cnidilide, 85, 270, 440
Cnidimine, 44, 190, 440
Cnidium monnieri, 44, 190, 375, 415, 440–441, 497–498, 522, 592, 610, 623
Cocain, 596
Cocculolidine, 45, 440
Cocculus diversifolius, 44, 97, 135, 349, 403, 458, 492, 513, 557, 572
Cocculus laurifolius, 45, 440–441, 483, 492, 507, 522, 570–571
Cocculus sarmentosus, 357
Cocculus trilobus, 367
Coclobine, 45, 441
Codeine, 1, 106, 217, 279, 441, 596
Codonopsis lanceolata, 45, 388, 405, 409, 414, 460, 505, 524, 561
Codonopsis pilosula, 45, 190, 329, 348, 410, 472, 477, 487, 512, 517, 559, 562, 565, 584, 621, 624, 631–632, 637
Codonopsis tangshen, 631
Coessidine, 441
Coix lachryma-jobi, 317, 329
Coixenolide, 45, 441
Coixol, 45, 441
Colchicine, 80, 86, 148, 440–441
Colchicine amide, 80, 441
Columbamine, 46, 48, 191–192, 247, 254–255, 273, 441, 596
Columbianetin, 44, 190, 441, 522
Colxol, 280, 596
Commelin, 45, 441
Commelina communis, 45, 388, 419, 441, 470
Commiphora molmol, 254
Commiphora myrrha, 45, 191, 253, 367, 408, 434, 464, 479, 507, 516, 546, 580, 583, 607, 626
Complanatine, 90, 210, 441
Conamine, 164, 441
Concuressine, 441
Condensed tannin, 441
Condurangin, 78, 441
Condurangogenins, 274, 596
Conessimine, 164, 441

Confertifolin, 108, 441
Cong, 12, 88, 155, 159, 342, 345, 351, 374
Cong Lan, 155, 345
Conhydrine, 160
Conicine, 160
Coniferaldehyde, 280, 596
Coniferyl alcohol, 29, 248, 441
Coniferyl cinnamate, 141, 297, 441
Conioselinum univittatum, 46, 351, 464
Conkurchine, 442
Convallamarin, 46, 115, 220, 254, 283, 442
Convallaria keiskei, 46, 254, 363, 442
Convallaria majalis, 254, 593, 596, 603
Convallarin, 115, 220, 283, 442
Convallatoxin, 9, 18, 46, 178, 237, 254, 442
Convallatoxol, 46, 254, 442, 596
Convalloside, 46, 254, 442, 596
Convolvulin, 254, 267, 596
Convolvulus arvensis, 46, 254, 382, 423, 428,
        494, 543
Convolvulus jajapa, 254, 596, 626
Conyza canadensis, 46, 191, 420, 434, 446,
        453–454, 464, 501–503, 509, 603, 613,
        633, 635
Copaene, 24, 245, 442
Coptis chinensis, 46, 191, 317, 421, 441–442, 493,
        504, 528, 573, 578, 590, 596, 640
Coptis groenlandica, 254, 611
Coptis japonica, 254, 356
Coptisine, 39, 46, 48, 92, 105, 144, 164, 187,
        191–192, 254–255, 279, 299, 442, 596
Corchorin, 47, 442
Corchoritin, 47, 442
Corchoroside A, 9, 64, 157, 178, 237, 261, 442
Corchorus capsularis, 47, 356, 404, 431, 442,
        478, 503, 524, 528, 560
Corchorus olitorius, 330, 373, 442, 455, 463,
        472, 477, 524, 562
Corchotoxin, 47, 442
Cordycepic acid, 151, 302, 442
Cordycepin, 47, 442
Cordyceps sinensis, 47, 308, 325–326, 328, 348,
        397, 403, 442, 450, 540, 549, 561
Coreximine, 48, 255, 442
Coriandrol, 47, 191, 443
Coriandrum sativum, 47, 161, 191, 384, 404, 426,
        432, 443–444, 448, 452, 458, 475,
        494, 501–502, 507, 521, 526, 532, 543,
        549, 554, 563, 566–567, 572, 585, 613,
        624, 635, 639
Corilagin, 129, 443
Coriolus versicolor, 47, 325–326, 538
Cornerin, 277, 596
Cornigerine, 80, 443

Cornin, 48, 93, 191, 255, 443
Cornus alba, 47, 255, 457, 494, 532, 544
Cornus canadensis, 255, 596, 626
Cornus florida, 255
Cornus kousa, 310, 379
Cornus macrophylla, 358
Cornus officinalis, 48, 191, 255
Cornus walteri, 48, 255, 361, 468, 503
Coroglaucigenin, 159, 443
Corosolic acid, 49, 316, 443
Cortenerin, 277, 596
Corticosteroids, 49, 443
Corycavine, 48, 255, 443
Corydalamine, 48, 192, 255, 443
Corydaline, 192, 255, 596
Corydalis, 48, 192, 255, 345, 347, 361, 385, 388,
        402, 407, 427, 441–443, 445, 449,
        451–452, 458, 489, 528, 541, 545, 554,
        557, 567, 596–597, 612, 624, 635
Corydalis ambigua, 48, 255, 407, 441–443, 445,
        449, 452, 458, 541, 567
Corydalis bungeana, 255, 347
Corydalis decumbens, 48, 255, 385, 427, 451
Corydalis incisa, 48, 345, 402, 442–443, 489,
        528, 545, 554, 557
Corydalis solida, 255
Corydalis ternata, 361
Corydalis yanhusuo, 192, 388, 596–597, 612,
        624, 635
Corydamine, 48, 255, 443
Coryfolin, 119, 224, 443
Corylifolinin, 119, 224, 443
Corylus avellana, 256, 591, 601–602, 633
Corylus heterophylla, 49, 256, 417, 422, 520,
        548, 569
Corylus mandshurica, 392
Corynantheine, 149, 301, 443
Corynoline, 48, 255, 443
Corynoloxin, 48, 255, 443
Corynoxeine, 149, 301, 443, 597
Corypalline, 299, 597
Cosin, 26, 183, 246, 443
Cosmosiin, 10, 40, 178, 443
Costene, 27, 444
Costol, 27, 171, 444
Costulactone, 27, 444
Costunolide, 66, 171, 262, 444
Costus specious, 49, 342, 398, 443, 457, 569
Coumaric acid, 221, 261, 597
Coumarin derivatives, 178, 298, 597

Coumarins, 5, 17–18, 47, 76, 115, 124, 128, 143,
        158, 165, 178, 181, 183, 189–190, 193,
        195, 198, 201–202, 208, 212–214, 231,
        242, 245–246, 256, 263, 269, 274,
        280, 287, 303, 317, 332, 444, 597
Coumaroylquinic acid, 33, 250, 444
Coumesterol, 95, 212, 444
Couvallatoxin, 157, 444
Crataegol acid, 141, 231, 444
Crataegolic acid, 38, 49, 119, 256, 444
Crataegus acid, 256, 597
Crataegus chlorusarca, 552
Crataegus cuneata, 49, 256, 428, 436, 439,
        443–444, 470, 485, 497, 509, 543, 573
Crataegus laevigata, 585–586, 594, 597, 600,
        603, 621, 624, 626–627, 634, 637
Creosol, 281, 597
Cresols, 275, 597
Crocetin, 49, 192, 444–445
Crocetin di-glucose ester, 49, 192, 444
Crocetin geniobiose glucose ester, 49, 192, 445
Crocine glycosides, 192, 597
Crocus sativus, 49, 192, 376, 422, 444–445, 506,
        590, 595, 597, 623, 627
Crotalaria mucronata, 50, 393, 515, 520, 546,
        573, 577–578
Crotin, 49–50, 192, 445
Croton cascarilloides, 50, 445, 533
Croton oil, 192, 597
Croton resin, 50, 192, 445
Croton tiglium, 192, 338, 597
Crotonic acid, 50, 192, 445
Crotonoside, 50, 192, 445
Crustecdysone, 170, 445
Crybulbine, 48, 192, 255, 445
Cryptone, 112, 219, 445
Cryptopine, 144, 299, 445
Cryptotaenen, 50, 192–193, 445
Cryptotaenia canadensis, 192, 388, 640
Cryptotaenia japonica, 50, 193, 357, 445, 491,
        495, 511–512, 531, 567, 570, 587, 597,
        607, 617, 622–623
Cryptotanshinone, 127, 291, 445
Cryptoxanthin, 77, 110, 143, 172, 205, 218, 281
Cryptoxan-thin-epoxide, 172, 445
Crytomeridiol, 93, 445
Crytoxanthin, 64, 199, 445
Cu Fei (Taiwan), 346
Cubebin, 112, 219, 445, 597
Cucumis melo, 50, 352, 445, 510, 561
Cucumis sativus, 50, 193, 356, 398, 415, 428,
        436, 445, 472–473, 491, 503, 508, 524,
        528, 549, 560, 613, 619–620, 631
Cucurbita maxima, 256, 613, 639

Cucurbita moschata, 51, 256, 368, 445, 561
Cucurbitacin B, 50, 100, 445
Cucurbitacin E, 50, 445
Cucurbitacins, 15, 42, 50, 180, 189, 193, 256,
        445, 597
Cucurbitine, 51, 256, 445
Cuercetin, 445
Cui Que, 56, 346
Cumaldehyde, 40, 160, 188, 445
Cumaric acid, 96, 213, 274, 446
Cumarin, 65, 446
Cumic alcohol, 154, 304, 446
Cuminic aldehyde, 41, 189, 446
Cumulene, 46, 191, 446
Cunninghamia lanceolata, 51, 379, 426, 430,
        438–439, 464, 501, 532, 536, 567
Curcasin, 82, 207, 446
Curcin, 82, 207, 446
Curcolone, 51, 193, 446
Curculigo capitulata, 51, 346, 429, 546, 564
Curculigo orchiodes, 525, 564
Curculigo stams, 386
Curcuma aromatica, 193, 391, 472, 572, 580,
        598, 601
Curcuma longa, 51, 193, 430, 446, 497, 556, 598,
        626, 637, 640
Curcuma pallida, 51, 349, 426, 438, 479, 546,
        578, 580
Curcuma phaeocoulis, 369
Curcuma zedoaria, 52, 384, 398, 446, 457, 539,
        580
Curcumin, 51–52, 193, 227, 446, 598
Curcuminoids diferuloylmethane, 156, 446
Curcumol, 51, 193, 446
Curdione, 51–52, 193, 446
Curmarin, 51, 193, 446
Curzenene, 51–52, 193, 446
Curzerenone, 52, 193, 446
Cuscuta australis, 52, 257, 348, 408, 432, 505,
        565
Cuscuta chinensis, 52, 194, 257, 383, 410, 421,
        423, 429, 436, 446, 561, 590, 603, 608
Cuscuta epithymum, 257
Cuscutalin, 52, 194, 257, 446
Cuscutin, 52, 194, 257, 446
Custeodysine, 104, 446
Cutins, 285, 598
Cyandidin-3-sophoroside, 77, 446
Cyanic acid, 99, 118, 223, 446
Cyanide, 103, 170, 223, 484, 598
Cyanidin, 68, 107, 118, 141–142, 200, 207, 217,
        223, 232, 262, 298, 447, 598
Cyanidin-3-gentiobioside, 136, 230, 447
Cyanidin-3-glucoside, 230, 447

Cyanidin-3-monogalactoside, 92, 447
Cyanidin-3-rutinoside, 33, 248, 447
Cyanin, 125, 226, 289, 447
Cyanogenic glycosides, 212, 257, 292, 598
Cyasterone, 10, 147, 238, 300, 447, 459
Cyathula prostrate, 52, 341, 460
Cycas revoluta, 52, 382, 410, 482, 559
Cyclanoline, 21, 139–140, 447
Cycloarternol, 4, 66, 447
Cycloastrangenol, 26, 183, 246, 447
Cycloencalenol, 96, 213, 447
Cyclomonerviol, 101, 447
Cyclomulberrochromene, 98, 214, 447
Cyclonerviol, 101, 447
Cycloprotobuxamines, 447
Cycloprotobuxines, 447
Cycloshikonin, 88, 209, 448
Cyclovirobuxine D, 31, 249
Cyclovivobuxine C, 31, 249, 448
Cydonia oblonga, 257, 586, 598, 602, 617, 621,
    633
Cydonia sinensis, 52, 257, 387, 439, 484, 507,
    565, 576
Cylindrin, 80, 89, 448
Cylon Rou Gui, 346
Cymarigenin, 157, 448
Cymarin, 18, 157, 242, 448, 598
Cymarol, 9, 157, 178, 237, 448
Cymarose, 173, 448
Cymbidium hyacinthinum, 53, 339, 464, 478, 515
Cymbopogon citratus, 53, 194, 257, 409, 422,
    431, 439, 452, 458, 467–468, 475, 493,
    502, 513, 519, 595–596
Cymbopogon distans, 53, 257, 464, 536
Cymbopogon goeringii, 392
Cymbopogon martinii, 314
Cymbopogon nardus, 386
Cymbopogon winterianus, 640
Cymbopogonol, 53, 194, 257, 448
Cymene, 9, 47, 99, 160–161, 167, 191, 215, 238,
    269, 275, 448, 598
Cynanchin, 53, 161, 448
Cynanchocerin, 53, 161, 448
Cynanchol, 53, 161, 448
Cynanchum atratum, 53, 340, 448
Cynanchum auriculatum, 318
Cynanchum japonicum, 53, 339
Cynanchum paniculatum, 53, 387, 452, 527, 553,
    570
Cynarin, 133, 293, 448
Cynaroside, 212, 598
Cynodon dactylon, 53, 382, 423
Cynoglossum divaricatum, 54, 258, 348, 539
Cynoglossum officinale, 258

Cynomorium coccineum, 54, 413, 423, 433, 451,
    518, 528, 562, 573
Cynomorium songarium, 380
Cyperoone, 54, 194, 258, 448
Cyperus brevifolius, 54, 314, 407–408, 423, 566
Cyperus difformis, 258, 373
Cyperus esculentus, 258, 602
Cyperus rotundus, 54, 194, 258, 386, 408,
    422–423, 431, 448, 462, 464, 484,
    523, 529, 536, 548, 556, 636
Cypripedin, 258, 598
Cypripedium calceolus, 258, 598, 604, 632
Cypripedium guttatum, 54, 258, 470, 533, 561,
    576
Cypripedium macranthum, 375
Cyrtomin, 55, 448
Cyrtomium falcatum, 55, 418, 448, 470, 491
Cysteic acid, 143, 172, 448
Cysteine, 143, 448
Cystine, 143, 159, 448
Cytisine, 137, 448
Cytisus scoparius, 55, 194, 359, 474, 553–554,
    559, 602, 629, 631
Cytochrome C, 125, 226, 449
Cytokinin, 73, 203, 449

D

Da Dou, 74, 346, 389
Da Fei Yang Cao, 67, 346
Da Feng Zi, 78, 346
Da Hua Tian Qing, 133, 346
Da Ji, 41, 66–67, 346
Da Ji Ru Zi Shu, 67, 346
Da Ma Ren, 34, 346
Da Qing, 27, 43, 81, 346
Da Qing Ye, 27, 346
Da Ri Jian Cao, 102, 346
Da Suan, 12, 346
Da Wan Hua, 32, 346
Da Xian Mao, 51, 346
Da Xian Ye Shu, 87, 346
Da Ye An, 65, 346
d-abscisin, 58, 196, 449
Daechu alkaloids, 156, 236, 451
Daemonorops draco, 55, 388, 411, 420, 439, 546
Daemonorops margaritae, 55, 356, 420, 459
Daidzein, 74, 95, 120, 137, 147, 203, 212, 224,
    234, 300, 451, 598
Daidzin, 120, 224, 451, 598
Dambonitol, 167, 173, 451
Dammaranedienol, 267, 598
Dan Gui, 17, 347
Dan Ye Mu Jing, 153, 347

Dan Ye Xiz Zhu, 109, 347
Dan Zhu Ye, 89, 347
Dao Dou, 34, 347
Daphne fortunei, 55, 258, 392, 397, 414, 475, 484, 579–580
Daphne genkwa, 195, 598, 616–617, 633
Daphne giraldii, 55, 394, 451
Daphne koreana, 55, 343
Daphne mezereum, 259
Daphnetin, 67, 153, 200, 451
Daphnetoxin, 195, 259, 598
Daphnidium myrrha, 55, 385
Daphnoretin, 30, 307, 451
d-apiose, 84, 449
Darutoside, 171, 451
d-asarinin, 168, 449
Dasycarponin, 299, 598
Datura alba, 56, 451, 485, 554
Datura innoxia, 195, 597, 603, 608, 633, 640
Datura suaveolens, 56, 372, 412–413, 418, 497
Datura tatula, 366
Daturodiol, 56, 195, 451
Daturolone, 56, 195, 451
Daucine, 56, 195, 451
Daucol, 56, 195, 451
Daucosterin, 5, 451
Daucosterol, 5, 8, 54, 56, 195, 237, 451
Daucus carota, 56, 161, 195, 336, 368, 396, 409, 417, 425, 430, 432–433, 439–440, 451, 475, 505–506, 516, 527, 532, 535, 542, 572, 593, 614, 619, 635, 638–639
Dauricine, 96–97, 275, 331, 451, 598
Dauricinoline, 96, 451
Dauricoline, 96
d-backuchiol, 119, 224, 449
d-borneol, 14, 155, 197, 236, 240, 449, 598
d-camphene, 7, 215, 598
d-camphor, 5, 14, 39–40, 118, 169, 176, 187–188, 223, 240, 449
d-carvone, 16, 35, 181, 186, 449
d-catechin, 4, 49, 117, 176, 256, 284, 449
d-catechol, 33, 115, 150, 235, 301, 449
d-corydaline, 48, 192, 255, 449
Deacetylfawcetine, 91, 211, 452
Deacetylo-leandrin, 167, 452
Deacylcynanchogenin, 53, 167, 452
Deacylmetaplexigenin, 53, 167, 452
Deanolic acid, 223, 599
Deaspidin, 60, 197, 452
Debilic acid, 21, 243, 452
Decalactone, 119, 223, 452
Decanal, 47, 53, 161, 191, 194, 257, 452
Decanol, 47, 161, 191, 452
Decanoylacetaldehyde, 78, 452

Decuroside, 17, 241, 452
Decursidin, 17, 108, 241, 280, 452
Decursin, 17, 158, 241
Decursind, 452
Decussatin, 141, 297, 452
Decylic aldehyde, 47, 161, 191, 452
Degueline, 143, 298, 452
Dehydroandrographolide, 15, 452
Dehydrocorydaline, 48, 192, 255, 452
Dehydrocorydalmine, 48, 192, 255, 452
Dehydroevodiamine, 67, 452
Dehydrofukinone, 181, 599
Dehydromatricaria, 46, 63, 191, 199, 453
Dehydromatricaria ester, 46, 191, 453
Dehydrosilybin, 134, 229, 453
Dehydrosoyasaponin, 303, 599
Delonix regia, 56, 349, 406, 479, 552
Delphinidin, 95, 118, 169, 180, 212, 223, 453, 599
Delphinidin-3,5-diglucoside, 173, 243, 453, 599
Delphinidin-3-monoglucoside, 136, 295, 453
Delta-3-carene, 153, 453
Delta-limonene, 219, 599
Delta-linalool, 182, 599
Demethyl-coclaurine, 277, 453
Demethyl-tubulosine, 11, 453
Dencichine, 105, 453
Dendrobine, 56, 453
Dendrobium nobile, 56, 63, 376, 453
Dendrolasin, 227, 599
Deng Qing Cao, 67
Deng Tai Ye, 13, 347
Deoxyandrograppholide, 15, 453
Deoxyelephantopin, 61, 453
Deoxypodophyllotoxin, 18, 60, 161, 242, 453
Deoxysantalin, 129, 227, 453
Deoxyschizandrin, 130, 228, 305, 453
Deoxytubulosine, 11, 454
Dephenyl methane-2-carboxylic acid, 46, 191, 454
Desacetylasperuloside, 102, 454
Desacetylmatricarin, 5, 176, 454
Desmodium gangeticum, 259, 584, 640
Desmodium microphyllum, 56, 259, 380, 494
Desmodium pulehellum, 57, 259, 369, 427, 520
Desmodium triforum, 57, 259, 372, 539, 556, 564
Desmodium triquetrum, 57, 259, 355
des-O-methyl-licariine, 63, 261, 454
Desoxypodophyllotoxin, 114, 454
Dextrose, 33, 78, 205, 248
d-*fen chone,* 169
d-fenchone, 69, 118, 201, 223, 449
d-fructose, 127, 449
d-galactose, 127, 134, 449
d-galacturonic acid, 94, 113, 220, 274, 282, 449

d-gallocatechol, 150, 235, 301, 449
d-glucaric acid, 73, 203, 450
d-glucose, 31, 132, 135, 295, 450
d-guereitol, 41, 189, 450
Di Dan Tou, 61, 347
Di Ding Zi Jing, 48, 347
Di Er Cao, 79, 347
Di Gu Pi, 90, 347
Di Huang, 58, 122, 347, 366
Di Jiao, 145, 347
Di Jin, 66, 107, 347
Di Jin Cao, 66, 347
Di Yu, 117, 347
Diacetate, 19, 67, 200, 454, 499–500
Diacetyl-atractylodiol, 26, 454
Diadzein, 224, 599
Diadzin-4,7-diglucoside, 120, 224, 454
Dianthrone glucosides, 186, 251, 599
Dianthronic heteroside, 36, 251, 454
Dianthus barbatus, 57, 259, 454, 464, 466
Dianthus caryophyllus, 259, 590, 601, 616
Dianthus chinensis, 161, 377, 420, 466, 513, 533,
    550
Dianthus saponin, 57, 259, 454
Dianthus superbus, 371
Diastase, 288, 599
Dibasic acids, 170, 454
Dibenzoylgagaimol, 167, 454
Dibilone, 21, 243, 454
Dicaffeoxylquinic acid, 154, 454
Dichrins, 57, 455
Dichroa cyanitis, 57, 399, 455
Dichroa febrifuga, 7, 343
Dichroidine, 57, 455
Dichroines, 57, 455
Dicoumarol, 95, 212–213, 274, 455, 599
Dicranopteris linearis, 57, 366, 404, 481, 521,
    544
Dictamine, 154, 304, 455
Dictamnin, 195, 599
Dictamnus albus, 57, 70, 195, 340, 418, 421, 437,
    455, 471, 490, 502, 523, 539, 542, 552,
    558, 571, 586, 599, 601
Dictamnus dasycarpus, 340
Dicumarol, 96, 213, 274, 455
Diethyl phthalate, 102, 455
Diffractaic acid, 149, 455
Digeneaside, 32, 455
Digicirin, 58, 196, 455
Digicoside, 58, 196, 455
Digifolein, 58, 196, 455
Digipurin, 58, 196, 455
Digitalia purpurea, 58, 196, 455, 476, 542, 562,
    592, 594–595, 599, 612, 625, 628

Digitalia sanguinalis, 366
Digitonin, 58, 196, 455
Digitoxigenin, 47, 58, 167, 196, 330, 455
Digitoxin, 58, 196, 305, 455, 599
Diglucoside, 5
Diglycoside, 142, 172, 232, 447, 455
Digoxin, 1, 196, 305, 599
Dihydro-artemisinin, 455
Dihydro-bigelovin, 456
Dihydro-carveol, 456
Dihydro-conessine, 164, 456
Dihydro-cyclonervilol, 456
Dihydro-foliamenthin, 456
Dihydro-harman, 456
Dihydro-N-methyl-isopelletierine, 293, 455
Dihydroxy methyl anthraquinone, 98, 276, 456
Dihydroxycoumarin, 28, 183, 456
Dihydroxyglutamic acid, 47, 255, 457
Dilactone, 222, 227, 600
Dillapiole, 16, 158, 181, 457
Dimeric indole alkaloids, 186, 600
Dimethoxyallylbenzene, 101, 457
Dimethy ether, 29, 457
Dimethyl thymohydroquinone, 65, 457
Ding Gong Teng, 64, 347
Ding Xian, 65, 141, 347
Diogenin, 49, 457
Diol, 51–52, 457
Diosbulbin, 58, 457
Diosbulbines, 58, 457
Dioscin, 58, 457
Dioscorea batatus, 58, 196, 378, 406, 449, 508,
    540
Dioscorea bulbifera, 58, 328, 356, 424, 429, 457,
    487, 552, 561, 564
Dioscorea cirrhosa, 58, 377, 515
Dioscorea nipponica, 58, 345, 398, 457, 571
Dioscorea opposita, 58, 196, 375, 415, 437, 477,
    501, 557, 572, 584, 600, 624, 628, 631
Dioscorecin, 58, 457
Dioscoretoxin, 58, 457
Diosgenin, 11, 25, 58, 106, 136, 147, 196, 217,
    230, 239, 295, 300, 457, 600
Diosgenin glycoside, 217, 457
Diosmetin-7-glucoside, 40, 457
Diosmin, 155, 304, 457
Diospyros chinensis, 59, 402, 425, 437, 556
Diospyros lotus, 377
Dioxybenzoic acid, 13, 240, 457
Dioxybenzol, 170
Dipalmiin, 148, 457
di-p-coumaroylmethane, 156, 454

Dipentene, 16, 47, 53, 63, 95, 99, 112, 153, 158, 169, 181, 189, 191, 194, 199, 215, 219, 257, 458, 600
Diploptene, 60, 197, 458
Dipsacus asper, 59, 259, 387, 406, 464, 584, 601
Dipsacus fullonum, 259, 609, 629
Disinomenine, 44, 97, 135, 275, 458
d-isochondrodendrine, 41, 189, 450
Disulphides, 201, 600
Diterpene acids, 243, 600
Diterpene jatrophone, 207, 600
Diterpenes, 73, 207, 246, 259, 262, 458, 600
Diterpenoids, 19, 243, 327, 458
Divaricoside, 140, 297, 458
dl-anabasine, 11, 458
dl-beheerine, 41, 189, 458
dl-borneol, 62, 197, 458
dl-catechol, 92, 458
dl-curine, 41, 189, 458
d-limonene, 9, 35, 62, 130, 142, 172, 186, 197, 228, 232, 238, 298, 450
dl-methylisopelletierine, 132, 293, 458
dl-N-noramepavine, 92, 458
dl-tetrahydropalmatine, 48, 192, 255, 458
dl-tetrandrine, 139, 458
d-lupaine, 90, 450
d-mannitol, 10, 47, 179, 238, 450
d-mannose, 134, 450
d-matrine, 137, 450
d-menthone, 130, 228, 450
d-N-methylpseudoephedrine, 63, 198, 260, 450
Do Xing Cao, 85, 341, 348
Dodecen-4-oic acid, 87, 271, 458
Dodium, 187, 600
Dodonaea viscosa, 59, 196, 343, 406, 477, 546, 564, 633
Dolichos lablab, 59, 260, 338, 415, 477, 487, 506, 572
Dolichos pruriens, 260
Domesticine, 99, 458
Don Gua, 28, 348
Dong Chong Xia Chao, 47, 348
Dong Kui Zi, 94, 348
Dong Ling Cao, 122, 348
Dong San Hu, 135–136, 348, 366
Dong Seng, 38, 45, 348, 367
Donoxime, 57, 259, 458
Dotriacontanol, 61, 459
Dou Kou, 13, 348
Dou Tu Si, 52, 348
Douminidine, 71, 202, 459
d-oxymatrine, 137, 450
d-pinene, 9, 238, 450
d-pseudoephedrine, 198, 260, 598

d-pseudoephedrinem, 63, 198, 260, 450
Draba nemorosa, 59, 382, 407
Draceana graminifolia, 59, 365, 459, 515, 522
Dracoalban, 55, 411, 459
Dracocephalum integrifolium, 59, 371, 464, 470
Dracoresene, 55, 411, 459
Dracoresinotannol, 55, 459
Dracorhodin, 59, 459
Dracorubin, 59, 459
d-raffinose, 127
Drosera anglica, 59, 439, 507
Drosera burmunni, 366
Drosera rotundifolia, 196, 600, 603, 618, 640
Drug interactions, 2
Dryobalanops aromatica, 59, 197, 364, 426, 430, 556, 567, 593, 598
Dryocrassin, 60, 197, 260, 459
Dryopteris crassirhizoma, 25, 60, 260, 352, 405, 458–459, 469–470
Dryopteris filix-mas, 260
Dryopteris laeta, 60, 367, 452, 458, 469, 524, 529, 546
d-sesamin, 168
d-sesamine, 5, 450
d-sophoranol, 137, 450
d-stachyose, 127, 450
d-terpineol, 150, 302, 450
d-tetrahydropalmatine, 48, 255, 451
d-tetrandrine, 140, 451
Du Jing Sha, 93, 348
Du Zhong, 65, 348
Duan Geng Ren Dong, 89, 348
Duan Geng Wu Jia, 5, 348
Duan Xue Liu, 44, 348
Duchesnea indica, 60, 70, 375, 429, 437, 461, 535, 578
Dui Ye Dou, 35, 348
Dulcilone, 131, 459
Dulciol, 131, 459
Dulcite, 65, 261, 459
Dulcitol, 95, 459
d-Usnic acid, 272, 598
d-verbascose, 127, 451
d-xylose, 113, 220, 282, 451
Dydimin, 44, 253, 459
Dysosma pleiantha (Hance) Woodson, 60, 161, 338, 369, 485, 494, 543

E

E Zhu, 349
Eauisetum palustre L., 162, 368
Ebelin lactone, 78, 459
Eburicoic acid, 117, 222, 459

Ecdysone, 104, 396, 459
Ecdysones cyasterone, 10, 238, 459
Ecdystecide, 459
Ecdysterone, 6, 10, 52, 95, 104, 114, 147, 177,
    212, 238, 300, 460
Ecgonine, 64, 199, 420, 460, 600
Echinocoside, 41, 460
Echinocystic acid, 45, 460
Echinopanacene, 103, 278, 460
Echinopanacol, 103, 278, 460
Echinops dahuricus, 60, 460
Echinops grijsii, 364
Echinopsine, 60, 460
Echitamidine, 13, 157, 180, 460
Echitamine, 13, 180, 460
Eclipta alba, 60, 197, 406, 460, 520, 585, 600,
    628
Eclipta erecta, 61, 367, 410, 421, 453, 464, 552,
    561, 563, 578
Eclipta prostrata, 334
Eclipta thermalis, 350
Ecliptine, 60–61, 197, 460, 600
Eicosanoic acid, 142, 460
Eicosenic acid, 30, 184, 460
Eikosanol, 87, 209, 460
Eissampeline, 460
Elaeagnus formosana, 61
Elaeagnus glabra, 61, 381, 410, 462, 470
Elaeagnus oldhumii, 61, 390, 416, 515, 558
Elaeagnus pungens, 61, 397, 428, 433, 436, 456,
    462, 480, 519, 567
Elaeagnus umbellata, 355
Elaeocarpid, 61, 460
Elcosanedicarboxylic acid, 460
Elegic acid, 460
Elemene, 9, 24, 238, 245, 460
Elemicin, 53, 194, 257, 460
Elephantin, 61, 461
Elephantopin, 61, 461
Elephantopus elatus, 61, 347, 460, 552
Elephantopus molis, 61, 365, 453, 459, 461–462,
    489, 505, 514, 532
Elephantopus scaber, 62, 381, 426
Elettaria cardamomum, 62, 197, 390, 409, 430,
    438–440, 450, 458, 475, 502–503, 516,
    524, 527–528, 535–536, 549, 566–567,
    569, 591, 593–594, 601, 608, 623,
    627, 635
Eleutherococcus senticosus, 62, 162, 198, 310,
    332, 336, 384, 552
Eleutherosides, 62, 198, 461, 600
Elixen, 265, 600
Ellagic acid, 65, 92, 141, 143, 200, 211, 232, 298,
    461

Ellagitannins, 284, 600
Elsholtzia argyi, 62, 461, 464, 472–473, 536, 566
Elsholtzia cristata, 79, 107, 386
Elsholtzia ketone, 62, 266, 461
Elsholtzia souliei, 386
Elsholtzianic acid, 62, 266, 461
Elskoliziaketon, 461
Emetic, 18, 56, 66, 71, 82, 93, 101, 123, 164, 167,
    202, 207, 215, 249, 277, 291, 600
Emetine, 11, 461
Emilia sonchifolia, 62, 162, 388, 393, 406, 431
Emilsin-like enzyme, 69, 461
Emmenagogue, 12, 20, 32–33, 46, 56–57, 62, 66,
    68–69, 71, 75, 85, 101, 113, 123–124,
    139, 143, 150, 154, 158, 160, 165, 170,
    179, 181, 192–193, 195, 197, 201–202,
    231, 248, 264, 269–270, 277, 287, 298,
    302, 304, 461
Emodin, 35, 60, 115–116, 123, 126–127, 179, 221,
    225, 227, 251, 262, 286, 290, 461, 600
Emodin methyl ester, 116, 221, 461
Emodin-monomethylether, 461
Emulsin, 256, 600
Entada phaseoloides, 62, 198, 352, 461, 628
Entageric acid, 62, 198, 461
Enzymes, 78, 196, 201, 205, 282, 310, 600
Ephedra distachya, 63, 198, 260, 365, 450, 461,
    497–498, 513, 522, 584, 598, 601, 611
Ephedra nevadensis, 260, 625
Ephedra sinica, 198, 602, 624, 628, 633, 640
Ephedrine, 63, 198, 306, 311, 461, 601
Epiafzalechin, 282, 601
Epialisol A, 11, 461
Epibrassicasterol, 101, 461
Epicatechin, 4, 61, 125, 176, 226, 282, 289, 462,
    601
Epicatechin gallate, 125, 226, 289, 462
Epicephalotaxin, 38, 462
Epicephalotaxine, 38, 462
Epidendrum monile, 63, 376
Epidendrum striatum, 63, 339
Epifriedelanol, 141, 298, 462
Epifriedelin, 44, 160, 190, 462
Epifriedelinol, 61, 65, 160, 261, 462
Epigallocatechin, 61, 125, 226, 282, 289, 462,
    601
Epiliguloxide, 166, 462
Epilobium amurense, 63, 260, 363
Epilobium angustifolium, 260, 582, 617, 633
Epilobium parviflorum, 260
Epimedium brevicorum, 63, 454, 462, 485–486,
    496, 507, 522, 538
Epimedium koreanum, 261, 360
Epimedium sagittatum, 261

Epimedoside A, 63, 261, 462
Epistephanine, 139, 462
Epiwilsonine, 38, 462
Epoxyquaine, 54, 194, 258, 462
Equisetonin, 63, 162, 199, 462
Equisetrin, 63, 162, 199, 462
Equisetum arvense, 63, 162, 199, 370, 384, 398, 417–418, 462, 474, 479, 481, 491–492, 495, 528, 538, 556, 601, 619, 630
Equisetum hyemale, 199, 367, 603, 622, 630, 632
Equisitine, 199, 601
Equistic acid, 162, 462
Er Cha, 4, 349
Eremophilene, 150, 302, 462
Ergostatetraen, 146, 462
Ergosterol, 15, 71, 83, 101, 116, 180, 269, 463
Ergosterol peroxide, 15, 180, 463
Ericolin, 39, 84, 170, 187, 208, 415, 463
Erigeron, 63, 162, 199, 342, 390, 431, 453, 463–464, 473, 481, 499, 501, 509, 512, 563, 566, 603, 613, 633, 635
Erigeron annuus, 390
Erigeron canadensis, 63, 162, 199, 342, 431, 453, 458, 463–464, 473, 481, 499, 501, 509, 512, 563, 566, 603, 613, 633, 635
Eriobotrya japonica, 64, 199, 369, 411, 432, 439, 445, 464, 501, 507, 533, 562, 565, 626, 640
Eriocaulon sieboldianum, 64, 361
Erisin, 116, 221, 283, 463
Erpinene, 150, 302, 463
Erucic acid, 27, 30, 34, 64, 134, 184, 229, 247, 250, 261, 463
Erycbelline, 64, 463
Erychroside, 64, 261, 463
Erycibe henryi, 64, 347, 463, 554
Erysimosol, 64, 261, 463
Erysimotoxin, 64, 261, 463
Erysimum amurense, 64, 261, 431, 442, 463, 481
Erysimum cheiranthoides, 380
Erysimum officinale, 261
Erythrina centaurium, 261, 629
Erythrina corallodendron, 64, 261, 372, 406
Erythrina indica, 353
Erythriside, 64, 261, 463
Escigenin, 9, 178, 463
Esculetin, 67, 155, 200, 250, 304, 463, 601
Esculetin dimethyl ether, 155, 304, 463
Esgoside, 98, 463
Estragol, 17, 155, 195, 241, 304, 465, 601
Estragole, 93, 154, 211, 245, 304, 465, 601
Estrogenic, 162, 233, 336, 465
Ethanol, 119, 223, 465
Ether oils, 5, 18, 76, 128, 143, 465

Ethoxychelerythrine, 92, 465
Ethoxysanguinarine, 92, 465
Ethyl alcohol, 64, 83, 199, 207, 465, 533
*Ethyl beta-*fructopyranoside, 125, 465
Ethyl cinnamate, 207, 601
Ethyl ester, 17, 465
Ethyl-p-methoxycinnamate, 601
Etoposide, 60, 466
Eucalptole, 197, 601
Eucalyptol, 23, 95, 182, 244, 261, 466, 601
Eucalyptus citriodora, 261, 592, 595, 597, 601, 604–605, 608, 632, 638
Eucalyptus robusta, 65, 261, 346, 438, 464, 473, 569
Euchresta japonicum, 65, 374, 396, 466, 505
Eucommia ulmoides, 65, 348, 467, 536
Eudesnol, 26, 183, 246, 466
Eugenal, 275, 601
Eugenia aromatica, 65, 402, 408, 422–423, 435, 461, 466, 483, 489, 494, 507, 524, 546, 572, 579
Eugenia caryophyllata, 200, 311, 601, 607, 630, 633
Eugenia ulmoides, 347
Eugenitin, 65, 200, 466
Eugenol, 13, 40–41, 57, 72, 93, 97, 99, 125, 153, 161, 188–189, 200, 203, 211, 215, 226, 231, 239, 259, 264, 289, 298, 321, 466, 512, 601
Eugianin, 141, 298, 466
Euonymus alatus, 65, 261, 322, 405, 459, 468, 472, 494, 543, 546, 556, 578
Euonymus atropurpureus, 262, 588, 593, 632–633
Eupafolin, 128, 262, 291, 466, 601
Eupafolin polysaccarides, 601
Eupaformonin, 66, 262, 466
Eupaformosanin, 66, 262, 330, 466
Euparin, 65, 466
Eupatol, 66, 163, 466
Eupatolide, 66, 262, 466
Eupatorium chinense, 65, 262, 394, 426, 446, 457, 466, 499, 502, 522, 547, 578
Eupatorium formosanum, 66, 369, 444, 466, 513, 529, 551, 556
Eupatorium odoratum, 66, 163, 349, 369, 396, 412, 416, 421, 435, 466, 492, 505, 523, 550, 555, 571
Eupatorium perfoliatum, 262, 600–601, 603, 630, 632, 640
Euphol, 66–67, 200, 466
Euphorbetin, 67, 200, 466
Euphorbia antiquorum, 66, 357, 408–409, 421, 447, 466

Euphorbia coraroides, 66, 427, 429, 467, 576
Euphorbia esula, 66, 392, 467, 480, 532, 552, 569
Euphorbia helioscopia, 347
Euphorbia hirta, 67, 200, 346, 436, 501, 544,
    547, 565, 584, 595, 603, 630, 635
Euphorbia humifusa, 67, 347, 430, 467, 473
Euphorbia kansui, 67, 371, 408, 495, 569
Euphorbia lasiocaula, 346
Euphorbia lathyrus, 67, 200, 387, 400, 403,
    425, 451, 463, 466–467, 499–500,
    601–602, 626
Euphorbia resinfera, 346
Euphorbia thymifolia, 386
Euphorbias, 66, 467
Euphorbiasteroid, 67, 200, 467
Euphorbine, 66, 467
Euphorbol, 67, 200, 467
Euphorbon, 66–67, 200, 467
Euphorbone, 200, 601
Euryale ferox, 67, 371, 540
Evocarpine, 67, 467
Evodia lepta, 67, 373, 410
Evodia rutaecarpa, 67, 385, 406, 467, 484, 502,
    517, 549
Evodiamine, 67, 467
Evodol, 67, 467
Evonymus alatus, 37, 68, 383
Eycinnuik, 467

F

Fabric, 149, 235, 467
Fagomine, 68, 467
Fagopyrin, 68, 467
Fagopyrum cymosum, 326–327
Fagopyrum esculentum, 68, 200, 262, 344, 428,
    445, 447, 470, 483, 501–502, 508,
    525, 533, 539, 544, 547, 549, 551, 573,
    577, 591
Fagopyrum tataricum, 68, 262, 361
Fagopyrum tutricum, 262, 627
Falcarindiol, 101, 329, 334, 467
Falvins, 13, 467
Falvonoids, 601
Fan Lu, 139, 349
Fan Mu Pen, 140, 349
Fan Qie, 90, 349
Fan Shi Lui, 119, 349
Fan Sui Xian, 14, 349
Fan Tian Hua, 149, 349
Fan Xie Ye, 35, 146, 349
Fan Xing, 144, 349
Fang Chi, 97, 349
Fang Feng, 76, 84, 171, 349, 369

Fang Ji, 44–45, 106, 140, 349, 357, 367, 382
Fangchinoline, 139, 467
Faradiol, 148, 234, 467
Farnesal, 53, 194, 257, 467
Farnesene, 9, 238, 467
Farnesiferols, 68, 201
Farnesol, 53, 113, 194, 257, 468
Farreol, 123–124, 287, 468
Fat, 68, 74, 104, 147, 149, 204, 215, 235, 468, 602
Fatty acids, 35, 72, 77, 86–87, 93, 107, 109, 118,
    120, 125, 137, 145, 152, 186, 203, 205,
    209, 218, 223–224, 226, 233, 236, 251,
    264, 271, 289, 296, 314, 330, 468, 602
Fatty oil, 84, 91, 143, 232, 281, 329, 468, 602
Fawcetimine, 91, 211, 468
Fawcetine, 91, 211, 468
Febrifugin, 79, 266, 468
Fei Jin Cao, 11
Fen Fang Ji, 349
Fenchone, 29, 145, 184, 201, 248, 282, 299, 468,
    602
Feng Dou Cai, 108, 349
Feng Huan Mu, 56, 349
Feng Lee, 349
Feng Lin Cao, 33, 350
Feng Lun Cai, 44, 350, 352
Feng Wei Cao, 119, 350
Feng Xian Hua, 80, 350
Feng Yang, 119, 350
Feriol, 123, 287, 468
Ferment, 288, 602
Fernadiene, 8, 178, 468
Fernene, 8, 60, 178, 260, 468
Fernenol, 80, 469
Ferric oxide, 256, 602
*Ferula assa-*foetida, 68, 201, 338, 417, 468–469,
    574, 597, 600, 604, 607, 626, 630
Ferulic acid, 17, 40, 68, 80, 91, 160, 163, 181, 188,
    201, 206, 211, 241–242, 265, 267, 280,
    469, 602
Feruloyl-p-coumaroylmethane, 469
Fetidine, 144, 299, 469
Fiber, 159, 602
Fibralactone, 469
Fibramine, 469
Fibraminine, 469
Fibraurea recisa, 68, 356, 469, 493, 528, 561
Ficus awkeotsang, 68, 338, 468, 472, 477, 479,
    540, 546
Ficus carica, 69, 201, 331, 384, 411, 421, 434,
    469, 477, 479, 505, 523, 529, 531, 542,
    544, 547, 549, 558, 572–573, 603,
    606, 639
Ficus inicrocarpa, 372

Ficus pumila, 69, 342, 499
Ficusin, 469
Filicenal, 469
Filicene, 469
Filicic acid, 25, 245, 469
Filicin, 469, 602
Filifolin, 163, 469
Filmarone, 60, 260, 469
Finitin, 469
Firmiana simplex, 69, 385, 421, 423–424, 437, 481, 505, 523, 549
Flagellaria indica, 69, 357, 406, 447, 461
Flavanones, 602
Flavaspidic acids, 469
Flavaspidin, 470
Flavocommelitin, 470
Flavon glucoside, 238, 470
Flavone, 10, 79, 89, 266, 470, 602
Flavonlignans, 251, 602
Flavonoid derivatives, 470
Flavonoid glycosides, 212, 225, 256, 603
Flavonoids, 14–15, 28, 49, 54, 61, 75, 100, 116, 126, 128, 138, 144–146, 183, 194, 200, 213–214, 221, 248, 256–258, 260, 266, 272, 274, 279, 281, 283, 290–292, 296, 299–300, 303, 322, 470–471, 603
Flavonols, 602–603
Flavoxanthin, 471
Flavoyadorinin, 471
Fluggein, 471
Fo Jia Cao, 132, 350
Fo Jia Cao (Taiwan), 350
Foeniculum officinale, 69
Foeniculum vulgare, 201, 386, 586, 602
Foetidin, 201, 604
Foliamenthin, 97, 214, 471
Folic acid, 210, 229, 604
Folinerin, 277, 604
Folinic acid, 17, 181, 241, 265, 471
Fon Xian Chi, 87, 350
Foo Shao Mai, 147, 350
Formonetin, 95, 212, 471
Formononetin, 26, 147, 183, 234, 246, 300, 471, 604
Forsythia suspensa, 69, 141, 201, 318, 362, 425, 456, 533, 549, 565, 604, 639
Forsythin, 201, 604
Fortunella crassifolia, 69, 359, 464, 473, 477, 531, 576
Fortunella japonica, 392
Fortunella margarita, 359
Fragaria indica, 70, 262, 375
Fragaria vesca, 263, 595, 615, 621, 633, 639
Fragarine, 57, 195, 471

Framine, 110, 471
Frangulin A, 286, 604
Frangulin B, 604
Fraxin, 70, 201, 263, 471
Fraxinella, 57, 70, 96, 213, 340, 471
Fraxinella dictamnus, 70, 340
Fraxinellone, 57, 195, 471
Fraxinus americana, 263, 636
Fraxinus bungeana, 70, 263, 403, 471
Fraxinus obovata, 372
Fraxinus ornus, 201, 597, 603, 633, 640
Fricdelin, 160
Friedelaun-3-ol, 66, 471
Friedelin, 44–45, 65, 89, 138, 141, 164, 190, 261, 298, 472
Fritillaria anheunensis, 70, 472, 529–530, 539, 542, 575
Fritillaria thunbergii, 341
Fritillaria verticillata, 201, 584, 621
Fritillarine, 70, 201, 472
Fritilline, 70, 201, 472
Fructose, 50, 68, 113, 119, 132, 193, 293, 472
Fu Ling, 117, 350
Fu Pen Zi, 126, 350
Fu Ping, 138, 350
Fu Rong Yie, 77, 350
Fu Shou Cao, 9, 157, 350
Fu Zi, 6, 170, 351, 388
Fuag Ji (Japanese), 351
Fukujusone, 157
Fulvoplumierin, 113, 169, 472
Fumaric acid, 129, 151, 291, 331, 472, 604
Fungal lysozyme, 71, 472
Fungiridal, 472
Furane, 62, 266, 472
Furanocoumarin, 73, 472
Furanodiene, 51–52, 193, 472
Furanodienone, 51–52, 193, 472
Furanoeremophilane, 166, 399
Furanoid, 245, 604
Furostanol saponins, 12, 472
Furylisobutyl ketone, 62, 266, 472
Furylmethyl ketone, 62, 266, 472
Furylpropyl ketone, 62, 266, 473
Fustin, 49, 73, 112, 170, 220, 473

G

Ga Song Xiang, 100, 351
Gagaminin, 167, 473
Galactan, 89, 121, 146, 210, 285, 473, 604
Galactitol, 36, 473
Galactomannan, 36, 251, 473
Galactose, 50, 69, 142, 170, 193, 282, 473, 604

Galangin, 13, 239–240, 473, 604
Galangol, 13, 239–240, 473, 604
Galanthamine, 77, 91, 166, 473
Galium aparine, 263, 609, 623, 633
Galium bungei, 70, 263, 379, 405, 542, 549
Galium spurium, 393
Galium verum, 202, 369, 584, 587–588, 603
Gallic acid, 4, 31, 48, 63, 65, 67, 72, 84, 93, 95,
　　　　112, 115, 120, 123–124, 161, 170, 191,
　　　　208, 220–221, 225, 247, 249, 255, 258,
　　　　260–261, 264, 286–287, 290, 298,
　　　　473, 604
Gallic acid derivatives, 260, 604
Gallocatechin, 125, 226, 282, 289, 473, 604
Gallotannic acid, 120, 124, 287, 473
Gallotannin, 216, 604
Galuteolin, 63, 162, 167, 199, 474
Gamatin, 116, 474
Gambirine, 4, 176, 474
Gamma-aminobutyric acid, 26, 183, 474
Gamma-dichroline, 474
Gamma-fernene, 8, 178, 474
Gamma-sitosterol, 102, 474
Gamma-terpinene, 474
Gan Cao, 74, 351
Gan Lan, 34, 111, 351
Gan Qi, 124, 351
Gan Su, 81, 351
Gang Ban Gui, 351
Ganoderma lucidum, 71, 325, 363, 403, 410, 463,
　　　　472, 525, 538, 540, 572
Gao Ben, 85, 351
Gao Liang Jiang, 13, 351
Gao Mu, 351
Gao Shan Liao, 104, 351
Gardenia angusta, 71, 202, 375, 408, 436, 474,
　　　　478, 508, 578, 605, 640
Gardenia jasminoides, 305
Gardenia maruba, 392
Gardenin, 71, 202, 474, 604
Gardenin crocin, 202, 604
Gardenoside, 71–72, 202, 474
Gastrodia elata, 71, 381, 474, 574, 576
Gastrodin, 71, 474
Gaultheria leucocarpa, 71, 263, 341, 512, 550
Gaultheria procumbens, 263, 616
Gaultherin, 283, 605
Ge Cong, 12, 351
Ge Gen, 104, 120, 351
Gedunin, 96, 213, 474
Gein, 72, 203, 264, 474
Gelatin, 29, 474
Gelidium amansii, 71, 263, 371, 404, 565
Gelidium cartilagineum, 263, 583, 617, 624

Gelsedine, 202, 605
Gelsemidine, 71, 202, 474
Gelsemine, 71, 202, 474, 605
Gelsemium sempervirens, 71, 202, 352, 459, 474,
　　　　496, 597, 605, 609, 633
Geniposide, 72, 202, 474, 605
Genisteine, 55, 74, 137, 147, 194, 474
Genkwadaphnin, 55, 195, 258, 475
Genkwanin, 24, 55, 195, 245, 258, 475
Gentialutine, 97, 214, 475
Gentiana algida, 72, 474–475, 486, 552
Gentiana dahurica, 72, 371, 475
Gentiana lutea, 202, 605
Gentiana macrophylla, 83
Gentiana scabra, 202
Gentiana squarrosa, 364
Gentianidine, 72, 202, 269, 475
Gentianindine, 202, 605
Gentianine, 72, 97, 202, 214, 269, 475, 605
Gentianol, 72, 202, 269, 475
Gentiatibetin, 97, 214, 475
Gentiatibetine, 97, 214, 475
Gentiopicrin, 72, 475
Gentiopicroside, 72, 202, 252, 321, 605
Gentisic acid, 80, 206, 261, 267, 475, 605
Geoside, 72, 203, 264, 475
Geranic, 53, 194, 257, 475
Geraniol, 39, 42, 47, 53, 56, 62, 99, 113, 125, 146,
　　　　153–154, 170, 187, 191, 194–195, 197,
　　　　215, 226, 233, 257, 289, 304, 475, 605
Geranium dahuricum, 72, 264, 473, 494, 543,
　　　　562, 564
Geranium eriostemon, 362
Geranium macrorrhizum, 264, 633
Geranylgeraniol, 87, 209, 475
Gerariol, 99, 276, 475
Germacrene, 83, 265, 475, 605
Germacrone, 123–124, 287
Germine, 150, 302, 475
Geum aleppicum, 72, 203, 264, 379, 466, 468,
　　　　470, 474–475, 601, 622, 630, 633
Geum urbanum, 264, 596, 640
Giang Huo, 101, 351
Gibberelin, 34, 476
Gibberelin A22, 34, 476
Gingerol, 155, 236, 476, 605
Ginkgetin, 73, 203, 476, 605
Ginkgo biloba, 3, 73, 203, 311, 321, 326, 389,
　　　　411, 424–425, 449–450, 458, 476, 491,
　　　　544, 549, 556, 582, 591, 593, 603, 605,
　　　　610, 629
Ginkgocide A, 203, 605
Ginkgocide B, 605
Ginkgocide C, 605

Ginkgocide J, 605
Ginkgocide M, 605
Ginkgol, 73, 203, 476
Ginkgolic acid, 73, 203, 476
Ginkgolides, 73, 203, 476
Ginnol, 73, 203, 476
Ginsenoside Ro, 476
Ginsenosides, 105, 216–217, 476, 605
Gitaloxigenin, 58, 196, 476
Gitaloxin, 58, 196, 476
Gitanin, 58, 196, 476
Gitoxigenin, 58, 167, 196
Gitoxin, 58, 196, 476
Givacoline, 181, 605
Glabralactone, 17, 158, 476
Glabridin, 264, 605
Glaucine, 144, 299, 476
Glechoma hederacea, 73, 203, 396, 409, 423,
        490–491, 497–498, 501–502, 510, 527,
        603, 605, 626, 628, 630, 633
Glechoma longituba, 359
Glechomine, 203, 605
Gleditsin, 73, 476
Glehnia littoralis, 73, 329–330, 341, 423, 486,
        525, 531, 538, 542, 561
Globulin, 19, 210, 476, 605
Globuline, 203, 606
Glucan, 11, 102, 165, 315, 328, 477
Glucides, 477
Glucobrassicin, 43, 81, 268, 477
Glucoevatromonoside, 47, 477
Glucofrangulin, 286, 606
Glucofrangulin A, 286, 606
Glucofrangulin B, 606
Glucokinin, 59, 260, 477
Glucominol, 12, 179, 477
Gluconasturin, 288, 606
Glucononitol, 153, 477
Glucopyranosides, 45, 477
Glucoquinone, 235, 606
Glucose, 49, 68, 74, 91, 100, 113, 127, 132, 142,
        170, 192, 201, 204, 211, 216, 264, 273,
        293, 298, 435, 445, 477, 530, 606
Glucoside apocynamarin, 242, 606
Glucoside asiaticoside, 37, 477
Glucoside morindin, 98, 276, 477
Glucosides, 5, 14–15, 18–19, 29, 34, 58, 65, 69,
        76, 128–129, 143, 186, 223, 251–252,
        279, 315, 320, 322, 330, 335, 487, 494,
        556, 599, 629
Glucosinolates, 225, 606
Glucovanillin, 27, 247, 477
Glucuronic acid, 158, 264, 477, 606
Glutamic acid, 88, 120, 210, 224, 477, 606

Glutamine, 58, 69, 172, 196, 201, 265, 477, 606
Gluten, 232, 606
Glutin-5-en-3-ol, 239, 477
Glutinol, 80, 478
Glycans, 198, 606
Glycine, 74, 143, 172, 203, 210, 346, 389, 437,
        451, 474, 478, 489, 520, 529, 540, 548,
        569, 592–593, 595, 597–598, 602, 606,
        609, 612, 616, 622, 624, 628, 636, 639
Glycine max, 74, 203, 346, 437, 451, 474, 489,
        520, 529, 540, 548, 569, 592–593, 595,
        597–598, 602, 606, 609, 612, 616, 622,
        624, 628, 636, 639
Glycine soja, 389
Glycoalkaloids, 134, 478
Glycogen, 53, 478
Glycolic acid, 25, 110, 182, 218, 281, 478
Glycorrhizin, 264, 606
Glycoside, 24–25, 37, 83, 108–109, 116, 135,
        144, 149, 163, 190, 217, 221, 223, 239,
        244–245, 249–250, 252, 266–267,
        276, 280, 283, 303, 315, 323, 425, 457,
        470, 478, 494–495, 543, 584, 602
Glycoside alkaloids, 135, 478
Glycoside nodakenin, 108, 280, 478
Glycoside pharbitin, 109, 267, 478
Glycoside rhaphantin, 116, 221, 478
Glycosides clerodendrin, 44, 190, 478
Glycosides G and K, 169, 478
Glycosides tribuloside, 146, 233, 478
Glycosine, 74, 478
Glycosmine, 74, 478
Glycosminine, 74, 478
Glycosmis cochinchinensis, 74, 387, 478, 558
Glycyrrhetinic acid, 264, 606
Glycyrrhiza, 3, 74, 204, 264, 319, 321, 328, 351,
        404, 422, 435, 470, 478, 490, 503–504,
        571, 582, 594, 603, 605–606, 610, 617,
        628, 632, 637
Glycyrrhiza glabra, 264, 319, 321, 582, 603,
        605–606, 617, 628, 632
Glycyrrhiza pallidiflora, 264
Glycyrrhiza uralensis, 3, 204, 328, 351, 594,
        610, 637
Glycyrrhizic acid, 74, 323, 478
Glycyrrhizin, 19, 176, 319, 321, 478, 606
Glypenosides, 75, 478
Gnaphalium affine, 74
Gnaphalium uliginosum, 204, 633, 640
Gobosterin, 84, 479
Gomphrena globosa, 75, 371, 410, 424, 479, 488,
        552
Gomphrenin, 75, 479
Gong Chong, 46, 351

Gong Lao Mu, 351
Gong Xian Teng, 94, 351
Gorlic acid, 78, 265, 479
Gossypetin, 123–124, 204–205, 287, 479, 606
Gossypin-3-sulfate, 274, 607
Gossypitrin, 63, 162, 199, 479
Gossypium herbaceum, 75, 204, 367, 400, 402,
        404, 479, 481–482, 511, 603, 607
Gossypol, 75, 204, 479, 607
Gou Gi, 90, 116, 352
Gou Gu, 94, 352
Gou Ji Guan Zhong, 153, 352
Gou Ma, 4, 352
Gou Mei, 99, 352
Gou Min, 71, 352
Gou Shi Cao, 133, 352
Gou Teng, 100, 149, 352
Gramme, 205, 607
Graveobiosides, 479
Grayanotoxin, 124, 170, 287, 479
Gu Dong Zi, 137
Gua Di, 50, 352
Gua Lou, 147, 352
Guaiacol, 69, 201, 479
Guaiaxulene, 69, 201, 479
Guaijaverin, 119, 479
Guan Ye Lean Qiao, 79, 352
Guan Zhong, 25, 55, 60, 104, 153, 352, 367, 371,
        394
Guang Feng Lun Cai, 44, 352
Guaridine, 183, 607
Guercitrin, 115, 479
Guggulsterol, 45, 191, 253, 479
Gui Hua, 93, 103, 125, 352, 366, 373
Gui Zhi, 40, 352
Guijaverin, 37, 479
Gum, 45, 51, 56, 68, 132, 142, 191, 200–201, 222,
        227, 231–232, 253–254, 257, 296, 298,
        479, 607
Guo Gang Long, 62, 352
Guo Ko Yi, 64, 352
Guo Tan Loan, 8, 352
Gurjuncene, 97, 214, 479
Guvacine, 20, 181, 479, 607
Guvacoline, 20, 181, 479
Gynostemma pentaphyllum, 75, 322–323, 360,
        528
Gynura bicolor, 75, 367, 470
Gynura japonica, 75, 373, 551
Gypsogenin, 149, 479

H

Haemanthidien, 75, 155

Haementhamine, 75, 479
Haemolytic sapogenin, 128, 479
Hai Dai, 83, 352
Hai Jin Sha Teng, 91, 353
Hai Tong Pi, 64, 353
Hai Zao, 129, 353
Hallucinogenic, 157, 291, 480
Han Cai, 125, 353
Han Mai Bin Cao, 134, 353
Han Xiao Hua, 97, 353
Han Xiou Cao, 97–98, 353
Han Xiou Cao (American), 353
Harman, 61, 77, 205, 480
Harmane, 146, 233, 480
Harmine, 146, 233, 480
Harmol, 77, 205, 480
Harpagide, 65, 131, 228, 238, 401, 583, 607
Harpagoside, 131, 228, 480, 607
Harringtonine, 38, 480
Hastanecine, 159, 480
Havonoids, 193, 607
Hayatidine, 41, 189, 480
Hayatine, 41, 189, 480
Hayatinine, 41, 189, 480
Hcn, 157
He Huan Pi, 4, 353
He Que She, 117, 353
He Shi, 35, 172, 353
He Shou Wu, 116, 353
He Ye, 100, 353
He Zi, 143, 353
Hectalactone, 119, 223, 480
Hedera acid, 480
Hedera helix, 265, 584, 590, 600, 605, 607, 609,
        633
Hedera rhombea, 75, 264, 343, 480, 524, 563
Hederacoside B, 265, 607
Hederacoside C, 265, 607
Hederagenin, 21, 243, 480
Hederasaponin B, 16, 241, 480
Hederin, 75, 264–265, 480, 607
Hedychium coronarium, 75, 314, 373, 396, 405,
        423, 430, 495, 533, 556
Hedyotis corymbosa, 76, 379, 426, 497, 502, 520
Hedyotis diffusa, 76, 315, 340, 403, 413, 464,
        487, 527
Hei Nan Pu Tao, 141, 353
Hei Shuo, 96, 353
Helenalin, 267, 607
Helenien, 142, 232
Helenin, 164, 480
Heliotropin, 296, 607
Helminthosporin, 36, 251, 480
Helveticoside, 47, 480

Helveticosol, 64, 261, 481
Hemerocallin, 164, 481
Hemigossypol, 75, 204, 481
Hemin, 77, 205, 481
Heneicosanic acid, 181, 607
Hentriacontane, 113, 220, 282, 481
Hepatica asiatica, 76, 265, 385
Hepatica nobilis, 265
Heptacosane, 12, 57, 164, 239, 481
Heptane, 113, 481
Heraclein, 265, 607
Heracleum dissectum, 76, 265, 369, 413, 432,
    444, 465, 484, 514, 526, 535, 556
Heracleum lanatum, 421, 579
Heracleum maximum, 265, 601, 606–607, 625,
    631
Herbacetrin, 63, 199, 481
Herbal medicine, 1–3, 308, 316–318, 323–325,
    332–334
Herclavin, 304, 607
Hernandezine, 144, 299, 481
Hernandine, 139, 481
Hernandoline, 139, 481
Hernandolinol, 139, 481
Herniarin, 35, 212, 481, 607
Hesperidin, 42, 44, 130, 253, 266, 481, 607
Heutriacontane, 69, 481
Hexacosanol, 131, 481
Hexadecenoic acid, 215, 607
Hexahydromatricaria, 63, 199, 481
Hexalactone, 119, 223, 481
Hexanoic acid, 119, 223, 482
Hexanol, 119, 223, 482
Hexenal, 170, 482
Hexylenaldehyde, 148, 301, 482
Hi Lu, 15, 353
Hi Tong, 113, 353
Hibifolin, 96, 274, 482
Hibiscus acid, 204–205, 608
Hibiscus chinensis, 76, 367, 551
Hibiscus mutabilis, 77, 350, 485, 543–544, 549
Hibiscus rhombifolius, 345
Hibiscus rosa-sinensis, 77, 393, 446, 520, 540,
    569, 589, 595, 606, 617, 634–635
Hibiscus sabdariffa, 77, 204, 363, 551–552, 577,
    586, 608, 615, 620
Hie Quin Cao, 40, 353
Hieracium pilosella, 265, 592, 597, 603, 637
Hieracium umbellatum, 77, 265, 374, 563, 576
Hierochloe odorata, 77, 205, 366, 441, 444, 597,
    612, 615
Hin Gu Cao, 354
Hinesol, 26, 183, 246, 482
Hippeastrum hybridum, 77, 377, 473, 506, 566

Hippophae rhamnoides, 77, 205, 311, 373,
    422, 445, 464, 468, 470, 480–481,
    491–492, 506, 538, 544, 555, 564,
    576–577, 593, 601–603, 625–626, 633,
    638–639
Hirsuteine, 149, 301, 482
Hirsutine, 149, 301, 482, 608
Hirsutrin, 75, 204, 482
Hispidulin, 128, 291, 482
Histamine, 152, 185, 207, 233, 235, 482, 608
Histidin, 159, 482
Homoarbutin, 39, 121, 187, 225, 482
Homoaromoline, 139, 482
Homolycorine, 91, 166, 482
Homoorentin, 141, 297, 482
Homoorientin, 68, 200, 262, 483
Homoplantagin, 113, 220, 282, 483
Homoplantaginin, 128, 291, 483
Homoserine, 26, 183, 246, 483
Homostephanoline, 139, 483
Homotrilobine, 45, 483
Hong Gin Tian, 123, 354
Hong Guan Yao, 25, 354
Hong Je Dan Hua, 113, 354
Hong Jua, 35, 354
Hong Ma Feng Shu, 82, 354
Hong Mao Dan, 100, 354
Hong Mei Xiao, 126, 354
Hong Men Lan, 103, 354
Hong Nan, 92, 354
Hong Pi, 140, 354
Hong Si Xian, 135, 354
Hong Teng, 129, 354
Hong Tu Cao, 29, 354
Hopadiene, 8, 178, 483
Hordenine, 205, 608
Hordeum vulgare, 78, 205, 366, 411, 454, 487,
    508, 540, 576, 607–608
Hormoharringtonine, 38, 483
Hosenkosides, 206, 608
Hou Po, 93, 354
Houttuynium, 78, 483
Houttynia cordata, 78, 391, 452, 464, 483, 491,
    544
Hovenia dulcis, 78, 392, 459, 483, 539
Hovenosides, 78, 483
Hoya carnosa, 78, 391, 441, 483, 535
Hoyin, 78, 483
Hu Chang, 115, 354
Hu Gu Xiao, 128, 354
Hu Hua Pi, 11, 355
Hu Huang Lain, 111, 355
Hu Ji Shang, 152, 355
Hu Li Wei, 149, 355

Hu Lu Cao, 57, 355
Hu Tao Ren, 82, 355
Hu Tin Chi, 61, 355, 381
Hu Tin Chi (Tiawan), 355
Hua Jiao, 154–155, 355, 374
Hua Qian Jin Teng, 140, 355
Hua Shan Seng, 111, 355
Huai Hua, 137, 355
Huai Niu Teng, 6, 355
Huan Hun Cao, 133, 355
Huang Bai, 93, 109, 120, 340, 355
Huang Bai Mu, 93, 355
Huang Gen Cao, 355
Huang Ging, 115, 355
Huang Gua, 50, 356
Huang Hua Jia Zhu Tao, 145, 356
Huang Hua Xuan Cao, 76, 356
Huang Lian, 46, 164, 356, 386
Huang Lu, 49, 161, 356
Huang Ma, 47, 63, 356
Huang Qin, 131, 356
Huang Qin (Taiwan), 356
Huang Shui Jia, 135, 356
Huang Teng, 55, 68, 356
Huang Wu Tien, 31, 356
Huang Yang (Taiwan), 356
Huang Yao Zi, 58, 356
Huang Zhi, 26, 356
Hui Hui Suan, 122, 356
Hui Mao Dou, 143, 356
Hui Qin, 111, 356
Humulene, 65, 83, 197, 200, 205, 207, 268, 483, 608
Humulus scandens, 78, 364, 417, 437, 505
Huo Ma Ren, 34, 356
Huo Qin Hua, 75, 357
Huo Tan Mo Cao, 116, 357
Huo Xiang, 9, 89, 114, 357
Huo Yu Jin, 66, 357
Huong Jing, 153, 357
Huperzine A, 79, 483
Hydnocarpus anthelmintica, 78, 265, 346, 435, 479, 483–484
Hydnocarpus kurzii, 265
Hydnocarpus oil, 78, 265, 483
Hydrangea arborescens, 266, 598, 603, 608, 628, 633
Hydrangea macrophylla, 79, 266, 387, 468, 483, 549
Hydrangeic acid, 79, 266, 483
Hydrangein, 266, 608
Hydrangenol, 79, 266, 483
Hydroagarofuran, 18, 242, 483
Hydrocarbons, 152, 227, 303, 312, 483, 630

Hydrociannic acid, 176, 179, 608
Hydrocinnamic acid, 18, 242, 483
Hydrocinnamic aldehyde, 41, 189, 484
Hydrocotylene, 153, 484
Hydrocotyline, 186, 608
Hydrocoumarin, 274, 608
Hydrocyanic acid, 22, 38, 52, 94, 109, 187, 212, 223, 242, 244, 257, 484, 608
Hydrogen cyanide, 103, 170, 484
Hydrojuglone, 207, 608
Hydroquinone, 79, 266, 484
Hydroxybenzoic acid, 143, 153, 261, 608
Hydroxycinnamic acid, 96, 136, 194, 213, 230, 257, 484, 608
Hydroxycoumarin, 274, 608
Hydroxyeephalotaxine, 38, 484
Hydroxyevodiamine, 67, 484
Hydroxygenkwanin, 55, 153, 195, 258, 484
Hydroxyleamptothecine, 484
Hydroxypeucedanin, 17, 241, 484
Hydroxyphenylethanol glycosides, 231, 608
Hylander, 162, 386, 484
Hymenocallis speciosa, 79, 378, 506
Hynocarpic acid, 265, 484
Hyoscine, 56, 79, 195, 206, 266, 484, 608
Hyoscyamine, 56, 79, 111, 131, 195, 206, 266, 485, 608
Hyoscyamus bohemicus, 79, 266, 363, 405–406, 437, 485, 515, 554
Hyoscyamus niger, 206, 266, 608, 637
Hyoscypierin, 79, 485
Hypaconitine, 6–7, 177, 485, 608
Hypaphorine, 4, 176, 485
Hypericin, 79, 206, 485, 609
Hypericum ascyron, 347
Hypericum attenuatum, 79, 360, 491, 543–544, 553
Hypericum chinensis, 485
Hypericum japonicum, 318
Hypericum perforatum, 206, 352, 494, 541, 573, 594, 603, 609, 625–627
Hypericum sumpsonii, 381
Hypericum triquetrifolium, 79, 485, 541
Hyperin, 39, 60, 79, 106, 108, 118, 126, 130, 148, 150, 161, 187, 227, 234–235, 290, 301, 485
Hyperoside, 33, 37, 49, 77, 107, 118, 150, 206, 223, 235, 247, 250, 256, 301, 485, 609
Hyperzia serrata, 79, 377, 483, 493
Hypoepistephanine, 139, 485
Hypoglycemic, 142, 173, 278, 485
Hypophyllanthin, 110, 485

Hypotensive, 12, 26, 31, 45, 56, 82, 94, 111, 120, 157, 160, 179, 183, 186, 224, 246, 256, 270, 272, 274, 281, 328, 485
Hyssopus ocymifolius, 79, 266, 386
Hyssopus officinalis, 266
Hystonin, 110, 218, 281, 485

I

Icariine, 63, 261, 485
Icarlin, 63, 261, 486
Ifflaionic acid, 131
Ilex aquifolium, 266, 610, 628, 637
Ilex chinensis, 79, 266, 376, 540, 563, 573
Ilex pubescens, 79, 266, 366, 399, 470, 484, 554, 578
Illicium lanacedatum, 80, 412, 518, 541, 556
Illicium verum, 80, 206, 338, 412, 550, 586, 616, 627
Ilungianins, 486
Imidazolylothylamine, 136, 295, 486
Impatiens balsamina, 80, 206, 267, 350, 382, 428, 469, 475, 500, 526, 554, 557, 582, 589, 608, 611, 621, 626, 628
Impatiens pallida, 267, 585, 590, 612
Imperata arundinaceae, 80, 339, 415, 417, 448, 469, 478, 557
Imperatorin, 16–17, 73, 116, 158, 171, 241, 486
India Bian Teng, 69, 357
Indian Ren Xian, 4, 357
Indian Tian Qing, 134, 357
Indican, 43, 81, 116, 268, 486
Indigo, 27, 81, 116, 268, 486
Indirubin, 27, 43, 486
Indo-brown, 27, 486
Indoid compounds, 72, 486
Indoxyl-5-ketogluconate, 81, 486
Indo-yellow, 486
Ineole, 486
Ingigo, 43, 486
Inlin, 35
Inokosterone, 6, 153, 177, 486
Inositol, 89, 137, 171, 185, 209, 218, 269, 272, 486, 609
Insecticidal, 37, 90, 123, 150, 157, 160, 277, 286, 302, 322, 486
Insulanoline, 106, 487
Insularine, 106, 139, 487
Insulin-*like peptide,* 214, 609
Insulins, 609
Inula britannica, 80, 164, 206, 267, 387, 425, 428, 436, 456, 470, 486–487, 491, 543–544, 565, 603, 622, 634, 637, 640

Inula helenium, 164, 267, 383, 405, 481, 598, 607, 609, 617, 626
Inula linariaefolia, 318
Inulicin, 80, 206, 267, 487
Inulin, 8, 20, 32–33, 41, 45, 84, 133, 142–143, 159, 164, 181–182, 185, 190, 229–230, 232, 234, 237, 259, 267, 294, 486–487, 609
Inusterols, 487
Invertase, 78, 113, 205, 487
Invertin, 97, 214, 487
Iodine, 58, 83, 115, 127, 129, 147, 179, 221, 245, 265, 269, 290, 315, 487, 609
Iphigenia indica, 80, 374, 441, 443, 517, 527
Ipolamiide, 165, 487
Ipomoea barbata, 81, 267, 372
Ipomoea cairica, 81, 267, 385, 423, 515
Ipomoea purga, 267, 596
Iridals, 247, 321, 609
Iridin, 81, 247, 267, 487, 609
Iridoid glycosides, 131, 191, 214, 228, 609
Iridoid valepotriates, 263, 609
Iridoidglycosides, 487
Iridoids, 202, 301, 303, 315, 609
Iridomyrmecin, 7, 177, 487, 609
Irigenin, 81, 247, 267, 487, 609
Iris aqyatuca, 81, 267, 372, 470, 487, 566
Iris buatatas, 351
Iris dichotoma, 375
Iris lactea, 81, 267, 384, 488
Iris pallasii, 81, 267, 365, 488
Iris pseudacorus, 585
Iris versicolor, 267, 585, 610, 617, 628, 637
Irisquinone, 81, 267, 488
Iron, 110, 117, 129, 157, 187, 208, 218, 292, 488, 609–610
Iron manganese, 208, 610
Iron oxide, 110, 218, 488
Isatan, 43, 81, 207, 268, 488
Isatan B, 43, 81, 268, 488
Isatis chinensis, 81, 268, 341, 438, 486, 488, 494, 505, 508, 543, 560
Isatis indigotica, 81, 268, 346, 477, 486, 488, 511
Isatis tinctoria, 207, 268
Iso-adiantone, 488
Iso-amaranthin, 488
Iso-amyl, 488
Iso-anthricin, 488
Iso-bergaten, 488
Iso-butyl, 162, 489
Iso-butyliso, 488
Iso-cerberin, 488
Iso-chaksine, 489
Iso-chlorogenic acid, 489

Iso-chondrodendrine, 106, 488–489
Iso-corynoline, 489
Iso-corypalmine, 489
Iso-cryptomerin, 489
Iso-cryptotanshinone, 489
Iso-cucurbitacin B, 100, 488
Iso-dexyelephantopin, 489
Iso-eugenitin, 489
Iso-eugenol, 489
Iso-fernene, 489
Iso-ferulic acid, 489
Iso-furanogermacrene, 489
Iso-ginketine, 489
Iso-harringtonine, 490
Iso-homoarbutin, 490
Iso-humulone, 490
Iso-hydrocarveol, 490
Iso-hyperine, 490
Iso-imperatorin, 490
Iso-indigo, 490
Iso-leucin, 490
Iso-leucine, 489–490
Iso-liensinine, 490
Iso-linderalactone, 490
Iso-linderoxide, 490
Iso-lineolone, 490
Iso-liquiritigenen, 490
Iso-liquiritigenin, 490
Iso-liquiritin, 490
Iso-lobelamine, 490
Iso-maculosindine, 490
Iso-mangiferin, 490
Iso-menthone, 490
Iso-*mesityl oxide,* 491
Iso-myristicin, 488, 491
Iso-neomatatabiol, 491
Iso-orientin, 491–492
Iso-patrinene, 491
Iso-paulownin, 491
Iso-pelletierine, 491
Iso-pentenic acid, 491
Iso-pimpinellin, 491
Iso-pinocamphone, 491
Iso-propylidere, 491
Iso-quercetin, 491
Iso-quercitrin, 491–492
Iso-quinoline, 491
Iso-ramanone, 491
Iso-rhamnetin, 491
Iso-rhynchophylline, 149, 301, 488, 492
Iso-ricinoleic acid, 492
Iso-sakuranetin, 492
Iso-sinomenine, 492
Iso-*steroidal alkaloids,* 492

Iso-tadeonal, 492
Iso-talatizidine, 492
Iso-tetrandrine, 492
Iso-thalidenzine, 492
Iso-thamnetin, 492
Iso-thiocyanates, 492
Iso-thujene, 492
Iso-trilobine, 492
Iso-valeraldehyde, 492
Iso-valeric acid, 493
Iso-vanihyperzine A, 493
Iso-vitexin, 493
Iso-xanthanol, 493
Izalpinin, 13, 239, 493

J

Jacoline, 229, 294, 611
Jaligonic acid, 111, 219, 493
Jambolin, 141, 298, 493
Japan Jin Fen Ju, 102, 357
Japan Liu Shan, 50, 357
Japan Mu Fang Ji, 357
Japan Mu Gua, 38, 357
Japan Niu Teng, 6, 357
Japan Nu Zhen, 357
Japan She Gen Cao, 103, 357
Japan Su, 36, 357
Jasminum grandiflorum, 268
Jasminum mesnyi, 81, 268, 391, 493, 508,
        563–564
Jasminum samba, 81, 268, 367, 401, 413, 420,
        471, 556
Jasmipierin, 81, 268, 493
Jatamansic acid, 100, 277, 493
Jatropha gospiifolia, 82, 207, 354, 415, 446, 503,
        516, 524, 528, 535, 560, 587, 598, 600,
        608, 610–611, 625–626, 633
Jatropha podagrica, 82, 315, 372, 414, 421, 423,
        470, 493, 505, 517, 543, 549, 567, 577
Jatrorhizine, 94, 144, 273, 299, 493
Je Koo Cai, 357
Je She, 357
Jervine, 76, 150, 302, 493
Ji Guan Hua, 37, 357
Ji Mu, 89, 357
Ji Xue Cao, 37, 358
Ji Xue Teng, 97, 138, 358
Jia Gou Ju, 95, 358
Jia Mu, 19, 358
Jia Zhu Tao, 101, 145, 167, 356, 358
Jian Xui Fuan Hou, 18, 358
Jian Zi Mu, 47, 358
Jiang Zhen Xiang, 7, 358

Jiao Cao, 150, 358
Jie Cai, 34, 358
Jie Geng, 33, 113, 358
Jie Gu Mu, 128, 358
Jin Cai, 152, 359
Jin Cao, 11, 24, 57, 66, 91, 347, 359, 372, 375
Jin Chuang Xian Cao, 10, 359
Jin Deng Long, 110, 359
Jin Gan, 359, 392
Jin Gi Er, 34, 359
Jin Gu Cao, 10, 90, 359
Jin He Huan, 4, 359
Jin Jia Dou, 109, 359
Jin Jing Zi, 125, 359
Jin Ju, 359
Jin Moa, 145, 359
Jin Qian Cao, 73, 359
Jin Qian Chao, 92, 359
Jin Que Hua, 55, 359
Jin Si Tao, 79, 360
Jin Tsan Jiu, 32, 360
Jin Xian Diao Wu Gui, 139, 360
Jin Yang Huo, 63, 360
Jin Yin Hua, 89, 360
Jing Mian Cao, 84, 360
Jing Tian, 132, 360
Jing Tian San Qi, 132, 360
Jiu Hong, 42, 360
Jiu Hua, 27, 39–40, 360, 389
Jiu Hua Teng, 27, 360
Jiu Jie Cha, 39, 360
Jiu Jie Mu, 119, 360
Jiu Lan, 155
Jiu Li Xiang, 99, 360
Jiu Pi, 42, 360
Joe Koo Lan, 75, 360
Ju Shi Cai, 137, 360
Juan Bai, 132–133, 360, 376
Jue Ming Zi, 35, 330, 361
Juglandic acid, 82, 493
Juglandin, 207, 611
Juglanin, 82, 207, 493
Juglans mandshurica, 82, 355, 408, 493
Juglans regia, 207, 608, 611, 633
Juglone, 207, 611
Jugone, 82, 207, 493
Jujuboside A, 156, 236, 493
Jujuboside B, 156, 236, 493
Jun Zi Lian, 44, 361
Juncus effusus, 82, 348, 414–415, 419, 493,
505–506, 526, 545, 571, 577
Juniperin, 269, 611
Juniperus communis, 269, 591, 598, 609, 611,
613, 623, 635

Juniperus rigida, 83, 207, 268, 383, 409, 412,
422, 426, 432–433, 440, 483, 501, 516,
527, 545, 566, 585, 591, 595, 600, 617,
626–627, 636
Justicia adhatoda, 269, 584, 640
Justicia gendarussa, 83, 269, 371
Jutrophine, 207, 611

K

Kadsura japonica, 83, 368, 475, 494
Kadsurarin A, 83, 494
Kadsuric acid, 83, 494
Kadsurin, 83, 494
Kaempferia galanga, 83, 207, 374, 426, 430, 438,
465, 591–593, 601, 618, 620
Kaempferin, 13, 35, 239, 251, 494
Kaempferitrin, 56, 72, 259, 264, 494
Kaempferol, 25–26, 32–33, 39, 46–47, 65, 79,
81, 106–107, 113, 116, 123–124, 134,
151, 161, 163, 165, 178, 183, 187, 200,
206–207, 245–246, 250, 254–255,
268, 272, 279, 286–287, 322, 400,
494, 611
Kaempferol derivatives, 206, 611
Kaempferol glucosides, 65, 494
Kaempferol trisaccharide, 134, 322, 494
Kaempferol-3,7-diglucoside, 63, 199, 495
Kaempferol-3-7-glycoside, 494
Kaempferol-3-galactoside Dacmpferol, 494
Kaempferol-3-glucoside, 125, 151, 303, 494
Kaempferol-3-glucosylgalactoside, 103, 494
Kaempferol-3-robinobioside, 495
Kaempferol-7-β-glycoside, 495
Kaempferol-7-shamnoside, 495
Kaempferol-*rhamno glucoside,* 230, 495
Kaempferol-rhamnoside, 102, 171, 224, 495
Kalosaponin, 165, 495
Kalotoxin, 165, 495
Kansuinine, 67, 495
Kanugin, 116, 495
Karabin, 101, 167, 495
Karanjin, 116, 495
Kaurene, 19, 114, 243, 495
Kaurene derivatives, 19, 243, 495
Kautschuk, 171
Ketone, 41, 50, 62, 75, 80, 150, 180, 189, 192–
193, 206, 266, 302, 412, 461, 472–473,
495, 512, 542, 616
Khellol, 40, 188, 495
Kiganen, 50, 192–193, 495
Kiganol, 50, 192–193, 495
Kino-tannic acid, 119, 495
Klob-Alisova, 495

Ko Cho Mo, 361
Kobusone, 54, 194, 258, 495
Koenigin, 276, 611
Kong Xin Lian Zi Cao, 361
Konokiol, 93, 496
Koo Jing Cao, 361
Korean Si Zhao Hua, 48, 361
Korean Yan Hu Suo, 48, 361
Korepimedoside A, 63, 261, 496
Korepimedoside B, 63, 261, 496
Koumine, 71, 202, 496
Kouminicine, 71, 202, 496
Kouminine, 71, 202, 496
Ku Dong Zi, 361
Ku Gua, 98, 361
Ku Lian Chi, 96, 361
Ku Lian Pi, 96, 361
Ku Seng, 137, 361
Ku Shu, 111, 361
Ku Zhi, 110, 361
Kuan Dong Hua, 148, 361
Kuei Chen Gao, 28, 361
Kui Shu Zi, 88, 361
Kukoamines, 90, 210, 496
Kulinone, 96, 213, 496
Kumatakenin, 183, 246, 611
Kumatakinin, 13, 239, 496
Kun Bu, 83, 361
Kuraridin, 137, 496
Kurrin, 111, 496
Kutkin, 111, 496

L

La Lian, 115, 362
Labdadien, 19, 498
Labenzyme, 41, 498
L-abrine, 4, 176, 496
Laburnine, 252, 612
Lacerol, 27, 43, 499
Lacnophyllum, 63, 199, 499
Lactone, 43, 78, 158, 203, 205, 224, 241, 253, 264, 284, 459, 499, 612
Lactone protoanemonin, 224, 241, 284, 612
Lactones atractylenolide II, 183, 246, 612
Lactones atractylenolide III, 612
Lactrile, 223, 612
Lactuca raddeana, 83, 269, 375, 435, 463, 507, 526, 529, 577
Lactuca serriola, 269, 597, 603, 612, 630
Lactucerin, 269, 612
Lactucopicrin, 269, 612
Laetrile, 223, 244, 612
Laggera alata, 83, 364, 410, 464, 470, 533

Lai Ye Sheng Ma, 7, 362
Lamiide, 165
Laminaria angusta, 83, 269, 352, 361, 405, 410, 429, 487, 499, 538
Laminaria digitata, 269, 609, 622, 624
Laminarin, 83, 269, 499
Laminine, 83, 269, 499
Lamiol, 165, 499
Lamioside, 165, 499
l-anagyrine, 137, 496
Lanatoside, 196, 612
Lang Ba Cao, 28, 362
Lao Huan Cao, 72, 362
Lao Jium Xiu, 362
Lapase, 499, 612
Lappa communis, 84, 369, 394, 415, 464, 468, 479, 487, 499, 546, 564
Lappaol, 20, 330, 499
Lappatin, 84, 499
Lappine, 84, 499
l-arabinose, 31, 94, 113, 220, 274, 282, 496
Larreagenin, 17, 499
l-asparagine, 217, 611
Latex, 69, 169, 171, 499
Lathyrol diacetate benzoate, 67, 200, 499
Lathyrol diacetate nicotinate, 67, 200, 500
Laudanine, 106, 217, 279, 500
Lauric acid, 42, 87, 96, 99, 129, 143, 181, 189, 213, 215, 271, 275, 500, 612
Laurifoline, 154, 304, 500
Laurotetanine, 36, 88, 500
Lavoxanthin, 133, 293, 500
Lawsone, 80, 84, 206, 208, 267, 500, 612
Lawsonia inermis, 84, 208, 392, 408, 423, 473, 500, 597, 603, 612, 618, 632–633
Laxogenin, 135, 230, 294, 500
l-baptifoline, 137, 496
l-beta-santonin, 23, 245, 496
l-borneol, 87, 271, 496
L-cadinene, 99, 276, 497
L-camphor, 76, 497
L-caryophyllene, 150, 302, 497
L-citronellol, 125, 226, 289, 497, 611
L-cocaine, 64, 199, 497
L-curcamene, 51, 193, 497
l-(d)-*isoleucine betaine,* 34, 185, 496
Lecithin, 116, 133, 203, 221, 229, 500, 612
Lecithine, 34, 250, 500
Lectin, 235, 250, 321, 604
Ledebouriella divaricata, 84, 349, 405, 464, 525
Ledum palustre, 84, 208, 383, 409, 415, 417, 423, 430, 492, 501, 516, 544, 549, 597, 603, 612, 618, 632–633
Lei Gong Teng, 147, 362

Lei Wan, 102, 362
Lemmaphyllum microphyllum, 84, 360, 449, 470, 506, 540, 546, 577
Lemna minor, 84, 208, 371, 588, 610, 614
Lenrosine, 36, 186, 500
Lenrosivine, 36, 186, 500
Leonaridine, 85, 269, 500
Leonticine, 192, 255, 612
Leonuride, 270, 612
Leonurin, 269–270, 612
Leonurinine, 85, 269, 500
Leonurus cardiaca, 270, 594–595, 612, 625, 628
Leonurus heterophyllus, 85, 269, 389, 468, 500, 576
Leonurus sibiricus, 85, 269, 340, 345, 464
l-ephedrin, 497
l-ephedrine, 63, 111, 198, 260, 497, 611
l-epicatechin, 49, 256, 497
l-epicatechol, 33, 150, 235, 301, 497
Lepidium apetalum, 85, 348, 492
Lepidium virginicum, 208, 341, 639
Leptostachyol acetate, 169
Lespedeza cuneata, 85, 389, 423, 470, 536
Leucaena leucocephala, 85, 390, 441, 500, 532
Leucaenine, 85, 308, 500
Leucanol, 85, 500
Leucine, 27, 58, 87–88, 103, 172, 196, 209, 218, 501, 612
Leucoanthocyanin, 33, 68, 200, 262
Leucocyanidin, 111, 501
Leucocyanidol, 67, 200, 501
Leucylphenylalanine anhydride, 85, 270, 501
Leurosine sulphate, 160, 501
Levidulinase, 14, 501
Leviduline, 14, 501
Levulin, 232, 613
Levulose, 64, 199, 501
l-homostarchydrine, 212, 611
l-hyoscyamine, 497
Li Chi, 88, 362
Li Chun Hua, 105, 362
Li Lu, 150, 362
Li Lu (Taiwan), 362
Li Shu, 121, 362
Li Tou Cao, 148, 362
Li Zhi Cao, 128, 362
Lian, 13, 15–16, 24, 39, 43–44, 46, 60–61, 69, 88, 96, 100–101, 114–115, 141, 144, 148, 158–159, 161–162, 164, 338–339, 341, 345, 356, 361–363, 365, 367, 369, 378–379, 382, 386, 390
Lian Qiao, 69, 141, 362
Lian Zi Xin, 100, 363
Liang Shi, 79, 363

Liao Ge Wang, 153, 363
Lichenin, 149, 501
Liderane, 87, 271, 501
Lie Xiang Du Juan, 123, 363
Liensinine, 100, 277, 501
Lignan, 229
Lignin, 30, 95, 121, 285, 335, 501
Lignoceric acid, 92, 134, 229, 501
Ligusticum chuanziang, 85, 270, 402, 440, 501, 519, 531, 567
Ligusticum jeholense, 85, 270, 437, 444, 522, 537
Ligusticum scoticum, 270, 601, 622
Ligustilide, 17, 85, 181, 241–242, 265, 270, 501, 613
Ligustrin, 231, 613
Lilacin, 231, 613
Lilium brownii, 86, 339, 441, 540
Lilium candidum, 271
Lilium japonicum, 270
Limonene, 16, 24, 29, 39, 42, 46–47, 51, 63, 73, 83–84, 87, 89, 92, 95, 101–102, 107, 112, 142, 146, 154, 158–160, 172, 181, 184, 186, 189, 191, 199, 203, 207–208, 210, 212–213, 219, 228, 232–233, 242, 245, 252, 266, 268–269, 271, 273, 275, 281, 304, 501–502, 613
Limonic acid, 272, 613
Limonin, 57, 67, 116, 195, 310, 502
Linalol, 62, 197, 502
Linalool, 9, 14, 23, 42, 46–47, 53, 65, 73, 76, 88–89, 95, 97, 99, 107, 113, 142, 146, 154–155, 161, 167, 172, 177, 182, 191, 194, 199, 203, 210, 212, 215, 232–233, 236, 238, 240, 244, 257, 304, 502, 613
Linalyl acetate, 146, 232–233, 502, 613
Linamarase, 68, 200, 262, 502
Linamarin, 87, 209, 502, 613
Linaracrine, 86, 208, 502
Linarezine, 86, 208, 502
Linaria vulgaris, 86, 208, 363, 613, 617, 632–633
Linarin, 86, 208, 503, 613
Lindera akoensis, 86, 271, 470, 502–503, 519, 529, 535
Lindera benzoin, 271
Lindera communis, 86, 271, 386
Lindera glauca, 87, 271, 341, 423, 426, 430, 433, 438, 468, 501
Lindera megaphylla, 87, 271, 346, 464
Lindera strychnifolia, 87, 271, 385, 489–490, 501, 503, 519
Linderalactone, 87, 271, 503
Linderene, 271, 503
Linderic acid, 87, 271, 503
Linderol, 87, 271, 503

Linderoxide, 87, 271, 503
Lindestrene, 87, 271, 503
Lindestreolide, 87, 271, 503
Lineolone, 157, 503
Ling, 21, 43, 46, 71, 116–117, 119, 122, 135, 137,
    146–147, 173, 317, 348, 350, 363, 365,
    383, 393
Ling Bi Long, 119, 363
Ling Lan, 46, 363
Ling Nan Huai, 137, 363
Ling Zhi, 71, 363
Linoleic acid, 5, 28, 34, 40, 42, 45, 50, 62, 82, 87,
    99, 134, 142, 179, 181, 183, 188–189,
    193, 197, 207, 209, 215, 225, 229–230,
    243, 250, 256, 271, 282, 503, 613
Linolein, 133, 148, 229, 503
Linolenic acid, 34, 48, 102, 172, 179, 209, 215,
    231, 250, 255, 312, 503, 614
Linoleyl acetate, 46, 191, 503
Linolic acid, 47, 503
Linositol, 171, 503
Linseed oil, 209, 614
Linum stelleroides, 87, 388, 418, 429, 436, 460,
    468, 475, 501–502, 504, 525, 561,
    574, 577
Linum usitatissimum, 209, 613–614, 617, 619, 631
Lipase, 148, 243, 301, 503
Lipids, 195, 314, 614
Liquidambar acerifolia, 87, 271, 350, 419, 430,
    438, 458, 496, 566
Liquidambar orientalis, 271, 595, 622, 637
Liquiritigenin, 74, 204, 264, 503
Liquiritin, 74, 204, 264, 504
Liquloxidol, 504
Liriodenine, 93, 100, 504
Liriope graminifolia, 88, 365, 515
Liriope spicata, 59, 103
Litchi chinensis, 88, 307, 362, 405, 410, 415, 417,
    423, 439, 477, 501, 506, 539, 555, 568,
    574, 576
Lithospermic acid, 209, 614
Lithospermin, 88, 209, 504
Lithospermum erythrorhizon, 88, 209, 393, 402,
    406, 448, 456, 504, 545, 556, 614
Litsea cubeba, 88, 374, 439, 500, 502
Liu Chun Yu, 86, 363
Liu Lan, 38, 363
Liu Shen Cao, 138, 363
Liu Xing Zi, 149, 363
Liu Ye Cai, 63, 363
Livistona chinensis, 88, 361
l-laudanidine, 299, 611
L-limonene, 95, 150, 217, 302, 497, 611
l-limonene, 95, 150, 217, 302, 497, 611

L-linalool, 41, 125, 142, 172, 189, 226, 232, 289,
    298, 497
l-linalool, 41, 125, 142, 172, 189, 226, 232, 289,
    298, 497
Llysine, 504
l-marmesin, 497
l-menthone, 73, 92, 203, 273, 497
l-methylcytisine, 137, 498
l-methylephedrine, 63, 198, 260, 498
l-norephedrine, 63, 198, 260, 498
Lo Han Song, 363
Lo Huang Zi, 363
Lo Sheng Kui, 363
Loan Mao Cao, 10, 363
Loan Now Xiang, 59, 364
Lobelanidine, 88, 209, 272, 504, 614
Lobelanine, 88, 209, 272, 504
Lobelia chinensis, 88, 209, 272, 341, 490, 504,
    602, 617, 632–633
Lobelia inflata, 272
Lobelia pulmonaria, 272, 587, 598, 624, 635
Lobelia pyramidalis, 373
Lobelia siphilitica, 272, 584
Lobelidiol, 272, 614
Lobeline, 88, 209, 272, 504, 614
Lochnerine, 160, 504
Loganin, 48, 89, 97, 168, 191, 209, 214, 231, 255,
    272, 504, 614
Loliolide, 95, 504
Long Dan, 72, 364
Long Kui, 136, 364
Longiceroside, 48, 191, 255, 504
Longifolene, 499
Lonicera acuminata, 89, 272, 372, 436, 486,
    504–505, 552, 563–564
Lonicera apodonta, 348
Lonicera brachypoda, 360
Lonicera caerulea, 272, 595, 609, 613, 615, 631
Lonicera chinensis, 372
Lonicera hypoglauca, 354
Lonicera japonica, 209, 628, 633, 640
Lonicerin, 89, 209, 272, 504
Lophanthus chinensis, 89, 357, 464
Lophanthus rugosus, 209, 640
Loranthus europaeus, 272, 322, 603, 611, 626
Loranthus parasiticus, 89, 272, 379, 419, 543,
    552
Loropetalum chinense, 89, 357, 470, 491, 544
Lotaustralin, 87, 209, 504
Lotusine, 100, 277, 504
Lou Lu, 60, 364
Lour Lu, 123, 364
l-perilla, 107, 217, 498
L-perilla-aiuehyde, 498

l-phellandrine, 41, 189, 498
l-phyllandrene, 219, 611
l-pinene, 44, 92, 190, 213, 273, 498
l-pinocamphone, 73, 92, 203, 273, 498
l-p-menthene, 125, 226, 289, 498
l-propenylsulforic acid, 498
l-pulegone, 73, 203, 498
l-rhamnose, 94, 113, 220, 274, 282, 498
l-sesamen, 62, 198, 498
l-stepharine, 96, 498
l-tetrahydropalmatine, 140, 498
Lu Cao, 57, 78, 355, 364
Lu Er Jin, 83, 364
Lu Gen, 25, 109, 364
Lu Teng, 97, 364
Lu Wen, 12, 364
Lu Xian, 14, 121, 364
Lu Xian Cao, 121, 364
Lu Yao, 135, 364
Lu Zhu, 25, 70, 364
Lucernol, 95, 212, 504
Luffa aegyptiaca, 89, 210, 401, 426, 433, 473,
        486, 502, 508, 510–511, 536, 552, 561,
        574, 576–577, 579, 604, 640
Luffa cylindrica, 98
Luffa faetida, 373
Lumicaerulic acid, 46, 191, 254, 504
Lun Ye Sha Seng, 364
Lunularic acid, 111, 504
Luo Bu Ma, 18, 364
Luo Fu Mu, 122, 364
Luo Han Guo, 98, 365
Luo Hua, 19, 32, 365
Luo Hua Zi Zhu, 32, 365
Luo Shi, 95, 106, 146, 173, 345, 365
Luo Ti Cao, 32, 365
Luo Xing Fu, 25, 365
Lupenone, 8, 12, 69, 237, 239, 505
Lupeol, 24, 61, 66, 69, 82, 113, 118, 152, 163, 169,
        201, 235, 402, 505
Lupeol acetate, 24, 61, 505
Lupeol palmitate, 82, 505
Lupin alkaloid, 65, 314, 505
Lupinidin, 90, 505
Lupinine, 90, 505
Lupinus luteus, 90, 391, 450, 505, 548
Lupulin, 78, 505
Lupulone, 78, 205, 505, 614
Lutein, 52, 143, 169, 172, 257, 505
Lutein-7-primveroside, 250, 614
Luteioic acid, 119, 505
Luteolin, 10, 18, 28, 45, 78, 89, 107, 110, 151, 159,
        169, 178, 183, 209, 212, 217–218, 233,
        238, 242, 272, 275, 281, 302, 505, 614

Luteolin-7-beta-D-glucopyranoside, 84, 208
Luteolin-7-glucoside, 82, 299, 505
Luteolin-7-rhamnoglucoside, 505
Luteoline, 24, 506
Luteolinidin, 27, 82, 506
Luteolinidin 5-glucoside, 27, 506
Luteolin-monoarabinoside, 506
Lutolin, 300, 614
Lycium barbarum, 90, 210, 273, 352, 432, 520,
        534, 577, 580, 591, 593, 595, 622, 625,
        639
Lycium chinense, 90, 210, 347, 403, 424, 439,
        496, 531, 538
Lycium pallidum, 273
Lycoclavanin, 91, 211, 506
Lycoclavanol, 91, 211, 506
Lycodoline, 91, 211, 506
Lycopene, 32, 49, 56, 77, 161, 185, 195, 205, 506
Lycoperiscon esculentum, 520, 591, 635
Lycopodine, 90–91, 210–211, 506, 614
Lycopodium annotinum, 90, 210, 375, 409, 441,
        506, 555, 569, 623, 637
Lycopodium cernum, 359
Lycopodium clavatum, 91, 211, 375, 409, 419,
        440, 452, 468–469, 506, 520, 574,
        599, 603, 614, 626, 637
Lycopose, 91, 273, 506
Lycopus fargesii, 91, 273, 375, 506
Lycopus lucidus, 377, 477, 545–546, 560
Lycopus obscurum, 391
Lycopus phlegmaria, 345
Lycopus veitchii, 391
Lycopus virginicus, 273, 592, 600, 622
Lycoramine, 77, 91, 166, 506
Lycorenine, 91, 166, 506
Lycoricidine, 91, 506
Lycoricidinol, 91, 506
Lycorin, 79, 506
Lycorine, 44, 77, 91, 99, 155, 166, 173, 215, 308,
        506
Lycoris aura, 91
Lycoris radiata, 377, 456, 482, 506, 541, 566
Lygodium japonicum, 91, 353, 468
Lyonia ovalifolia, 91, 368, 412, 506
Lyoniols, 91, 506
Lysimachia barystachys, 92, 273, 359, 396, 464,
        497–498, 502
Lysimachia vulgaris, 273, 590, 603, 628, 633
Lysine, 59, 88, 103, 143, 159, 172, 208, 260, 506,
        614
Lythrum salicaria, 92, 211, 371, 436, 447, 461,
        508, 525, 550, 564, 577, 628, 633,
        636, 639
l-zingiberene, 236, 612

M

Ma An Teng, 365
Ma Bian Cao, 151, 365
Ma Bo, 365
Ma Chi Xian, 117, 365
Ma Dou Ling, 21, 365
Ma Dou Ling (Taiwan), 365
Ma Huang, 63, 323, 365
Ma Lan Zi, 81, 365
Ma Qian Zi, 140, 365
Ma Wei Lian, 144, 365
Ma Xian Gao, 107, 365
Machilus thunbergii, 92, 354, 458, 501, 517, 543, 546
Macleaya cordata, 92, 342, 407, 426, 435, 442, 465, 481, 526, 541, 550
Maclurin, 98, 214, 507
Macrephyllic acid, 114, 507
Macrocarpium officinalis, 93
Macrophylline, 133, 293, 507
Madasiatic acids, 186, 614
Madecassic, 186, 614
Madecassoside, 37, 186, 507
Maesa japonica, 93, 348, 507
Maesa perlarius, 93, 373, 406
Maesa tenera, 93, 373
Maesaguinone, 93, 507
Magnesium, 117, 187, 222, 507, 614
Magnocurarine, 93, 211, 507, 614
Magnolia biloba, 93, 387, 412, 438–439, 465–466, 507, 550
Magnolia coco, 93, 389
Magnolia fortunei, 351
Magnolia japonica, 355
Magnolia officinalis, 93, 211
Magnolol, 93, 507
Mahonia aquifolium, 273, 319, 590, 596, 615, 620, 624
Mahonia japonica, 94, 273, 352, 421, 493, 556
Mai Dong, 102, 365
Mai Men Dong, 88
Mai Ya, 78, 366
Mairin, 65, 200, 507
Makisterones, 114, 507
Makulor, 45, 191, 253, 507
Malic acid, 38, 47–48, 52, 59, 64, 83, 93–94, 104, 118–119, 124, 151, 153, 161, 180, 187, 191, 196, 199, 212, 223, 226, 235, 255, 257, 263, 269, 272, 278, 288–289, 291, 303, 507, 615
Mallol, 282, 615
Mallorepine, 94, 322, 508
Mallotinin, 94, 508

Mallotus japonicus, 94, 389, 406, 546, 564
Mallotus paniculatus, 94, 338, 410
Mallotus repandus, 94, 322, 351, 421, 508, 546
Malonic acid, 177, 615
Maltase, 68, 136, 200, 262, 295, 508
Maltose, 78, 205, 508
Maluidin glucoside, 141, 298, 508
Malva chinensis, 94, 274, 348, 449, 496, 498
Malva rotundifolia, 274
Malva sylvestris, 586, 603, 607, 615, 617, 633
Malvidin, 92, 95, 211–212, 508
Malvin, 92, 211, 274, 508, 615
Man Jiang Hong, 27, 366
Man Shan Hong, 123, 366
Man Ti Xian, 366
Man Tu Luo, 56, 366
Mandelonitrile, 118, 223, 508
Mang Ji, 57, 366
Manganese, 117, 208, 610
Mangiferin, 15, 305, 508
Manihot esculenta, 94, 212, 378, 484, 598
Mannan, 58, 196, 508
Manneotetrose, 81, 207, 268, 508
Mannit, 71, 202, 508
Mannitol, 131, 137, 145, 151, 171, 225, 302, 508, 615
Mannosan, 89, 146, 210, 508
Mannose, 14, 50, 81, 193, 268, 508
Mao Di Huang, 58, 366
Mao Dong Qing, 79, 366
Mao Dong San Hu, 135, 366
Mao Gao Cai, 59, 366
Mao Guo Yan Ming Cao, 121, 366
Mao Liang, 61, 122, 366
Mao Xian, 51, 77, 366
Mao Zhua Chao, 122, 366
Mar Dong, 103, 366
Margaric acid, 129, 181, 508, 615
Markogenin, 15, 508
Marmesin, 16, 171, 241, 509
Marsdenia condurango, 274, 596, 623, 640
Marsdenia tenacissima, 94, 274, 382, 509, 511, 552–553
Marsdeoreophisides, 94, 274, 509
Martaicaria ester, 509
Maslinic acid, 38, 49, 120, 187, 256, 509
Masperuloside, 98, 276, 509
Massoilactone, 205, 615
Mastic acid, 220, 615
Masticin, 220, 615
Masticinic acid, 112, 220, 509
Masticonic acid, 112, 220, 509
Masticoresene, 112, 220, 509
Matabic acid, 508

Matai-resinol, 20, 181, 509
Matairesinoside, 146, 173, 509
Matatabic acid, 7, 177, 509, 615
Matatabiether, 7, 177, 509, 615
Matatabistic acid, 7, 177, 509, 615
Matricaria, 46, 63, 94, 199, 212, 388, 412–413,
    419, 488–489, 507, 509, 564, 569, 578,
    585–586, 589, 597–598, 603, 607, 614,
    618, 633, 637
Matricaria chamomilla, 94, 212, 388, 412–413,
    419, 488–489, 507, 564, 569, 578,
    585–586, 589, 597–598, 603, 607, 614,
    618, 633, 637
Matrine, 137, 306, 509
Matteuccia struthiopteris, 95, 212, 312, 315, 358,
    418, 423, 428–429, 436, 460, 469,
    526–528, 538, 540, 542, 561, 574, 588,
    591–592, 594, 619–620, 624, 632, 638
Matteucinol, 168, 509
Maysin, 236, 615
Maytanacine, 95, 509
Maytanbutine, 95, 509
Maytanprine, 95, 509
Maytansine, 95, 509
Maytansinol, 95, 509
Maytanvalinc, 95, 510
Maytenus diversifolia, 95, 345, 399, 459, 504,
    509, 562–563
Maytenus serrata, 95, 366, 509–510
Meconic acid, 217, 615
Meconine, 106, 217, 279, 510
Medicagemic acid, 95, 212, 510
Medicago falcata, 95, 368, 430, 440, 444, 451,
    453, 455, 471, 502, 504, 508, 510, 513,
    516, 525, 532, 553, 570
Medicago sativa, 212, 584, 597, 610–611, 624,
    631, 639
Mei Deng Mu, 95, 366
Mei Gui Hua, 125, 366
Mei Hua Cao, 106, 366
Mei Li Cao, 39, 367
Mei She Chao, 32, 367
Melaleuca leucadendra, 95, 213, 340, 419–420,
    425, 429, 438, 458, 466, 473, 491, 497,
    501, 510, 520, 524, 536, 544, 556, 567,
    573–574, 585, 591, 595, 635
Melaleucin, 95, 510
Melasma arvense, 96, 353, 418, 515
Melia azedarach, 213, 361, 471, 510, 560, 603,
    615, 633, 637
Melia japonica, 96, 345, 419, 427, 447, 474, 496,
    500, 510, 513, 520–521, 528, 570, 574
Meliacins, 213, 615
Melialactone, 96, 213, 510

Melianotriol, 96, 213, 510
Meliatin, 97, 214, 510
Melibiase, 136, 295, 510
Melilotic acid, 96, 213, 274, 510
Melilotoside, 96, 213, 274, 510
Melilotus alba, 96, 213, 274, 341, 399, 422, 428,
    436, 444, 446, 455, 484, 510, 554, 572,
    597, 599, 603, 626, 633, 640
Melilotus arvensis, 274
Melilotus officinalis, 274
Melochia corchonfolia, 96, 389, 482, 510, 570
Melocorin, 96, 274, 510
Melosatin D, 275, 615
Melotoxin, 50, 510
Melovinone, 275, 322, 615
Menisnine, 41, 189, 510
Menisperine, 96, 510
Menispermine, 97, 275, 510
Menispermum canadense, 275, 583, 599, 615,
    618, 635, 638
Menispermum dauricum, 96–97, 275, 349, 374,
    389, 403, 451–452, 458, 498, 507, 510,
    557, 560–561, 567
Menispermum palmatum, 275, 590, 597, 601,
    612–613, 617, 620, 622–623, 625
Mentha arvensis, 97, 213, 275, 342, 510–511, 592,
    613, 615–616, 635
Mentha pulegium, 275, 610, 625
Mentha spicata, 275, 593, 595, 598, 610, 614, 617,
    623, 635
Menthene, 125, 226, 289, 510
Menthiafolin, 97, 214, 510
Menthol, 73, 89, 97, 176, 203, 210, 213, 275, 510,
    615
Menthone, 89, 97, 210, 213, 228, 275, 511, 616
Menthyl acetate, 97, 213, 511, 616
Menyanthes trifoliata, 97, 214, 378, 409, 416,
    425, 428, 437, 456, 471, 475, 479, 487,
    504, 510–511, 554, 561, 570, 597, 603,
    609, 622, 632–633, 637
Menyanthin, 97, 214, 511
Meoglucobrassicin, 81, 268, 511
Mesaconitine, 6–7, 177, 511, 616
Mesityl oxide, 50, 193, 511
Mesoinositol, 44, 152, 190, 235, 511, 616
Metaphanine, 139, 511
Metaplexis japonica, 167, 335, 363, 420, 452,
    454, 473, 491, 531, 553, 573
Metaploxigenin, 94, 274, 511
Methanethiol, 25, 182, 511
Methanolic, 98, 276, 511
Methioine, 103, 511
Methlxanthines, 616
Methoxyl-camptothecine, 511

Methoxylhemigosipol, 75, 204, 511
Methyl acetyl-isocupressate, 511
Methyl anthranilate, 42, 512
Methyl caffeate, 33, 250, 512
Methyl chavicol, 206, 245, 616
Methyl cinnamate, 13, 239, 512
Methyl communate, 19, 512
Methyl eugenol, 97, 512
Methyl isobutyl ketone, 50, 192–193, 512
Methyl isocupressate, 19, 512
Methyl nigakinone, 111, 281, 512
Methyl palmitate, 45, 190, 512
Methyl salicylate, 161, 187, 199, 247, 259, 283, 616
Methyl vanillin, 163
Methylacetic acid, 63, 199
Methylamine, 177, 616
Methylanthranilate, 42, 99, 276
Methylchavicol, 9, 69, 201, 238, 512, 616
Methyl-corypalline, 100, 277, 512
Methylcytisine, 36, 137, 188, 252, 616
Methylene-bishydroxy-coumarin, 513
Methylephedrine, 63, 198, 260
Methylethylacetic ester, 97, 513
Methylheptenol, 53, 194, 257, 513
Methylheptenone, 53, 155, 194, 236, 257, 513
Methylisopelletierine, 120, 224, 513
Methylkulonate, 96, 213, 513
Methyl-laurate, 103, 512
Methyllcaconitine, 56, 513
Methyl-n-amyl ketone, 41, 189, 512
Methyl-n-propyl ketone, 180, 616
Methyl-pelletierine, 224, 512
Methyl-swertianin, 297, 512
Mezerein, 195, 259, 616
Mi Hou Tao, 7, 157, 367
Mi Meng Hua, 367
Mian Bao Shu, 24, 367
Mian Hua Gen, 75, 367
Mian Ma Guan Zhong, 60, 367
Mian Zi Soo, 75, 367
Michelabine, 97, 513
Michelenolide, 66, 262, 513
Michelia alba, 97, 340, 401, 502, 512–513, 526, 550, 573
Mi-hem erocallin, 164, 513
Milletia reticulata, 97, 358, 412, 548
Milletia taiwaniana, 364
Miltirone, 127, 291, 513
Mimosa arborea, 97, 276
Mimosa hostilis, 276
Mimosa invisa, 98, 276, 353, 514
Mimosa pudica, 214, 353, 619
Min Dong Seng, 38, 367

Mineral elements, 104, 215, 513, 616
Minerals, 24, 157, 218, 230, 250, 295–296, 513, 616
Miniatine, 44, 514
Minosine, 98, 214, 276, 514
Mitraphylline, 4, 176, 514
Mo Han Lian, 61, 162, 367
Mo Ja Chao, 63, 367
Mo Li Hua, 81, 367
Mo Yao, 45, 367
Mo Yue, 14, 367
Molephantin, 61, 514
Momordica charantia, 98, 214, 361, 414, 561, 594, 602, 609, 617
Momordica cylindrica, 98, 373
Momordica grosvenori, 98, 365
Monocrotalines, 50, 514
Monoterpene, 5, 18, 76, 115, 128, 143, 158, 221, 235, 514, 616
Monoterpene glycosides, 235, 616
Monoterpenoid glycosides, 216, 616
Monotropin, 121, 225, 514
Moo Tune, 10, 367
Mophilenolide, 166, 399
Morin, 98, 214, 514
Morinda citrifolia, 98, 276, 326, 357, 406, 409, 456, 477, 509, 514, 521, 549, 559, 564
Morinda didyma, 276, 616, 639
Morinda parvifolia, 98, 276, 387, 406, 511, 514
Morindadiol, 98, 276, 514
Morindaparvin-A, 98, 276, 514
Morindin, 98, 276, 477, 616
Mormordicine, 214, 617
Mormordin, 214, 617
Morolic acid, 8, 514
Morphine, 44, 106, 217, 279, 514, 617
Morroniside, 48, 168, 191, 255, 514
Morus alba, 98, 214, 373, 397, 447, 456, 507, 514–515, 586, 588, 603, 638–639
Motephantinin, 61, 514
Mu Dan Pi, 104, 367
Mu Er Cao, 75, 367
Mu Fang Ji, 45, 357, 367
Mu Gui, 103, 367
Mu Jin, 76, 367
Mu Jing, 152–153, 347, 368
Mu Jing Chi, 153, 368
Mu Lan, 93, 368
Mu Tong, 21, 42, 368
Mu Tou Hui, 107, 368
Mu Xiang, 21, 27, 164, 171, 368, 371, 383
Mu Xu, 95, 368
Mu Yu Ma, 30, 368
Mu Ziang, 130, 368

Mucic acid, 109, 514
Mucilage, 10, 20, 53, 59, 79, 88, 107, 115, 142, 176–179, 181, 185–187, 189–190, 195, 204–205, 208–209, 217, 219–222, 234–238, 240, 251–253, 257, 259–260, 263–264, 267, 274, 278, 282–283, 299–301, 303, 515, 617
Mucronatine, 50, 515
Mucronatinine, 50, 515
Mucus, 58, 139, 176, 178, 209, 214, 223, 230–231, 265, 273, 280, 515
Mukorosside, 129, 515
Mulberrin, 98, 214, 515
Mulberrochromene, 98, 214, 515
Munjistin, 126, 289, 515
Muramine, 105, 279, 515
Muricatin A, 81, 267, 515
Murraya koenigii, 276, 611, 633, 640
Murraya paniculata, 99, 276, 360, 400, 422, 425, 432, 475, 497, 512, 525
Musa paradisiaca, 99, 386, 522, 555
Musaenoide, 96, 515
Muscarine, 34, 185, 515
Muslinic acid, 61, 515
Mustard oil, 30, 184, 229, 515, 617
Mustard-oil glycosides, 184, 617
Mutaxanthin, 110, 218, 281, 515
Myoinositol, 216, 617
Myrcene, 24, 45, 56, 62, 83–84, 95, 102, 116, 146, 150, 154, 160–161, 168–169, 187, 191, 195, 197, 207–208, 212, 233, 245, 253, 268, 275, 302, 304, 516, 617
Myrica cerifera, 277, 603, 622, 626, 633, 637
Myrica rubra, 99, 276, 352, 516
Myricetin, 49, 99, 123–124, 141, 145, 161, 276, 282, 287, 298–299, 400, 516, 617
Myricetol, 161, 516
Myricitrin, 49, 516
Myricyl, 35, 138, 251, 267, 516, 617
Myricyl alcohol, 35, 251, 267, 516, 617
Myriogynine, 37, 516
Myristic acid, 29, 42, 45, 82, 87, 99, 133, 143, 152, 172, 181, 184, 189, 204–205, 207, 211, 215, 235, 271, 275, 516, 617
Myristica fragrans, 99, 215, 372, 411, 430, 448, 458, 466, 475, 481, 489, 500, 502–503, 516, 524, 529, 536, 546, 550, 560, 567, 598, 612–613, 619, 631
Myristicin, 99, 193, 215, 516, 617
Myristoleic acid, 618
Myrocin, 30, 145, 184, 516
Myrosinase, 34, 134, 145, 229, 233, 250, 516
Myrtenol, 150, 302, 516
Myrtocyan, 301, 618

N

Na Yang Shan, 19, 368
Naginataketone, 162, 518
Nan Gua Zi, 51, 368
Nan He Chi, 56, 368
Nan Tian Zhu, 99, 368
Nan Wu Wei Zi, 83, 368
Nan Zhu, 91, 368
Nandazurine, 99, 518
Nandina domestica, 99, 368, 421, 458, 518
Nandinine, 99, 518
Napelline, 177, 618
Naphthalene glycosides, 186, 251, 618
Naphthaquinone, 113, 518
Naphthopyrones, 35, 251, 518
Narcissus tazetta, 99, 114, 215, 378, 506, 518, 566, 583, 612
Narcitine, 99, 215, 518
Narcotine, 106, 217, 279, 518, 618
Nardostachys grandiflora, 277
Nardostachys jatamansi, 100, 277, 351, 464, 493, 556
Naringenin-4β-O-pyranogluoside, 518
Naringin, 42, 310, 518
Nasunin, 136, 295, 518
Nauclea rhynchophylla, 100, 352, 492, 547
Naucleoside, 518
n-*butyl allophanate*, 45, 190, 517
n-butyl-2-*ethyl butyl phthalate*, 102
n-butyldenephthalide, 242, 618
n-caprylaldehyde, 103, 278, 517
n-coumaric, 212, 250, 618
N'-desmethyldauricine, 618
n-dodecanol, 181, 618
Neferine, 100, 277, 518
Negundoside, 153, 518
Nei Don Zi, 86–87, 369
Nelumbium nelumbo, 100, 277, 353, 363, 453, 490, 501, 504, 512, 518
Nelumbium nuciferum, 100, 277, 362, 413, 456, 504, 517, 522, 539, 548
Nelumbium officinale, 277
Neo-allicin, 12, 179, 518
Neoalsomitra integrifoliola, 100, 341, 432, 445, 488
Neoandrographolide, 15, 518
Neoanisatin, 80, 518
Neoboschnialactone, 41, 518
Neocarthamin, 35, 185, 519
Neochlorogenic acid, 61, 519
Neocnidilide, 85, 270, 519
Neocryuptomerin, 114, 519
Neogitogenin, 15, 519

Neoglucobrassicin, 43, 519
Neoherculin, 304, 618
Neohesperidin, 116, 519
Neo-lignans, 93, 518
Neolinarin, 86, 208, 519
Neolinderalactone, 87, 271, 519
Neoline, 177, 618
Neomatabiol, 7, 177, 519, 618
Neo-nepetalactone, 7, 177, 518, 618
Neothujic acid, 282, 618
Neotigogenin, 135, 230, 294, 519
Neoxanthin, 143, 169, 519
Nepenthes raffsiana, 100, 393, 410, 413, 470, 533
Nephelium lappaceum, 354
Nephelium longana, 100, 372, 477, 552, 562, 564–565, 576
Nepodin, 126–127, 227, 290, 519, 619
Neriin, 277, 619
Nerinine, 155, 173, 519
Nerioderin, 101, 519
Neriodin, 101, 519
Neriodorin, 101, 519
Neriororin, 519
Nerium indicum, 101, 167, 277, 358, 451–452, 455, 476, 519, 523
Nerium oleander, 277, 596, 604, 619, 625, 627
Nerol, 53, 103, 125, 194, 226, 257, 289, 519
Nerolidiol, 95, 520
Nerolidol, 14, 76, 240, 520
Nervilia purpurea, 101, 389, 447, 456, 463, 520, 561
Nervisterol, 101, 520
Nevadersin, 92, 520
n-formyl-N-deacetylcolchine, 80, 517
n-hentriacontane, 11, 517
n-hexacosane, 82, 517
Niacin, 6, 14, 19, 28, 30, 34, 36, 49, 74, 77, 108, 141, 155, 163, 167, 171, 183, 185, 203–204, 231, 236, 240, 251, 256, 288, 520, 619
Nicotelline, 101, 215, 520
Nicotiana tabacum, 101, 215, 388, 520–521, 584, 619, 640
Nicotimine, 101, 215, 520
Nicotine, 60–61, 91, 101, 162, 197, 199, 211, 215, 520, 619
Nicotinic acid, 17, 37, 90, 136, 181, 195, 210, 234, 241, 265, 273, 301, 520, 619
Nigakihemiacetal A, 111, 281, 520
Nigakilactone A, 111, 281, 520
Nigakinone, 111, 281, 512, 520
Nigerine (N,N-dimethyltryptamine), 214, 619
Nilgirine, 50, 520
Nimbin, 96, 213, 520

Nimbolins, 96, 213, 521
Ning Meng Sian Mao, 53, 369
Niranthin, 110, 521
Nirtetralin, 110, 521
Nishindaside, 521
Nishindine, 153, 521
Nitidine, 155, 304, 521
Nitroacronycine, 7, 521
Nitryl-glycoside, 243, 521, 619
Niu Bang Chi, 20, 369
Niu Fang Feng, 76, 369
Niu Zi Qie, 136, 369
n-*methyl anabasine,* 293, 618
n-methyl anthranilic acid, 67, 517
n-methylcoclaurine, 100, 277, 517
n-methylisococlaurine, 100, 277, 517
n-methyl-isopelletierine, 132, 293, 517
n-methylmorpholine, 36, 251, 517
n-methyltyramine, 42, 190, 517
n,n-*dimethyltryptamine oxide,* 110, 517
n-nonyl, 142, 172, 232, 298, 512, 517
n-*nonyl aldehyde,* 142, 172, 232, 298, 517
n-norarmepavine, 92, 517
Nobiletin, 42, 164, 190, 453, 521
Nocoteine, 521
Nodakenetin, 17, 108, 158, 171, 241, 280, 521
Nodakenin, 17, 108, 158, 241, 280, 478, 521
Nonacosane, 57, 118, 125, 160, 226, 289, 521
Nonadecanoid, 181, 619
Nonalactone, 119, 223, 521
Nonanal, 47, 191, 521
Nonyl aldehyde, 125, 226, 289, 521
Nonylic aldehyde, 41, 189, 521
Norargemonine, 299, 619
Nordamnacanthal, 98, 276, 521
Nordracorubin, 59, 522
Norepinephrine, 99, 117, 222, 522
Noreugenin, 8, 522
Noricariin, 63, 261, 522
Norkurarinone, 137, 522
Normenisarine, 45, 522
Nornuciferine, 522
Norpseudoephedrine, 63, 522
nor-rubrofusarin, 35, 251, 521
Norsecurinine, 132, 522
Nortracheloside, 146, 173, 522
Nothosmyrnium japonicum, 101, 351, 457, 464, 522
Nothosmyrnol, 85, 101, 270, 522
Notoptero, 101, 522
Notopterygium incisium, 101, 351, 464, 467, 490, 502, 522
Nou Me, 103, 369
Novacine, 140, 522

n-pentadecane, 207, 618
n-phenylethyl alcohol, 125, 226, 289, 517
n-tetradecanol, 181, 618
n-trans-coumaroyltyramine, 618
n-trans-feruloyltyramine, 233, 618
n-triacontane, 518
Nu Zhen Zi, 86, 369
Nuciferine, 100, 167, 277, 522
Nudicaulin, 105, 279, 522
Nuphar japonicum, 101, 370, 522, 524, 528, 558
Nupharamine, 101, 522
Nupharine, 278, 619
Nuzhenide, 86, 208, 270, 522
Nymphaea alba, 278, 619, 626, 636
Nymphaea tetragona, 101, 277, 378, 410
Nymphaeine, 278, 619
Nymphoides peltata, 335, 397, 492, 549

O

o-acetylcolumbianetin, 44, 190, 522
Obacunone, 109, 218, 523
Obakinone, 57, 195, 523
Obtusifolin, 35, 251, 523
Obtusin, 35, 251, 523
Ocimene, 142, 232, 523, 619
Octacosane, 69, 201, 523
Octacosanol, 69, 275, 523, 619
Octadecatetraenic acid, 231, 619
Octalactone, 119, 223, 523
Octanol, 9, 238, 523
Octopinic acid, 107, 523
o-cumaric acid, 65, 522
Odine, 292, 523
Odoilin, 523
Odoratin, 66, 523
Oenathera biennis, 375
Oenathera javanica, 378
Oenathera odorata, 372
Oenathera terythrosepala, 346
O-*isovaleryl columbianetin,* 44, 190, 522
Okinalein, 120, 224, 284, 523
Okinalin, 120, 224, 284, 523
Olcanolic acid, 160, 523
Oldenlandia chrysotricha, 102, 376, 425, 474, 524, 561, 573
Oldenlandia diffusa, 102, 340, 418, 423, 454, 523, 528
Oldenlandoside, 102, 523
Oleandomycin, 277, 619
Oleandrin, 101, 277, 523, 619
Oleandrose, 101, 277, 523
Oleanene derivatives, 25, 523
Oleanic acid, 524

Oleanolic acid, 16, 19, 21, 32, 43, 45, 65, 86, 95, 102, 105, 118, 141, 152, 158, 169, 185, 200, 208, 216, 223, 235, 241, 243, 253, 270, 298, 402, 523–524
Oleic acid, 11, 30, 34, 42, 45, 47, 50, 62, 75, 82, 87, 99, 101, 142, 179, 186, 189, 193, 197, 207, 209, 215, 229, 243, 250, 256, 264, 271, 282, 308, 524, 619
Olein, 125, 133, 172, 226, 229, 524
Olein acid, 125, 133, 229, 524
Oleoresin, 10, 60, 186, 197, 524
Oleostearic acid, 179, 619
Oleyl alcohol, 160, 524
Oligopeptides, 223, 312, 619
Oligosaccharides, 9, 148, 178, 234, 524
Olitoriside, 47, 524
Omphalia lapidescens, 102, 315, 362, 477
Onjisaponin A, 114, 282, 524
Onjisaponin B, 114, 282, 525
Ononitol, 95, 212, 525
O-nornuciferine, 100, 277, 522
Onychium japonicum, 102, 357, 495
Ophelic acid, 37, 252, 525
Ophioglossum japonicus, 102, 365, 423
Ophioglossum thermale, 168, 370
Ophioglossum vulgatum, 103, 390
Ophiopogenins, 103, 525
Ophiopogon gracilus, 103, 365
Ophiopogon japonicus, 103, 366, 388, 423, 494, 525, 538, 562
Ophiorrhiza japonica, 103, 357, 399, 406, 423, 484, 546, 563
Oplopanax elatus, 103, 278, 323, 345, 460, 464, 470, 517, 525
Oplopanax horridus, 278, 319, 623, 630
Oplopanaxosides, 103, 278, 525
Orchis latifolia, 103, 278, 354, 406
Orchis mascula, 278, 617
Organic acids, 71, 84, 92, 116, 124, 128, 147, 226, 233, 288, 290, 296, 525, 620
Oridonin, 121–122, 525
Orientin, 68, 87, 92, 108, 116, 153, 165, 200, 209, 211, 262, 525
Orientoside, 108, 525
Oripavine, 217, 279, 620
Orobanchin, 168, 525
Orthomethylcoumaric aldehyde, 40, 189, 525
Oryza sativa, 103, 369, 410, 490, 501, 511, 533, 569, 572, 574
Osalic acid, 525
Osmane, 103, 525
Osmanthus fragrans, 103, 352, 367, 423, 512–513, 519, 525
Osmunda japonica, 104, 394, 446, 459–460, 538

Osthol, 17, 99, 158, 276, 525
Oxalate, 5–6, 51, 66, 104, 133, 161, 225, 229, 278,
    429, 525, 592
Oxalic acid, 47, 82–83, 153, 161, 191, 235, 269,
    291, 620
Oxalis acetosela, 278
Oxalis corriculaza, 104, 278, 373, 429, 439, 507,
    525, 565, 577
Oxoushinsunine, 97, 526
Oxyberberine, 247, 273, 620
Oxycanthine, 28, 144, 247, 526
Oxychelerythrine, 155, 304, 526
Oxylenzoic acid, 248, 620
Oxylysin, 27, 247, 526
Oxymatrine, 137, 526
Oxymethyl anthraquinone, 35–36, 186, 251, 526
Oxynitidine, 155, 304, 526
Oxypeucedanine, 16, 241, 526
Oxypurpureine, 144, 299, 526
Oxyria digyna, 104, 215, 351, 513, 540, 546, 588,
    593, 602, 616, 624, 627
Oxyristic acid, 111, 219, 526
Oxysanguinarine, 92, 526

P

Pa Jiao Lian, 60, 369
Pachymarose, 117, 222, 527
Pachymic acid, 117, 222, 527, 620
Pachyrhizus thunbergianus, 104, 351
Paeonidin, 125, 226, 289, 527
Paeonin, 53, 104, 151, 216, 303, 527
Paeonol, 53, 104, 216, 527, 620
Paeonoside, 104, 216, 528
Pai Lan (Taiwan), 369
Pai Qian Chao, 57, 369
Pai Qian Shu, 110, 369
Palamatine, 28, 247, 528
Palderoside, 102, 528
Paliloleic acid, 620
Pallidine, 48, 255, 528
Palmaline, 191, 254, 528
Palmatine, 32, 46, 68, 94, 109, 144, 218, 250,
    299, 528
Palmitic acid, 5, 10–11, 28–29, 40, 42, 45, 47,
    50, 54, 62, 82, 95–96, 101, 129, 137,
    142–143, 171, 177, 179, 181, 183–184,
    189, 193, 197, 204–205, 207, 212–213,
    216, 220, 230, 233, 243, 275, 282, 308,
    528, 620
Palustridine, 162, 528
Palustrine, 63, 162, 199, 528
Pan Chan Teng, 36, 369

Panax ginseng, 105, 216, 332, 372, 422, 524, 528,
    542, 545, 559, 572, 624
Panax japonicum, 105, 393, 435, 476, 552
Panax notoginseng, 105, 382, 453, 470, 528
Panax quinquefolium, 105, 217, 311, 379, 476,
    535, 583, 605, 620
Panax zingiberensis, 105, 373, 415
Panaxadiol, 75, 105, 528
Panaxosides, 216–217, 620
Panaxynol, 105, 216, 329, 334, 528
Paniculatincoumurrayin, 99, 276, 528
Pantothenic acid, 74, 203, 210, 529, 620
Papain, 69, 201, 529
Papaver amurense, 105, 279, 391, 411, 442, 515,
    522
Papaver bracteatum, 279
Papaver rhoeaes, 105, 217, 279, 362, 547, 620,
    635
Papaver somniferum, 106, 217, 279, 391, 400,
    421, 423, 441, 489, 500, 507, 510, 514,
    518, 529, 541, 562, 568, 580, 584, 596,
    615, 617–618, 626, 632, 640
Papaverine, 106, 217, 279, 529, 620
Paraaspidin, 60, 197, 529
Paracyclea insularis, 106, 382, 487–488
Paracyclea ochiaiana, 106, 382, 487
Paradin, 217, 621
Parasorbic acid, 137, 296, 529
Parechites adnascens, 106, 365
Paridol, 217, 621
Parietaria judaica, 279, 603, 634
Parietaria micrantha, 106, 279, 371, 540
Parietin, 116, 221, 529
Parinaric acid, 206, 621
Paris polyphylla, 106, 331, 393, 409, 457
Paris quadrifolia, 217, 457, 595, 611, 621
Paristyphnine, 217, 621
Parnassia palustris, 106, 366, 485, 494
Parthenocissus tricuspidata, 107, 347, 447, 468,
    507, 523
Parthenolide, 66, 262, 529
Patchoulenone, 54, 194, 258, 529
Patrinene, 168, 529
Patrinia heterophylla, 107, 368
Patrinia scabiosaefolia, 107,[1]465
Patrinia villosa, 168, 335, 340, 504, 514, 529, 575
Patrinoside, 107, 529
Patuletin, 142, 172, 232, 529
Paulownin, 168, 529
Paulownioside, 168, 529
p-coumaric acid, 80, 160, 206, 234, 267, 300, 620
p-cymene, 39, 41, 56, 62, 73, 83, 92, 145–146,
    169, 187, 189, 195, 197, 203, 207, 233,
    268, 273, 299

Pectic, 37, 83, 165, 269, 318, 529
Pectic acid, 37, 529
Pectic compound, 83, 269, 529
Pectins, 99, 113, 142, 215, 256, 529
Pectolinarigenin, 86, 208, 529
Pectolinarin, 41, 86, 208, 529
Pedicularis palustris, 279, 584, 609, 622
Pedicularis resupinata, 107, 279, 365, 407, 421,
     425, 436, 494
Peganine, 86, 208, 529
Pei Lan, 66, 369
Peimidine, 70, 201, 529
Peimilidine, 70, 201, 530
Peimine, 70, 201, 530, 621
Peiminine, 70, 201, 530
Peimisine, 70, 201, 530
Peiniphine, 70, 201, 530
Pelargonidin glycosides, 180, 621
Pelargonin, 104, 216, 530
Pelletierene alkaloids, 224, 621
Pelletierine, 120, 224, 530
Peltatin, 60, 530
Peltatoside, 167
Pencordin, 108, 280, 530
Peng Lai Teng, 107, 369
Peng Wo Mao, 369
Peng Zi Cao, 369
Penicillin, 171, 530
Pentacosane, 234, 621
Pentagallotannin, 216, 621
Pentagalloyl glucoside, 216, 621
Penta-m-digalloyl-glucose, 530
Pentane, 245, 621
Pentanoic acid, 119, 223, 530
Pentosan, 4, 121, 133–134, 229, 285, 530
Pentosane, 69, 531
Pentose, 4, 237, 531
Pepsin, 69, 201, 531
Peptides, 90, 210, 531
Peraksine, 122, 285, 531
Peregrinine, 216, 621
Pergularin, 9, 167, 178, 237, 531
Pericalline, 36, 186, 531
Pericamylus formosanus, 107, 369, 515, 518
Perilla arguta, 394
Perilla frutescens, 107, 169, 217, 317, 341, 383,
     386, 396, 399–400, 403, 405, 408,
     413–415, 422, 447, 460, 489, 497–498,
     502, 505, 531, 584–586, 595, 602, 611,
     621, 624, 630
Perilladehyde, 107, 531
Perillaketon, 531
Perillanin chloride, 217, 621
Peripalloside, 18, 531

Periploca sepium, 107, 169, 386, 394, 399,
     431–432, 478, 531, 539, 561
Periplocin, 107, 169, 531
Periplocymarin, 169, 531
Perividine, 36, 186, 531
Perivine, 36, 186, 531
Perlolyrine, 45, 85, 270, 329, 531
Persicaria amphibia, 107, 419, 485, 494, 505,
     543–544
Persicaria hydropiper, 108, 441, 485, 492, 501,
     531, 537, 544, 546, 563
Persicaria orientalis, 108, 525, 536, 577
Persicarin, 102, 108, 115, 168, 221, 531
Peruvosides, 145, 233, 531
Pesticidal for maggots, 170, 531
Petasalbin, 166, 399
Petasites japonicus, 108, 349, 417, 422–423, 520,
     548, 569
Petroselaidic acid, 621
Petroselenic acid, 73, 531
Petroselic acid, 50, 192–193, 531
Petroselidinic acid, 73, 531
Petroselinic acid, 56, 168, 172, 195, 532, 621
Petunidin, 95, 141, 212, 298, 532
Petunidin glucoside, 141, 298, 532
Peucedanum decursivum, 108, 280, 370, 478
Peucedanum formosanum, 108, 370, 413, 444,
     532
Peucedanum graveolens, 280, 593, 597, 603,
     637, 640
Peucedanum japonicum, 108, 350, 452, 521, 530,
     542, 549, 572
Peucedanum rubricaule, 317
Peucedanum terebinthaceum, 478, 497, 521
Peuformosin, 108, 280, 532
Phanol, 290, 621
Phantomolin, 61, 532
Pharbilic acid, 109, 267, 532
Pharbitis diversifolia, 109, 371, 476, 478, 532
Pharbitis hederacea, 81
Phaseolus angularis, 109, 359, 408, 422, 468,
     495, 548, 576
Phaseolus vulgaris, 218, 584, 588, 609, 632, 637
Phasin, 66, 532
Phasine, 66, 532
Phellandrene, 27, 40, 47, 130, 150, 154–155, 158,
     161, 169, 189, 191, 228, 236, 292, 302,
     304, 396, 532
Phellandrine, 51, 532
Phellodendrine, 109, 218, 532
Phellodendron amurense, 109, 218, 355, 421, 431,
     523, 528, 532, 590, 611, 630, 632
Phellopterin, 16, 171, 241, 532
Phene, 130, 228, 292, 532

Phenethylamine, 47, 255, 532
Phenol, 227, 229, 258, 275, 622
Phenolic acid, 124, 199, 228, 245, 287, 532, 622
Phenolic compounds, 85, 310, 329, 532
Phenolic derivatives, 153, 532
Phenolic glycosides, 203, 234, 264, 300, 622
Phenols, 54, 75, 83, 100, 122, 149, 222, 269, 277,
    284–285, 533, 578, 622
Phenyl ethyl, 64, 199, 288, 533, 622
Phenyl ethyl alcohol pentosans, 64, 199
Phenylacetic acid, 125, 226, 289, 533
Phenylalanine, 103, 159, 533
Phenylethylalcohol, 161
Phenylprepyl cinnamate, 271, 622
Phenylpropane derivatives, 25, 157, 329, 533
Phenyl-*propanoid glycosides,* 279, 622
Phenylpropanoids, 183, 622
Phenylpropyl alcohol, 40, 189, 533
Phenytheptatriyne, 28, 533
Phetidine, 144, 299, 533
Phillyrin, 69, 141, 201, 231, 533
Phlobaphene, 148, 284, 301, 533
Phorbol, 50, 192, 533
Phosphatase, 134, 229, 533
Phosphates, 117, 533
Phosphatides, 68, 200, 262, 533
Phosphatidylcholine, 144, 533
Phosphatidyl-ethanolamine, 533
Phosphatidyl-inositol, 144, 533
Phosphatidyl-serine, 144, 533
Phosphorus, 157, 533
Photinia serrulata, 109, 376, 484, 564
Phototoxic, 28, 333, 534
Phragmites australis, 280, 582, 588, 596, 606,
    624, 630, 632, 636, 638
Phragmites communis, 25, 109, 218, 364, 417,
    540, 593, 597, 616, 624
Phthalide, 85, 326, 398, 534
Phthatate, 534
p-hydroxyacetophenone, 133, 294, 527
p-hydroxybenzoic, 95, 124, 170, 212, 287, 527,
    620
p-hydroxybenzoic acid, 124, 170, 287, 527
p-hydroxycinnamic acid, 527
Phyllandrene, 189, 242, 622
Phyllanthin, 110, 534
Phyllanthine, 110, 534
Phyllanthus emblica, 109, 319–320, 389, 409,
    435, 577
Phyllanthus niruri, 393
Phyllanthus simplex, 109, 315, 347, 557
Phyllanthus urinaria, 110, 390, 485, 521, 534
Phyllanthus virgatus, 110, 385, 410
Phyllantidine, 110, 131, 534

Phyllitis scolopendrium, 384
Phyllodium pulchellum, 110, 369, 427, 471, 517,
    534
Phyllostachys bambusoide, 110, 344, 410, 420,
    429, 488, 538, 556
Phyllostachys nigra, 218
Phylteralin, 110, 534
Physalein, 90, 210, 273, 534
Physalin, 110, 210, 218, 273, 281, 534, 622
Physalin A, 110, 218, 281, 534
Physalin B, 534
Physalin C, 534
Physalis alkekengi, 110, 218, 281, 359, 418, 445,
    478, 485, 505, 534, 569, 580, 584, 603,
    622, 632, 639
Physalis angulata, 110, 281, 361
Physalis franchetti, 281, 623, 638
Physalis pubescene, 321
Physanols, 110, 218, 281, 534
Physcim-l-gluco-rhamnoside, 110, 534
Physcion, 35, 123, 126–127, 225, 227, 251, 286,
    290, 534, 622
Physochlaina infundibularis, 111, 355, 485, 554
Physoxanthin, 110, 218, 281, 534
Phytic acid, 58, 196, 203, 534, 622
Phytin, 133, 195, 229, 535, 622
Phytoene, 192, 623
Phytoestrogens, 147, 234, 300, 535
Phytolacca acinosa, 111, 219, 375, 493, 526, 535,
    552, 612, 617, 626, 637
Phytolacca americana, 219, 610
Phytolaccatoxin, 111, 219, 535
Phytolacine, 111, 219, 535
Phytosterindigitonid, 78, 535
Phytosterine, 86, 208
Phytosterols, 11, 35, 105, 141, 179, 185–186, 217,
    225, 228, 230–231, 234, 274, 295,
    300, 623
Phytotoxin, 82, 207, 535
Pi Jiang, 13, 369
Pi Pa Yie, 64, 369
Picein, 160, 535
Picralinal, 13, 180, 535
Picrasma excelsa, 281, 584, 597, 626, 629, 639
Picrasma quassioides, 111, 281, 361, 397–399,
    512, 520, 535, 543
Picrasmin, 111, 281, 535
Picrinine, 13, 180, 535
Picrorhiza kurroa, 111, 355, 404, 496, 535, 574
Picrorhizin, 111, 535
Pienen acid, 535
Pigments, 234, 323, 623
Pileostegia viburnoides, 111, 344, 401, 501, 543
Pimaradene, 19, 243, 535

Pimela alba, 111, 351
Pimpinella anisum, 281, 583, 586, 597, 602–603, 624
Pimpinella thellungiana, 111, 281, 315, 356, 486
Pin Di Mu, 20, 369
Pin Peng Cao (Japan), 370
Pinacene, 169, 535
Pinecamphene, 266, 623
Pinellia ternata, 111, 341, 410, 437, 497
Pinene acid, 153
Pinicolic acid, 117, 222, 536
Pinipicrin, 29, 184, 536
Pinitol, 85, 112, 219, 281, 536
Pinnatin, 116, 536
Pinocarveol, 40, 188, 536
Pinus albicaulis, 282, 586, 591, 604, 615, 626, 634, 638–639
Pinus bungeana, 112, 281, 379, 465, 502, 536
Pinus madshurica, 169, 380, 396, 399, 404, 458, 499, 527, 535, 545, 549
Pinus sylvestris, 219, 585, 591, 599
Piper cubeba, 112, 219, 419, 428, 430, 432, 438, 445, 458, 536, 549, 567, 597, 623, 626
Piper longum, 112, 219, 341, 536, 578, 611, 624, 640
Piper nigrum, 112, 219, 433, 435, 445, 456, 536, 594, 623
Piperamine, 112, 219, 536
Piperidine, 132, 219, 293, 517, 623
Piperine, 112, 219, 536, 623
Piperitone, 53, 154, 257, 304, 536
Piperonal, 112, 163, 219, 536
Pistacia lentiscus, 112, 220, 372, 469, 473, 509, 543, 585, 590, 615, 634
Pittosporum tobira, 113, 353, 456, 481
Plantagin, 113, 220, 282, 536
Plantago asiatica, 113, 220, 282, 343, 418, 449, 451, 481, 483, 496, 498, 536, 573, 602, 613, 619–620
Plantago major, 220, 589, 593–594, 617, 634
Plantago psyllium, 282
Plantasan, 113, 220, 282, 536
Plantenolic acid, 113, 220, 282, 536
Plastoquinone, 108, 536
Plastoquinone-9, 116, 536
Platycladus occidentalis, 282, 582–583, 594, 601–602, 604, 617, 623–624
Platycladus orientalis, 29, 113, 145, 282, 343
Platycodigenic acids, 113, 537
Platycodigenin, 33, 113, 537
Platycodon autumnalis, 113, 398, 425, 537, 539, 559
Platycodonin, 113, 537
Platycodosides, 113, 537

Platyconin, 113, 537
Pleridine, 537
Plumbago zeylanica, 113, 339, 398, 472, 477, 479, 487, 501, 515, 518, 525, 537, 558
Plumeria rubra, 113, 169, 354, 357, 402, 404–405, 434, 440, 465, 468, 472, 475, 494–495, 502, 505, 529, 533, 537, 543
p-lumicolchicine, 80, 527
Plumieric acid, 113, 169, 537
Plumieride, 113, 537
p-methoxybenzylacetone, 19, 243, 527
p-methoxycinnamaldehyde, 9, 238, 527
p-methoxylcinnamic acid, 131, 228, 527
p-methoxystyrene, 207, 620
Po Po Na, 151, 370
Po Shu, 37, 370
Po Yen, 124, 370
Podocarpene, 114, 537
Podocarpus macrophyllus, 114, 363, 428, 430, 460, 482, 495, 507, 519, 536–538, 570
Podophylium peltatum, 338, 454, 470, 537
Podophyllotoxin, 60, 114, 161, 282, 537, 623
Podophyllotoxin type lignans, 623
Podototarin, 114, 537
Pogostemon cablin, 114, 220, 357, 465, 592, 630
Polyacetylene, 35, 85, 537
Polyanthus narcissus, 99, 114, 378
Polydatin, 115, 537
Polygala japonica, 114, 282, 380, 552, 566
Polygala senega, 283, 320, 616, 622–623, 632
Polygala tenuifolia, 114, 282, 392, 525
Polygala vulgaris, 283, 605, 617, 637, 640
Polygalacic acid, 113, 136, 230, 537
Polygalic acid, 230, 623
Polygalitol, 283, 623
Polygodiol, 108, 537
Polygonatum chinense, 115, 283, 355, 423, 442, 515, 561
Polygonatum multiflorum, 283
Polygonatum odoratum, 220, 317, 358, 603, 628, 638
Polygonin, 115, 537
Polygonone, 108, 537
Polygonum aviculare, 115, 221, 342, 419, 428, 436, 449, 564, 577, 603, 630, 634
Polygonum bistorta, 115, 221, 316, 342, 413, 432, 444, 465, 473, 484, 487, 514, 526, 535, 540, 544, 556, 612
Polygonum cuspidatum, 115, 354, 470, 477, 537
Polygonum hydropiper, 115, 221, 492, 531, 546, 563, 587, 595
Polygonum lapidosa, 29
Polygonum multifolrum, 116, 221, 353, 437, 461, 478, 500, 529, 547, 623

Polygonum orientale, 116, 378, 491, 493, 525, 577
Polygonum perfoliatum, 116, 351, 410, 438, 486, 525, 540
Polygonum tinctorium, 341
Polyine, 73, 329, 537
Polynes, 623
Polypeptides, 185, 233, 623
Polyphenolic acids, 263, 623
Polyphenols, 77, 145, 168, 185, 205, 211, 221, 325, 538, 623
Polyporus umbellatus, 116, 393, 425, 463, 540
Polysaccharides, 5–6, 12, 45, 47, 63, 71, 73, 103, 124, 147, 177–179, 185–186, 190–191, 198, 216–217, 233, 235, 254, 261–263, 269, 272, 287, 318, 325, 329–330, 538, 583, 624
Polysaccharopeptide, 47, 326
Polythienyls, 142, 232, 538
Ponasterone, 95, 104, 114, 170, 212, 538
Ponasterone A, 95, 104, 212, 538
Poncirin, 42, 116, 538
Poncirus trifoliata, 116, 352, 421, 430, 439, 474, 486, 502, 516, 519, 538
Pongamia pinnata, 116, 378, 420, 474, 494–495, 536, 538
Pongapin, 116, 538
Ponicidine, 122, 538
Populin, 116, 221, 283, 538, 624
Populnin, 63, 162, 199, 538
Populus alba, 116, 221, 283, 390, 420, 463, 538, 550, 564, 570, 595, 597, 602–603, 635
Populus balsamifera, 283, 622
Populus tremuloides, 283, 624, 628, 634
Poria cocos, 117, 222, 350, 459, 527, 536, 572, 590, 620
Porphyrins, 212, 624
Portulaca grandiflora, 117
Portulaca oleracea, 117, 222, 365, 433, 459, 522, 539, 576, 592, 617
Portulaca pilosa, 117, 353, 410, 429, 488, 507–508, 533, 539, 564, 573
Portulal, 117, 538
Potassium, 30, 54, 57, 78, 110, 117, 157, 159, 184, 187, 218, 222, 258–259, 538–539, 624
Potassium hydroxide, 110, 218, 538
Potassium malate, 78, 538
Potassium myronate, 30, 184, 539
Potassium nitrate, 54, 78, 258, 539
Potassium oxide, 57, 259, 539
Potassium salts, 117, 222, 539
Potassium sodium, 117, 539
Potentilla anserina, 284, 595, 603
Potentilla bifurca, 117, 284, 383, 449
Potentilla erecoa, 284, 594, 600, 622, 634
Poterium officinale, 117, 222
Precatorine, 4, 176, 539
Pregnenes, 107, 539
Preskinnianine, 57, 195, 539
Primula sieboldii, 118, 284, 390, 403, 539, 541
Primula veris, 284, 640
Primula vulgaris, 222, 603, 622, 634, 637, 640
Primulagenin A, 118, 222, 284, 539
Proanthocyanidins, 256, 319, 624
Procurcumenol, 51–52, 193, 539
Procyanidins, 256, 282, 624
Proesapanin A, 31, 249, 539
Progestoron, 196, 624
Proline, 88, 539
Prometaphanine, 139, 539
Pronuciferine, 100, 277, 539
Propeimin, 70, 201, 539
Propionic acid, 539
Prosapogenin, 113, 539
Protease, 68, 113, 200, 262, 539
Protein (TAP29), 147, 233, 540
Proteinase, 71, 78, 205, 540
Proteins, 191, 219, 230, 247, 540
Protoalkaloids, 198, 624
Protoalnulin, 239, 624
Protoanemonin, 15, 32, 120, 122, 180, 224, 241, 249, 253, 284–285, 540, 612, 624
Protoanemonoid compound, 237, 624
Protoberberine alkaloids, 247, 273, 624
Protocatechetic acid, 248, 624
Protocatechuic, 41, 79, 95, 106, 115, 124, 170, 212, 221, 266, 279, 287, 540, 624
Protocatechuic acid, 41, 79, 106, 115, 124, 170, 221, 266, 287, 540
Protoescigenine, 9, 178, 540
Protohypericin, 79, 541
Proto-isoerubosides, 12, 179, 540
Protopine, 39, 48, 92, 144, 187, 192, 255, 299, 541, 624
Protoprimulagenin A, 118, 222, 284, 541
Protostemonine, 139, 541
Protostephanine, 139, 541
Protoveratrine, 76, 150, 541
Provitamin A, 205, 232, 625
Prudomenin, 118, 223, 541
Prunasin, 118, 223, 541
Prunelin, 169, 541
Prunella vulgaris, 118, 169, 223, 385, 428, 447, 449, 453, 485, 524, 549, 573, 589, 591–592, 599, 628, 634, 637–639
Prunus americana, 244
Prunus domestica, 118, 223, 391, 439, 598

Prunus mume, 118, 223, 312, 384, 411, 507, 541, 562, 565, 588, 590, 595, 598, 612, 615, 624, 635
Prunus padus, 118, 344, 424, 468, 485, 505, 521
Prunus persica, 119, 223, 382, 401, 420, 452, 465, 480, 482, 501, 521, 523, 530, 564, 601
Prussic acid, 223, 625
Pseudoaconitine, 6, 541
Pseudoanisatin, 80, 541
Pseudobrucine, 140, 541
Pseudoconhydrine, 160
Pseudocuramine, 277, 625
Pseudoephedrine, 260, 625
Pseudohypericin, 79, 206, 541, 625
Pseudojervine, 76, 150, 302, 541
Pseudolycorine, 91, 541
Pseudomorphine, 106, 217, 279, 541
Pseudopelletierine, 120, 224, 541
Pseudopurpurin, 126, 289, 541
Pseudostellaria heterophylla, 119, 380, 472, 560
Pseudostrychnine, 140, 541
Psidium guajava, 119, 349, 411, 414–415, 419, 444, 479, 505
Psoralea corylifolia, 119, 224, 342, 412, 443, 449, 488, 542, 589, 610, 625
Psoralen, 57, 69, 73, 119, 171, 195, 201, 224, 265, 542, 625
Psoralidin, 119, 224, 542
Psoraline, 224, 625
Psychotria rubra, 119, 360, 406
Psychotria serpens, 363
Psychotrine, 11, 542
Psyllic acid, 210, 273, 625
Psyllostearyl alcohol, 163, 542
Pteridium aquilinum, 170, 418, 445, 469, 482, 484, 491, 537–538, 563–564
Pteris cretica, 119, 350, 469, 560, 564
Pterocarya stenoptera, 119, 350, 495, 540, 542, 550
Pterosterone, 95, 212, 542
p-terpinene, 145, 299, 527
p-tyrosol, 123, 286, 527
Pu Gong Ying, 142–143, 370
Pu Gong Ying (Taiwan), 370
Pu Gong Ying (Western), 370
Pu Huang, 148, 370
Pueraria lobata, 120, 224, 451, 454, 468, 489, 495, 542, 548, 579, 598–599, 610
Pueraria montana, 120, 374, 403, 417, 427, 477
Pueraria thunbergiana, 104, 224, 351, 598, 625, 632
Pueraria vittata, 385
Puerarin, 120, 224, 542, 625

Pulegone, 275, 625
Pulin, 232, 625
Pulsatilla ambigua, 120, 284, 340, 412, 523, 540, 545, 552
Pulsatilla chinensis, 224, 326, 572, 586, 612, 624–625
Pulsatilla vulgaris, 284, 612, 634, 637, 640
Pulsatoside, 224, 625
Punica granatum, 120, 224, 376, 491, 512–513, 530, 541, 600, 621, 637
Puqiedinone, 70, 317, 542
Purgative oil, 207, 625
Purine, 185, 625
Purpurea-glycosides A, 196, 625
Purpurea-glycosides B, 625
Purpureal glycosides, 58, 196, 542
Purpurin, 70, 125–126, 202, 263, 289, 542, 625
Purulent, 58, 171, 196, 542
Putrescine, 42, 105, 216, 542
p-vinylguaiacol, 76, 527
p-vinylphenol, 76, 527
Pyrethrins, 188, 225, 625
Pyrethrum cinerariifolium, 120, 225, 344, 403, 437, 465, 595, 613, 620, 625, 630
Pyrocaledol, 127, 290, 542
Pyrocatechic tannin, 184, 542
Pyrocatechine acid, 119, 542
Pyrogallol, 122, 270, 285, 542, 625
Pyrogallol tannin, 285, 542
Pyrola decorata, 121, 364, 415, 435, 482, 490, 514
Pyrola rotundifolia, 225, 587, 594, 603, 630
Pyromeconic acid, 542
Pyrrolidine, 56, 195, 542
Pyrrolizidine, 229, 234, 294, 312, 625
Pyrrolizidine alkaloids, 234, 625
Pyrrosia adnascens, 121, 377, 411, 415, 471, 564–565
Pyrrosia lingua, 121, 377, 458, 490
Pyrryl-*methyl ketone,* 542

Q

Qian Cao, 73, 126, 359, 370
Qian Hu, 17, 108, 370
Qian Hu (Taiwan), 370
Qian Jin Teng, 139–140, 355, 370
Qian Jin Zi, 67, 371
Qian Li Guang, 133, 371
Qian Li Guang (European), 371
Qian Qu Cai, 92, 371
Qian Ri Hong, 75, 371
Qian Shi, 67, 371
Qiang Cao, 106, 371

Qianhucocumarin, 108, 280, 542
Qin Cai, 18, 371
Qin Jiu, 83, 371
Qing Feng Teng, 135, 371
Qing Guo, 22, 371
Qing Mu Xiang, 21, 371
Qing Ping, 84, 371
Qing Ye Dan, 141, 371
Qiong Zhi, 71, 371
Qiu Jiang, 156, 371
Qluconic acid, 543
Qu Mai, 57, 371
Quan Yuan Guan Zhong, 55, 371
Quao Ye Ging Lan, 59, 371
Quassin, 111, 179, 238, 281, 543, 625
Quassinoid, 281
Querbrachitol, 152, 176, 235, 543, 626
Quercetagetin, 142, 232, 543
Quercetagetrin, 142, 232, 543
Quercetin glucoside, 143, 298, 543
Quercetin-3-β-glycoside-7-β-glycoside, 163
Quercetin-3-galactoside, 126, 227, 290, 544
Quercetin-4-glucoside, 77, 544
Quercetin-monomethylether, 142, 544
Quercetol, 125, 289, 544
Quercimeritrin, 77, 95, 107–108, 115, 221, 544
Quercitin, 37, 69, 77, 201, 205–206, 235,
        255–256, 544
Quercitol, 47, 67–68, 152, 161, 200, 235, 255,
        262, 544
Quercitrin, 29, 57, 67, 73, 78–79, 89, 108, 130,
        136, 184, 200, 203, 206, 230, 544
Quercus acutissima, 121, 285, 362, 434, 473, 501,
        530, 540
Quercus robur, 285, 598, 632, 634
Queretin, 26
Quian Niu, 371–372
Quin Pi, 372
Quinene, 84, 208
Quinic acid, 136, 230, 544
Quinochalone, 35, 185, 544
Quinonic substance, 93, 544
Quinonoid, 88, 209, 545
Quinoric acid, 8, 545
Quinquenosides, 105, 216, 545
Quisqualic acid, 121, 545
Quisqualis grandiflora, 121

R

Rabdosia lasiocarpus, 121, 366, 525, 549, 566
Rabdosia rubescens, 122, 348, 465, 538, 564
Racemic acid, 153, 235, 545
Raddanoside, 16, 241, 545

Raddeanin A, 16, 241, 545
Raffinose, 91, 167, 273
Ramalic acid, 149, 545
Ranunculin, 120, 122, 180, 285, 545, 626
Ranunculus chinensis, 122, 285, 356, 412, 540,
        545
Ranunculus ficaria, 285
Ranunculus japonicus, 122, 285, 366
Ranunculus occidentalis, 285, 586
Ranunculus sceleratus, 122, 285, 376, 400, 542,
        555
Ranunculus ternatus, 122, 285, 366, 532, 564,
        578
Raphanin, 122, 225, 545
Raphanol, 288, 626
Raphanolide, 288, 626
Raphanus sativus, 122, 225, 342, 545, 588, 606,
        608, 638–639
Rauvolfia serpentina, 286
r-cadinene, 83, 207, 268, 545
Rebaudiosides, 140, 545
Rebixanthin, 32, 185, 545
Rehmannia chinensis, 122, 347, 429, 433, 538,
        545
Rehmannia glutinosa, 225, 305, 591, 615, 623,
        626, 632
Rehmannin, 122, 225, 545, 626
Reliculine, 48, 255, 545
Ren Dong, 89, 348, 354, 372
Ren Seng, 105, 372
Renifolin, 39, 187, 545
Repandusinic acids, 94, 545
Repandusinin, 94, 546
Rerpinenol, 154, 304, 546
Rescinnamine, 122, 286, 626
Reserpine, 41, 122, 285–286, 546
Resin, 24, 32, 37, 45, 48, 50–51, 55, 60, 65, 68,
        74, 78, 84, 87, 91, 93–94, 103, 112,
        124, 135, 147, 153, 161, 170, 176, 178,
        180–182, 185, 188, 191–193, 197,
        199–201, 204–205, 207, 213, 217, 219,
        228–230, 232–233, 235–237, 239,
        243, 247, 253–255, 261, 267, 269, 271,
        273–274, 278, 287, 289, 292, 294
Resin albaspidin, 197, 546
Resinous oil, 124, 287, 546
Reticuline, 92, 546
Retinol, 104, 215, 546, 627
Retroresine, 50, 546
r-glutamyl-valyl-glutamic acid, 82, 545
Rhamnazin, 108, 115, 221, 546
Rhamnetin, 65, 200, 546
Rhamnetin-3-0-beta-D-galactoside, 250, 627
Rhamnocitrin, 13, 26, 183, 239, 246, 547

Rhamnodiastase, 123, 286, 547
Rhamnose, 33, 67–69, 200–201, 220, 262, 547
Rhamnus catharticus, 286, 587, 595, 600, 604,
    606, 621–622, 639
Rhamnus davurica, 123, 286, 377, 407, 438, 461,
    494, 547
Rhaponticum uniflorum, 123
Rhein, 35, 116, 123, 164, 186, 221, 225, 251, 286,
    547, 627
Rhein anthrones, 225, 286, 627
Rhein aurantioobtusin, 251, 547
Rhein chrysarobin, 35, 186, 251, 547
Rhein monoglucoside, 35, 251, 547
Rheum koreanum, 286
Rheum officinale, 123, 225
Rheum tanguticum, 286
Rhodea japonica, 123, 384, 547
Rhodexins, 547
Rhodiola elongata, 123, 286, 354, 470, 527, 547
Rhodiola rosea, 287, 320, 602, 627
Rhodioloside, 123, 286–287, 547, 627
Rhododendron anthopogon, 123, 287, 363, 479,
    552
Rhododendron dauricum, 123, 287, 413, 420,
    468, 470, 475, 485, 490, 516, 547, 554
Rhododendron maximum, 287
Rhododendron micranthum, 170, 475
Rhododendron molle, 124, 170, 287, 338, 368,
    531, 547
Rhododendron mucronatum, 124, 287, 340, 400,
    419, 444, 456, 468, 479, 527, 532, 540,
    563, 574
Rhododendron sinensis, 27, 124, 287, 389, 412,
    574
Rhodotoxin, 123, 287, 547
Rhoeadine, 105–106, 217, 279, 547
Rhoeagenine, 105, 217, 279, 547
Rhomotoxin, 124, 287, 306, 547
Rhus chinensis, 124, 170, 287, 334, 384, 388, 473,
    530, 538, 546
Rhus radicans, 288, 582, 636, 638
Rhus semialata, 124, 287, 370, 564
Rhus verniciflua, 124, 287
Rhymohydroquinone, 65, 547
Rhynchophylline, 100, 149, 301, 547
Rhyncophylline, 301, 627
Ri Jian Ca Sha Jiang Cao, 372
Ribes lacustre, 587
Ribes mandshurica, 124, 288, 374, 439, 507, 525
Ribes nigrum, 288, 587, 595, 621, 634, 639
Ricin, 226, 627
Ricinine, 125, 226, 548, 627
Ricinoleic acid, 226, 627
Ricinolein, 125, 226, 548

Ricinus communis, 125, 226, 342, 433, 449, 492,
    524, 548, 560, 612, 627
r-linolenic acid, 139, 231, 545
Robinin, 109, 120, 122, 218, 224, 285, 548
Ron Yen Raw, 100, 372
Rong Cai, 81, 372
Rong Shu, 372
Rorifamide, 125, 288, 548
Rorifone, 125, 288, 548
Rorippa indica, 125, 288, 353, 409, 477, 548
Rorippa nasturtium-aquaticum, 288, 599,
    601–602, 606, 619, 622, 626, 639
Rosa acicularis, 125, 226, 289, 359, 433, 462,
    465, 470, 473, 511, 567, 571, 577, 605,
    611, 615, 621
Rosa canina, 289, 596, 638–639
Rosa chinensis, 125, 289, 389, 465
Rosa davurica, 317
Rosa multiflora, 125, 289
Rosa rugosa, 125, 226, 289, 366, 420, 439, 447,
    466, 475, 497–498, 510, 521, 527, 533,
    548, 595, 639
Rosagenin, 277, 627
Rosenoxide, 125, 226, 289, 548
Rotenone, 97, 143, 298, 548, 627
Rotundone, 54, 194, 258, 548
Rotunol, 54, 194, 258, 548
Rou Dau Kou, 99, 372
Rovidine, 36, 186, 548
Roxburghine D, 4, 176, 548
r-terpinene, 169, 545
Ru Xiang, 112, 372
Ruberythric acid, 289, 627
Rubescensine B, 122, 548
Rubescensins, 121, 549
Rubia akane, 125, 289, 355, 406, 542, 549
Rubia chinensis, 126, 289, 370, 515, 541, 549
Rubia tinctorum, 289, 585, 587–588, 592, 609,
    625–627
Rubiadin-I-*methyl ether,* 276, 549
Rubichloric acid, 98, 276, 549
Rubierythrinic acid, 126, 289, 549
Rubijervine, 150, 302, 549
Rubricauloside, 108, 549
Rubrierythrinic acid, 70, 125, 202, 263, 289, 549
Rubus chamaemorus, 290, 582, 589, 628, 636
Rubus coreanus, 126, 226, 290, 350, 424, 429,
    436, 562, 573, 620, 634, 639
Rubus fruiticosus, 290
Rubus idaeus, 290, 586, 603–604, 621
Rubus parvifolius, 126, 290, 354, 471
Rui Ye Ren Dong, 89, 372
Rumex acetosa, 126, 290, 380, 437–438, 461,
    485, 519, 534, 544, 575, 577

Rumex acetosella, 227, 595, 622, 632
Rumex crispus, 127, 227, 290, 388, 587, 619–620, 634, 640
Rumex obtusifolia, 290
Rumex patientia, 127, 383, 413, 461, 558
Rutaecarpine, 67, 549
Rutin, 4, 35, 41, 47, 50, 68–69, 73, 77, 79, 82, 118, 137, 143, 148, 157, 163, 167, 178, 193, 200–201, 203, 206, 223, 231, 234, 237, 245, 256, 262, 266, 277, 292, 298, 549, 627

S

Sabinen, 184, 627
Sabinene, 62, 84, 112, 154, 169, 197, 207–208, 219, 304, 549, 627
Saccharase, 136, 295, 549
Saccharides, 47, 549
Saccharose, 146, 549
Safflomin A, 35, 185
Safflower yellow, 35
Safranal, 192, 627
Safrole, 25, 80, 93, 99, 181–182, 188, 206, 211, 215, 550, 627
Sagittardia sagittifolia, 127, 417, 449–451, 477, 576
Saikosaponins, 31, 184, 550
Saikosides, 184, 628
Salicarin, 92, 211, 550, 628
Salicifoline, 93, 97, 211, 550
Salicin, 116, 221, 283, 291, 420, 550, 628
Salicinase, 116, 221, 283, 550
Salicortin, 116, 221, 283, 291, 550, 628
Salicylates, 188, 296, 628
Salicylic acid, 32, 71, 119, 131, 171, 185, 188, 209, 230, 234, 263, 266–267, 272, 290, 300, 550, 628
Salicylic compounds, 303, 628
Salidroside, 150, 235, 301, 550
Saligenin glucoside, 127, 290, 550
Salireposide, 116, 221, 283, 550
Salix alba, 291, 603, 606, 622, 628, 634, 636
Salix babylonica, 127, 290, 363, 487, 542, 550, 552
Salix discolor, 628
Salsola collina, 127, 393, 424, 550
Salsolidine, 127, 550
Salsoline, 127, 550
Saluianin, 127, 227, 291, 550
Salvia chinensis, 127, 291, 376, 554
Salvia clevelandii, 291
Salvia coccinea, 127, 227, 291, 393, 550, 635
Salvia miltiorrhiza, 326, 331

Salvia plebeia, 128, 291, 362, 465–466, 471, 482–483
Salvigenin, 66, 550
Salvin, 291, 628
Salviol, 127, 291, 550
Sambucus formosana, 128, 292, 354, 407
Sambucus nigra, 292, 598, 603, 622, 632, 637–639
Sambucus racemosa, 292, 589, 606, 627, 634
San Bai Cao, 130, 372
San Dian Jin Cao, 57, 372
San Hai Ton, 372
San Hu Ci Tong, 64, 372
San Hu Shu, 56, 372
San Hu You Tong, 82, 372
San Jian Shan, 38, 372
San Long Zhi, 131, 373
San Qi, 75, 105, 132, 360, 373
San Se Xian, 14, 373
San Ya Ko, 67, 373
San Ye Wu Jia, 5, 373
Sang Gen Bai Pi, 98, 373
Sang Zhi, 98, 373
Sanguinarine, 92, 106, 164, 217, 279, 550
Sanguisorba officinalis, 128, 227
Sanguisorbic acid, 222, 227, 628
Sanguisorbins, 551
Sanjoinines, 156, 236, 551
Sansevieria trifosciate, 128, 350, 401, 525
Sanshol, 154, 551
Santalic acid, 129, 227, 551
Santalin, 129, 227, 551
Santalols, 227, 628
Santalone, 129, 227, 551
Santalum album, 129, 227, 380, 409, 423, 431, 453, 492, 551, 557, 563, 566, 570, 598–599, 628, 630
Santamarine, 66, 262, 551
Santene, 129, 227, 551
Santhophylls, 183, 628
Sapindus mukorossi, 129, 384, 515, 552
Sapium discolor, 374
Sapium sebiferum, 129, 384
Sapogenins, 31, 184, 306, 551
Saponaretin, 68, 77, 144, 147, 200, 204, 234, 262, 299, 551
Saponaria officinalis, 129, 228
Saponarin, 76, 129, 228, 551
Saponartin-4-0-glucoside, 551

Saponin, 6, 8, 11, 13, 20–21, 31–32, 37, 39, 42, 56–57, 61, 66, 73–75, 77, 89, 105, 115, 129, 135–136, 144, 147, 151, 154, 177–180, 182–183, 187, 198, 204, 230, 233, 238, 243–244, 246–247, 249, 253, 259, 264, 294, 322–323, 335, 435, 454, 553, 561

Saponin akebin, 42, 238, 244, 253, 553

Saponin alpha-methylester, 553

Saponin beta-methylester, 553

Sarcandra glabra, 129, 331, 377, 465, 472, 477, 562

Sarcostin, 53, 94, 167, 274, 553

Sargassum officinalis, 292

Sargassum pallidum, 129, 292, 353, 405, 488, 523, 538

Sargentodoxa cuneata, 129, 354, 402

Sarmentosin, 132, 553

Sarmentoslin, 132, 293, 553

Sarmutoside, 140, 297, 553

Sarolactone, 79, 553

Sarothamine, 55, 194, 553

Sarracine, 133, 293, 553

Sarsapic acid, 230, 295, 629

Sarsasapogenin, 15, 25, 553

Sativol, 95, 212, 553

Saturated acids, 152, 303, 553

Saururus chinensis, 130, 372, 410, 419, 485, 491, 544

Saussarine, 629

Saussurea japonica, 130, 292, 368, 532, 553

Saussurea lappa, 27, 171, 228, 292, 368, 587, 626, 629–630, 635

Savinin, 5, 553

Scabioside, 259, 629

Schisandra arisanensis, 130, 553

Schisandra chinensis, 130, 228, 385, 553, 613, 623, 639

Schisantherins, 553

Schizandrer, 130, 228, 553

Schizandrin, 130, 228, 553

Schizandrol, 130, 228, 553

Schizonepeta multifida, 130

Schizonepeta tenuifolia, 228, 315, 613, 616

Sciadopitysin, 114, 203, 553, 629

Scolopendrium subspinipes, 384

Scopalia dulcis, 131, 410, 424, 459, 481, 486, 508, 550, 554, 564, 567

Scopanol, 131, 554

Scoparoside, 194, 629

Scoplin, 111, 554

Scopolamine, 56, 111, 131, 195, 554

Scopoletin, 16, 24, 28, 32, 35, 64, 79–80, 96, 101, 111, 151, 157, 183, 206, 213, 241, 245, 249, 266–267, 274, 281, 303, 554, 629

Scopoline, 64, 101, 554

Scopolomine, 79, 554

Scordinins, 179, 629

Scoulerine, 48, 255, 554

Scrophularia buergeriana, 131, 387, 401, 465, 471, 480, 487, 527, 554

Scrophularia ningpoensis, 228, 583, 589, 603, 607, 622

Scrophularin, 131, 228, 554

Scutellarein heteroside, 145, 299, 554

Scutellaria baicalensis, 131, 229, 292, 312, 327, 401, 419, 424, 558, 578, 589, 591–592, 629, 632, 634, 640

Scutellaria formosana, 131, 292, 356, 421

Scutellaria lateriflora, 293

Scutellarin, 127, 229, 291, 293, 554, 629

Se Gua, 89, 373

Sebiferic acid, 129, 554

Secalose, 27, 247, 554

Secoiridoid glucosides, 252, 320, 629

Secoiridoids, 261, 629

Secologanin, 97, 214, 554

Securinega suffruticosa, 131, 390, 407, 456, 534, 555

Securinega virosa, 132, 340, 471, 575

Securinine, 131, 554

Securinol, 131, 555

Securitinine, 131, 555

Sedacrine, 293, 629

Sedacryptine, 293, 629

Sedinine, 293, 629

Sedocaulin, 132, 293, 555

Sedocitrin, 132, 293, 555

Sedoheptose, 132, 293

Sedoheptulosan, 168, 555

Sedoheptulose, 132, 293

Sedum acre, 293, 602, 618, 629

Sedum aizoon, 132, 293, 415, 555

Sedum erythrostichum, 132, 293, 360, 553

Sedum formosanum, 132, 293, 411, 571

Sedum lineare, 132, 293, 350, 472, 477, 555

Sedum sarmentosum, 132, 293, 345, 455–456, 458, 517, 553

Selaginella involvens, 132, 376, 406, 450, 570

Selaginella tamarisina, 133, 360, 410, 414, 482, 489, 559

Selenium, 179, 629

Selinene, 150, 302, 555

Semiaquilegia adoxoides, 133, 381

Sempervirine, 71, 202, 555

Senecio argunensis, 133, 293, 371, 436–437, 448, 500, 507, 553
Senecio aureus, 294, 626
Senecio campestris, 133, 352
Senecio cannabifolius, 133, 355, 415, 527
Senecio scandens, 171, 371, 414, 530, 542, 562
Senecio vulgaris, 133, 229, 312, 487, 555, 611, 625, 629, 636, 640
Senecionine, 133, 229, 294, 555, 629
Seneciphyline, 229, 294, 629
Sennoside A, 629
Sennosides, 35, 123, 186, 225, 251, 286, 555
Sequiterpine, 163, 555
Seratonin, 122, 285, 555
Serine, 88, 143, 210, 555, 629
Serotonin, 77, 99, 205, 235, 280, 555, 630
Serpentine, 122, 285, 555
Serratenediol, 90, 210, 555
Sesamin, 133, 229, 555
Sesamol, 133, 229, 555
Sesamum indicum, 133, 229, 312–313, 429, 436–437, 500, 503, 516, 524, 528, 530, 535, 555, 560, 576, 604, 613, 619, 622, 624, 638–639
Sesbinia grandiflora, 133
Sesbinia javanica, 134, 382, 449–450
Sesbinia sesbin, 134, 357, 478, 494, 552, 561, 571
Sesquijasmine, 81, 268, 555
Sesquilignans, 556
Sesquiterpene, 5, 13, 18, 22, 29, 51, 59, 65, 73, 76, 81, 95, 100, 115, 128, 143, 154, 157, 176, 181, 184, 188, 193, 197, 203, 218, 225, 227, 239–240, 248, 251, 253, 262, 264, 268–269, 277–278, 318, 321, 556, 630
Sesquiterpene alcohol, 29, 184, 556
Sesquiterpene alkaloids, 65, 556
Sesquiterpene glucosides, 5, 18, 76, 128, 143, 556
Sesquiterpene hydrocarbons, 227, 630
Sesquiterpene ketones, 157, 556
Sesquiterpene lactone, 203, 253, 264
Sesquiterpenes patchoulol, 220, 630
Sesquiterpenoid, 201, 630
s-guaiazulene, 24, 245, 549
Sha Cao, 54, 373
Sha Cha Hua, 33, 373
Sha Gen Cai, 88, 373
Sha Gui Hua, 93, 373
Sha Gui Hua (Taiwan), 373
Sha Hong Fan Cao, 29, 373
Sha Ji, 77, 373
Sha Jiang Cao, 104, 372–373
Sha Ma, 47, 373
Sha Seng, 8, 153, 343, 364, 374, 385

Sha Yuan Zi, 26, 374
Shan, 7, 13, 19, 23, 38, 48–51, 57–58, 65, 71, 74–75, 77, 79–80, 83, 88, 90–91, 93, 96, 101, 104, 108, 111, 120, 123–124, 129, 136–137, 143, 147–148, 151, 153–154, 159, 317, 324, 327, 343, 345, 351, 355, 357, 366, 368, 372, 374–375, 377, 387, 391, 393
Shan Ci Ko, 374
Shan Cong Zi, 88, 374
Shan Dou Gen, 96, 137, 374
Shan Ge, 120, 374
Shan Guo, 23, 374
Shan Hua Jiao, 154, 374
Shan Jiang, 13, 374
Shan Jiu, 129, 374
Shan Liu Jiu, 77, 374
Shan Ma Zi, 124, 374
Shan Na, 83, 374
Shan Pu Tao, 153, 374
Shan Teng Zi, 151, 374
Shan Wo Ju, 83, 375
Shan Yan Cao, 136, 375
Shan Yao, 58, 375
Shan Ye Man Shi Song, 90, 375
Shan Ye Shi Song, 375
Shan Zha, 49, 375
Shan Zhi Ma, 375
Shan Zhu Yu, 48, 93, 375
Shang Han Cao, 375
Shang Lu, 111, 375
Shanzhiside, 71, 556
Shao Ci Wu Jia, 5, 375
Shao Lan, 54, 375
She Cheung Zi, 44, 375
She Gan, 28, 81, 375
She Ma, 78, 375
She Mei, 60, 70, 375
Shen Jin Cao, 91, 375
Shen Liu, 142, 375
Sheng Hong Yu, 9, 375
Sheng Jiang, 155, 376
Sheng Ma, 7, 40, 362, 376
Sheng Teng, 31, 41, 376, 385
Shi Cao, 5, 102, 133, 157, 340, 352, 376, 388
Shi Da Chuan, 102, 376
Shi Diao Lan, 92, 376
Shi Dou, 56, 63, 376
Shi Hong Hua, 49, 376
Shi Ji Qing, 79, 376
Shi Jian Chuan, 127, 376
Shi Jiun Zi, 121, 376
Shi Juan Bai, 132, 376
Shi Li, 11, 376

Shi Liu Pi, 120, 376
Shi Long Nei, 122, 376
Shi Luo, 16, 158, 376
Shi Mu Zi, 149, 376
Shi Nan Ye, 109, 376
Shi Shan, 79, 377
Shi Shang Bai, 132, 377
Shi Sheng Yu, 29, 377
Shi Song, 90–91, 345, 375, 377, 391
Shi Suan, 77, 91, 377
Shi Suan Hua, 77, 377
Shi Wei, 121, 377
Shi Wu Tou, 37, 377
Shi Zhu Yu, 154, 377
Shi Zi, 59, 377
Shibuol, 59, 556
Shikimic acid, 73, 80, 200, 203, 556, 630
Shikonin, 22, 88, 209, 556
Shionon, 25, 183, 556
Shishonin, 217, 630
Shobakunine, 556
Shogaols, 236, 630
Shong Jie Fong, 129, 377
Shu Gi, 20, 377
Shu Li, 123, 377
Shu Liang, 58, 377
Shu Long, 121, 377
Shu Qu Cao, 74, 378
Shu Shu, 94, 378
Shu Yu, 58, 378
Shuang Mian Ci, 155, 378
Shui Cai, 97, 378
Shui Fei Ji, 134, 378
Shui Gui Jiao, 79, 378
Shui Hong Cao, 116, 378
Shui Huang Pi, 116, 378
Shui Jin, 102, 378
Shui Ku Shi, 151, 378
Shui Lian, 101, 378
Shui Man Chin, 151, 378
Shui Shai, 99, 114, 378
Shui Shai Gen, 114, 378
Shui Su, 138, 378
Shui Tuan Hua, 8, 379
Shui Xian Cao, 76, 102, 379
Shui Yang Mei, 8, 72, 379
Shui Yang Mei Gen, 8, 379
Shui Yu, 137, 379
Shun, 379
Shyobunones, 157, 556
Si Gua, 98, 379
Si Yang Bai Hua Cai, 379
Si Yang Seng, 379
Si Ye Lian, 39, 379

Si Ye Lu, 70, 379
Si Zhao Hua, 47–48, 361, 379
Siang Si Zi, 4, 379
Siegesbeckia orientallis, 171, 404, 451, 488
Sien Feng Cao, 28, 379
Silene jenisseensis, 134, 294, 353, 400, 465
Silene ocaulis, 294, 582, 600, 631
Silibinin, 229, 313, 630
Silica, 110, 218, 556
Silicates, 199, 630
Silicic acid, 57, 199, 221, 234, 259, 300, 556, 630
Silybin, 134, 229, 557
Silybinomer, 134, 229, 557
Silybum marianum, 134, 229, 313, 378, 453, 557,
    603, 630
Silyckristin, 134, 229, 557
Silydiamin, 134, 229, 557
Silymarin, 134, 229, 251, 557, 630
Silymarin polyacetylenes, 251, 630
Simiarenol, 80, 557
Simplexine, 109, 315, 557
Sinactine, 44, 135, 557
Sinalbine, 134, 229, 557
Sinapic acid, 30, 80, 184, 206, 267, 557
Sinapine, 30, 184, 557
Sinapis alba, 134, 229, 339, 415, 463, 501, 503,
    516, 533, 557, 617
Sinapyl aldehyde, 129, 227, 557
Sinigrin, 30, 145, 184, 233, 557
Sinigroside, 34, 250, 557
Sinoacutine, 44, 135, 557
Sinoccatine, 48, 255, 557
Sinodiosgenin, 58, 196, 557
Sinomenine, 44, 96–97, 135, 275, 557
Sinomenium acutum, 135, 351, 371, 403, 458,
    492, 507, 557, 572
Sinoside, 140, 297, 557
Sinostroside, 140, 297, 557
Sioimperatorin, 17, 241, 557
Sioquercitrin, 150, 235, 301, 557
Siosakuranetin, 44, 253, 558
Sitosterol, 24, 37, 61, 69, 101, 120, 148, 157,
    164–165, 172–173, 186, 201, 230, 233,
    260, 301, 308, 396, 558, 630
Sitosteryl glucopyranosid, 61, 558
Skimmianine, 57, 74, 154–155, 195, 304, 558
Slliptinone, 113, 558
Sloeemodin, 127, 558
Smilacin, 135, 230, 294, 558
Smilacina japonica, 135, 294, 364
Smilacina stellata, 294
Smilax aristolochiifolia, 295, 591

Smilax china, 135, 230, 294, 383, 500, 519, 546, 552, 558, 564, 569, 616, 623, 626, 629, 631
Solamargine, 135, 230, 295, 558, 631
Solanidine, 136, 295, 558
Solanigrines, 136, 230, 295, 558
Solanine, 136, 230, 295, 558, 631
Solanocapsin, 295, 558
Solanocapsine, 135–136, 295, 558
Solanocarpine, 296, 631
Solanum aculeatissimum, 135, 230, 295, 387, 424, 558, 593, 624, 638–639
Solanum biflorum, 135, 295
Solanum capsicastrum, 135, 295, 366, 558
Solanum dulcamara, 296, 631, 634
Solanum incanum, 135, 295, 356, 406, 424, 450, 558, 573
Solanum indicum, 136, 295, 369, 457, 508, 510, 549, 558
Solanum lyratum, 136, 295, 340, 403, 416, 437, 453, 486, 518, 556, 571
Solanum nigrum, 136, 230, 295, 364, 520, 548, 552, 558, 577, 594, 600, 613, 620, 630–631, 636
Solanum pseudo-capsicum, 136, 295, 348, 558
Solanum tuberosum, 296, 589, 616, 639
Solanum verbascifolium, 136, 295, 375
Solanum xanthocarpum, 296, 631
Solasodine, 135–136, 230, 295–296, 558, 631
Solasurine, 135, 230, 295, 558
Soldulcamaridine, 296, 631
Solidago canadensis, 136, 230, 428, 544, 593, 623, 634
Solidago dahurica, 136, 390, 418, 428, 436, 447, 471, 484, 495, 537, 544
Solidago virgaurea, 171, 230, 495, 544, 609, 628–629
Somalin, 157, 559
Sonchus arvensis, 137, 171, 334, 360, 369, 486, 499, 528, 560, 565
Sonchus oleraceus, 171, 361, 396, 417, 520, 548, 569
Song Ji Shang, 89, 379
Song Lo, 379
Song Ta, 112, 379
Song Ye Mo, 117, 379
Sophora alopecurosides, 361, 570
Sophora flavescens, 137
Sophora japonica, 137, 231, 317, 355, 474, 549, 559
Sophora subprostrata, 137, 374, 411, 451, 509, 513, 526, 559
Sophora tomatosa, 137, 363, 448
Sophorabioside, 137, 231, 559

Sophoradin, 137, 559
Sophoradiol, 137, 231, 559
Sophoranochromene, 137, 559
Sophoranone, 137, 559
Sophoricoside, 137, 231, 559
Soranjudiol, 98, 276, 559
Sorbitol, 272, 631
Sorbose, 170, 559
Sorbus alnifolia, 137, 296, 379, 465, 468, 471, 489, 529
Sorbus aucuparia, 296, 620–621, 634, 639
Soyasapogenol, 163, 559
Sparteine, 39, 55, 187, 194, 559, 631
Spatholobus suberectus, 138, 358, 472, 565
Spathulenol, 24, 245, 559
Spemine, 105, 216, 559
Spermindine, 105, 216, 559
Sphenomeris chusana, 138, 384, 532, 559
Sphenone A, 138, 559
Sphondin, 265, 631
Spilanthes acmella, 138, 363, 382, 407, 421, 516, 558–559, 562
Spilanthol, 138, 559
Spinacetin, 172, 559
Spinacia oleracea, 172, 370, 396, 400, 432, 436, 529, 559, 579
Spinasterol, 45, 172, 190, 294, 396, 559, 631
Spiraea salicifolia, 138, 296, 387, 406, 432, 471, 577
Spiraea ulmaria, 296, 603, 607, 628, 634, 638
Spirodela polyrhiza, 138, 350, 414
Spirostanol saponins, 106, 331, 559
Spongesterol, 17, 241, 560
Springic acid, 560
Squalene, 4, 143, 176, 315, 560
Stachydrine, 39–40, 42, 85, 120, 136, 138, 157, 188, 212, 225, 252, 269, 295–297, 560, 631
Stachydrine chloride, 138, 296, 560
Stachyose, 81, 91, 169, 207, 268, 273, 560
Stachys chinensis, 138, 296, 378, 560
Stachys officinalis, 297
Starch, 67, 119, 137, 148, 230, 295–296, 560, 631
Stauntonia hexaphylla, 138, 389, 560
Stauntonin, 138, 560
Stearic acid, 11, 28, 42, 45, 47, 50, 82, 96, 99, 137, 143, 161, 179, 181, 183, 189, 193, 207, 209, 213, 215, 230, 233, 560, 631
Stearin acid, 125, 133, 229, 560
Stellaria alsine, 139, 381
Stellaria media, 139, 231, 349, 523, 545, 597, 603, 614, 619, 637, 639
Stemona japonica, 139, 338, 492, 541, 560
Stemona tuberosa, 347

Stemonidine, 139, 560
Stemonine, 139, 560
Stephania cepharantha, 139, 360, 420, 434, 482, 492
Stephania hernendifolia, 139, 398–399, 458, 467, 481, 489, 561
Stephania japonica, 139, 370, 447, 462, 480, 483, 485, 487, 511, 539, 541, 560–561
Stephania sinica, 140, 355, 498, 560, 572
Stephania tetrandraq, 140, 451
Stephanine, 139, 560
Stephanoline, 139, 560
Stepharine, 97, 140, 275, 560
Stepharotine, 140, 560
Stephisoferuline, 139, 561
Stepholidine, 96, 561
Stepinonine, 139, 561
Steponine, 139, 561
Stereoisomer, 20, 181, 561
Sterins, 190, 631
Steroid alkaloid glycosides, 135, 323, 561
Steroid glycosides, 107, 134, 561
Steroidal, 12, 15, 25, 115, 196, 230, 239, 249, 252, 295–296, 302, 561, 631
Steroidal alkaloids, 249, 296, 302, 631
Steroidal saponins, 12, 15, 196, 230, 239, 252, 295–296, 631
Steroids, 82, 205, 315, 328, 631
Sterol, 42, 50–51, 68, 75, 89, 98, 203, 224, 228, 561
Stevia rebaudiana, 140, 381, 418, 545, 561
Steviolbioside, 140, 561
Stevioside, 140, 561
Sticinic acid, 209, 632
Stictic acid, 209, 632
Stigmast-4-3-one, 235, 632
Stigmast-7-enol, 97, 214, 561
Stigmasterol, 8, 11, 24, 27, 41, 43, 45, 52, 58, 61, 73, 87, 95, 101–103, 106, 126, 130, 138, 141, 190, 194, 209, 212, 217, 225–226, 229, 231, 235, 239, 257, 275, 279, 290, 293, 398, 561–562, 632
Streptomycin, 171, 562
Strophalloside, 18, 562
Strophanthus divaricatus, 140, 297, 388, 434, 458, 553, 557
Strophanthus gratus, 297, 593
Strophantidin, 18, 242, 562
Strospeside, 58, 196, 562
Strumaroside, 154, 562
Strychnine, 111, 140, 231, 562, 632
Strychnos nux-vomica, 140, 231, 331, 349, 427, 514, 562, 594, 609, 614, 632
Strychnos pierriana, 140, 365, 522, 541, 578

Stylopine, 39, 187, 562
Styracin, 141, 297, 562
Styrax benzoin, 297, 589, 595, 632, 638
Styrax suberifolus, 140, 297, 354
Styrax tonkinensis, 141, 297, 338, 409, 419, 441, 562, 574
Su Cao, 114, 380
Su Mu, 31, 380
Su Yang, 54, 380
Suan Cheng, 42, 380
Suan Mo, 126, 380
Suan Zao, 156, 380
Suan Zao Ren, 156, 380
Suberins, 285, 632
Succinacid, 562
Succinic acid, 17, 54, 64, 72, 95, 118, 129, 148, 181, 223, 241, 264–265, 562
Sucrose, 45, 64, 100, 190, 199, 562
Sugars, 208, 217–218, 236, 632
Sui Me Jie, 56
Sui Mi Jie, 34, 380
Sulfuretin, 161, 562
Sumaresinolic acid, 141, 297, 562, 632
Sumaresinolic acid esters, 297, 632
Sun Cha, 380
Suo Lou Zi, 380
Swertia chirata, 297, 584–585, 602, 609, 640
Swertia diluta, 141, 297, 371, 524
Swertiamarin, 141, 252, 297, 321, 562, 632
Swertifrancheside, 141, 297, 562
Swertisin, 141, 297, 562
Sworoside, 48, 191, 255, 563
Synephrine, 42, 190, 563
Syringa dilatata, 141, 362, 399, 533, 551, 563
Syringa suspensa, 231, 608, 613
Syringa vulgaris, 313
Syringaldehyde, 280, 632
Syringaresinol, 5, 563
Syringen, 563
Syringic acid, 10, 95, 124, 170, 179, 238, 261, 287, 563, 632
Syringic aldehyde, 129, 227, 563
Syringin, 81, 89, 141, 168, 209, 231, 268, 272, 563

T

Tadeonal, 108, 115, 221, 563
Taeniafuge, 170, 563
Tagetes erecta, 142, 232, 298, 344, 410, 450, 497, 517, 563, 613
Tagetes minuta, 298, 597, 604, 606–607, 621, 627, 634

Tagetes patula, 142, 172, 232, 361, 383, 447, 455, 480, 502, 523, 529, 538, 543, 601, 613, 619, 632, 635

Tagetone, 142, 172, 232, 298, 563, 632

Tai Huang, 123, 127, 380, 383

Tai Zi Shen, 119, 380

Talatisamine, 6–7, 177, 563

Talictrine, 144, 299, 563

Tamarindus indicus, 142, 363, 414, 421, 424, 460, 473, 477, 503, 515, 524, 528, 564, 573, 576, 579

Tamarix juniperina, 142, 375, 544

Tan Seng, 127, 380

Tan Xian, 129, 380

Tang Jie, 64, 380

Tang Song Cao, 144, 381

Tang Song Cao (Taiwan), 381

Tangeratin, 42, 190, 563

Tangshenoside, 190, 632

Tannates, 103, 563

Tannic acid, 4–5, 31, 47–48, 61, 63, 75, 77, 79, 93, 122–123, 145, 153, 162, 185, 191, 199, 224–225, 227, 249, 255, 258, 264–266, 286, 290, 563–564, 632

Tanshinol, 127, 291, 564

Tanshinone, 127, 291, 564

Taraxacerin, 143, 565

Taraxacin, 143, 232, 565, 634

Taraxacum formosanum, 142, 370, 487, 529, 565

Taraxacum mongolicum, 143, 172, 370, 405, 413, 421, 424, 445, 448, 478, 500, 504–505, 516, 519, 528, 555, 560, 565, 575, 580

Taraxacum officinale, 143, 232

Taraxanthin, 52, 148, 172, 234, 257, 565

Taraxasterol, 80, 137, 142–143, 171, 206, 232, 267, 565, 634

Taraxasteryl acetate, 565

Taraxasteryl palmitate, 69, 80, 565

Taraxerol, 5, 12, 45, 67, 143, 172, 190, 200, 232, 239, 300, 565, 634

Taraxerone, 8, 67, 138, 200, 237, 565

Taraxeryl acetate, 45, 190, 565

Taraxol, 143, 172, 565

Taraxsteryl acetate, 41

Tartalic acid, 255, 634

Tartaric acid, 38, 48, 52, 64, 93, 100, 104, 118, 121, 137, 153, 187, 191, 199, 223, 235, 255–257, 278, 565

Tartrate, 159, 565

Taspine, 36, 252, 565

Taurine, 71, 263, 565

Taxettin, 155, 565

Taxifolin, 112, 220, 565

Taxinine E, 143, 298, 566

Taxol, 143, 298, 320, 566, 634

Taxus brevifolia, 320

Taxus cuspidata, 143, 298, 393, 397, 401, 419, 434, 566, 580

Tazettine, 77, 91, 99, 166, 215, 566

Tectoridin, 28, 81, 247, 566, 634

Ten Min Qing, 381

Teng Hu Tin Chi, 61, 381

Tenuidine, 114, 282, 566

Tenuifolin, 114, 282, 566

Tephrosia purpurea, 143, 298, 356, 452, 543, 548–549, 566

Tephrosia virginiana, 298, 599, 627

Tephrosin, 143, 298, 566, 634

Teresantalic, 129, 227, 566

Teresantalol, 129, 227, 566

Terminalia chebula, 143, 232, 353, 435, 461, 468, 564, 587, 594, 627, 634

Terpene, 14, 19, 23, 62, 87, 240, 243, 266, 271, 566

Terpeneol, 63, 199, 566

Terpenoids, 165, 178, 183, 200, 208, 213, 221, 275, 310, 322, 635

Terpenylacetate, 153, 566

Terpinen-4-ol, 146, 233, 566

Terpinene, 47, 62, 83, 160, 191, 197, 207, 268–269, 275, 566, 635

Terpineol, 51, 59, 62, 99, 112, 182, 188, 191, 197, 199, 215, 219, 567, 635

Terpinol, 95, 213, 567

Terpinolene, 50, 102, 150, 168, 191–193, 302, 567

Terrestriamide, 233, 635

Teteracylic acid, 567

Tetraacetylbrazilin, 31, 249, 567

Tetradecen-4-oic acid, 87, 271, 567

Tetragonia tetragonoides, 144, 349, 403, 437, 533, 567, 571

Tetragonin, 144, 567

Tetrahydro-cannabinols, 185, 635

Tetramethylpyrazine, 82, 85, 270, 315, 567

Tetramethylpyrazinesteroids, 567

Tetrandrine, 96–97, 275, 567, 635

Tetulinic acid, 131, 567

Thalfoetidine, 144, 299, 567

Thalicarpine, 144, 299, 567, 635

Thalictrum aquilegifolium, 144, 299, 381, 407, 420–421, 445, 469, 471, 476, 481, 492–493, 526, 528, 533, 541, 546, 567–568

Thalictrum dasycarpum, 299, 322, 591, 597–598, 611, 619, 635–636

Thalictrum foetidum, 144, 381, 431, 470, 507, 528, 552

Thalictrum ichangense, 144, 299, 365, 551, 563

Thalictrum occidentale, 597
Thalidasine, 144, 299, 568, 635
Thalidezine, 144, 299, 568
Thalisopavine, 299, 635
Thalphinine, 144, 299, 568
Thalpine, 144, 299, 568
Thamnolic, 272, 635
Thea assamica, 145, 343, 428, 538, 564, 568, 579
Thea sinensis, 33
Theasaponin, 33, 568
Thebaine, 106, 217, 279, 568, 635
Thelic simidine, 144, 299, 568
Theobromine, 145, 185, 568
Theophylline, 145, 185, 568, 635
Theronine, 88, 568
Thesium chinense, 145, 339, 471, 508
Theveside, 145, 233, 568
Thevetai peruviana, 145, 233, 356, 531, 568, 575, 587, 593, 620, 627, 631
Thevetin A, 145, 233, 568
Thevetin B, 568
Theviridoside, 145, 233, 568
Thiamine, 6, 11, 13–14, 19, 28, 30, 34, 36, 49, 74, 77, 90, 108, 155, 158, 163, 167, 171, 179, 182–183, 185, 195, 204–205, 210, 223, 236, 239–240, 251, 256, 569, 635
Thiophenes, 232, 635
Thlaspi arvense, 145, 233, 359, 465, 468, 516, 557, 583, 585, 592, 603, 608, 623, 637
Threonin, 247, 569
Threonine, 27, 103, 172, 569
Thuja chinensis, 145, 299, 343
Thuja koraiensis, 410, 416, 433, 468, 482, 516, 536, 543, 569
Thuja occidentalis, 299, 603, 617, 634–635, 640
Thuja orientalis, 29, 113
Thujone, 29, 62, 145, 182, 227, 253, 282, 291, 299, 569, 635
Thujyl alcohol, 245, 635
Thymine, 162, 569
Thymol, 65, 145, 233, 261, 299–300, 569, 636
Thymus amurensis, 145, 299, 347, 409, 414, 426, 432, 505, 527, 554, 569, 573, 578, 580
Thymus capitatus, 300, 614
Thymus vulgaris, 146, 233, 396, 409, 426, 430, 439, 475, 502, 516, 566, 572, 587, 593, 614, 629, 634, 636
Tian Bao Cao, 79, 381
Tian Cai, 24, 381
Tian Hua Fen, 147, 381
Tian Ja Cai, 62, 381
Tian Jiu, 140, 381
Tian Kui Zi, 133, 381
Tian Ma, 71, 381

Tian Men Dong, 25, 381
Tian Nan Xing, 21, 381
Tian Peng Cao, 139, 381
Tian Qi, 105, 382
Tian Qing, 133–134, 346, 357, 382
Tian We Cao, 138, 382
Tian Xuan Hua, 46, 382
Tian Zhu Cao, 131, 382
Tie Dao Mu, 36, 382
Tie Shu, 52, 382
Tie Xian Cai, 4, 382
Tie Xian Cao, 53, 382
Tie Xian Jiu, 8, 382
Tie Xian Lian, 43, 382
Tienmulilmine, 150, 302, 569
Tienmulilminine, 150, 302, 569
Tiger nut oil, 194, 258, 636
Tiglic acid, 17, 94, 212, 569
Tigloidine, 110, 218, 281, 569
Tigonenin, 230, 636
Tilia amurensis, 146, 394, 465, 471
Tilia cordata, 300, 592, 603, 623, 634, 636, 638
Tilia mandshurica, 300
Tiliadine, 300, 636
Ting Li Zi, 59, 161, 382
Tinnevelly senna, 146, 349
Tinnins, 207, 229, 242, 263, 278, 636
Tirucallol, 67, 569
Tithymalin, 66, 569
Tocopherol, 32, 172, 185, 203, 290, 396, 569, 636
Tohogenol, 90, 210, 569
Tomentogenin, 53, 570
Tong Guan Teng, 94, 382
Toosendanin, 96, 213, 570
Toralacton, 35, 251, 570
Torilis japonica (Houtt.) DC Prodr, 172
Torosachrysone, 36, 251, 305, 570
Totarol, 114, 570
Tou Gu Cao, 80, 382
Tou Ren, 119, 382
Toxicodendrol, 288, 636
Tracheloside, 146, 173, 570
Trachelospermum jasminoides, 106, 146, 173, 365, 396, 415, 448, 451, 509, 522, 570
Trachitin, 165, 570
Trachycarpus excelsa Wendl, 173
Trans-5-hexadecenoic acid, 299, 636
Trans-aconitic acid, 7, 25, 237, 328, 570, 636
Trans-beta-ocimene, 50, 193, 570
Trans-caryophyllene, 24, 245, 570
Trapa bicornis Osbeck, 173, 363
Trapa bispinosa, 146, 363, 411, 424, 456, 540
Trapa manshurica, 173, 363, 396–397
Trehalose, 132–133, 570

Tremulacin, 116, 221, 283, 570
Triacetonamine, 176, 636
Triacylglycerols, 211, 636
Triandrin, 291, 636
Triboline, 45, 570
Tribulus terrestsis, 146, 233, 345, 418, 478, 480,
         618, 629, 635–636
Tribulusamide A, 233, 313, 636
Tribulusamide B, 636
Trichosanthes kirilowii, 147, 233, 318, 352, 381,
         525, 538, 540, 546, 552, 570, 636
Trichosanthin, 147, 233, 318, 570
Tricin, 95, 212, 280, 570, 636
Tricycloekasantal, 129, 227, 570
Trifolin, 33, 165, 250, 570
Trifolioside, 97, 214, 570
Trifolirhizin, 137, 570
Trifolium incarnatum, 300, 603
Trifolium pratense, 147, 234, 300, 343, 451, 471,
         474, 487, 535, 592, 620, 622, 628,
         630, 634
Triglyceride, 88, 211, 570
Trigonella foenum-graecum, 147, 385, 491,
         551–552, 571, 577, 595, 597, 613–614,
         619–620, 636
Trigonelline, 4, 27, 34, 57, 121, 136, 144, 147, 157,
         176, 185, 195, 234, 295, 571, 636
Trigonotis peduncularis, 173, 344, 453
Trihydric alcohol, 163, 571
Trihydroxytriptolide, 147, 571
Trillarin, 147, 300, 571
Trillin, 58, 147, 300–301, 571, 636
Trillium camschatcense, 147, 300, 389, 447, 457,
         460, 571
Trillium erectum, 301, 602, 627, 629, 633, 636,
         640
Trilobamine, 45, 571
Tripchilorolide, 147, 571
Tripdiolide, 147, 571
Tripdioltonide, 147, 571
Tripeptide, 82, 571
Triptein, 147, 571
Tripterygium hypoglaucum, 147, 372, 394, 406,
         571
Tripterygium wilfordii, 147, 324, 327, 362, 434,
         571, 578
Triptolide, 147, 571
Triptolidenol, 147, 571
Triptonide, 147, 571
Triptophenolide, 147, 571
Triterpene acid, 271, 571, 637
Triterpene glycosides, 5, 134, 188, 329, 571, 637
Triterpene saponins, 204, 637

Triterpenes, 11, 24, 132, 176, 181, 185, 197, 206,
         211, 245–246, 260, 277, 280, 292–293,
         306, 328, 571, 637
Triterpenoid bitters, 213, 637
Triterpenoid saponins, 31, 177, 182, 184, 186,
         190, 198, 219, 222, 231, 241, 283–284,
         303, 322, 571, 637
Triterpenoids, 42, 45, 71, 120, 162, 214, 224, 238,
         244, 253, 266–267, 303, 306, 322,
         326, 414, 572, 637
Triticum vulgare, 147, 350, 431, 468, 540,
         576–577
Tropane alkaloids, 195, 206, 266, 637
Tryptanthrin, 27, 43, 572
Tryptophane, 59, 103, 260, 572
Tsudzuic acid, 87, 271, 572
Tu Er Cao, 29, 382
Tu Fang Ji, 106, 382
Tu Fang Ji (Taiwan), 382
Tu Gu Ling, 135, 383
Tu Hung Hua, 32, 383
Tu Niu Teng, 6, 383
Tu Ren Shen, 142, 383
Tu Si Zi, 52, 383
Tu Soon, 83, 383
Tu Tai Huang, 127, 383
Tu Wei Cao, 149, 383
Tu Xian, 84, 383
Tubulosine, 11, 572
Tuduranine, 44, 135, 140, 572
Tulipa edulis, 80, 148, 374, 406, 441, 560
Tulipa gesneriana, 391
Turmerone, 52, 193, 572, 637
Tussilago farfara, 148, 234, 361, 465, 467, 485,
         549, 552, 564–565, 604, 609, 615, 617,
         623, 625, 632, 634
Tymol, 146, 233, 572
Typha angustata, 148, 370, 410, 492, 524
Typha angustifolia, 234, 611, 621, 623
Typhonium divaricatum, 148, 362
Typhonium giganteum, 148, 503, 562, 572–574
Tyramine, 152, 185, 233, 572, 637
Tyrosine, 58–59, 69, 74, 103, 148, 196, 201, 203,
         218, 260, 572, 637

U

Ucarvone, 25, 182, 572
Ugenol, 65, 572
Ulmus campestris, 148, 301, 391, 427, 431, 482,
         503, 533, 535, 558
Ulmus rubra, 301, 591–592, 595, 617, 621, 634

Umbelliferone, 17, 24, 28, 32, 35, 47, 56, 96, 108, 157–158, 161, 163, 183, 195, 212–213, 241, 245, 249, 265, 274, 280, 572, 637
Umbelliprenin, 16–17, 158, 181, 241, 572
Uncaria gambir, 301, 594, 597, 608, 611, 619, 627
Uncaria hirsuta, 149, 301, 352, 443, 482, 488, 547
Unsaturated acids, 152, 303, 572
Uracil, 17, 148, 181, 241, 265, 573
Uraria crinita, 149, 355, 525, 551, 577
Uraria lagopodiodes, 383
Urbenine, 46, 191, 254, 573
Urea, 36, 117, 251, 573
Urease, 34, 68–69, 200–201, 262, 573
Urena procumbens, 149, 349, 410, 470, 533
Uric acid, 178, 637
Uronic acid, 4, 142, 237, 573
Ursolic acid, 17, 39, 44, 49, 54, 79, 86, 95, 101–102, 113, 118, 120, 126, 135, 145, 168–169, 187, 208, 220, 223, 225–226, 253, 256, 266, 268, 270, 282, 290, 295, 299, 637
Ursone, 150, 235, 301, 573
Urtica angustifolia, 149, 391, 400, 406, 417, 431, 436, 467–468, 540
Urtica laetevirens, 149, 376
Urtica tenacissima, 30
Urtica urens, 235, 588, 591, 595, 606, 608, 624, 626, 630, 632
Urushiol, 124, 287–288, 546, 573, 638
Usaramine, 50, 573
Ushinsunine, 97, 573
Usigtoercin, 79, 573
Usnea diffracta, 149, 362, 419, 455, 501, 545, 573
Usnea longissima, 379
Usnic acid, 149, 573
Utendin, 167, 573
Uvaol, 103, 573
Uzarigenin, 159, 574

V

Vaccaria pyramidata, 363
Vaccaria segetalis, 129, 149, 383, 479, 574
Vaccarin, 149, 574
Vaccaroside, 149, 574
Vaccinium bracteatum, 150, 301, 384, 397, 400, 415, 419, 449, 485, 497, 550, 557, 564, 573
Vaccinium macrocarpon, 301
Vaccinium myrtilloides, 301, 587, 609, 618, 634
Vaccinium myrtillus, 313
Vaccinium vitis-idaea, 235, 587, 608, 638–639
Valepotriates, 263, 302, 609, 638

Valeraldehyde, 95, 574
Valerenone, 150, 302, 574
Valeriana alternifolia, 150, 302, 358, 408, 425–426, 428, 430, 435, 450, 462–463, 493, 497, 516, 532, 536, 542, 555, 567, 574
Valeriana officinalis, 302
Valerianic acid, 96, 180, 213, 574, 638
Valerianol, 150, 302, 574
Valeric acid, 89, 210, 242, 574, 638
Valine, 87–88, 103, 148, 172, 209, 574
Vallarine, 37, 574
Valtrate, 302, 638
Vanillic acid, 10, 91, 95, 111, 124, 160, 170, 179, 211, 238, 261, 280, 287, 574, 638
Vanillin, 68, 141, 163, 180, 201, 296–297, 300, 574
Vanillyl alcohol, 71, 574
Vellarin, 186, 638
Vellosimine, 122, 285, 574
Venoterpine, 33, 574
Veratrine alkaloids, 124, 287, 574
Veratrum dahuricum, 150, 302, 362, 475, 493, 541, 549, 569, 580
Veratrum formosanum, 150, 406, 541, 574
Veratrum viride, 302, 584, 594, 631
Verbena officinalis, 151, 235
Verbenalin, 151, 191, 235, 255, 574, 638
Verbenalol, 151, 235, 575
Veronicastroside, 151, 302, 575
Verticine, 70, 201, 575
Vervenin, 235, 638
Viburnito, 275, 638
Viburnum opulus, 303, 597, 608, 627, 634
Viburnum sargenti, 151, 303, 374, 403, 421, 424, 428, 436, 439, 465, 489, 494, 507, 527, 554
Vicia faba, 151, 342, 425, 472, 494
Villoside, 168, 575
Vilmorrianines, 6, 575
Vinblastine, 36, 186, 575, 638
Vincristine, 36, 186, 575
Vindoline, 160, 575
Vindolinine, 36, 160, 186, 575
Viola acuminata, 152, 303, 359, 405, 434, 483, 553, 572
Viola tricolor, 303, 617, 628–629, 634, 638
Violaxanthin, 32, 126, 143, 169, 185, 227, 290, 575
Violin, 303, 638
Viroallosecurinine, 132, 575
Virosecurinin, 132, 575
Virosine, 132, 575
Viscin, 235, 638
Viscotoxin, 235, 638

Viscum album, 152, 235, 355, 402, 404, 406, 421,
        437, 468, 471, 482, 505, 511, 516, 524,
        543–544, 572, 577, 583–584, 592–593,
        595, 603–604, 613, 624, 627, 638
Vitamin A, 5, 34, 42, 71, 85, 90, 117, 133, 158,
        164, 182, 185, 210, 220, 222, 229, 269,
        282–283, 576, 638
Vitamin B, 7, 17, 34, 74, 78, 117, 127, 133, 142,
        157, 177, 195, 205, 222, 229, 247, 576,
        638
Vitamin *B* complex, 195, 247, 638
Vitamin C, 7, 10, 29, 38, 52, 54, 66, 69, 77–78,
        90, 104, 109–110, 126, 138, 177–178,
        187, 191, 195, 205, 208, 210, 218,
        226–227, 231, 248, 250, 257–258,
        263, 265, 273, 276, 278, 282, 286,
        289–290, 296, 302, 320, 576, 639
Vitamin E, 17, 77, 83, 115, 181, 185, 205, 221,
        241, 269, 309, 326, 577, 639
Vitamin G, 577
Vitamin K, 10, 178, 577, 639
Vitamin P, 201, 639
Vitamins, 14–15, 82, 84, 88–89, 100, 109, 125,
        143, 147, 152, 180, 190, 201, 203, 205,
        210, 212, 214, 218, 223, 225–226,
        228–230, 232, 235–236, 240, 256,
        281, 288–289, 292, 296, 301, 577, 639
Vitex, 147, 152–153, 234, 303, 321, 347, 357, 368,
        404, 418, 422, 430, 433, 438, 440, 453,
        458, 465–466, 470, 475, 477, 484,
        487, 491, 499, 505, 518, 521, 525, 532,
        535–536, 564, 566, 577–578, 584, 589,
        594–595, 603, 609, 639
Vitex agnus-castus, 321
Vitex chinensis, 152, 303, 368, 422, 433
Vitex labrusca, 303, 584, 589, 594–595, 603,
        609, 639
Vitex nequndo, 153, 357, 404, 430, 438, 440, 453,
        458, 465–466, 475, 477, 484, 487, 491,
        499, 518, 521, 525, 535, 564
Vitex rotundifolia, 347
Vitex trifolia, 153, 368, 418, 433, 470, 505, 536,
        566, 577–578
Vitexicarpin, 153, 577
Vitexin, 50, 68, 77, 82, 87, 92, 108, 116, 126, 149,
        155, 200, 204, 209, 211, 227, 262, 290,
        304, 577, 639
Vitexin-7-glucoside, 147, 577
Viticine, 303, 639
Vitis amurensis, 153, 374, 507, 526, 545, 565
Vitis vinifera, 235, 603, 615–616, 619–620, 631,
        634
Vitrexin-4-O-xyloside, 578
Vitricine, 153, 578

Volatile carbonyl, 122, 285, 578
Volatile oil, 7, 31, 39, 51, 60, 65, 71, 93–94, 99,
        112, 165, 176, 183¦ 187, 193, 196–199,
        201–202, 204, 206, 209, 211–213,
        215, 219, 222, 227–229, 231, 235–236,
        239–245, 248–249, 251, 257, 259–260,
        262–264, 269, 274, 276, 280, 283–285,
        301, 316, 329, 578, 640
Volatile phenols, 122, 285, 578
Vomicine, 140, 578

W

Wahlenbergia marginata, 153, 385, 406
Wan Shou Jiu, 142, 383
Wang Bu Liu Xing, 129, 149, 383
Wang Jiang Nan, 36, 383
Wax, 86, 217, 287, 299, 640
Wedolactone, 61, 453, 578
Wei Ling Cai, 117, 383
Wei Ling Xian, 43, 317, 383
Wei Mao, 37, 65, 68, 383
Wei Sui Xian, 13, 384
Wei Yan Xian, 36, 384
Wikestroemia indica, 153, 363, 402, 451, 484,
        578
Wikstroemin, 153, 578
Wilfordine, 65, 147, 322, 578
Wilsonine, 38, 578
Wisteria brachybotrys, 322
Wisteria sinensis, 153, 303, 394, 478, 546
Withanolides, 195, 640
Wo Seng, 18, 384
Wo Zu, 384
Wognoside, 131, 229, 292, 578
Wogonin, 131, 229, 292–293, 327, 578, 640
Won Nian Qing, 123, 384
Woodorien, 153, 578
Woodwardia japonica, 153, 352, 486, 578
Worenine, 46, 191, 254, 578, 640
Wu An, 384
Wu Bei Zi, 124, 384
Wu Fan Shu, 150, 384
Wu Gan, 81, 384
Wu Hua Go, 69, 384
Wu Huan Shu, 129, 384
Wu Jia Pi, 5, 162, 384
Wu Jiu, 129, 384
Wu Ju, 138, 384
Wu Ma, 133, 384
Wu Mai, 118, 384
Wu Ru Ba, 147, 385
Wu Song Ju, 385
Wu Tao, 6, 385

Wu Tong, 44, 69, 94, 160, 344, 385, 389
Wu Wei Zi, 83, 130, 368, 385
Wu Wei Zi (Taiwan), 385
Wu Yao, 55, 87, 385
Wu Zhao Jin Long, 81, 385
Wu Zhu Yu, 67, 385

X

Xamthophyll, 579
Xanthanol, 154, 579
Xanthine, 145, 185, 579
Xanthinin, 154, 579
Xanthium chinense, 154, 342, 428, 454, 493, 562, 567, 579
Xanthium sibiricum, 173
Xanthoagathin, 133, 579
Xanthones, 181, 252, 280, 297, 321, 640
Xanthophyllepoxyl, 32, 249, 579
Xanthophylls, 172, 248, 640
Xanthoplanine, 155, 304, 579
Xanthorin, 36, 251, 579
Xanthotoxine, 16, 241, 579
Xanthoxylene, 155, 304, 579
Xanthoxylin, 129, 579
Xanthoxylinin, 154, 579
Xanthoxylum piperitum, 154, 345, 421, 439–440, 465, 475, 502, 532, 551–552, 556, 579
Xanthumin, 154, 579
Xi Sheng Teng, 41, 385
Xi Shin, 76, 385
Xi Shu, 33, 385
Xi Xin, 25, 385
Xi Yang Bai Hua Cai, 43
Xi Ye Sha Seng, 153, 385
Xi Yc Zhu Chi Cao, 110, 385
Xia Ku Chao, 118, 385
Xia Tian Wu, 48, 385
Xian, 4, 8, 10, 13–14, 22, 29, 36, 43, 51–53, 57, 65, 70, 76–77, 80, 84, 86–87, 94, 102, 107, 117, 121, 129, 135, 138–139, 141, 148, 161, 163–164, 171, 317, 338, 340–341, 343, 346–347, 349–351, 354, 357, 359–360, 364–366, 373, 379–380, 382–387, 391–392
Xian He Cao, 10, 386
Xian Mao, 51, 53, 161, 346, 386
Xian Xia Hua, 386
Xian Ye Shu, 86–87, 346, 386
Xiang Fu, 54, 386
Xiang Jia Pi, 107, 169, 386
Xiang Jiao, 99, 386
Xiang Si Shu, 4, 386
Xiang Tian Huang, 43, 386

Xiang Xu, 62, 79, 107, 386
Xiao Fei Yang Cao, 67, 386
Xiao Hui Xiang, 69, 386
Xiao Ji, 34, 37, 41, 387
Xiao Shan Ju, 74, 387
Xiao Ye Yang Jiao Teng, 98, 387
Xiao Yeh, 28, 387
Xiao Ying Qie, 135, 387
Xin Yi, 93, 387
Xing Ren, 118, 387
Xiu Qiu, 79, 387
Xiu Xian Jiu, 138, 387
Xu Chang Qing, 53, 387
Xu Duan, 59, 387
Xu Sui Zi, 67, 387
Xuan Fu Hua, 80, 164, 387
Xuan Mu Gua, 38, 52, 387
Xuan Seng, 131, 387
Xue Jian Chou, 19, 388
Xue Jie, 55, 388
Xue Shang Yi Zhi Hao, 6, 388
Xylan, 210, 640
Xylose, 82, 89, 142, 185, 210, 220, 579, 640

Y

Ya Dan Zi, 30, 388
Ya Er Qin, 50, 388
Ya Ma, 87, 388
Yan Cao, 101, 136, 375, 388
Yan Hu Suo, 48, 361, 388
Yan Jie Cao, 103, 388
Yang Gan Jiu, 94, 388
Yang Guo Nau, 140, 388
Yang Lu, 17, 45, 388
Yang Lu Kui, 17, 388
Yang Ti Gen, 127, 388
Yang Yu Lan, 93, 389
Yang Zhi Zu, 27, 124, 389
Yao Jiu Hua, 389
Yatanoside, 30, 579
Ye Bai He, 50, 389
Ye Da Dou, 74, 389
Ye Dou Gen, 96, 389
Ye Gan Zi, 109, 389
Ye Guan Men, 85, 389
Ye He Hua, 389
Ye Lu Kui, 96, 389
Ye Mu Gua, 138, 389
Ye Wo, 389
Ye Wu Tong, 94, 389
Ye Yen Me, 27, 389
Yejuhualactone, 39, 252, 579
Yen Lin Cao, 389

Yen Xing, 389
Yeu Je Hua, 389
Yi Dian Hong, 101, 389
Yi Mu Cao, 85, 340, 389
Yi Nian Pon, 390
Yi Wu, 61, 390
Yi Ye Chan, 131, 390
Yi Yi, 45, 390
Yi Zhi, 6, 13, 62, 103, 136, 162, 166, 171, 346,
    370, 388, 390
Yi Zhi Huang Hua, 136, 171, 390
Yi Zhi Jian, 103, 166, 346, 390
Yi Zhi Zi, 62, 390
Yie Huang Hua, 390
Yie Mian Hua, 390
Yie Pu Tao Teng, 390
Yie Xiz Zhu, 390
Yin Bai Yang, 116, 390
Yin Cao, 118, 390
Yin Chen, 23, 390
Yin He Huan, 85, 390
Yin Lian Hua, 16, 390
Ying Chun Hua, 81, 391
Ying Su, 105, 391
Ylangene, 65, 200, 579
Yohimbine, 286, 640
You Tong, 11, 82, 372, 391
Y-sitosterol, 10, 579
Yu Bai, 148, 391
Yu Bai Pi, 148, 391
Yu Dei Mei, 78, 391
Yu Jin, 51, 66, 357, 391
Yu Jin Xian, 391
Yu Lee Ren, 118, 391
Yu Ma, 30, 149, 368, 391
Yu Ma Gen, 30, 391
Yu Mei Ku, 106, 391
Yu Mi Xu, 155, 391
Yu Shan Dou, 90, 391
Yu Shan Shi Song, 91, 391
Yu Xing Cao, 78, 391
Yu Ye Jin Hua, 99, 391
Yuan Bai, 391
Yuan Hua, 55, 392
Yuan Jin Gan, 392
Yuan Xi Huang San, 12, 392
Yuan Zhi, 114, 392
Yuanhuacine, 55, 195, 258, 579
Yuanhuadine, 55, 195, 258
Yuanhuatine, 55, 195, 258, 580
Yue Tao, 13, 392
Yun Shi, 31, 392
Yun Xian Cao, 53, 392
Yunnanxana, 143, 298, 580

Z

Zanthaline, 106, 217, 279, 580
Zanthoxylum ailanthoides, 154, 304, 377, 455,
    491, 500, 507, 512
Zanthoxylum americanum, 304, 588, 594, 607,
    618, 627, 634
Zanthoxylum bungeanum, 154, 304, 374, 409,
    420, 423, 428, 446, 465, 502, 516, 536,
    546, 549
Zanthoxylum nitidum, 155, 304, 378, 400, 516,
    521, 526, 558, 577
Zanthoxylum schinifolium, 155, 304, 355, 421,
    440, 463, 465, 532, 579
Zao Ci, 73, 392
Zao Zhui, 20, 392
Ze Lan, 20, 392
Ze Qi, 66, 392
Ze Xie, 11, 392
Zea mays, 155, 236, 391, 417, 422, 431, 520, 548,
    569, 601–602, 615, 617, 627, 629,
    634–635
Zeaxanthin, 90, 110, 143, 172, 210, 218, 273, 281,
    580
Zederone, 51–52, 193, 580
Zedoarin, 51, 580
Zephyranthes candida, 155, 173, 345, 351, 479,
    506, 519, 565–566
Zephyranthes carinata, 155, 349, 406
Zerumbone, 52, 156, 580
Zerumbone epoxide, 156, 580
Zhang Shu, 26, 392
Zhe Gu Cai, 32, 392
Zhen, 7, 49, 70, 86, 357–358, 369, 392
Zhi, 6, 13, 15–16, 20, 25–27, 40, 45, 62, 71, 78,
    84, 98, 101, 103, 110, 114, 124, 128,
    131, 136, 162, 164, 166, 171, 334–335,
    340, 345–346, 352, 356, 361–363,
    370–371, 373, 375, 388–390, 392–393
Zhi Bei Zi, 78, 392
Zhi Jia Hua, 84, 392
Zhi Jin Niu, 20, 393
Zhi Mu, 15, 393
Zhi Wen, 25, 393
Zhow Sho, 393
Zhu Cao, 32, 88, 131, 382, 393–394
Zhu Chun Hua, 127, 393
Zhu Je Seng, 105, 393
Zhu Jin, 393
Zhu Ling, 116, 393
Zhu Long Cao, 100, 393
Zhu Mao Chao, 127, 393
Zhu Shan, 143, 393
Zhu Shi Tou, 50, 393

Zhu Wei, 33, 393
Zhu Ye Lan, 24, 393
Zhu Yin Yin, 393
Zhu Zi Cao, 110, 393
Zi Bei Cao, 62, 393
Zi Cao, 22, 70, 110, 361, 369, 393
Zi Duan, 146, 394
Zi Jin Pi, 147, 394
Zi Kee Guan Zhong, 104, 394
Zi Lan, 65, 394
Zi Su, 107, 394
Zi Teng, 153, 394
Zi Wei Hua, 28, 394
Zi Yu, 128, 222, 227, 394, 580
Zi *Yu glucoside* I, 128, 222, 227, 580
Zi *Yu glucoside* II, 128, 222, 227, 580

Zi Zhu Cao, 32, 394
Ziang Jia Pi, 394
Zingiber officinale, 155, 236
Zingiber zerumbet, 156, 314, 371, 398, 408, 446, 454, 469, 580
Zingiberen, 193, 640
Zingiberene, 13, 51–52, 145, 155, 193, 236, 239, 299, 580
Zingiberol, 13, 155, 236, 239, 580
Ziziphus jujuba, 156, 236, 380, 425, 451, 493, 551–552, 603, 617, 629, 632, 639
Zong Lu, 146, 173, 394
Zong Shi, 84
Zu Si Ma, 55, 394
Zuo Yie He Cao, 49, 394
Zygadenine, 150, 302, 580

9 780367 384968